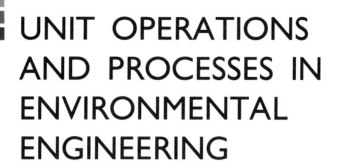

UNIT OPERATIONS AND PROCESSES IN ENVIRONMENTAL ENGINEERING

PWS Series in Engineering

Anderson, *Thermodynamics*

Askeland, *The Science and Engineering of Materials*, Third Edition

Baliga, *Power Semiconductor Devices*

Bolluyt/Stewart/Oladipupo, *Modeling for Design Using SilverScreen*

Borse, *FORTRAN 77 and Numerical Methods for Engineers*, Second Edition

Clements, *68000 Family Assembly Language*

Clements, *Microprocessor Systems Design*, Second Edition

Clements, *Principles of Computer Hardware*, Second Edition

Das, *Principles of Foundation Engineering*, Second Edition

Das, *Principles of Geotechnical Engineering*, Second Edition

Das, *Principles of Soil Dynamics*

Duff/Ross, *Freehand Sketching for Engineering Design*

Duff/Ross, *Mastering 3D Studio*

El-Wakil/Askeland, *Materials Science and Engineering Lab Manual*

Feldman/Valdez-Flores, *Applied Probability and Stochastic Processes*

Fleischer, *Introduction to Engineering Economy*

Forsythe, *FORTRAN 90: Contemporary Computing for Scientists and Engineers with Spreadsheets*

Garber/Hoel, *Traffic and Highway Engineering*, Second Edition

Gere/Timoshenko, *Mechanics of Materials*, Third Edition

Glover/Sarma, *Power System Analysis and Design*, Second Edition

Hayter, *Probability and Statistics for Engineers and Scientists*

Howell, *Introduction to AutoCAD Designer 1.1*

Janna, *Design of Fluid Thermal Systems*

Janna, *Introduction to Fluid Mechanics*, Third Edition

Kassimali, *Structural Analysis*

Keedy, *An Introduction to CAD Using CADKEY 5 and 6*, Third Edition

Keedy/Teske, *Engineering Design Using CADKEY 5 and 6*

Knight, *The Finite Element Method in Mechanical Design*

Knight, *A Finite Element Method Primer for Mechanical Design*

Liley, *Tables and Charts for Thermodynamics*

Logan, *A First Course in the Finite Element Method*, Second Edition

McDonald, *Continuum Mechanics*

McGill/King, *Engineering Mechanics: An Introduction to Dynamics*, Third Edition

McGill/King, *Engineering Mechanics: Statics*, Third Edition

McGill/King, *Engineering Mechanics: Statics and An Introduction to Dynamics*, Third Edition

Meissner, *FORTRAN 90*

Raines, *Software for Mechanics of Materials*

Ray, *Environmental Engineering*

Reed-Hil/Abbaschian, *Physical Metallurgy Principles*, Third Edition

Reynolds/Richards, *Unit Operations and Processes in Environmental Engineering*, Second Edition

Russ, *CD-ROM for Materials Science*

Russ, *Materials Science: A Multimedia Approach*

Sack, *Matrix Structural Analysis*

Schmidt/Wong, *Fundamentals of Surveying*, Third Edition

Segui, *Fundamentals of Structural Steel Design*

Segui, *LRFD Steel Design*

Shen/Kong, *Applied Electromagnetism*, Third Edition

Stewart/Bolluyt/Oladipupo, *Modeling for Design using AutoCAD Release 12*

Stewart/Bolluyt/Oladipupo, *Modeling for Design using AutoCAD Release 13*

Sule, *Manufacturing Facilities*, Second Edition

Vardeman, *Statistics for Engineering Problem Solving*

Weinman, *FORTRAN for Scientists and Engineers*

Weinman, *VAX FORTRAN*, Second Edition

Wempner, *Mechanics of Solids*

Woolsey/Kim/Curtis, *VizAbility CD-Rom and Handbook*

Wolff, *Spreadsheet Applications in Geotechnical Engineering*

Zirkel/Berlinger, *Understanding FORTRAN 77 and 90*

UNIT OPERATIONS AND PROCESSES IN ENVIRONMENTAL ENGINEERING

Second Edition

Tom D. Reynolds, Ph.D., P.E.
Professor Emeritus of Civil Engineering
Texas A & M University

Paul A. Richards, Ph.D., P.E.
Professor of Civil Engineering
University of Southwestern Louisiana

PWS Publishing Company

I(T)P   An International Thomson Publishing Company

Boston • Albany • Bonn • Cincinnati • Detroit • London • Madrid • Melbourne • Mexico City • New York • Paris
San Francisco • Singapore • Tokyo • Toronto • Washington

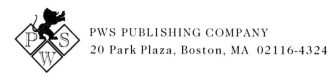

PWS PUBLISHING COMPANY

20 Park Plaza, Boston, MA 02116-4324

This book is printed on recycled, acid-free paper.

International Thomson Publishing
The trademark ITP is used under license.

For more information, contact:

PWS Publishing Co.
20 Park Plaza
Boston, MA 02116

International Thomson Publishing Europe
Berkshire House 168–173
High Holborn
London WC1V 7AA
England

Thomas Nelson Australia
102 Dodds Street
South Melbourne, 3205
Victoria, Australia

Nelson Canada
1120 Birchmount Road
Scarborough, Ontario
Canada M1K 5G4

International Thomson Editores
Campos Eliseos 385, Piso 7
Col. Polanco
11560 México D.F., Mexico

International Thomson Publishing GmbH
Königswinterer Strasse 418
53227 Bonn, Germany

International Thomson Publishing Asia
221 Henderson Road
#05-10 Henderson Building
Singapore 0315

International Thomson Publishing Japan
Hirakawacho Kyowa Building, 31
2-2-1 Hirakawacho
Chiyoda-ku, Tokyo 102
Japan

Library of Congress Cataloging-in-Publication Data

Reynolds, Tom D.
    Unit operations and processes in environmental engineering / Tom D. Reynolds, Paul A. Richards. — 2nd ed.
      p.   cm.
    Includes bibliographical references and index.
    ISBN 0-534-94884-7
    1. Water — Purification. 2. Sewage — Purification. 3. Water — Purification — Problems, exercises, etc.
  4. Sewage — Purification — Problems, exercises, etc. I. Richards, Paul A. II. Title.
TD430.R48 1995
628.1′62 — dc20
                                                                     95-38371
                                                                     CIP

Sponsoring Editor: Jonathan Plant
Production Coordinator: Robine Andrau
Marketing Development Manager: Nathan Wilbur
Manufacturing Coordinator: Wendy Kilborn
Interior Designer: Robine Andrau
Interior Illustrator: Carl Brown

Typesetter: Best-set Typesetter Ltd., Hong Kong
Cover Designer: Julia Gecha
Cover Printer: Coral Graphic Services, Inc.
Text Printer/Binder: R.R. Donnelley & Sons/
  Crawfordsville

Printed and bound in the United States of America
96 97 98 99 — 10 9 8 7 6 5 4 3 2

Contents

v

9

SEDIMENTATION   219

10

FILTRATION   284

14

MEMBRANE PROCESSES 395

15

ACTIVATED SLUDGE 411

20

AEROBIC DIGESTION 610

21

SOLIDS HANDLING 629

22

LAND TREATMENT OF MUNICIPAL WASTEWATER AND SLUDGES 666

Preface

Unit Operations and Processes in Environmental Engineering, Second Edition is for an advanced undergraduate course or a graduate-level course in environmental engineering. It has been written primarily for the civil/environmental engineering curriculum; however, it may be used in other fields, such as chemical engineering, where water and wastewater treatment are taught. The book presents water and wastewater treatment under a single cover, using the unit operation and process approach. The fundamentals of each operation and process are first presented in each chapter; then the application in the water treatment and wastewater treatment fields is given. Some operations or processes, such as activated sludge, are used in only one field and are presented accordingly. The undergraduate civil engineer's background in chemistry, mathematics, and fluid mechanics, and a prior introductory course in environmental engineering, are adequate preparation for use of the text. In nearly all the chapters, numerous example problems are given along with practice problem sets at the ends of the chapters, since an engineering discipline is most rapidly learned by problem solving. The book is oriented toward engineering design based on fundamentals.

Chapters 1 through 5 present background material on chemical concepts, biological concepts, mass balances, flow models, reactor fundamentals, water quantities, water quality, wastewater quantities, and wastewater quality. The instructor may vary the depth of coverage of these chapters according to the background of his or her class. Chapter 6 should then be covered because it has typical flowsheets of water, wastewater, and advanced wastewater treatment plants. After Chapter 6, the chapters need not be covered in sequential order; each chapter is written to be complete within itself. This presentation allows the instructor to select chapters or parts of the chapters in any sequence desired. If the instructor wants to cover water treatment in the first part of a semester and wastewater treatment in the latter part, he or she may choose the sequence of the chapters or parts of the chapters accordingly. The material coverage for a water treatment approach, a wastewater treatment approach, and an advanced wastewater treatment approach would include the following chapters or parts of chapters and major topics.

Water Treatment

Chapter	Topic
7	Screening, Presedimentation, Aeration, Adsorption, Prechlorination
8	Coagulation and Flocculation
9	Sedimentation
10	Filtration
12	Adsorption
24	Disinfection
13	Ion Exchange
14	Membrane Processes
21	Solids Handling
23	Iron and Manganese Removal, Volatile Contaminant Removal from Groundwater

Wastewater Treatment

Chapter	Topic
7	Screening and Shredding, Grit Removal, Flow Equalization, Quality Equalization
9	Sedimentation
15	Activated Sludge
16	Oxygen Transfer and Mixing
17	Trickling Filters and Rotary Biological Contactors
18	Stabilization Ponds and Aerated Lagoons
24	Disinfection
19	Anaerobic Digestion
20	Aerobic Digestion
21	Solids Handling
22	Land Treatment of Municipal Wastewater and Sludges
23	Flotation, Oil Separation, Septic Tank, and Soil Absorption Systems

Advanced Wastewater Treatment

Chapter	Topic
8	Coagulation and Flocculation
9	Sedimentation
11	Ammonia Stripping
10	Filtration
12	Adsorption
14	Membrane Process
13	Ion Exchange
11	Ammonia Removal by Ion Exchange, Breakpoint Chlorination, and Biological Means
23	Anaerobic Contact Process, Submerged Anaerobic Filter, Biological Phosphorus Removal, Aquatic Systems

If the instructor wants to cover the chapters based on the fundamental principle involved — that is, physical, chemical, or biological — he or she may choose the chapters accordingly. The material coverage for this approach would include the following chapters or parts of chapters and major topics.

Physical Treatment

Chapter	Topic
7	Screening, Presedimentation, Aeration, Adsorption, Screening and Shredding, Grit Removal, Flow Equalization, Quality Equalization
9	Sedimentation
10	Filtration
11	Ammonia Stripping
12	Adsorption
14	Dialysis and Reverse Osmosis
16	Oxygen Transfer and Mixing
23	Volatile Contaminant Removal from Groundwaters, Flotation, Oil Separation, Septic Tanks
24	UV Disinfection

Chemical Treatment

Chapter	Topic
7	Neutralization
8	Coagulation and Flocculation
13	Ion Exchange
14	Electrodialysis
24	Disinfection
11	Ammonia Removal by Ion Exchange and Breakpoint Chlorination
24	Chlorine and Ozone Disinfection

Biological Treatment

Chapter	Topic
15	Activated Sludge
17	Trickling Filters and Rotary Biological Contactors
18	Stabilization Ponds and Aerated Lagoons
19	Anaerobic Digestion
20	Aerobic Digestion
11	Ammonia Removal by Nitrification–Denitrification
22	Land Treatment of Municipal Wastewaters and Sludges
23	Anaerobic Contact Process, Submerged Anaerobic Filter, Biological Phosphorus Removal, Aquatic Systems, Septic Tank and Soil Absorption Systems

Solids handling is best covered as a separate subject because it involves both physical and chemical treatments.

A third approach to using the book is to cover the topics in the order of their occurrence — from the most widely used treatment topics first to the least common treatment topics last. In using this approach, the instructor may choose the sequence of topics to his or her own liking.

The book contains more material than can be conveniently covered in a one-semester, three-credit-hour course; thus, the instructor may choose to limit the depth of coverages of some topics. However, the material that is not covered in class can serve as a valuable reference source for the student.

Based on feedback from users of the first edition, all homework problems have been reviewed. Some problems have been deleted, other problems have been revised, and numerous new problems have been added. Over 60 percent of the homework problems in the second edition are new. There are over 300 homework problems in the second edition, compared to about 100 in the first edition.

The glossary includes most terms that are covered in an introductory environmental engineering course. The Appendix contains tables that give common conversion factors, atomic numbers and weights, properties of water, dissolved oxygen concentrations at various temperatures, chloride concentrations and atmospheric pressures, and three diagrams for laboratory biological treatment reactors, along with instructions.

Although the book is intended as a textbook, it will be useful as a reference book for practicing engineers and for engineers who wish to do self-study. The examples in the metric system (International System of Units, abbreviated SI) will help practicing engineers, as well as students, to use this dimensional system.

A Solutions Manual, which includes full worked-out solutions to the homework problems, is also available for instructors.

The authors appreciate the efforts of Dana L. Montet in preparing the Solutions Manual. The authors also gratefully acknowledge the thorough reading of the manuscript and the meaningful suggestions by the following reviewers:

Frederic C. Blanc
Northeastern University

Robert P. Carnahan
University of South Florida

Ronald A. Chadderton
Villanova University

John D. Dietz
University of Central Florida

Austin Grogan
University of Central Florida

Richard O. Mines, Jr.
University of South Florida

Vito Punzi
Villanova University

Aarne Vesilind
Duke University

Irvine W. Wei
Northeastern University

Tom D. Reynolds
Paul A. Richards

CHEMICAL CONCEPTS

I

The purpose of this chapter is to review basic inorganic chemistry and present an introduction to physical chemistry and organic chemistry as related to the field of water and wastewater treatment.

INORGANIC CHEMISTRY

The basic chemical identities that compose all substances are the elements. The various elements vary from each other in atomic weight, size, and chemical properties. The symbols for the elements are used in writing chemical formulas and equations. The **atomic weight** is the weight of an element relative to carbon-12, which has an atomic weight of 12. The **valence** is the combining power of an element relative to that of the hydrogen atom, which has an assigned value of 1. Thus, an element with +3 valence can replace three hydrogen atoms in a compound, and an element with −3 valence can react with three hydrogen atoms. The **equivalent weight of an element** is equal to its atomic weight divided by the valence. For example, the equivalent weight of calcium (Ca^{+2}) is 40.0 gm divided by the valence of +2, to give 20.0 gm.

The **molecular weight** of a compound equals the sum of the atomic weights of the combined elements and is usually expressed in grams. The **equivalent weight of a compound** is that weight of the compound which contains 1 gram atom of available hydrogen or its chemical equivalent. Equivalent weight also equals the molecular weight/Z, where Z is a positive integer whose value depends upon the chemical context. For acids, Z is equal to the moles of H^+ obtainable from 1 mole of the acid. For H_2SO_4, $Z = 2$, so the equivalent weight is $[(1 \times 2) + 32 + (16 \times 4)/2] = 49.0$ gm. For bases, the value of Z is equal to the number of moles of H^+ with which 1 mole of the base reacts. Thus for $Ca(OH)_2$, $Z = 2$. For salts, Z is the oxidation or valence state of the ions. For $BaCl_2$, $Z = 2$ because Ba^{+2} is equivalent to 2 moles of H^+. For $Al_2(SO_4)_2$, $Z = 6$ because $2Al^{+3}$ is equivalent to 6 moles of H^+. For oxidizing or reducing agents, Z is equal to the change in oxidation or valence state. For $K_2Cr_2O_7$ in acid conditions, it undergoes a change in valence state of 3×2 or $Z = 6$.

In a large number of different molecules, certain groupings of atoms act together as a unit and are termed **radicals**. Radicals are given special names for identity. Some common radicals encountered in water and wastewater engineering are the hydroxyl (OH^-), bicarbonate (HCO_3^-), carbonate (CO_3^{-2}), orthophosphate (PO_4^{-3}), sulfate (SO_4^{-2}), ammonium (NH_4^+), nitrite (NO_2^-), nitrate (NO_3^-), and hypochlorite (OCl^-) ions. The hydroxyl group

1

(OH$^-$) is one of the most frequent that occurs. Radicals themselves are not compounds, but they join with other elements to form compounds.

Concentration Units

Usually, the concentration of ions or chemicals in solution is expressed as the weight of the element or compound in milligrams per liter of water, abbreviated mg/ℓ. Sometimes, the term **parts per million** (ppm) is used rather than mg/ℓ. If the specific gravity is unity, conversion gives $(1 \text{ mg}/\ell)$ $(\text{m}\ell/1000 \text{ mg})$ $(\ell/1000 \text{ m}\ell)$ or $1:10^6$. Thus 1 mg/ℓ equals to 1 part per million. Usually, the specific gravity is so close to 1 that this conversion can be assumed. Chemical dosages are usually expressed as pounds per million gallons (kilograms per million liters); however, older books may refer to grains per gallon. One grain per gallon equals to 17.1 mg/ℓ.

The concentrations of elements are usually stated in milligrams of that particular element, and it is usually understood that the solution contains the stated number of milligrams. For example, a water that contains 36.2 mg/ℓ of sodium contains 36.2 mg of sodium ion (Na$^+$) by weight per liter. In special cases, the concentration expressed is not for a specific element. For example, hardness that is due to Ca^{+2} and Mg^{+2} is frequently stated in unit weights of CaCO$_3$. This facilitates computations involving hardness by using a single term rather than two terms, one for Ca^{+2} and one for Mg^{+2}. The alkalinity of water may be due to one or more of the following ions: HCO$_3^-$, CO$_3^{-2}$, and OH$^-$. These also are frequently expressed as mg/ℓ as CaCO$_3$, to simplify certain computations. All nitrogen compounds — that is, ammonia, nitrite, and nitrate — are usually expressed as mg/ℓ nitrogen — that is, mg/ℓ-N. Also, all phosphates are usually expressed as mg/ℓ phosphorus — that is, mg/ℓ-P. Sometimes concentrations are expressed as **milliequivalents per liter** (meq/ℓ). These are obtained by dividing mg/ℓ by the milliequivalent weight — that is, the equivalent weight expressed as milligrams.

In order to standardize tests so that common procedures are used, the book titled *Standard Methods for the Examination of Water and Wastewater* (APHA, 1992) is used for all analyses done routinely.

Hydrogen Ion Concentration

Pure water (H$_2$O) dissociates to a slight degree to yield a concentration of hydrogen ions, H$^+$, equal to 10^{-7} mole/ℓ. The dissociation is represented by

$$H_2O \rightleftharpoons H^+ + OH^- \tag{1.1}$$

Since water dissociates to produce one hydroxyl ion, OH$^-$, for each hydrogen ion, H$^+$, there are also 10^{-7} mole/ℓ of hydroxyl produced. Since for pure water the amount of hydroxyl ions is equal to the amount of hydrogen ions, pure water is considered neutral. The acidic or basic nature of water is related to the concentration of hydrogen ions and is expressed by the symbol, pH, where

$$pH = \log \frac{1}{[H^+]} \tag{1.2}$$

Since the logarithm of 1 over 10^{-7} is 7, the pH at neutrality is 7. The slight ionization of water is represented by

$$[H^+][OH^-] = K_w = 10^{-14} \tag{1.3}$$

where K_w is the ionization constant for pure water. When an acid is added to water, it ionizes in the water and the hydrogen ion concentration increases, resulting in a lower pH. Consequently, the hydroxyl ion concentration decreases. For example, if an acid is added to increase the $[H^+]$ to 10^{-2}, the pH is 2 and the $[OH^-]$ must decrease to 10^{-12}, since

$$[10^{-2}][10^{-12}] = 10^{-14} \tag{1.4}$$

Likewise, if a base is added to increase the $[OH^-]$ to 10^{-3}, the $[H^+]$ decreases to 10^{-11} and the pH is 11. It is important to remember that the $[H^+]$ and $[OH^-]$ can never be reduced to zero, no matter how acidic or basic the solution may be. The pH scale ranges from 0 to 14, where pH 0 to 7 is acidic and pH 7 to 14 is basic or

Frequently, the basic side is referred to as alkaline. The pH may be measured by an electronic electrode or pH indicators that exhibit significant color change. Adjustment of pH is used to optimize coagulation, softening, and disinfection reactions and is used in corrosion control. In wastewater treatment, the pH must be maintained in a favorable range for optimum biological activity.

Chemical Equilibria

Most chemical reactions are reversible to some degree, and the concentrations of reactants and products determine the final equilibrium. Consider the equation

$$a\mathrm{A} + b\mathrm{B} \rightleftharpoons c\mathrm{C} + d\mathrm{D} \tag{1.5}$$

where

    A, B = reactants

    C, D = products

An increase in A or B shifts the reaction to the right, and an increase in C or D shifts it to the left. The relationship may be expressed by the mass-action equation

$$\frac{[C]^c[D]^d}{[A]^a[B]^b} = K \tag{1.6}$$

where

    [ ] = molar concentrations

    K = equilibrium constant

Strong acids and bases in dilute solutions approach 100% ionization, while weak acids and bases are poorly ionized. An equilibrium relationship that is very important in water chemistry is the carbonic acid-bicarbonate-carbonate system. Carbonic acid dissociates as follows:

$$H_2CO_3 \rightleftharpoons H^+ + HCO_3^- \tag{1.7}$$

and for this system, the mass action equation is

$$\frac{[H^+][HCO_3^-]}{[H_2CO_3]} = K_1 = 4.45 \times 10^{-7} \text{ at } 25°C \tag{1.8}$$

The bicarbonate ion $[HCO_3^-]$ dissociates as follows:

$$HCO_3^- \rightleftharpoons H^+ + CO_3^{-2} \tag{1.9}$$

and for this system, the mass action equation is

$$\frac{[H^+][CO_3^{-2}]}{[HCO_3]} = K_2 = 4.69 \times 10^{-11} \text{ at } 25°C \tag{1.10}$$

The previous discussion is for **homogenous chemical equilibria**, wherein all reactants and products occur in the same physical state. **Heterogeneous equilibria** occur between substances in two or more states. For example, at pH greater than 10, solid calcium carbonate in water reaches a stability with calcium and carbonate ions in solution since the dissociation is

$$CaCO_3 \rightleftharpoons Ca^{+2} + CO_3^{-2} \tag{1.11}$$

Equilibrium between crystals of a compound in the solid state and its ions in solution may be treated mathematically as if the equilibrium were homogenous in nature. For Eq. (1.11), the equilibrium equation according to mass action is

$$\frac{[Ca^{+2}][CO_3^{-2}]}{[CaCO_3]} = K \tag{1.12}$$

The concentration of a solid substance can be treated as a constant K_s in the mass action equation; therefore, $[CaCO_3]$ may be treated as K_s to give

$$[Ca^{+2}][CO_3^{-2}] = K_sK = K_{sp} = 5 \times 10^{-9} \text{ at } 25°C \tag{1.13}$$

The constant, K_{sp}, is termed the **solubility product constant**.

If the product of the ionic molar concentrations is less than the solubility product constant, K_{sp}, the solution is undersaturated. Conversely, if the product is greater than the solubility product, K_{sp}, the solution is supersaturated. For this case, crystals form and precipitation progresses until the ionic concentrations are reduced to those of a saturated solution. From Eq. (1.13), the solubility of calcium carbonate is about $7 \, mg/\ell$.

Ways to Shift Equilibria   Chemical reactions in water or wastewater treatment rely on shifting equilibria to bring about desired changes or desired results. The most common ways to shift equilibria are the production of insoluble substances, weakly ionized compounds, or gaseous end products, and oxidation and reduction reactions.

An example of production of an insoluble substance is in water softening, where slaked lime, $Ca(OH)_2$, is added to remove Ca^{+2}.

$$Ca^{+2} + 2HCO_3^- + Ca(OH)_2 \rightleftharpoons 2CaCO_3 \downarrow + 2CO_2 \qquad \textbf{(1.14)}$$

Other reactions occur in softening, but this one illustrates the production of an insoluble compound. The solubility product constant of the $CaCO_3$ is 5×10^{-9} at 25°C.

An example of production of a poorly ionized compound is the neutralization of acid or caustic wastes. Many textile mill wastewaters have NaOH present and are neutralized by using H_2SO_4 as follows:

$$2H^+ + SO_4^{-2} + 2Na^+ + 2OH^- \rightarrow 2H_2O + 2Na^+ + SO_4^{-2} \qquad \textbf{(1.15)}$$

The combination of $[H^+]$ and $[OH^-]$ produces water that is poorly ionized.

Production of a gaseous product is illustrated by adding chlorine at the breakpoint as follows:

$$3Cl_2 + 2NH_3 \rightarrow N_2 \uparrow + 6H^+ + 6Cl^- \qquad \textbf{(1.16)}$$

Here, the gaseous nitrogen produced, N_2, causes the reaction to go to completion.

An example of using oxidation-reduction reactions to complete the reaction is the adding of chlorine to cyanide ion, CN^-, in electroplating wastes, to give

$$5Cl_2 + 2CN^- + 8OH^- \rightarrow 10Cl^- + 2CO_2 + N_2 \uparrow + 4H_2O \qquad \textbf{(1.17)}$$

This reaction goes to completion because chlorine is reduced while the cyanide ion is oxidized to CO_2 and H_2O. Also, this reaction goes to completion because a gaseous product, N_2, is formed.

pH and Alkalinity Alkalinity is a measure of the capacity of a water to neutralize acids — that is, to absorb hydrogen ions without significant pH changes. It is determined in the laboratory by titrating the water with a standardized sulfuric acid solution, usually 0.02N H_2SO_4. The three chemical forms that cause alkalinity in natural waters are bicarbonate, carbonate, and hydroxyl ions: HCO_3^-, CO_3^{-2}, and OH^-. The CO_3^{-2}, HCO_3^-, and OH^- are related to the CO_2 and H^+ ion concentration (or pH) as shown by the following equilibrium reactions (Sawyer *et al.*, 1994),

$$CO_2 + H_2O \rightleftharpoons H_2CO_3 \rightleftharpoons HCO_3^- + H^+ \qquad \textbf{(1.18)}$$

$$M(HCO_3)_2 \rightleftharpoons M^{+2} + 2HCO_3^- \qquad \textbf{(1.19)}$$

$$HCO_3^- \rightleftharpoons CO_3^{-2} + H^+ \qquad \textbf{(1.20)}$$

$$CO_3^{-2} + H_2O \rightleftharpoons HCO_3^- + OH^- \qquad \textbf{(1.21)}$$

From these equations, it can be seen that CO_2 and the three forms of alkalinity are all part of one system that exists in equilibrium since all equations have HCO_3^-. A change in the concentration of any one member of the system will cause a shift in the equilibrium, cause a change in concentration of the other ions, and result in a change in pH. Conversely, a change in pH will shift the equilibrium and result in a change in concentration of the other ions. Figure 4.6 shows the relationship between CO_2 and the three forms of alkalinity in a water with a total alkalinity of $100\,mg/\ell$ as $CaCO_3$ over the pH range commonly found in water and wastewater engineering. As can be seen, for most natural waters the pH is from 5 to 8.5; thus the HCO_3^- content represents the major portion of the alkalinity.

Substances that offer resistance to a pH change are referred to as **buffers**. Since the pH range for natural waters and wastewaters is from about 5 to 8.5, the main buffering system is the bicarbonate-carbonate system. If an acid is added, a portion of the H^+ ions combine with HCO_3^- to form un-ionized H_2CO_3, and only the H^+ remaining free affect the pH. If a base is added, then some of the OH^- ions react with HCO_3^- to form CO_3^{-2}, and only the OH^- remaining free affect the pH. Both chemical reactions and biological processes depend upon the natural buffering action to control pH changes. In coagulation, when the alkalinity naturally present is insufficient, calcium hydroxide, $Ca(OH)_2$, and sodium carbonate, Na_2CO_3, can be added to furnish alkalinity to react with the coagulant.

PHYSICAL CHEMISTRY

The main areas of physical chemistry that are utilized in water and wastewater engineering are concerned with chemical kinetics, gas laws, and colloidal dispersions.

Chemical Kinetics

Chemical kinetics is concerned with the rate at which reactions occur. The types of reactions discussed in this section are irreversible reactions occurring in one phase wherein the reactants are dispersed uniformly throughout the liquid. In irreversible reactions, the stoichiometric combination of the reactants leads to almost complete conversion to products. Most reactions in water and wastewater engineering are sufficiently irreversible to allow this assumption for the kinetic interpretation of experimental data. Of the reactions discussed in this text, the first-order kinetics occur the most frequently in water and wastewater engineering.

Zero-order reactions occur at a rate that is independent of the concentration of the reactant or product, so the disappearance of the reactants is linear with time. Consider the following conversion of a single reactant into a single product.

$$A(reactant) \rightarrow B(product) \qquad (1.22)$$

If C represents the concentration of A, the rate at which A disappears with respect to time is

$$-\frac{dC}{dt} = k = -r \qquad (1.23)$$

where

$$\frac{dC}{dt} = r = \text{rate of change in the concentration of A with time, mg/\ell\text{-day or hour}}$$

k = reaction constant, mg/ℓ-day or hour

Rearranging Eq. (1.23) for integration gives

$$\int_{C_0}^{C} dC = -k \int_0^t dt \tag{1.24}$$

where

C = concentration of A at any time, t, mg/ℓ

C_0 = initial concentration of A, mg/ℓ

t = time, days or hours

Integration gives

$$C \Big]_{C_0}^{C} = -kt \tag{1.25}$$

or

$$C - C_0 = -kt \tag{1.26}$$

or

$$C = C_0 - kt \tag{1.27}$$

Equation (1.27) is a linear equation of the type $y = b + mx$, so it will plot a straight line on arithmetic graph paper, as shown in Figure 1.1. If the data plots a straight line, the reaction is zero-order and the slope of the line is equal to k.

First-order reactions occur at a rate that is proportional to the concentration of one reactant. Again consider the conversion of a single reactant to a single product.

$$A(\text{reactant}) \rightarrow P(\text{product})$$

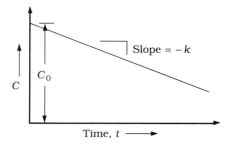

FIGURE 1.1 *Plot of Data for a Zero-Order Reaction*

If C represents the concentration of A at any time, t, the disappearance of A is represented by

$$-\frac{dC}{dt} = kC = -r \tag{1.28}$$

where

$$\frac{dC}{dt} = r = \text{rate of change in the concentration of A with time, mg/}\ell\text{-day or hour}$$

k = rate constant, day^{-1} or hour^{-1}

C = concentration of A at any time, mg/ℓ

Rearranging Eq. (1.27) for integration gives

$$\int_{C_0}^{C} \frac{dC}{C} = -k \int_{0}^{t} dt \tag{1.29}$$

which gives

$$\ln C \Big]_{C_0}^{C} = -kt \tag{1.30}$$

or

$$\ln C - \ln C_0 = -kt \tag{1.31}$$

which may be rearranged to

$$\ln C = \ln C_0 - kt \tag{1.32}$$

This is a linear equation of the type $y = b + mx$, so a plot of first-order data will give a straight line on logarithm paper, as shown in Figure 1.2. If the data plot a straight line, the reaction is first-order and the slope of the line is equal to k. Another common form for Eq. (1.32) is

$$\frac{C}{C_0} = e^{-kt} \tag{1.33}$$

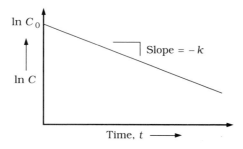

FIGURE I.2   *Plot of Data for a First-Order Reaction*

Second-order reactions proceed at a rate proportional to the second power of a single reactant being converted to a single product. Again, consider the conversion of a single reactant to a single product.

$$2A(\text{reactant}) \rightarrow P(\text{product})$$

If C represents the concentration of A at any time, t, the disappearane of A is represented by

$$-\frac{dC}{dt} = kC^2 = -r \tag{1.34}$$

where

$\dfrac{dC}{dt} = r$ = rate of change in concentration of A with time, mg/ℓ-day

k = rate constant, ℓ/mg-day

C = concentration of A at any time, mg/ℓ.

Integration of Eq. (1.34) and rearranging gives

$$\frac{1}{C} = \frac{1}{C_0} + kt \tag{1.35}$$

This equation is a linear equation of the type $y = b + mx$. Thus, a plot of second-order data will give a straight line on arithmetic paper, as shown in Figure 1.3. If the data plot a straight line as shown, the reaction is second-order and the slope of the line is equal to k.

Most chemical reactions increase with an increase in temperature, provided that the higher temperature does not affect the reactant or catalyst. The basic relationship between temperature and the reaction-rate constant of chemical reactions was derived by Arrhenius in 1889. When his original equation is modified and simplified, the common expression for temperature dependence on rate constants becomes

$$k_2 = k_1\theta^{T_2 - T_1} \tag{1.36}$$

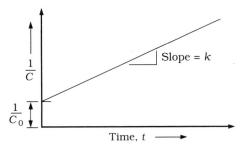

FIGURE 1.3 *Plot of Data for a Second-Order Reaction*

where

k_1 = rate constant at temperature T_1
k_2 = rate constant at temperature T_2
θ = temperature coefficient, dimensionless
T_1, T_2 = temperatures, °C

This equation allows for the value of a rate constant for a given temperature to be converted to another temperature. The temperature coefficient, θ, is relatively constant within the narrow ranges of temperatures usually occurring in water or wastewater treatment processes. A common value of θ is 1.072, which doubles the reaction rate for a 10°C temperature increase.

Gas Laws   The gas laws, particularly their influence on the solution or removal of gases from liquids, are of significance to the environmental engineer.

The **general gas law** gives the relationship between the volume, pressure, and temperature of a gas at two different conditions as follows:

$$\frac{P_1 V_1}{T_1} = \frac{P_2 V_2}{T_2} \tag{1.37}$$

where

P_1, P_2 = absolute gas pressures
V_1, V_2 = gas volumes
T_1, T_2 = absolute gas temperature, °R, °K

The general gas law is particularly useful in sludge gas computations and diffused air computation in aeration calculations.

Dalton's law states that in a mixture of gases, such as air, each gas exerts a pressure independent of the others. The partial pressure of each gas is proportional to the percent by volume of that gas in the mixture — that is, the partial pressure is equal to the pressure the gas would exert if it were the sole occupant of the volume available to the mixture.

Henry's law states that the weight of any gas that will dissolve in a given volume of a liquid, at a constant temperature, is directly proportional to the pressure the gas exerts above the liquid. In equation form, it is

$$C_s = (\text{Constant})(p_g) \tag{1.38}$$

where

C_s = concentration of a gas dissolved in the liquid at equilibrium
Constant = Henry's law constant for the gas at the given temperature
p_g = partial pressure of the gas above the liquid

A knowledge of Dalton's and Henry's laws is useful in solving problems involving gas transfer into and out of liquids. As an example, the Henry's law constant for oxygen in water at 20°C is 43.8 mg/ℓ-atm. Since air contains

20.9% by volume of oxygen, the partial pressure of oxygen in air, according to Dalton's law, is 0.209 atm when the total air pressure is 1 atm. Therefore, the equilibrium concentration of oxygen in water at 20°C and in the presence of 1 atm of air would be $43.8 \times 0.209 = 9.15 \, mg/\ell$. At elevation 2000 ft (610 m), the total atmospheric pressure is 706 mm Hg. Since 1 atm is 760 mm Hg, the total atmospheric pressure is 706/760 or 0.929 atm. Thus the oxygen solubility would be $43.8 \times 0.209 \times 0.929 = 8.50 \, mg/\ell$.

In environmental engineering, most of the problems related to the transfer of a gas into a liquid involve the addition of oxygen by aeration to maintain aerobic conditions for microorganisms. The removal of gases from liquids is also accomplished by aeration devices of one type or another. Usually the process involves gas transfer at or near atmospheric pressure from air bubbles passed through the liquid, liquid drops falling through air, or thin films of liquid flowing over surfaces exposed to air. Although Henry's law is for equilibrium, it serves to show how far a gas-liquid system is from equilibrium. The rate of solution of oxygen is proportional to the difference between the equilibrium concentration as given by Henry's law and the actual concentration in the liquid, or,

$$\frac{dC}{dt} \propto (C_s - C_a) \tag{1.39}$$

where

$\dfrac{dC}{dt}$ = rate of solution of oxygen

C_s = equilibrium concentration of oxygen

C_a = actual concentration of oxygen

This concept is the basis for engineering calculations in aerobic methods of waste treatment, such as the activated sludge process, and in the evaluation of the reaeration capacity of lakes and streams.

The removal from water of undesirable gases such as carbon dioxide and hydrogen sulfide is usually accomplished by some method of aeration. For gas removal, C_a is greater than C_s, and the rate of removal is proportional to the difference $C_a - C_s$.

Colloidal Dispersions Many of the suspended materials that water treatment plants remove from surface waters are colloids. Also, many of the suspended materials that wastewater treatment plants remove from wastewaters are colloids. Colloids are fine materials suspended in solids, liquids, and gases, but in water and wastewater engineering, the dispersions in a liquid, such as water, are most important. The colloid material may be a solid, such as clay, organic debris, or waste materials, or may be a liquid, such as oils secreted by algae in natural waters or oils present in wastewaters. Colloidal particles range in size from about 1 millimicron (mμ), or 10^{-6} mm, to 1 micron (μ), or 10^{-3} mm, a micron being 10^{-3} mm, or 10^{-6} m. They are not visible individually to the unaided or naked eye. Colloids may be defined as ultrafine dispersions or suspensions. Since colloids are so small, their surface-area-to-mass ratio is very high. A cubic centimeter of 10-$m\mu$ colloids has a surface area of $600\,m^2$, or 1/7 of an acre. As a result of the large surface area, surface phenomena predominate and control the behavior of colloidal dispersions or suspensions. The mass of colloids is so small that gravitational effects are negligible. Some of the properties of colloids that are of importance are as follows:

1. Colloids are electrically charged and may be positive or negative. Colloids such as clays, algae, and bacteria are negatively charged in the pH range that occurs in natural waters or wastewaters. Since like charges repel, the particles cannot come close enough together to flocculate or agglomerate into larger particles. A discussion of the electrostatic field around colloids is presented in the chapter on coagulation and flocculation.

2. Colloids, due to their small size, reflect light, and the measurement of turbidity of a water sample is based on this principle.

3. Colloids have tremendous surface areas and, of course, have great adsorptive powers. Adsorption is preferential, some ions being chosen while others are not.

4. Colloids, due to their large particle size, do not pass through semipermeable membranes and are removed by reverse osmosis. In reverse osmosis, a water under pressure passes through a semipermeable membrane and leaves colloidal and some ionic material behind.

ORGANIC CHEMISTRY

Organic chemistry is concerned with carbon compounds. Many of these are foods or substrates for bacteria and other microorganisms, whereas some are foods for higher forms of life such as humans. Most herbicides, pesticides, and hazardous materials are synthetic organic chemicals. All organic compounds contain carbon atoms connected together in chain or ring structures with other elements attached. The bonding between atoms is covalent bonding of outer-shell electrons (that is $C\overset{\cdot}{\underset{\cdot}{\times}}C$), and the bonding is abbreviated as C—C for purposes of simplicity. Typical chain or ring structures are

$$
\begin{array}{ccccc}
| & | & | & | & | \\
-C\!-\!C\!-\!C\!-\!C\!-\!C- \\
| & | & | & | & |
\end{array}
\quad\text{Straight-chain structure}
$$

$$
\begin{array}{ccccc}
 & & | & & \\
 & & -C- & & \\
| & | & | & | & | \\
-C\!-\!C\!-\!C\!-\!C\!-\!C- \\
| & | & | & | & | \\
 & -C- & & & \\
 & | & & &
\end{array}
\quad\text{Branched-chain structure}
$$

Cyclic structure

The major components of organic compounds are carbon, hydrogen, and oxygen, and minor elements include nitrogen, phosphorus, sulfur, and certain metals. Each carbon has four connecting bonds. Certain organics are derived from nature, such as plant fibers and animal tissues. Some organics, such as rubber, plastics, anti-freeze, and the like, are produced synthetically. Certain organics are produced from fermentation processes, such as antibiotic production. In contrast to inorganic compounds, organic compounds are usually combustible, high in molecular weight, only slightly soluble in water, and a source of food for animal life and microbial life.

The properties of organic compounds are determined by the chain length, or structure, by the degree of unsaturation or multiple bonding (C=C, C≡C) in the carbon chain, and by the groups attached to the carbon chain. Simple substitutions with hydrogen are referred to as **hydrocarbons**, other functional groups are the hydroxyl (—OH) to form **alcohols**; the aldehyde
$$-\overset{O}{\overset{\|}{C}}-H$$
to form **aldehydes**; the carboxylic (—COOH) to form **carboxylic acids**; the ketone
$$-\overset{O}{\overset{\|}{C}}-$$
to form **ketones**; the ether (—O—) to form **ethers**; and the amine (—NH_2) to form **amines**. Molecules based on simple straight chains or branched straight chains are referred to as **aliphatic**. Other groupings include **aromatic** compounds based on the benzene skeletons (C_6H_6) in a cyclic stabilized form (⬡) and heterocyclic compounds based on ring structures substituted with atoms other than carbon, commonly oxygen, nitrogen, and sulfur.

Food and foodstuffs are mainly organic compounds or organic materials, and living organisms are also mainly organic compounds or organic materials. The principal constituents of naturally synthesized organic materials are carbon (C), hydrogen (H) and also present in small amounts, less than 5%, are nitrogen (N), phosphorus (P), and sulfur (S). These five elements (termed major elements) usually amount to 99% of the chemical composition of organic materials. Some inorganic ions (termed minor elements), such as sodium (Na^+), potassium (K^+), calcium (Ca^{+2}), magnesium (Mg^{+2}), and chloride (Cl^-), are required for biological growth. Elements (termed trace elements) required in small amounts for biological growth are iron (Fe), copper (Cu), zinc (Zn), bromine (Br), and iodine (I). In municipal wastewater treatment, the most important classes of organic molecules are certain large molecules, such as carbohydrates, proteins, and lipids (fats or oils) or their degradation products.

Carbohydrates   **Carbohydrates** are composed of carbon, hydrogen, and oxygen in the molecular form or equation $C_nH_{2n}O_n$. Glucose ($C_6H_{12}O_6$) is a simple carbohydrate referred to as a **monosaccharide**. When two monosaccharides are joined together they form a **disaccharide**, a common one being sucrose, found as common sugar. Organic materials are used for a food and an energy source for microbes. These simple carbohydrates are easily decomposed in the presence of oxygen to yield water (H_2O) and carbon dioxide (CO_2) and to release energy that can be used by microbes. For example, glucose degrades as follows:

$$C_6H_{12}O_6 + 6CO_2 \rightarrow 6CO_2 + 6H_2O + \text{energy}$$

The formation of monosaccharides into disaccharides can be continued to produce **polysaccharides**, which are usually more resistant to degradation by microbes. Polysaccharides can be divided into two groups: (1) readily degradable starches found in potatoes, corn, wheat, rye, rice, and other edible plants and (2) poorly degradable materials such as cellulose found in wood, cotton, paper, and other similar plant tissues. Cellulose degrades biologically much more slowly than starches.

Proteins   **Proteins** are long strings of amino acids containing carbon, hydrogen, oxygen, nitrogen, and phosphorus. Amino acids have the amino group ($-NH_2$) in their structure, and a hydrogen from this group can be replaced to form long strings of amino acids. Protein forms an essential part of all living tissue and is necessary in the diet for all higher animals. A small section of a protein is shown below,

$$\begin{array}{ccccc} H & & O & H & CH_3 \\ | & & \| & | & | & \diagup O \\ -N-CH_2- & C- & N- & CH- & C \end{array}$$

Proteins always contain the peptide linkage

$$\begin{array}{cc} O & H \\ \| & | \\ -C- & N- \end{array}$$

In the presence of oxygen, micoorganisms break proteins down to amino acids, which are further degraded to ammonia (NH_3), carbon dioxide (CO_2), and water (H_2O).

Lipids   A **lipid** is a fat or oil that is only slightly soluble in water. Fats are solids or
(Fats and Oils)   semisolids at room temperatures, whereas oils are liquids. These compounds include a long hydrocarbon structure, which usually has a more ionic group (such as $-COOH$) at one end. The $-COOH$ is ionic in that the hydrogen

can be ionized in water to change the group to —COO^-. Most organic compounds do not readily ionize. Animal fats and oils are mainly saturated carbon bonds ($-\overset{|}{C}-\overset{|}{C}-$), whereas plant products contain some unsaturated bonds ($-\overset{|}{C}-\overset{|}{C}-$). These structures form water-soluble layers or group together to form emulsions. Because of their limited solubility, fats and oils are only slowly degraded by microbes. In the presence of oxygen, fats and oils decompose into carbon dioxide (CO_2) and water (H_2O).

Miscellaneous Compounds

Soaps are salts of long-chain carboxylic acids, such as sodium palmitate, $CH_3(CH_2)_{13}CH_2\overset{\displaystyle O}{\overset{\|}{C}}-O^-Na^+$. They are slowly degraded by microbes. Urea is an amide whose structural equation is $NH_2-\overset{\displaystyle O}{\overset{\|}{C}}-NH_2$. It is found in human urine and is slowly degraded by microbes.

Chlorinated hydrocarbons, such as endrin, are some of the most useful pesticides used to control insects in agricultural practice. Chlorinated hydrocarbons are highly toxic compounds that may appear in surface runoff from treated areas or in percolation into groundwaters. They often appear in wastewaters from manufacturing plants. Usually, chlorinated pesticides are resistant to biological degradation and may appear for months or years in soils and water, and some may tend to be concentrated in aquatic plants and animals. The chlorinated hydrocarbons used as herbicides include 2,4-D and 2,4,5-TP, which are used mainly in agricultural practices. Compounds called chloroacetamides, such as alachlor and metolachor, are also used as herbicides. Other chlorinated pesticides of significance are DDT, dieldrin, lindane, aldrin, chlorodane, toxaphene, heptachlor, methoxychlor, and DDD, which have been used as insecticides or fungicides. All chlorinated pesticides are considered to be significant because of their slow biodegradability and their toxicity to humans and life in the environment. For this reason, many of these compounds (such as DDT, dieldrin, and aldrin) are now banned in the United States, and the use of many other chlorinated pesticides has been greatly restricted.

Other pesticides used are organic compounds containing phosphorus, organic carbamate compounds, and organic s-triazine compounds. Organic compounds containing phosphorus include parathion and malathion. Parathion has been an effective insecticide against the fruit fly, but it has high toxicity to mammals. Malathion is a widely used insecticide that has low toxicity to mammals. Organic carbamate compounds include IPC, which is used as a herbicide and appears to have low toxicity to mammals. The organic s-triazines are of three types, which are chloro s-triazines, methylthio s-triazines, and methoxy s-triazines. Two of the most common chloro s-triazines are atrazine and cyanazine. Both of these are herbicides, and atriazine has been found to be very resistant to biodegradation.

**PRIORITY
POLLUTANTS**

Waterborne chemicals (both inorganic and organic) in the environment are of concern, and the Environmental Protection Agency (EPA) lists these as **priority pollutants**. Priority pollutants are selected on the basis of their known or suspected carcinogenicity, mutagenicity, teratogenicity, or high acute toxicity. These are of importance in a municipal water supply, since they must be monitored in the treated water and must be lower than the allowable level for drinking waters. They are also important in a municipal or industrial wastewater treatment plant effluent, since they must be monitored and must be below the allowable level of the plant permit.

The **toxic inorganic chemicals** include heavy metals, such as barium, cadmium, chromium, lead, and mercury, and the nonmetals, such as arsenic and selenium. All of these are very toxic to humans at very low concentrations. The **toxic organic chemicals** include such chemicals as endrin, lindane, methoxychlor, toxaphene, 2,4-D, 2,4,5-TP, benzene, para-dichlorobenzene, carbon tetrachloride, 1,2-dichloroethane, 1,1,1-trichloroethane, 1,1-dichloroethylene, trichloroethylene, vinyl chloride, toluene, and xylenes. All of these compounds are very toxic to humans at very low concentrations. Endrin, lindane, methoxychlor, and toxaphene are insecticides, whereas, 2,4-D and 2,4,5-TP are herbicides. Benzene, para-dichlorobenzene, carbon tetrachloride, 1,2-dichloroethane, 1,2-dichloroethane, 1,1,1-trichloroethane, 1,1-dichloroethylene, trichloroethylene, and vinyl chloride are synthetic organic chemicals commonly classified as **volatile organic chemicals** (VOCs). Toluene and xylenes are synthetic organic chemicals commonly classified as **synthetic organic chemicals** (SOCs). Both classes of these synthetic organic chemicals are used in a wide variety of industrial and commercial applications. It must be kept in mind that the previously mentioned chemicals are typical examples of the toxic inorganic and organic chemicals on the priority pollutant listing. The present listing has over 150 chemicals, and the listing is continuously being reviewed and updated. Thus the reader should contact the Environmental Protection Agency for a current listing.

REFERENCES

American Public Health Association (APHA). 1992. *Standard Methods for the Examination of Water and Wastewater*. 16th ed. American Public Health Association (APHA), American Water Works Association (AWWA), and Water Environment Federation (WEF).

Barnes, D.; Bliss, P. J.; Gould, B. W.; and Valletine, H. R. 1981. *Water and Wastewater Engineering Systems*. London: Pittman Press.

Brey, W. S., Jr. 1958. *Principles of Physical Chemistry*. New York: Appleton-Century-Crofts.

Hammer, M. J. 1986. *Water and Wastewater Technology*. 2nd ed. New York: Wiley.

McGauhey, P. H. 1968. *Engineering Management of Water Quality*. New York: McGraw-Hill.

McKinney, R. E. 1962. *Microbiology for Sanitary Engineers*. New York: McGraw-Hill.

O'Connor, R. 1974. *Fundamentals of Chemistry*. New York: Harper & Row.

Sawyer, C. N., and McCarty, P. L. 1978. *Chemistry for Environmental Engineering*. 4th ed. New York: McGraw-Hill.

Sawyer, C. N.; McCarty, P. L.; and Parkin, G. F. 1994. *Chemistry for Environmental Engineering*. 5th ed. New York: McGraw-Hill.

Viessman, W., and Hammer, M. J. 1993. *Water Supply and Pollution Control*. 5th ed. New York: Harper & Row.

Wertheim, E., and Jeskey, H. 1956. *Introductory Organic Chemistry*. 3rd ed. New York: McGraw-Hill.

PROBLEMS

1.1 What are the atomic weight and the equivalent weight of Na^{+1} and Ca^{+2}? If a water sample has $102 \, mg/\ell$ of Na^{+1} and $68 \, mg/\ell$ of Ca^{+2}, how many milliequivalents of each are present?

1.2 What are the molecular weight and the equivalent weight of $CaCl_2$? If a water sample contains $168 \, mg/\ell$ of $CaCl_2$, how many milliequivalents are present?

1.3 If a water sample contains 134 parts per million of Na^{+1}, how many milligrams per liter are present?

1.4 If a water has a pH of 7.6, how many moles per liter of hydrogen ion, H^+, and hydroxyl ion, OH^-, are present?

1.5 In water coagulation with alum (aluminum sulfate) and bicarbonate alkalinity, the simplified reaction is

$Al_2(SO_4)_3 \cdot 14H_2O + 6HCO_3^-$
$\rightarrow 2Al(OH)_3 \downarrow + 3SO_4^{-2} + 14H_2O + 6CO_2$

Why does this reaction go to completion?

1.6 What are the radicals that cause alkalinity?

1.7 A water has an hardness of $185 \, mg/\ell$ as $CaCO_3$. What is the hardness expressed as meq/ℓ?

1.8 A water has an alkalinity of $225 \, mg/\ell$ as $CaCO_3$. What is the alkalinity expressed as meq/ℓ?

1.9 Give the reaction order for the following rate equations

a. $\dfrac{dC}{dt} = kC^2$ **b.** $\dfrac{dC}{dt} = k$ **c.** $\dfrac{dC}{dt} = kC$

1.10 A laboratory kinetic study was done for the reaction

$$A \rightarrow Products$$

and the following data were obtained.

TIME, hour	CONCENTRATION OF A, mg/ℓ
0	135
1	98
2	71
4	37
6	19
8	10

Determine:
a. The reaction order.   **b.** The rate constant, k.

1.11 A laboratory kinetic study was done for the reaction

$$A \rightarrow Products$$

and the following data were obtained.

TIME, hour	CONCENTRATION OF A, mg/ℓ
0	135
1	53
2	33
3	24
4	19
5	15
6	13
7	11
8	10

Determine:
a. The reaction order.   **b.** The rate constant, k.

1.12   A laboratory study was done for the reaction

$$A \rightarrow Products$$

and the following data were obtained.

TIME, hours	CONCENTRATION OF A, mg/ℓ
0	135
1	119
2	104
3	88
4	73
6	41
8	10

Determine:

a. The reaction order.
b. The rate constant, k.

1.13   A water at El 1500 ft (457 m) is at 20°C. The atmospheric pressure at El 1000 ft (305 m) is 733 mm Hg and at El 2000 ft (610 m) is 706 mm Hg. Determine the saturation dissolved-oxygen concentration in mg/ℓ.

1.14   The high-purity oxygen process used in some activated sludge wastewater treatment plants uses almost pure oxygen gas as an oxygen source instead of air. If the gas used is 80% oxygen and 20% nitrogen and is at 1 atm pressure, what is the equilibrium oxygen concentration at 20°C?

1.15   An activated sludge wastewater treatment plant for a municipality has anaerobic digesters with sludge gas utilization. The anaerobic digesters produce 110,000 ft^3 (3114 m^3) of sludge gas at 10 in. (254 mm) of water column pressure (relative pressure) and 95°F (35°C). The gas is to be compressed and stored in a gas dome and then used to fuel internal combustion engines that drive generators and produce electricity for the plant. If the compressed gas is at 30 psig (207 kPa gage) pressure and 95°F (35°C), how many ft^3 (m^3) will it occupy?

1.16   An activated sludge wastewater treatment plant for a municipality has anaerobic digesters with sludge gas utilization. The anaerobic digesters produce 162,000 ft^3/day (4590 m^3/day) of sludge gas at 10 in. (254 mm) of water column pressure (relative pressure) and 95°F (35°C). The gas is to be compressed and stored in a gas dome and used to fuel internal combustion engines that drive generators and produce electricity for the plant. If the compressed gas is at 35 psig (241 kPa gage) pressure and 95°F (35°C), how many ft^3/day (m^3/day) will it occupy?

1.17   A stream has a saturation dissolved oxygen content of 9.0 mg/ℓ and an actual dissolved oxygen content of 4.0 mg/ℓ. If the actual dissolved oxygen was 6.0 mg/ℓ, what is the ratio of the two driving forces for reoxygenation?

2
BIOLOGICAL CONCEPTS

Numerous microorganisms are important in the field of water and wastewater treatment. Thus, an introductory survey of microorganisms, such as bacteria, protozoa, fungi, algae and viruses, is presented in this chapter.

In the treatment of decomposable organic wastewaters, microorganisms — that is, microbes — play an important role, and most of the species found in water and wastewater are harmless to humans. However, a number of microorganisms are pathogenic and are responsible for a variety of waterborne diseases, and their presence poses a health problem. Microbes are too small to be seen by the unaided human eye, and the viruses, bacteria, protozoa, fungi, and algae are in this category. With higher forms of life, it is convenient to classify organisms as plants or animals. Plants have rigid cell walls, are photosynthetic, and are nonmotile — that is, they cannot move independently. Animals have flexible cell walls, require organic food materials, and are capable of self-movement. The application of such standards of differentiation to microorganisms is difficult because of the simple structures of their cells. Based on cell structure, microorganisms are divided into two classes. (1) **Prokaryotes** are small, simple, one-cell structures, <5 microns (μ), with a primitive nuclear area consisting of one chromosome. Reproduction is normally by binary fission: a simple division of the parent cell into two daughter cells. All bacteria, both single-celled and multicellular, are prokaryotes, as are blue-green algae. (2) **Eukaryotes** are larger cells, >20 microns (μ), with a more complex structure, and each cell contains a distinct membrane-bound nucleus with many chromosomes. They may be single-celled or multicellular, reproduction may be asexual or sexual, and quite complex life cycles may be found. This class of microorganisms includes fungi, algae (except blue-green), and protozoa. A further group of microbes, the viruses, do not readily fit into either of the previous classes of cells and are thus considered separately.

BACTERIA

Single-celled bacteria may be classified by the shape of their cells as shown in Figure 2.1. A **bacillus** is rod-shaped, a **coccus** is spherical in shape, and a **spirillum** is spiral-shaped. They may occur as single cells, as clusters of cells, or in chains made up of cells, although some other groupings may occur. Cell shape and cell grouping are characteristics of each individual species. For example, *E. coli* cells are rod-shaped and occur as single cells. Rod-shaped cells are usually from about 1.0 micron (μ) in diameter and about 3 to 5μ long. Spherical cells are about 0.2 to 2μ in diameter. Spiral-shaped

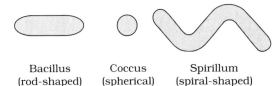

FIGURE 2.1
*Shapes of Single-
Celled Bacteria*

Bacillus
(rod-shaped)

Coccus
(spherical)

Spirillum
(spiral-shaped)

cells are about 0.3 to 5μ in diameter and 6 to 15μ long. Single-celled bacteria are microscopic organisms, usually nonphotosynthetic, that normally multiply by binary fission. In binary fission, the parent cell divides and breaks apart into two daughter cells as shown in Figure 2.2. The division time, also called the **fission** or **generation time**, varies with the species and environmental conditions, but usually it is about 20 minutes. Bacterial cells must have their food in soluble form, and they may be capable of movement (motile) or incapable of movement (nonmotile), depending on the species. Figure 2.3 shows a schematic drawing of a rod-shaped bacterial cell. The cytoplasm consists of a colloidal suspension of carbohydrates, proteins, and other complex organic compounds. Biochemical reactions, such as the synthesis of proteins, essential to the life processes of the cell take place in the cytoplasm. The nuclear area is well defined in some species and poorly defined in other species. The nuclear area is responsible for the hereditary characteristics of the cell that control reproduction of cell components. The cell membrane controls the flow of material into and out of the cell. The cell wall, which is semirigid, protects the cell interior and also retains the cytoplasm in a definite volume and shape. The slime layer or capsule protects the cell from drying if it is placed in a dry environment and also serves as a reserve food. The flagella are hair-like organelles for movement and are present only in cells that are motile. Enzymes (not shown in Figure 2.3) are organic catalysts that catalyze necessary biochemical reactions. Some enzymes are within the cell (endocellular), whereas others are secreted

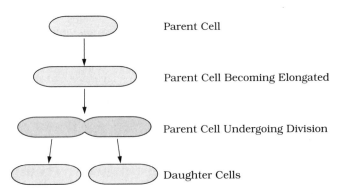

Parent Cell

Parent Cell Becoming Elongated

Parent Cell Undergoing Division

Daughter Cells

FIGURE 2.2 *Reproduction of a Rod-Shaped Bacterial Cell by Binary Fission*

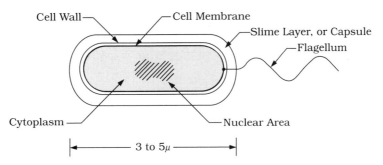

FIGURE 2.3 *Schematic Drawing of a Rod-Shaped Bacterial Cell*

to the outside of the cell (exocellular). Exocellular enzymes break down large molecules, such as proteins and starches, into smaller ones that can pass through the cell wall and cell membrane. When bacterial cells die, they lyse or break apart, thus releasing their cell contents.

Most of the single-celled bacteria used in wastewater treatment are soil microorganisms, and very few are of enteric origin. The majority of the bacterial species, with the exception of some groups such as the nitrifying bacteria, are saprophytic heterotrophs because they require nonliving pre-formed organic materials as substrates. The nitrifying bacteria, which convert the ammonium ion to nitrite and the nitrite ion to nitrate, are autotrophs because they use carbon dioxide as their carbon source instead of pre-formed organic substances. The microbes present consist of both aerobes and facultative anaerobes, since they use free molecular oxygen in their respiration process.

Growth Phases   The growth of microorganisms is primarily related to the number of viable cells present and the amount of substrate or limiting nutrient present, in addition to other environmental factors. Many of the concepts involved in the continuous growth of microbes, as in the activated sludge process, are illustrated by growth relationships of batch cultures.

First, consider the case of a single species of bacteria inoculated in (that is, added to) a medium containing a substrate and all substances required for growth. The growth that will occur is shown in Figures 2.4 and 2.5. In Figure 2.4 the growth is measured by the number of viable cells, N. There are distinct phases to this curve, which are usually designated as (1) lag, (2) log growth, (3) declining growth, (4) maximum stationary, (5) increasing death, and (6) log death. Frobisher *et al.*, (1974) describe the following physiological changes that occur in the various phases.

In the beginning of the **lag phase**, the microbes are becoming adjusted to their new environment. Although this phase is not completely understood, it is known that the organisms are recovering from transplant injuries and are absorbing water and substrate and secreting exocellular enzymes to

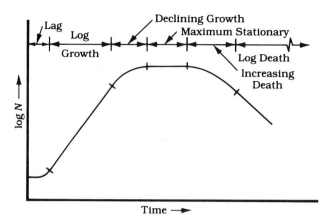

FIGURE 2.4 *Microbial Growth Phases Based on the Number of Viable Cells,* **N**

break down large substrate molecules. There is no increase in the number of viable cells at the start of the lag phase; however, toward the end of the phase, metabolic activity begins and the cells begin to increase. During the latter part of the lag phase, the division time gradually decreases, cell size diminishes, and the fission rate reaches a maximum determined by the species and growth conditions. The duration of the lag phase is greatly dependent upon the age of the inoculum culture and the amount of inoculum. If the parent culture is young and biologically active, the lag phase will be extremely short. Also, if the inoculum is relatively large, the lag phase will be minimized. A large inoculum will introduce minute but effective amounts of enzymes, certain essential nutrients, and growth factors from the parent culture, or it will quickly synthesize or release these due to the large number of cells in the inoculum.

During the **log growth phase**, the rate of fission is the maximum possible and the average size of the cells is at its minimum for the species. The cell wall and membranes are the thinnest during this phase, and the metabolic activities are at the maximum rate. The cells are physiologically young, biologically active, and more vulnerable to deleterious influences than mature, less active, cells. It is a phase of exponential growth, and the log of the number of viable cells versus time is a straight-line relationship. The fission time or generation time depends upon the species and the nutrient and environmental conditions and may vary from a few minutes to as much as several days. Usually, however, it is from 10 to 60 minutes.

Toward the end of the log growth phase, the cells begin to encounter difficulties such as the depletion of the substrate or an essential nutrient and the accumulation of toxic end products, which may have reached an inhibitory level. This is the beginning of the **declining growth phase**, in which the rate of fission begins to decline and the microorganisms die in increasing numbers so that the increase in the number of viable cells is at a slower rate. In most cases in wastewater treatment, the declining growth phase is due to a depletion of substrate and not due to an accumulation of toxic end products.

Eventually, the number of cells dying equals the number of cells being produced, and the culture is in the **maximum stationary phase**, in which the population of viable cells is at a relatively constant value. The time required to reach this phase depends primarily on the species, the concentration of the microbes, the composition of the medium, and the temperature. As the environment becomes more and more adverse to microbial growth, the **increasing death phase** begins, in which the cells reproduce more slowly and the rate at which the cells die exceeds the growth rate. Finally, the increasing death phase progresses into the **log death phase**.

If the cell growth is measured as the total mass of viable cells produced instead of the number of viable cells, the growth curve will be as shown in Figure 2.5. The slope of a tangent to the curve at any time represents the rate of cell production based upon mass. Immediately after inoculation, the cells begin to absorb water and substrate, and the cell mass starts to increase. Thus the **log growth phase** based on mass includes both the lag and log phase based on viable cell numbers. As the cells begin to multiply, the log growth phase progresses, and the rate of cell mass production increases.

When the substrate or an essential nutrient or other factor becomes limiting, or if there is an inhibitory level of accumulated toxic end products, the rate of cell mass production begins to decrease. This condition represents the end of the log growth phase and the beginning of the **declining growth phase**. As environmental conditions become more unfavorable for cell growth, the decrease in growth becomes more pronounced. Finally, when the rate at which the cell mass is being produced equals the rate at which the mass is decreasing, the curve reaches a maximum value. This represents the

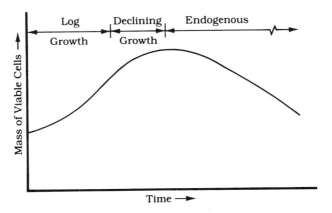

FIGURE 2.5 *Microbial Growth Phases Based on the Mass of Viable Cells*

end of the declining growth phase and the beginning of the endogenous phase.

In the **endogenous phase**, the microbes utilize as food their own stored food materials and protoplasm in addition to a portion of the dead cells in the environment. The net effect of the endogenous phase is a decrease in total cell mass with respect to time. The rate at which the cell mass is endogenously decayed is a relatively constant value of about 5 to 20% per day. The endogenous degradation rate is continuous; it exists in the log growth and declining growth phases but is masked by the much larger rate of growth (Wilner *et al.*, 1954).

Nutrition and Respiration   Microbes may be classified according to the source of their energy and carbon requirements (Frobisher *et al.*, 1974). Microbes that chemically oxidize exogenic inorganic substances for their energy needs are termed **chemolithotrophs**, whereas those that oxidize exogenic organic substances are termed **chemoorganotrophs**. Those that use light as an energy source but require some inorganic substances for respiration are termed **photolithotrophs**, and those using light but requiring some organic substances for respiration are **photoorganotrophs**. Microbes that use inorganic substances as their carbon source, such as carbon dioxide, are termed **autotrophs**, and those using organic substances as a carbon source are **heterotrophs**. Most microbes used in the activated sludge process use organic materials for both energy (that is, respiration) and cell synthesis; thus they are chemoorganotrophic and are heterotrophs. The nitrifying bacteria, *Nitrosomonas* and *Nitrobacter*, use the chemical oxidation of an inorganic nitrogen compound for their energy source and use carbon dioxide for synthesis; thus they are chemolithotrophic and are autotrophs.

The essential elements and substances required for nutrition are classified as (1) major elements, (2) minor elements, (3) trace elements, and (4) growth factors. The major elements are carbon, hydrogen, oxygen, nitrogen, and phosphorus. The minor elements are principally sulfur, potassium, sodium, magnesium, calcium, and chlorine. The trace elements mainly consist of iron, manganese, cobalt, copper, boron, zinc, molybdenum, and aluminum. The growth factors are essential substances required in trace amounts that the cells themselves cannot synthesize. These principally include vitamins, essential amino acids, and precursors for the synthesis of essential amino acids or other required synthesized compounds. It should be understood that not all the minor elements, trace elements, and growth factors previously listed are required for all microbes because the requirements vary for each species.

In municipal wastewaters, all the essential elements and growth factors are present in adequate amounts if no significant industrial wastewaters enter the collection system. In certain industrial wastewaters, there may be a deficiency of nitrogen and phosphorus, and these must be added so that the $N : BOD_5$ ratio is at least $1 : 32$ and the $P : BOD_5$ ratio is at least $1 : 150$ (Sawyer, 1956). For industrial wastewaters, the minor elements, trace elements, and growth factors are usually present in sufficient amounts in the carriage water.

Microorganisms may be classified according to their respiration requirements as **aerobes**, **anaerobes**, and **facultatlve anaerobes** (Frobisher, et al. 1974). Respiration produces the energy for their life processes and essentially consists of enzymatically removing hydrogen atoms from a hydrogen donor substance, which, as a result, is oxidized. The hydrogen atoms are united with a hydrogen acceptor, which is constantly reduced. In aerobic respiration, the hydrogen acceptor is molecular oxygen and the end product is water. In anaerobic respiration, the hydrogen acceptor may be combined oxygen in the form of radicals, such as the carbonate, nitrate, and sulfate ions, or it may be an organic compound. The respective end products for these acceptors are methane, ammonia, hydrogen sulfide, or a reduced organic compound. Facultative anaerobes use aerobic respiration if molecular oxygen is in the environment; if it is absent, they employ anaerobic respiration. In the activated sludge process, the microbes consist of both aerobes and facultative anaerobes.

Environmental Factors Affecting Microbial Activity

Environmental factors may be broadly classified as physical, chemical, and biological according to their nature. The effects of environmental factors upon microbial activity are an important consideration, because (1) it is desirable to maintain a biological treatment culture at its optimum activity, (2) environmental factors are important in evaluating the feasibility of treating certain organic wastewaters by a biological treatment process, and (3) environmental factors are important in disinfecting drinking waters and effluents from biological treatment processes.

Some of the major physical factors affecting microbial activity are (1) temperature, (2) osmotic pressure, (3) the presence of molecular oxygen, (4) the presence of liquid water, and (5) ultraviolet radiation (uv). Since the temperature within a microbial cell is virtually equal to the temperature of the environment, an increase in temperature increases the microbial activity up to that point where the cells are killed. A temperature increase of 10°C approximately doubles the microbial activity. Microbes may be classified according to their optimum temperature range as psychrophils, mesophils, and thermophils, which have the respective optimum temperature ranges of 0 to 10°C, 10 to 45°C, and 45 to 75°C. Above 75°C, microbes are rapidly killed if the contact time is sufficient. The majority of microbes utilized in the activated sludge process are psychrophils and mesophils, although there are always some thermophils present. The osmotic pressure, which is dependent upon the salt concentration in the environment, must be within a certain range, because microbes feed by osmosis. Most microbes are not affected by the salt content if it is between 500 to 35,000 mg/ℓ. Hydrostatic pressure within the range encountered in the activated sludge process does not affect the microbial activity. Molecular or dissolved oxygen must be present for aerobes and for facultative anaerobes when using aerobic respiration. Usually, a concentration of about 2.0 mg/ℓ is used as a design value for an aerobic biological reactor. Microorganisms must have liquid water available for their feeding, and in the activated sludge process this is always the case. However, oxidation or holding ponds for seasonal wastewaters, such as cannery wastes, may freeze throughout their depth in the winter in severe climates, and if this occurs, microbial action ceases. Ultraviolet light is toxic to microorganisms at low intensities. In lakes and reservoirs, ultraviolet light from the sun assists in killing microbes, but it has negligible effects on the activated sludge process. High-intensity ultraviolet lights have been used to disinfect drinking waters on passenger ships and effluents from municipal wastewater treatment plants.

The major chemical factors affecting microbial activity are (1) pH, (2) the presence of certain acids and bases, (3) the presence of oxidizing and reducing agents, (4) the presence of heavy metal salts and ions, and (5) the presence of certain chemicals. The microbes utilized in the activated sludge process, like the majority of all microbes, thrive best at a neutral pH range of 6.5 to 9.0. Since carbon dioxide is one of the end products from aerobic bio-oxidation, the carbonate-bicarbonate buffering system will be established in the biological reactor, and this will assist in maintaining a neutral pH. However, certain industrial wastewaters may have such a low or high pH that neutralization is required prior to treatment.

Certain acids, such as benzoic acid, and certain bases, such as ammonium hydroxide, are toxic to microbes if present in sufficient concentrations; and at high concentrations all acids and bases are toxic. Strong oxidizing or reducing reagents are toxic to microorganisms at relatively low concentrations. All of the halogens (chlorine, fluorine, bromine, and iodine) and their salts are very toxic if the halogen has a valence above its lowest valence state because, in this form, it is a strong oxidizing agent.

Heavy metal salts and heavy metal ions are toxic in relatively low concentrations. The toxicity of a metallic ion, in general, increases with an increase in the atomic weight. The heavy metals that may be encountered in wastewaters are usually mercury, arsenic, lead, chromium, zinc, cadmium, copper, barium, and nickel. Silver sometimes is found because it is used in photographic and silver plating operations, but as a consequence of its value, it is usually recovered. Generally, mercury, arsenic, and lead are the most toxic of the heavy metals.

Certain industrial chemicals such as organic acids, alcohols, ethers, aldehydes, phenol, chlorophenol, cresols, and dyes are toxic if present in sufficient concentrations. Soaps and detergents are toxic in relatively high concentrations; however, in the concentrations usually found in wastewaters, they are usually not toxic. Antibiotics produced by pharmaceutical fermentations are toxic in minute concentrations. Greases, if present in significant amounts, will coat the microbes in activated sludge and interfere with their aerobic respiration. Cyanide compounds and the cyanide ion (CN^{-1}) are toxic at very low concentrations. Certain ammonia compounds and the ammonium ion (NH_4^{+1}) can be toxic at moderate concentrations.

The effect of all the previously discussed chemical agents and compounds is primarily a function of the concentration, temperature, and contact time.

As the contact time increases, and usually as the temperature increases, the relative toxicity increases. At a constant contact time and temperature, the effect of concentration upon microbial activity is as depicted below:

Concentration

$\longrightarrow$

No Effect	Stimulates	Inhibits	Kills

At relatively low concentrations there is no effect; however, as the concentration increases, the chemical agent or compound stimulates microbial growth. A further increase in concentration causes inhibition, and an increase beyond this range results in a killing effect or toxicity to the microorganisms. For example, phenol in sufficient concentrations is a good disinfectant. However, if a phenolic wastewater has a phenol concentration less than about 500 mg/ℓ, it readily stimulates microbial activity and is bio-oxidized by the activated sludge process, providing it is an acclimated culture. Also, petrochemical wastewaters containing organic compounds such as organic acids, alcohols, ethers, and aldehydes are readily bio-oxidized if the concentrations are dilute. Even pharmaceutical wastes from the production of antibiotics may be treated by the activated sludge process if the concentrations are extremely dilute.

Biological factors affecting microbial activity, such as in the activated sludge process, are usually a result of undesirable microorganisms in the mixed culture. For example, excessive growth of the filamentous genus *Sphaerotilis* creates a sludge having poor settling characteristics, making proper final clarification difficult. Figure 2.6(a) and 2.6(b) shows photographs made through a light microscope of two activated sludge cultures. Figure 2.6(a) shows a normal activated sludge having agglomerated floc made up of mostly single-celled bacteria. The agglomerating substance is gummy material from the attrition of slime layers from bacterial cells. Figure 2.6(b) shows a poorly settling activated sludge culture having wiry, filamentous, multicellular growths which prevent the sludge from settling and compacting properly in the final clarifier.

The waterborne pathogenic bacteria are single-celled bacteria. The previously discussed principles concerning bacteria also apply to the pathogenic bacteria. When municipal wastewater is treated by biological processes, the pathogens rapidly die, because they cannot compete with the soil microbes for food materials in the existing environment. The pathogens are accustomed to 98°F (35°C) or body temperature, to the presence of bile salts, some blood, and other environmental factors. Although they rapidly die, some will be present in the effluent; thus, disinfection is usually provided.

Actinomycetes   Actinomycetes are multicellular, filamentous, rod-shaped bacteria. Nearly all are aerobic and are prokaryotes. They occur widely in soil and water, and their end products can cause taste and odor problems in water supplies.

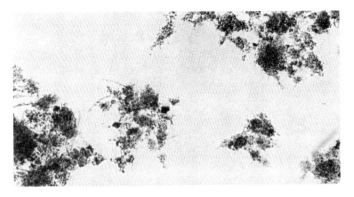

(a) *Photomicrograph of an Activated Sludge Culture with Excellent Settling Characteristics (100×). Note agglomerated floc.*

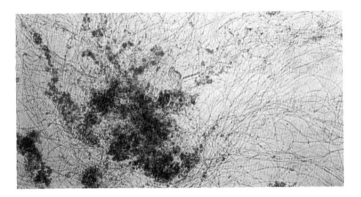

(b) *Photomicrograph of an Activated Sludge Culture with Extremely Poor Settling Characteristics (430×). Note wiry, filamentous growths.*

FIGURE 2.6 *Photographs Made Through a Light Microscope of an Activated Sludge Culture*

OTHER MICROBES AND SOME LARGER ORGANISMS

Protozoa   Protozoa are single-celled microbes without cell walls and are eukaryotes. Most are aerobic chemoheterotrophs and are usually classified on the basis of their motility. Those of concern in water and wastewater treatment include amoebas, flagellates, and free-swimming and stalked ciliates. In aerobic biological treatment processes, such as the activated sludge process, protozoa become numerous. They are useful in obtaining a clear effluent since they feed mainly on single-celled bacteria. The single-celled bacteria are essential to utilize organic wastes as food materials, but if they are not agglomerated into floc particles, they cause turbidity. Several protozoa are pathogenic and can cause waterborne disease. *Endamoeba histolytica* causes amoebic dysentery, and *Giardia lamblia* causes giardiasis.

Fungi   Fungi are aerobic, multicellular, nonphotosynthetic chemoheterotrophs, and they are eukaryotes. These microbes use organic matter as a substrate and reproduce by asexual spores. They have metabolic requirements similar to those of bacteria, but they require less nitrogen and can grow at lower pH values. Fungi are larger than bacteria but have poor settling characteristics. They are important in composting and are significant in some fixed-film processes, such as rotating biological contactors. When relatively large segments of biomass shear off the contactors, they can be settled. In suspended culture processes, such as the activated sludge process, fungi are unwanted because their presence causes a sludge to have poor settling characteristics.

Algae   Algae are single-celled and multicellular photosynthetic microbes that are commonly classified by their color as green, yellow to golden brown, greenish tan to golden brown, blue green, brown (marine), or red (marine). By photosynthesis, algae convert CO_2, H_2O, and nutrients into dissolved oxygen, O_2, other end products, and new algal cells. Algae require adequate amounts of nitrogen and phosphorus to grow effectively, and these are usually present as nitrates and phosphates. If these nutrients are available in streams and lakes, large amounts of algae, known as algal blooms, can develop. These blooms can easily be seen by the unaided human eye. This process, by which nutrients can accumulate in a water body in unwanted growths of algae, is termed **eutrophication**. The eutrophication of some surface waters has become a major problem in recent years because large amounts of nutrients enter mainly as a result of agricultural fertilizers and the discharge of municipal effluents. In water treatment, algae present in a water supply source, such as a river, lake, or reservoir, can be a problem since some algae produce oily end products that can produce tastes and odors. In wastewater treatment, algae are important in oxidation or stabilization ponds, where the oxygen they produce provides part of that required by the bacteria in stabilizing the organic wastes.

Viruses   Viruses are smaller than bacteria. They reproduce only if they can colonize in a living cell, and, having done so, they multiply very rapidly. Viruses are responsible for several human waterborne diseases, such as infectious hepatitis and poliomyelitis, thus there is concern to minimize their transmission by water or wastewater. The determination of the types and numbers of viruses is a difficult procedure that is normally not done on a routine basis.

Other Organisms In addition to microorganisms, more complex larger organisms — many microscopic and many visible to the unaided human eye — are found in natural waters. These include rotifers and crustaceans, which are normally found in relatively clean waters with significant dissolved oxygen and low pollutant levels. Rotifers are multicellular animals with flexuous bodies and hairlike cilia on the head to rake in food materials and provide motility. Rotifers feed on bacteria and small organic particles and sometimes are found in activated sludges. Crustaceans are hard-shelled multicellular animals, such as the *Daphnia* and *Cyclops*, which feed on bacteria and algae. Both rotifers and crustaceans are important food supplies for fish. Worms and insect larvae are normal inhabitants in trickling-filter biological slimes treating municipal wastewaters. Nematode worms have been found in activated sludges, where they feed on other microorganisms and on small particles of organic matter not used by other microorganisms. Two common organisms used in stream pollution studies are the worm *Turbifix* and the midge fly larva. *Turbifix* is found in polluted streams, and the midge fly larvae are found as stream recovery begins.

PATHOGENS AND WATERBORNE COMMUNICABLE DISEASES

A communicable disease is one that is readily spread from person to person. Certain bacteria, protozoa, and viruses may cause a waterborne communicable disease — that is, a disease transmitted by improperly treated drinking water. In the United States, the most common waterborne diseases that can occur and their causative agents are as follows (AWWA, 1990): (1) Bacterial pathogens: typhoid fever, *Salmonella typhi*; paratyphoid fever, *Salmonella paratyphi*; bacillary dysentery, *Shigella* genus; and salmonellosis, *Salmonella* genus. (2) Protozoa pathogens: Amoebic dysentery, *Endamoeba histolytica*, and giardiasis, *Giardia lamblia*. (3) Virus pathogens: poliomyelitis, polioviruses, and infectious hepatitis, Hepatitis A virus. For a more complete listing of all waterborne diseases that have occurred in the United States, the reader is referred to the publication *Water Quality and Treatment* (AWWA, 1990). It should be kept in mind that not all waterborne communicable diseases are spread by water alone. Most of these diseases can also be spread by other means, such as contaminated milk and foods.

BACTERIAL ANALYSES

Some microorganisms are **pathogenic** — that is, they are capable of producing diseases in humans — but most species are nonpathogenic. It is important to ensure that drinking water supplies (and, to a lesser extent, wastewater effluents) are free of pathogenic microbes. Unfortunately, there are numerous pathogenic species and each requires a specific, difficult analysis, so determination of their numbers is impractical on a routine basis. Routine microbial analysis of waters and wastewaters depends upon detecting indicator organisms, such as the coliforms. Coliforms are species that exist in large numbers in the intestines of humans and other warm-blooded animals, so they are present in large numbers in municipal wastewaters. The coliforms mainly consist of *Escherichia coli (E. coli)* and *Aerobacter aerogenes (A.*

aerogenes). These species are normally not considered pathogenic in waters; however, a strain of *E. coli* has caused a waterborne disease, a gastroenteritis. The detection of coliforms suggests that human fecal contamination of the water has occurred and that enteric pathogens (those of human intestinal origin) could be present. *E. coli* exists in large numbers in the human intestines, and it is present in such high concentrations that, even after dilution, routine analysis is possible. Unfortunately, *E. coli* is present in the intestines of all warm-blooded animals; it is, therefore, found to a lesser extent in soil. Also, although *A. aerogenes* may occur in the intestines of all warm-blooded animals, it is most commonly found in soil. Thus a surface water sample may have moderate numbers of coliforms due to animals living in the watershed or due to runoff over soil. Their presence in large numbers in a raw water sample is indicative of human contamination.

The presence of coliforms is detected by growth of visible colonies on plates of a suitable solid medium or from the fermentation of a lactose broth or lauryl tryptose broth in fermentation tubes.

Membrane Filter

As bacteria multiply on a solid medium, their accumulated mass increases until a colony is produced large enough to be seen by the unaided eye. A special filter, the millipore filter, can be used to filter a known volume of water, and its pores are small enough to filter out bacteria. The volume of water used is usually 100 mℓ and is passed through a sterile membrane filter that traps the bacteria. The filter is transferred to an agar plate, which consists of an absorbent pad saturated with a gelatin-like solid medium with a special dye in the bottom of a shallow glass dish. The glass cover is placed on the dish, and the plate is incubated for 24 hours at 35°C (98°F). After incubation the cover is removed, and the visible coliform colonies can be counted since the dye present causes the coliforms to be pink or purple or to have a green metallic sheen. Noncoliform colonies will be clear. Then, because the number of coliform colonies and the volume of sample used are known, it is possible to determine the number of coliforms per 100 ml of water sample. This estimate of the number of coliforms per 100 ml of sample is referred to as the most probable number (MPN).

Multiple-Tube Fermentation

Coliforms have the ability to ferment — that is, to use lactose sugar as a substrate and produce carbon dioxide gas — within 48 hours. Only a few other, rare species have this ability. A series of diluted samples can be added to a series of tubes containing lactose broth or lauryl tryptose broth. Each tube has a small inverted tube that is also filled with the liquid medium. Usually three dilutions are used. Each dilution uses three or five tubes. The inoculated tubes are incubated for 48 hours at 35°C (98°F) and the presence of gas (carbon dioxide) in the inverted entrapment tube characterizes a positive tube. Knowing the number of positive tubes, one can use statistical tables to estimate the number of coliforms per 100 mℓ in the original water sample. This statistically derived number of coliforms per 100 mℓ is referred to as the MPN. Since there are several other species that can ferment lactose and produce carbon dioxide, this test is preliminary, and the MPN is referred to as the presumptive test MPN. The positive tubes can be used to inoculate lactose broth tubes containing a green-colored dye and bile salts. These tubes can be incubated for 48 hours at 35°C (98°F). The dye and the bile salts are toxic to noncoliforms, so positive tubes can be assumed to be due to coliforms. From the number of positive tubes, the confirmed MPN can be obtained.

Fecal Coliform Test

This test is more specific for *E. coli* than the multiple-tube method. For it, lactose tubes are inoculated from positive presumptive tubes. The tubes are incubated for 24 hours ± 2 hours at 44.5°C (112°F). Positive tubes exhibit the production of gas. Then, once the number of positive tubes is known, statistical tables can be used to get a fecal MPN. Since this test is more specific for *E. coli*, and *E. coli* is frequently of human intestinal origin, this test is more indicative of fecal origin.

Microscopic Examination

Sometimes, operators of activated sludge treatment plants examine a drop of activated sludge under the light microscope, which magnifies from about 200 to 1400 times. This is mainly for the detection of undesirable filamentous forms or for the presence of protozoa. Filamentous forms can indicate possible problems in sludge settling, and protozoa indicate a well-populated activated sludge culture.

BIOCHEMICAL KINETICS, GROWTH KINETICS, AND TEMPERATURE EFFECTS

Two basic approaches are used to determine the rate of substrate removal that occurs in biological processes, and both approaches utilize relationships applied in the fermentation industries. The first approach uses a modification of chemical kinetics (Eckenfelder, 1966), and its equations can be derived from the Michaelis-Menten relationship for substrate utilization. The second approach uses the Monod relationship for microbial growth kinetics (McCarty and Lawrence, 1970). If a laboratory treatability study is performed for a particular wastewater to obtain the kinetic parameters for both approaches and an activated sludge plant is designed to give the required effluent quality, both approaches will give essentially the same detention time (that is, reaction time), θ, for the reactor basin.

The first approach employs a formulation that has been developed for industrial fermentations. It has been found that most enzyme-catalyzed reactions involving a single substrate are zero order with respect to the substrate at relatively high substrate concentrations and are first order with respect to the substrate at relatively low substrate concentrations. When this occurs in a stirred laboratory vessel having no inflow or outflow (that is, a batch reactor), the substrate utilization is zero order for a period after the inoculation and later becomes first order. An explanation of this occurrence is that at high substrate concentrations, the surface of the enzymes is saturated with substrate; thus the reaction is independent of the substrate concentration. At relatively low substrate concentrations, the portion of the surface of the enzymes that is covered with the substrate is proportional to the substrate concentration. This phenomenon is explained by the Michaelis-Menten concept.

The Michaelis-Menten equation gives the relationship between the rate of fermentation product production, dP/dt, and the substrate concentration, S. The rate of product production may also be expressed as the specific rate of product production, $(1/X)(dP/dt)$, where X is the cell mass concentration (Aiba *et al.*, 1965). Since the specific rate of substrate utilization, $(1/X)(dS/dt)$, is proportional to the specific rate of product production (Aiba *et al.*, 1965), it may be incorporated into the Michaelis-Menten formulation as follows:

$$\frac{1}{X}\frac{dS}{dt} = k_s \left(\frac{S}{K_m + S} \right) \qquad (2.1)$$

where

$(1/X)(dS/dt)$ = specific rate of substrate utilization, mass/(mass microbes) × (time)

dS/dt = rate of substrate utilization, mass/(volume)(time)

k_s = maximum rate of substrate utilization, mass/(mass microbes)(time)

K_m = substrate concentration when the rate of utilization is half the maximum rate, mass/volume

S = substrate concentration, mass/volume

Figure 2.7 shows the rate of substrate utilization versus the substrate concentration for the Michaelis-Menten relationship. When the rate is $\frac{1}{2}k_s$, the value of S is K_m. The relationship of Eq. (2.1) may be used to relate the rate of

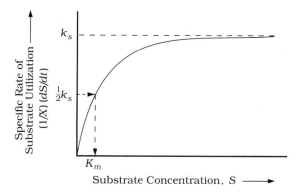

FIGURE 2.7 *Relationship between the Specific Rate of Substrate Utilization, (1/X)(dS/dt), and the Substrate Concentration, S*

substrate utilization to the substrate and cell concentrations in various biological processes.

In examining Eq. (2.1), note that there are two limiting cases. If S is relatively large, K_m may be neglected. S cancels out in the numerator and denominator, and the reaction is zero order in substrate (Brey, 1958). Thus

$$\frac{1}{X}\frac{dS}{dt} = k_s = k = K \tag{2.2}$$

where k or K is the rate constant for a zero-order reaction. If, on the other hand, S is relatively small, it may be neglected in the denominator, and the reaction is first order in substrate (Brey, 1958), or

$$\frac{1}{X}\frac{dS}{dt} = \frac{k_s}{K_m}(S) = kS = KS \tag{2.3}$$

where k or K is the rate constant for a first-order reaction in substrate. Equations (2.2) and (2.3) represent the specific rate of substrate utilization, $(1/X)(dS/dt)$. The equations for the specific rate of substrate utilization are modifications of zero-order and first-order chemical equations with a term added for the cell mass, X. These equations have been used by Professor W. W. Eckenfelder, Jr. (1966, 1970, 1980, 1989) for the design of various biological treatment processes. To summarize the approach using the Michaelis-Menten concept, high substrate concentrations yield zero order in substrate utilization, and low substrate concentrations yield first order in substrate utilization.

The Monod equation (1949) is similar to the Michaelis-Menten equation except that it relates growth to substrate concentration and is

$$\mu = \mu_{\max}\left(\frac{S}{K_s + S}\right) \tag{2.4}$$

where

μ = growth rate constant, time^{-1}

$\mu_{\max}$ = maximum value of the growth rate constant, time^{-1}

S = substrate concentration in solution, mass/volume

K_s = substrate concentration when the growth rate constant is half the maximum rate constant, mass/volume

Monod observed that the microbial growth that occurs is represented by

$$\frac{dX}{dt} = \mu X \tag{2.5}$$

where

dX/dt = rate of cell production, number/time or mass/time

X = number or mass of microbes present

μ = growth rate constant, time^{-1}

Monod found the relationship between the growth rate constant, μ, and the substrate concentration, S, to be as shown in Figure 2.8. When the growth rate constant is $\frac{1}{2}\mu_{max}$, the value of S is K_s. At high substrate concentrations, the growth rate constant is μ_{max}. At lower substrate concentrations, it is as shown in Eq. (2.4) and Figure 2.8. The similarities between the Michaelis-Menten and the Monod relationships can be seen by comparing Figures 2.7 and 2.8 and Eqs. (2.1) and (2.4). The Monod equation has also been used to derive design equations for biological treatment processes.

Chemical reactions generally occur faster as the temperature is increased, and biochemical reactions act similarly. Since microbial cells and their cell contents are at the same temperature as their environment, and since metabolic reactions are biochemical-enzyme-catalyzed reactions, an increase in temperature generally increases the rate of reaction. A simplified equation based on the Arrhenius temperature relationship (Brey, 1958) is

$$\frac{k_{T_2}}{k_{T_1}} = \theta^{T_2 - T_1} \tag{2.6}$$

where

k_{T_1} = reaction rate constant at temperature T_1, °C

k_{T_2} = reaction rate constant at temperature T_2, °C

θ = temperature correction coefficient

T_1 = temperature, °C

T_2 = temperature, °C

Generally, an increase of 10°C temperature doubles the reaction rate, and for this case, θ is 1.072. For biological treatment processes, θ ranges from 1.01 to 1.09, with 1.03 being typical.

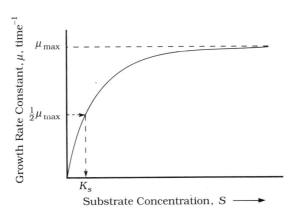

FIGURE 2.8 *Relationship between the Growth Rate Constant, μ, and the Substrate Concentration, S*

REFERENCES

Aiba, S.; Humphery, A. E.; and Millis, N. F. 1965. *Biochemical Engineering*. New York: Academic Press.

Aiba, S.; Humphrey, A. E.; and Millis, N. F. 1973. *Biochemical Engineering*. 2nd ed. New York: Academic Press.

Alexander, M. 1961. *Soil Microbiology*. New York: Wiley.

American Water Works Association (AWWA). 1990. *Water Quality and Treatment*. 4th ed. New York: McGraw-Hill.

Barnes, D.; Bliss, P. J.; Gould, B. W.; and Valletine, H. R. 1981. *Water and Wastewater Engineering Systems*. London: Pittman Press.

Black, J. G. 1993. *Microbiology: Principles and Applications*. Englewood Cliffs, New Jersey: Prentice-Hall.

Brey, W. S., Jr. 1958. *Principles of Physical Chemistry*. New York: Appleton-Century-Crofts.

Brock, T. D., and Madigen, M. T. 1991. *Biology of Microorganisms*. Englewood Cliffs, New Jersey: Prentice-Hall.

Castellan, G. W. 1964. *Physical Chemistry*. Reading, Mass: Addison-Wesley.

Eckenfelder, W. W., Jr. 1966. *Industrial Water Pollution Control*. New York: McGraw-Hill.

Eckenfelder, W. W., Jr. 1970. *Water Quality Engineering for Practicing Engineers*. New York: Barnes and Noble.

Eckenfelder, W. W., Jr. 1980. *Principles of Water Quality Management*. Boston: CBI.

Eckenfelder, W. W., Jr. 1989. *Industrial Water Pollution Control*. 2nd ed. New York: McGraw-Hill.

Environmental Protection Agency (EPA). 1983. *Guidance Manual for POTW Pretreatment Program Development*. Washington, D.C.

Frobisher, M.; Hinsdill, R. D.; Crabtree, K. T.; and Goodheart, C. R. 1974. *Fundamentals of Microbiology*. Philadelphia: Saunders.

Gaudy, A., and Gaudy, E. 1980. *Microbiology for Environmental Scientists and Engineers*. New York: McGraw-Hill.

Gaudy, A., and Gaudy, E. 1988. *Elements of Bioenvironmental Engineering*. San Jose, Calif.: Engineering Press.

McCarty, P. L., and Lawrence, A. W. 1970. Unified Basis for Biological Treatment Plant Design and Operation. *Jour. SED*, No. SA3:757.

McKinney, R. W. 1962. *Microbiology for Sanitary Engineers*. New York: McGraw-Hill.

Metcalf and Eddy, Inc. 1991. *Wastewater Engineering: Treatment, Disposal and Reuse*. 3rd ed. New York: McGraw-Hill.

Monod, J. 1949. The Growth of Bacterial Cultures. *Ann. Rev. Microbiol*. 3:371.

Novotny, V.; Imhoff, K. R.; Olthof, M.; and Krenkel, P. A. 1989. *Karl Imhoff's Handbook of Urban Drainage and Wastewater Disposal*. New York: Wiley.

Reynolds, T. D. 1982. *Unit Operations and Processes in Environmental Engineering*. Boston: PWS.

Stanier, R. Y.; Doudoroff, M.; and Adelberg, E. A. 1963. *The Microbial World*. Englewood Cliffs, N.J.: Prentice-Hall.

Water Pollution Control Federation (WPCF). 1990. *Wastewater Biology. The Microlife*. Washington, D.C.

PROBLEMS

2.1 An aerated laboratory vessel is filled with a mixture of substrate and microbial cells. At time zero, the substrate concentration (S) is $150\,mg/\ell$ and the cell concentration (X) is $1500\,mg/\ell$. The biochemical reaction is pseudo-first order, and the rate constant K is $0.40\,\ell/(gm\ cells)(hr)$. After 6 hours of aeration, the substrate concentration (S) is $4\,mg/\ell$, and the cell concentration (X) is $1590\,mg/\ell$. Using the specific rate of substrate utilization equation, $(1/X)(dS/dt) = KS$, determine:

a. The rate of substrate utilization, dS/dt, at time zero in $mg/(\ell)(hr)$.

b. The rate of substrate utilization, dS/dt, at time equals 6 hours in $mg/(\ell)(hr)$.

2.2 For Problem 2.1, $\mu_{max} = 3.0\ day^{-1}$ and $K_s = 60\,mg/\ell$. Using Monod's equation, determine:

a. The growth rate constant, μ, at time zero.

b. The rate of cell growth, dX/dt, at time zero in $mg/(\ell)(hr)$.

c. The growth rate constant, μ, at time equals 6 hours.

d. The rate of cell growth, dX/dt, at time equals 6 hours in mg/(ℓ)(hr).

2.3 A laboratory study has been done at 23°C, and the reaction is pseudo-first order and $K = 0.45$ ℓ/(gm cells)(hr). The temperature correction coefficient θ equals 1.03. Determine:

a. The maximum rate constant, K, if the maximum operating temperature is 27°C.

b. The minimum rate constant, K, if the minimum operating temperature is 17°C.

3

MASS BALANCES, FLOW MODELS, AND REACTORS

In the study of unit operations and processes employed in water and waste-water treatment plants, the concepts of mass balances, flow models, and reactors are useful. For physical operations, mass balances and flow models are helpful in understanding and analyzing the operation. For chemical and biochemical processes, not only are mass balances and flow models important, but also the rate at which the process occurs and the amount of conversion required determine the size of the treatment facilities that must be provided. The material presented in this chapter is intended to serve as an introduction to operation and process analysis and to the text presented in future chapters.

MASS BALANCES

The fundamental approach to show the change occurring in a vessel or some type of container, such as a tank, is the mass-balance analysis. For non-nuclear processes, mass cannot be created or destroyed. Thus the mass that accumulates is equal to the mass flow entering minus the mass converted minus the mass flow leaving, or,

$$[\text{Accumulation}] = [\text{Input}] - \begin{bmatrix} \text{Decrease due} \\ \text{to reaction} \end{bmatrix} - [\text{Output}] \quad (3.1)$$

If the system has no reaction occurring, such as a unit operation, there is no decrease due to reaction, and the mass-balance equation becomes

$$[\text{Accumulation}] = [\text{Input}] - [\text{Output}] \quad (3.2)$$

Consider the reactor shown in Figure 3.1. An envelope is drawn to show the system boundaries that must be established so that all the flows of mass into and out of the system can be identified. The mass balance on the Component A being studied is

$$dC_1 V = QC_0 dt - Vr dt - \underline{Q}C_1 dt \quad (3.3)$$

where

dC_1 = change of Component A in the vessel, mass/volume

V = vessel volume, volume

Q = flowrate, volume/time

C_0 = initial concentration, mass/volume

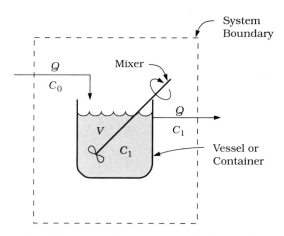

FIGURE 3.1 *Sketch of a Reactor for the Application of Mass-Balance Analysis*

C_1 = final concentration, mass/volume

dt = time increment, time

r = rate at which Component A reacts, mass/volume-time

In a mass-balance equation, it is advisable first to check to make sure all units cancel properly and the answer is in the proper units. For Eq. (3.3), the units would be

$$(\text{mass/volume})(\text{volume})$$
$$= (\text{volume/time})(\text{mass/volume})(\text{time})$$
$$- (\text{volume})(\text{mass/volume-time})(\text{time})$$
$$- (\text{volume/time})(\text{mass/volume})(\text{time})$$

or

$$\text{mass} = \text{mass} - \text{mass} - \text{mass}$$

or

$$\text{mass} = \text{mass}$$

which are the correct units. Dividing Eq. (3.3) by dt gives

$$\frac{dC_1}{dt}V = QC_0 - Vr - QC_1 \tag{3.4}$$

The term dC_1/dt is the change in C_1 in the reactor from time zero to some later time, such as from day zero to the next successive day. If the reaction is first order, $-r = kC_1$ and the balance becomes

$$\frac{dC_1}{dt}V = QC_0 - VkC_1 - QC_1 \tag{3.5}$$

In studying Eq. (3.4) and Eq. (3.5), it can be seen that the term $dC_1/dt \cdot V$ represents the rate of accumulation of Component A in the system. The term QC_0 represents the rate of flow of Component A into the system. The terms Vr and VkC_1 represent the rate of change in Component A in the system, and QC_1 represents the rate of flow of Component A out of the system.

Dividing Eq. (3.5) by V yields

$$\frac{dC_1}{dt} = \frac{Q}{V}C_0 - kC_1 - \frac{Q}{V}C_1 \tag{3.6}$$

Most unit processes are designed for steady state, since this simplifies the system. For steady state, $dC_1/dt = 0$; thus, if at time zero $C_1 = 10\,\text{mg}/\ell$, and at the next successive day $C_1 = 10\,\text{mg}/\ell$, then $dC_1/dt = (10\,\text{mg}/\ell - 10\,\text{mg}/\ell)/(1\ \text{day}) = 0$. If at time zero $C_1 = 10\,\text{mg}/\ell$, and at the next successive day $C_1 = 5\,\text{mg}/\ell$, then $dC_1/dt = (10\,\text{mg}/\ell - 5\,\text{mg}/\ell)/(1\ \text{day}) = 5\,\text{mg}/\ell\text{-day}$ and the system is not at steady state. Assuming steady state for Eq. (3.6), $dC_1/dt = 0$ and the detention time in the reactor, $\theta = V/Q$. Incorporating these in Eq. (3.6) gives

$$0 = \frac{C_0}{\theta} - kC_1 - \frac{C_1}{\theta} \tag{3.7}$$

from which

$$0 = C_0 - C_1(1 + k\theta) \tag{3.8}$$

Rearranging gives

$$C_1 = \frac{C_0}{1 + k\theta} \tag{3.9}$$

From this equation the concentration in the effluent, C_1, can be computed if C_0, k, and θ are known.

EXAMPLE 3.1   *Mass Balance*

If a reactor, as shown in Figure 3.1, has $C_0 = 150\,\text{mg}/\ell$ and $k = 2.0\,\text{hr}^{-1}$, make a plot of the effluent concentration, C_1, versus θ for θ values up to 10 hours. Also, determine the effluent concentration, C_1, if the reaction time is 5.5 hours.

SOLUTION   The equation for C_1 is

$$C_1 = \frac{150\,\text{mg}/\ell}{1 + (2.0\,\text{hr}^{-1})\theta} \tag{3.10}$$

Substitution for various values of θ gives

θ	C_1
0	150
1	50
2	30
3	21
4	17
5	14
6	12
7	10
8	8.8
9	7.9
10	7.1

A plot of the effluent concentration, C_1, versus θ is shown in Figure 3.2. From the figure, at $\theta = 5.5\,\text{hr}$, $C_1 = 13\,\text{mg}/\ell$. From Eq. (3.9), $C_1 = C_0/(1 + k\theta)$ or $C_1 = 150\,\text{mg}/\ell/(1 + 2.0\,\text{hr}^{-1} \times 5.5\,\text{hr})$. Thus $C_1 = \boxed{13\,\text{mg}/\ell.}$

As previously mentioned, for a mass balance it is advisable to check to make sure all units cancel properly and the answer is in the proper units. In Eq. (3.9) the units, when substituted, give

$$\text{mg}/\ell = \frac{\text{mg}/\ell}{1 + (\text{hr}^{-1})(\text{hr})} \tag{3.11}$$

$$\text{mg}/\ell = \text{mg}/\ell \tag{3.12}$$

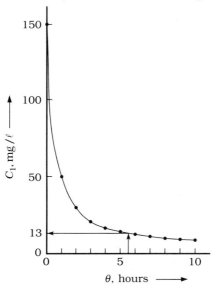

FIGURE 3.2 *Graph for Example 3.1*

FLOW MODELS

In most environmental engineering unit operations and processes, the residence time and flowrate are significant. Flow models are useful in evaluating the effects of residence time and flowrate on a system. The flow regime in continuous-flow systems may approach either **ideal plug flow** (piston flow) or **ideal completely mixed flow.** In the ideal plug-flow regime, the elements of fluid that enter the system at the same time flow through it with the same velocity and leave at the same time. Also, there is no longitudinal mixing of the fluid elements as they move through the system. Plug flow is approached in systems that have large length-to-width ratios. For instance, if a tube has a length-to-width ratio of 50:1, the flow regime will approach plug flow if the velocity is not excessive. In the ideal completely mixed flow regime, the elements of fluid that enter the system are immediately dispersed uniformly throughout the system. Also, the properties of the fluid at all locations within the system are the same and are identical to the properties of the discharge or effluent stream. The completely mixed model is approached in mixing vessels that are circular, square, or slightly rectangular in plan view. It should be realized that plug flow and completely mixed flow are limiting cases and that actual flow regimes will range in a broad spectrum between these ideal models. The intermediate case is usually referred to as **plug flow with dispersion, dispersed plug flow, plug flow with longitudinal mixing, intermediate-mixed flow,** or **arbitrary flow.**

The residence-time distribution of the fluid elements that compose the effluent characterizes each flow model. Tracer inputs to the system make it possible to measure the residence-time distribution (RTD) curve for the system being studied. Such tracer inputs may be pulse inputs (slug release of tracer) or step inputs (continuous release of tracer).

Plug Flow

Consider the plug-flow model shown in Figure 3.3 that has a volume, V, and an influent and effluent flowrate, Q. Assume there is a white dye in the inflow and in the model. If a pulse or slug of red dye of mass S is released in the influent, as in Figure 3.3(b), it will move into the model and form a red band or piston at Point 1, as in Figure 3.3(c). As the flow continues, the red band or piston moves through the model, as in Figure 3.3(d). Eventually, the red band of dye reaches Point 2, as in Figure 3.3(e), and appears in the discharge flow. The residence-time distribution (RTD) curve is a plot of C_1/C_0 versus $t/\bar{t}$, where

t = time after dye release

$\bar{t}$ = mean residence time = V/Q

C_1 = concentration of red dye at Point 2 at time t

C_0 = concentration of red dye released, S/V

Figure 3.4 shows the RTD curve for the slug release of tracer or dye in the described plug-flow model. The red dye makes its appearance in the effluent when the time after release, t, is equal to the mean residence time, $\bar{t}$, or $t/\bar{t} = 1.0$.

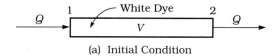

(a) Initial Condition

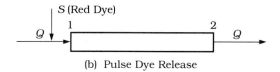

(b) Pulse Dye Release

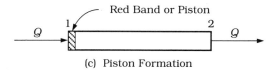

(c) Piston Formation

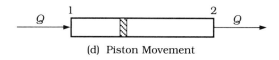

(d) Piston Movement

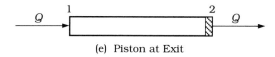

(e) Piston at Exit

FIGURE 3.3 *Plug Flow with Slug Tracer Input*

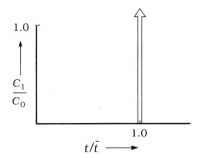

FIGURE 3.4 *Tracer Curve for Plug-Flow Model with a Slug of Tracer Input*

Completely Mixed Flow   Consider the completely mixed model shown in Figure 3.5(a) that has a volume, V, and an influent and effluent flowrate, Q. The model has continuous mixing or stirring, as indicated by the propeller. Assume there is a white dye in the inflow and in the model. If a pulse or slug of red dye of mass S is released in the influent, as in Figure 3.5(b), it will quickly move into the model and immediately be mixed throughout the system at a uniform concentration, C_0, as in Figure 3.5(c). The immediate concentration, C_0, is equal to S/V, and the time after release, t, is considered to be zero. Since the properties in the model are identical to the properties of the discharge stream, C_0 equals C_1 at time zero. As the inflow continues, the red dye concentration in the model, C_1, is diluted, as in Figure 3.5(d). Eventually, the red dye is completely diluted from the system, and the model is at its original condition, as in Figure 3.5(e). The curve for the

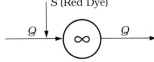

(a) Initial Condition

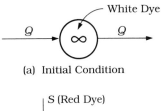

(b) Pulse Dye Release

(c) Immediate Dye Mixing

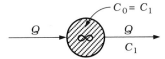

(d) Dye Dilution

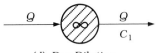

(e) Original Condition

FIGURE 3.5 *Completely Mixed Flow with a Slug of Tracer Input*

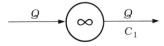

tracer input can be solved by mass balances. According to the equation for a mass balance,

$$[\text{Accumulation}] = [\text{Input}] - \begin{bmatrix} \text{Decrease} \\ \text{due to} \\ \text{reaction} \end{bmatrix} - [\text{Output}] \qquad (3.1)$$

After $t = 0$, the input term is zero, and since there is no reaction, the change due to reaction is zero and Eq. (3.1) becomes

$$dC_1 V = 0 - 0 - C_1 Q dt \qquad (3.13)$$

Rearranging for integration yields

$$\int_{C_0}^{C_1} \frac{dC_1}{C_1} = -\frac{Q}{V} \int_0^t dt \qquad (3.14)$$

and integration gives

$$\frac{C_1}{C_0} = e^{-(Q/V)t} = e^{-t/\bar{t}} \qquad (3.15)$$

Equation (3.15) is the equation for the tracer curve from a slug of dye input, and Figure 3.6 shows the curve plot.

Plug Flow with Dispersion The third flow model is plug flow with dispersion or plug flow with longitudinal mixing. This model is an intermediate between plug flow and completely mixed flow, and sometimes it is referred to as intermediate-mixed flow. It represents a case of plug flow where there is dispersion or mixing of individual elements of fluid along the longitudinal axis. Figure 3.7(a) shows a slug release of red dye in this type of model, and the front formed is not straight because of axial mixing. As the red dye moves through the model, as in Figure 3.7(b), the mixing tends to lengthen the zone of red dye, Z.

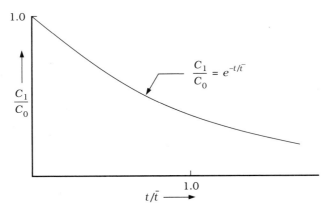

FIGURE 3.6 *Tracer Curve for a Completely Mixed Model with a Slug of Tracer Input*

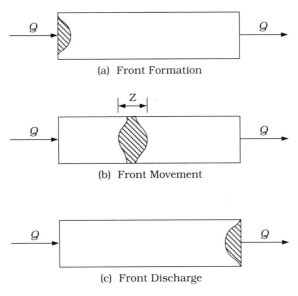

(a) Front Formation

(b) Front Movement

(c) Front Discharge

FIGURE 3.7 *Plug Flow with Dispersion and a Slug Input of Tracer*

Finally, the red dye is discharged from the system as in Figure 3.7(c). The tracer curve for the slug release of tracer is shown in Figure 3.8. Thus it can be seen from the curve that the average residence time is less than the theoretical time. The exact shape of the curve cannot be determined except by tracer studies on a pilot or full-scale tank or reactor basin. This is because the curve shapes are a function of the geometry and relative dimensions of the tank, the mixing intensity, the dispersion coefficient, and other variables.

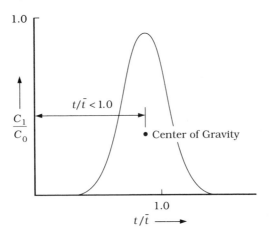

FIGURE 3.8 *Tracer Curve for Plug Flow with Dispersion and a Slug of Tracer Input*

Many tanks or reactor basins used in environmental engineering unit operations and processes have flow regimes that are plug flow with dispersion. Usually these are designed as plug-flow systems, and a correction term is used to simulate the dispersed plug-flow condition. The design of dispersed plug-flow reactors is presented later in this chapter.

In summary, there are three basic flow models: (1) plug flow, (2) completely mixed flow, and (3) plug flow with dispersion. Plug flow occurs in tanks or conduits that have a large length-to-width or length-to-diameter ratio, such as 50:1, if the velocities are not excessive. Completely mixed flow occurs in tanks or basins that are circular, square, or slightly rectangular. Plug flow with dispersion is an intermediate flow regime between plug flow and completely mixed flow. Flow models are useful in studies involving unit operations and processes such as settling tanks, activated sludge reactor basins, aerated lagoons, oxidation ponds, and high-rate anaerobic digesters.

REACTORS

Reactions may be classified in numerous ways, one of which is whether heat is liberated (**exothermic**) or absorbed (**endothermic**) during the reaction. Since a change in temperature will cause a change in the rate of reaction, temperature is important in reactor design. Fortunately, in environmental engineering most solutions are dilute, and any temperature change due to reaction is usually negligible. Temperature effects from the surroundings, however, may significantly influence the rate of reaction. Another classification of reactions is according to the number and types of phases involved, the phases being gas, liquid, and solid. A reaction is **homogeneous** if it occurs in one phase alone and **heterogeneous** if it requires the presence of at least two phases. Occasionally, this classification is not clear-cut; and an example of this is the enzyme-substrate reactions involved in biological systems. In biochemical reactions, the enzymes act as organic catalysts. Enzymes are large-molecular-weight proteins of colloidal size; thus, enzyme-containing solutions represent a hazy region between homogeneous and heterogeneous systems. How one classifies a borderline case, such as an enzyme-catalyzed reaction, depends upon which description one thinks is more useful for the reaction. In reactor design, enzyme-catalyzed reactions are usually treated as homogeneous systems to simplify the design computations. Reactions also may be classified as **catalytic** or **noncatalytic**. A catalytic reaction is one whose rate is changed by materials — the catalysts — that are neither reactants nor products. The catalysts need not be present in large amounts, and they either hinder or accelerate a reaction. During a reaction, a catalyst is not changed or is changed very slowly. Enzyme and microbial reactions are special cases of catalytic reactions.

The rate of reaction is affected by many variables. In homogeneous systems, the temperature, pressure, and composition are the main variables. In heterogeneous systems, since more than one phase is involved, the number and types of variables become more complex. In both homogeneous and heterogeneous reactions, if the overall reaction consists of a number of steps, the slowest step will have the greatest effect and will be controlling. In most reactions involved in water and wastewater engineering, the solutions are dilute, and heat effects from the reaction are negligible. The volume of the fluid is usually the same as the volume of the reactor; thus, the reactor is of constant volume. The reactor is usually under constant pressure; thus, pressure is not an important variable. Also, the density of the influent is considered the same as that of the effluent.

Reactors may be operated in either a batchwise or a continuous-flow manner. In a **batch reactor**, the reactor is charged with the reactants, the contents are well mixed and left to react, and then the resulting mixture is discharged. The capital cost of a batch reactor is usually less than that of a continuous-flow reactor, but the operating costs are much greater. It is usually limited to small installations. In a continuous-flow reactor, the feed to the reactor and the discharge from it are continuous. The three types of continuous-flow reactors are the plug-flow reactor, the dispersed plug-flow reactor, and the completely mixed or continuously stirred tank reactor (CSTR). The continuous-flow reactors may be operated as either steady-state or unsteady-state systems. In most cases, they are considered steady-state processes in which the feed stream flowrate and its composition are constant with respect to time. Frequently, reaction kinetics are studied in the laboratory under batchwise conditions to determine the rate constant, k. The application of the kinetic constant, k, to the design of a continuous-flow reactor, however, involves no changes in kinetic principles; thus this is valid.

The **plug-flow reactor** ideally has the geometric shape of a long tube or tank and has continuous flow in which there is movement of one or all of the reactants in the axial direction. The reactants enter at the upstream end of the reactor, and the products leave at the downstream end. Ideally, there is no induced mixing between elements of fluid along the direction of flow. The composition of the reacting fluid changes in the direction of flow. Long and narrow rectangular activated sludge reactor basins approach plug flow and are considered plug-flow reactors.

The **dispersed plug-flow reactor** has a flow regime that is plug flow with dispersion. This type of reactor has a smaller mean residence time and less conversion than a plug-flow reactor. Short and wide activated sludge reactor basins are examples of dispersed plug-flow reactors.

The **completely mixed or continuously stirred tank reactor (CSTR)** consists of a stirred tank that has a feed stream of the reactants and a discharge stream of partially reacted materials. These tanks are usually round, square, or slightly rectangular in plan view, and it is necessary to provide sufficient stirring. The stirring of a CSTR is extremely important, and it is assumed that the fluid in the reactor is perfectly mixed — that is, the contents are uniform throughout the reactor volume. As a result of the mixing, the composition of the discharge stream is the same as that of the contents in the reactor. Frequently, several CSTRs in series are employed to improve their conversion or performance. Round, square, or slightly rectangular activated sludge reactor basins are examples of completely mixed or continuously stirred tank reactors.

The previous discussions on the plug-flow, dispersed plug-flow, and completely mixed or continuously stirred tank reactors have been for homogeneous reactions, and for these reactions, the reactors contain only the reacting fluids. There are other reactors that are used for heterogeneous reactions — that is, those employing more than one phase, such as a solid and a liquid. The **fixed-bed reactor** or **packed-bed reactor** is a reactor packed with particles of a solid catalyst; such reactors employ gasses or liquids contacted with the solid phase. An ion exchange column used for the softening or demineralization of water is an example of a fixed-bed reactor. The **moving-bed reactor** has a fluid that passes upward through a moving bed of the solids. The solid phase is fed to the top of the bed, flows downward, and is removed at the bottom. A countercurrent ion exchange column used for the softening or demineralization of water is an example of a moving-bed reactor. The **fluidized-bed reactor** is used for a solid and a fluid, either a liquid or a gas. In these reactors, the solid material is in the form of fine particles contained in a bed. The reacting stream flows upward through the bed at a velocity sufficient to lift the particles but not great enough to carry them out of the bed. A fluidized-bed incinerator used for the incineration of organic wastewater sludges is an example of a fluidized-bed reactor. The heterogeneous reactors are special cases, and in this chapter, only homogeneous reactors will be presented.

The fundamental equations for the design of a homogeneous reactor are the **rate equation** and the **equation for the law of conservation of mass**. The rate equation for Species A for the various reaction orders is

$$-\frac{dC_A}{dt} = kC_A^n = -r_A \tag{3.16}$$

where n is the reaction order and r_A is the rate of conversion of Species A. The law of conversion of mass, which has been previously presented, is

$$[\text{Accumulation}] = [\text{Input}] - \left[\begin{array}{c}\text{Decrease due} \\ \text{to reaction}\end{array}\right] - [\text{Output}] \tag{3.1}$$

If the component is increasing due to the reaction, the reaction term in Eq. (3.1) will have the sign opposite to that shown.

In environmental engineering applications, solutions are usually dilute; therefore, there is no change in density due to the reaction, and the reactor has a constant volume.

Batch Reactors   A batch reactor is shown in Figure 3.9. In batchwise operation, the batch type of reactor is charged with the reactants, the contents are well mixed for a sufficient reaction time for the desired conversion to occur, and then the resultant mixture is discharged. This is an unsteady-state process where the

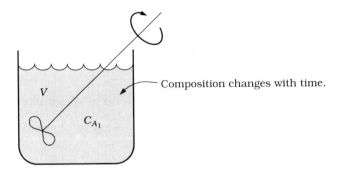

FIGURE 3.9   *Batch Reactor*

composition changes with time; however, at any instant, the composition throughout the reactor is uniform. The component undergoing conversion due to the reaction is designated A. Since the composition is uniform throughout the reactor, at any given time, a mass balance for A can be made for the entire reactor. The input and output terms are zero; thus, the mass-balance equation, Eq. (3.1), becomes

$$[\text{Accumulation}] = -\begin{bmatrix}\text{Decrease due} \\ \text{to reaction}\end{bmatrix} \qquad (3.17)$$

Consider a batch reactor with a **first-order reaction**. The accumulation term is dC_A/dt and the reaction term is kC_A; thus the mass-balance equation, Eq. (3.17), becomes

$$\frac{dC_A}{dt} = -kC_A \qquad (3.18)$$

which is the same as Eq. (3.16) where $n = 1$. Rearranging the equation for integration gives

$$\int_{C_{A_0}}^{C_{A_1}} \frac{dC_A}{C_A} = -k\int_0^\theta dt \qquad (3.19)$$

Integration and rearranging gives

$$\frac{C_{A_1}}{C_{A_0}} = e^{-k\theta} \qquad (3.20)$$

where θ is the reaction time.

EXAMPLE 3.2   *Batch Reactors*

A batch reactor is to be designed for a first-order reaction, and the required removal or conversion of A is 90%. If the rate constant, k, is $0.35\,\text{hr}^{-1}$, what is the required residence time in hours for the batch reactor?

SOLUTION If X_A is the fraction of conversion, the value of C_A is

$$C_{A_1} = C_{A_0}(1 - X_A) = C_{A_0}(1 - 0.90) = 0.10C_{A_0}$$

The performance equation is

$$\frac{C_{A_1}}{C_{A_0}} = e^{-k\theta}$$

Thus,

$$\frac{0.10C_{A_0}}{C_{A_0}} = e^{-(0.35)\theta}$$

from which $\theta = \boxed{6.58 \text{ hours.}}$

In order for a batch reactor to perform as designed, the mixing must be sufficient to make the contents of the reactor uniform. Some small activated sludge reactor basins are batch reactors.

Plug-Flow Reactors A plug-flow reactor operated at steady state is shown in Figure 3.10. It has an idealized flow regime since the fluid elements pass through the reactor in a piston-like manner. For steady state, the feed and its composition, the discharge and its composition, and the composition at any point in the reactor are constant with respect to time. The plug-flow regime for the reactor is an idealized flow regime, since there is negligible dispersion of fluid elements along the direction of flow.

In the plug-flow reactor, the composition of the fluid varies from point to point along the flow path. Therefore, the mass balance for a reaction component must be made for a differential volume, dV_r, as shown in Figure 3.10. The mass-balance equation, Eq. (3.1), for steady state becomes

$$[\text{Input}] = [\text{Output}] + \begin{bmatrix} \text{Decrease due} \\ \text{to reaction} \end{bmatrix} \qquad (3.21)$$

The material balance for Component A in the differential reactor volume, dV_r, in Figure 3.10 is therefore

$$QC_A = Q(C_A - dC_A) + r_A dV_r \qquad (3.22)$$

from which

$$QdC_A = r_A dV_r \qquad (3.23)$$

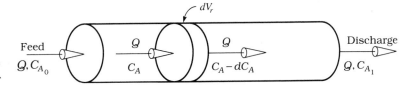

FIGURE 3.10
Plug-Flow Reactor

where r_A is the reaction rate. Equation (3.23) is the general design equation for plug-flow reactors.

Consider a plug-flow reactor with a **first-order reaction**. The reaction term, r_A, equals $-kC_A$ according to Eq. (3.16). Substituting for r_A in Eq. (3.23) gives

$$QdC_A = -kC_A dV_r \qquad (3.24)$$

Rearranging for integration results in

$$\int_{C_{A_0}}^{C_{A_1}} \frac{dC_A}{C_A} = -\frac{k}{Q} \int_0^{V_r} dV_r \qquad (3.25)$$

Integration yields

$$\ln\left(\frac{C_{A_1}}{C_{A_0}}\right) = -\frac{k}{Q} V_r \qquad (3.26)$$

Since $V_r = Q\theta$, where θ = mean residence time of the fluid, substituting for V_r in Eq. (3.26) yields

$$\ln\left(\frac{C_{A_1}}{C_{A_0}}\right) = -k\theta \qquad (3.27)$$

or

$$\frac{C_{A_1}}{C_{A_0}} = e^{-k\theta} \qquad (3.28)$$

Since Eq. (3.28) for a plug-flow reactor is the same as Eq. (3.20) for a batch reactor, a plug-flow reactor will have the same performance as a batch reactor if the reaction or residence times are the same.

EXAMPLE 3.3 *Plug-Flow Reactors*

A plug-flow reactor is to be designed for a first-order reaction, and the required conversion of A (X_A) is 90%. If the rate constant, k, is $0.35\,\mathrm{hr}^{-1}$, what is the required residence time in hours?

SOLUTION
$$C_{A_1} = C_{A_0}(1 - X_A) = C_{A_0}(1 - 0.90) = 0.10 C_{A_0}$$

The performance equation is

$$\frac{C_{A_1}}{C_{A_0}} = e^{-k\theta}$$

Thus,

$$\frac{0.10 C_{A_1}}{C_{A_0}} = e^{-(0.35)\theta}$$

from which $\theta = \boxed{6.58 \text{ hours.}}$

It can be noted that the reaction time for the plug-flow reactor in Example 3.3 is the same as the reaction time for the batch reactor in Example 3.2. Thus, the performance of a plug-flow reactor will be the same as that of a batch reactor if the reaction times are the same.

The plug-flow reactor is an ideal reactor, since the flow regime is assumed to be plug flow. In a real reactor, plug flow cannot be perfectly attained because of dispersion of fluid elements along the flow path or axis of the reactor. For rectangular activated sludge reactor basins, the turbulence created by the aeration system creates velocities normal to the direction of flow and also velocities in the direction of flow. The velocities in the direction of flow, in particular, create longitudinal dispersion. Usually, activated sludge reactor basins will approach plug flow if the basins are long and narrow and have a length to width ratio of 50:1 or more. For shorter rectangular reactor basins, the flow regime will be dispersed plug flow, and the reactors are dispersed plug-flow reactor basins.

Completely Mixed Reactors or Continuously Stirred Tank Reactors

The completely mixed or continuously stirred tank reactor (CSTR) is shown in Figure 3.11. The reactor consists of a stirred tank into which there is a feed of reacting material entering and from which there is a discharge of partially reacted material. Upon entering the reactor, the feed fluid is almost instantaneously mixed with the fluid already present, and the reactor contents are uniform throughout the entire reactor volume. As a result of the mixing, the composition of the discharge stream is the same as that of the contents in the reactor. The mixing in a CSTR is extremely important, and it is assumed that the fluid in the reactor is perfectly mixed — that is, the contents are uniform throughout the reactor volume. In practice, perfect mixing can be obtained if the mixing is sufficient and the liquid is not too viscous. If the mixing is inadequate, there will be bulk streaming between the inlet and the outlet, and the composition of the reactor contents will not be uniform. The flow regime in the completely mixed reactor or CSTR is completely mixed flow. The round, square, or slightly rectangular reactor basins used for the activated sludge process are examples of completely mixed reactors or CSTRs.

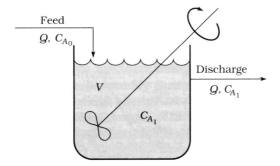

FIGURE 3.11 *Completely Mixed Reactor or Continuously Stirred Tank Reactor (CSTR)*

For the CSTR the mass-balance equation, Eq. (3.1), for the reactor is

$$[\text{Accumulation}] = [\text{Input}] - \begin{bmatrix} \text{Decrease due} \\ \text{to reaction} \end{bmatrix} - [\text{Output}] \qquad (3.1)$$

Consider a CSTR with a **first-order reaction**. The rate of reaction is kC_{A_1}, and substituting the designations into the mass-balance equation gives

$$dC_{A_1}V = C_{A_0}Q\,dt - VkC_{A_1}\,dt - C_{A_1}Q\,dt \qquad (3.29)$$

Rearrangement yields

$$\frac{dC_{A_1}}{dt} = C_{A_0}\frac{Q}{V} - kC_{A_1} - C_{A_1}\frac{Q}{V} \qquad (3.30)$$

For steady state, $dC_{A_1}/dt = 0$, and since $\theta = V/Q$, Eq. (3.30) becomes

$$0 = \frac{C_{A_0}}{\theta} - kC_{A_1} - \frac{C_{A_1}}{\theta} \qquad (3.31)$$

Eq. (3.31) may be rearranged to give

$$\frac{C_{A_1}}{C_{A_0}} = \frac{1}{1 + k\theta} \qquad (3.23)$$

If the time for a given conversion is desired, Eq. (3.32) may be rearranged to give

$$\theta = \frac{C_{A_0} - C_{A_1}}{kC_{A_1}} \qquad (3.33)$$

EXAMPLE 3.4   *Completely Mixed Reactors*

A CSTR is to be designed for a first-order reaction, and the required conversion of A is 90%. If the rate constant, k, is $0.35\,\text{hr}^{-1}$, what is the required residence time in hours? How does this compare to the time for the plug-flow reactor in Example 3.3?

SOLUTION
$$C_{A_1} = C_{A_0}(1 - X_A) = C_{A_0}(1 - 0.90) = 0.10C_{A_0}$$

The performance equation is

$$\theta = \frac{C_{A_0} - C_{A_1}}{kC_{A_1}}$$

Thus,

$$\theta = \frac{C_{A_0} - 0.10C_{A_0}}{(0.35\,\text{hr}^{-1})(0.10C_{A_0})}$$
$$= \boxed{25.7\,\text{hours.}}$$

The residence time for the batch and plug-flow reactors was 6.58 hr; therefore, the CSTR required a much longer residence time.

Thus, for a first-order reaction, the residence time, θ, for a CSTR will be larger than for a batch or plug-flow reactor. In order to improve the performance of the CSTR, it is possible to use several in a series instead of one large CSTR. Figure 3.12 shows N number of CSTRs in series. The figure shows the reactors to be equal in volume; if desired, however, they may be of different volumes.

The algebraic solution for a series of CSTRs becomes complex if there are many reactors involved having different reaction times. A graphic solution can easily be made on any number of reactors in a series. All that is required is a graph showing the rate, r_A, versus concentration, C_A, as shown in Figure 3.13. The value on the y-axis is $-r$, but since $-r = kC_A^n$, it appears in the positive y-axis direction. The term $-r$ means that Species A is decreasing in concentration with respect to time. The proof of the graphic solution is given in the following discussion. The mass-balance equation, Eq. (3.1), for a steady-state CSTR may be rearranged to

$$[\text{Output}] = [\text{Input}] - \begin{bmatrix} \text{Decrease due} \\ \text{to reaction} \end{bmatrix} \tag{3.34}$$

Substituting for the variables in Eq. (3.34) for the first CSTR gives

$$C_{A_1}Q = C_{A_0}Q - r_{A_1}V \tag{3.35}$$

Dividing by Q gives

$$C_{A_1} = C_{A_0} - r_{A_1}\frac{V}{Q} \tag{3.36}$$

Since $V/Q = \theta$, Eq. (3.36) may be rearranged to

$$C_{A_1} = C_{A_0} - \theta r_{A_1} \tag{3.37}$$

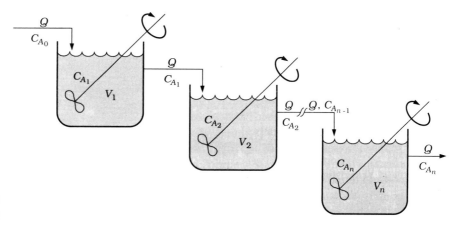

FIGURE 3.12
A Series of CSTRs

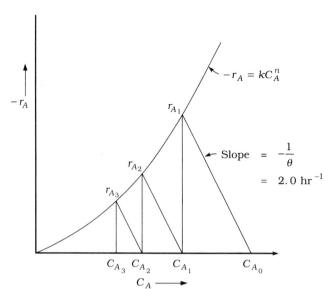

FIGURE 3.13   *Graphic Solution for a Series of Three CSTRs*

On the graph of r_A versus C_A, Figure 3.13, a line drawn through the point $(C_{A_0}, 0)$ with a slope $-1/\theta$ will intersect the rate curve at r_{A_1} and have an abscissa of C_{A_1}. It follows that

$$\text{Slope} = -\frac{1}{\theta} = \frac{-r_{A_1}}{C_{A_1} - C_{A_0}} \tag{3.38}$$

Cross-multiplying gives

$$C_{A_1} - C_{A_0} = -\theta r_{A_1} \tag{3.39}$$

or

$$C_{A_1} = C_{A_0} - \theta r_{A_1} \tag{3.40}$$

It can be seen that Eq. (3.40) is identical to Eq. (3.37), which proves the graphic solution. Equation (3.37) and Eq. (3.38) may be generalized for a reactor (number i) in a series to give

$$C_{A_i} = C_{A_{i-1}} - \theta r_{A_i} \quad \text{or} \quad -\frac{1}{\theta} = \frac{-r_{A_i}}{C_{A_i} - C_{A_{i-1}}} \tag{3.41}$$

Suppose there are three CSTRs in a series and the rate-versus-concentration curve is as shown in Figure 3.13. Assume the residence time of each reactor is 0.50 hr. It is desired to know the rate in each of the reactors and the concentration of Species A in each reactor and its effluent. Since the residence times are equal, the slopes of all the lines will be equal to $-1/\theta$, or $-1/0.50\,\text{hr}^{-1} = -2.0\,\text{hr}^{-1}$. The value of C_{A_0} is located and a line is drawn with a slope of $-2.0\,\text{hr}^{-1}$; and it intersects the curve at r_{A_1} and C_{A_1}. This is the rate and concentration in the first reactor. Similar construction for the other two reactors gives r_{A_2}, C_{A_2}, r_{A_3}, and C_{A_3}.

EXAMPLE 3.5 *Series of CSTRs*

It is desired to compare the fraction of conversion for a series of 10 CSTRs to the conversion for a plug-flow reactor having the same residence time. The kinetics of the first-order reaction A $\rightarrow$ B have been studied, and the reaction rate, k, is $0.35\,\text{hr}^{-1}$ and the initial concentration of A is $150\,\text{mg}/\ell$.

SOLUTION First, it is necessary to construct the rate versus concentration curve. The rate for a first-order reaction is

$$-\frac{dC_{A_1}}{dt} = -r_A = kC_{A_1}$$

The tabulations are

C_{A_1}, mg/ℓ	$-r_A = kC_{A_1}$, mg/ℓ-hr
20	7
40	14
60	21
80	28
100	35
120	42
140	49
160	56

The graph of $-r_A$ versus C_{A_1} is shown in Figure 3.14. The slope for the line construction is arbitrarily chosen as $1.50\,\text{hr}^{-1}$, and the triangle diagram for 10 reactors is constructed as shown in Figure 3.14.

The discharge from the 10th CSTR has $18\,\text{mg}/\ell$ C_A, and the residence time is $1/1.50\,\text{hr}$, or $0.667\,\text{hr}$ per reactor. The total residence time for 10 CSTRs is $(0.667)10 = 6.67\,\text{hr}$. For the plug-flow reactor, the performance equation is

$$\frac{C_{A_1}}{C_{A_0}} = e^{-k\theta}$$

which may be rearranged to

$$C_{A_1} = C_{A_0} e^{-k\theta}$$

Thus,

$$C_{A_1} = (150)e^{-(0.35)(6.67)} = 14.5\,\text{mg}/\ell$$

The conversion for the series of 10 CSTRs is

$$X_A = (150 - 18)(1/150)100\% = 88.0\%$$

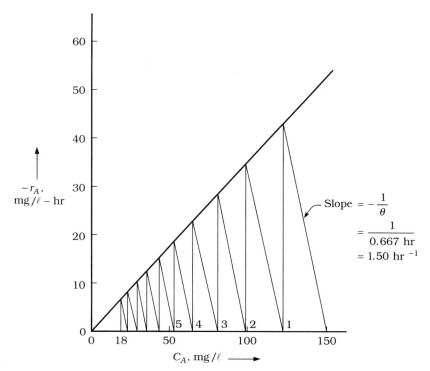

FIGURE 3.14
Graph for
Example 3.5

The conversion for the plug-flow reactor is

$$X_A = (150 - 14.5)(1/150)100\% = 90.3\%$$

Thus the performance for the series of 10 CSTRs is 88.0/90.3 or $\boxed{97.5\ \%}$ of that for the plug-flow reactor. Therefore, the performance of 10 CSTRs in series approaches that of a plug-flow reactor that has a residence time equal to the total residence time of the series of CSTRs.

For an algebraic solution for a series of CSTRs having the same residence time, θ, assume for the first stage,

$$\frac{C_{A_1}}{C_{A_0}} = \frac{1}{1 + k\theta} \tag{3.42}$$

or

$$C_{A_1} = \left(\frac{1}{1 + k\theta}\right)C_{A_0} \tag{3.43}$$

For the second stage,

$$\frac{C_{A_2}}{C_{A_1}} = \frac{1}{1 + k\theta} \tag{3.44}$$

or

$$C_{A_2} = \left(\frac{1}{1 + k\theta}\right) C_{A_1} \tag{3.45}$$

Combining Eqs. (3.43) and (3.45) gives

$$C_{A_2} = \left(\frac{1}{1 + k\theta}\right)\left(\frac{1}{1 + k\theta}\right) C_{A_0} \tag{3.46}$$

or

$$\frac{C_{A_2}}{C_{A_0}} = \left(\frac{1}{1 + k\theta}\right)^2 \tag{3.47}$$

Generalizing for n stages gives

$$\frac{C_{A_n}}{C_{A_0}} = \left(\frac{1}{1 + k\theta}\right)^n \tag{3.48}$$

where

C_{A_n} = effluent concentration of Species A from reactor number n

C_{A_0} = influent concentration of Species A in the influent to the first reactor

k = rate constant

θ = detention time of each reactor in the series — that is, $\theta = V/Q$, where V is the volume of each reactor and Q is the flowrate

n = number of reactors in series

EXAMPLE 3.6 *Series of CSTRs*

Repeat Example 3.5 for the algebraic solution of 10 CSTRs. The values $k = 0.35\,\text{hr}^{-1}$, $C_{A_0} = 150\,\text{mg}/\ell$, and $\theta = 0.667\,\text{hr}$ for each reactor.

SOLUTION The generalized equation is

$$C_{A_n} = \left(\frac{1}{1 + k\theta}\right)^n C_{A_0}$$

Substituting gives

$$C_{A_n} = \left[\frac{1}{1 + (0.35)(0.667)}\right]^n \cdot 150$$
$$= 0.81734^n \cdot 150$$

Substituting gives $n = 1$, $C_{A_1} = 122\,\text{mg}/\ell$; $n = 2$, $C_{A_2} = 99$; $n = 3$, $C_{A_3} = 80$; $n = 4$, $C_{A_4} = 65$; $n = 5$, $C_{A_5} = 53$; $n = 6$, $C_{A_6} = 43$; $n = 7$, $C_{A_7} = 35$; $n = 8$, $C_{A_8} = 28$; $n = 9$, $C_{A_9} = 23$; and $n = 10$, $C_{A_{10}} = 18\,\text{mg}/\ell$.

Since the effluent $C_{A_{10}} = 18\,\text{mg}/\ell$ is the same as in Example 3.5, the performance of 10 CSTR is the same as that found in Example 3.5.

The previous discussion on the continuously stirred tank reactor (CSTR) has shown that for a first-order reaction, the residence time for a CSTR will be greater than for a plug-flow reactor. In order to improve the performance of a single CSTR, it is possible to use a series of the reactors with the total volume equal to that for a single reactor. A series of 10 CSTRs will have a performance approaching that of a plug-flow reactor with a residence time equal to the total residence time of the 10 CSTRs.

Dispersed Plug-Flow Reactors   The flow regime in these reactors is between plug-flow and completely mixed flow and is termed plug-flow with dispersion or dispersed plug flow. The flow pattern is considered a nonideal-flow regime. Danckwerts (1953) developed the fundamental approach to understanding nonideal flow in reactors, and further developments by researchers have contributed to the present state of knowledge.

The performance equation developed by Wehner and Wilhelm (1956) for the dispersed plug-flow reactor is

$$\frac{C_{A_1}}{C_{A_0}} = \frac{4i \exp[1/2(vL/D)]}{(1+i)^2 \exp[i/2(vL/D)] - (1-i)^2 \exp[-i/2(vL/D)]} \qquad (3.49)$$

Where

i = $[1 + 4k\theta(D/vL)]^{0.5}$, dimensionless

v = velocity in the axial direction, length/time

L = reactor length, length

D = longitudinal or axial dispersion coefficient, (length)^2/time

k = rate constant, time^{-1}

θ = reaction time or residence time, time

The dispersed model will range from plug flow at one extreme to completely mixed flow at the other extreme, depending on the intensity of turbulence or mixing. Consequently, the reactor volume for a dispersed plug-flow reactor will lie between those calculated for plug flow and completely mixed flow. The parameter D, the longitudinal or axial dispersion coefficient, is related to the degree of mixing that occurs as flow goes through the reactor. The flow is characterized by the dimensionless group D/vL, which is referred to as the dispersion number, d, or,

$$d = \frac{D}{vL} \qquad (3.50)$$

where

d = dispersion number, dimensionless

D = longitudinal or axial dispersion coefficient, (length)^2/time

v = axial velocity, length/time

L = reactor length, length

The dispersion number, d, measures the degree of axial dispersion. Thus when $D/vL \to 0$, there is negligible dispersion and the flow is plug flow. On the other hand, when $D/vL \to \infty$, there is considerable dispersion and the flow is completely mixed. The dispersion number, $d = D/vL$, will affect the performance of a dispersed plug-flow reactor.

Figure 3.15 gives a graphical comparison of the performance of the dispersed plug-flow reactor to that of the ideal plug-flow reactor for a first-order reaction (Levelspiel and Bischoff, 1959 and 1961). It was prepared by using Eq. (3.49) and the performance equation for an ideal plug-flow reactor. This figure relates the effluent concentration, C_{A_1}, the influent concentration, C_{A_0}, the dispersion number, d, where $d = D/vL$, and the ratio of dispersed plug-flow reaction time, θ_{dpf}, to the plug-flow reaction time, θ_{pf}. Since the volume is proportional to the reaction time, it follows that $\theta_{dpf}/\theta_{pf} = V_{dpf}/V_{pf}$. From the graph, it can be seen that the dispersion number, d, varies from $d = D/vL = 0$ (for plug flow) to $d = D/vL = \infty$ (for completely mixed flow).

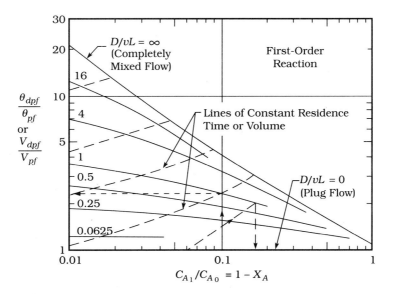

FIGURE 3.15 *Comparison between Dispersed Plug-flow and Plug-Flow Reactors for a First-Order Reaction of the Type $A \to$ Products*

Adapted with permission from "Backmixing in the Design of Chemical Reactors" by O. Levenspiel and K. B. Bischoff in *Industrial and Engineering Chemistry* 51, no. 12 (December 1959): 1431; and from "Reaction Rate Constant May Modify the Effects of Backmixing" by O. Levenspiel and K. B. Bischoff in *Industrial and Engineering Chemistry* 53, no. 4 (April 1961):313. Copyright 1959, 1961 American Chemistry Society.

EXAMPLE 3.7 *Dispersed Plug-Flow Reactors*

A first-order dispersed plug-flow reactor is to have 90% conversion of Species A, and using the performance equation for a plug-flow reactor gives a residence time of 6.58 hours. If the dispersion number $d = 1.0$, determine the reactor residence time for the dispersed plug-flow reactor.

SOLUTION The value of C_{A_1}/C_{A_0} is

$$\frac{C_{A_1}}{C_{A_0}} = 1 - X_A = (1 - 0.90) = 0.10$$

From Figure 3.15, the y-axis intercept for $C_{A_1}/C_{A_0} = 0.10$ and $d = D/vL = 1.0$ is 2.2. Thus,

$$\theta_{\text{dpf}} = 2.2\theta_{\text{pf}} = 2.2(6.58) = \boxed{14.5 \text{ hours.}}$$

EXAMPLE 3.8 *Dispersed Plug-Flow Reactors*

A first-order reaction has a rate constant, k, of $0.35 \, \text{hr}^{-1}$, and a plug-flow reactor has a residence time of 8 hours. What is the fraction of Species A remaining for the plug-flow reactor? If the dispersion number $d = 1.0$, what fraction of A will remain for a dispersed plug-flow reactor?

SOLUTION The performance of a plug-flow reactor is given by

$$\frac{C_{A_1}}{C_{A_0}} = e^{-k\theta} = e^{-(0.35)8} = \boxed{0.061}$$

Using $C_{A_1}/C_{A_0} = 0.061$ in Figure 3.15 and following an interpolated line for a constant residence time until the D/vL line of 1.0 is intersected gives an x-axis value of 0.18. Thus,

$$\frac{C_{A_1}}{C_{A_0}} = \boxed{0.18}$$

A comparison of a CSTR and a plug-flow reactor can be made using Figure 3.15. It can seen that the size of the CSTR relative to that of the plug-flow reactor is a function of the fractional conversion. For example, if $X_A = 0.90$, then $1 - X_A = 1 - 0.90 = 0.10 = C_{A_1}/C_{A_0}$. From Figure 3.15, a CSTR would have a volume 3.9 times that of a plug-flow reactor for a first-order reaction. If $X_A = 0.80$, then $1 - X_A = 1 - 0.80 = 0.20 = C_{A_1}/C_{A_0}$. From Figure 3.15, a CSTR would have a volume 2.6 times that of a plug-flow reactor for a first-order reaction.

In conventional plug-flow activated sludge reactor basins, the dispersion number, d, is from 0 to 0.2. For dispersed plug-flow activated sludge reactor basins, the dispersion number, d, is from 0.2 to 4.0. For completely mixed activated sludge reactor basins with mechanical aerators, the dispersion number, d, is from 4.0 to ∞. For oxidation ponds, the dispersion number, d, usually is from 0.1 to 2.0 (Metcalf and Eddy, 1979).

The dispersion number, $d = D/vL$, for any reactor can be determined by slug tracer injections in the feed stream and observation of the tracer concentrations versus time in the effluent stream. The procedures for the data analysis are presented in the textbook by O. Levenspiel (1972).

In review of the previous discussion on the dispersed plug-flow reactor, it can be seen that it represents a reactor in which axial dispersion is occurring. For a given conversion, the dispersed plug-flow reactor will have a residence or reaction time greater than that of the plug-flow reactor. The design procedure is first to determine the reaction time required for a plug-flow reactor, θ_{pf}, and then to use Figure 3.15 to determine the ratio θ_{dpf}/θ_{pf}. From this, the reaction time for the dispersed plug-flow reactor, θ_{dpf}, can be determined. The dispersed plug-flow reactor is between a plug-flow reactor and a completely mixed reactor.

In summary, there are four types of reactors: (1) batch reactors, (2) plug-flow reactors, (3) completely mixed reactors (CSTRs), and (4) dispersed plug-flow reactors. For a first-order reaction, the residence times of batch and plug-flow reactors are the same. For a first-order reaction, the residence times of the dispersed plug-flow and completely mixed reactors are larger than those of batch and plug-flow reactors. The residence time of the completely mixed reactor is the largest. Even so, the completely mixed reactor offers many advantages that make it widely used, as shown in subsequent chapters of this book. For further study on reactors, the reader is referred to the textbook by O. Levenspiel (1972) entitled *Chemical Reaction Engineering*.

REFERENCES

Bischoff, K. B. 1966. Optimal Continuous Fermentation Reactor Design. *Can. Jour. Chem. Eng.* 44, 281.

Brey, W. S., Jr. 1958. *Principles of Physical Chemistry*. New York: Appleton-Century-Crofts.

Castellan, G. W. 1964. *Physical Chemistry*. Reading, Mass: Addison-Wesley.

Cholette, A.; Blanchet, J.; and Cloutier, L. 1960. Performance of Flow Reactors at Various Levels of Mixing. *Can. Jour. Chem. Eng.* 38:1.

Danckwerts, P. V. 1953. Continuous Flow Systems. *Chem. Eng. Sci.* 2, no. 1:1.

Denbigh, K. G. 1961. Instantaneous and Overall Reaction Yields. *Chem. Eng. Sci.* 14:25.

Denbigh, K. G. 1944. Velocity and Yield in Continuous Reaction Systems. *Trans. Faraday Soc.* 40:352.

Denbigh, K. G., and Turner, J. C. R. 1971. *Chemical Reactor Theory*. 2nd ed. Cambridge, England: Cambridge University Press.

Douglas, J. N., and Bischoff, K. B. 1964. Variable Density Effects and Axial Dispersion in Chemical Reactors. *Ind. Eng. Chem. Process Design Develop.* 3, no. 2:130.

Eldridge, J. W., and Piret, E. L. 1959. Continuous-Flow Stirred-Tank Reactor Systems. *Chem. Eng. Prog.* 55, no. 2:44.

Greenhalgh, R. E.; Johnson, R. L.; and Nott, H. D. 1959. Mixing in Continuous Reactors. *Chem. Eng. Prog.* 55, no. 2:44.

Harrell, J. E., and Perona, J. J. 1968. Mixing of Fluids in Tanks of Large Length-to-Diameter Ratio by Recirculation. *Ind. Eng. Chem. Process Design Develop.* 7, no. 3:464.

Jones, R. W. 1951. A General Graphical Analysis of Continuous Reactions in Series of Agitated Vessels. *Chem. Eng. Prog.* 47, no. 1:46.

Lessells, G. A. 1957. Evaluate Kinetic Data Rapidly. 1957. *Chem. Eng.* 64:251.

Levenspiel, O. 1958. Longitudinal Mixing of Fluids Flowing in Circular Pipes. *Ind. Eng. Chem.* 50, no. 3:343.

Levenspiel, O. 1972. *Chemical Reaction Engineering.* 2nd ed. New York: John Wiley.

Levenspiel, O., and Bischoff, K. B. 1959. Backmixing in the Design of Chemical Reactors. *Ind. Eng. Chem.* 51, no. 12:1431.

Levenspiel, O., and Bischoff, K. B. 1961. Reaction Rate Constant May Modify the Effects of Backmixing. *Ind. Eng. Chem.* 53, no. 4:313.

Levenspiel, O., and Chatlynne, C. Y. 1970. Tracer Curves and the Residence Time Distribution. *Chem. Eng. Sci.* 25:1611.

Levenspiel, O., and Smith, W. K. 1957. Notes on the Diffusion-type Model for the Longitudinal Mixing of Fluids in Flow. *Chem. Eng. Sci.* 6:227.

Levenspiel, O., and Turner, J. C. R. 1970. The Interpretation of Residence-Time Experiments. *Chem. Eng. Sci.* 25:1605.

MacDonald, R. W., and Piret, E. L. 1951. Continuous Flow Stirred Tank Reactor Systems. *Chem. Eng. Prog.* 47, no. 7:363.

Metcalf and Eddy, Inc. 1991. *Wastewater Engineering: Treatment, Disposal and Reuse.* 3rd ed. New York: McGraw-Hill.

Reynolds, T. D. 1982. *Unit Operations and Processes in Environmental Engineering.* Boston: PWS Publishing.

Smith, J. M. 1970. *Chemical Engineering Kinetics.* 2nd ed. New York: McGraw-Hill.

Szepe, S., and Levenspiel, O. 1964. Optimization Backmix Reactors in Series for a Single Reaction. *Ind. Eng. Chem. Process Design Develop.* 3, no. 3:214.

Trambouze, P. J., and Piret, E. L. 1959. Continuous Stirred Tank Reactors. *A. I. Ch. E. Jour.* 5, no. 3:384.

Wehner, J. F., and Wilhelm, R. H. 1956. Boundary Conditions of Flow Reactor. *Chem. Eng. Sci.* 6:89.

PROBLEMS

3.1 During a chemical reaction, the following concentrations of Species A at the various times were observed.

TIME, hours	CONCENTRATION OF A, mg/ℓ
0	50.8
7.5	32.0
15	19.7
22.5	12.3
30	7.6

Determine:

a. The reaction order and rate constant, k.

b. The concentration of A after 20 hr.

3.2 During a chemical reaction, the following concentrations of Species A at the various times were observed.

TIME, hours	CONCENTRATION OF A, mg/ℓ
0	125
5	38.5
10	23.3
15	16.1
20	12.5

Determine:

a. The reaction order and rate constant, k.

b. The concentration of A after 18 hours.

3.3 A plug-flow chemical reactor has an influent flow with a concentration of 150 mg/ℓ of A and a flowrate of 100 gal/min (380 ℓ/min). The reaction is first order, the rate equation is

$$-\frac{dC_A}{dt} = kC_A = -r_A$$

and the rate constant is 0.40 hr^{-1}.

Determine:

a. The required detention time and the volume of the reactor if the effluent contains 20 mg/ℓ of A. Express volume as gallons and liters.

b. A plot of percent C_A removed or converted versus detention time.

3.4 A completely mixed chemical reactor has an influent flow with a concentration of 150 mg/ℓ of A and a flowrate of 100 gal/min (380 ℓ/min). The reaction is first order, the rate equation is

$$-\frac{dC_A}{dt} = kC_A = -r_A$$

and the rate constant is 0.40 hr^{-1}.
Determine:
a. The required detention time and the volume of the reactor if the effluent contains 20 mg/ℓ of A. Express volume as gallons and liters.
b. A plot of percent C_A removed or converted versus detention time.
c. How many times larger a completely mixed reactor must be than a plug-flow reactor for 80% removal or conversion.
d. How many times larger a completely mixed reactor must be than a plug-flow reactor for 90% removal or conversion.

3.5 A dispersed plug-flow chemical reactor has an influent flow with a concentration of 150 mg/ℓ of A and a flowrate of 100 gal/min (380 ℓ/min). The reaction is first order, the rate equation is

$$-\frac{dC_A}{dt} = kC_A = -r_A$$

and the rate constant is 0.40 hr^{-1}. The axial or longitudinal dispersion number, D/vL, is 1.0.
Determine:
a. The required detention time and the required volume if the removal or conversion of A is 90%. Express volume as gallons and liters.
b. The required detention time and reactor volume if the removal or conversion of A is 80%. Express volume as gallons and liters.

3.6 A plug-flow reactor has a conversion of A of 90%, and the reaction is first order. If a dispersed plug-flow reactor has the same volume and has a dispersion number, D/vL, of 1.0, what will be the percent conversion?

3.7 Three CSTRs of equal volume are to be used in series. The influent flow has a concentration of 150 mg/ℓ of A, and the flowrate is 100 gal/min (380 ℓ/min). The reaction is first order, the rate equation is

$$-\frac{dC_A}{dt} = kC_A = -r_A$$

and the rate constant is 0.40 hr^{-1}.
Determine:
a. A graph of $-r_A$ on the y-axis versus C_A on the x-axis on arithmetic paper.
b. The mean residence time and volume for each reactor if the removal or conversion of A is 90%. What are the concentration of A and the rate of conversion in each reactor? Express volume as gallons and liters.
c. If a single CSTR has a mean residence time equal to the total of the mean residence times for the three CSTRs in series in part (b), what will the effluent concentration of A be? What will the percent removal or conversion be? For parts (b) and (c), use a graphical solution using the graph from part (a).

3.8 Three CSTRs are to be used in series. The second reactor has a volume twice that of the first and third reactors. The influent flow has a concentration of 150 mg/ℓ of A, and the flowrate is 100 gal/min (380 ℓ/min). The reaction is first order, the rate equation is

$$-\frac{dC_A}{dt} = kC_A = -r_A$$

and the rate constant is 0.40 hr^{-1}.
Determine:
a. A graph of $-r_A$ on the y-axis versus C_A on the x-axis on arithmetic paper.
b. The mean residence time and volume of each reactor if the removal or conversion of A is 90%. What are the concentration of A and the rate of conversion in each reactor? use a graphic solution using the graph from part (a). Express volume as gallons and liters.

3.9 Three CSTRs of equal volume are to be used in series. The influent flow has a concentration of 150 mg/ℓ of A, and the flowrate is 100 gal/min (380 ℓ/min). The reaction is second order, the rate equation is

$$-\frac{dC_A}{dt} = kC_A^2 = -r_A$$

and the rate constant is 0.0070 ℓ/mg-hr.
Determine:
a. The graph of $-r_A$ on the y-axis versus C_A on the x-axis on arithmetic paper.
b. The mean residence time and volume of each reactor if the removal or conversion of A is 80%. What are the concentration of A and the rate of conversion in each reactor? Use a graphical solution employing the graph from part (a). Express volume as gallons and liters.

3.10 Three CSTRs are to the used in series. The second reactor has a volume twice that of the first and third reactors. The influent flow has a concentration of 150 mg/ℓ of A, and the flowrate is 100 gal/min (380 ℓ/min). The reaction is second order, the rate equation is

$$-\frac{dC_A}{dt} = kC_A^2 = -r_A$$

and the rate constant is 0.0070 ℓ/mg-hr.
Determine:

a. A graph of $-r_A$ on the *y*-axis versus C_A on the *x*-axis on arithmetic paper.

b. The mean residence time and volume of each reactor if the removal or conversion of A is 80%. What are the concentration of A and the rate of conversion in each reactor? Use a graphic solution employing the graph from part (a). Express volume as gallons and liters.

3.11 Three CSTRs of equal volume are to be used in series. The influent flow has a concentration of 150 mg/ℓ of A, and the flow rate is 100 gal/min (380 ℓ/min). The reaction is first order, the rate equation is

$$-\frac{dC_A}{dt} = kC_A = -r_A$$

and the reaction constant is 0.40 hr^{-1}.
Determine using the algebraic equation:

a. The mean residence time and volume for each reactor if the conversion of A is 90%. Express volume as gallons and liters.

b. The concentration of A in each reactor and its effluent.

c. The total mean residence time and total volume for the three reactors. Express volume as gallons and liters.

d. The mean residence time and volume for a single CSTR having the same conversion of A. Express volume as gallons and liters.

4

WATER QUANTITIES AND WATER QUALITY

The design of a water treatment plant requires a knowledge of the quantity or flowrate of the water to be provided, the quality of the untreated or raw water, and the quality required for the product water. The quantity or flowrate to be provided determines the size of the various unit operations and processes to be provided and also the hydraulic design of the plant. The quality of the untreated water and the quality of the product water determine which unit operations and processes are to be provided for the plant.

WATER QUANTITIES

Determining the quantity of water to be provided for a treatment plant requires knowing the population to be served and the per capita water usage — that is, the water demand. It is common to express water consumption or usage in terms of gallons or liters used per capita per day. This figure is determined by dividing the total daily consumption by the total population in the city. Usually, the average daily consumption over a period of one year is used. Fluctuations in flow are usually expressed as percentage ratios of maximum or minimum monthly, weekly, daily, or hourly rates to their respective average values.

Most water supply systems include relatively large structures, such as a dam, reservoir, and transmission system, such as a canal, aqueduct, or pipeline, that require considerable time to build and are not easily expanded. They also include, below city streets or in the right-of-ways of streets, distribution lines that require interrupting traffic when being constructed. Consequently, the main components are designed to meet the city's needs for a reasonable period of time. The **design period** is the number of years for which the system or component is to be adequate. The **design population** is the number of persons to be served. The **design flows** are the rates of consumption for the residential, commercial, industrial, and public areas of the city. The **design area** is the area to be served by the system or component for the residential, commercial, and industrial districts and public areas. The **design capacity** is the capacity of the water treatment plant and its component structures, which for a surface water system would include such components as the intake structure and the low-lift pumps located in or near the intake structure, the conduits or canal to the treatment plant, the treatment plant, the storage reservoirs or clearwells, and the high service pumps that pump the finished water to the distribution system.

In selecting the **design period** for a water treatment plant and its components, several factors must be kept in mind, the main ones being (1) the expected population growth and the development of residential, commerical, and industrial areas, (2) the ease or difficulty of expanding or adding to existing or planned works, (3) the economic life of the existing or planned works, and (4) the rate of interest to be paid on the municipal bonds. Long design periods are justified where the population growth is small, stagewise expansion is difficult, the economic life is long, and the interest rate is low. Structures that are difficult and costly to enlarge, such as dams, intake structures, and conduits or canals to the treatment plant, usually have design periods of 25 to 50 years. Structures that are easily expanded, such as well fields, distribution systems, and water treatment plants, usually have design periods of 10 to 25 years.

Design Populations   Before a water treatment plant or one of its components can be designed, it is necessary to determine the design population — that is, the future population that must be served. This requires not only the design period but also a population forecast for the future.

The U.S. Census Bureau is the main source of past population figures, since the Bureau makes population counts and publishes reports concerning its findings every 10 years. Census Bureau figures are called decennial counts. In order to find the present population, if the year is not a decennial one, an engineer may determine the ratios of the last census population to the number of school enrollments and utility connections. Then, knowing the present school enrollments and present utility connections, the engineer can estimate the present population. The engineer may use a number of utility services, such as water, gas, electric, and telephone connections. Usually, an engineer will compute the present population in as many ways as possible and then use the average value. For North American cities, the ratio of the population to school enrollment is about $5:1$; the ratio of the population to the water, gas, and electric connections is about $3:1$; and the ratio of the population to telephone connections is about $4:1$.

Engineers may use several methods to make population forecasts. Sometimes they use several methods and then apply their judgment to select the most appropriate method. In making a population forecast, it is useful to be familiar with the economic base of the city, its trade territory, whether or not its industries are expanding, the development of the surrounding rural areas, and the city's transportation facilities, such as railroad, highway, and marine facilities. Unusual events, such as the discovery of a new oil field or the sudden development of a new industry, can make a population forecast invalid and require rapid development of its water supply system or of components of that system.

For the **logistic method**, the population versus time plots an S-shaped curve on arithmetic paper, as shown in Figure 4.1, or a straight line on logistic paper, as shown in Figure 4.2. In order to see if this method is

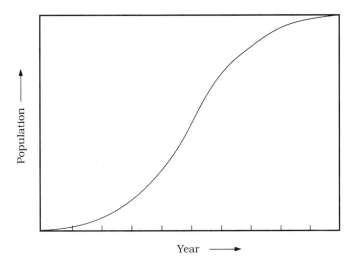

FIGURE 4.1 *Plot of the Logistic Method on Arithmetic Paper*

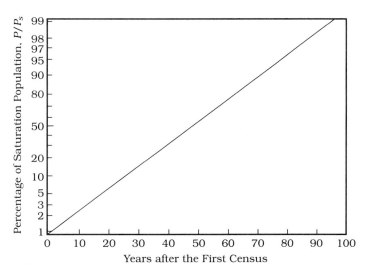

FIGURE 4.2 *Plot of the Logistic Method on Logistic Graph Paper*

applicable, the population as a percent of the saturation population is plotted versus time on logistic paper; and if a straight line results, the method is applicable. The equation for the logistic plot is

$$P = \frac{P_s}{1 + e^{a + b\Delta t}} \qquad (4.1)$$

where

P = population for a future year

P_s = saturation population

a, b = data constants

Δt = future time period, years

To fit the curve, three successive years, represented by t_0, t_1, and t_2, are chosen that are equidistant from each other in the period of time. The years should be selected so that one is near the earliest recorded population for the city, one is near the middle, and one is near the end of the available record of time. The fitted curve will pass through the values of P_0, P_1, P_2, which occur at year periods of t_0, t_1, t_2. The number of years between t_0 and t_1 and between t_1 and t_2 is designated as n. The values of P_s, a, and b are determined from the following equations:

$$P_s = \frac{2P_0P_1P_2 - P_1^2(P_0 + P_2)}{P_0P_1 - P_1^2} \tag{4.2}$$

$$a = \ln\frac{P_s - P_0}{P_0} \tag{4.3}$$

$$b = \frac{1}{n}\ln\frac{P_0(P_s - P_1)}{P_1(P_s - P_0)} \tag{4.4}$$

Once P_s, a, and b have been determined, they may be substituted into Eq. (4.1) to give the estimated population, P, for any time period, Δt, beyond the base year where P_0 occurs. The logistic method is based on the fact that populations will grow until they reach a saturation population that is established by the limit of economic opportunity. All populations, without regard to size, tend to grow according to the S-shaped curve or logistic curve. The curve starts with a low rate of growth, followed by a high rate, and then by a progressively lower rate until the saturation population is reached.

For the **constant percentage method**, the rate of growth is proportional to the population and is represented by

$$\frac{dP}{dt} = KP \tag{4.5}$$

where

dP/dt = rate of growth

K = rate constant

P = population

Integration of this equation yields

$$\ln P = \ln P_0 + K\Delta t \tag{4.6}$$

In order to see if this method is applicable, the population versus time is plotted on semi-log paper; and if the method applies, the data plot a straight line. The value of the constant K may be determined from the slope of the line. The constant K may also be determined from the equation

$$K = \frac{\ln P - \ln P_0}{\Delta t} \tag{4.7}$$

If this method applies, it approximates the early era of a logistic curve — that is, the beginning of an S-shaped growth curve.

For the **arithmetic method**, the rate of growth is equal to a constant and is represented by

$$\frac{dP}{dt} = K \tag{4.8}$$

where

dP/dt = growth rate

K　　= constant

Integration of Eq. (4.8) gives

$$\frac{\Delta P}{\Delta t} = K \tag{4.9}$$

Also, integration of Eq. (4.8) gives

$$P = P_0 + K\Delta t \tag{4.10}$$

If this method is applicable, a plot of population versus time on arithmetic paper will yield a straight line with a slope equal to K. If this method applies, it indicates a middle era of an S-shaped growth curve, which approximates a straight line.

For the **declining-growth method**, the rate of growth is represented by

$$\frac{dP}{dt} = K(P_s - P) \tag{4.11}$$

where

dP/dt = growth rate

K　　= constant

P_s　　= saturation population

P　　= population at any time, t

This method, like the logistic method, is based on the city having a limited saturated population that represents the maximum economic development. The value of the saturation population, P_s, may be determined the same way as described for the logistic method or by assuming a maximum area of growth and a maximum population density. The constant, K, may be determined for 10-year increments using the equation

$$K = -\frac{1}{n} \cdot \ln \frac{P_s - P}{P_s - P_0} \qquad \textbf{(4.12)}$$

where n equals the time interval between successive censuses. By solving for several K values, an average K may be determined for calculations. Integration of Eq. (4.12) and rearranging give

$$P = P_0 + (P_s - P_0)(1 - e^{K\Delta t}) \qquad \textbf{(4.13)}$$

where P equals the future population and all other variables are as previously defined. If this method applies, it approximates the latter era of an S-shaped growth curve.

For the **percentage method** the population projections by state or federal agencies, such as the Bureau of Census are used, and it assumes that the city being studied will grow in proportion to the growth of the state or nation — that is, the city's population will be a determined percent of the state or national growth, and the percent will remain constant in the future.

For the **curvilinear method**, a graphical projection is made of the past populations of similar but larger cities, as shown in Figure 4.3. The cities selected must be similar to, but larger than, the city being studied. The cities selected for comparison should be similar in as many characteristics as possible, such as economic base, geographical proximity, access to transportation systems, and so on. As shown in Figure 4.3, the city being studied is City X. Its population is plotted up to the last year of decennial census by using census records. The population is extended up to the present year — that is the reference population — using data on the school enrollments and utility connections for the last year of decennial census and for the present year. City A reached the present population about 65 years ago, and its population is plotted from then to about 26 years into the future at the intersection of the present population and the present year. Cities B, C, and D are plotted likewise, and from these plots, the average populations were determined for the future decennial years. A smooth curve is plotted through these points by eye to give the projected, dashed-line curve for City X. If care is used in selecting the cities to be compared, this method will give reasonable results. Because of its accuracy, this is the most commonly used method of making population projections. In comparing this method to the S-shaped growth curve, it can be seen that this method encompasses any era on the S-shaped curve.

Water Usage Water is used for domestic, commercial, industrial, and public purposes in addition to that lost to leakage and wastage. Table 4.1 shows the average annual demand for numerous North American cities and counties. Table 4.2 shows the range in usage for the North American continent as well as average values.

Water consumption depends upon the amount of water used for various

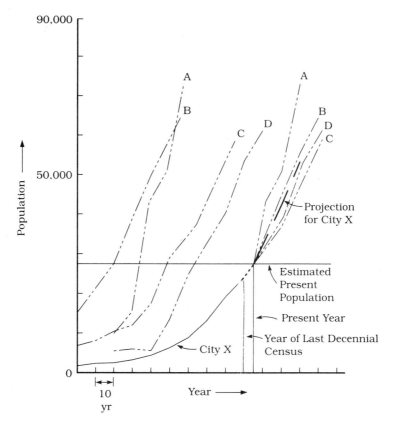

FIGURE 4.3 *Graphical Comparison with Similar, Larger Cities*

purposes, as shown in Table 4.2. **Domestic use** includes water furnished to houses, apartments, and other dwellings, and it ranges from 15 to 70 gal/cap-day (60 to 265 ℓ/cap-day). A breakdown of domestic consumption shows the following usage (Durfor and Becker, 1964): flushing toilets, 41%; washing and bathing, 37%; kitchen use, 6%; drinking water, 5%; washing clothes, 4%; general household cleaning, 3%; watering lawns and gardens, 3%; and washing the family car or cars, 1%. The amount of water used varies with the standards of living, high standards of living having higher demands. Domestic usage may be 50/150 or 33% of the total consumption for the average city, but for small cities the domestic usage will be a greater percentage.

Commercial use includes water used for commercial enterprises, such as retailing and wholesaling of various goods and the performance of various services. Commercial endeavors include businesses, laundries, hospitals, etc., and one of the main uses is for air conditioning. **Industrial use** is the use of water in producing a product, and it consists of process water, boiler

TABLE 4.1 *Average Annual Water Demands of Selected Cities and Districts in the United States*

CITY OR DISTRICT	AVERAGE ANNUAL WATER DEMAND	
	gpd/cap	ℓpd/cap
Baltimore, Md.	160	606
Berkeley, Calif.	76	288
Boston, Mass.	145	549
Grand Rapids, Mich.	178	674
Greenville County, S.C.	110	416
Hagerstown, Md.	100	378
Jefferson County, Ala.	102	386
Johnson County, Kan.	70	265
Lancaster County, Neb.	167	632
Las Vegas, Nev.	410	1551
Little Rock, Ark.	50	189
Los Angeles, Calif.	185	700
Peoria, Ill.	90	341
Memphis, Tenn.	125	473
Orlando, Fla.	150	568
Rapid City, S.D.	122	462
Santa Monica, Calif.	137	519
Shreveport, La.	125	473
Wyoming, Mich.	150	568
Average Values	153	580

Adapted from *Design and Construction of Sanitary and Storm Sewers*, Water Pollution Control Federation Manual of Practice No. 9. Copyright © 1969 by American Society of Civil Engineers and Water Pollution Control Federation. Reprinted by permission.

TABLE 4.2 *Water Usage for the North American Continent*

CLASS OF USAGE	QUANTITY, gpd/cap		QUANTITY, ℓpd/cap	
	Range	Average	Range	Average
Domestic	15–70	50	60–265	189
Commercial and industrial	10–100	65	40–380	246
Public	5–20	10	20–75	38
Leakage and waste	10–40	25	40–150	95
	40–230	150	160–870	568

Adapted from *Water and Wastewater Engineering*, Vol. 1, *Water Supply and Wastewater Removal* by G. M. Fair, J. C. Geyer, and D. A. Okun. Copyright © 1966 by John Wiley & Sons, Inc. Reprinted by permission of John Wiley & Sons, Inc.

makeup water, and cooling water. Industrial endeavors include paper mills, breweries, steel mills, chemical plants, refineries, fermentation industries, textile mills, meat packing, milk processing, etc. The water used for commercial and industrial purposes varies from about 2200 to 7640 gal/acre-day (20,600 to 71,400 ℓ/ha-day) (Davis and Sorensen, 1969). These figures are given as approximations, since, for a particular industrial or commercial area, a special study should be made to determine usage per unit of ground area. In many cases, industries will develop their own water supplies, particularly if groundwater is available, and will be independent of the city's system. Industries often require large amounts of water; the actual amount depends upon the extent of the manufacturing and the type of industry. Typical values for various industries are reported in Table 4.3. Some industries develop their own water supplies and place little if any demand on the municipal water supply system. Extensive reuse of water by industries will result in demands lower than those shown in Table 4.3.

Public uses are for public buildings (such as city halls, fire stations, jails, and schools), sprinkling lawns for public buildings, flushing streets, and fire fighting. The total amount of water used annually for fighting fires is relatively small, but the flowrate at which it is used is quite high for short durations. Water used for public uses is generally not paid for. Public usage is from 5 to 20 gal/cap-day (20 to 75 ℓ/cap-day).

Leakage and wastage, or water unaccounted for, is due to meter slippage, leaks in the city water mains, and unauthorized connections. Leakage and wastage is from 10 to 40 gal/cap-day (40 to 150 ℓ/cap-day).

TABLE 4.3 *Water Usage for Selected Industries*

PRODUCT	UNIT OF PRODUCTION	TYPICAL WATER USE gal/unit	TYPICAL WATER USE ℓ/unit
Beet sugar	Ton of beets	25,000	95,000
Beverage alcohol	Gallon	80	300
Beer	Barrel	470	1,800
Meat	2000 lb. live weight	5,000	19,000
Canned vegetables	Case	250	950
Cotton goods	Ton	48,000	182,000
Woolens	Ton	140,000	585,000
Leather (tanned)	Ton	16,000	61,000
Paper	Ton	39,000	148,000
Paper pulp	Ton	110,000	416,000
Oil refining	Barrel of crude oil	770	29,000
Steel	Ton	35,000	132,000
Electricity, steam	KW-hr	80	300

Adapted from *Engineering Management of Water Quality* by P. H. McGauhey. Copyright © 1968 by McGraw-Hill, Inc.; and from *Water Resources Engineering* by R. H. Linsley and J. B. Franzini, 3rd ed. Copyright © 1979 by McGraw-Hill, Inc. Reprinted by permission.

Variations in water consumption occur throughout the day, week, and season of the year. The average annual water consumption, as previously mentioned, is the total water consumption for the year divided by the number of days per year and the population of the city. There will be a variation in water demand throughout the day, the week, and the month. The variations are a characteristic of each city, and they must be determined by studying water consumption or water pumpage records for the city being investigated.

Figure 4.4 shows the hourly variation throughout a given day for a large city, and it can be seen that the hourly demand is low throughout the night and reaches a maximum in the morning and in the afternoon. For Figure 4.4, the average hourly consumption is for a given day, such as the maximum day or any particular day, and it is not necessarily the average hourly consumption for the entire year. For a day in the year, representing the average day, the hourly demand will be the average hourly demand for the year. The shape of the characteristic curve will depend upon the city or area being studied. A considerable flow during the night represents loss due to leakage and wastage, unless there is a large industrial demand during the night. The increase in demand during the morning represents the beginning of the daily activities of the consumers. If there is a large industrial demand throughout the day and night, the characteristic curve will be less pronounced than that shown in Figure 4.4. During the week, the minimum consumption

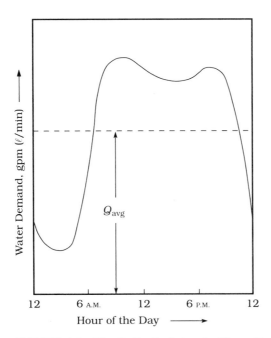

FIGURE 4.4 *The Daily Variation in Water Demand for a Typical City*

is on Sunday and the maximum consumption is usually on Monday. During the fall and winter months, the demand is below average, and during the spring and summer months, the demand is above average due to the warmer weather. Usually, the maximum consumption in North America is during July or August. The smaller the community, the greater the variation in demand. If the demand curve in Figure 4.4 is for the maximum day, the average flow will be the capacity of the water treatment plant or plants. The area above the curve represents the amount of stored water that must empty into the system to meet the demand. The area below the curve represents the excess treated water used to refill storage.

Table 4.4 shows the variations in demand expressed as a ratio of the maximum or minimum to the average or mean. For the average city with a demand of 150 gal/cap-day (570 ℓ/cap-day), the maximum day would be about 150 × 1.5 or 225 gal/cap-day (570 × 1.5 or 855 ℓ/cap-day). The maximum hourly demand would be about 150/24 × 2.5 or 15.6 gal/cap-hr (570/24 × 2.5 or 59 ℓ/cap-hr). The minimum hourly demand would be about 150/24 × 0.40 or 2.5 gal/cap-hr (570/24 × 0.40 or 9.5 ℓ/cap-hr). Due to the wide variation in demand among the various cities or districts shown in Table 4.1, the only way to determine the average annual demand, as well as the other demands, for a particular city or district is by studying their past pumping records. By studying their past-12-month pumping records, it is possible to determine the average annual demand (average day demand), the maximum day demand, the maximum hourly demand, and the minimum hourly demand. These demands must be increased to obtain design demands at the end of the design period.

As a city grows, the **future water consumption** increases. For North American cities, the per capita consumption of water has increased by about one-tenth of their population increase. If in 1980 a city's population was 50,000, and for the year 2000 the population is expected to be 75,000, the growth is (75,000 − 50,000)/50,000 × 100%, or 50%. If in 1980 the city's water consumption was 150 gal/cap-day (570 ℓ/cap-day), the demand in the year 2000 would be 150[100% + (0.1%)(50)] or 158 gal/cap-day (570 × 105% or 599 ℓ/cap-day).

TABLE 4.4 *Variations in Demand for North American Cities*

RATIO	NORMAL RANGE	AVERAGE
Maximum day : average day	(From 1.2 to 2.0) : 1	1.5 : 1
Maximum hour : average hour	(From 2.0 to 3.0) : 1	2.5 : 1
Minimum hour : average hour	(From 0.25 to 0.50) : 1	0.40 : 1

Adapted from *Water and Wastewater Engineering*, Vol. 1, *Water Supply and Wastewater Removal* by G. M. Fair, J. C. Geyer, and D. A. Okun. Copyright © 1966 by John Wiley & Sons, Inc. Reprinted by permission of John Wiley & Sons, Inc.

Design Capacities   The **river** used as a water supply must have a flow equal to or greater than the flow on the maximum day. If a **well field** is used, the flow must be equal to that on the maximum day. The **low-lift pumps**, located in intake structures on a river or a reservoir, must have a capacity equal to the average flow on the maximum day plus a reserve of usually 10 to 33% of the pump capacity. The **treatment plant**, which is usually a rapid sand filter plant for a surface water, must have a capacity equal to the average flow on the maximum day — that is, the flow on the maximum day. The **service reservoirs** at a treatment plant must have sufficient volume so that the treatment plant can operate at the average flow on the maximum day, the excess flow during the high demand during the day being provided by the reservoir. The **high service pumps**, which pump from reservoirs for treated water at the treatment plant, must have a capacity equal to the maximum hourly flow plus a reserve of usually 10 to 33%. The **distribution system** must have a capacity equal to the maximum hourly flow or the average hourly flow on the maximum day plus the fire demand, this being referred to as the coincident draft plus fire demand. One of these flow conditions will, of course, control. The **ground storage reservoirs** at the treatment plant and at other locations in the city's network system must have enough storage to furnish the peak hourly flow on the maximum day or the average hourly flow on the maximum day plus the fire demand. Usually, this is about equal to the amount of water used on the maximum day or about 150 gal/cap (about 570 ℓ/cap). Ground storage reservoirs within a distribution system may be located on elevated terrain, such as hills, or if located on relatively flat terrain, they will require booster pump stations to furnish the required flow at an adequate pressure. The **elevated storage** required for fire demand is about 55 gal/cap (210 ℓ/cap) or more.

WATER QUALITY   Pure water is a colorless, tasteless, and odorless compound containing hydrogen and oxygen with the chemical formula H_2O. It is a universal solvent, and as a result, it dissolves many natural and human-made substances as it travels through the environment. In addition, it suspends some fine solids, both natural and human-made. The dissolved and suspended materials are termed **impurities**, and they gain entry into water as a result of the hydrologic cycle.

Impurities in Water   In falling, rain water picks up dust particles, plant pollens, dissolved gases such as oxygen, nitrogen, carbon dioxide, and sulfur dioxide, and other chemical substances. Thus rain water is not pure but contains some impurities collected as it falls. Surface runoff carries anything it can transport or dissolve, such as silt and clay, vegetation fragments, bacteria, microscopic life forms, waste products, and weathered minerals from rock strata. Surface runoff is the main contribution to streams, rivers, and lakes, along with subsurface runoff, groundwater flow, and wastewater flows. In lakes and reservoirs, some suspended materials, such as silt and clay, and other impurities, may settle out. However, microbes, plants, and animal life may grow and also die, thus contributing impurities to the water.

The rainfall, which percolates down into the groundwater zone, dissolves some minerals in strata through which it passes. The extent of the dissolving depends upon the quantity of flow, the composition of the strata, and the acidity of the water. Thus groundwater usually contains more dissolved minerals than surface water. The usual storage strata for groundwaters are termed aquifers, and they usually have significant porosity and permeability, such as sedimentary formations of sands and gravels. Limestones may yield some groundwater from caverns in the formations. Sandstones and shales may yield groundwater because of their bedding planes, joints, and porosities. Igneous formations, such as granite and basalt, and metamorphic formations, such as slate and marble, usually are poor sources of groundwater because of limited joints, crevices, and cleavage planes.

The common impurities imparted to water as it passes through nature are classified as **dissolved** and **suspended**. The main dissolved impurities are cations, anions, organic compounds such as vegetative dyes, and gases such as oxygen, carbon dioxide, and hydrogen sulfide. The main suspended impurities are silts, clays, organic and inorganic colloids, and living organisms, such as bacteria, viruses, protozoa, and algae. Most bacteria, viruses, and protozoa present are harmless to humans, but some may cause disease. Certain algae cause disagreeable tastes and odors, in addition to turbidity. Certain salts in high concentrations cause hardness, tastes, and corrosiveness. Some gases cause corrosiveness, acidity, and odors. Normally, the chemicals produced by the decomposition of organic matter are not present in sufficient concentrations to cause any effects. Surface waters, because of their overland travel, usually contain appreciable amounts of suspended material as compared to groundwaters. Groundwaters, since they undergo natural filtration while passing through sand formations, usually have negligible amounts of suspended material, but because of their contact with certain strata, such as limestone, they may have high dissolved salt contents.

Sampling and Units

Samples collected must be representative to yield valid test results. Groundwaters are usually very uniform in quality throughout the year, whereas surface waters exhibit wide variation in quality with time. This variation in surface water quality requires that samples be taken over a period of time to be representative. The collecting procedures, preservation procedures, and laboratory analyses, such as dissolved gases and temperature, must be done in the field to be meaningful. Bacterial samples must be carefully preserved and analyzed as soon as possible to be valid.

The most commonly used unit to express the concentration of an impurity is milligrams per liter (mg/ℓ), which is also equal to grams per cubic meter (g/m^3). Coliform bacterial analyses, however, are expressed as the number of organisms per $100\,m\ell$. The **most probable number (MPN)** is the statistical number that represents the number of coliforms per $100\,m\ell$.

Sometimes concentrations are expressed on a weight basis as parts per million (ppm), the relationship between ppm and mg/ℓ being

$$\text{ppm} = \frac{\text{mg}/\ell}{\text{specific gravity of liquid}} \tag{4.14}$$

It can be seen that, if the specific gravity is 1.00, which is the usual case, 1 ppm = 1 mg/ℓ. Another expression frequently used is the number of milliequivalents per liter (meq/ℓ). The weight of a milliequivalent is

$$\text{meq weight} = \frac{\text{milligram atomic or molecular weight}}{Z} \tag{4.15}$$

where, for most compounds, Z is the number of replaceable hydrogen atoms or their equivalent. Thus Z is the valence involved. For oxidation and reduction reactions, Z is the change in valence. The number of milliequivalents (meq) is

$$\text{Number of milliequivalents (meq)} = \frac{\text{mg weight}}{\text{meq weight}} \tag{4.16}$$

Frequently, chemical analyses, such as alkalinity and hardness, are expressed in terms of calcium carbonate $CaCO_3$, which has 50 mg/meq.

Water Quality Characteristics   The quality characteristics of a water may be classified as physical, chemical, or biological, according to the nature of the constituent or impurity, as shown in Figure 4.5. Common physical characteristics are suspended solids, color, taste, odor, and temperature. Common chemical characteristics are pH, cation and anion concentrations, alkalinity, acidity, hardness, dissolved solids, conductivity, and priority pollutants. Common biological chacteristics are coliform and viral counts.

Physical Characteristics   The **total solids** are the solids remaining after evaporating a sample and drying the residue. The **suspended solids** are the solids that may be removed by filtration of the sample in the laboratory. The suspended solids in a water are mainly due to silt, clay, and algal content; however, if the filter pore size is very small, the filtration procedure will also

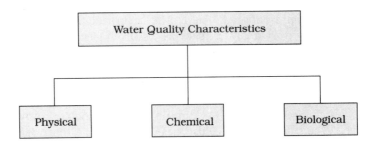

FIGURE 4.5 *Water Quality Characteristics*

remove colloidal materials such as bacteria and organic debris. The **dissolved solids** are the difference between the total solids and the suspended solids. Thus they represent the salt content of the water — that is, the sum of the cations and anions present.

Turbidity decreases the clarity of a water and is due to fine suspended materials, which may range in size from colloidal to coarse dispersions. It is usually due to colloidal materials such as bacteria, silt, clay, algae, and organic debris. Turbidity interferes with the passage of light through a sample. It is measured by a nephelometric instrument, which measures the degree of reflected light or the reduction in light transmitted through the sample as compared to a standard.

Color is due to fine suspended materials, including colloids, or to dissolved materials. Some surface waters may have color as a result of organic materials leached from organic debris, such as leaves, needles of conifers, and wood in various stages of decomposition. Much of this color is due to colloidal and dissolved materials. Some streams and rivers that drain areas of red clay soils may be reddish during flood stages because of colloidal clays in suspension that cause turbidity. Some surface waters and groundwaters may have color due to the presence of certain metallic ions, such as iron and manganese. Waters with iron will be brownish to reddish, whereas waters with manganese will be brownish to blackish. The color is due to fine suspended precipitates of iron and manganese oxide. **Apparent color** is due to fine suspended materials, most of which is colloidal, such as turbidity, whereas **true color** is due to dissolved substances.

Tastes and odors in surface waters are usually due to organic compounds from algal growths or other microbial growths; decomposing organic matter such as leaves, weeds, or grasses; industrial organic chemicals; synthetic detergents; or agricultural chemicals. In addition, chlorine may react with certain organic compounds to produce products that cause tastes and odors. **Tastes and odors in groundwaters** are usually due to dissolved gases, such as hydrogen sulfide or methane, hydrogen sulfide being the most common cause. Hydrogen sulfide has a distinct rotten egg odor and is very offensive. Both tastes and odors are measured by diluting the sample until the taste or odor is no longer detectable by the analyst. In the threshold odor test, a water sample is diluted until the odor is no longer detectable. The **threshold odor number (TON)** is the total diluted volume divided by the sample volume. For example, if a 20-mℓ sample is diluted with 40 mℓ of distilled water and the odor is not detectable, then the threshold odor number (TON) is (20 + 40)/20 or 3. The test for tastes is the flavor threshold test, which is done similarly to the threshold odor test. If an analyst dilutes a 100-mℓ sample with 100 mℓ of distilled water and the flavor is barely detectable, then the **flavor threshold number (FTN)** is (100 + 100)/100 or 2. Since odor and taste detection varies with the individual person, several analysts usually perform these tests. For details of these and other tests for tastes and odors, the reader is referred to *Standard Methods for Examining Water and Wastewater*.

Temperature of a water is important in terms of the treatments required and its use. Groundwater temperatures vary with the depth and the characteristics of the aquifer. Generally, the temperature of a groundwater is relatively constant throughout the year. Surface water temperatures of a stream or river vary according to the source of the water and its prior use. The temperature of surface waters in reservoirs varies with the depth; the deeper the water, the lower the temperature. Generally, the temperature of a surface water varies significantly throughout the year due to seasonal climatic changes.

Chemical Characteristics   The pH of a water is a measure of its acidic, neutral, or basic nature. The pH is the logarithm of the reciprocal of the hydrogen ion concentration in moles per liter, or

$$pH = \log \frac{1}{[H^+]} \tag{4.17}$$

where $[H^+]$ is the hydrogen ion concentration in moles per liter. For pure water at 24°C, the H^+ ion and OH^- ion concentrations are equal to 10^{-7} mole per liter; thus, pure water has a pH of 7 and is neutral. Waters with a pH less than 7 are acidic, and waters with a pH greater than 7 are basic. The pH is usually determined by a potentiometer, which measures the electrical potential exhibited by the H^+ ion, or by color-indicating dyes. The pH values of natural waters range from 5 to 8.5 and are acceptable except from a corrosive standpoint. The concept of pH is important in chemical coagulation, water softening, disinfection, and corrosion control.

The major **cations** in water — those found in appreciable concentrations — are Ca^{+2}, Mg^{+2}, Na^+, and K^+. The minor cations — those found in smaller amounts — are mainly Fe^{+2} and Mn^{+2}. The major **anions** in water are HCO_3^-, SO_4^{-2}, and Cl^-. The minor anions are F^- and NO_3^-. If a water analysis is correct, the sum of the cations and the sum of the anions, when expressed as equivalents per liter or milliequivalents per liter, must be equal according to the principle of electroneutrality. The main source of cations and anions in surface waters is the weathering of soils, minerals, and rocks in the watershed. The main source in groundwaters is the slow dissolving of soils, minerals, and rock formations as the water passes through the strata. If NH_4^+ and NO_3^- are present in significant amounts, it is indicative of organic matter being present, which means possible waste contamination. The analyses of two surface waters and two well waters are shown in Table 4.5. One surface water, Number (1), and one well water, Number (3), come from a limestone and dolomite plateau area. Since these rocks are essentially $CaCO_3$ and $CaMg(CO_3)_2$, the waters are high in Ca^{+2}, Mg^{+2}, and HCO_3^- due to weathering and dissolution.

The sum of the cation and anion concentrations represents the salt content or all of the **dissolved solids**. The dissolved solids are the residue remaining after evaporation of a filtered sample. Certain cations and anions, as well as the dissolved solids, are important in evaluating a water for domestic purposes.

TABLE 4.5 *Analyses of Typical Waters*

	CONCENTRATION, mg/ℓ			
	Surface Water		Groundwater	
CONSTITUENT	(1)	(2)	(3)	(4)
Calcium (Ca^{+2})	51	11	90	9.2
Magnesium (Mg^{+2})	17	3.5	38	2.6
Sodium (Na^+)	21	22	9.7	202
Potassium (K^+)	4.0	3.3	5.2	5.0
Iron (Fe^{+2})	0.08	0.12	0.10	0.02
Bicarbonate (HCO_3^-)	181	22	383	410
Sulfate (SO_4^{-2})	30	19	39	2
Chloride (Cl^-)	48	38	36	98
Fluoride (F^-)	0.2	0.6	0.2	1.4
Nitrate (NO_3^-)	0.2	0.2	2.2	0
Silica (SiO_2)	8.8	16	16	16
Total dissolved solids	271	142	432	538
Total hardness as $CaCO_3$	197	42	380	34
pH	7.1	6.7	7.0	7.6

(1) Colorado River at Marble Falls, Texas, U.S. Geol. Survey Water Supply Paper 1069, 1949.

(2) Neches River at Port Arthur, Texas, U.S. Geol. Survey Water Supply Paper 1047, 1948.

(3) Groundwater at Stephenville, Texas, U.S. Geol. Survey Water Supply Paper 1069, 1949.

(4) Groundwater at Houston, Texas, U.S. Geol. Survey Water Supply Paper 1047, 1948.

The **alkalinity** of a water is its ability to neutralize an acid and is due to its CO_3^{-2}, HCO_3^-, and OH^- content. The CO_3^{-2}, HCO_3^-, and OH^- are related to the CO_2 and H^+ ion concentration (or pH) as shown by the following equilibrium reactions (Sawyer *et al.*, 1994):

$$CO_2 + H_2O \rightleftharpoons H_2CO_3 \rightleftharpoons HCO_3^- + H^+ \qquad \textbf{(4.18)}$$

$$M(HCO_3)_2 \rightleftharpoons M^{+2} + 2HCO_3^- \qquad \textbf{(4.19)}$$

$$HCO_3^- \rightleftharpoons CO_3^{-2} + H^+ \qquad \textbf{(4.20)}$$

$$CO_3^{-2} + H_2O \rightleftharpoons HCO_3^- + OH^- \qquad \textbf{(4.21)}$$

From these equations, it can be seen that CO_2 and the three forms of alkalinity are all part of one system that exists in equilibrium since all equations have HCO_3^-. A change in the concentration of any one member of the system will cause a shift in the equilibrium and cause a change in concentration of the other ions and result in a change in pH. Conversely, a change in pH will shift the equilibrium and result in a change in concentration of the other ions. Figure 4.6 shows the relationship between CO_2 and the three forms of alkalinity in a water with a total alkalinity of 100 mg/ℓ as

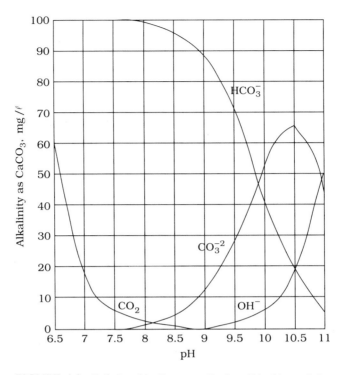

FIGURE 4.6 *Relationship Between Carbon Dioxide and the Three Forms of Alkalinity at Various pH Levels (values calculated for water with a total alkalinity of 100 mg/ℓ at 25°C)*

Adapted from *Chemistry for Environmental Engineering* by C. N. Sawyer, P. L. McCarty, and G. F. Parkin, 4th ed. Copyright © 1994 by McGraw-Hill, Inc. Reprinted by permission.

$CaCO_3$ over the pH range commonly found in water and wastewater engineering. As can be seen, for most natural waters the pH is from 5 to 8.5; thus, the HCO_3^- content represents the major portion of the alkalinity except for cases such as surface waters with a high algal content. Algae use CO_2 in their biochemical processes, and as a result the pH is increased. The alkalinity forms change as follows (Sawyer *et al.*, 1994):

$$2HCO_3^- \rightleftharpoons CO_3^{-2} + H_2O + CO_2 \qquad \textbf{(4.22)}$$

$$CO_3^{-2} + H_2O \rightleftharpoons 2OH^- + CO_2 \qquad \textbf{(4.23)}$$

As the CO_2 is used by the algae, these reactions shift to the right, and the OH^- created by the second equation causes an increase in the pH. Algae have been found to increase the pH of a water up to pH 11. Alkalinity of a water is determined by titrating the water with an acid down to pH 8.3 and then down to pH 4.5. Textbooks, such as that by C. N. Sawyer, P. L. McCarty, and G. F. Parkin (1994), can be used to determine the relative amounts of HCO_3^-, CO_3^{-2}, and OH^- using the data from titrations. Usually alkalinity is expressed in terms of milliequivalents per liter or equivalent $CaCO_3$. Knowing the alkalinity of a water is important in coagulation, in lime-soda softening, and in evaluating the buffering capacity of a water.

The **acidity** of a water is due to the presence of CO_2 or mineral acids, CO_2 being the most common cause. The biochemical action of both aerobic and anaerobic soil bacteria releases CO_2, and some groundwaters have CO_2 contents as high as 30 to $50\,mg/\ell$ because of their passage through soil. Acid waters are corrosive, and waters with a high CO_2 content consume considerable lime in lime-soda softening.

The **hardness** of a water is due to the presence of polyvalent metallic ions, principally Ca^{+2} and Mg^{+2}. Hard waters require considerable amounts of soap before a lather can be produced, and they also produce scale in hot water pipes, heaters, boilers, and other units where the temperature of the water is increased appreciably. When a hard water is heated, scale-producing calcium carbonate is formed, as shown by the equation

$$Ca^{+2} + 2HCO_3^- \rightarrow \underset{\downarrow}{CaCO_3} + CO_2 \uparrow + H_2O \qquad \textbf{(4.24)}$$

Hardness of waters varies considerably from place to place. In general, however, groundwaters are harder than surface waters. Hardness can be expected in regions where large amounts of limestone are found, since water with carbon dioxide will dissolve limestone, releasing the calcium ion. Hardness is measured in terms of milliequivalents per liter or equivalent $CaCO_3$, and the degree of hardness is shown in Table 4.6. Knowing the hardness of a water is important in evaluating its use as a domestic or industrial water supply. The hardness must be known in determining the amount of chemicals required for lime-soda softening and in the design of ion exchange softening units.

The **conductivity** of a water is determined by measuring the electrical resistance between two electrodes and comparing it to the resistance of a standard solution of KCl at 25°C. The conductivity of a particular water can be used to estimate its dissolved solids concentration. For most waters, the dissolved solids in mg/ℓ is equal to the conductivity in micromhos per centimeter (1 micromhos/cm = 1 microsiemen/cm) at 25°C times 0.55 to

TABLE 4.6 *Hardness of Waters*

HARDNESS, mg/ℓ AS $CaCO_3$	DEGREE OF HARDNESS
1–75	Soft
75–150	Moderately hard
150–300	Hard
300 or more	Very hard

Adapted from *Chemistry for Environmental Engineering* by C. N. Sawyer, P. L. McCarty, and G. F. Parkin, 4th ed. Copyright © 1994 by McGraw-Hill, Inc. Reprinted by permission.

0.90. The exact value of the coefficient depends upon the types of salts present.

The **dissolved gases** that occur in natural waters are O_2, CO_2, H_2S, NH_3, N_2, and CH_4. Of these gases, the most common are O_2, CO_2, and H_2S. Oxygen is corrosive in water; however, the use of cement-lined cast iron pipe or other types of pipes minimizes this problem. Dissolved oxygen is usually present in surface waters, except those that are badly polluted. In groundwaters, it is usually absent. Carbon dioxide is very corrosive in water, with most surface waters having moderate amounts and groundwaters having high concentrations. Carbon dioxide is an end product of bacterial action, and groundwaters absorb it while percolating through soil strata containing soil bacteria. Usually, groundwaters are aerated to release excess CO_2. The gas H_2S is not only corrosive but also has an offensive, rotten egg odor, even in very small concentrations. It is a product of anaerobic bacterial action, and it is most common in groundwaters that have passed through lignite beds or strata containing sulfates or iron pyrites. Usually, it is removed by aeration. Ammonia is released by bacterial decomposition of organic matter, and its presence in appreciable amounts is indicative of pollution. Dissolved nitrogen is nearly always present; however, it is inert and has no detrimental effects. Methane is a product of anaerobic bacterial action, and it occurs in groundwaters that have passed through lignite beds or have passed through strata containing natural gas. It may cause tastes and odors in groundwaters and is usually removed by aeration.

The **priority pollutants** (both inorganic and organic) may be present in a water to be used for a municipal water supply, and determining which of these must be monitored in the untreated and treated water is important. If present in the treated water in amounts greater than allowed, they must be reduced, which is usually done at their source, or by adding treatments, or by a combination of these two alternatives. If priority pollutants are present in a new water supply in excessive amounts, frequently it is possible to find an alternate water supply that does not have this problem.

Biological Characteristics   Microbes are commonly found in surface waters, but they are usually absent from groundwaters due to the natural filtration that has occurred. Microbes present in water include bacteria, protozoa, fungi, algae, and viruses. Viruses are classified according to the host they infect. Most bacteria in water are benefical to humans, but some bacteria present in polluted waters are pathogenic and can cause waterborne diseases.

The coliform bacteria, which consist primarily of the species *Escherichia coli* and *Aerobacter aerogenes*, are indicator organisms that live in soil and in the intestines of humans and other warm-blooded animals. Their presence in a water, especially in large numbers, indicates that the water may have been polluted by human intestinal discharges. They usually do not multiply in a water but are steadily dying at a logarithmic rate. Coliforms are important not only because they indicate possible pollution, but also because their absence usually indicates an absence of pathogens.

The routine laboratory bacterial test for the coliforms can be done with a minimum amount of training; however, an individual test is required for each pathogenic organism, and the individual tests require considerable skill. Another reason why the coliforms are used for the routine test for bacterial water quality is that they are much more numerous than pathogens. Each person excretes about 82 gm of feces per day, which contain from 5,000,000 to 500,000,000 coliforms per gram of feces (Steel and McGhee, 1979).

The ratio of the number of pathogens to the number of coliforms varies according to the disease case rate. For typhoid fever in the United States, there is about one typhoid organism per million coliforms or less. The routine coliform bacterial test is for *Escherichia coli* and *Aerobacter aerogenes*; however, the fecal coliform test is for *Escherichia coli*, which is more prevalent in the intestines of humans and other warm-blooded animals than in soil. Certain single-celled bacteria, protozoa, and viruses may cause a waterborne communicable disease — that is, a disease transmitted by improperly treated drinking water.

Actinomycetes, fungi, and algae are microbes that produce end products that may cause tastes and odors in water supplies. Of these, algae are the most common problem. A more complete discussion on coliforms, waterborne diseases, pathogenic bacteria, protozoa, viruses, and microbes such as actinomycetes, fungi, and algae is presented in Chapter 2.

Drinking Water Standards

Drinking water standards are water quality parameters established for public water supplies by regulatory authorities to define the limiting concentrations of various constituents. Limiting concentrations are those concentrations that can be tolerated for the intended use. Drinking water standards are revised periodically, and the limiting allowable concentrations of a given constitutent may change over a period of time. In the United States, the first drinking water standards were established in 1914 to protect the health of the traveling public and to assist in enforcing interstate quarantine regulations. These standards have undergone several revisions, and the latest revision was in 1974 when Congress passed the Safe Drinking Water Act (PL 93-523). The act gives the Environmental Protection Agency (EPA) the power to set the limiting levels of contaminants permitted in drinking waters and to enforce these limits if the individual states fail to comply. In 1986 an important amendment to the Safe Drinking Water Act was passed and it is concerned primarily with toxic substances in drinking waters. As a result of the Safe Drinking Water Act (PL 93-523), the drinking water standards in the United States are established by the Environmental Protection Agency. These standards are continuously being reviewed and updated; thus, the reader should contact the EPA for a current listing. Drinking water standards are classified as primary and secondary standards. Primary standards are health related, whereas secondary standards are for aesthetic considerations. The following discussion on standards is based on the Environmental Protection Agency standards for drinking waters.

Primary standards include such parameters as coliform count, turbidity, toxic inorganic chemicals, toxic organic chemicals, and radionuclides. The **coliform count** is for an average of less than 1 coliform per 100 mℓ per month, which is an average MPN of less than 1 per month. The **turbidity** is health related since turbidity is an indirect measure of the microbial count. The standard for turbidity is 0.5 turbidity unit.

The **toxic inorganic chemicals** include such chemicals as arsenic, barium, cadmium, chromium, lead, mercury, selenium, fluoride, and nitrate. All of these, except fluoride and nitrate, are very toxic to humans at very low concentrations. The fluoride ion standard is from 0.7 to 2.4 mg/ℓ; the exact value within this range is dependent upon the annual average of the maximum daily air temperature for the particular community. If this temperature is 79.3 to 90.5°F (26.3 to 32.5°C), the standard is 0.7 mg/ℓ, whereas if this temperature is 40.0 to 53.7°F (4.4 to 12.1°C), the standard is 1.2 mg/ℓ. For temperatures between these ranges, the standard varies from 0.8 to 1.1 mg/ℓ. The reason this standard varies with temperature is that people drink more water during the summer season in warm climates than they do in cooler climates. A fluoride ion content equal to or less than the recommended standard insures that discolored ("mottled") teeth do not develop when the permanent teeth are formed. A small amount of fluoride ion in a water is desirable, because it promotes the formation of hard, decay-resistant permanent teeth. Many cities using a water with a negligible amount of fluoride add fluoride to the water for this reason. However, the amount added must be lower than the recommended standards. The nitrate ion standard is equal to or less than 10 mg/ℓ as N. A nitrate ion content less than the recommended standard insures that the disease methemoglobinemia does not occur, particularly among infants. In this disease, nitrate is reduced to nitrite in the digestive tract. This nitrite enters the blood and oxidizes hemoglobin in the blood and impairs the oxygen-carrying capacity of the blood.

The **toxic organic chemicals** include such chemicals as endrin, lindane, methoxychlor, toxaphene, 2,4-D and 2,4,5-TP, benzene, para-dichlorobenzene, carbon tetrachloride, 1,2-dichloroethane, 1,1,1-trichloroethane, 1,1-dichloroethylene, trichloroethylene, vinyl chloride, and trihalomethanes. All of these compounds, with the exception of trihalomethanes, are very toxic to humans at very low concentrations. Endrin, lindane, methoxychlor, and toxaphene are insecticides, whereas 2,4-D and 2,4,5-TP are herbicides. Benzene, para-dichlorobenzene, carbon tetrachloride, 1,2-dichloroethane, 1,1,1-trichloroethane, 1,1-dichloroethylene, trichloroethylene, and vinyl chloride are industrial organic chemicals commonly classified as volatile organic chemicals (vocs). Insecticides, herbicides, and the volatile organic chemicals have been found in some surface waters and groundwaters in the environment. The trihalomethanes result from the chlorination of drinking waters containing trace amounts of organic materials, such as humic acid. Since commerical chlorine contains some bromine and naturally occurring waters contain trace amounts of bromine, some of the compounds resulting from chlorination are bromine compounds. The trihalomethanes include trichloromethane (chloroform), dibromochloromethane, dichlorobromomethane, and bromoform. Usually trichloromethane (chloroform) amounts to about 75% of the trihalomethane concentration. Trichloromethane (chloroform) and dibromochloromethane have been found to be carcinogenic in test animals. Also, some epidemiological studies on chlorinated drinking waters have indicated an increased mortality from a variety of cancers as the trihalomethane concentration increased. As a result, the trihalomethane standard has been set at $0.10 \, \text{mg}/\ell$.

The most common **radionuclides** in the environment are naturally occurring, but since human-made radionuclides exist, the standard includes both types. Radionuclide atoms decay to release three forms of energy (radioactivity): alpha, beta, and gamma radiation. Human-made radionuclides come mainly from weapon testing, from radiopharmaceuticals, and from the processing of nuclear fuel elements. Most naturally occurring radionuclides have alpha emissions, whereas most human-made radionuclides lack alpha emissions and have mainly beta and gamma emissions. The standards include radium-226, radium-228, gross alpha particle activity, beta particle activity, radon, and uranium.

Secondary standards include such parameters as color, foaming agents, odor, chloride ion, sulfate ion, dissolved solids, iron, manganese, pH, copper, and zinc. A drinking water with less than 15 color units and with less than $0.5 \, \text{mg}/\ell$ foaming agents will be extremely clear. A water with a threshold odor number less than 3 will be essentially odorless. A chloride ion (Cl^-) greater than $250 \, \text{mg}/\ell$ creates a salty taste. A sulfate ion (SO_4^{-2}) greater than $250 \, \text{mg}/\ell$ and dissolved solids greater than $500 \, \text{mg}/\ell$ can produce a laxative effect, particularly among consumers not accustomed to the water. An iron content less than $0.3 \, \text{mg}/\ell$ prevents red staining of laundered white fabric, whereas a manganese content less than $0.05 \, \text{mg}/\ell$ prevents brown or black staining of laundered white fabric. A water with a pH between 6.5 to 8.5 will not be corrosive or scale forming. To avoid tastes due to metallic ions, the copper ion should be less than $1 \, \text{mg}/\ell$ and the zinc ion should be less than $5 \, \text{mg}/\ell$.

In summary, the primary standards, which are enforceable, insure a safe drinking water. Most of the toxic chemicals (both inorganic and organic) are priority pollutants. The secondary standards, which are nonenforceable, insure a pleasant drinking water that is satisfactory for other domestic uses. Although the secondary standards are nonenforceable, most cities make considerable effort to meet these because they want to satisfy their consumers. The drinking water standards are continuously being reviewed and updated; thus, the reader should contact the Environmental Protection Agency for the current listing.

Water Sources and Characteristics

Most of the world population use surface or groundwaters as a source of public water supplies. The most common surface water sources are streams, rivers, lakes, and impounded reservoirs. The most common groundwater sources are pumped wells or flowing artesian wells. About 50% of the public water supplies in the United States use surface water, and the main users are medium- to large-sized cities. The remaining 50% use groundwaters, and the main users are small- to medium-sized cities. Typical groundwaters have low coliform counts, low total bacterial counts, low turbidity, low color, pleasant taste, low odor, moderate to high dissolved solids, low radioactivity, low dissolved oxygen, and moderate to high carbon dioxide contents. In contrast, typical surface waters have moderate to high coliform counts, high total bacterial counts, moderate to high turbidity, variable color, variable taste, variable odor, low to moderate dissolved solids, variable radioactivity, variable dissolved oxygen, and variable carbon dioxide contents (Zilly, 1975).

In general, groundwaters have higher quality than surface waters, and the quality is quite uniform throughout the year. Frequently, groundwater has such a high quality that only disinfection is required for drinking water purposes to safeguard against subsequent contamination in the distribution system. The uniform characteristics of groundwater make it easy to treat throughout the year. A disadvantage of groundwaters is that many have high dissolved solids, calcium, magnesium, iron, manganese, sulfate, and chloride contents. The removal of dissolved solids, sulfates, and chlorides is difficult and expensive since distillation or ion exchange is required. Groundwaters are usually not as abundant as surface waters, but because of their high quality and the rather simple treatments required, they are used by many small- to medium-sized cities. Due to the high quality of groundwaters, most Third World countries are in the process of providing well waters for their rural communities.

Surface waters have a high variation in quality with respect to time. Their variability in quality means that surface waters require more flexibility in treatment than groundwaters. Surface water is usually more abundant than groundwater, and most major cities in the world use surface waters as supplies.

REFERENCES

Al-Tayla, M. A.; Ahmad, S.; and Middlebrooks, E. J. 1978. *Water Supply Engineering Design*. Ann Arbor, Mich: Ann Arbor Science.

American Water Works Association (AWWA). 1990. *Water Quality and Treatment*. New York: McGraw-Hill.

American Water Works Association (AWWA). 1990. *Water Treatment Plant Design*. New York: AWWA.

Benefield, L. D.; Judkins, J. F.; and Weand, B. L. 1982. *Process Chemistry for Water and Wastewater Treatment*. New York: Prentice-Hall, Inc.

Betz. 1962. *Betz Handbook of Industrial Water Conditioning*. Trevose, Pa.

Camp, T. R. 1963. *Water and Its Impurities*. New York: Reinhold.

Davis, C. V., and Sorensen, K. E. 1969. *Handbook of Applied Hydraulics*. 3rd ed. New York: McGraw-Hill.

Davis, M. L., and Cornwell, D. A. 1991. *Introduction to Environmental Engineering*. 2nd ed. New York: McGraw-Hill.

Dunfor, C. N., and Becker, E. 1964. Public Water Supplies of the 100 Largest Cities in the United States. U. S. Geological Survey Water Supply Paper 1812. Washington, D.C.: U.S. Geological Survey.

Fair, G. M.; Geyer, J. C.; and Okun, D. A. 1966. *Water and Wastewater Engineering.* vols. 1 and 2. New York: Wiley.

Hammer, M. J. 1986. *Water and Wastewater Technology.* 2nd ed. New York: Wiley.

Linsley, R. K., and Franzini, J. B. 1979. *Water Resources Engineering.* 3rd ed. New York: McGraw-Hill.

McGauhey, P. H. 1968. *Engineering Management of Water Quality.* New York: McGraw-Hill.

McGhee, T. J. 1991. *Water Supply and Sewerage.* 6th ed. New York: McGraw-Hill

McKinney, R. E. 1962. *Microbiology for Sanitary Engineers.* New York: McGraw-Hill.

Montgomery Consulting Engineers, Inc. 1985. *Water Treatment Principles and Design.* New York: Wiley.

Nalco. 1979. *The Nalco Water Handbook.* New York: McGraw-Hill.

Nordell, E. 1968. *Water Treatment for Industrial and Other Uses.* New York: Reinhold.

Reynolds, T. D. 1982. *Unit Operations and Processes in Environmental Engineering.* Boston: PWS.

Salvato, J. A. 1992. *Environmental Engineering and Sanitation.* 4th ed. New York: Wiley-Interscience.

Sawyer, C. N., and McCarty, P. L. 1978. *Chemistry for Environmental Engineering.* 3rd ed. New York: McGraw-Hill.

Sawyer, C. N.; McCarty, P. L.; and Parkin, G. F. 1994. *Chemistry for Environmental Engineering.* 4th ed. New York: McGraw-Hill.

Steel, E. W., and McGhee, T. S. 1979. *Water Supply and Sewerage.* 5th ed. New York: McGraw-Hill.

Tchobanoglous, G., and Schroader, E. D. 1985. *Water Quality.* Reading, Mass.: Addison-Wesley.

Viessman, W., and Hammer, M. J. 1993. *Water Supply and Pollution Control.* 5th ed. New York: Harper & Row.

Water Pollution Control Federation (WPCF). 1969. *Design and Construction of Sanitary and Storm Sewers.* WPCF Manual of Practice No. 9, Washington, D.C.

Wolff, J. B. 1957. Forecasting Residential Requirements. *Jour. AWWA* 49, no. 3, pp. 225–235.

Zilly, R. G. 1975. *Handbook of Environmental Civil Engineering.* New York: Reinhold.

PROBLEMS

4.1 It is desired to know the population of City X for 1996, 2006, 2011, 2016, and 2021. The economic background, climate, and so on are similar to those of the cities in the list that follows.

a. Estimate the 1996 population.

Note: The following statistics concerning utility connections, school enrollment, and telephone connections have been compiled.

	1990	1996
Utility connections	7,152	8,633
School enrollment	4,478	5,398
Telephone connections	5,589	6,748

b. Estimate the future populations by the curvilinear rate of growth.

Bureau of Census Records

YEAR	CITY X	CITY A	CITY B	CITY C	CITY D
1890	1,683	—	6,908	14,445	—
1900	2,029	—	8,069	20,686	—
1910	2,216	5,155	10,400	26,425	9,957
1920	3,291	5,713	12,085	38,500	15,494
1930	4,102	5,036	17,113	52,848	43,132
1940	6,390	13,766	28,256	55,982	51,497
1950	8,908	24,502	38,968	84,706	74,246
1960	12,622	40,050	51,230	97,808	137,969
1970	17,878	45,547	57,770	95,326	144,396
1980	23,583	62,762	70,508	101,261	149,230
1990	28,209	80,215	84,628	108,215	156,815

4.2 It is desired to know the population of City X for 1996, 2006, 2011, 2016, and 2021.

a. Estimate the 1996 population using the data on school enrollment and utility connections given in Problem 4.1.

b. Estimate the future populations using the constant percentage method.

4.3   A city has a present population of 53,678, and the average water consumption is 162 gal/cap-day. Determine the future estimated consumption if the 10-year projected population is 65,300.

4.4   A surface water sample has been evaporated until dry in a crucible dish. The dry (tare) weight of the crucible is 44.6420 gm. The weight of the residue plus the crucible is 44.6484 gm. Determine the total solids (both dissolved and suspended) in the sample if the sample volume was 10 mℓ.

4.5   For the surface water in Problem 4.4, a sample has been filtered through a crucible and filter mat. The dry (tare) weight of the crucible and mat was 17.5482 gm. The filtered residue plus the crucible and mat weighed 17.5504 gm. If the sample was 100 mℓ, determine:
a. The suspended solids in mg/ℓ.
b. The dissolved solids in mg/ℓ.

4.6   For the Colorado River water analysis given in Table 4.5, Column (1), determine:
a. The calcium, magnesium, and iron hardness expressed as meq/ℓ and as mg/ℓ equivalent $CaCO_3$.
b. The total hardness as meq/ℓ and as mg/ℓ equivalent $CaCO_3$.
c. The alkalinity as meq/ℓ and as mg/ℓ equivalent $CaCO_3$.
d. The nitrate as nitrogen (N).
e. The hydrogen and hydroxyl ion concentration in moles/ℓ.

4.7   Repeat Problem 4.6 for the Neches River water analysis given in Table 4.5, Column (2).

4.8   Repeat Problem 4.6 for the Stephenville groundwater analysis given in Table 4.5, Column (3).

4.9   Repeat Problem 4.6 for the Houston groundwater analysis given in Table 4.5, Column (4).

4.10   Classify the waters listed in Table 4.5 according to their hardness using the criteria listed in Table 4.6.

4.11   In the water analyses listed in Table 4.5, only the bicarbonate alkalinity is given. Why are the carbonate and hydroxyl alkalinities omitted from the analysis?

5

WASTEWATER QUANTITIES AND WASTEWATER QUALITY

The design of a wastewater treatment plant requires a knowledge of the quantity or flowrate of the wastewater, the quality of the untreated or raw wastewater, and the quality required for the effluent. The quantity or flowrate to be provided determines the size of the various unit operations and processes and also the hydraulic design of the plant. The quality of the untreated or raw wastewater and the quality required for the effluent determine which unit operations and processes are to be provided for the plant.

WASTEWATER QUANTITIES

The purpose of the **sanitary sewer system** is to receive liquid wastes from dwellings, buildings, institutions, and other entities and transport them to the wastewater treatment plant without creating offensive conditions or health hazards. The system consists of the collection pipes or conduits and appurtenances, such as oil and grease traps, manholes, inverted siphons, and pump stations, where necessary.

Sewage or wastewater is the liquid transported by a sewer. The term *sewage* is an older term; the most common modern term is *wastewater*. **Sanitary sewage** or **domestic wastewater** is the liquid wastewater from sanitary facilities within a building. **Industrial wastewater** is the liquid wastewater from an industry, such as a paper-making plant or a brewery. **Storm sewage** or **stormwater** is the storm runoff that occurs from rainfall. **Infiltration** is the groundwater or rainfall seepage that enters sanitary sewers through cracks in pipe joints and manholes, service connections, and defective pipes. **Inflow** is relatively unpolluted water that enters the sewer through such sources as manhole covers, roof downspouts, yard drains, foundation drains, and cooling-water discharge from air conditioners and industries. Inflow through manhole covers is almost unavoidable; however, these other sources can be avoided. Most cities have ordinances that prohibit connections that allow these sources of inflow to occur. Such sources should discharge into sewers that carry stormwater. Proper surface drainage will minimize inflow through manhole covers. Infiltration is maximum during storm events and may be termed **peak infiltration** or **storm infiltration** to differentiate it from **dry-weather flow infiltration**, which occurs during dry weather.

93

A **sewer** is a conduit, usually a pipe, that carries wastewater or stormwater. A **sanitary sewer** carries sanitary sewage or domestic wastewater and, in some cases, any industrial wastewater from the area it serves. A **storm sewer** carries rainfall runoff or stormwater from the area it serves. A **combined sewer** carries sanitary sewage and rainfall runoff or stormwater and, in some cases, industrial wastewater from the area is serves. A combined sewer system is not considered good practice and is usually limited to the older cities in the United States. A **separate sewer system** has both sanitary and storm sewers and is considered good, modern engineering practice. In a combined sewer system, the dry-weather flow is discharged to the wastewater treatment plant, and overflow structures are provided that enable excess flows to be diverted to the receiving body of water, usually without any treatment. The excess flows consist of a mixture of sanitary sewage and stormwater. The Environmental Protection Agency will not approve the construction of a wastewater treatment plant that serves a combined sewer system; thus, new systems are usually separate sewers.

When domestic wastewater temperature is very warm, travel times in the sewers are long, and sulfates are present in significant amounts, corrosion of the crowns or inside tops of concrete sewers has occurred. This is particularly true for the Southern states, where sewer grades are unusually flat and air temperatures, particularly during the summer, are high. As the wastewater flows, bacterial action begins and some of the organic materials are used as bacterial food substances or substrates. Oxygen will usually be present in fresh wastewater; however, it is consumed by aerobic bacterial action faster than it can diffuse from the air into the surface of the wastewater. After a certain period of time, such as several hours or more, all the dissolved oxygen is depleted and anaerobic bacterial action begins in the wastewater.

Sulfates, in anaerobic action, are converted to hydrogen sulfide, as shown in Figure 5.1. The reduction of SO_4^{-2} to H_2S is done by sulfur bacteria, such as the genus *Desulfovibrio*, which is found in municipal wastewater. As the hydrogen sulfide is produced, some diffuses from the wastewater into the air above the flow. It has a very foul, rotten egg odor and is detectable in very low concentrations. Once it is in the air, it does no damage if it is well ventilated and the crown is dry. If poor ventilation exists, water condensation occurs at the crown area of the pipe. Some of the hydrogen sulfide is absorbed by the crown moisture, and sulfur bacteria, such as the bacterium *Thiobaccillus*, convert the hydrogen sulfide to a dilute sulfuric acid solution by the simplified reaction $H_2S + 2O_2 \rightarrow H_2SO_4$. *Thiobaccillus* is present in municipal wastewater and probably infects the crown during high flows. The dilute H_2SO_4 attacks the concrete in the crown area by the simplified reaction $H_2SO_4 + CaCO_3 \rightarrow 2H^+ + CO_3^{-2} + Ca^{+2} + SO_4^{-2}$. The corrosion due to this action can be very serious, and many affected concrete sewers have collapsed.

Corrosion by bacteria-generated sulfuric acid occurs also in cast iron and steel pipes, as well as in other acid-soluble materials. For these pipes, an

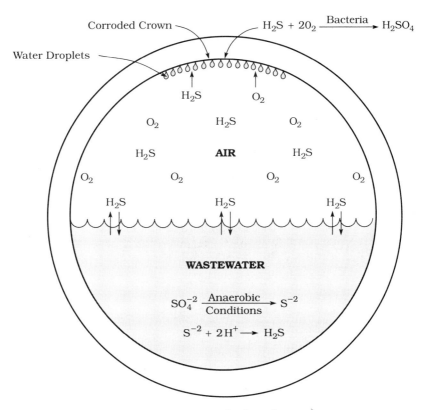

FIGURE 5.1 *Anaerobic Conditions in a Sanitary Sewer*

Adapted from *Chemistry for Environmental Engineering* by C. N. Sawyer, P. L. McCarty, and G. F. Parkin, 4th. ed. Copyright © 1994 by McGraw-Hill, Inc. Reprinted by permission.

inside protective coating is required. The main reason why clay pipe is such a good sewer conduit for sanitary sewers is that it is inert and resistive to acid attack. Sewer conditions that cause hydrogen sulfide generation are (1) flat grades creating low velocities and long travel times, (2) warm wastewater temperatures, (3) high sulfate ion concentration in the wastewater, and (4) high organic content in the wastewater — that is, a high five-day biochemical oxygen demand, BOD_5. In the Southern states that border the Gulf of Mexico and the Atlantic Ocean, flat grades and warm temperatures are common, and hydrogen sulfide generation is a problem in these areas.

The capacity of a proposed sewer system is determined by the estimated requirements of the community at the end of the **design period**. Laterals and branch sewers are designed on the assumption that the sub-areas they are to serve are fully developed. Sewer lines and pumping stations are difficult to enlarge; thus they have design periods from 25 to 50 years. Treatment plants are easier to expand and have design periods from 10 to 25 years.

Wastewater Sources   Municipal wastewater consists of liquid wastes from **domestic, commercial, and industrial sources**. The **amount of municipal wastewater** will be related to the water demand, since this is the source of the carriage water. An estimate of such wastes is related to water consumption studies, either at the present date or sometime in the future. Table 5.1 gives the wastewater flow from numerous cities or districts along with the wastewater flow as a percentage of the water demand. The portion of water consumed that will reach the sewage system depends upon local conditions. Water used for boilers, industrial plants, and lawn irrigation may or may not reach the sanitary sewers. Some industries may have their own water supplies and discharge wastewater into the municipal collection system. The relationship between water demand and wastewater flow varies from city to city, but wastewater flow is usually from 50 to 100% of the water demand. From Table 5.1, it can be seen that the only way to determine accurately the wastewater flow for a city is from past gauging records, since the flow varies appreciably from city to city. Frequently, it is assumed that the wastewater flow, including an allowance for dry-weather infiltration, equals the average water con-

TABLE 5.1   *Average Annual Wastewater Flows of Selected Cities and Districts in the United States*

	AVERAGE ANNUAL WASTEWATER FLOW		PERCENTAGE OF THE WATER DEMAND
CITY OR DISTRICT	gal/cap-d	ℓ/cap-d	
Baltimore, Md.	100	379	63
Berkeley, Calif.	68	257	89
Boston, Mass.	140	530	97
Grand Rapids, Mich.	190	719	84
Greenville, S.C.	150	568	73
Hagerstown, Md.	100	379	100
Jefferson County, Ala.	100	379	98
Johnson County, Kan.	60	227	86
Lancaster County, Neb.	92	348	55
Las Vegas, Nev.	209	791	51
Little Rock, Ark.	50	189	100
Los Angeles, Calif.	85	322	46
Peoria, Ill.	75	284	83
Memphis, Tenn.	100	378	80
Orlando, Fla.	70	265	47
Rapid City, S.D.	121	458	99
Santa Monica, Calif.	92	348	67
Wyoming, Mich.	82	310	56
Averages	100	396	76

Adapted from *Design and Construction of Sanitary and Storm Sewers*, Water Pollution Control Federation Manual of Practice No. 9. Copyright © 1969 by American Society of Civil Engineers and Water Pollution Control Federation. Reprinted by permission.

sumption. Some authorities (GLUMRB, 1990) recommend that the design flow not be less than 100 gal/cap-d (380 ℓ/cap-d). This includes nominal infiltration. Higher design flows are recommended when justified by gauging. Most cities in the United States have wastewater treatment plants with flow recording devices, and daily and weekly flow record charts are available to determine the average daily flow.

As with water demand, a **variation in wastewater flowrate** occurs daily, weekly, and monthly. Figure 5.2 is a hydrograph showing the flow variation throughout the day for a typical city. During the night, the main flow is from industries. An increase in wastewater flow occurs once daily activity begins around 6 A.M., and a peak occurs shortly before noon lunchtime and after evening dinnertime. Each city will have its own characteristic hydrograph, which must be determined from flow gauging records. A weekly variation also occurs, and the maximum day is usually Monday and the minimum is usually Sunday. A monthly variation in flow occurs, and usually the maximum monthly flow occurs during the summer season. Usually the minimum monthly flow occurs during the winter season.

The wastewater flow does not vary as much as the water demand. This is because of the storage volume in the sewers and the time it takes the flow to reach the point of gauging. In one city, the maximum monthly water demand was 1.34 times the average monthly water demand, and the maximum monthly wastewater flow was 1.08 times the average monthly wastewater flow. The minimum monthly water demand was 0.85 times the average monthly water demand, and the minimum monthly wastewater flow was 0.86 times the average monthly wastewater flow.

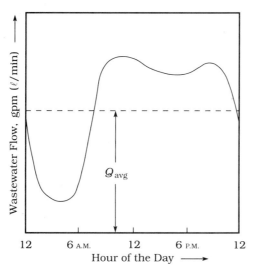

FIGURE 5.2 *The Daily Variation in Wastewater Flow for a Typical City*

The hour of the day when the daily peak occurs depends upon the type of district served and the flow time. For residential areas, the peak will occur at about 9 A.M. if the flow is gauged near the area served. If the flow is gauged at considerable distance downstream, the peak will be at a later hour due to the travel time involved. In commerical and industrial districts, water is used throughout the working hours, and the peak is less pronounced than in residential areas. For small residential areas, the peak may be as much as 500% of the average flowrate for the day. For commerical areas, the peak may be as much as 150% of the average flowrate, and for industrial areas, the peak is usually less than 150% of the average flowrate.

The ratio of the peak hourly flow to the average hourly flow for the day has been investigated by numerous researchers. The most widely accepted formula is that recommended by Babbitt (1958), which is

$$\frac{Q_P}{Q_A} = \frac{5}{P^{0.2}} \tag{5.1}$$

where

Q_P = peak hourly flow

Q_A = average hourly flow

P = tributary population, thousands

For populations less than 1000 persons, the maximum value from this equation is taken as $Q_P/Q_A = 5.0$. The equation is for North American cities and represents normal variations in the dry-weather flow. The maximum hourly flow will be the maximum flow from the previous equation plus an allowance for storm infiltration. Fire demand does not enter into sewage calculations, because it drains into storm sewers. Figure 5.3 shows the relationship among the peak, average, and minimum flows for municipal wastewater according to the investigation by Babbitt (1958). Numerous other researchers have found similar results. The previous formula and this graph are for municipal wastewater flows during dry-weather conditions and, in design, allowances must be made for infiltration, commercial flows, and industrial flows. Usually these additional flows tend to decrease the maximum-to-average ratio and to increase the minimum-to-average ratio.

Infiltration and inflow, as previously mentioned, occur in sanitary sewer systems. Since infiltration will be low during dry weather, the dry-weather flow may be considered to be domestic wastewater, commercial and industrial wastewater, and a nominal amount of infiltration and inflow. In wet weather, high infiltration occurs since the portions of the soil become saturated with water, and also high inflow occurs. Some sewers are located below the permanent groundwater table, and for such cases, significant infiltration will occur in dry weather. Sewers near streams are particularly susceptible to high infiltration.

The amount of infiltration to be expected will depend upon the care with which the system is constructed, the height of the groundwater table,

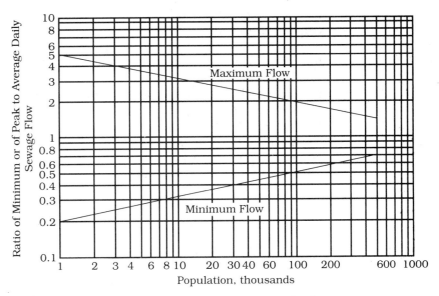

FIGURE 5.3 *Ratio of Extreme Flows to Average Daily Flow for Municipal Wastewaters in Various Areas of the United States*

Adapted from *Design and Construction of Sanitary and Storm Sewers*, Water Pollution Control Federation Manual of Practice No. 9. Copyright © 1969 by American Society of Civil Engineers and Water Pollution Control Federation. Reprinted by permission.

and the type of soil. Special plastic-type lock joints are available for clay and concrete pipe, and these minimize infiltration. Expansive clay soils, in particular, contribute to high infiltration, since the drying and swelling action will pull pipe joints apart. A very pervious soil, such as sand, permits water to percolate to the sewer, where it will travel along the pipe until it reaches a joint where it can enter. Since construction practice, soil types, and construction conditions vary, the amount of infiltration in different sewer systems varies considerably. Large pipes have more joint lengths, but joints in large sewers are usually of better workmanship. Infiltration in old systems, where an appreciable portion of the system is below groundwater and cement mortar joints were used, may be 60,000 gal/mi-day (142 m^3/km-d) or greater. Present specifications usually limit infiltration to 500 gal/mi-day-inch (45 m^3/km-d-mm) of pipe diameter.

Since sewers deteriorate, engineers are usually conservative in estimating infiltration rates for design. It is common to design for the peak hourly wastewater flow plus 30,000 gal/mile-day (71 m^3/km-d) of sewer and house connections. This is where most of the system is above the permanent groundwater table. If most of the system is below the permanent groundwater table, 60,000 gal/mi-day (142 m^3/km-d) is usually used for the design infiltration. Metcalf and Eddy, Inc. (1981) have found that the peak infiltration in gal/acre-day (m³/ha-d) decreases as the tributary area increases, as shown in Figure 5.4. Since rainfall intensity decreases as the drainage area increases, it follows that peak infiltration should decrease as the tributary area increases. In the absence of field measurements, these curves are considered conservative. Curve A is for old sewers, and Curve B is for old and new sewers. The choice between Curve A and Curve B for old sewers depends upon present and expected future conditions of the sewers, the elevation of the groundwater table, and the method of joint construction. For example, if the sewer joints are cement mortar and the groundwater is above the sewers, Curve A should be used. If the sewer joints are cement mortar and the groundwater is below the sewers, Curve B should be used. Curve B is also for new sewers with pipe joints sealed with compression gaskets or rubber or rubber-like materials. For high-groundwater conditions, the peak infiltration varies from 1900 gal/acre-day (17.8 m³/ha-d) to 5200 gal/acre-day (48.5 m³/ha-d). For low-groundwater conditions, the peak infiltration varies from 350 gal/acre-day (3.3 m³/ha-d) to 1500 gal/acre-day (14.0 m³/ha-d).

Commerical and industrial contributions are usually estimated on the basis of gal/acre-day (ℓ/ha-d). In smaller cities where no significant commerical or industrial wastewater flow occurs, allowances for commerical and

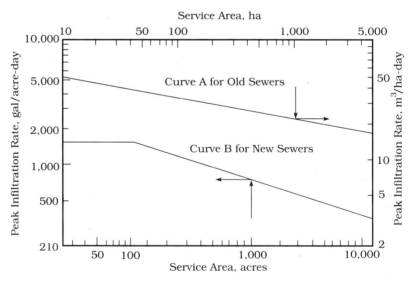

FIGURE 5.4 *Peak Infiltration Allowances for Sanitary Sewers*

Adapted from *Wastewater Engineering: Collection and Pumping of Wastewater* by Metcalf and Eddy, Inc. Copyright © 1981 by McGraw-Hill, Inc. Reprinted by permission.

industrial flows can be included in the per capita flows for the domestic wastewater. Commerical and industrial areas in large cities are larger and more completely developed. Large commerical areas that attract workers and customers from relatively long distances away will have more fixtures with greater usage and thus generate a greater flow per acre (hectare). The number of stories in commerical buildings also affects wastewater flows. Allowances for commerical flows vary from city to city and may range from 2000 to 90,000 gal/acre-day (18.7 to 842 m³/ha-d) with an average value of 28,200 gal/acre-day (278 m³/ha-d) (WPCF, 1969). Industrial wastewater quantities may vary from 2000 to 250,000 gal/acre-day (18.7 to 2340 m³/ha-d) with an average value of 49,800 gal/acre-day (465 m³/ha-d) (WPCF, 1969). The type of industry served, the size of the plant, the type of supervision, the degree of water reuse, the method of on-site treatment (if any), and the future growth are all factors that must be considered for industrial contributions. If the water requirements or demands for the industries are known, wastewater flow projections can be based on water flow projections. For industries without reuse programs, about 85 to 95% of the water used will probably become wastewater. For large industries with reuse programs, individual estimates must be made. The average domestic or sanitary wastewater contributed from industrial activities may vary from 8 to 25 gal/cap-day (30 to 95 ℓ/cap-d).

Design Flows The **minimum flowrate** is important in designing pipes and channels in a wastewater treatment plant that carry wastewater with suspended solids. The minimum velocity required to keep organic solids in suspension is about 1.0 ft/sec (0.3 m/s) and to keep silt and fine sand in suspension is about 2.0 ft/sec (0.6 m/s).

The **design flow rate** is usually assumed to be the average daily flow at the end of the design period. Usually the average daily flow is considered to be the average daily flow for a continuous 12-month period. The design flowrate is used for determining the organic loading to the treatment plant and for sizing the primary, secondary, and (where required) tertiary treatment units and the sludge treatment and handling facilities.

The **maximum wet-weather flowrate** is the peak hourly flow plus flow due to the maximum storm infiltration and inflow. The maximum wet-weather flow determines the hydraulic capacity of the treatment plant and the collection system.

An analysis of the flow gauging records at the existing wastewater treatment plant is the most accurate way to determine the average annual wastewater flow in gal/cap-day (ℓ/cap-d), to determine the maximum and minimum flow rates, and to estimate the storm infiltration and inflow. A study of the flowrates can determine the ratios Q_P/Q_A, and $Q_{Minimum}/Q_A$, and these can be compared to the values given in Figure 5.3. The Q_P/Q_A and $Q_{Minimum}/Q_A$ ratios must be adjusted to represent values to be expected at the end of the design period when the population will be larger.

WASTEWATER QUALITY

The degree of treatment of a wastewater depends upon its quality characteristics and the required effluent quality characteristics. The characteristics may be classified as physical, biological, and chemical, according to their nature, as shown in Figure 5.5. There are numerous tests that can be done at a wastewater treatment plant, but usually only the tests required to meet the plant's discharge permit and the tests required to insure proper plant operation are performed. In addition to the impurities in the water supply, a wastewater will contain impurities as a result of its use. For a municipal wastewater, the added impurities are a result of domestic, commercial, and industrial use. The characteristics of typical, untreated municipal wastewater in the United States are shown in Table 5.2. These characteristics are for an average annual wastewater flow of about 100 gal/cap-day (380 ℓ/cap-d). The data in Table 5.2 are intended to serve only as a guide and not as a basis for design. The characteristics are discussed in the following text.

Physical Characteristics

The main physical characteristics of a wastewater, such as a municipal wastewater, are turbidity, color, odor, total solids (both suspended and dissolved), and temperature.

The **turbidity** of a wastewater is mainly due to suspended organic solids, which range in size from colloidal to coarse suspensions. Municipal wastewater is about 99.95% water, but the organic solids present have a pronounced effect in that they exert a biochemical oxygen demand.

The **color** of fresh municipal wastewater is a light tan if the wastewater is less than 2 to 6 hours old. Stale wastewater, wastewater over 6 hours old, has undergone some biochemical oxidation in the collection system and is usually a grey color. Extremely stale wastewater has undergone extensive biochemical oxidation under anaerobic conditions and will have a dark grey or black color. The blackening is due to the production of various sulfides, in particular ferrous sulfide. Hydrogen sulfide, which is produced under anaerobic conditions, reacts with small amounts of the ferrous ion, producing ferrous sulfide, which is black.

The **odor** of fresh municipal wastewater is a soapy or oily odor and usually is not offensive. Stale municipal wastewater has undergone extensive anaerobic bacterial action, which produces offensive compounds, such as

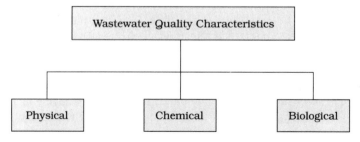

FIGURE 5.5 *Wastewater Quality Characteristics*

TABLE 5.2 *Typical Characteristics of Untreated Municipal Wastewater in the United States*

CONSTITUENT	CONCENTRATION		
	Strong	Medium	Weak
Solids, total:	1250	800	450
Dissolved, total:	890	560	350
Fixed	295	295	185
Volatile	595	265	165
Suspended, total:	360	240	100
Fixed	145	75	25
Volatile	215	165	75
Settleable solids, mℓ/ℓ	7	5	3
Biochemical oxygen demand, 5-day, 20°C (BOD_5, 20°C)	400	200	100
Total organic carbon (TOC)	290	145	75
Chemical oxygen demand (COD)	910	455	230
Nitrogen (total as N):	75	40	16
Organic	40	20	8
Free ammonia	35	20	8
Nitrites	0	0	0
Nitrates	0	0	0
Phosphorus (total as P):	15	8	4
Organic	5	3	1
Inorganic	10	5	3
Chlorides[a]	83	42	21
Alkalinity (as $CaCO_3$)[a]	200	100	50
Grease	40	20	5

All values except settleable solids are expressed in mg/ℓ.
$1 \, mg/\ell = 1 \, g/m^3$
[a] Values should be increased by the amount in the domestic water supply.
Adapted from *Water Supply and Sewerage* by E. W. Steel, 4th ed. Copyright © 1960 by McGraw-Hill Book Company, Inc.; and from *Wastewater Engineering: Treatment, Disposal and Reuse* by Metcalf and Eddy, Inc., 3rd ed. Copyright © 1991 by McGraw-Hill, Inc. Reproduced with permission of McGraw-Hill, Inc.

hydrogen sulfide, mercaptans, indol, and skatol. Hydrogen sulfide is the principal odor-producing compound and has a distinctive rotten egg odor. Municipal wastewater from medium-sized or larger cities is usually stale when it arrives at the wastewater treatment plant.

The **total solids** of a municipal wastewater consist of both the dissolved and the suspended solids. The **suspended solids** are those separated by filtering a sample and drying and weighing the residue. The **dissolved solids** consist of those remaining after evaporating the filtered sample, which leaves a salt residue. The **volatile solids** are those burned off when the filtered residue or evaporated residue is ignited at 550°C. Since there are both suspended and dissolved solids, there will be **volatile suspended solids** and **volatile dissolved soldis**. The volatile solids are an approximate measure of the organic materials, since most organic materials are heat sensitive. The main organic materials in a municipal wastewater are proteins, carbohydrates, and fats (or greases); their degradation products; and microbial cells. The major portion of the volatile suspended solids are organic materials. The **fixed solids** are those remaining after ignition; thus there are **fixed suspended solids** and **fixed dissolved solids**. The major portion of the fixed dissolved solids consists of the minerals or salts present in the water supply in addition to the salts added during usage, such as the preparation and eating of foods. The **settleable solids** are those that can be removed by sedimentation using a 1-liter conical glass container (referred to as an Imhoff cone) and settling for a given period of time, usually equal to the detention time in the primary settling tanks. The settleable solids are reported in $m\ell/\ell$. For the usual municipal wastewater, about 50% of the total solids are suspended solids, and about 60 to 65% of the suspended solids are settleable.

The **temperature** of a municipal wastewater is usually higher than that of the water supply due to heating of water — for instance, in domestic use water is heated for washing and bathing. The average wastewater temperature is usually slightly higher than the average monthly air temperature during the winter and is usually slightly lower than the average monthly air temperature during the summer. The temperature of a wastewater is important since the biological processes used in treating wastewaters are temperature dependent.

The turbidity, color, odor, total solids, suspended and dissolved solids, volatile and fixed solids, settleable solids, and temperature of **industrial wastewaters** (both organic and inorganic) will vary appreciably according to the source of the wastewater.

Chemical Characteristics The main chemical characteristics of a wastewater, such as a municipal wastewater, are the chemical oxygen demand (COD), total organic carbon (TOC), various forms of nitrogen, various forms of phosphorus, chloride ion, sulfate ion, alkalinity, pH, heavy metal ions, trace elements, and priority pollutants.

The **chemical oxygen demand (COD)** is a measure of the organic materials in a wastewater in terms of the oxygen required to oxidize the organic materials chemically. It is measured by boiling and refluxing the sample with a strong oxidizing agent and determining the amount of oxidizing agent used.

The **total organic carbon (TOC)** is a measure of the organic materials in a wastewater in terms of the amount of carbon in the organic materials. It is measured by removing the carbon dioxide present, combusting the sample, and measuring the amount of carbon in the carbon dioxide evolved.

The **organic nitrogen** is the amount of nitrogen present in organic compounds, such as proteins and urea and their degradation products. The **ammonia nitrogen** is the amount of nitrogen present in the form of ammonia (NH_3) or the ammonia ion (NH_4^+). The **nitrite nitrogen** is the amount of nitrogen in the form of the nitrite ion (NO_2^-). The **nitrate nitrogen** is the amount of nitrogen present in the form of nitrates (NO_3^-). The ammonia, nitrite, and nitrate nitrogens in municipal wastewaters originate in the form of organic compounds, such as proteins and urea or their degradation products, and evolve from the bacterial decomposition of these materials.

The **organic phosphorus** is the amount of phosphorus present in organic compounds, such as proteins and their degradation products. The **inorganic phosphorus** is the amount of phosphorus contained in inorganic compounds, such as the phosphate ion (PO_4^{-3}). Most of the inorganic phosphorus is from the bacterial decomposition of organic phosphorus compounds.

Tests such as those for the **chloride ion** (Cl^-), the **sulfate ion** (SO_4^{-2}), the **alkalinity**, and the **pH** are used mainly in assessing the suitability of reusing the treated wastewater. Knowing the alkalinity and pH is frequently useful in controlling some treatment processes. The pH of untreated wastewater must be between 6.5 and 9 in order for the wastewater to be treated by biochemical processes, since these require a limited pH range.

The **grease** content is used in controlling some biological processes, since a grease content greater than 5 to 7% of the activated sludge dry weight will coat the activated sludge floc and interfere with proper oxygen transfer.

The amounts of **heavy metal ions**, such as mercury (Hg), arsenic (As), lead (Pb), zinc (Zn), cadmium (Cd), copper (Cu), nickel (Ni), chromium (Cr), and silver (Ag), are important in assessing the treatability of a wastewater, since above certain concentrations, these ions inhibit biological processes.

The amount of **trace elements**, many of which are heavy metals, such as zinc, copper, cobalt, and iron, are rarely analyzed in municipal wastewaters, but these are essential trace elements for biological processes. In municipal wastewaters, the concentrations in the carriage water are adequate for treatment.

The **priority pollutants** (both inorganic and organic) present in a municipal or industrial wastewater are of importance in determining which of these pollutants must be monitored in the plant effluent. If present in the plant effluent in greater amounts than allowed, they must be reduced, which is usually done at their source of origin. These pollutants are discussed in Chapter 1.

Organic industrial wastewaters generally have COD and TOC concentrations much higher than municipal wastewaters. The amounts of organic nitrogen, ammonia nitrogen, nitrite nitrogen, nitrate nitrogen, organic phosphorus, inorganic phosphorus, and heavy metal ions, as well as the alkalinity, grease content, and pH, vary according to the source of the wastewater. Frequently, the pH requires adjustment, and nitrogen and phosphorus have to be added for the wastewater to be treated by biological processes. Usually the trace elements in the carriage water are adequate for biological processes. **Inorganic industrial wastewaters**, such as metal-plating wastewaters from copper, zinc, and nickel plating, usually have very low

COD and TOC concentrations but high concentrations of the metallic ion that is plated. These wastewaters are usually treated by chemical processes, such as the ion exchange process.

Biological Characteristics

The main biological characteristics of a wastewater, such as a municipal wastewater, are the biochemical oxygen demand (a measurement of the organic materials), the oxygen required for the nitrification of organic and ammonia nitrogen, and the microbial population of the wastewater.

The **biochemical oxygen demand (BOD)** is the amount of oxygen required by microbes, mainly bacteria, in the stabilization of organic materials under aerobic conditions. Since the amount of oxygen varies with the length of time and the temperature, the standard test is for five days at 20°C, and the value it yields is referred to as the **five-day biochemical oxygen demand (BOD$_5$)**. For typical North American wastewater, the BOD$_5$ is about 200 mg/ℓ. The amount of oxygen required by nitrifying bacteria in the conversion of organic and ammonia nitrogen to nitrates is frequently called the **nitrogenous oxygen demand**.

Municipal wastewater will contain microbial life such as bacteria, protozoa, fungi, viruses, algae, rotifers, and, sometimes, nematodes, the last two being worm-like organisms. The bacterial population is relatively high and may range from 100,000 to 100,000,000 cells per milliliter. Much of the bacterial population are soil organisms that enter by infiltration and are important in biological treatment processes. The coliforms, which are mainly the species *Escherichia coli* and *Aerobacter aerogenes*, are found in the intestines of humans and other warm-blooded animals and in soil and will be numerous. Pathogens, or disease-causing organisms, mainly bacteria, will also be present. Tests for the coliform population or count are sometimes done to evaluate coliform removal by certain treatment plant operations or processes and to determine the suitability of discharging the treated wastewater to downstream water intakes, recreational waters, or shellfish-producing waters.

Organic industrial wastewaters generally have much higher BOD$_5$ concentrations than municipal wastewaters. The nitrogen demand varies according to the source of the wastewater. Generally, organic industrial wastewaters have low microbial populations. Inorganic industrial wastewaters, such as wastewaters from metal-plating industries, usually have very low BOD$_5$ concentrations, very low nitrogen demands, and negligible microbial populations.

MEASUREMENT OF WASTE ORGANIC MATERIALS

The measurement of the concentration of waste organic materials in a wastewater is important in designing the plant and in the control and operation of the plant.

Biochemical Oxygen Demand (BOD)

Acclimated microorganisms, mainly soil bacteria, readily use organic material in a wastewater as substrates under aerobic conditions. They require molecular oxygen in their respiration and produce end products, such as carbon dioxide and water. The **biochemical oxygen demand**, as previously mentioned, is the amount of oxygen used by microbes in the stabilization of a decomposable organic waste material under aerobic conditions. If the amount of oxygen used versus time is plotted, it will produce a curve similar to Figure 5.6. Up to about 10 to 12 days, the demand will be the oxygen required for oxidation of carbonaceous materials as shown by

$$\text{Organic matter} + O_2 \xrightarrow[\text{microbes}]{\text{Aerobic}} CO_2 + H_2O + NH_3 + \text{new cells}$$
$$+ \text{ energy} + \text{other end products} \quad (5.2)$$

After about 10 to 12 days, the NH_3 will be oxidized to nitrites, and then to nitrates, by the reactions

$$2NH_3 + 3O_2 \xrightarrow[\substack{\text{bacteria} \\ (\textit{Nitrosomonas})}]{\text{Nitrite}} 2NO_2^- + 2H^+ + 2H_2O + \text{new cells} + \text{energy}$$
$$+ \text{ other end products} \quad (5.3)$$

$$2NO_2^- + O_2 + 2H^+ \xrightarrow[\substack{\text{bacteria} \\ (\textit{Nitrobacter})}]{\text{Nitrate}} 2NO_3^- + 2H^+ + \text{new cells} + \text{energy}$$
$$+ \text{ other end products} \quad (5.4)$$

Thus there are two distinct curves or stages — one for the breakdown of carbonaceous material and one for the nitrification of ammonia. In order to

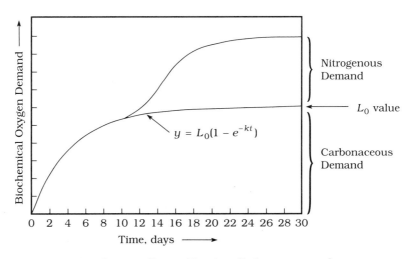

FIGURE 5.6 *The* BOD *Curves Showing Carbonaceous and Nitrogenous Demand*

provide a standard test, a time of five days at 20°C was selected and is referred to as the **five-day biochemical oxygen demand** (BOD₅).

For the carbonaceous or first stage, a plot of the organic material oxidized and a plot of that remaining are shown in Figure 5.7. Studies have shown that if the seed microorganisms are acclimated to the waste, then the removal of organic material is a pseudo-first-order reaction:

$$-\frac{dC}{dt} = kC \tag{5.5}$$

where

dC/dt = rate of removal of organic material

k = rate constant to the base e

C = concentration of organic material remaining at time t

Rearranging Eq. (5.5) for integration gives

$$\int_{C_0}^{C} \frac{dC}{C} = -k \int_{0}^{t} dt \tag{5.6}$$

where C_0 = concentration of organic material initially. Thus,

$$\ln C \Big]_{C_0}^{C} = -kt \tag{5.7}$$

or

$$\frac{C}{C_0} = e^{-kt} \tag{5.8}$$

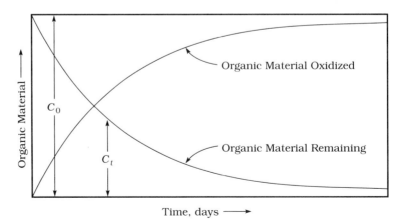

FIGURE 5.7 *First-Stage or Carbonaceous Demand*

where

C = oxidizable organic material remaining at time t

C_0 = total oxidizable organic material originally present

t = test duration

Since the amount of organic material used is proportional to the amount of oxygen required, $C \propto L$, where L is the BOD remaining at time t. Similarly, $C_0 \propto L_0$; thus Eq. (5.8) may be expressed as

$$\frac{L}{L_0} = e^{-k_1 t} = 10^{-K_1 t} \tag{5.9}$$

where

L = BOD remaining at time t

L_0 = ultimate first-stage or carbonaceous BOD

k_1 = rate constant to the base e

t = test duration

K_1 = rate constant to the base 10

Thus, from Eq. (5.9), $L = L_0 e^{-k_1 t}$ and $L = L_0 10^{-K_1 t}$. The BOD exerted up to a given time, y, is $y = L_0 - L_0 e^{-k_1 t}$ or $y = L_0 - L_0 10^{-K_1 t}$, or

$$y = L_0(1 - e^{-k_1 t}) \tag{5.10}$$

$$y = L_0(1 - 10^{-K_1 t}) \tag{5.11}$$

where y is the BOD exerted up to time t. The relationship between K_1 and k_1 is K_1 (base 10) $= 0.434 k_1$ (base e). For typical municipal wastewaters, k_1 (base e) varies from about 0.23 to 0.70 day^{-1} at 20°C, the average value being 0.39 day^{-1}. A typical value for k_1 (base e) for municipal wastewaters is 0.39 day^{-1} (Eckenfelder, 1989). For typical treated municipal effluents, k_1 (base e) varies from 0.14 to 0.28 day^{-1} at 20°C, a typical value being 0.23 day^{-1} (Eckenfelder, 1989).

EXAMPLE 5.1   BOD Test

A wastewater has a BOD$_5$ of 200 mg/ℓ, and the k_1 value is 0.34 day^{-1}. Determine the ultimate first-state BOD, L_0.

SOLUTION   From the previous equations,

$$y = L_0(1 - e^{-k_1 t})$$

or

$$200 = L_0[1 - e^{-(0.34)5}]$$

from which $L_0 = \boxed{245 \text{ mg}/\ell.}$

Chemical Oxygen Demand (COD)

The BOD_5 test requires five days duration; and in many cases, such as treatment plant control, effluent monitoring, and industrial wastewater treatment, it is desirable to have a rapid test. The **chemical oxygen demand** (COD) test consists of oxidizing the sample with a strong oxidizing agent, potassium dichromate, and determining the amount of potassium dichromate used. Then the oxygen required in the chemical oxidation can be determined. The test consists of boiling and refluxing the vapor evolved for 2 hours and measuring the potassium dichromate used by titrating with ferrous ammonium sulfate. The total test time is about 4 or 5 hours. The test will not measure the oxygen required for nitrification; thus, it is a measure of only the organic material. For a particular wastewater or effluent, a correlation graph of BOD_5 values versus COD values from numerous samples can be made, as shown in Figure 5.8, and the BOD_5 value can then be estimated if the COD value is known. A disadvantage of the COD test is that it does not differentiate between the biologically oxidizable and the biologically resistant organic materials — that is, the nondegradable COD. The nondegradable COD can be determined from the correlation graph as shown in Figure 5.8. When the BOD_5 value is zero, the line intersects the y-axis, and the intercept COD value is the nondegradable COD. The biodegradable COD equals the total COD minus the nondegradable COD. For the industrial waste-

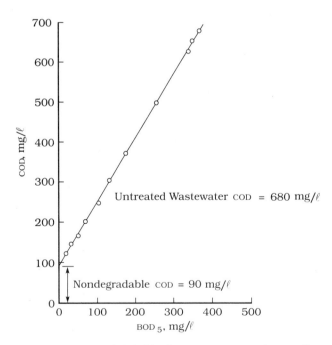

FIGURE 5.8 *Relationship between BOD_5 and COD for a Soluble Organic Industrial Wastewater*

water shown in Figure 5.8, the untreated wastewater had a COD of $680\,mg/\ell$ and the nondegradable COD was $90\,mg/\ell$. Thus the untreated wastewater had a biodegradable COD of $680 - 90 = 590\,mg/\ell$. The nondegradable COD is the COD not degradable within the wastewater treatment plant, and it appears in the plant effluent. Once the effluent is discharged into the receiving body of water, most of these materials are slowly degraded by microorganisms.

Total Organic Carbon (TOC)

The **total organic carbon (TOC)** test is a measure of the organic material in terms of the organic carbon content. This test, like the COD test, is used in treatment plant control and effluent monitoring. The sample is acidified and stripped of CO_2; then it is ignited in a combustion chamber, and the amount of carbon in the CO_2 evolved is measured by an infrared analyzer. Standards must be made to convert the infrared reading to organic carbon. The test requires only 15 to 20 minutes once a standard is prepared. The test is rapid, but the equipment is expensive and laboratory skill is required. For a particular wastewater or effluent, a correlation graph of BOD_5 values versus TOC values from numerous samples can be made, and the BOD_5 value can then be estimated if the TOC value is known.

Theoretical Oxygen Demand (TOD)

The **theoretical oxygen demand (TOD)** is the amount of oxygen theoretically required to convert the elements in the organic material to stable end products. The chemical equation or empirical equation for the waste material must be known, and a balanced chemical equation must be developed to convert the elements to stable end products. Since the most common elements in a waste material are carbon, hydrogen, oxygen, and nitrogen, the stable end products are CO_2, H_2O, and NO_3^-.

Sample Collection

The concentrations of the quality parameters, such as the BOD_5 and suspended solids, vary in approximate proportion to the wastewater flowrate. Thus, for a sample to be representative, it must be a composited sample made up of a series of flow-weighted grab samples.

EFFLUENT REQUIREMENTS

The most common method to dispose of treated effluents is their discharge to receiving water bodies, such as streams, lakes, and estuaries. To maintain a certain quality in the receiving body of water, two methods of control have been used. The first is the setting of receiving-water standards, and the second is the setting of effluent standards. With receiving-water standards, the degree of treatment required by the dischargers must be increased as additional dischargers obtain discharge privileges. The greatest problem with receiving-water standards is maintaining equity among the various dischargers. In order to avoid this problem, most countries, including the United States, have chosen to set effluent standards or issue effluent permits that protect the quality of the receiving water bodies. Effluent standards are easy to enforce since the dischargers must provide records to demonstrate compliance with the standards.

REFERENCES

American Public Health Association (APHA), American Water Works Association (AWWA), and Water Environment Federation (WEF). 1992. *Standard Methods for the Examination of Water and Wastewater.* 18th ed. Washington, D.C.

American Concrete Pipe Association. 1980. *Concrete Pipe Design Manual.* Vienna, Virginia.

Babbitt, H. E. 1958. *Sewerage and Sewage Treatment.* 8th ed. New York: Wiley.

Barnes, D.; Bliss, P. J.; Gould, B. W.; and Vallentine, H. R. 1981. *Water and Wastewater Engineering Systems.* London: Pitman Books.

Camp, T. R. 1963. *Water and Its Impurities.* New York: Reinhold.

Davis, C. V., and Sorensen, K. E. 1969. *Handbook of Applied Hydraulics.* 3rd ed. New York: McGraw-Hill.

Eckenfelder, W. W., Jr. 1966. *Industrial Water Pollution Control.* New York: McGraw-Hill.

Eckenfelder, W. W., Jr. 1989. *Industrial Water Pollution Control.* 2nd ed. New York: McGraw-Hill.

Eckenfelder, W. W., Jr., and O'Connor, D. J. 1961. *Biological Waste Treatment.* New York: Pergammon Press.

Environmental Protection Agency (EPA). 1974. *Sulfide Control in Sanitary Sewerage Systems.* EPA Process Design Manual. Washington, D.C.

Fair, G. M.; Geyer, J. C.; and Okun, D. A. 1968. *Water Supply and Wastewater Engineering.* vols. 1 and 2. New York: Wiley.

Great Lakes–Upper Mississippi River Board of State Sanitary Engineers (GLUMRB). 1978. *Recommended Standards for Sewage Works.* Albany, N.Y.

Great Lakes–Upper Mississippi River Board of State Public Health and Environmental Managers (GLUMRB). 1990. *Recommended Standards for Wastewater Facilities.* Albany, N.Y.

Gifft, H. M. 1945. *Waterworks and Sewerage.* vol. 92, p. 175.

Hammer, M. J. 1986. *Water and Wastewater Technology.* 2nd ed. New York: Wiley.

Harman, W. G. 1918. *Engineering News-Record.* vol. 80, p. 1234.

McGauhey, P. H. 1968. *Engineering Management of Water Quality.* New York: McGraw-Hill.

McGhee, T. J. 1991. *Water Supply and Sewerage.* 6th ed. New York: McGraw-Hill.

Metcalf and Eddy, Inc. 1981. *Wastewater Engineering: Collection and Pumping of Wastewater.* New York: McGraw-Hill.

Metcalf and Eddy, Inc. 1972. *Wastewater Engineering: Collection, Treatment and Disposal.* New York: McGraw-Hill.

Metcalf and Eddy, Inc. 1979. *Wastewater Engineering: Treatment, Disposal and Reuse.* 2nd ed. New York: McGraw-Hill.

Metcalf and Eddy, Inc. 1991. *Wastewater Engineering: Treatment, Disposal and Reuse.* 3rd ed. New York: McGraw-Hill.

National Clay Pipe Institute. 1978. *Clay Pipe Engineering Manual.* Washington, D.C.

Novotny, V.; Imhoff, K. R.; Olthof, M.; and Krenkel, P. A. 1989. *Karl Imhoff's Handbook of Urban Drainage and Wastewater Disposal.* New York: Wiley.

Peavy, H. S.; Rowe, D. R.; and Tchobanoglous, G. 1985. *Environmental Engineering.* New York: McGraw-Hill.

Sawyer, C. N., and McCarty, P. L. 1978. *Chemistry for Environmental Engineering.* 3rd ed. New York: McGraw-Hill.

Sawyer, C. N.; McCarty, P. L.; and Parkin, G. F. 1994. *Chemistry for Environmental Engineering.* 4th ed. New York: McGraw-Hill.

Steel, E. W., and McGhee, T. J. 1979. *Water Supply and Sewage.* 5th ed. New York: McGraw-Hill.

Tchobanoglous, G., and Schroader, E. D. 1985. *Water Quality.* Reading, Mass.: Addison-Wesley.

Viessman, W., and Hammer, M. J. 1993. *Water Supply and Pollution Control.* 5th ed. New York: Harper & Row.

Water Pollution Control Federation (WPCF). 1969. *Design and Construction of Sanitary and Storm Sewers.* WCPF Manual of Practice No. 9. Washington, D.C.

Water Pollution Control Federation (WPCF). 1982. *Gravity Sanitary Sewer Design and Construction.* WPCF Manual of Practice No. FD-5. Washington, D.C.

PROBLEMS

5.1 A municipal wastewater has a BOD$_5$ of 200 mg/ℓ, and k_1 (base e) is 0.39 day^{-1}. Determine:
a. The ultimate first-stage BOD in mg/ℓ, L_0. 233
b. The ratio y/L_0 as a fraction and as a percent. 858

5.2 A municipal effluent has a BOD$_5$ of 10 mg/ℓ, and k_1 (base e) is 0.23 day^{-1}. Determine:
a. The ultimate first-stage BOD in mg/ℓ, L_0.
b. The ratio, y/L_0 as a fraction and as a percent.

5.3 A municipal wastewater treatment plant effluent sample was filtered through a crucible and filter mat. The crucible plus the mat had a dry (tare) weight of 17.8216 gm. After filtration, the crucible mat and residue weighed 17.8374 gm. After weighing, the crucible, mat, and residue were ignited at 600°C. The crucible, mat, and ash weighed 17.8258 gm. If the sample was 50 mℓ, determine the suspended solids, the volatile solids, the fixed solids, and the percent volatile suspended solids.

5.4. A municipal wastewater treatment plant effluent sample was filtered through a crucible and filter mat. The crucible plus the mat had a dry (tare) weight of 17.2814 gm. After filtration, the crucible, mat, and residue weighed 17.2827 gm. If the sample volume was 50 mℓ, determine the suspended solids.

6

WATER AND WASTEWATER TREATMENT PLANTS

The unit operations and processes used in environmental engineering may be classified as physical, chemical, or biological treatments according to their functional principle. In the strict sense, a unit operation is a physical treatment and a unit process is a chemical or biological treatment; however, the terms **unit operations** and **unit processes** are frequently used interchangeably. Some typical unit operations are sedimentation, flotation, and granular bed filtration. Some typical unit processes are coagulation, flocculation, carbon adsorption, ion exchange, chlorination, activated sludge, trickling filters, aerobic digestion, and anaerobic digestion. Many unit operations and processes, such as coagulation, flocculation, and sedimentation, are used in both water and wastewater treatment. Consequently, these operations or processes may be studied by investigating their fundamentals and then studying their application in water and wastewater treatment. Some unit operations or processes are limited to one field — that is, water or wastewater treatment. However, these may still be studied by investigating their fundamentals and then their application. This approach to studying the various unit operations and processes is presented in this text.

The degree to which a water must be treated depends on the raw water quality and the desired quality of the finished water. Likewise, the degree to which a wastewater must be treated depends on the raw wastewater quality and the required effluent quality. Since the degree of treatment determines the number and types of unit operations and processes to be used, there are numerous flowsheets employed in water treatment and, in particular, in wastewater treatment. In order to illustrate the integration of unit operations and unit processes in the overall plant design, the following discussion presents some of the most common flowsheets used in water and wastewater treatment and, also, a brief description of the unit operations and unit processes involved.

WATER TREATMENT PLANTS

The most common treatment plants for surface waters are rapid sand filtration plants and lime-soda softening plants. Groundwaters usually have a much better quality than surface waters; consequently, for groundwaters the most common plants are gas stripping and chlorination plants and softening plants, either lime-soda or ion exchange type. Slow sand filtration plants are sometimes used for surface waters.

From a bacteriological standpoint, the degree to which a water must be treated to obtain drinking water depends upon the coliform count of the raw water, as shown in Table 6.1. As previously discussed in a prior chapter, the MPN (the most probable number) represents the statistical number of coliforms per 100 mℓ. In the United States, only isolated watersheds will have surface waters in Groups 1 and 2. Most surface waters will be in Group 3. In the United States, most groundwater will be in Group 1.

The flowsheet for a rapid sand filtration plant, shown in Figure 6.1, consists of coarse and fine screens, chemical coagulation, flocculation, sedimentation, granular media filtration, and chlorination. The coarse screens remove large debris, whereas the finer traveling screens remove

TABLE 6.1 *A Classification of Waters by Concentration of Coliform Bacteria and Treatment Required to Render the Water of Safe Sanitary Quality[a]*

GROUP NUMBER	MAXIMUM PERMISSIBLE AVERAGE MPN TOTAL COLIFORM BACTERIA PER MONTH[b,c]	TREATMENT REQUIRED
1	MPN not more than 1.0 coliform bacteria	None for protected underground water, but, at the minimum, chlorination for surface water
2	MPN not more than 50	Simple chlorination or equivalent
3	MPN not more than 5000, and this MPN exceeded in not more than 20% of samples	Rapid sand filtration (including coagulation) or its equivalent plus continuous chlorination
4	MPN greater than 5000 in more than 20% of samples and not exceeding 20,000 in more than 5% of the samples	Auxiliary treatment such as presedimentation or prechlorination or its equivalent (either separately or combined) or presedimentation for 30 days or more plus rapid sand filtration and chlorination
5	MPN exceeds Group 4	Prolonged storage or equivalent to bring within Groups 1 to 4

[a] Physical characteristics, inorganic and organic chemicals, and radioactivity concentrations in the raw water and ease of removal by the proposed treatment must also be taken into consideration.

[b] The MPN (most probable number) is the number of coliforms per 100-mℓ sample.

[c] Fecal coliforms not to exceed 20% of total coliform organisms. The monthly geometric mean of the MPN for Group 2 may be less than 100 and, for Groups 3 and 4, less than 20,000/100 mℓ with the indicated treatment. (The total coliform density of 20,000/100 mℓ may be exceeded if fecal coliform do not exceed 2000/100 mℓ monthly geometric mean.) Complete treatment for Group 2 water is recommended.

Source: U.S. Public Health Service.

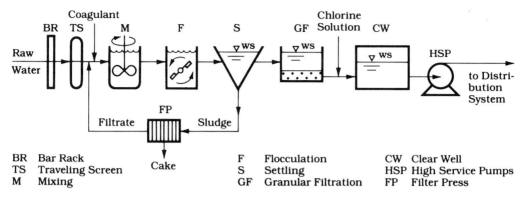

BR   Bar Rack
TS   Traveling Screen
M   Mixing

F   Flocculation
S   Settling
GF   Granular Filtration

CW   Clear Well
HSP   High Service Pumps
FP   Filter Press

FIGURE 6.1   *Rapid Sand Filtration Plant*

smaller debris. Chemical coagulation and flocculation produce a precipitate or floc that enmeshes most of the colloidal solids. Most of the floc is removed in the settling basins. The granular media filters remove most of the fine nonsettling floc, and disinfection kills any pathogenic organisms present. After disinfection, the water is stored in the clear well and is pumped to the distribution system by the high service pumps. The plant capacity is equal to the average flow on the day of maximum demand, Q_{avg}, as shown by Figure 6.2. The clear well, as shown in Figure 6.2, provides storage so that the plant may operate at a constant rate on the day of

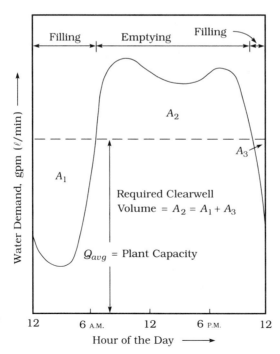

FIGURE 6.2
Required Clearwell Volume

maximum demand; that is, when the hourly demand is greater than the hourly water production, the water required is provided from storage in the clear well. The solids handling system, for disposal of the sludge from the clarifier, consists of a filter press to dewater the sludge. The cake is disposed of by sanitary landfill, although other solids handling systems may be used.

The flowsheet for a lime-soda softening plant, shown in Figure 6.3, consists of coarse and fine screens, chemical precipitation, flocculation, sedimentation, recarbonation, granular media filtration, and disinfection. The addition of slaked lime (calcium hydroxide) and soda ash (sodium carbonate) precipitates the calcium and magnesium ions as calcium carbonate and magnesium hydroxide. During flocculation, the precipitate or floc formed enmeshes most of the colloidal solids, and most of the floc is removed by the settling basins. Recarbonation by carbon dioxide lowers the pH and stabilizes the water so that further precipitation does not occur. The granular media filters remove most of the fine nonsettling floc, and disinfection kills any pathogenic organisms present. Lime-soda softening, in addition to removing hardness produces coagulation and settling of colloidal solids, although sometimes a coagulant is added to improve performance. The solids handling system consists of centrifugation and disposal of the dewatered cake by sanitary landfill; however, other solids handling systems may be used. Sometimes the lime-soda softening process uses two flocculators and two settling basins. For this case, the flowsheet will have the first flocculator, the first settling basin, the second flocculator, and then the second settling basin. This is done when the chemicals are added and flocculated in a stagewise manner. Also, dual units are used when stabilization is accomplished in two stages.

Frequently, groundwaters are of such high quality that the only

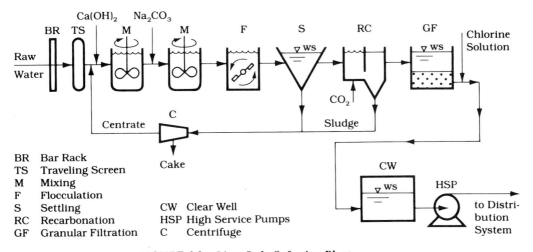

FIGURE 6.3 *Lime-Soda Softening Plant*

treatments required are gas stripping, to remove gases such as carbon dioxide when they are present in supersaturated amounts, and chlorination, to provide a residual in the distribution system as shown in Figure 6.4. The supply and treatment plant for such a water would consist of the well, a gas stripping unit, disinfection facilities, a ground storage reservoir, and a high service pump station. If a groundwater has sufficient hardness, softening by the lime-soda process or by the ion exchange process is required. Also, if a groundwater has sufficient iron or manganese content, special treatment is required for their removal.

Slow sand filtration plants consist of screening, plain sedimentation for 30 days or more, slow sand filtration, and disinfection. Slow sand filters operate at about 1/20 of the hydraulic rate of rapid sand filters; consequently, the area of the filter beds must be about 20 times as large as rapid sand filter beds. Also, 30 days or more plain settling requires a large area. Most slow sand filter plants are existing plants that have never been converted to rapid sand filtration plants. In the United States, there has been some interest in using slow sand filters for treating waters from isolated and extremely clear lakes or reservoirs where disinfection is the only treatment presently being done.

In recent years, it has been found that certain toxic substances, such as certain synthetic organic chemicals and heavy metal ions, are present in some waters, particularly surface waters, although they have also been found in groundwaters. Thus plants must monitor for these to provide a safe drinking water. Since very little removal occurs in the usual rapid sand filtration plants, particularly for organic chemicals, the most practical solution consists of eliminating these substances at their sources. This is done by

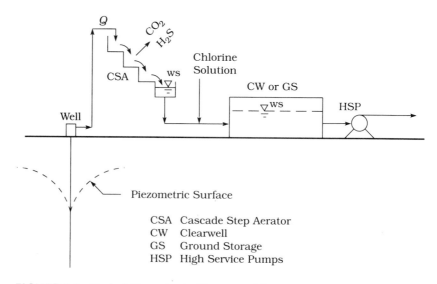

FIGURE 6.4 *Typical Groundwater Treatment Plant*

regulatory agencies, on both the state and national levels. Where occurrence is possible, activated carbon adsorption after filtration should be provided.

Industries usually require process water for manufacturing their products, boiler feed water for their boilers, and cooling water for their condensers. Usually, process water is provided by the previously described flowsheets; however, some industries may require specialized treatments such as demineralization for process water. Boiler feed water must be demineralized if high pressure boilers are used. Demineralization is usually accomplished by cation and anion exchangers operated on the hydrogen and hydroxyl cycles. Cooling water is usually drinking water quality except in those cases in which surface, well, or salt waters are used without treatment.

WASTEWATER TREATMENT PLANTS

The most common municipal wastewater treatment plants are primary and secondary treatment plants, tertiary treatment plants, and physical-chemical treatment plants.

Primary treatment consists of removing a substantial amount of the suspended solids from a wastewater. The collected solids must be treated, in most cases, followed by proper disposal. Secondary treatment consists of bio-oxidizing the remaining organic suspended solids and the organic dissolved solids. The flowsheet of a conventional activated sludge plant, shown in Figure 6.5 consists of screening, grit removal, primary clarification,

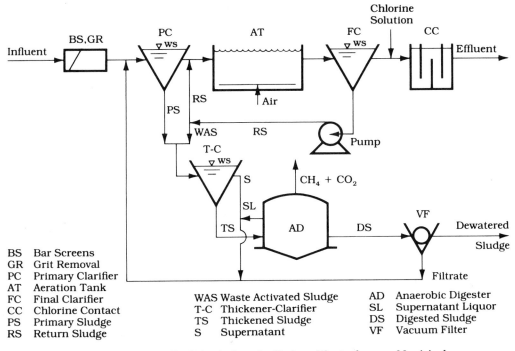

BS	Bar Screens
GR	Grit Removal
PC	Primary Clarifier
AT	Aeration Tank
FC	Final Clarifier
CC	Chlorine Contact
PS	Primary Sludge
RS	Return Sludge

WAS	Waste Activated Sludge
T-C	Thickener-Clarifier
TS	Thickened Sludge
S	Supernatant

AD	Anaerobic Digester
SL	Supernatant Liquor
DS	Digested Sludge
VF	Vacuum Filter

FIGURE 6.5 *Activated Sludge Plant for a Municipal Wastewater*

activated sludge treatment, and chlorination. The coarse solids are removed by screening, and the sand and silt are removed by the grit removal system. Primary clarification removes as many suspended solids as possible, and the primary effluent is mixed with the return activated sludge. The mixed liquor then flows to the aeration tank. Bio-oxidation of most of the remaining organic matter occurs in the aeration tank, and the final clarifier removes the biological solids, which are returned to mix with the incoming primary effluent. The effluent from the final clarifier is disinfected to kill pathogenic organisms and then discharged to the receiving body of water. The primary clarifier sludge and the waste activated sludge (in other words, the excess activated sludge produced by the microbial solids, which has to be wasted from the system) are mixed together and then thickened to increase the solids content. The thickened sludge is sent to the anaerobic digester for bio-oxidation of the organic solids. The digested sludge is dewatered by vacuum filtration, and the dewatered sludge is disposed of in a sanitary landfill or by land disposal. Minor flows, such as the thickener supernatant, the anaerobic digester supernatant, and the vacuum filter filtrate, are returned to the head of the plant. Although the solids handling system shown consists of thickening, anaerobic digestion, and vacuum filtration, other solids handling systems, such as aerobic digestion and centrifugation, are used. This flowsheet gives about 85 to 95% five-day biochemical oxygen demand (BOD_5) and suspended solids removal.

Tertiary treatment of a secondary effluent consists of providing further treatment to increase the quality of the effluent. The flowsheet for a tertiary treatment plant is shown in Figure 6.6. It consists of lime coagulation, flocculation, sedimentation, ammonia stripping, recarbonation, sedimentation, multimedia filtration, carbon adsorption, and breakpoint chlorination. The coagulant used is quicklime (calcium oxide), which is reacted with water to produce slaked lime (calcium hydroxide), which is added ahead of the mixing basin. Lime coagulation, flocculation, and sedimentation at a high pH removes most of the suspended solids and phosphorus. Ammonia stripping at a high pH removes most of the ammonia. Recarbonation is provided to lower the pH and stabilize the wastewater. The settling basin downline from the recarbonation basin removes the calcium carbonate precipitated by recarbonation. Multimedia filtration removes most of the nonsettling floc, and carbon adsorption removes most of the remaining dissolved organic compounds. In addition to disinfection, breakpoint chlorination chemically oxidizes the remaining ammonia to nitrogen gas and the remaining organic matter to other end products. The solids handling system shown permits recovery of the quicklime coagulant, thus reducing the lime requirements and the amount of lime sludge to be disposed of. The coagulant is recovered by lime recalcination, in which the calcium carbonate precipitate in the sludge is heated at a high temperature to produce the coagulant, calcium oxide. In addition to coagulant recovery, the organic solids in the sludge are incinerated. The flowsheet in Figure 6.6 will produce an effluent

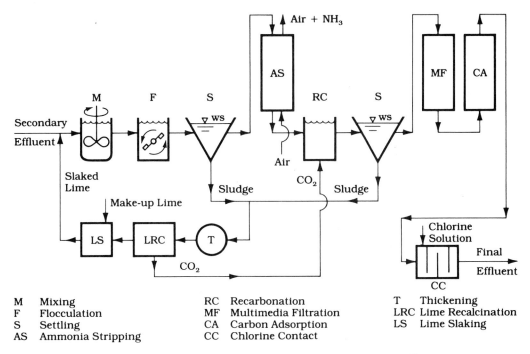

FIGURE 6.6 *Tertiary Treatment of a Secondary Effluent by Physical-Chemical Methods*

approaching drinking water quality when treating municipal secondary effluents.

The flowsheet of a physical-chemical treatment plant for raw municipal wastewaters is shown in Figure 6.7. It consists of lime coagulation, flocculation, sedimentation, recarbonation, sedimentation, multimedia filtration, carbon adsorption, and breakpoint chlorination. Lime coagulation, flocculation, and sedimentation remove most of the suspended solids, phosphorus, and organic nitrogen. Recarbonation lowers the pH and stabilizes the wastewater. The sedimentation basin downline from the recarbonation unit removes most of the calcium carbonate precipitated by recarbonation. Multimedia filtration removes most of the fine, nonsettling floc, and carbon adsorption removes most of the remaining organic compounds. In addition to disinfection, breakpoint chlorination chemically oxidizes the remaining ammonia to nitrogen gas and the remaining organic materials to other end products. The solids handling system shown permits recovery of the quicklime coagulant and reduces the amount of lime sludge to the disposed of, in addition to incinerating the organic solids in the lime sludge. The flowsheet shown removes from 96 to 99% of the BOD_5 and suspended solids when treating municipal wastewaters.

Industrial wastewaters may be broadly classified as organic and inorganic

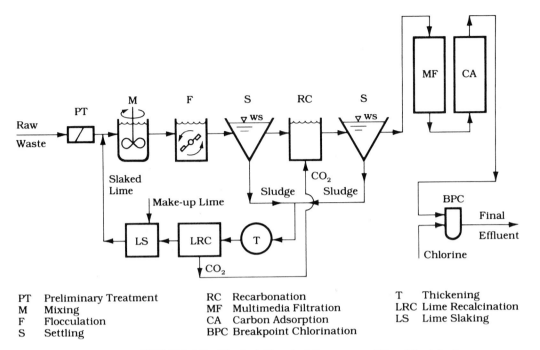

PT   Preliminary Treatment
M   Mixing
F   Flocculation
S   Settling

RC   Recarbonation
MF   Multimedia Filtration
CA   Carbon Adsorption
BPC Breakpoint Chlorination

T   Thickening
LRC Lime Recalcination
LS   Lime Slaking

FIGURE 6.7   *Physical-Chemical Treatment of Raw Municipal Wastewater*

wastewaters. In the United States, the total daily BOD_5 contribution from organic industrial wastewaters is several times the BOD_5 contribution from municipal wastewaters. The main industries, in order of magnitude, are (1) pulp, paper, and paper board (cardboard); (2) fermentation for brewed and distilled alcoholic beverages, yeast production, and antibiotic production; (3) sugar beet processing; (4) slaughter house and meat packing; (5) textile manufacture and textile dying; (6) canning and freezing; (7) petroleum refining and petrochemical production; and (8) miscellaneous, such as tanning leather, dairy processing, food processing (coffee, rice, fish, pickles, soft drink bottling, and baking), cornstarch production, and soap and detergent manufacturing. Even for biodegradable organic industrial waste-waters, the quality parameters and the flowrates vary appreciably between the numerous industries in this category. Therefore, a wide variation exists in the various flowsheets employed. For organic industrial wastewaters with low to moderate suspended solids content, the flowsheet shown in Figure 6.8 is frequently used. It consists of a completely mixed activated sludge unit and an aerobic digestion system for treating the waste activated sludge. A completely mixed activated sludge unit has a reactor basin that is usually square or circular in plan view. The influent wastewater, on entering, is spread throughout the reactor basin volume in a very short period of time. The digested sludge from the aerobic digestion system is chemically con-

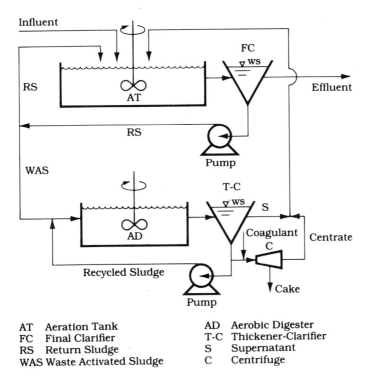

FIGURE 6.8 *Completely Mixed Activated Sludge Plant for an Industrial Wastewater*

ditioned to release water and is then dewatered by centrifugation. The cake is usually disposed of by sanitary landfill or land disposal, and the centrate is returned to the head of the plant. For organic industrial wastewaters with low to moderate suspended solids content in which the organic content varies appreciably during the day, a constant level equalization basin may be required, as shown in Figure 6.9. This flowsheet consists of an equalization basin, a dispersed plug-flow activated sludge unit, and an incinerator for disposal of the waste activated sludge. A dispersed plug-flow activated sludge unit has a reactor basin that is usually rectangular in plan view and that has significant longitudinal dispersion of fluid elements throughout the length of the basin. The waste activated sludge is chemically conditioned and is then dewatered by a filter press. The sludge cake is incinerated and the ash disposed of in a sanitary landfill. The mixing system for the completely mixed reactor and aerobic digester in Figure 6.8 and for the equalization basin in Figure 6.9 is shown schematically as a propeller. In actual basins, the mixing is usually provided by the aeration system. If appreciable land area is available, organic industrial wastewaters with low to moderate suspended solids content may be treated by the aerated lagoon flowsheet shown in Figure 6.10. It consists of an aerated lagoon, which is essentially an

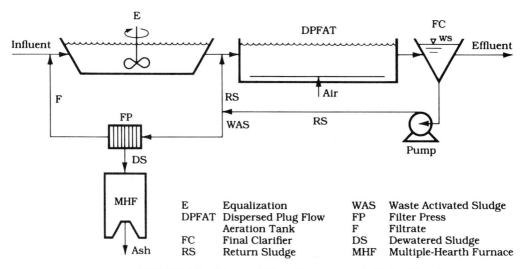

FIGURE 6.9 *Dispersed Plug-Flow Activated Sludge Plant for an Industrial Wastewater*

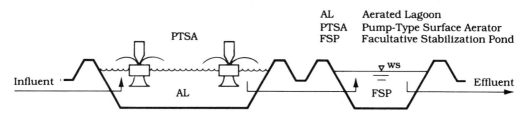

FIGURE 6.10 *Aerated Lagoon System for an Industrial Wastewater*

activated sludge unit without recycle, and a facultative stabilization pond, which serves as a final clarifier. The biological solids that settle in the stabilization pond undergo anaerobic decomposition on the pond bottom, and the solids are usually removed once every several years. Frequently, a settling tank with sludge rakes for continuous sludge removal is used for final clarification; the removed sludge is treated to stabilize it.

The main industries having inorganic industrial wastewaters are metal-plating industries, such as cadmium, zinc, chromium, copper, nickel, and tin plating, and metal industries, such as steel-mill and iron foundries. For inorganic industrial wastewaters, the unit operations and processes used for treatment depend on the wastewater characteristics. For instance, industrial wastewaters from plating industries, which contain heavy metallic ions such as copper, zinc, and cadmium, may be treated by ion exchange. To locate flowsheets used for a particular industrial wastewater, the reader is referred to the various texts and literature on industrial waste treatment.

**DESIGN
FLOWRATES
AND
PARAMETERS**

Water and wastewater treatment plants for municipalities are usually designed for a flowrate that will occur from 10 to 25 years in the future. This requires a population projection for the design period and also an estimate of the water demand or wastewater flow per capita for the future period. The most accurate way to determine the water demand per capita is by studying long-term water pumpage records (at least 12 months' duration). Pumping records will also allow determination of the minimum, average, and maximum day demands in addition to the maximum hourly demand. Usually, in the United States, the average annual water demand for municipalities is from about 100 to 200 gal/cap-day (380 to 760 ℓ/cap-d) Frequently, the existing demand is increased slightly to obtain the demand per capita in the future, since the demand per capita increases as the population increases. The percent increase is usually taken as 0.1 times the percent increase in population. Thus, if the existing demand is 145 gal/cap-day (549 ℓ/cap-d), and the population is expected to increase 80% in the future, the future demand would be (145)[100% + (0.1)80%] or 157 gal/cap-day [(549)(100% + 0.10 × 80%) = 593 ℓ/cap-d]. As previously mentioned, the design capacity of a water treatment plant is the average flow on the day of maximum demand. Long-term analyses on the water supply will give quality parameters, such as pH, turbidity, color, odor, taste, hardness, alkalinity, dissolved solids, and coliform count. Short-term analyses on parameters subject to variation, such as the turbidity of a surface water, should be used with caution.

The most accurate way to determine the wastewater flow per capita is by studying long-term flow records. Usually, for municipalities with separate sewer systems in the United States, the wastewater flow is approximately equal to the water demand. The future flowrate per capita is increased over the present flowrate in a manner similar to what is done for future water demands. Also, flow records will allow determination of the ratio of the peak hourly flow to the average hourly flow, the ratio of the minimum hourly flow to the average hourly flow, and the amount of storm water infiltration. The design capacity of a wastewater treatment plant is usually considered to be the maximum daily flow during wet-weather conditions. Long-term analyses of the wastewater will give quality parameters, such as the five-day biochemical oxygen demand (BOD$_5$), suspended solids concentration, chemical oxygen demand (COD), organic nitrogen, ammonia nitrogen, nitrite nitrogen, nitrate nitrogen, grease content, and pH. For separate sewer systems in the United States, the average BOD$_5$ is about 200 mg/ℓ and the average suspended solids concentration is about 240 mg/ℓ. The wastewater flow is usually from 60 to 140 gal/cap-day (227 to 530 ℓ/cap-d), with 100 gal/cap-day (380 ℓ/cap-d) being typical. Arbitrarily assumed design parameters, such as the population equivalent of 0.17 lb BOD$_5$/cap-day (0.077 kg/cap-d) and 0.20 lb suspended solids/cap-day (0.091 kg/cap-d), should be used with caution and should be substantiated by several months of analyses and flow gauging.

Water and wastewater treatment plants for industries are usually designed to accommodate water demands and wastewater flows for a future flow projection of 5 to 10 years, and the plants are designed to facilitate future expansions. This is done because industries usually prefer stagewise expansions; these require less initial expense than plants for longer design periods. Also, the uncertainty of the future market for their product necessitates prudent capital expenditures. Water demands for an industry are usually expressed in gallons (liters) per production unit. For example, for a pulp and paper mill it would be gallons used per ton (liters per tonne) of paper produced. Knowing the number of production units expected in the future, engineers can determine the future water demand.

In designing a wastewater treatment plant for an industry, it is desirable to have an industrial waste survey done for the industry. In a waste survey, a flowsheet for the industry is developed that shows, in sequential order, the operations used in producing the product. All sources of water use and wastewater generation are located, and the flowrate of each is determined. Wastewater streams are sampled and analyzed to give quality parameters such as BOD_5 and suspended solids. The final result of the survey is a flow and material balance diagram showing the flowrates and quality parameters for each waste stream and the outfall sewer. In studying the diagram, places are located where reuse and usable byproduct recovery is feasible, where a reduction in water usage is possible, where beneficial process modifications can be done, and where segregation of flows is desirable or necessary. An example of reuse is the reusing of washwaters instead of using them on a once-through basis. An example of usable byproduct recovery is the recovery of oil from waste streams within a refinery by the use of gravity-type oil-water separation tanks. The oil that floats to the top of the tanks is skimmed off, and the recovered oil is sent back for processing. An example of a beneficial process modification is the substitution of low-BOD_5 detergents for high-BOD_5 soaps in the wash operations done in a woolen textile mill. Flow segregation should be done for streams having a low BOD_5 concentration, inert materials, or characteristics that make them incompatible with the main wastewater flow. Some segregated streams may be eliminated from the main wastewater flow and not require any treatment. Other segregated flows may require separate treatment from the main wastewater flow, or, in some cases, they may be pretreated and the pretreated stream sent back to the main wastewater flow. For instance, an acidic stream with a significant BOD_5 concentration may be neutralized and then remixed with the main wastewater flow. Once the flows are adjusted for reuse, usable byproduct recovery, flow reduction, beneficial process modifications, and flow segregation, the final flow in gallons (liters) per day for the outfall sewer and the final quality parameters, such as pounds (kilograms) of BOD_5 and pounds (kilograms) of suspended solids per day, can be determined. These can be correlated with production units to give gallons (liters) of wastewater, pounds (kilograms) of BOD_5, and pounds (kilograms) of suspended solids per production unit. Then, when the number of production units in the future is known, the design parameters and design flow for the industrial wastewater treatment plant can be determined. In most cases, an industrial waste survey significantly reduces the amount of wastewater flow to be treated, thus decreasing the costs of treatment.

Drinking water standards for water treatment plants and effluent standards for wastewater treatment plants are established by governmental regulations. Primary drinking water standards for public water supplies are based on the National Safe Drinking Water Act, Public Law 93-523, which was signed into law on December 16, 1974. Public Law 93-523 gives the Environmental Protection Agency (EPA) the responsibility for establishing primary drinking water standards relating to health. Secondary regulations relating to taste, odor, and appearance of drinking water have been recommended by the EPA, and most states have adopted these or a modification of them. An amendment to the Safe Drinking Water Act was passed and signed into law on July 2, 1986, to cover toxic substances in drinking waters.

Comprehensive federal water pollution control legislation was enacted on October 18, 1972 under Public Law 92-500. This law requires the EPA to establish effluent guidelines from which individual states issue discharge permits and impose specific effluent limitations. Under Public Law 92-500, municipalities and industries are required to have secondary treatment. An amendment to Public Law 92-500 was passed in 1987 and is mainly concerned with toxic substances in effluents. In many of the areas of the United States where populations are dense and/or industrial wastewaters are significant, effluent permits may require more than secondary treatment — that is, tertiary treatment beyond disinfection for municipalities and tertiary treatment by industries.

REFERENCES

American Water Works Association. 1990. *Water Quality and Treatment*. 4th ed. New York: AWWA.

American Water Works Association. 1990. *Water Treatment Plant Design*. 2nd ed. New York: AWWA.

Eckenfelder, W. W., Jr. 1989. *Industrial Water Pollution Control*. 2nd ed. New York: McGraw-Hill.

McGhee, T. J. 1991. *Water Supply and Sewerage*. 6th ed. New York: McGraw-Hill.

Metcalf and Eddy, Inc. 1991. *Wastewater Engineering: Treatment, Disposal and Reuse*. 3rd ed. New York: McGraw-Hill.

Reynolds, T. D. 1982. *Unit Operations and Processes in Environmental Engineering*. Boston: PWS Publishing Company.

United States Public Health Service. 1969. *Manual for Evaluating Public Drinking Water Supplies*. Cincinnati, Ohio.

Viessman, W., and Hammer, M. J. 1993. *Water Supply and Pollution Control*. 5th ed. New York: Harper & Row.

Water Pollution Control Federation (WPCF). 1977. *Wastewater Treatment Plant Design*. WPCF Manual of Practice. Washington, D.C.

7

PRELIMINARY UNIT OPERATIONS AND PROCESSES

In the treatment of waters and wastewaters, preliminary treatments may be required to remove certain objectionable impurities or to make the water or wastewater more amenable to subsequent treatments.

WATER TREATMENT

In the treatment of certain surface waters, preliminary treatments such as screening, presedimentation, aeration, adsorption, and prechlorination may be required. In the treatment of certain groundwaters, preliminary treatments such as aeration may be required.

Screening

Coarse bar racks and fine traveling racks are employed at intake structures on reservoirs and rivers. Coarse bar screen racks usually have clear spaces up to 3 in. (75 mm) between the bars and are used to prevent the entry of large debris, such as logs, into the intake structure. A trolley and hoist must be provided to remove any logs that hang against the bar rack. Fine traveling screens located behind the bar racks, as shown in Figure 7.1, usually have openings of about 3/8 to 1/2 in. (10 to 13 mm) and are used to prevent the entry of small debris, such as sticks, bark, leaves, and fish.

Presedimentation

Some river waters have high turbidities and coliform counts that may require presedimentation prior to other treatments. Two types of presedimentation are usually used. The first type is for river waters, which usually have low to moderate turbidities and coliform counts but which on occasions of high rainfall intensities, have high turbidities and coliform counts. The second type is for river waters consistently having high turbidities and coliform counts. The first type of presedimentation requires plain sedimentation in reinforced concrete tanks with sludge rakes for sludge removal. Also reinforced concrete tanks with steeply sloped, hopper-shaped bottoms and gravity sludge removal have been used. Both types of tanks may be circular, rectangular, or square in plan view. The detention time for these basins is from 0.5 to 1.0 hour, and an overflow rate or surface loading of 1000 to 3000 gal/day-ft^2 (40.7 to 122 m^3/day-m^2) is usually used. These basins are recommended for river waters occasionally having turbidities greater

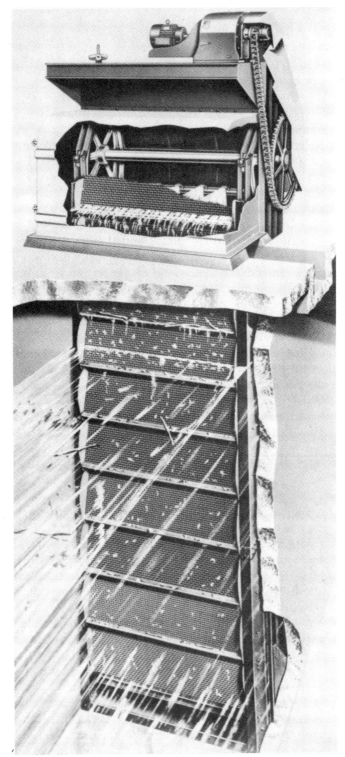

FIGURE 7.1
Traveling Water Screen
Courtesy of Envirex, Inc.

than 10,000 turbidity units, and turbidity removals from 65 to 80% can be expected (Degremont, Inc., 1973).

The second type of presedimentation requires plain sedimentation in large basins having extremely long detention times and is used for river waters having turbidities greater than 10,000 turbidity units. The Mississippi River and many of its tributaries commonly have turbidities from 10,000 to 40,000 turbidity units. In addition, some of these rivers may have coliform counts greater than 5000/100 mℓ. Much of the turbidity will settle and be removed by plain sedimentation basins, and it is desirable to remove as much of this turbidity as possible to avoid overloading the treatment plant. The detention times are usually 30 to 60 days. Usually these basins are constructed of earthen dikes, and they may or may not have continuous sludge removal facilities. The dikes usually have slope protection consisting of reinforced concrete slabs or riprap, such as crushed stone, to avoid erosion and unwanted growths of aquatic plants, which could cause tastes and odor problems. Self-purification occurs by plain sedimentation and, in the case of bacterial removal, by settling to a limited degree and by the action of ultraviolet light in the sun's rays. Some bacteria adhere to silt and other particles and settle along with the particles. Additional removal occurs by the disinfecting action of ultraviolet light from the sun. If the river water has a coliform count greater than 5000/100 mℓ, presedimentation for 30 to 60 days by plain sedimentation is recommended. Removal of 80 to 90% of bacteria and viruses can be expected after 30 days' storage (Amirtharajah, 1986).

Aeration   Aeration may be used for gas stripping (degasification) to remove unwanted gases, such as carbon dioxide and hydrogen sulfide, and iron and manganese. Groundwaters, in particular, may require aeration to remove these contaminants. If present in the free ion form, iron and frequently manganese may be oxidized to insoluble compounds that may be removed by coagulation (if required), sedimentation, and filtration. Usually, aeration is accomplished by cascades, multiple-tray aerators, spray nozzles, or diffused compressed air tanks.

A **cascade** usually is a flight of three or four concrete or metal steps over which the water tumbles as a thin sheet. For metal cascades, some types have low weirs at the periphery of the steps. The cascade is of little use in removing odors due to algae but may reduce the carbon dioxide by 20 to 45% (Steel and McGhee, 1979), and the head required is from 3 to 10 ft. (0.9 to 3 m). The cascade should be in a screened enclosure to preclude the entry of birds and insects, such as midge flies.

A **multiple-tray aerator** consists of a series of horizontal traps, each containing 8 to 12 in. (200 to 300 mm) of medium, the medium being ceramic balls 2 to 6 in. (50 to 150 mm) in diameter, slag, or stones. The medium is supported by perforated plates, slots, or screen trays. Some tray aerators have no medium and depend upon the perforated plates, slots, or screen trays. Usually three trays are employed and are spaced 1.5 ft. (0.5 m) apart. Application of the water is usually by a shallow pan with a perforated bottom above the top tray. The perforations are usually 3/16 to 1/2 in. (5 to 13 mm) in diameter and are spaced at 3 in. (75 mm) on center. Sometimes the water is applied by spray nozzles spraying directly on the top tray. The hydraulic loading is from 1 to 5 gpm/ft² (40 to 200 ℓ/min-m²), depending upon the concentration and the kind of gas being removed. Low hydraulic loadings and media 1 1/2 to 2 1/2 in. (38 to 64 mm) in diameter are used when the gas is highly concentrated. The removal of carbon dioxide can be estimated as follows (AWWA, 1990):

$$\frac{C}{C_0} = e^{-kn} \tag{7.1}$$

where

$\quad C$ = effluent concentration, mg/ℓ
$\quad C_0$ = influent concentration, mg/ℓ
$\quad k$ = rate constant (base e)
$\quad n$ = number of trays

Typically, k is from 0.28 to 0.37 for carbon dioxide, the higher values being for good tray ventilation. Multiple-tray aerators should be in screened enclosures to prevent entry by birds and insects. The flow regime in a cascade or multiple-tray aerator is similar to a series of completely mixed tanks. For instance, three trays would be similar to three completely mixed tanks in series.

EXAMPLE 7.1   *Carbon Dioxide Removal*

A groundwater containing 8 mg/ℓ of carbon dioxide is to be degasified using a multiple-tray aerator with three trays. Two tray aerators in parallel are to be used for flexibility in operation if one aerator is inoperative. The design population is 5000 persons, and the maximum day demand is 150 gal/cap-day. The k value is 0.33, and the hydraulic loading is 3 gpm/ft². Determine:

1. The carbon dioxide content of the product water.
2. The size of the trays if the length-to-width ratio is 2:1 and the trays are made to 1-in. increments.

SOLUTION   The performance equation is

$$\frac{C}{C_0} = e^{-kn}$$

Thus,

$$C = (8)e^{-0.33 \times 3} = (8)(0.3716) = 2.98 \text{ mg/ℓ or } \boxed{3.0 \text{ mg/ℓ}}$$

The flow to each aerator is

$$Q = (5000 \text{ persons})\left(\frac{150 \text{ gal}}{\text{cap-day}}\right)\left(\frac{\text{day}}{1440 \text{ min}}\right)\left(\frac{1}{2}\right) = 260 \text{ gpm}$$

The area is

$$A = (260\,\text{gpm})\left(\frac{\text{ft}^2}{3\,\text{gpm}}\right) = 86.67\,\text{ft}^2$$

$L = 2W$; thus,

$$(W)(2W) = 86.67\,\text{ft}^2$$

$$W = 6.58\,\text{ft or } \boxed{6\,\text{ft-7 in.}}$$

$$L = (2)(6.58) = 13.17\,\text{ft or } \boxed{13\,\text{ft-3 in.}}$$

EXAMPLE 7.1 SI *Carbon Dioxide Removal*

A groundwater containing $8\,\text{mg}/\ell$ of carbon dioxide is to be degasified using a multiple-tray aerator with three trays. Two tray aerators in parallel are to be used for flexibility in operation if one aerator is inoperative. The design population is 5000 persons, and the maximum day demand is $570\,\ell/\text{cap-d}$. The k value is 0.33, and the hydraulic loading is $400\,\ell/\text{min-m}^2$. Determine:

1. The carbon dioxide content of the product water.
2. The size of the trays if the length-to-width ratio is $2:1$.

SOLUTION The performance equation is

$$\frac{C}{C_0} = e^{-kn}$$

Thus,

$$C = (8)e^{-0.33 \times 3} = (8)(0.3716) = 2.98\,\text{mg}/\ell \text{ or } \boxed{3.0\,\text{mg}/\ell}$$

The flow to each aerator is

$$Q = (5000)\left(\frac{570\,\ell}{\text{cap-d}}\right)\left(\frac{\text{d}}{1440\,\text{min}}\right)\left(\frac{1}{2}\right) = 990\,\ell/\text{min}$$

The area is

$$A = \left(\frac{990\,\ell}{\text{min}}\right)\left(\frac{\text{min}^2\text{-m}}{400\,\ell}\right) = 2.47\,\text{m}^2$$

$L = 2W$; therefore,

$$(W)(2W) = 2.47\,\text{m}^2$$

$$W = \boxed{1.57\,\text{m}}$$

$$L = (2)(1.57) = \boxed{3.14\,\text{m}}$$

Spray nozzles are sometimes used for aeration; however, they require considerable head and so much space that housed enclosures are difficult. Operation in cold weather may be impossible. Carbon dioxide removal may be as high as 90% with pressure heads from 10 to 20 psig (70 to 140 kilopascals). The amount of water discharged per nozzle will depend upon the nozzle design and head used. A typical nozzle with a 1-in. (24-mm) orifice will discharge about 70 gpm (265 ℓ/min) to a height of 7 ft (2.1 m) at a 10-psig (70-kilopascals) pressure.

Diffused compressed air tanks are sometimes used for aeration. Normally these tanks are from 9 to 15 ft (2.7 to 4.6 m) in depth, are from 10 to 30 ft (3.0 to 9.1 m) in width, and have a length sufficient to provide for a contact time of 5 to 30 min. The air is supplied by diffusers such as those used in wastewater treatment by the activated sludge process. The diffusers are placed along the bottom of one wall to give a spiral roll to the water, and the air flow is from 0.005 to 0.2 ft^3/gal (0.0374 to 1.50 m^3/m^3) of water (Steel and McGhee, 1979). This type of aerator has an advantage over other types of aerators since it has a long contact time, eliminates freezing, and allows the unit to be used for mixing and flocculation at a rapid sand filtration plant or an iron and manganese removal plant.

Adsorption   Activated carbon is a universal adsorbent since it adsorbs nearly all organic compounds causing taste, odor, or color problems; halogens; hydrogen sulfide, iron, and manganese ions; and numerous other dissolved substances. In water treatment it is useful for removing organic compounds that cause taste, odor, or color. Activated carbon is made from a variety of organic materials, such as ground fruit pits, coconut shells, or sawdust. It is heated in a closed retort to a high temperature and "activated" by injecting steam. The particles have a large surface-area-to-volume ratio, and the surface area has slight positive and negative charges. This allows it to adsorb organic compounds having slightly polar charges, which nearly all organic compounds have. It also adsorbs hydrogen sulfide, iron, and manganese ions, along with other substances. It is available as a powder or in granular form. At many water filtration plants, the powdered carbon is added at the intake structure or ahead of the chemical mixing basins. It adsorbs most taste-, odor-, or color-causing organic compounds and is removed with the sludge from the settling basins. When used in this manner, it is not recovered and is disposed of with the sludge. Dosing is accomplished with dry chemical feeders equipped with a special device using compressed air to prevent uncontrolled flow through the feeder. Feeders are in separate rooms because of carbon dust problems, and to control carbon dusts, hoods should be used where the sacks of carbon are added to the hopper. Dosages range from 2 to 70 lb/million gallons (0.24 to 8.40 gm/m^3) and are typically 10 to 15 lb/million gallons (1.20 to 1.80 gm/m^3). The required dosage is frequently determined by the threshold odor number of the finished water. Granular activated carbon is regenerated, is used as a postfiltration treatment, and is discussed in a subsequent chapter.

Prechlorination   Many filtration plants chlorinate before any other treatments, and the chlorine solution is added at the suctions of the raw water pumps or at the mixing basins. Frequently ammonia is added at the same locations so that chloramines are formed, and this helps prevent the formation of trihalomethanes. Also, adding ammonia may prevent tastes and odors that can be formed from chlorine reacting with algal end products in the raw water supply. Prechlorination may prevent odors and taste compounds from being produced by bacterial action in the settling basin sludge. Also, prechlorination may prevent algal growths on the filter media, which can cause tastes and odors. Chloramines have a slower disinfecting action than free chlorine; however, ample time is given since many hours are required in the sedimentation basins and clearwell storage reservoirs. The chlorine residual prior to filtration is usually 0.1 to 0.5 mg/ℓ. Some surface waters may have an MPN greater than 5000/100 mℓ, and these require that both prechlorination and postchlorination be practiced if presedimentation for 30 days or more is not provided.

WASTEWATER TREATMENT   In the treatment of municipal and industrial wastewaters, preliminary treatments such as screening or shredding, grit removal, flow equalization, quality equalization, and neutralization may be required. For municipal wastewaters, screening or shredding and grit removal are always required for good plant performance. Some small-package municipal plants do not have grit removal; however, they experience problems due to grit accumulations in aeration tanks and digesters.

Screening and Shredding   Screens and shredders are used in municipal wastewater treatment, and sometimes in industrial wastewater treatment, for the removal or shredding of coarse solids. Typical coarse solids in municipal wastewaters are pieces of wood, plastic materials, and rags. Although hand-cleaned screens were formerly used, almost all screens presently installed are mechanically cleaned screens, as shown in Figure 7.2, and the screened solids are dumped into a receptacle located behind the screen for storage prior to disposal. Usually, disposal is by sanitary landfill or incineration. For manually cleaned bar screens, the bar spacing is usually from 1 to 2 in. (25 to 50 mm), and the bars are mounted at a 30 to 75° angle to the horizontal, 30 to 45° being typical. For mechanically cleaned screens, the bar spacing is usually 1/2 to 1 1/2 in. (12 to 38 mm), and the bars are at a 45 to 90° angle to the horizontal, 60° being typical. Usually, the mechanical rakes that clean a screen are operated by a time clock or by floats that activate the drive motor when the head loss across the screen is greater than about 2 in. (50 mm). The approach channel for a manually or mechanically cleaned bar screen should be straight for several feet (0.6 m) ahead of the screen to give uniform flow across the screen, and the approach velocity should be at least 1.5 ft/sec (0.46 m/s) to avoid grit deposition. The velocity through the bars should be less than 2 ft/sec (0.62 m/s) at design flow and not more than 3 ft/sec (0.91 m/s) at maximum flow. At least two screens should be provided for flexibility in

FIGURE 7.2
Mechanically Cleaned
Bar Screen
Courtesy of
Envirex, Inc.

operation since one may be used as a standby unit. At small plants, it is common to have one mechanical screen and one manual screen. The manual screen is used only when the mechanical screen is inoperative. At medium- to large-size plants, two or more mechanical screens are provided. Stop-gate slots should be provided ahead of and behind each screen so that the unit may be dewatered for maintenance. The head loss through bar screens may be determined from

$$h_L = \frac{(V_b^2 - V_a^2)}{2g} \cdot \frac{1}{0.7} \qquad (7.2)$$

where

h_L = head loss, ft (m)

V_a = approach velocity, ft/sec (m/s)

V_b = velocity through the bar openings, ft/sec (m/s)

g = acceleration due to gravity

The amount of screenings removed by bar screens depends upon the bar spacing. For ½-in. (13-mm) spacing, a typical value is $8\,\text{ft}^3$ per million gallons ($60\,\text{m}^3$ per million cubic meters) treated (wpcf, 1977). For 1½-in. (38 mm) spacing, a typical value is $1.5\,\text{ft}^3$ per million gallons ($11.2\,\text{m}^3$ per million cubic meters) treated (wpcf, 1977).

EXAMPLE 7.2   *Bar Screen*

A mechanical bar screen is to be used in an approach channel with a maximum velocity of 2.1 ft/sec. The bars are ½ in. thick and have 1½-in. clear openings. Determine:

    1. The velocity between the bars.

    2. The head loss in feet.

SOLUTION   Assume W is the channel width and D is the channel depth. The net area is $(W)(D)[1.5/(1.5 + 0.5)]$ or $0.75WD$. The channel area is WD, so using the continuity equation for flow gives

$$V_a A_a = V_b A_b$$

or

$$V_b = \frac{V_a A_a}{A_b}$$

$$= (2.1)\left(\frac{WD}{0.75WD}\right) = \boxed{2.80\,\text{ft/sec}}$$

The head loss is given by

$$h_L = \frac{V_b^2 - V_a^2}{2g} \cdot \frac{1}{0.7}$$

or

$$h_L = \frac{2.80^2 - 2.1^2}{2g} \cdot \frac{1}{0.7} = \boxed{0.08\,\text{ft}}$$

EXAMPLE 7.2 SI   *Bar Screen*

A mechanical bar screen is to be used in an approach channel with a maximum velocity of 0.64 m/s. The bars are 10 mm thick, and the openings are 30 mm wide. Determine:

    1. The velocity between the bars.

    2. The head loss in meters.

SOLUTION   Assume W is the channel width and D is the channel depth. The net area is $(W)(D)[30/(30 + 10)]$ or $0.75WD$. The channel area is WD, so using the continuity equation for flow gives

$$V_a A_a = V_b A_b$$

or

$$V_b = \frac{V_a A_a}{A_b}$$

$$= (0.64)\left(\frac{WD}{0.75WD}\right) = \boxed{0.85\,\text{m/s}}$$

The head loss is given by

$$h_L = \frac{V_b^2 - V_a^2}{2g} \cdot \frac{1}{0.7}$$

or

$$h_L = \frac{0.85^2 - 0.64^2}{2g} \cdot \frac{1}{0.7} = \boxed{0.023\,\text{m}}$$

The purpose of shredding coarse solids is to reduce their size so they will be removed by subsequent treatment operations, such as primary clarification, where both floating and settling solids are removed. Shredding may be accomplished by grinders, barminutors, and comminutors. Grinders shred the solids removed by a mechanically cleaned bar screen, and the shredded solids are returned to the wastewater flow downstream from the screen. The barminutor is a bar screen with a shredder that moves up and down the screen. The rotating cutter blades pass through the bar spaces, thus cutting the accumulated coarse solids. The comminutor, as shown in Figure 7.3, consists of a rotating slotted cylinder through which the entire wastewater flow passes. Solids too large to pass through the slots are cut by blades as the cylinder rotates, thus reducing their size until they pass through the slot openings. Figure 7.4 shows a comminutor installation employing two units, which provides for flexibility since one unit can remain in service while the other is inoperative during maintenance.

Grit Removal   Grit removal is usually limited to municipal wastewaters; however, some industrial wastewaters have grit. Grit consists of sand and silt particles that enter by infiltration, cinders, fragments of egg shells, bone chips, coffee grounds, and some shredded garbage if home garbage grinders are prevalent. For the most part, it is sand and silt. Grit is removed as much as possible to prevent wear on pumps; accumulations in aeration tanks, clarifiers, and digesters; and the clogging of sludge piping. Grit removal is usually done by horizontal-velocity settling chambers, diffused air chambers, and square settling chambers.

Typical horizontal-velocity grit settling chambers are shown in Figures 7.5, 7.6, and 7.7. The horizontal velocity in these chambers is controlled by

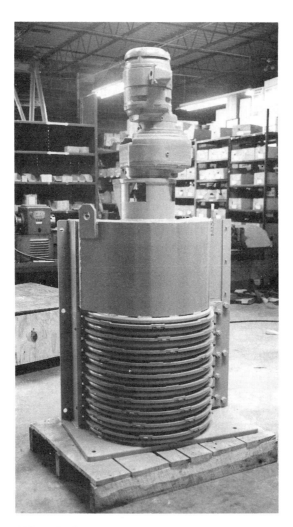

FIGURE 7.3 *Comminutor Unit*
Courtesy of Infilco Degremont, Inc.

either a Parshall flume or a proportional weir located at the downstream end of the chamber. In Figure 7.5, the mechanical grit conveyor buckets, which are chain driven, can be seen. In Figure 7.6, the Parshall flume, which is basically an open-channel venturi section, is shown at the downstream end of the chamber. In Figure 7.7, the proportional weir is shown at the downstream end of the chamber. Figure 7.8 shows the details of the Parshall flume, and Figure 7.9 shows the details of the proportional weir. Although both types of controls are used, the Parshall flume is more widely used mainly because it has less head loss than the proportional weir. The Parshall flume or proportional weir will probably be used as a flow measuring device as well, since in plant operation, it is useful to know the plant flowrate.

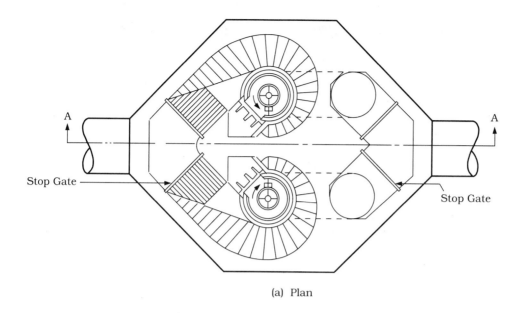

(a) Plan

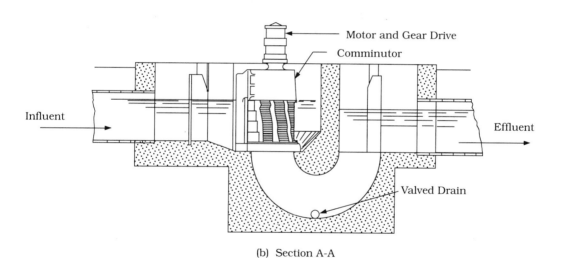

(b) Section A-A

FIGURE 7.4 *Comminutor Installation*
Courtesy of Chicago Pump Company.

Also, measuring records allow the per capita flow to be determined for future plant expansions. In the horizontal-velocity settling chambers, a constant horizontal velocity is maintained at all discharges by proper selection of the chamber cross-sectional geometry for both types of downstream flow control devices. The horizontal velocity must be adequate to keep organic solids in suspension, while not being sufficient to scour settled grit along the

FIGURE 7.5 *Two Horizontal-Velocity Grit Settling Chambers. The channel cross-sections are trapezoidal and grit conveyor buckets are mounted on chains driven by sprocket wheels and electric motors. Submerged conveyor buckets rake the settled grit toward the upstream end, where it is raised vertically and deposited into receptacles. View is toward the upstream end.*
Courtesy of Envirex, Inc.

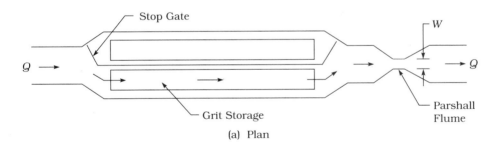

(a) Plan

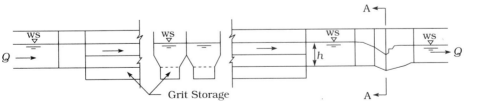

(b) Profile and Channel Cross Section

FIGURE 7.6 *A Horizontal-Velocity Grit Settling Chamber with a Parshall Flume Control Section*

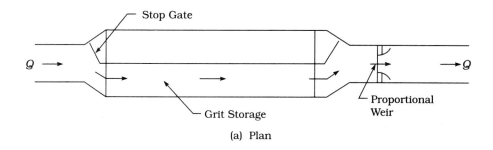

(a) Plan

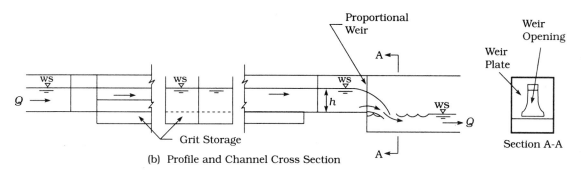

(b) Profile and Channel Cross Section

FIGURE 7.7 *A Horizontal-Velocity Grit Settling Chamber with a Proportional Weir Control Section*

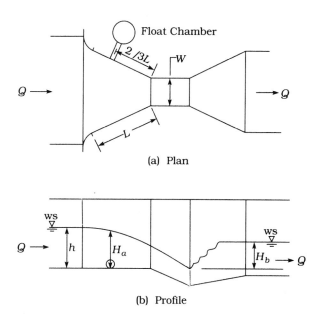

(a) Plan

(b) Profile

FIGURE 7.8 *The Parshall Flume*

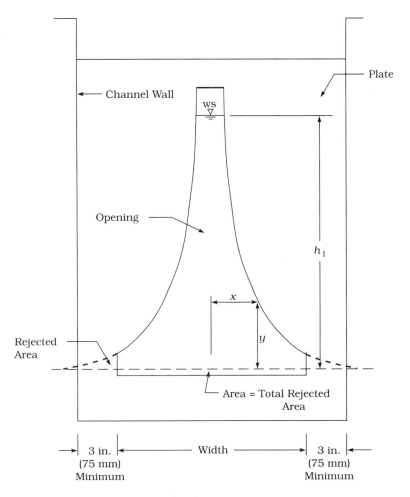

FIGURE 7.9 *The Proportional Weir*

bottom of the chamber. Both research and practice have shown that a range from 0.75 to 1.25 ft/sec (0.23 to 0.38 m/s) and a design value of 1.0 ft/sec (0.30 m/s) are commonly used. T. R. Camp (1942) showed that the ideal chamber cross-sectional area is a parabola if a Parshall flume is used as the control section. Since a parabolic section would be difficult to construct, in practice a combination of rectangular and trapezoidal areas is used to approximate a parabola as shown by Figure 7.10. Camp also showed that the ideal chamber cross-sectional area is a rectangle if a proportional weir is used as the control section. A commonly used discharge equation for Parshall flumes with free flow — that is, no backwater effects — is

$$Q = CWH_a^{1.5} \tag{7.3}$$

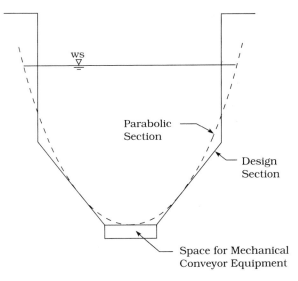

FIGURE 7.10 *Ideal Parabolic Cross Section and Design Cross Section for Chamber with a Parshall Flume*

where

Q = discharge, ft³/sec (m³/s)

C = 4.1 for USCS units and 2.26 for SI units

W = throat width, ft (m)

H_a = upstream depth, ft (m)

Textbooks such as Chow (1959) and French (1985) give standard dimensions for Parshall flumes. Equation (7.3) is for a free-flowing flume — that is, one with no backwater effects — and the discharge is a function of the upstream head, H_a. If a flume is not free-flowing and has backwater effects, the discharge is a function of the difference between the upstream and downstream heads, $H_a - H_b$. For throat widths less than 12 in. (300 mm), free flow occurs if H_b/H_a is 0.60 or less. For throat widths greater than 12 in. (300 mm), free flow occurs if H_b/H_a is 0.70 or less. The discharge equation for a proportional weir is

$$Q = Ckh_1 \qquad (7.4)$$

where

Q = discharge, ft³/sec (m³/s)

C = 7.82 for USCS units and 4.32 for SI units

k = weir coefficient

h_1 = head, ft (m)

The head, h, is usually about 3 in. (75 mm) greater than h_1. For the proportional weir, once Q and h_1 have been determined, the weir coefficient, k, may be determined from these equations. The x and y values, shown in Figure 7.9, are related by

$$x = \frac{k}{2\sqrt{y}} \tag{7.5}$$

where

x = horizontal distance from the y-axis, ft (m)

k = weir coefficient

y = vertical distance, ft (m)

Once the cross-sectional area has been designed, the theoretical length of the chamber must be determined. The design quartz sand particle is 0.2 mm in diameter, has a specific gravity of 2.65, and the settling velocity has been found to be 0.0689 ft/sec (21 mm/s). All sand particles 0.2 mm in diameter or larger are removed. Figure 7.11 shows the critical trajectory of the design sand particle in a horizontal-velocity grit chamber. If the design particle enters just below the top of the water surface, the time required for it to settle to the bottom at the exit of the chamber is

$$t = \frac{h}{V_s} \tag{7.6}$$

where

V_s = settling velocity

t = settling time

The horizontal distance that the particle travels during settling is the theoretical chamber length

$$L = Vt \tag{7.7}$$

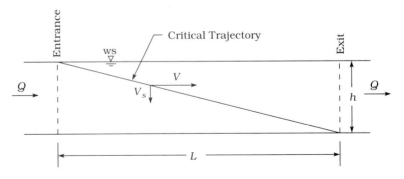

FIGURE 7.11 *Critical Trajectory of the Design Particle in a Horizontal-Velocity Grit Removal Chamber*

where

L = theoretical chamber length

V = horizontal velocity

In design, a length of 1.2 to 1.5 times the theoretical length is used to even out turbulence at the entrance of the chamber. For example, if h = 2.0 ft (0.61 m), then t = (2.0 ft) ÷ (0.0689 ft/sec) = 29 sec (t = 0.61 m ÷ 0.021 m/s = 29 s). Thus L = (1.0 ft/sec)(29 sec) = 29 ft (L = 0.30 m/s × 29 s = 8.70 m). If the actual length is 1.35 times the theoretical length, then actual length L = (1.35)(29 ft) = 39.2 ft (L = 1.35 × 8.70 m = 11.8 m).

For ease of construction, rectangular cross sections usually have a depth of 1.0 to 1.5 times the width. For parabolic cross sections, the depth is usually 0.60 to 1.5 times the top width. The volume provided for grit storage in the bottom of a chamber is from 6 to 12 in. (150 to 300 mm), depending on the size of the chamber and the equipment available.

Frequently, grit from horizontal-velocity grit chambers is washed in special devices to free it from any organic particles that may have been collected. Any organic particles removed from a washing operation are returned to the wastewater flow downstream from the chamber. The flow regime in horizontal-velocity grit chambers approaches plug flow.

EXAMPLE 7.3 *Grit Chamber with a Parshall Flume*

A municipal wastewater treatment plant has an average flow of 4.50 million gallons a day (MGD), and two horizontal-velocity grit chambers with a Parshall flume are to be designed. One chamber is to be operated as a standby. The peak flow is 2.34 times the average flow, or (2.34)(4.50) = 10.53 MGD or 16.3 ft³/sec. The Parshall flume has a throat width of 6 in., the horizontal velocity is 1.0 ft/sec, and the settling velocity is 0.0689 ft/sec. The head losses from the end of the grit chamber to the entrance of the Parshall flume are negligible because of the low velocity. Determine:

1. The head on the flume, ft.
2. The parabolic cross section and the design cross section.
3. The settling time, sec.
4. The theoretical chamber length, ft.
5. The design chamber length if the design length is 1.35 times the theoretical length and is an increment of 1 in.

SOLUTION The discharge equation for a 6-in. flume is

$$Q = 4.1WH_a^{1.5} = (4.1)(6/12)H_a^{1.5} = 2.05H_a^{1.5}$$

or

$$H_a = \left(\frac{Q}{2.05}\right)^{1/1.5} = \left(\frac{Q}{2.05}\right)^{0.6667}$$

$$= \left(\frac{16.3}{2.05}\right)^{0.6667} = \boxed{3.98 \text{ ft}}$$

A plan of the flume with a 6-in. throat width is drawn (not shown) according to the standard dimensions given by Chow (1959) and French (1985), and the channel width at the float tube is 1.08 ft. Thus the velocity at H_a is $(16.3)/(1.08 \times 3.98) = 3.79$ ft/sec. Writing Bernoulli's energy equation between the water surface in the downstream end of the grit chamber and the point where H_a is measured gives

$$\frac{V_1^2}{2g} + h = \frac{V_2^2}{2g} + H_a + h_L \text{ (entrance to the flume)}$$

or

$$\frac{(1.0 \text{ ft/sec})^2}{2g} + h = \frac{(3.79 \text{ ft/sec})^2}{2g} + 3.98 \text{ ft}$$
$$+ 0.10 \left[\frac{(3.79 \text{ ft/sec})^2}{2g} - \frac{(1.0 \text{ ft/sec})^2}{2g} \right]$$

or

$$0.02 \text{ ft} + h = 0.22 \text{ ft} + 3.98 \text{ ft} + 0.02 \text{ ft}$$

Thus,

$$h = 4.20 \text{ ft}$$

The cross-sectional area is given by

$$A = \frac{Q}{V}$$

or

$$A = \frac{16.3 \text{ ft}^3/\text{sec}}{1.0 \text{ ft/sec}} = 16.3 \text{ ft}^2$$

The top width for a parabola is

$$w = \frac{3}{2} \cdot \frac{A}{h}$$

or

$$w = \frac{3}{2} \cdot \frac{16.3 \text{ ft}^2}{4.20 \text{ ft}} = 5.82 \text{ ft}$$

For a head, H_a, of 0.50 ft, the discharge is

$$Q = (2.05)0.5^{1.5} = 0.725 \text{ ft}^3/\text{sec}$$

The velocity at H_a is $(0.725)/(1.08 \times 0.5) = 1.34$ ft/sec. Rewriting Bernoulli's energy equation as before gives $h = 0.51$ ft. The cross-sectional area of the wastewater in the chamber is

$$A = \frac{0.725 \text{ ft}^3/\text{sec}}{1.0 \text{ ft/sec}} = 0.725 \text{ ft}^2$$

The top width is

$$w = \frac{3}{2} \cdot \frac{0.725\,\text{ft}^2}{0.51\,\text{ft}} = 2.13\,\text{ft}$$

In a like manner, other heads are assumed, and the Q, h, A, and w values are determined and are shown in Table 7.1. The depth values, h, and the top width values, w, are plotted, and a smooth curve is drawn as shown in Figure 7.12. In order to select the design cross section, the width of the channel for the conveyor buckets must be known. A manufacturer's representative is contacted and several widths are obtained, which, in this range, are 15, 18, 21, and 24 in. After each of these is tried, the value of 21 in. (1 ft-9 in.) is selected because the design cross section fits the chamber the best.

TABLE 7.1 *Tabulations for Examples 7.3 and 7.3 SI*

H_a		h		Q		A		w	
ft	(m)	ft	(m)	ft³/sec	(m³/s)	ft³	(m³)	ft	(m)
0.5	(0.15)	0.51	(0.153)	0.725	(0.0197)	0.725	(0.0657)	2.13	(0.644)
1.0	(0.30)	1.04	(0.312)	2.05	(0.0557)	2.05	(0.186)	2.96	(0.894)
2.0	(0.60)	2.10	(0.630)	5.80	(0.158)	5.80	(0.525)	4.14	(1.25)
3.0	(0.90)	3.17	(0.948)	10.7	(0.289)	10.7	(0.963)	5.06	(1.52)
3.98	(1.23)	4.20	(1.30)	16.3	(0.461)	16.3	(1.54)	5.82	(1.78)

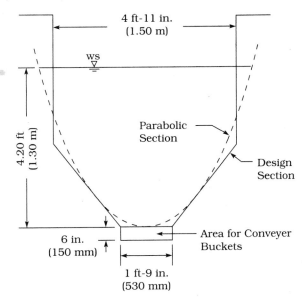

FIGURE 7.12 *Chamber Cross-Sectional Area for Examples 7.3 and 7.3 SI*

The design cross section is shown in Figure 7.12 by solid lines. When the design cross section is chosen, the side slopes should be relatively steep (45° or greater), and the areas above and below the parabola should be well balanced. The top width is made with an increment of 1 in. The settling time is given by

$$t = \frac{h}{V_s}$$

For a design settling velocity of 0.0689 ft/sec and a depth, h, of 4.20 ft, the time is

$$t = \frac{4.20 \, \text{ft}}{0.0689 \, \text{ft/sec}} = \boxed{61.0 \, \text{sec}}$$

The theoretical chamber length is

$$L = Vt$$

For a 1.0 ft/sec horizontal velocity, the theoretical length is

$$L = \left(\frac{1.0 \, \text{ft}}{\text{sec}}\right) 61.0 \, \text{sec} = \boxed{61.0 \, \text{ft}}$$

The design chamber length is

$$L = (1.35)(61.0 \, \text{ft}) = 82.4 \, \text{ft or} \boxed{82 \, \text{ft-5 in.}}$$

EXAMPLE 7.3 SI   *Grit Chamber with a Parshall Flume*

A municipal wastewater treatment plant has an average flow of 0.197 m³/s, and two horizontal-velocity grit chambers with a Parshall flume are to be designed. One chamber is to be operated as a standby. The peak flow is 2.34 times the average, or (2.34)(0.197) = 0.461 m³/s. The flume has a throat width of 150 mm, the horizontal velocity is 0.30 m/s, and the settling velocity is 21 mm/s. The head losses from the end of the grit chamber to the entrance of the Parshall flume are negligible because of the low velocity. Determine:

1. The head on the flume, m.
2. The parabolic cross section and the design cross section.
3. The settling time, s.
4. The theoretical chamber length, m.
5. The design chamber length if the design length is 1.35 times the theoretical length.

SOLUTION   The discharge equation for a 150-mm flume is

$$Q = 2.26WH_a^{1.5} = (2.26)(150/1000)H_a^{1.5} = 0.339H_a^{1.5}$$

or

$$H_a = \left(\frac{Q}{0.339}\right)^{1/1.5} = \left(\frac{Q}{0.339}\right)^{0.6667}$$

or

$$H_a = \left(\frac{0.461}{0.339}\right)^{0.6667} = \boxed{1.23 \, \text{m}}$$

A plan of the flume with a 150-mm throat width is drawn (not shown) according to the standard dimensions given by Chow (1959) and French (1985), and the channel width at the float tube is 330 mm. Thus the velocity at H_a is $0.461/(0.330 \times 1.23) = 1.14$ m/s. Writing Bernoulli's energy equation between the water surface in the downstream end of the grit chamber and the point where H_a is measured gives

$$\frac{V_1^2}{2g} + h = \frac{V_2^2}{2g} + H_a + h_L \quad \text{(entrance to flume)}$$

or

$$\frac{(0.3 \, \text{m/s})^2}{2g} + h = \frac{(1.14 \, \text{m/s})^2}{2g} + 1.23 \, \text{m} + 0.10\left[\frac{(1.14 \, \text{m/s})^2}{2g} - \frac{(0.3 \, \text{m/s})^2}{2g}\right]$$

or

$$0.005 \, \text{m} + h = 0.066 \, \text{m} + 1.23 \, \text{m} + 0.006 \, \text{m}$$

Thus,

$$h = 1.30 \, \text{m}$$

The cross-sectional area is given by

$$A = \frac{Q}{V} = \frac{0.461 \, \text{m}^3/\text{s}}{0.30 \, \text{m/s}} = 1.54 \, \text{m}^2$$

The top width for a parabola is

$$w = \frac{3}{2} \cdot \frac{A}{h}$$

or

$$w = \frac{3}{2} \cdot \frac{1.54 \, \text{m}^2}{1.30 \, \text{m}} = 1.78 \, \text{m}$$

For a head, H_a, of 0.15 m, the discharge is

$$Q = (0.339)(0.15)^{1.5} = 0.0197 \, \text{m}^3/\text{s}$$

The velocity at H_a is $(0.0197)/(0.330 \times 0.15) = 0.398\,\text{m/s}$. Rewriting Bernoulli's energy equation as before gives $h = 0.153\,\text{m}$. The cross-sectional area of the wastewater in the chamber is

$$A = \frac{0.0197\,\text{m}^3/\text{s}}{0.30\,\text{m/s}} = 0.0657\,\text{m}^2$$

The top width is

$$w = \frac{3}{2} \cdot \frac{0.0657\,\text{m}^2}{0.153\,\text{m}} = 0.644\,\text{m}$$

In a like manner, other heads are assumed and the Q, h, A, and w values are determined and are shown in Table 7.1. The depth values, h, and top width values, w, are plotted, and a smooth curve is drawn as shown in Figure 7.12. In order to select the design cross section, the width of the channel for the conveyor buckets must be known. A manufacturer's representative is contacted and several widths are obtained, which, in this range, are 380, 460, 530, and 610 mm. After each of these is tried, the value of 530 mm is selected because the design cross section fits the chamber best. The design cross section is shown in Figure 7.12 by solid lines. When the design cross section is chosen, the side slopes should be relatively steep (45° or greater), and the areas above and below the parabola should be well balanced. The settling time is given by

$$t = \frac{h}{V_s}$$

For a design settling velocity of 21 mm/s (0.021 m/s) and a depth, h, of 1.30 m, the time is

$$t = \frac{1.30\,\text{m}}{0.021\,\text{m/s}} = \boxed{61.9\,\text{s}}$$

The theoretical length is

$$L = Vt$$

For a 0.30 m/s horizontal velocity, the theoretical length is

$$L = \left(\frac{0.30\,\text{m}}{\text{s}}\right)(61.9\,\text{s}) = \boxed{18.6\,\text{m}}$$

The design chamber length is

$$L = (1.35)(18.6\,\text{m}) = \boxed{25.1\,\text{m}}$$

The design of a horizontal-velocity grit settling chamber with a proportional weir is similar to the design of a chamber using a Parshall flume except for determining the cross-sectional area. The discharge is divided by the horizontal velocity of 1.0 ft/sec (0.30 m/s) to obtain the cross-sectional area. For the proportional weir, the ideal chamber cross section is a rectangle. Once the desired depth-to-width ratio and the width of the available conveyor buckets are known, the chamber cross-sectional dimensions are determined, thus giving the head, h. As previously mentioned, h is normally about 3 in. (75 mm) greater than h_1. Since $Q = Ckh_1$, the weir coefficient, k, can be determined, and since $x = k/(2\sqrt{y})$, the x and y values can be determined for making the weir plate. The chamber length, L, is determined in the same manner as for a chamber with a Parshall flume.

Grit is usually removed by mechanical rakes and temporarily discarded into receptacles. If horizontal-velocity chambers are used, at least two are provided, one of which is mechanically cleaned, and one serves as a standby. It is desirable to have all chambers mechanically cleaned.

Aerated grit chambers may be rectangular, as shown in Figures 7.13 and 7.14, square, as shown in Figure 7.15, or circular. The rectangular tanks are usually for medium- to large-sized treatment plants, whereas the square and round tanks are usually for small- to medium-sized treatment plants. In rectangular aerated grit chambers, as shown in Figures 7.13 and 7.14, the flow has a spiral or helical roll due to the flow entering at one end of the tank and due to the air rising from the diffusers along one wall. The suspended particles have two to three helical rolls at peak flow, as they pass

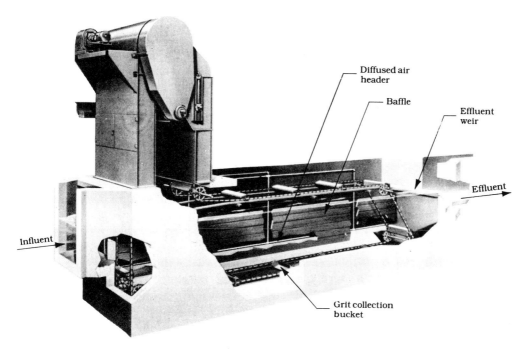

FIGURE 7.13 *Aerated Grit Chamber*
Courtesy of Envirex, Inc.

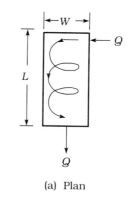

(a) Plan

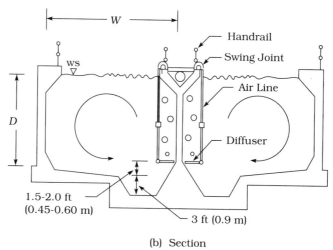

(b) Section

FIGURE 7.14 *Aerated Grit Chamber with Spiral-Roll Flow*

Adapted from *Wastewater Engineering: Treatment, Disposal and Reuse*, 3rd ed. by Metcalf & Eddy, Inc. Copyright © 1991 by McGraw-Hill, Inc. Reprinted by permission.

through the length of the tank. The velocity of roll governs the size of the particles of a given specific gravity that will be removed. If the velocity is too low, organic particles will be removed along with the grit. The air flowrate is adjustable, and once it is properly adjusted, almost 100% of the grit will be removed and it will be well washed. In the aerated grit chamber shown in Figure 7.13, the settled grit is removed by chain-driven buckets which rake the grit to the upstream end of the tank, where it is lifted vertically and deposited into a storage receptacle. In the aerated grit chamber shown in Figure 7.14, the entering wastewater is introduced at the top of the chamber so that it is in the direction of the roll. The grit that moves along the tank bottom collects in the hopper and is removed by scraper buckets to one end of the tank. From there it is elevated by buckets, screw conveyors, grab

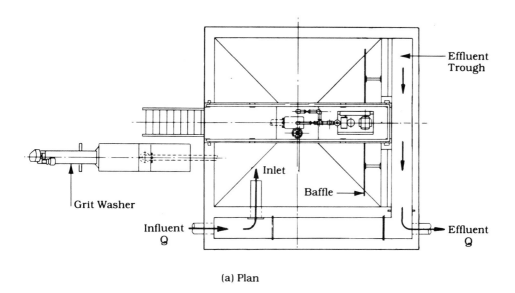

(a) Plan

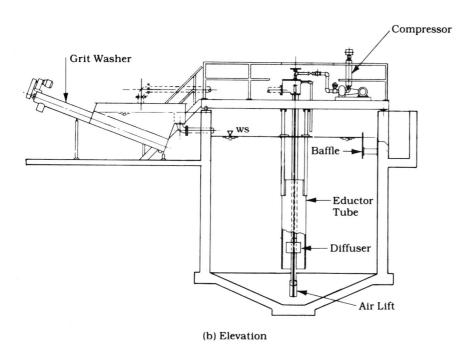

(b) Elevation

FIGURE 7.15 *Aerated Grit Removal Chamber*
Courtesy of Walker Process Equipment; Division of McNish
Corporation.

TABLE 7.2 *Criteria for Aerated Spiral-Roll Grit Chambers*

	VALUE	
ITEM	**Range**	**Typical**
Dimensions:		
Depth, ft (m)	7–16 (2.1–4.9)	
Length, ft (m)	25–65 (7.6–19.8)	
Width, ft (m)	8–23 (2.4–7.0)	
Width-depth ratio	1:1–5:1	1.5:1
Length-width ratio	3:1–5:1	4:1
Detention time at peak flow, min	2–5	3
Air supply, ft^3/min-ft of length (m^3/min-m of length)	2.0–5.0 (0.18–0.46)	
Grit quantities, ft^3/MG (m^3/10^{3}m^3)	0.5–27 (0.004–0.200)	2.0 (0.015)

Adapted from *Wastewater Engineering*, 3rd ed. by Metcalf & Eddy, Inc. Copyright © 1991 by McGraw-Hill, Inc. Reprinted by permission.

buckets, air lifts, or jet pumps. The detention time is based on the peak flowrate, ranges from 2 to 5 min, and is usually 3 min. Typical design criteria for rectangular aerated grit chambers are shown in Table 7.2. In a square aerated grit chamber, as shown in Figure 7.15, the flow enters near a corner of the tank and imparts rotational flow to the contents. At the same time, the diffused air flow creates a vertical roll, and the grit settles out while the organic particles remain in suspension and leave with the effluent. The bottom is sloped and terminates in a hopper or sump, where the grit is removed either by air lifts or mechanical means. The detention time is based on the peak flowrate and is from 1 to 5 min, 3 min being typical. To determine head losses through a tank, expansion in depth caused by air must be considered. This is usually about 3 in. (75 mm). The flow regime in square or round aerated grit removal chambers is completely mixed flow and, in rectangular aerated grit removal chambers, it is dispersed plug flow. The diffused-air grit removal chambers are very popular, because the tanks are compact, the grit is cleaner, and the dissolved oxygen imparted to the wastewater creates aerobic conditions.

EXAMPLE 7.4 *Aerated Grit Chamber*

A municipal wastewater treatment plant has an average flow of 10.8 MGD, and two aerated grit chambers are to be designed. Dual units are to be provided for flexibility in operation, since one could be used should the other require temporary maintenance. During normal operation, both chambers would be used. Each chamber is to be a rectangular tank with spiral-roll flow. The peak flow is 2.29 times the average flow. The tank width is 1.5 times the depth, and the length is 4.0 times the width. Determine:

1. The theoretical dimensions of the tanks, in feet, if the detention time is 3 min.
2. The total air flow in ft^3/min, if 3.2 ft^3/min per foot of tank length is to be provided.

SOLUTION The peak flow is $(2.29)(10.8) = 24.7$ MGD, or 38.2 ft^3/sec. The total tank volume is

$$V = Qt$$

where V = tank volume, Q = flowrate, and t = detention time. Thus,

$$V = \left(\frac{38.2\,\text{ft}^3}{\text{sec}}\right)(3\,\text{min})\left(\frac{60\,\text{sec}}{\text{min}}\right) = 6876\,\text{ft}^3$$

or, for each tank, the volume is $(6876)(1/2) = 3438$ ft^3. Since the length, L, is 4 times the width, W, and the depth, D, is $W/1.5$, it follows that

$$(4\,W)(W)(W/1.5) = 3438\,\text{ft}^3$$

or

$$W = \boxed{10.9\,\text{ft}}$$

$$L = (10.9)(4) = \boxed{43.6\,\text{ft}}$$

$$D = (10.9)(1/1.5) = \boxed{7.3\,\text{ft}}$$

The total air flow is

$$\text{ft}^3/\text{min} = (3.2\,\text{ft}^3/\text{min-ft})(43.6\,\text{ft}) = \boxed{140\,\text{ft}^3/\text{min}}$$

EXAMPLE 7.4 SI *Aerated Grit Chamber*

A municipal wastewater treatment plant has an average flow of 0.473 m^3/s, and two aerated grit chambers, are to be designed. Dual units are to be provided for flexibility in operation, since one could be used should the other require temporary maintenance. During normal operation, both chambers would be used. Each chamber is to be a rectangular tank with spiral-roll flow. The peak flow is 2.29 times the average flow. The tank width is 1.5 times the depth, and the length is 4.0 times the width. Determine:

1. The theoretical dimensions of the tanks, in meters, if the detention time is 3 min.
2. The total air flow in m^3/min, if 0.3 m^3/min per meter of tank length is to be provided.

SOLUTION   The peak flow is $(2.29)(0.473) = 1.083 \, \mathrm{m^3/s}$. The total tank volume is

$$V = Qt$$

where V = tank volume, Q = flowrate, and t = detention time. Thus,

$$V = \left(\frac{1.083 \, \mathrm{m^3}}{\mathrm{s}}\right)(3 \, \mathrm{min})\left(\frac{60 \, \mathrm{s}}{\mathrm{min}}\right) = 194.94 \, \mathrm{m^3}$$

or, for each tank, the volume is $(194.94)(1/2) = 97.47 \, \mathrm{m^3}$. Since the length, L, is 4 times the width, W, and the depth, D, is $W/1.5$, it follows that

$$(4\,W)(W)(W/1.5) = 94.47 \, \mathrm{m^3}$$

or

$$W = 3.28 \text{ or } \boxed{3.3 \, \mathrm{m}}$$

$$L = (3.3)(4) = \boxed{13.2 \, \mathrm{m}}$$

$$D = (3.3/1.5) = \boxed{2.2 \, \mathrm{m}}$$

The total air flow is

$$\mathrm{m^3/min} = (0.3 \, \mathrm{m^3/min\text{-}m})(13.2 \, \mathrm{m}) = \boxed{3.96 \, \mathrm{m^3/min}}$$

Square grit chambers, as shown in Figure 7.16, are also used and two units are usually provided for flexibility in operation. The horizontal velocity in these at maximum flow is 1.0 ft/sec (0.30 m/s), and this results in lower velocities at lower flows; thus, some organic solids are removed with the grit. The grit is raked to a sump, and there it is moved upward on an incline by rakes. As the solids are raked up the incline, much of the organic solids are separated by washing and flow back to the basin, and by this method, a cleaner grit is produced. The flow regime in square grit removal tanks is dispersed plug flow.

The grit removed at a plant serving separate sanitary sewers will be from 0.5 to 27 ft³ per million gallons (4 to 200 m³ per million cubic meters) treated (Metcalf & Eddy, Inc., 1991). A typical value is 2 ft³ per million gallons (15 m³ per million cubic meters). The grit removed at a treatment plant is usually disposed of by burial or sanitary landfill.

Flow Equalization   The equalization of flow to give a relatively constant flowrate to a wastewater treatment plant is applicable to both municipal and industrial wastewaters. Although the following discussion is directed toward municipal wastewater treatment plants, the principles employed are also applicable to industrial wastewater treatment plants.

The use of flow equalization basins after preliminary treatment (that is, screening and grit removal) provides a relatively constant flowrate to the subsequent treatment operations and processes; thus it enhances the degree

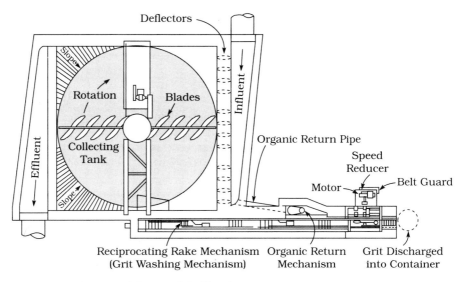

FIGURE 7.16 *Typical Square Grit Chamber*
Courtesy of Dorr-Oliver, Inc.

of treatment. Not only does equalization dampen the daily variation in flowrate, but it also dampens the variation in the concentration of BOD_5, suspended solids, and so on, throughout the day. Flow equalization can significantly improve the performance of an existing plant, and in the case of a new plant design, it will reduce the required size of the downstream treatment facilities. Equalization is feasible for dry-weather flows in separate sanitary sewers and sometimes for storm infiltration flows. The flow regime in equalization basins is completely mixed flow.

Equalization basins, as shown in Figure 7.17, may be either in-line facilities or side-line facilities. For the in-line equalization basin, as shown in Figure 7.17(a), the entire wastewater flow is pumped at a relatively constant rate to the downstream wastewater treatment facilities. In the side-line equalization basin, as shown in Figure 7.17(b), the flow during the day that is in excess of the average hourly flow overflows to the equalization basin, and when the influent flowrate becomes less than the average flow, the wastewater is pumped from the basin to the downstream treatments. The in-line system provides greater dampening of the concentration of BOD_5, suspended solids, and so on, than is attained in the side-line system. The equalization basin will have a fluctuating water level, as shown in Figure 7.18, and aeration must be provided to keep the solids in suspension and maintain aerobic conditions. Usually, the fluctuating volume is from 10 to 25% of the average daily dry-weather flow and may be determined from a flow hydrograph of the influent flow to the plant, as shown in Figure 7.19. Although basins may be constructed with earthen sides and bottoms, concrete-lined sides and bottoms are better engineering practice because erosion is eliminated.

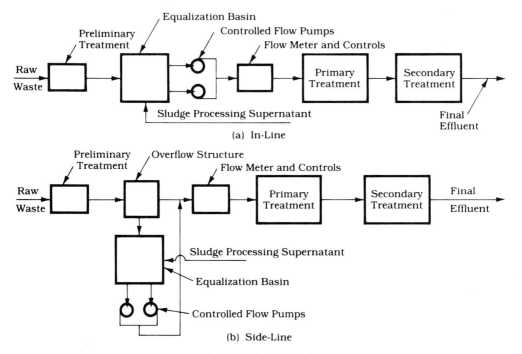

FIGURE 7.17 *Equalization Basins*

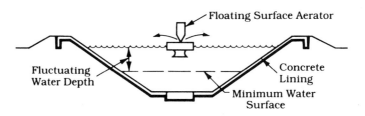

FIGURE 7.18 *Section through Equalization Basin*

For in-line basins, approximately 10 to 20% of the BOD_5 entering is stabilized in the basin (EPA, 1974a). At an existing wastewater treatment plant, the use of flow equalization has increased the removal of suspended solids in the primary settling units from 23 to 47% (EPA, 1974a). Also, the performance of biological treatment significantly benefits from equalization, because shock loads are minimized and the flowrate approaches steady state, which is beneficial to biological units and final clarifiers. At a new activated sludge plant employing side-line flow equalization and multimedia filtration as tertiary treatment, the final effluent is reported to have a BOD_5 less than 4 mg/ℓ and suspended solids less than 5 mg/ℓ (EPA, 1974a). The volume required for the fluctuating water depth depicted in Figure 7.18 may

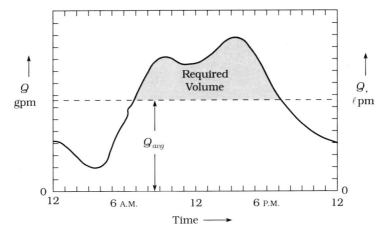

FIGURE 7.19 *Fluctuating Volume Determined by Hydrograph*

be determined using a hydrograph of the hourly flowrate throughout the day, as shown in Figure 7.19. The average hourly flowrate is determined, and a horizontal line representing this flowrate is drawn on the hydrograph. The area between the horizontal line depicting the average flow and the curve depicting the hourly flow exceeding the average represents the volume required for the fluctuating water depth once proper unit conversions have been made. Details for the design of flow equalization basins are given in appropriate publications (EPA, 1974a, 1974b).

EXAMPLE 7.5   *Flow Equalization Basin*

An in-line flow equalization basin is to be designed for a wastewater treatment plant. From plant records, a compilation has been made that gives the average flowrate versus hour of the day as shown in Table 7.3. Determine the fluctuating volume required for the basin.

SOLUTION   A plot of the hourly flowrate versus hour is shown in Figure 7.20. From the previous data, the average hourly flow for the day is 1765 gpm. Plotting the average flow on Figure 7.20 and determining the area above the average flow gives the required fluctuating volume as

$$\text{Fluctuating volume} = \boxed{298{,}000\,\text{gal}}$$

EXAMPLE 7.5 SI   *Flow Equalization Basin*

An in-line flow equalization basin is to be designed for a wastewater treatment plant. From plant records, a compilation has been made that gives the average flowrate versus hour of the day as shown in Table 7.3. Determine the fluctuating volume required for the basin.

TABLE 7.3 *Flowrate versus Hour for Examples 7.5 and 7.5 SI*

	FLOWRATE			FLOWRATE	
HOUR	**gpm**	**(ℓpm)**	**HOUR**	**gpm**	**(ℓpm)**
12 A.M.	1740	(6590)	1 P.M.	2330	(8820)
1	1630	(6170)	2	2290	(8670)
2	1390	(5260)	3	2220	(8400)
3	1180	(4470)	4	2220	(8400)
4	1040	(3940)	5	2150	(8140)
5	910	(3440)	6	2010	(7610)
6	910	(3440)	7	1940	(7340)
7	920	(3480)	8	1940	(7340)
8	1530	(5790)	9	1880	(7120)
9	2080	(7870)	10	1630	(6170)
10	2270	(8590)	11	1560	(5900)
11	2330	(8820)	12 A.M.	1740	(6590)
12 noon	2330	(8820)			

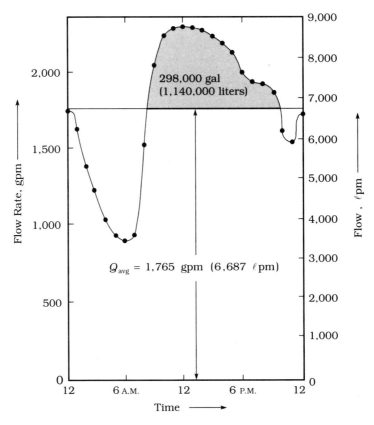

FIGURE 7.20 *Graph for Examples 7.5 and 7.5 SI*

SOLUTION   A plot of the hourly flowrate versus hour is shown in Figure 7.20. From the data shown in Table 7.3, the average hourly flow for the day is 6687 ℓpm. Plotting the average flow on Figure 7.20 and determining the area gives the fluctuating volume as

$$\text{Fluctuating volume} = \boxed{1{,}140{,}000 \text{ liters}}$$

Quality Equalization   Quality equalization consists of using a constant level equalization basin ahead of a municipal or industrial wastewater treatment plant to dampen the variation in the concentration of BOD_5, suspended solids, and so on, throughout the day. Usually, quality equalization basins are employed when the variation in the organic concentration of 4-hour composite samples exceeds 4:1 during a day (EPA, 1974a). The use of quality equalization not only will increase the performance of an existing plant but also will give better treatment by a proposed plant. Equalization tanks are similar to the one shown in Figure 7.18 except that the wastewater is at a constant level. Aeration must be provided to keep suspended solids in suspension and to maintain aerobic conditions in the basin. The flow regime in constant level equalization basins is completely mixed flow.

Neutralization   Frequently, industrial wastewaters may be acidic or basic and may require neutralization prior to subsequent treatments or release to a municipal sanitary sewer system. If the downstream treatment is a biological process, the wastewater should have a pH between about 6.5 and 9.0 to avoid inhibition. Sometimes it is feasible to mix an acidic waste stream with a basic waste stream and then use a constant level equalization basin as a neutralization tank.

   Acidic wastewaters may be neutralized by passage through limestone beds, by the addition of slaked lime, $Ca(OH)_2$, caustic soda, $NaOH$, or soda ash, Na_2CO_3. Limestone beds may be of the upflow or downflow type; however, the upflow type is the most common. Limestone beds should not be used if the sulfuric acid content is greater than 0.6%, because the $CaSO_4$ produced will be deposited on the crushed limestone and, as a result, effective neutralization will cease. Also, metallic ions such as Al^{+3} and Fe^{+3}, if present in sufficient amounts, form hydroxide precipitates that will coat the crushed limestone and reduce neutralization. Upflow beds are the most common because the products of the reaction, such as CO_2, are removed more effectively than in downflow beds. Laboratory- or pilot-scale studies should be made before limestone beds are used. Acidic wastewaters may be neutralized with slaked lime, $Ca(OH)_2$, and usually two or three agitated vessels are used in series, as shown in Figure 7.21. Each vessel has a pH sensor that controls the slaked lime feed rate. Since slaked lime is less expensive than other bases or soda ash, it is the most commonly used chemical for acidic neutralization. Caustic soda, $NaOH$, or soda ash, Na_2CO_3, may be used in a similar manner.

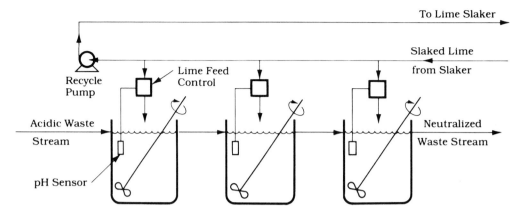

FIGURE 7.21   *Three-Stage Lime Neutralization*

Alkaline wastewaters may be neutralized with a strong mineral acid, such as H_2SO_4 or HCl, or with CO_2. Usually, if a source of CO_2 is not available, neutralization is done with H_2SO_4 because it is cheaper than HCl. The reaction with mineral acids is rapid, and agitated vessels are used with pH sensors that control the acid feed rate. Neutralization of alkaline wastewaters with CO_2 usually consists of bubbling CO_2 from a perforated pipe grid in the bottom of the neutralization tank, thus creating carbonic acid, H_2CO_3, which reacts with the alkaline substances. Frequently, flue gas is available as a source of CO_2, which makes the neutralization process more economical.

REFERENCES

American Society of Civil Engineers (ASCE). 1990. *Water Treatment Plant Design*. 2nd ed. New York: McGraw-Hill.

American Water Works Association (AWWA). 1990. *Water Quality and Treatment*. 4th ed. New York: McGraw-Hill.

Amirtharajah, A. 1986. Variance Analyses and Criteria for Treatment Regulations. *Jour.* AWWA 3:34.

AWARE, Inc. 1974. *Process Design Techniques for Industrial Waste Treatment*. Edited by C. E. Adams, Jr., and W. W. Eckenfelder, Jr. Nashville, Tenn.: Enviro Press.

Barnes, D.; Bliss, P. J.; Gould, B. W.; and Vallentine, H. R. 1981. *Water and Wastewater Engineering Systems*. London: Pitman Books Limited.

Benefield, L. D.; Judkins, J. F.; and Parr, A. D. 1984. *Treatment Plant Hydraulics for Environmental Engineers*. Englewood Cliffs, N.J.: Prentice-Hall.

Camp, T. R. 1942. Grit Chamber Design. *Journal of Sewage Works* 14:368.

Chow, V. T. 1959. *Open-Channel Hydraulics*. New York: McGraw-Hill.

Davis, C. V., and Sorensen, K. E. 1969. *Handbook of Applied Hydraulics*. 3rd ed. New York: McGraw-Hill.

Degremont, Inc. 1973. *Water Treatment Handbook*. Rueil-Malmaison, France.

Eckenfelder, W. W., Jr. 1989. *Industrial Water Pollution Control*. 2nd ed. New York: McGraw-Hill.

Eckenfelder, W. W., Jr. 1980. *Principles of Water Quality Management*. Boston: CBI Publishing.

Eckenfelder, W. W., Jr. 1970. *Water Quality Engi-*

neering for Practicing Engineers. New York: Barnes & Noble.

Eckenfelder, W. W., Jr., and Ford, D. L. 1970. *Water Pollution Control.* Austin, Tex.: Pemberton Press.

Eckenfelder, W. W., Jr.; Patoczka, J.; and Watkin, A. T. 1992. Wastewater Treatment. In *Environmental Engineering in the Process Plant.* Edited by N. P. Chopey. From *Chemical Engineering.* New York: McGraw-Hill.

Environmental Protection Agency (EPA). 1974a. Flow Equalization. EPA Technology Transfer Seminar Publication. Washington, D.C.

Environmental Protection Agency (EPA). 1974b. *Upgrading Existing Wastewater Treatment Plants.* EPA Process Design Manual. Washington, D.C.

Fair, G. M.; Geyer, J. C.; and Okun, D. A. 1968. *Water and Wastewater Engineering.* Vol. 2. New York: Wiley.

French, R. H. 1985. *Open-Channel Hydraulics.* New York: McGraw-Hill.

Great Lakes–Upper Mississippi Board of State Sanitary Engineers. 1978. *Recommended Standards for Sewage Works.* Ten state standards. Albany, N.Y.

Horan, N. J. 1990. *Biological Wastewater Treatment Systems, Theory and Operation.* Chichester, England: Wiley.

McGhee, T. J. 1991. *Water Supply and Sewerage.* 6th ed. New York: McGraw-Hill.

Metcalf and Eddy, Inc. 1972. *Wastewater Engineering: Collection, Treatment and Disposal.* New York: McGraw-Hill.

Metcalf and Eddy, Inc. 1979. *Wastewater Engineering: Treatment, Disposal and Reuse.* 2nd ed. New York: McGraw-Hill.

Metcalf and Eddy, Inc. 1991. *Wastewater Engineering: Treatment, Disposal and Reuse.* 3rd ed. New York: McGraw-Hill.

Montgomery, J. M., Consulting Engineers, Inc.

1985. *Water Treatment Principles and Design.* New York: Wiley.

Novotny, V.; Imhoff, K. R.; Olthof, M.; and Krenkel, P. A. 1989. *Karl Imhoff's Handbook of Urban Drainage and Wastewater Disposal.* New York: Wiley.

Ramalho, R. S. 1977. *Introduction to Wastewater Treatment Processes.* New York: Academic Press.

Reynolds, T. D. 1982. *Unit Operations and Processes in Environmental Engineering.* Boston: PWS.

Salvato, J. A. 1992. *Environmental Engineering and Sanitation.* 4th ed. New York: Wiley.

Sanks, R. L. 1978. *Water Treatment Plant Design.* Ann Arbor, Mich.: Ann Arbor Science Publishers.

Schroeder, E. D. 1977. *Water and Wastewater Treatment.* New York: McGraw-Hill.

Steel, E. W., and McGhee, T. J. 1979. *Water Supply and Sewerage.* 5th ed. New York: McGraw-Hill.

Sundstrom, D. W., and Klei, H. E. 1979. *Wastewater Treatment.* Englewood Cliffs, N.J.: Prentice-Hall.

Tchobanoglous, G., and Schroeder, E. D. 1985. *Water Quality.* Reading, Mass.: Addison-Wesley.

Tebbutt, T. H. Y. 1992. *Principles of Water Quality Control.* 4th ed. Oxford, England: Pergamon Press Limited.

Viessman, W., and Hammer, M. J. 1993. *Water Supply and Pollution Control.* 5th ed. New York: Harper & Row.

Water Environmental Federation (WEF). 1992. *Design of Municipal Wastewater Treatment Plants.* Vols. 1 and 2. WEF Manual of Practice No. 8. Alexandria, Va.

Water Pollution Control Federation (WPCF). 1977. *Wastewater Treatment Plant Design.* WPCF Manual of Practice No. 8. Washington, D.C.

Zipparro, V. J. and Hasen, H. 1993. *Davis' Handbook of Applied Hydraulics.* 4th ed. New York: McGraw-Hill.

PROBLEMS

7.1 A mechanically cleaned bar screen has bars $\frac{3}{8}$ in. (5 mm) thick and $1\frac{1}{4}$-in. (30-mm) clear spaces between the bars. If the velocity through the bars is 3 ft/sec (0.90 m/s), determine the approach velocity, in ft/sec (m/s), and the head loss through the screen, in ft (m). Use both USCS and SI units.

7.2 A municipal wastewater treatment plant has an average flow of 3.95 MGD, and two horizontal-velocity grit chambers with a Parshall flue are to be designed. One chamber is to be operated as a standby. The peak flow is 2.20 times the average flow. The Parshall flume has a throat width of 6 in. The horizontal velocity is 1.0 ft/sec, and the settling rate is 0.0689 ft/sec.

Determine:

a. The head on the flume, ft.

b. The parabolic cross section and the design cross section if the channel for the conveyor buckets is 18 in. wide.

c. The settling time, sec.

d. The theoretical chamber length, ft.

e. The design chamber length, if the design length is 1.35 times the theoretical length, ft, and, for ease of construction, is an increment of 1 in.

f. The head and the horizontal velocity for the average flow, ft, ft/sec.

g. The head and the horizontal velocity for the minimum flow if the minimum flow is one-half the average flow, ft, ft/sec.

7.3 A municipal wastewater treatment plant has an average flow of 14,950,000 ℓ/d (14.95 MLD), and two horizontal-velocity grit chambers are to be designed. The peak flow is 2.20 times the average flow. The Parshall flume has a throat width of 150 mm. The horizontal velocity is 0.30 m/s, and the settling rate is 21 mm/s.

Determine:

a. The head on the flume, m.

b. The parabolic cross section and the design cross section if the channel for the conveyor buckets is 450 mm wide.

c. The settling time, s.

d. The theoretical chamber length, m.

e. The design chamber length, if the design length is 1.35 times the theoretical length, m.

f. The head and horizontal velocity for the average flow, m, m/s.

g. The head and the horizontal velocity for the minimum flow, if the minimum flow is one-half the average flow, m, m/s.

7.4 A municipal wastewater treatment plant has an average flow of 2.62 MGD, and two horizontal-velocity grit chambers with a proportional weir are to be designed. One chamber will operate as a standby. The peak flow is 2.51 times the average flow. The channel cross-sectional area is to have a depth about 1.25 times the width. The horizontal velocity is 1.0 ft/sec, and the settling velocity is 0.0689 ft/sec. The head, h, is 3 in. greater than h_1.

Determine:

a. The channel width and depth, ft, if the channel width is to be an increment of 3 in. to accommodate the collector equipment.

b. The design of the proportional weir.

c. The settling time, sec.

d. The theoretical chamber length, ft.

e. The design chamber length, ft, if the design length is 1.35 times the theoretical length and, for ease of construction, is an increment of 1 in.

f. The head and the horizontal velocity for the average flow, ft, ft/sec.

g. The head and the horizontal velocity for the minimum flow if the minimum flow is one-half the average flow, ft, ft/sec.

7.5 A municipal wastewater treatment plant has an average flow of 9,920,000 ℓ/d (9.92 MLD), and two horizontal-velocity grit chambers with a proportional weir are to be designed. One chamber will operate as a standby. The peak flow is 2.51 times the average flow. The channel is to have a depth 1.25 times the width. The horizontal velocity is 0.30 m/s, and the settling velocity is 21 mm/sec. The head, h, is 75 mm greater than h_1.

Determine:

a. The channel width and depth, m.

b. The design of the proportional weir.

c. The settling time, s.

d. The theoretical chamber length, m.

e. The design chamber length if the design length is 1.35 times the theoretical length, m.

f. The head and the horizontal velocity for the average flow, m, m/s.

g. The head and the horizontal velocity for the minimum flow if the minimum flow is one-half the average flow, m, m/s.

7.6 A municipal wastewater treatment plant has an average flow of 15.6 MGD, and two aerated grit chambers are to be designed. Dual units are to be provided for flexibility in operation, since one could be used should the other require temporary maintenance. During normal operation, both chambers will be used. Each chamber is to be a rectangular tank with spiral-roll flow. The peak flow is 1.91 times the average flow. The tank width is 1.5 times the depth, and the length is 4.0 times the width.

Determine:

a. The theoretical dimensions of each tank if the detention time is 3 min, ft.

b. The total air flow, if 3.2 ft^3/min per foot of tank length is to be provided, ft^3/min.

7.7 A municipal wastewater treatment plant has an average flow of 59.0 MLD, and two aerated grit chambers are to be designed. Dual units are to be provided for flexibility in operation, since one could be used should the other require temporary maintenance. During normal operation, both chambers will be used. Each chamber is to be a rectangular tank with spiral-roll flow. The peak flow is 1.91 times the average flow. The tank width is 1.5 times the depth, and the length is 4.0 times the width. Determine:

a. The theoretical dimensions of each tank, m, if the detention time is 3 min.

b. The total air flow, if $0.30\,m^3/min$-m of length is provided, m^3/min.

7.8 A municipal wastewater treatment plant treats 12.3 MGD. The screenings amount to $1.5\,ft^3$ per million gallons treated, and the grit amounts to $4.0\,ft^3$ per million gallons treated.

Determine:

a. The volume of screenings removed per day, ft^3.

b. The volume of grit removed per day, ft^3.

c. The time in days required to fill a 4 ft × 6 ft × 6 ft solid waste storage container used for temporary storage.

7.9 A municipal wastewater treatment plant treats 45.4 million liters per day (MLD). The screenings amount to $11.2\,m^3$ per million cubic meters treated, and the grit amounts to $30\,m^3$ per million cubic meters treated.

Determine:

a. The volume of screenings removed per day, m^3.

b. The volume of grit removed per day, m^3.

c. The time in days required to fill a $1.2\,m \times 1.8\,m \times 1.8\,m$ solid waste storage container used for temporary storage.

7.10 An in-line equalization basin is to be designed for a wastewater treatment plant. From plant records, a compilation has been made that gives the hourly flowrate versus hour of the day as follows:

HOUR	MGD	MLD	HOUR	MGD	MLD
12 A.M.	3.0	11.4	1 P.M.	6.6	25.0
1	2.2	8.3	2	6.6	25.0
2	2.1	7.9	3	7.1	26.9
3	2.0	7.6	4	7.6	28.8
4	2.4	9.1	5	8.7	32.9
5	2.8	10.6	6	9.9	37.5
6	3.6	13.6	7	9.9	37.5
7	5.5	20.8	8	8.4	31.8
8	6.6	25.0	9	6.1	23.1
9	7.4	28.0	10	4.2	15.9
10	7.6	28.8	11	3.5	13.2
11	7.1	26.9	12 A.M.	3.0	11.4
12 noon	6.8	25.7			

Determine the fluctuating volume required for the basin in gallons and cubic feet.

7.11 Repeat Problem 7.10 for SI units. The volume should be in cubic meters.

8

COAGULATION AND FLOCCULATION

Coagulation and flocculation consist of adding a floc-forming chemical reagent to a water or wastewater to enmesh or combine with nonsettleable colloidal solids and slow-settling suspended solids to produce a rapid-settling floc. The floc is subsequently removed in most cases by sedimentation. **Coagulation** is the addition and rapid mixing of a coagulant, the resulting destabilization of the colloidal and fine suspended solids, and the initial aggregation of the destabilized particles. **Flocculation** is the slow stirring or gentle agitation to aggregate the destabilized particles and form a rapid-settling floc. Figure 8.1 shows a paddle-wheel flocculation basin at a rapid sand water filtration plant. Figure 8.2 shows a turbine-type flocculation basin at a tertiary wastewater treatment plant.

In water treatment, the principal use of coagulation and flocculation is to agglomerate solids prior to sedimentation and rapid sand filtration. In municipal wastewater treatment, coagulation and flocculation are used to agglomerate solids in the physical-chemical treatment of raw wastewaters and primary or secondary effluents. In industrial waste treatment, coagulation is employed to coalesce solids in wastewaters that have an appreciable suspended solids content. In water treatment, the principal coagulants used are aluminum and iron salts, although polyelectrolytes are employed to some extent. In wastewater treatment, aluminum and iron salts, lime, and polyelectrolytes are used. Chemical precipitation, which is closely related to chemical coagulation, consists of the precipitation of unwanted ions from a water or wastewater. In the coagulation of municipal wastewaters, not only does coagulation of solids occur, but also the chemical precipitation of much of the phosphate ion takes place.

THEORY OF COAGULATION

A portion of the dispersed solids in surface waters and wastewaters are nonsettleable suspended materials that have a particle size ranging from 0.1 millimicron (10^{-7} mm) to 100 microns (10^{-1} mm). Since colloids have a particle size ranging from one millimicron (10^{-6} mm) to one micron (10^{-3} mm), a significant fraction of the nonsettleable matter is colloidal particulates. The supracolloidal fraction has a particle size ranging from one micron (10^{-3} mm) to 100 microns (10^{-1} mm). Many of the particles in this fraction have certain colloidal characteristics such as negligible settling velocities.

FIGURE 8.1 *A Three-Compartment Paddle-Wheel Floccula-tion Basin at a Water Treatment Plant. Note that the flow is parallel to the paddle-wheel shafts, as shown in Figure 8.20. The flow passes through the orifices in the walls. Tapered flocculation is provided by varying the size of the compartments and the number and length of the blades on the paddle wheels.*

Colloidal Characteristics

Colloidal dispersions are classified according to the dispersed phase and the dispersed medium. The principal systems involved in both water and waste-water treatment are solids dispersed in liquids (sols) and liquids dispersed in liquids (emulsions). When suspended in water, organic matter, such as microbes, and inorganic matter, such as clays, are examples of a system consisting of solids dispersed in a liquid. An oil dispersed in water is an example of a liquid dispersed in a liquid. In water and wastewater treatment, solids dispersed in liquids (sols) are of particular interest. An important feature of a solid colloid dispersed in water is that the solid particles will not

FIGURE 8.2 *A Turbine-Type Flocculation Basin at a Tertiary Wastewater Treatment Plant. Note baffle fence for outlet flow.*

settle by the force of gravity. When a solid colloid stays in suspension and does not settle, the system is in a stable condition.

Colloids have an extremely large surface area per unit volume of the particles — that is, a large specific surface area. Because of the large surface area, colloids tend to adsorb substances, such as water molecules and ions, from the surrounding water. Also, colloids develop or have an electrostatic charge relative to the surrounding water.

Colloidal solids in water may be classified as **hydrophilic** or **hydrophobic** according to their affinity for water. Hydrophilic colloids have an affinity for water because of the existence of water-soluble groups on the colloidal surface. Some of the principal groups are the amino, carboxyl, sulfonic, and hydroxyl. Since these groups are water-soluble, they promote hydration and cause a water layer or film to collect and surround the hydrophilic colloid. Frequently, this water layer or film is termed the **water of hydration** or **bound water**. Usually organic colloids, such as proteins and their degradation products, are hydrophilic. Hydrophobic colloids have little, if any, affinity for water; as a result, they do not have any significant water film or water of hydration. Usually inorganic colloids, such as clays, are hydrophobic.

Colloidal particles have electrostatic forces that are important in maintaining a dispersion of the colloid. The surface of a colloidal particle tends to acquire an electrostatic charge due to the ionization of surface groups and the adsorption of ions from the surrounding solution. Also, colloidal minerals, such as clays, have an electrostatic charge due to the ionic deficit within the mineral lattice. Hydrophilic colloids, such as proteinaceous materials and microbes, have charges due to the ionization of such groups as the amino ($-NH_2$) and the carboxyl ($-COOH$), which are located on the colloidal surface. When the pH is at the isoelectric point, the net or overall charge is zero since the amino group is ionized ($-NH_3^+$) and also the carboxyl group is ionized ($-COO^-$). At a pH below the isoelectric point, the carboxyl group is not ionized ($-COOH$), and the colloid is positively charged as a result of the ionized amino group ($-NH_3^+$). At a pH above the isoelectric point, the amino group loses a hydrogen, producing a neutral group ($-NH_2$), and the colloid is negatively charged because of the ionized carboxyl group ($-COO^-$). In general, most naturally occurring hydrophilic colloids, such as proteinaceous matter and microbes, have a negative charge if the pH is at or above the neutral range. Some colloidal materials, such as oil droplets and some other chemically inert substances, will preferentially adsorb negative ions, particularly the hydroxyl ion, from their surrounding solution and become negatively charged. Colloidal minerals, such as clays, have more nonmetallic atoms than metallic atoms within their crystalline structure, resulting in a net negative charge. Usually most naturally occurring hydrophobic colloids, such as clays, are inorganic materials and have a negative charge. The sign and magnitude of the charge of a colloid will depend on the type of colloidal matter and on the characteristics of the surrounding solution.

In most colloidal systems, the colloids are maintained in suspension (in other words, stabilized) as a result of the electrostatic forces of the colloids themselves. Since most naturally occurring colloids are negatively charged and like charges are repulsive, the colloids remain in suspension because of the action of the repulsive forces.

A negative colloidal particle will attract to its surface ions of the opposite charge — counterions — from the surrounding water, as depicted in Figure 8.3. The compact layer of counterions is frequently termed the **fixed layer**; outside the fixed layer is the **diffused layer**. Both layers will contain positive and negative charged ions; however, there will be a much larger number of positive ions than negative ions. The two layers represent the region surrounding the particle where there is an electrostatic potential due to the particle, as illustrated in Figure 8.3. The concentration of the counterions is greatest at the particle surface; it decreases to that of the bulk solution at the outer boundary of the diffused layer. The shear plane or shear surface surrounding the particle encloses the volume of water (in other words, the bound water or water envelope) that moves with the particle. The **zeta**

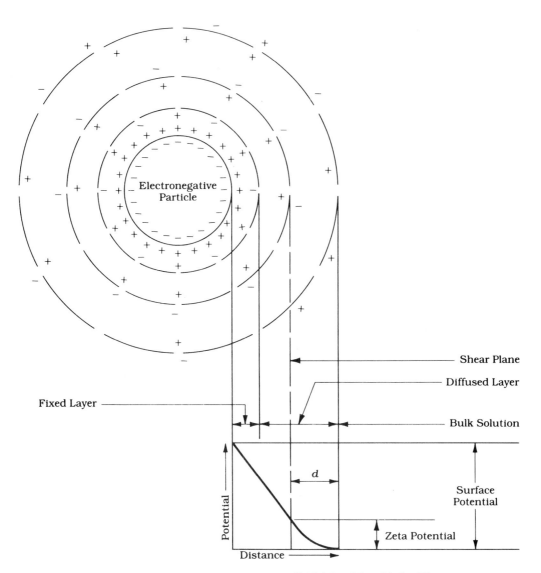

FIGURE 8.3 *A Negative Colloidal Particle with Its Electrostatic Field*

Adapted with permission from "A Theory of Coagulation Useful for Design" by C. P. Priesing. In *Industrial and Engineering Chemistry* 54, no. 8 (August 1962): 38. Copyright 1962, American Chemical Society.

potential is the electrostatic potential at the shear surface, as shown in Figure 8.3. This potential is usually related to the stability of a colloidal suspension.

A colloidal suspension is stable if the particles remain in suspension and do not coagulate. The colloidal stability depends on the relative magnitude of the forces of attraction and the forces of repulsion. The forces of attraction are due to van der Waals forces, which are effective only in the immediate neighborhood of the colloidal particle. The forces of repulsion are due to the electrostatic forces of the colloidal dispersion. The magnitude of these forces is measured by the zeta potential, which is

$$\zeta = \frac{4\pi q d}{D} \qquad (8.1)$$

where

ζ = zeta potential

q = charge per unit area

d = thickness of the layer surrounding the shear surface through which the charge is effective, as shown in Figure 8.3

D = dielectric constant of the liquid

Thus the zeta potential measures the charge of the colloidal particle, and it is dependent on the distance through which the charge is effective. It follows that the greater the zeta potential, the greater are the repulsion forces between the colloids and, therefore, the more stable is the colloidal suspension. Also, the presence of a bound water layer and its thickness affects colloidal stability, since this layer prevents the particles from coming into close contact.

Hydrophilic colloids have a shear surface at the outer boundary of the bound water layer. Hydrophobic colloids have a shear surface near the outer boundary of the fixed layer.

Coagulation of Colloids (Destabilization)   When a coagulant is added to a water or wastewater, destabilization of the colloids occurs and a coagulant floc is formed. Since the chemistry involved is very complex, the following discussion is intended to be introductory and will illustrate the interactions known to occur. These interactions are (1) the reduction of the zeta potential to a degree where the attractive van der Waals forces and the agitation provided cause the particles to coalesce; (2) the aggregation of particles by interparticulate bridging between reactive groups on the colloids; and (3) the enmeshment of particles in the precipitate floc that is formed. Figure 8.4 shows the interparticulate forces acting on a colloidal particle. The repulsive forces are due to the electrostatic zeta potential, and the attractive forces are due to van der Waals forces acting between the particles. The net resultant force is attractive out to the distance x. Beyond this point the net resultant force is repulsive.

When a coagulant salt is added to a water, it dissociates, and the metallic ion undergoes hydrolysis and creates positively charged hydroxo-metallic ion complexes. Usually a coagulant is an aluminum salt, such as $Al_2(SO_4)_3$, or an iron salt, such as $Fe_2(SO_4)_3$. Numerous species of hydroxo-metallic complexes are formed because the complexes, which are hydrolysis

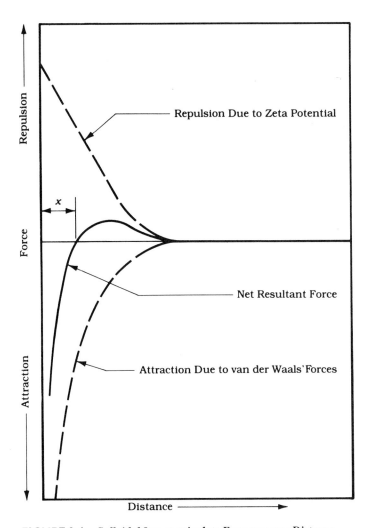

FIGURE 8.4 *Colloidal Interparticulate Forces versus Distance*

products, tend to polymerize (Stumm and Morgan, 1962; Stumm and O'Melia, 1968). The generalized expression for these complexes is $Me_q(OH)_p^{z+}$. For an aluminum salt, some of the resulting polymers are $Al_6(OH)_{15}^{+3}$, $Al_7(OH)_{17}^{+4}$, $Al_8(OH)_{20}^{+4}$, and $Al_{13}(OH)_{34}^{+5}$. For an iron salt, some of the resulting polymers are $Fe_2(OH)_2^{+4}$ and $Fe_2(OH)_4^{+5}$. The hydroxo-metallic complexes are polyvalent, possess high positive charges, and are adsorbed to the surface of the negative colloids. This results in a reduction of the zeta potential to a level where the colloids are destabilized. The destabilized particles, along with their adsorbed hydroxo-metallic complexes, aggregate by interparticulate attraction due to van der Waals' forces. These forces are aided by the gentle agitation of the water. In the aggregation process, the agitation is very important since it causes the destabilized particles to come in close vicinity or collide and then coalesce.

The aggregation of the destabilized particles also occurs by interparticulate bridging involving chemical interactions between reactive groups on the destabilized particles. The agitation of the water is also important in this type of aggregation, since it causes interparticulate contacts.

The dosages of coagulant salts used in coagulating waters and wastewaters are usually in appreciable excess of the amount required to produce the necessary positive hydroxo-metallic complexes. The excess complexes continue to polymerize until they form an insoluble metallic hydroxide, $Al(OH)_3$ or $Fe(OH)_3$, and the solution will be supersaturated with the hydroxide. In the formation of the metallic hydroxide, there is enmeshing of the negative colloids with the precipitate as it forms. This enmeshment type of coagulation is sometimes referred to as **precipitate** or **sweep coagulation**.

Originally it was thought that the zeta potential reduction was caused by the adsorption of the metallic ions from the coagulant salt. However, it is now known that the principal action is the adsorption of the highly positively charged hydroxo-metallic complexes. The species of polyvalent metallic ion complexes are much more effective in coagulating a colloidal dispersion than are monovalent complexes; thus, polyvalent metallic salts are always used in coagulation.

For dilute colloidal suspensions, the rate of coagulation may be extremely slow because of the low particulate concentrations, which cause an inadequate number of interparticulate contacts. In many water and wastewater treatment plants, it has been found that recycling a small portion of the settled sludge, before or after rapid mixing, maintains the colloidal concentration at a level where rapid coagulation and flocculation occur. For dilute suspensions, relatively large coagulant dosages may cause restabilization of the colloids. When this occurs, the negatively charged colloids become positively charged; this is believed to be due to existing positively charged reactive sites on the colloidal surfaces.

The coagulation of colloids by organic polymers occurs by a chemical interaction or bridging. The polymers have ionizable groups such as carboxyl, amino, and sulfonic, and these groups bind with reactive sites or groups on the surfaces of the colloids. In this manner several colloids may be bound to a single polymer molecule to form a bridging structure. Bridging between particles is optimum when the colloids are about one-half covered with adsorbed segments of the polymers.

Frequently in coagulation, the terms *electrokinetic*, *perikinetic*, and *orthokinetic coagulation* are encountered. **Electrokinetic coagulation** refers to coagulation that is a result of zeta potential reduction. **Perikinetic coagulation** refers to coagulation in which the interparticle contacts result from Brownian movement. In **orthokinetic coagulation** the interparticle contacts are caused by fluid motion due to agitation.

Proper coagulation and flocculation of a water are very important in the settling of the coagulated water, because the settling velocity is proportional to the square of the particle diameter. Thus the production of large floc particles results in rapid settling.

COAGULANTS

The most widely used coagulants in water treatment are aluminum sulfate and iron salts. Aluminum sulfate (filter alum) is employed more frequently than iron salts because it is usually cheaper. Iron salts have an advantage over filter alum because they are effective over a wider pH range. In the lime-soda softening process, the lime serves as a coagulant since it produces a heavy floc or precipitate consisting of calcium carbonate and magnesium hydroxide. This precipitate has coagulating and flocculating properties. The most widely used coagulants in wastewater treatment are filter alum and lime. Sometimes coagulant aids, such as recycled sludge or polyelectrolytes, are required to produce a rapid-settling floc. The principal factors affecting the coagulation and flocculation of water or wastewater are turbidity, suspended solids, temperature, pH, cationic and anionic composition and concentration, duration and degree of agitation during coagulation and flocculation, dosage and nature of the coagulant, and, if required, the coagulant aid. The selection of a coagulant requires the use of laboratory or pilot plant coagulation studies, since a given water or wastewater may show optimum coagulation results for a particular coagulant. Usually laboratory studies using the jar test are adequate for selecting a coagulant for a water treatment plant, whereas laboratory and frequently pilot studies are required for wastewaters. As shown in the previous section, coagulation chemistry is complex, and theoretical chemical equations to determine the amount of the metallic hydroxides produced give only approximate results. In this section, the common coagulants, their chemical reactions with alkalinity, and their characteristics are discussed.

Aluminum Sulfate

Sufficient alkalinity must be present in the water to react with the aluminum sulfate to produce the hydroxide floc. Usually, for the pH ranges involved, the alkalinity is in the form of the bicarbonate ion. The simplified chemical reaction to produce the floc is

$$Al_2(SO_4)_3 \cdot 14H_2O + 3Ca(HCO_3)_2 \rightarrow$$
$$\underline{2Al(OH)_3} \downarrow + 3CaSO_4 + 14H_2O + 6CO_2 \quad \textbf{(8.2)}$$

Certain waters may not have sufficient alkalinity to react with the alum, so alkalinity must be added. Usually alkalinity in the form of the hydroxide ion is added by the addition of calcium hydroxide (slaked or hydrated lime). The coagulation reaction with calcium hydroxide is

$$Al_2(SO_4)_3 \cdot 14H_2O + 3Ca(OH)_2 \rightarrow \underline{2Al(OH)_3} \downarrow + 3CaSO_4 + 14H_2O \quad \textbf{(8.3)}$$

Alkalinity may also be added in the form of the carbonate ion by the addition of sodium carbonate (soda ash). Most waters have sufficient alkalinity, so no chemical needs to be added other than aluminum sulfate. The optimum pH range for alum is from about 4.5 to 8.0, since aluminum

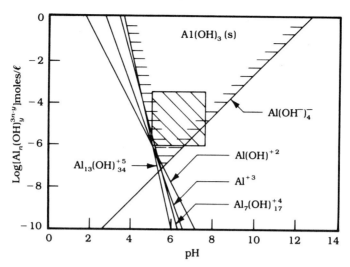

FIGURE 8.5 *Solubility of Aluminum Hydroxide. (Shaded area is the usual operating region used in water treatment.)*
Adapted from *Journal of American Water Works Association*, 60, no. 5 (May 1968), by permission. Copyright 1968, the American Water Works Association.

hydroxide is relatively insoluble within this range, as shown in Figure 8.5. Also illustrated in this figure is the usual range of aluminum hydroxide concentrations for dosages used in water treatment. These usually produce an oversaturated solution of aluminum hydroxide. Aluminum sulfate is available in dry or liquid form; however, the dry form is more common. The dry chemical may be in granular, powdered, or lump form, the granular being the most widely used. The granules, which are 15 to 22% Al_2O_3, contain approximately 14 waters of crystallization, weigh from 60 to 63 lb/ft^3 (960 to 1010 kg/m^3), and may be dry fed. The dry chemical may be shipped in bags, barrels, or bulk (carload). The liquid form is 50% alum and is shipped by tank car or tank truck.

Ferrous Sulfate Ferrous sulfate requires alkalinity in the form of the hydroxide ion in order to produce a rapid reaction. Consequently, slaked or hydrated lime, $Ca(OH)_2$, is usually added to raise the pH to a level where the ferrous ions are precipitated as ferric hydroxide. This reaction is an oxidation-reduction reaction requiring some dissolved oxygen in the water. In the coagulation reaction, the oxygen is reduced and the ferrous ion is oxidized to the ferric state, where it precipitates as ferric hydroxide. The simplified chemical reaction is

$$2FeSO_4 \cdot 7H_2O + 2Ca(OH)_2 + \tfrac{1}{2}O_2 \rightarrow \underline{2Fe(OH)_3} \downarrow + 2CaSO_4 + 13H_2O$$
$$(8.4)$$

For this reaction to occur, the pH must be raised to about 9.5, and sometimes stabilization is required for the excess lime employed. Ferrous sulfate and lime coagulation is usually more expensive than alum. In general, the precipitate formed, ferric hydroxide, is a dense, quick-settling floc. Ferrous sulfate is available in dry or liquid form; however, the dry form is more common. The dry chemical may be granules or lumps, the granules being the more widely used. The granules, which are 55% $FeSO_4$, contain seven waters of crystallization, weigh from 63 to 66 lb/ft^3 (1010 to 1060 kg/m^3), and are usually dry fed. The dry chemical may be shipped in bags, barrels, or bulk.

Chlorinated copper as treatment is another method to use ferrous sulfate. In this process ferrous sulfate is reacted with chlorine, and the ferrous ion is oxidized to the ferric ion as follows:

$$3FeSO_4 \cdot 7H_2O + 1.5Cl_2 \rightarrow Fe_2(SO_4)_3 + FeCl_3 + 21H_2O \quad \textbf{(8.5)}$$

This reaction occurs at a pH as low as about 4.0. The products, ferric sulfate and ferric chloride, are very effective coagulants, and they are discussed in the next two sections.

Ferric Sulfate   The simplified reaction of ferric sulfate with natural bicarbonate alkalinity to form ferric hydroxide is

$$Fe_2(SO_4)_3 + 3Ca(HCO_3)_2 \rightarrow \underline{2Fe(OH)_3} \downarrow + 3CaSO_4 + 6CO_2 \quad \textbf{(8.6)}$$

The reaction usually produces a dense, rapid-settling floc. If the natural alkalinity is insufficient for the reaction, slaked lime may be used instead. The optimum pH range for ferric sulfate is from about 4 to 12, since ferric hydroxide is relatively insoluble within this range, as shown in Figure 8.6. Also illustrated in Figure 8.6 is the usual range of ferric hydroxide concentrations for dosages used in water treatment; these usually produce an oversaturated solution of ferric hydroxide. Ferric sulfate is available in dry form as granules or as a powder, the granules being the more common. The granules are 90 to 94% $Fe_2(SO_4)_3$, contain nine waters of crystallization, and weigh from 70 to 72 lb/ft^3 (1120 to 1155 kg/m^3). The chemical is usually dry fed. The dry chemical may be shipped in bags or barrels.

Ferric Chloride   The simplified reaction of ferric chloride with natural bicarbonate alkalinity to form ferric hydroxide is

$$2FeCl_3 + 3Ca(HCO_3)_2 \rightarrow \underline{2Fe(OH)_3} \downarrow + 3CaSO_4 + 6CO_2 \quad \textbf{(8.7)}$$

If the natural alkalinity is insufficient for the reaction, slaked lime may be added to form the hydroxide, as given by the equation

$$2FeCl_3 + 3Ca(OH)_2 \rightarrow \underline{2Fe(OH)_3} \downarrow + 3CaCl_2 \quad \textbf{(8.8)}$$

The optimum pH range for ferric chloride is the same as for ferric sulfate, which is from about 4 to 12. The floc formed is generally a dense, rapid-settling floc. Ferric chloride is available in dry or liquid form. The dry chemical may be in powder or lump form, the lump form being the more

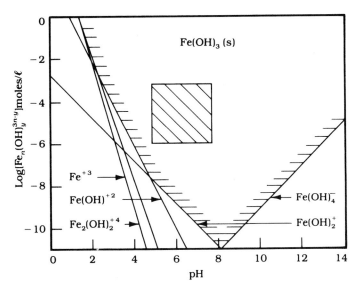

FIGURE 8.6 *Solubility of Ferric Hydroxide. (Shaded area is the usual operating region used in water treatment.)*

Adapted from *Journal of American Water Works Association* 60, no. 5 (May 1968), by permission. Copyright 1968, the American Water Works Association.

common. The lumps, which are 59 to 61% $FeCl_3$, contain six waters of crystallization and weigh from 60 to $64\,lb/ft^3$ (960 to $1026\,kg/m^3$). The lumps are very hydroscopic and are usually solution fed. Upon absorbing water, they decompose to yield hydrochloric acid. The powdered or anhydrous form is 98% $FeCl_3$, contains no water of crystallization, and weighs from 85 to $90\,lb/ft^3$ (1360 to $1440\,kg/m^3$). The liquid form is 37 to 47% $FeCl_3$. The dry form is shipped in barrels, the solution form in bulk.

Lime High lime treatment is frequently used in wastewater treatment, and the lime may be in the form of slaked lime or hydrated lime. Slaked lime (milk of lime), $Ca(OH)_2$, is produced by reacting quicklime, CaO, with water in lime-slaking equipment. Quicklime is available in the dry form as granules or lumps, the lump form being the more common. Quicklime lumps are usually 70 to 96% CaO and weigh from 55 to $70\,lb/ft^3$ (880 to $1120\,kg/m^3$). Quicklime is shipped in bags, barrels, or bulk. If the amount of lime required is rather small, it may be desirable to use hydrated lime, $Ca(OH)_2$, since the feeders are less expensive than slaking equipment. Hydrated lime is from 82 to 99% $Ca(OH)_2$, and it is available as a powder in two different densities. Light hydrated lime weighs from 24 to $48\,lb/ft^3$ (385 to $770\,kg/m^3$), whereas dense hydrated lime weighs from 40 to $70\,lb/ft^3$ (640 to $1120\,kg/m^3$). It is available in bags, barrels, or bulk.

COAGULANT AIDS

In both water and wastewater treatment, coagulant aids are sometimes used to produce a quick-forming, dense, rapid-settling floc and to insure optimum coagulation.

Alkalinity addition is required to aid coagulation if the natural alkalinity is insufficient to produce a good floc. Lime is usually used, and it may be fed as slaked lime (milk of lime) or hydrated lime. Soda ash (sodium carbonate) is used to a lesser extent than lime since it is more expensive than quicklime. Soda ash is available as a powder and may be dry fed. The powder is 99.4% Na_2CO_3 and is available in three specific weights: 23, 35, and 65 lb/ft^3 (369, 560, and 1040 kg/m^3). It comes in bags, barrels, or bulk.

Polyelectrolytes are also used to obtain optimum coagulation, and these aids may be classified according to their ionic characteristics. There are anionic (negatively charged), cationic (positively charged), and poly-ampholites, which have both positively and negatively charged groups. Polyelectrolytes may be of natural origin such as starch or polysaccharide gums, or they may be synthetic in origin. Most polyelectrolytes used in water and wastewater treatment are synthetic organic chemicals. When used as coagulant aids, they assist in coagulation, principally by chemical bridging or interaction between reactive groups on the polyelectrolyte and the floc. In some cases, polyelectrolytes may be used as the sole coagulant, with no other chemicals required. Polyelectrolytes are frequently in powder form and may require specific procedures to prepare aqueous solutions for feeding. Usually the dosage is less than about 0.3 mg/ℓ. Activated silica is an inorganic, anionic polyelectrolyte made from sodium silicate. It has been used to some degree as a coagulant aid.

Turbidity addition, such as recycling some chemically precipitated sludge ahead of the mixing or flocculation basins, is occasionally required to furnish sufficient particulate concentrations for rapid coagulation. Adequate particulate concentrations provide sufficient interparticulate collisions to yield optimum coagulation. Clays are sometimes used for turbidity addition instead of recycled sludge.

Adjustment of pH is required if the pH of the coagulated water does not fall within the pH range for minimum solubility of the metallic hydroxide. Increasing the pH is usually done by addition of lime; pH reduction is usually accomplished by the addition of a mineral acid, such as sulfuric acid.

JAR TESTS

The laboratory technique of the jar test is usually used to determine the proper coagulant and coagulant aid, if needed, and the chemical dosages required for the coagulation of a particular water. In this test, samples of the water are poured into a series of glass beakers, and various dosages of the coagulant and coagulant aid are added to the beakers. The contents are rapidly stirred to simulate rapid mixing and then gently stirred to simulate flocculation. After a given time, the stirring is ceased and the floc formed is allowed to settle. The most important aspects to note are the time for floc formation, the floc size, its settling characteristics, the percent turbidity and color removed, and the final pH of the coagulated and settled water. The chemical dosage determined from the procedure gives an estimate of the dosage required for the treatment plant. A detailed outline of the procedure for the jar test may be found in numerous publications (Black *et al.*, 1957; Black *et al.*, 1969; Camp, 1968, 1952). Camp (1968) developed a modification of the jar test that gives the *G* and *GT* parameters needed for the design of rapid-mixing and flocculation basins.

CHEMICAL FEEDERS

Chemical feeders may be classified as solution- or dry-feed type according to the manner of measuring the chemical dosage. It is preferable to use dry feeders because less equipment and labor are required than for solution feeders. For solution feeders, a solution of a known concentration of the chemical is prepared in a storage tank. The feed consists of metering the solution by a metering pump as it is fed to the mixing basins. Frequently, two storage tanks are used so that one may be charged while the other is on-line. Solution feeding is not as desirable as dry feeding; however, some chemicals, such as ferric chloride, which is hydroscopic, must be fed in solution. Dry feeders measure the chemical dosage by metering the dry chemical as it is transferred from a storage bin into a dissolving chamber. A minor stream of water dissolves the chemical; then the solution is fed to the mixing basins. The measurement of the dry chemical feed rate may be done by gravimetric or volumetric methods; numerous types of equipment are available. Thus the dry-feed system consists of a hopper, a proportioning mechanism (gravimetric or volumetric), the dissolving chamber, and the chemical feed lines to the mixing basins. If a large volume of chemical is stored, usually the bin is directly above the feeder, and the hopper is mounted on the bottom of the storage bin.

Quicklime slakers are usually dry feeders, as shown in Figure 8.7. They may be one of two types: the paste (pug-mill) type and the slurry or detention type. For the pug-mill type, the dry feeder feeds the quicklime to a pug mill, where the quicklime and water are agitated to form a very thick slurry or paste of 30 to 40% calcium hydroxide. The paste is then diluted with water to form a slurry of about 10% calcium hydroxide (slaked lime or milk of lime), and this slurry is fed to the rapid-mix basins. For the detention type, the dry feeder feeds the quicklime to a slaking chamber, where the quicklime and water are agitated to form a 16 to 20% slurry of calcium hydroxide. This slurry is diluted to about 10% calcium hydroxide and then fed to the rapid-mix basins. The efficiency of the slaker depends, in part, on the temperature of the quicklime and water mixture under agitation. Although heat is evolved during slaking, some slakers require the temperature in the slaking chamber to be up to 170°F (76.7°C), which may necessitate preheating of the water fed to the slaker, particularly during startup. Approximately 500 Btu of heat are evolved per pound of calcium oxide slaked (1.16×10^6 J/kg), and the time required for slaking depends partially on the quality of the quicklime. The slaking time usually ranges from several minutes to as long as one hour.

FIGURE 8.7 *Lime Slaker at a Lime-Soda Softening Plant. The calcium oxide moves from the hopper onto an endless, horizontal moving belt mounted on scales in the box with the BIF emblem. The calcium oxide then falls from the belt into the dissolving box below. The inert materials are removed by moving buckets and fall into the wheelbarrow.*

RAPID MIXING AND FLOC-CULATION

In the rapid-mix basins, intense mixing or agitation is required to disperse the chemicals uniformly throughout the basin and to allow adequate contact between the coagulant and the suspended particles. By the time the water leaves the rapid-mix basins, the coagulation process has progressed sufficiently to form microfloc.

In the flocculation basins, the fine microfloc begins to agglomerate into larger floc particles. This aggregation process (flocculation) is dependent on the duration and amount of gentle agitation applied to the water. By the time the water leaves the flocculation basins, the floc has agglomerated into large, dense, rapid-settling floc particles.

The types of devices usually used to furnish the agitation required in both rapid mixing and flocculation may be generally classified as (1) mechanical agitators, such as paddles, (2) pneumatic agitators, and (3) baffle basins. The mechanical type is the most common.

A rational approach to evaluate mixing and to design basins employing mixing has been developed by T. R. Camp (1955). Camp realized that rapid mixing and flocculation are basically mixing operations and, consequently, are governed by the same principles and require similar design parameters. According to his research, the degree of mixing is based on the power imparted to the water, which he measured by the **velocity gradient**. The velocity gradient of two fluid particles which are 0.05 ft (0.01524 m) apart and have a velocity relative to each other of 2.0 fps (feet per second) (0.6096 m/s) is 2.0 divided by 0.05, or 40 fps/ft (0.6096 m/s ÷ 0.01524 m = 40 mps/m). The equation for the velocity gradient for mechanical or pneumatic agitation is

$$G = \sqrt{\frac{W}{\mu}} = \sqrt{\frac{P}{\mu V}} \qquad (8.9)$$

where

G = velocity gradient, fps/ft or sec^{-1} (mps/m or s^{-1})

W = power imparted to the water per unit volume of the basin, ft-lb/sec-ft^3 (N-m/s-m^3)

P = power imparted to the water, ft-lb/sec (N-m/s)

V = basin volume, ft^3 (m^3)

μ = absolute viscosity of the water, lb-force-sec/ft^2 (at 50°F, μ = 2.73 × 10^{-5} lb-sec/ft^2) (at 10°C, μ = 0.00131 N-s/m^2).

The velocity gradient for baffle basins is given by

$$G = \sqrt{\frac{\gamma h_L}{\mu T}} \qquad (8.10)$$

where

γ = specific weight of water

h_L = head loss due to friction, turbulence, and so on

T = detention time

The rate of particulate collisions is proportional to the velocity gradient, G; therefore, the gradient must be sufficient to furnish the desired rate of particulate collisions. The velocity gradient is also related to the shear forces in the water; thus, large velocity gradients produce appreciable shear forces. If the velocity gradient is too great, excessive shear forces will result and prevent the desired floc formation. The total number of particle collisions is proportional to the product of the velocity gradient, G, and the detention time, T. Thus the value of GT is of importance in design. Figure 8.8 depicts the relationship between the velocity gradient, the water temperature, and the power imparted to the water per unit volume.

Camp (1968) developed a jar-test procedure using beakers with and without inside baffles. From mixing and flocculation tests using his

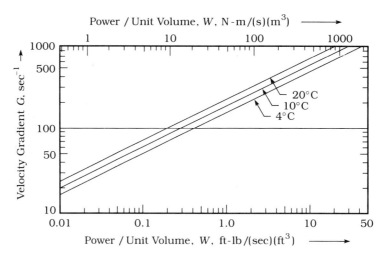

FIGURE 8.8 *Mixing Power Requirements*

Adapted from *Water Treatment Plant Design*, by permission. Copyright 1969, the American Water Works Association.

technique, it is possible to determine the optimum G and T values for a particular coagulant and a given water or wastewater.

Rapid Mixing   Although power for rapid mixing may be imparted to the water by mechanical agitation, pneumatic agitation, and baffle basins, the power required for each method must be the same if the mixing is to be at the same intensity.

Mechanical agitation is the most common method for rapid mixing since it is reliable, very effective, and extremely flexible in operation. Usually rapid mixing employs vertical-shaft rotary mixing devices such as turbine impellers, paddle impellers, or, in some cases, propellers. All of the rotary mixing devices impart motion to the water in addition to turbulence. Types of rapid-mixing chambers or basins are shown in Figure 8.9 (a)–(f). The in-line mixer is the most compact method and is increasing in popularity. Since the optimum velocity gradient may vary with respect to time, it is desirable to have equipment with variable-speed drives. A speed variation of 1:4 is commonly used. Numerous variable drive devices are commercially available. If only one chemical is added, a mixing basin with only one compartment may be used. If, however, more than one chemical is required, sequential application and dispersion of each chemical is desirable, necessitating multiple compartments. Mechanical mixing basins are not affected to any extent by variation in the flowrate and have low head losses. Typical rapid-mixing basins have detention times and velocity gradients as shown in Table 8.1 (AWWA, 1969).

Detention times from 20 to 60 sec are generally used, although some mixing basins have had detention times as small as 10 sec or as long as 2 to 5 min. To obtain high velocity gradients, such as 700 to 1000 fps/ft (700 to

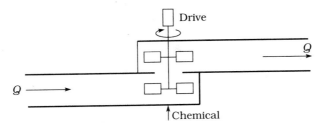

(a) In-Line Mixer

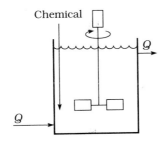

(b) Turbine Chamber

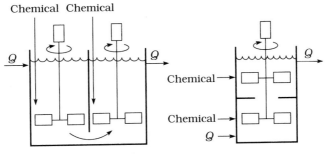

(c) Double Compartment Turbine Chamber

(d) Double Compartment Turbine Chamber

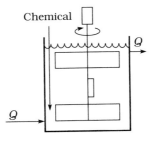

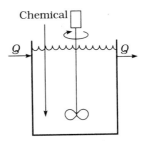

FIGURE 8.9
Rapid-Mixing Devices

(e) Paddle Chamber

(f) Propeller Chamber

TABLE 8.1 *Detention Times and Velocity Gradients of Rapid-Mixing Basins*

DETENTION TIME (sec)	G (fps/ft or sec^{-1}; mps/m or s^{-1})
20	1000
30	900
40	790
50 or more	700

1000 mps/m), requires relatively high mixing power levels. Single compartment mixing basins are usually circular or square in plan view, and the fluid depth is 1.0 to 1.25 times the basin diameter or width. Tanks may be baffled or unbaffled; however, small baffles are desirable because they minimize vortexing and rotational flow.

Rotary mixing devices may be classified as turbines, paddle impellers, or propellers according to McCabe and Smith (1976). The types of **turbine impellers**, as shown in Figure 8.10, are the straight blade, vaned disc, curved blade, and shrouded curved blade with a stationary diffuser, with the vaned disc being the most widely used. The stationary vanes of the shrouded turbine prevent rotational flow. The impeller blades may be pitched or vertical, but the vertical are the most common. The diameter of the impeller is usually from 30 to 50% of the tank diameter or width, and the impeller is usually mounted one impeller diameter above the tank bottom. The speeds range from 10 to 150 rpm, and the flow is radially outward from the turbine. It divides at the tank wall, giving a flow pattern as shown in Figure 8.11(a). Small baffles extending into the tank a distance of 0.10 times the tank width

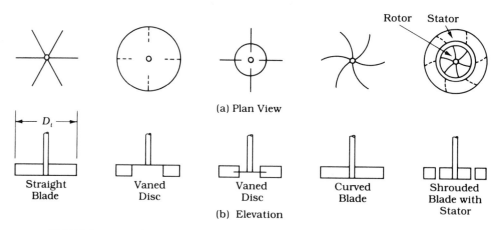

FIGURE 8.10 *Types of Turbine Impellers*

Adapted from *Unit Operations of Chemical Engineering* by W. L. McCabe and J. C. Smith. Copyright © 1976 by McGraw-Hill, Inc. Reprinted by permission.

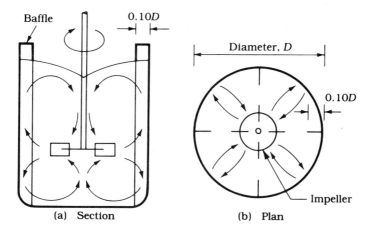

FIGURE 8.11 *Flow Regime in a Turbine-Impeller Tank*

Adapted from "Mixing — Present Theory and Practice: Parts I and II" by J. H. Rushton and J. Y. Oldshue. In *Chemical Engineering Progress* 46, no. 4 (April 1953):161; and 49, no. 5 (May 1953):267. Reprinted by permission.

or diameter will minimize vortexing and rotational flow and, consequently, cause more power to be imparted to the liquid. This results in greater turbulence, which is desirable for agitation. Turbines are the most effective of all the mechanical agitation or mixing devices because they produce high shear, turbulence, and velocity gradients.

Paddle impellers usually have two or four blades. The blades may be pitched or vertical, the vertical type being the more common. The diameter of a paddle impeller is usually from 50 to 80% of the tank diameter or width, and the width of a paddle is usually $\frac{1}{6}$ to $\frac{1}{10}$ of the diameter. The paddles usually are mounted one-half of a paddle diameter above the tank bottom, as shown in Figure 8.12(a). The flow regime for a two-blade paddle, which is similar to that for the turbine impeller, is depicted in Figure 8.13. The paddle speeds range from 20 to 150 rpm, and baffling is required to minimize vortexing and rotational flow except at very slow speeds. The paddle is not as efficient as the turbine type since it does not produce as much turbulence and shear forces.

The **propeller impeller**, shown in Figure 8.14, may have either two or three blades, and the blades are pitched to impart axial flow to the liquid. The rotation of a propeller traces out a helix in the liquid, and the pitch is defined as the distance the liquid moves axially during one revolution, divided by the propeller diameter. Usually the pitch is 1.0 or 2.0, and the maximum propeller diameter is about 18 in. The axial flow, as depicted in Figure 8.15(a), strikes the bottom of the tank and divides and imparts a flow regime as shown. For deep tanks, two propellers may be mounted on the same shaft and may produce liquid motion in the same direction or in opposite directions. The propeller speed is ordinarily from 400 to 1750 rpm,

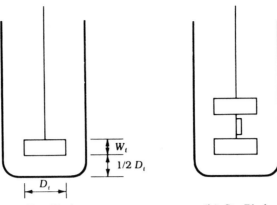

FIGURE 8.12 *Types of Paddle Impellers*

(a) Two Blades

(b) Six Blades

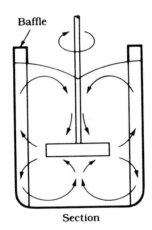

FIGURE 8.13 *Flow Regime in a Paddle-Impeller Tank*

Section

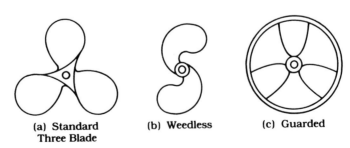

(a) Standard Three Blade

(b) Weedless

(c) Guarded

FIGURE 8.14 *Types of Propeller Impellers*

Adapted from *Unit Operations of Chemical Engineering* by W. L. McCabe and J. C. Smith. Copyright © 1976 by McGraw-Hill, Inc. Reprinted by permission.

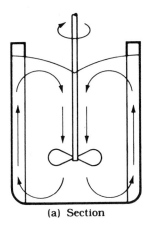 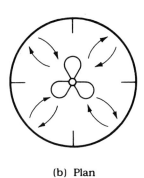

(a) Section   (b) Plan

FIGURE 8.15 *Flow Regime in a Propeller-Impeller Tank*

Adapted from "Mixing — Present Theory and Practice: Parts I and II," by J. H. Rushton and J. Y. Oldshue. In *Chemical Engineering Progress* 46, no. 4 (April 1953):161; and 49, no. 5 (May 1953):267.

and baffling is required in large tanks to minimize vortexing and rotational flow. In small tanks, the propeller may be mounted off-center to avoid rotational flow. Propeller agitators are very effective in large tanks because of the high velocities imparted to the liquid.

The power imparted to the liquid by various impellers may be determined using relationships developed by Rushton (Rushton, 1952; Rushton *et al.*, 1947; Rushton *et al.*, 1950; Rushton and Oldshue, 1953; Rushton and Mahoney, 1954) for impellers employed in chemical process industries. For **turbulent flow** ($N_{Re} > 10,000$), the power imparted by an impeller in a baffled tank is given by the following equation:

$$P = K_T n^3 D_i^5 \rho \tag{8.11}$$

where

P = power, ft-lb/sec (N-m/s)

K_T = impeller constant for turbulent flow

n = rotational speed, rps

D_i = impeller diameter, ft (m)

ρ = density of the liquid, $\rho = \gamma/g_c$

γ = specific weight of the liquid, lb/ft³ (N/m³)

g_c = acceleration due to gravity, 32.17 ft/sec² (9.806 m/s²)

If the flow is **laminar** ($N_{Re} < 10$ to 20), the power imparted by an impeller in either a baffled or an unbaffled tank is given by

$$P = K_L n^2 D_i^3 \mu \tag{8.12}$$

where

K_L = impeller constant for laminar flow

μ = absolute viscosity of the liquid, lb-force-sec/ft^2 (N-s/m^2)

The Reynolds number for impellers is given by

$$N_{\text{Re}} = \frac{D_i^2 n \rho}{\mu} \qquad\qquad \textbf{(8.13)}$$

where

N_{Re} = Reynolds number, dimensionless

For laminar flow, the power imparted in a tank is independent of the presence of baffles. In turbulent flow, however, the power imparted in an unbaffled tank may be as low as one-sixth the power imparted in the same tank with baffles. Values of the impeller constants, K_T and K_L, for various types of impellers, are given in Table 8.2 for circular tanks having four baffles. For turbulent flow, it has been found that the power required for agitation in a baffled vertical square tank is the same as in a baffled vertical circular tank having a diameter equal to the width of the square tank. In an unbaffled square tank, the power imparted is about 75% of that imparted in a baffled square or a baffled circular tank. Also, two straight blade turbines mounted one turbine diameter apart on the same shaft impart about 1.9

TABLE 8.2   *Values of Constants K_L and K_T in Eqs. (8.11) and (8.12) for Baffled Tanks Having Four Baffles at Tank Wall, with Width Equal to 10% of the Tank Diameter*

TYPE OF IMPELLER	K_L	K_T
Propeller, pitch of 1, 3 blades	41.0	0.32
Propeller, pitch of 2, 3 blades	43.5	1.00
Turbine, 4 flat blades, vaned disc	60.0	5.31
Turbine, 6 flat blades, vaned disc	65.0	5.75
Turbine, 6 curved blades	70.0	4.80
Fan turbine, 6 blades at 45°	70.0	1.65
Shrouded turbine, 6 curved blades	97.5	1.08
Shrouded turbine, with stator, no baffles	172.5	1.12
Flat paddles, 2 blades (single paddle), $D_i/W_i = 4$	43.0	2.25
Flat paddles, 2 blades, $D_i/W_i = 6$	36.5	1.70
Flat paddles, 2 blades, $D_i/W_i = 8$	33.0	1.15
Flat paddles, 4 blades, $D_i/W_i = 6$	49.0	2.75
Flat paddles, 6 blades, $D_i/W_i = 6$	71.0	3.82

From: (1) "Mixing of Liquids in Chemical Processing" by J. H. Rushton. In *Industrial and Engineering Chemistry* 44, no. 2 (December 1952): 2931, copyright 1952, American Chemical Society; (2) "Mixing — Present Theory and Practice" by J. H. Rushton and J. Y. Oldshue. In *Chemical Engineering Progress* 46, no. 4 (April 1953): 161; and (3) *Unit Operations of Chemical Engineering* by W. L. McCabe, J. C. Smith, and P. Harriott. 5th ed. Copyright © 1993 by McGraw-Hill, Inc. Reprinted by permission.

times as much power as one turbine alone. In nearly all cases of rapid mixing for coagulation, the flow regime is well within the turbulent range.

Pneumatic mixing basins employ tanks and aeration devices somewhat similar to those used in the activated sludge process, as depicted in Figure 8.16. The detention times and velocity gradients are of the same magnitude and range as those used for mechanical rapid mixing. Variation of the velocity gradient may be obtained by varying the air flowrate. Pneumatic mixing is not affected to any extent by variations in the influent flowrate, and the hydraulic head losses are relatively small. By selecting the design velocity gradient, G, it is possible to determine the power required by Eq. (8.9) or Figure 8.8. The basin volume, V, may be determined from the flowrate and detention time, T. The air flowrate to impart the desired power to the water may then be determined by the equation

$$P = C_1 G_a \log\left(\frac{h + C_2}{C_2}\right) \tag{8.14}$$

where

P = power, ft-lb/sec (N-m/s)

C_1 = 81.5 for USCS units and 3904 for SI units

G_a = air flowrate at operating temperature and pressure, cfm (m^3/min)

h = depth to the diffusers, ft (m)

C_2 = 34 for USCS units and 10.4 for SI units

The baffle-type mixing basins, as depicted in Figure 8.17, depend on hydraulic turbulence to furnish the desired velocity gradient. The velocity gradient imparted to the water is given by Eq. (8.10), and the volume is determined from the flowrate and the detention time, T. The head loss usually varies from 1 to 3 ft (0.3 to 0.9 m), and these basins have very little short circuiting. Baffle basins are not suitable where there is a wide variation in the flowrate, and it is not possible to vary the velocity gradient to any extent. Because of these disadvantages, baffle basins are presently not widely used. In both mechanical and pneumatic mixing, varying the power imparted results in a variation of the velocity gradient.

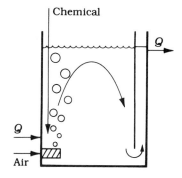

FIGURE 8.16
Pneumatic Rapid Mixing

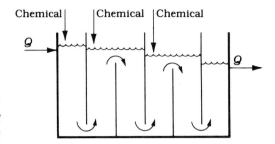

FIGURE 8.17
Baffle Basin Rapid Mixing

EXAMPLE 8.1 *Rapid Mixing*

A square rapid-mixing basin, with a depth of water equal to 1.25 times the width, is to be designed for a flow of 2.0 million gallons per day (MGD). The velocity gradient is to be 790 fps/ft, the detention time is 40 sec, the operating temperature is 50°F (10°C), and the turbine shaft speed is 100 rpm. Determine:

1. The basin dimensions if increments of 1 in. are used.

2. The horsepower required.

3. The impeller diameter if a vane-disc impeller with six flat blades is employed and the tank has four vertical baffles (one on each tank wall). The impeller diameter is to be 30 to 50% of the tank width.

4. The impeller diameter if no vertical baffles are used.

5. The air required if pneumatic mixing is employed and the diffusers are 0.5 ft above the tank bottom.

SOLUTION The volume is

$$V = \frac{2.0 \times 10^6 \,\text{gal}}{1440 \,\text{min}} \left| \frac{1 \,\text{min}}{60 \,\text{sec}} \right| \frac{1 \,\text{ft}^3}{7.48 \,\text{gal}} \left| \frac{40 \,\text{sec}}{} \right.$$
$$= 124 \,\text{ft}^3$$

The dimensions are given by

$$(W)(W)(1.25W) = 124 \,\text{ft}^3 \quad \text{or} \quad W = 4.63 \,\text{ft or } 4 \,\text{ft-8 in.}$$

From this, $H = (1.25)(4.67 \,\text{ft}) = 5.84 \,\text{ft or } 5 \,\text{ft-10 in.}$

$$\boxed{\text{Use } W = 4 \,\text{ft-8 in.;} \ H = 5 \,\text{ft-10 in.}}$$

$$\text{Volume} = (4.67)^2(5.83) = 127 \,\text{ft}^3$$

Equation (8.9) may be rearranged to give the power imparted as $W = G^2\mu$ or $W = (790/\text{sec})^2(2.73 \times 10^{-5} \,\text{lb-sec/ft}^2) = 17.04 \,\text{ft-lb/sec-ft}^3$. The value of the total power, P, is

$$P = \frac{17.04 \,\text{ft-lb}}{\text{sec-ft}^3} \left| \frac{127 \,\text{ft}^3}{} \right. = 2164 \,\text{ft-lb/sec}$$

$$P = \frac{2164\,\text{ft-lb}}{\text{sec}} \left| \frac{\text{sec}}{550\,\text{ft-lb}} \right| \text{hp} = \boxed{3.93\,\text{hp}}$$

The impeller speed, n, is 100 rpm, or 100/60 or 1.667 rps. Assume turbulent flow; thus, from Table 8.2, $K_T = 5.75$.

The power equation for USCS units is

$$P = \frac{K_T n^3 D_i^5 \gamma}{g_c}$$

Rearranging the power equation to determine the impeller diameter gives

$$D_i = \left(\frac{Pg_c}{K_T n^3 \gamma}\right)^{1/5}$$

$$= \left[\frac{2164\,\text{ft-lb}}{\text{sec}} \left| \frac{32.17\,\text{ft}}{\text{sec}^2} \right| \frac{1}{5.75} \left| \frac{1}{(1.667\,\text{rps})^3} \right| \frac{\text{ft}^3}{62.4\,\text{lb}} \right]^{1/5}$$

$$= \boxed{2.11\,\text{ft}}$$

$$D_i/W = 2.11/4.67 = 0.452 \text{ or } 45.2\%$$

Check on the Reynolds number, N_{Re}:

$$N_{\text{Re}} = \frac{D_i^2 n \rho}{\mu} = \frac{D_i^2 n \gamma}{\mu g_c}$$

$$= \frac{(2.11\,\text{ft})^2}{} \left| \frac{1.667\,\text{rps}}{} \right| \frac{62.4\,\text{lb}}{\text{ft}^3} \left| \frac{\text{ft}^2}{2.73 \times 10^{-5}\,\text{lb-sec}} \right| \frac{\text{sec}^2}{32.17\,\text{ft}}$$

$$= 527{,}000 >>> 10{,}000 \qquad \text{Thus the flow regime is turbulent.}$$

If no vertical baffles are used, the power imparted is 75% of that for a baffled tank. Therefore, to impart the same power requires a larger impeller. The value K_T is $0.75(5.75) = 4.31$. The impeller diameter is given by

$$D_i = \left(\frac{Pg_c}{K_T n^3 \gamma}\right)^{1/5}$$

$$= \left[\frac{2164\,\text{ft-lb}}{\text{sec}} \left| \frac{32.17\,\text{ft}}{\text{sec}^2} \right| \frac{1}{4.31} \left| \frac{1}{(1.667\,\text{rps})^3} \right| \frac{\text{ft}^3}{62.4\,\text{lb}} \right]^{1/5}$$

$$D_i = \boxed{2.24\,\text{ft}}$$

$$D_i/W = 2.24/4.67 = 0.480 = 48.0\%$$

If pneumatic mixing is used, the power equation for diffused air for USCS units may be rearranged to yield

$$G_a = \frac{P/81.5}{\log\left(\dfrac{h + 34}{34}\right)}$$

and $h = 5.83\,\text{ft} - 0.50\,\text{ft} = 5.33\,\text{ft}$.

Thus,

$$G_a = \frac{2164/81.5}{\log\left(\dfrac{5.33 + 34}{34}\right)}$$

$$G_a = \boxed{420 \, \text{cfm}}$$

The air flow required is at operating conditions.

EXAMPLE 8.1 SI   *Rapid Mixing*

A square rapid-mixing basin, with a depth of water equal to 1.25 times the width, is to be designed for a flow of $7570 \, \text{m}^3/\text{d}$. The velocity gradient is to be 790 mps/m, the detention time is 40 s, the operating temperature is 10°C, and the turbine shaft speed is 100 rpm. Determine:

1. The basin dimensions.

2. The power required.

3. The impeller diameter if a vane-disc impeller with six flat blades is employed and the tank has four vertical baffles (one on each tank wall). The impeller diameter is to be 30 to 50% of the tank width.

4. The impeller diameter if no vertical baffles are used.

5. The air required if pneumatic mixing is employed and the diffusers are 0.15 m above the tank bottom.

SOLUTION   The volume is

$$V = \frac{7570 \, \text{m}^3}{1440 \, \text{min}} \Big| \frac{\text{min}}{60 \, \text{s}} \Big| \frac{40 \, \text{s}}{}$$

$$= 3.50 \, \text{m}^3$$

The dimensions are given by

$$(W)(W)(1.25\,W) = 3.50 \, \text{m}^3 \quad \text{or} \quad W = 1.41 \, \text{m}$$

From this, $H = (1.25)(1.41 \, \text{m}) = 1.76 \, \text{m}$.

$$\boxed{\text{Use } W = 1.41 \, \text{m}; \; H = 1.76 \, \text{m.}}$$

Equation (8.9) may be rearranged to give the power imparted as $W = G^2\mu$ or $W = (790/\text{s})^2(0.00131 \, \text{N-s/m}^2)(\text{m/m}) = 818 \, \text{N-m/s-m}^3$. The value of the total power, P, is

$$P = \frac{818 \, \text{N-m/s}}{\text{m}^3} \Big| \frac{3.50 \, \text{m}^3}{} = \boxed{2863 \, \text{N-m/s} = 2863 \, \text{J/s} = 2863 \, \text{W}}$$

The impeller speed, n, is 100 rpm, or 100/60 or 1.667 rps. Assume turbulent flow; thus, from Table 8.2, $K_T = 5.75$. The power equation for SI units is

$$P = K_T n^3 D_i^5 \rho$$

Rearranging the power equation to determine the impeller diameter gives

$$D_i = \left(\frac{P}{K_T n^3 \rho}\right)^{1/5}$$

$$= \left[\frac{2863 \text{ N-m}}{\text{s}} \middle| \frac{}{5.75} \middle| \frac{}{(1.667 \text{ rps})^3} \middle| \frac{\text{m}^3}{999.7 \text{ kg}} \middle| \frac{\text{kg-m}}{\text{N-s}^2}\right]^{1/5}$$

$$= \boxed{0.640 \text{ m}}$$

$D_i/W = 0.640/1.41 = 0.454$ or 45.4%

Check on the Reynolds number, N_{Re}:

$$N_{\text{Re}} = \frac{D_i^2 n \rho}{\mu}$$

$$= \frac{(0.640 \text{ m})^2 \middle| 1.667 \text{ rps} \middle| 999.7 \text{ kg} \middle| \frac{\text{m}^2}{\text{m}^3} \middle| \frac{\text{N-s}^2}{0.00131 \text{ N-s}} \middle| \frac{}{\text{kg-m}}}{}$$

$$= 521{,}000 >>> 10{,}000 \qquad \text{Thus the flow regime is turbulent.}$$

If no vertical baffles are used, the power imparted is 75% of that for a baffled tank. Therefore, to impart the same power requires a larger impeller. The value K_T is 0.75(5.75) = 4.31. The impeller diameter is given by

$$D_i = \left(\frac{P}{K_T n^3 \rho}\right)^{1/5}$$

$$= \left[\frac{2863 \text{ N-m}}{\text{s}} \middle| \frac{}{4.31} \middle| \frac{}{(1.667 \text{ rps})^3} \middle| \frac{\text{m}^3}{999.7 \text{ kg}} \middle| \frac{\text{kg-m}}{\text{N-s}^2}\right]^{1/5}$$

$$= \boxed{0.678 \text{ m}}$$

$D_i/W = 0.678/1.41 = 0.481 = 48.1\%$

If pneumatic mixing is used, the power equation for diffused air for SI units may be rearranged to yield

$$G_a = \frac{P/3904}{\log\left(\dfrac{h + 10.4}{10.4}\right)}$$

and $h = 1.76 \text{ m} - 0.15 \text{ m} = 1.61 \text{ m}$.
  Thus,

$$G_a = \frac{2863/3904}{\log\left(\dfrac{1.61 + 10.4}{10.4}\right)}$$

$$G_a = \boxed{11.7 \text{ m}^3/\text{min}}$$

The air flow is at operating conditions.

Flocculation   The power required for the gentle agitation or stirring of the water during flocculation may be imparted by mechanical or pneumatic agitation, with mechanical agitation being the most common. Formerly, baffle basins were used for flocculation; however, since the available range of G and GT values is limited, these are not employed at present to any extent. Most mechanical agitators are paddle wheels, as shown in Figure 8.18, although turbines and propellers are also used.

The degree of completion of the flocculation process depends on the relative ease and rate by which the small microfloc aggregate into large floc particles and on the total number of particulate collisions during flocculation. Thus the degree of completion is dependent on the floc characteristics, the velocity gradient, G, and the value of GT (Culp and Culp, 1978). The magnitude of the dimensionless parameter, GT, is related to the total number of collisions during aggregation in the flocculation process. A high GT value indicates a large number of collisions during aggregation. A more accurate parameter is GCT, where C is the ratio of the floc volume to the total water volume being flocculated. If the velocity gradient in flocculation is too great, the shear forces will prevent the formation of a large floc. If the velocity gradient is insufficient, adequate interparticulate collisions will not occur and a proper floc will not be formed. If the water is difficult to coagulate, the floc will be fragile and a final velocity gradient less than 5 fps/ft (mps/m) may be required. If, however, the water coagulates readily, a high-strength floc usually results and the final velocity gradient may be as large as 10 fps/ft (mps/m) (AWWA, 1969, 1990).

Flocculation basins are frequently designed to provide for tapered flocculation in which the flow is subjected to decreasing G values as it passes through the flocculation basin. This produces a rapid buildup of small, dense floc, which subsequently aggregates at lower G values into larger, dense, rapidly settling floc particles. Tapered flocculation is usually accomplished by providing a high G value during the first third of the flocculation period, a lower G value during the next third, and a much lower G value during the last third. For example, a typical series of G values could be 50, 20, and 10 fps/ft (mps/m). Although many basins are designed that do not have tapered flocculation, optimum flocculation usually necessitates its use.

Typical arrangements for flocculators employing paddle wheels on horizontal shafts are shown in Figures 8.19 and 8.20. In the cross-flow pattern, Figure 8.19(a), the paddle-wheel shafts are mounted at right angles to the overall water flow. In the axial-flow pattern, Figure 8.20(a), the paddle-wheel shafts are parallel to the flow. At least three consecutive compartments, as shown in Figures 8.19 and 8.20, are required to minimize short circuiting. The partitions between compartments are usually wood baffle fences made of horizontal wood slats with either spacings between the slats for passage of the flow or concrete walls with orifices. The wood baffle fences are more flexible since different slat spacings may be used. Multiple compartments, in addition to minimizing short circuiting, facilitate tapered flocculation design. For the cross-flow pattern, tapered flocculation may

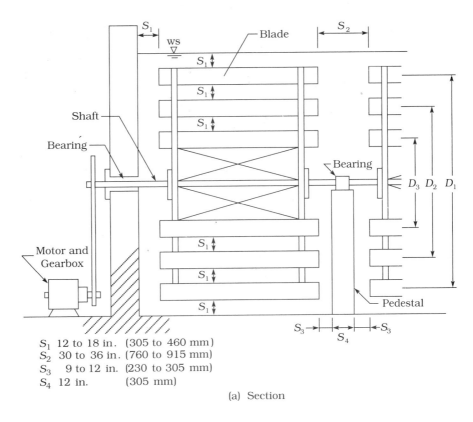

S_1 12 to 18 in. (305 to 460 mm)
S_2 30 to 36 in. (760 to 915 mm)
S_3 9 to 12 in. (230 to 305 mm)
S_4 12 in.   (305 mm)

(a) Section

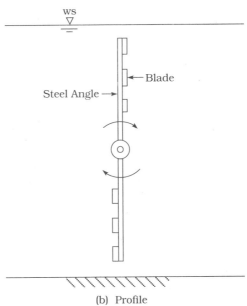

(b) Profile

FIGURE 8.18 *Horizontal-Shaft Flocculation Paddle Wheels*

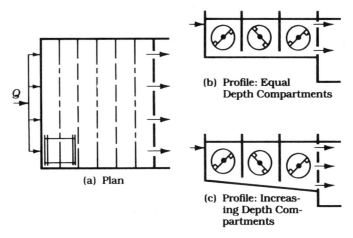

FIGURE 8.19 *Horizontal-Shaft Paddle-Wheel Flocculator (Cross-Flow Pattern)*

Adapted from *Water Treatment Plant Design*, by permission. Copyright 1969, the American Water Works Association.

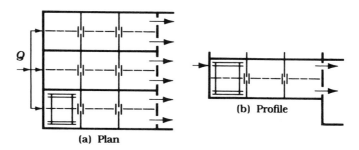

FIGURE 8.20 *Horizontal-Shaft Paddle-Wheel Flocculator (Axial-Flow Pattern)*

Adapted from *Water Treatment Plant Design*, by permission. Copyright 1969, the American Water Works Association.

be provided by varying the paddle size, the number of paddles, and the diameter of the paddle wheels on the various horizontal shafts. Also, it can be obtained by varying the rotational speed of the various horizontal shafts. For the axial-flow pattern, tapered flocculation may be obtained by varying the paddle size and the number of paddles on each paddle wheel that have a common horizontal shaft. All mechanical flocculation devices should be equipped with variable-speed drives having a range up to 1:4 to meet variations in the quality of the feed water. If the compartments are separated by concrete walls with orifices, the orifices should have circular deflector plates immediately upstream and downstream to minimize short circuiting.

The velocity gradient through each orifice should not exceed the gradient in the compartment immediately upstream. The velocity gradient may be estimated using the head loss, h, from the orifice equation $Q = 0.60A \times \sqrt{2gh}$, where Q is in cfs (m³/s), A is the area in ft² (m²), and h is in ft (m). The velocity and the time of passage through the orifice, T, may be computed, and the velocity gradient may be determined from Eq. (8.10), which is $G = (\gamma h_L/\mu T)^{1/2}$, where $h_L = h$ for an orifice.

Vertical-shaft devices, such as the paddle wheels shown in Figures 8.21(a) and (b), are sometimes used. A typical flocculator layout employing these units is shown in Figure 8.21(c). It should be noted from the layout that the compartments are arranged in series to minimize short circuiting and also to facilitate tapered flocculation design.

The power imparted to the water by paddle wheels may be determined using Newton's law for the drag force exerted by a submerged object moving in a liquid. The drag force for a paddle-wheel blade is given by

$$F_D = C_D A \rho \frac{v^2}{2} \qquad (8.15)$$

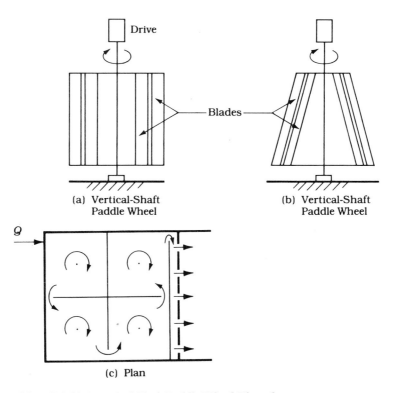

(a) Vertical-Shaft Paddle Wheel

(b) Vertical-Shaft Paddle Wheel

(c) Plan

FIGURE 8.21 *Vertical-Shaft Paddle-Wheel Flocculator*

Adapted from *Water Treatment Plant Design*, by permission. Copyright 1969, the American Water Works Association.

where

F_D = drag force of the paddle, lb (N)

C_D = coefficient of drag

A = paddle-blade area at right angle to the direction of movement, ft² (m²)

ρ = density of the water, $\rho = \gamma/g_c$

v = velocity of the paddle blade relative to the water, fps (mps)

Since the power is equal to the force times the velocity, the power is given by

$$P = F_D v = C_D A \rho \frac{v^3}{2} = C_D A \frac{\gamma}{g_c} \frac{v^3}{2} \qquad (8.16)$$

where

$$P = \text{power, ft-lb/sec (N-m/s)}$$

The drag coefficient depends basically on the geometry of the paddle. Values for various paddle dimensions are given in Table 8.3 (Rouse, 1950).

Practice has shown that the peripheral velocity of the paddle blades should range from 0.3 to 3.0 fps (0.09 to 0.91 mps), and the velocity of a paddle blade relative to the water is approximately three-fourths the peripheral blade velocity. Also, the total paddle-blade area on a horizontal shaft should not exceed 15 to 20% of the total basin cross-sectional area. The area should be at least 15% to insure adequate mixing and should be equal to or less than 20% to avoid excessive rotational flow.

In the design of a horizontal-shaft paddle-wheel flocculator, usually a design velocity gradient, G, and detention time, T, are determined from laboratory or pilot plant studies, and the GT value is computed to insure that it is within an acceptable GT range. Then the flowrate and detention time are used to compute the basin volume, V. Basin dimensions are calculated by knowing the volume of the basin and the number of horizontal shafts selected. If tapered flocculation is used, the G values for each compartment must be determined from laboratory or pilot plant studies. Then, a paddle-wheel design is assumed and the peripheral speeds of the paddles to impart the desired power are computed. If the resulting peripheral speeds

TABLE 8.3 *Drag Coefficient Values for Various Paddle Dimensions*

LENGTH-WIDTH RATIO	C_D
5	1.20
20	1.50
∞	1.90

are excessive, larger paddles or more paddles are assumed, and the peripheral speeds are redetermined to see if they are satisfactory.

Other mechanical agitators such as the walking-beam flocculator, turbines, and propellers have been used occasionally; however, the paddle wheels are by far the most widely used.

Pneumatic flocculation has been used, at times, in lieu of mechanical agitation. When the design G value is known, the required power to be imparted per unit basin volume, P/V, may be determined from Eq. (8.9). The design GT value is used to compute the detention time, T, and the basin volume is determined from the flowrate. The air flow, G_a, needed to impart the desired power may then be determined using Eq. (8.14).

The velocity of flocculated water, in orifices or ports in baffle fences and in conduits (when employed) that lead from the flocculator, should not be greater than 0.5 to 1.0 fps (0.15 to 0.30 mps) to avoid shearing apart the floc.

EXAMPLE 8.2 *Flocculation*

A cross-flow, horizontal-shaft, paddle-wheel flocculation basin is to be designed for a flow of 6.5 MGD, a mean velocity gradient of $26.7 \, \text{sec}^{-1}$ (at 50°F), and a detention time of 45 min. The GT value should be from 50,000 to 100,000. Tapered flocculation is to be provided, and three compartments of equal depth in series are to be used, as shown in Figure 8.19(b). The G values determined from laboratory tests for the three compartments are $G_1 = 50 \, \text{sec}^{-1}$, $G_2 = 20 \, \text{sec}^{-1}$, and $G_3 = 10 \, \text{sec}^{-1}$. These give an average G value of $26.7 \, \text{sec}^{-1}$. The compartments are to be separated by slotted, redwood baffle fences, and the floor of the basin is level. The basin should be 50 ft in width to adjoin the settling basin. The speed of the blades relative to the water is three-quarters of the peripheral blade speed. Determine:

1. The GT value.
2. The basin dimensions if increments of 1 in. are used.
3. The paddle-wheel design.
4. The power to be imparted to the water in each compartment.
5. The rotational speed of each horizontal shaft in rpm.
6. The rotational speed range if $1:4$ variable-speed drives are employed.
7. The peripheral speed of the outside paddle blades in fps.

SOLUTION The GT value is

$$GT = \frac{26.7}{\text{sec}} \left| \frac{45 \, \text{min}}{} \right| \frac{60 \, \text{sec}}{\text{min}}$$

$$= \boxed{72,100}$$

Since the GT value is between 50,000 and 100,000, the detention time of 45 min is satisfactory.

Basin volume, V, is given by

$$V = \frac{6.5 \times 10^6\,\text{gal}}{24\,\text{hr}} \left| \frac{\text{hr}}{60\,\text{min}} \right| \frac{45\,\text{min}}{} \left| \frac{\text{ft}^3}{7.48\,\text{gal}} \right.$$

$$= 27{,}156\,\text{ft}^3$$

Profile area $= 27{,}156\,\text{ft}^3/50\,\text{ft} = 543\,\text{ft}^2$.

Assume compartments are square in profile, and x is the compartment width and depth. Thus,

$$(3x)(x) = 543\,\text{ft}^2 \quad \text{or} \quad x = 13.46 \quad \text{or} \quad 13\,\text{ft-6 in.}$$
$$3x = 3(13\,\text{ft-6 in.}) = 40\,\text{ft-6 in.}$$

$$\boxed{\text{Use width} = \text{depth} = 13\,\text{ft-6 in.;} \quad \text{length} = 40\,\text{ft-6 in.}}$$

$$\text{Volume} = (13.5)(40.5)(50.0) = 27{,}338\,\text{ft}^3$$

Assume a paddle-wheel design as shown in Figure 8.18(a), with $D_1 = 11.0\,\text{ft}$, $D_2 = 8.0\,\text{ft}$, and $D_3 = 5.0\,\text{ft}$. Use four paddlewheels per shaft, and assume the blades are $6\,\text{in.} \times 10\,\text{ft}$. The space between blades is 12 in.

$$\text{Blade area per shaft} = (0.5\,\text{ft})(10\,\text{ft})(6)(4)$$
$$= 120\,\text{ft}^2$$

$$\text{Percent of cross-sectional area} = \frac{120}{} \left| \frac{}{50} \right| \frac{100}{13.5}$$

$$= 17.8\%$$

Since this is between 15 and 20%, make the trial design using the assumed paddle-wheel design. The power, P, is given by

$$G = \sqrt{\frac{P}{\mu V}} \quad \text{or} \quad P = \mu G^2 V$$

Power for first compartment, P, is given by

$$P = \frac{2.73 \times 10^{-5}\,\text{lb-sec}}{\text{ft}^2} \left| \frac{(50)^2}{\text{sec}^2} \right| \frac{27{,}338\,\text{ft}^3}{3}$$

$$= \boxed{622\,\text{ft-lb/sec} = 1.13\,\text{hp}}$$

Power per wheel $= (622\,\text{ft-lb/sec})\,1/4 = 156\,\text{ft-lb/sec}$, and

$$P = C_D A \frac{\gamma}{g_c} \frac{v^3}{2} = C_D A \left(\frac{62.4}{32.17}\right)\left(\frac{v^3}{2}\right) = 0.97 C_D A v^3$$

The length-width ratio is $10/0.5 = 20$; thus $C_D = 1.50$.

The blade velocity relative to water is

$$v = (\text{rps})\left(\frac{\pi D}{\text{rev}}\right)0.75$$

Thus,

$$v_1 = (\text{rps})(\pi)(11.0\,\text{ft})0.75 = 25.92\,(\text{rps}).$$

In a like manner, $v_2 = 18.85\,(\text{rps})$ and $v_3 = 11.78\,(\text{rps})$

The power per wheel is

$$P = 0.97C_D A_1 v_1^3 + 0.97C_D A_2 v_2^3 + 0.97C_D A_3 v_3^3$$

Since $A_1 = A_2 = A_3$ and all C_D values are equal,

$$P = (0.97C_D)(A_1)[(25.92)^3(\text{rps})^3 \\ + (18.85)^3(\text{rps})^3 + (11.78)^3(\text{rps})^3]$$

or

$$156 = (0.97)(1.5)(0.5)(10)(2)(17{,}414 + 6698 + 1635)(\text{rps})^3$$

From this,

$$\text{rps} = 0.075 \text{ and } \boxed{\text{rpm} = (0.075)60 = 4.50}$$

Peripheral speed $= (\pi D)\text{rps}$; thus, for the outside blade,

$$v_1' = (\pi)(11.0\,\text{ft})(0.075) = \boxed{2.59\,\text{fps}}$$

Since $v_1' < 3.0\,\text{fps}$, the design of the paddle wheels is satisfactory. The maximum rotational speed is 4.50 rpm; therefore, the minimum rotational speed is $(1/4) \times 4.50 = 1.13$ rpm. Thus,

$$\boxed{\text{Rotational speed for } 1{:}4 \text{ drive} = 1.13 \text{ to } 4.50\,\text{rpm}}$$

In a like manner, the power, rotational speed, peripheral blade speed, and rotational speed range are computed for the second and third compartments. Following is a summary of the values.

First compartment:

$$\boxed{\begin{array}{l} P = 622\,\text{ft-lb/sec} = 1.13\,\text{hp} \\ \text{rpm} = 4.50; \text{range} = 1.13 \text{ to } 4.50\,\text{rpm} \\ \text{fps} = 2.59 \end{array}}$$

Second compartment:

$$\boxed{\begin{array}{l} P = 99.5\,\text{ft-lb/sec} = 0.18\,\text{hp} \\ \text{rpm} = 2.40; \text{range} = 0.60 \text{ to } 2.40\,\text{rpm} \\ \text{fps} = 1.38 \end{array}}$$

Third compartment:

$$\boxed{\begin{array}{l} P = 24.9\,\text{ft-lb/sec} = 0.045\,\text{hp} \\ \text{rpm} = 1.53; \text{range} = 0.38 \text{ to } 1.53\,\text{rpm} \\ \text{fps} = 0.88 \end{array}}$$

EXAMPLE 8.2 SI   *Flocculation*

A cross-flow, horizontal shaft, paddle-wheel flocculation basin is to be designed for a flow of 25,000 m^3/d, a mean velocity gradient of 26.7 s^{-1} (at 10°C), and a detention time of 45 min. The GT value should be from 50,000 to 100,000. Tapered flocculation is to be provided, and three compartments of equal depth in series are to be used, as shown in Figure 8.19(b). The G values determined from laboratory tests for the three compartments are G_1 = 50 s^{-1}, G_2 = 20 s^{-1}, and G_3 = 10 s^{-1}. These give an average G value of 26.7 s^{-1}. The compartments are to be separated by slotted, redwood baffle fences, and the floor of the basin is level. The basin should be 15.0 m in width to adjoin the settling basin. The speed of the blades relative to the water is three-quarters of the peripheral blade speed. Determine:

1. The GT value.
2. The basin dimensions.
3. The paddle-wheel design.
4. The power to be imparted to the water in each compartment.
5. The rotational speed of each horizontal shaft in rpm.
6. The rotational speed range if 1 : 4 variable-speed drives are employed.
7. The peripheral speed of the outside paddle blades in m/s.

SOLUTION   The GT value is

$$GT = \frac{26.7}{s} \left| \frac{45\,min}{} \right| \frac{60\,s}{min}$$

$$= \boxed{72,100}$$

Since the GT value is between 50,000 and 100,000, the detention time of 45 min is satisfactory. Basin volume, V, is given by

$$V = \frac{25,000\,m^3}{24\,h} \left| \frac{h}{60\,min} \right| \frac{45\,min}{}$$

$$= 781\,m^3$$

Profile area = 781 m^3/15.0 m = 52.1 m^2.

Assume compartments are square in profile, and x is the compartment width and depth. Thus,

$$(3x)(x) = 52.1\,m^2 \qquad x^2 = 17.37\,m^2 \qquad x = 4.17\,m$$

$$3x = 3(4.17\,m) = 12.51\,m$$

$$\boxed{\text{Use width} = \text{depth} = 4.17\,m, \text{length} = 12.51\,m}$$

$$\text{Volume} = (4.17\,m)(12.51\,m)(15.0\,m) = 783\,m^3$$

Assume a paddle-wheel design as shown in Figure 8.18(a) with $D_1 = 3.35\,\text{m}$, $D_2 = 2.44\,\text{m}$, and $D_3 = 1.52\,\text{m}$. Use four paddle wheels per shaft, and assume the blades are 15 cm wide and 3.00 m long.

$$\text{Blade area per shaft} = (0.15\,\text{m})(3.00\,\text{m})(6)(4)$$
$$= 10.8\,\text{m}^2$$
$$\text{Percent of cross-sectional area} = \frac{10.8}{15}\left|\;\right.\frac{100}{4.17}$$
$$= 17.3\%$$

Since this is between 15 to 20%, make the trial design using the assumed paddle-wheel design. The power, P, is given by

$$G = \sqrt{\frac{P}{\mu V}} \quad \text{or} \quad P = \mu G^2 V$$

Absolute viscosity, μ, at 10°C is 0.00131 N-s/m²; thus the power for the first compartment is

$$P = \frac{0.00131\,\text{N-s}}{\text{m}^2}\left|\;\right.\frac{(50)^2}{\text{s}^2}\left|\;\right.\frac{783\,\text{m}^3}{3}$$
$$= \boxed{855\,\text{N-m/s} = 855\,\text{J/s} = 855\,\text{W}}$$

Power per wheel $= (855\,\text{N-m/s})1/4 = 214\,\text{N-m/s}$

$$P = C_D A \rho \frac{v^3}{2}$$

The length-width ratio is $3.0/0.15 = 20$; thus $C_D = 1.50$.
The blade velocity relative to water is

$$v = (\text{rps})\left(\frac{\pi D}{\text{rev}}\right)0.75$$

Thus,

$$v_1 = (\text{rps})(\pi)(3.35\,\text{m})(0.75) = 7.893\,(\text{rps})$$

In a like manner, $v_2 = 5.749\,(\text{rps})$ and $v_3 = 3.581\,(\text{rps})$.
The power per wheel is

$$P = C_D A_1 \rho \frac{v_1^3}{2} + C_D A_2 \rho \frac{v_2^3}{2} + C_D A_3 \rho \frac{v_3^3}{2}$$

Since $A_1 = A_2 = A_3$, and all C_D values are equal,

$$P = C_D A_1 \frac{\rho}{2}(v_1^3 + v_2^3 + v_3^3)$$

or

$$214\,\text{N-m/s} = (1.5)(0.15\,\text{m})(3.0\,\text{m})(2)(999.7\,\text{kg/m}^3)(1/2)[(7.893)^3(\text{rps})^3$$
$$+ (5.749)^3(\text{rps})^3 + (3.581)^3(\text{rps})^3]\text{m}^3/\text{s}^3$$

$$214\,\text{N-m/s} = (674.8\,\text{kg-m}^2/\text{s}^3)(1\,\text{N-s/kg-m})(491.7 + 190.0 + 45.9)(\text{rps})^3$$

From this,

$$\text{rps} = 0.076 \text{ and } \boxed{\text{rpm} = (0.076)60 = 4.55}$$

Peripheral speed $= (\pi D)\text{rps}$; thus, for the outside blade,

$$v_1' = (\pi)(3.35\,\text{m})(0.076) = \boxed{0.80\,\text{m/s}}$$

Since $v_1' < 0.91\,\text{m/s}$, the design of the paddle wheels is satisfactory. The maximum rotational speed is 4.55 rpm; therefore, the minimum rotational speed $= (1/4)4.55 = 1.14\,\text{rpm}$. Thus,

$$\boxed{\text{Rotational speed for 1:4 drive} = 1.14 \text{ to } 4.55\,\text{rpm}}$$

In a like manner, the power, rotational speed, peripheral blade speed, and rotational speed range are computed for the second and third compartments. Following is a summary of the values.

First compartment:

$$\boxed{\begin{array}{l} P = 855\,\text{N-m/s} = 855\,\text{J/s} = 855\,\text{W} \\ \text{rpm} = 4.55; \text{ range} = 1.14 \text{ to } 4.55\,\text{rpm} \\ \text{m/s} = 0.80 \end{array}}$$

Second compartment:

$$\boxed{\begin{array}{l} P = 137\,\text{N-m/s} = 137\,\text{J/s} = 137\,\text{W} \\ \text{rpm} = 2.46; \text{ range} = 0.62 \text{ to } 2.46\,\text{rpm} \\ \text{m/s} = 0.43 \end{array}}$$

Third compartment:

$$\boxed{\begin{array}{l} P = 34.2\,\text{N-m/s} = 34.2\,\text{J/s} = 34.2\,\text{W} \\ \text{rpm} = 1.56; \text{ range} = 0.39 \text{ to } 1.56\,\text{rpm} \\ \text{m/s} = 0.27 \end{array}}$$

Solids-Contact Units

Solids-contact units, shown in Figures 8.22(a) and (b), 8.23, and 8.24, are frequently called upflow clarifiers. These units combine mixing, flocculation, and sedimentation into a single structural unit. These are designed to maintain a large volume of flocculated solids within the system, which enhances the flocculation of incoming solids since there are more interparticulate collisions. The volume of the solids in the contact zone may vary from 5 to 50% of the zone volume, depending on the particular use. The volume of

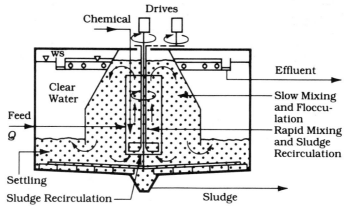

(a) Slurry-Recirculation Type

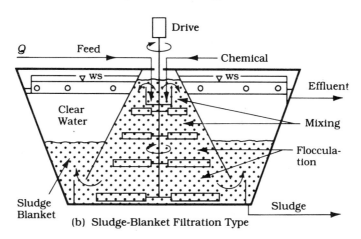

FIGURE 8.22
Solids-Contact Units

(b) Sludge-Blanket Filtration Type

the solids is taken as the percent volume after 30 min settling in a test cylinder. Solids-contact units are of two basic designs: (1) the slurry-recirculation type and (2) the sludge-blanket filtration type.

In the slurry-recirculation type, as shown in Figure 8.22(a), the large floc mass is maintained by recycling an appreciable portion of the floc through a center compartment by means of a pitched blade impeller. The mixing and flocculation are accomplished in the compartment surrounding the recycle center well. The flocculated water passes under the skirt upward into the clarification zone; however, it passes only through the top of the sludge blanket. After passing through the clarification section, the water leaves by an effluent launderer.

In the sludge-blanket filtration type, as shown in Figure 8.22(b), mixing and flocculation are achieved in a center compartment. The necessary agitation and gentle stirring are provided by pitched blade impellers. The flocculated water, on leaving the flocculation section, passes upward through the sludge blanket to obtain floc removal by contact with flocculated solids

FIGURE 8.23 *Solids-Contact Unit at a Water Treatment Plant. Unit is empty so that painting and maintenance can be done.*

in the blanket. The water continues to flow upward through the clarification section into the effluent launder.

The main advantage of the contact-solids units over conventional mixing, flocculation, and clarification units is their reduced size. Consequently, the units are more compact and occupy less land space. The units are best suited to treat a feed water that has a relatively constant quality with respect to time. Because of the short mixing, flocculation, and clarification detention times, it is difficult to treat a feed water that has a wide variation in quality. They are frequently used to coagulate lake waters and soften well waters. The basic designs shown in Figure 8.22(a) and (b) are available from numerous manufacturers.

LIME-SODA SOFTENING

Hardness may be defined as the ability of a water to consume excessive amounts of soap prior to forming a lather and to produce scale in hot water heaters, boilers, or other units in which the temperature of the water is significantly increased. It is due to the presence of polyvalent metallic ions, principally calcium and magnesium. Calcium and magnesium ions will react with soap to form insoluble organic salts that are present as scum on the

FIGURE 8.24 *Solids-Contact Unit in Operation. Some floc is visible in the water.*

water surface. Once all the calcium and magnesium ions have been precipitated, a lather can be formed. Calcium and magnesium may be removed by the lime-soda softening process, in which the unwanted ions are precipitated by adding slaked lime and soda ash. These reagents produce a voluminous precipitate of calcium carbonate and magnesium hydroxide, which acts by sweep coagulation, thus enmeshing the suspended particles as the precipitate is formed. Occasionally a coagulant, such as ferrous sulfate, is added along with the slaked lime and soda ash to aid in the coagulation and flocculation process. The total hardness (TH) is the sum of the calcium and magnesium ion concentrations. It is usually expressed as meq/ℓ or mg/ℓ of equivalent $CaCO_3$. Alkalinity is due to the bicarbonate, carbonate, and hydroxyl ion concentrations. For natural waters with a pH less than 9, the alkalinity (Alk) is the bicarbonate and carbonate alkalinities, since the hydroxyl ion concentration is negligible below this pH. Alkalinity is also expressed as meq/ℓ or mg/ℓ equivalent $CaCO_3$. The carbonate hardness (CH) is that part of the total hardness that is chemically equivalent to the bicarbonate and carbonate alkalinities. The noncarbonate hardness (NCH) is equal to the total hardness minus the carbonate hardness. The amount of lime and soda ash required in softening depends on the concentration of the total hardness (TH), the carbonate hardness (CH), the noncarbonate hardness (NCH), the magnesium ion, and the carbon dioxide.

If the total hardness (TH) is greater than the alkalinity (Alk) when measured as meq/ℓ or mg/ℓ $CaCO_3$, then

$$\text{Carbonate hardness (CH)} = \text{alkalinity (Alk)}$$

and

$$\text{Noncarbonate hardness (NCH)} = \text{total hardness (TH)} \\ - \text{carbonate hardness (CH)}$$

If the total hardness (TH) is less than the alkalinity (Alk), then

$$\text{Carbonate hardness (CH)} = \text{total hardness (TH)}$$

and

$$\text{Noncarbonate hardness (NCH)} = 0$$

For the pH ranges that occur in natural waters, the alkalinity is usually in the form of the bicarbonate ion, HCO_3^{-1}. In fact, at pH 7.5 or below, the bicarbonate ion is essentially all the alkalinity.

As slaked lime is added to a water, it will react with any carbon dioxide present as follows:

$$Ca(OH)_2 + CO_2 \rightarrow \underline{CaCO_3} \downarrow + H_2O \qquad \textbf{(8.17)}$$

The lime reacts with carbonate hardness as shown by

$$Ca(OH)_2 + Ca(HCO_3)_2 \rightarrow \underline{2CaCO_3} \downarrow + 2H_2O \qquad \textbf{(8.18)}$$

and

$$Ca(OH)_2 + Mg(HCO_3)_2 \rightarrow MgCO_3 + \underline{CaCO_3} \downarrow + 2H_2O \qquad \textbf{(8.19)}$$

The product magnesium carbonate in Eq. (8.19) is soluble, but more lime will remove it by the reaction

$$Ca(OH)_2 + MgCO_3 \rightarrow \underline{CaCO_3} \downarrow + \underline{Mg(OH)_2} \downarrow \qquad \textbf{(8.20)}$$

Also, magnesium noncarbonate hardness, shown as magnesium sulfate, is removed, as additional lime is added, by the reaction

$$Ca(OH)_2 + MgSO_4 \rightarrow CaSO_4 + \underline{Mg(OH)_2} \downarrow \qquad \textbf{(8.21)}$$

Although in Eq. (8.21) the magnesium is precipitated, an equivalent amount of calcium that has been added remains in solution. The water will now contain only the original calcium noncarbonate hardness and the calcium noncarbonate hardness produced from Eq. (8.21), which equals the magnesium noncarbonate hardness. The removal of this calcium may be accomplished by soda ash by the reaction

$$Na_2CO_3 + CaSO_4 \rightarrow Na_2SO_4 + \underline{CaCO_3} \downarrow \qquad \textbf{(8.22)}$$

To precipitate calcium carbonate requires a pH of about 9.5; and to precipitate magnesium hydroxide requires a pH of about 10.8, which necessitates adding an excess of about $1.25 \, meq/\ell$ of lime to raise the pH.

From the previous discussion, the meq/ℓ of lime required is equal to the sum of the carbon dioxide, carbonate hardness, and magnesium ion concentrations when expressed in meq/ℓ, plus $1.25 \, meq/\ell$ excess to raise the pH. The meq/ℓ of soda ash required is equal to the noncarbonate hardness expressed as meq/ℓ.

After softening of the water, it will contain the excess lime and the magnesium hydroxide and calcium carbonate that did not precipitate or settle. The excess lime and magnesium hydroxide are stabilized by carbon dioxide, which lowers the pH to about 9.5, resulting in

$$CO_2 + Ca(OH)_2 \rightarrow \underline{CaCO_3} \downarrow + H_2O \qquad (8.23)$$

and

$$CO_2 + Mg(OH)_2 \rightarrow MgCO_3 + H_2O \qquad (8.24)$$

Further stabilization to about pH 8.5 will stabilize the calcium carbonate by the reaction

$$CO_2 + CaCO_3 + H_2O \rightarrow Ca(HCO_3)_2 \qquad (8.25)$$

It is not possible to remove all of the hardness from a water because of the slight solubility of calcium carbonate and magnesium hydroxide and, also, the presence of some of the precipitate as a very fine, nonsettling floc. Once stabilization has been accomplished, the fine precipitate dissolves back into solution. In actual practice, the lime-soda process will soften a water to about 50 to $80 \, mg/\ell$ residual hardness as calcium carbonate.

EXAMPLES 8.3 AND 8.3 SI

Lime-Soda Softening

A water that is to be softened by the lime-soda process has the following carbon dioxide, calcium, magnesium, and bicarbonate concentrations:

CO_2	$8 \, mg/\ell$
Ca^{+2}	$65 \, mg/\ell$
Mg^{+2}	$32 \, mg/\ell$
HCO_3^-	$260 \, mg/\ell$

Determine the calcium oxide and soda ash required per million gallons (million liters) if the purities are 85 and 95%, respectively. Use meq/ℓ as the basis of computations for both the alkalinity and the hardness, instead of mg/ℓ as $CaCO_3$.

SOLUTION

The weights of 1 meq of the ions are $Ca^{+2} = 40/2 = 20 \, mg$, $Mg^{+2} = 24/2 = 12 \, mg$, and $HCO_3^- = 61/1 = 61 \, mg$. One meq of carbon dioxide weighs $22 \, mg$. Thus the total hardness (TH) is $65/20 + 32/12$ or $3.25 + 2.67 =$

5.92 meq/ℓ. The alkalinity (Alk) is 260/61 = 4.26 meq/ℓ. Since the total hardness (TH) is greater than the alkalinity (Alk), the carbonate hardness (CH) is 4.26 meq/ℓ and the noncarbonate hardness (NCH) is 5.92 − 4.26 = 1.66 meq/ℓ. The magnesium ion content is 2.67 meq/ℓ, and the carbon dioxide is 8/22 or 0.36 meq/ℓ. Thus the lime required in meq/ℓ for the various reactants, and the excess to raise the pH, are as follows:

CO_2	0.36 meq/ℓ
CH	4.26 meq/ℓ
Mg^{+2}	2.67 meq/ℓ
Excess	1.25 meq/ℓ
Total =	8.54 meq/ℓ

The soda ash required for the NCH is 1.66 meq/ℓ. One meq of CaO is (40 + 16)1/2 = 28 mg, and one meq of Na_2CO_3 is [2(23) + 12 + 3(16)]1/2 = 53 mg. The lime required is (8.54)28 = 239.0 mg/ℓ, and the soda ash is (1.66)53 = 88.0 mg/ℓ. Thus the lime per million gallons is

$$CaO = \frac{239}{10^6} \left| \frac{10^6 \, gal}{MG} \right| \frac{8.34 \, lb}{gal} \left| \frac{}{0.85} \right. = \boxed{2345 \, lb/MG}$$

The soda ash per million gallons is

$$Na_2CO_3 = \frac{88}{10^6} \left| \frac{10^6 \, gal}{MG} \right| \frac{8.34 \, lb}{gal} \left| \frac{}{0.95} \right. = \boxed{773 \, lb/MG}$$

The lime per million liters is

$$CaO = \frac{239 \, mg}{\ell} \left| \frac{gm}{1000 \, mg} \right| \frac{kg}{1000 \, gm} \left| \frac{10^6 \, \ell}{ML} \right. = \boxed{239 \, kg/ML}$$

The soda ash per million liters is

$$Na_2CO_3 = \frac{88 \, mg}{\ell} \left| \frac{gm}{1000 \, mg} \right| \frac{kg}{1000 \, gm} \left| \frac{10^6 \, \ell}{ML} \right. = \boxed{88 \, kg/ML}$$

COAGULATION AND FLOCCULATION IN WATER TREATMENT

All of the previously discussed coagulants have been used in water treatment, with aluminum sulfate and ferrous sulfate and lime being the most common. The coagulant dosage depends on the turbidity and the relative ease by which coagulation occurs, and the usual chemical dosages vary from 5 to 90 mg/ℓ. In addition to the previously discussed coagulant aids, activated silica and turbidity addition by means of clay have been used. Polyelectrolyte coagulant aids, if required, are usually added after rapid mixing.

In the softening of water by the lime-soda process, the precipitates formed, calcium carbonate and magnesium hydroxide, enmesh particulates and remove considerable turbidity. Lime and soda ash used in this manner act as coagulants and usually yield a large, dense, fast-settling floc. Frequently, optimum floc settling characteristics are produced by using a coagulant, such as ferrous sulfate, in conjunction with these chemicals.

In water coagulation and water softening, the rapid-mixing basins usually have detention times from 20 to 60 sec and G values from 1000 to 700 sec^{-1}, as described in the previous section on rapid mixing. Flash mixers, such as impellers inside pipe lines, are sometimes used, and employ high G values as previously mentioned and detention times from 1 to 5 sec. For flash mixers, the volume is considered as the flowrate times the detention time. The flocculators used in water coagulation and water softening are usually of the paddle-wheel type employing horizontal shafts and a cross-flow pattern. One source (Camp, 1955) found flocculation times to be usually from 20 to 60 min, G values to be from 35 to 70 sec^{-1}, and GT values to be from 48,000 to 210,000. A subsequent source (ASCE and AWWA, 1990) reports that GT values can be as low as 30,000. Another source (Montgomery, 1985) reports flocculation times from 15 to 45 min and G values from 10 to 75 sec^{-1}. For coagulation of river waters, the minimum flocculation time is considered to be 20 min and G values are from 10 to 50 sec^{-1}. For coagulation of reservoir waters, the minimum flocculation time is considered to be 30 min and G values are from 10 to 75 sec^{-1}. Since river waters are generally more turbid than reservoir waters, they flocculate more easily, and shorter detention times and lower G values are used. For lime-soda softening, the minimum detention flocculation time is considered to be 30 min and G values are from 10 to 50 sec^{-1}. If iron salts are used for coagulation, the G value should not exceed 50 sec^{-1} since these salts form a large and dense floc. If cationic polymers are used as coagulants, the G values should be 50% greater than those previously mentioned. The number of compartments in a flocculation basin is usually three. Almost all coagulation and lime-soda softening plants now use tapered flocculation, and the range in G values previously given is the range that occurs across the flocculator. The second compartment usually has a G value about 40% of that of the first compartment. This means that if the range in G values is 10 to 50 sec^{-1} and tapered flocculation is used, then the G values for the first, second, and third compartments are 50, 20 (50 $\times$ 0.40), and 10 sec^{-1}. The average G value is $(50 + 20 + 10)1/3 = 26.7$ sec^{-1} if the compartments are of equal size. If the compartments differ in size, then the average G value is $(50V_1 + 20V_2 + 10V_3)/(V_1 + V_2 + V_3)$, where the various V values are the volumes of the respective compartments.

Direct filtration, which is sometimes used, is rapid sand filtration without sedimentation, and in this process, the flowsheet includes mixing, flocculation, and rapid sand filtration. The filter removes all of the floc formed; consequently, this process is limited to waters that are extremely clear (less than 10 turbidity units) throughout the year. For criteria for flocculation basins for direct filtration, the reader is referred to the books by J. M. Montgomery (1985) and the ASCE and AWWA (1990).

The solids-contact units have been used successfully for feed waters with a relatively constant quality with respect to time; thus, they have been successfully used in the coagulation and the softening of lake waters and the softening of well waters. For coagulation, the solids concentration in the contact zone is usually 5% by volume; the G values vary from 75 to $175\,sec^{-1}$ and the GT values from 125,000 to 150,000. For water softening, using a 10% solids concentration by volume results in G values from 130 to $200\,sec^{-1}$ and GT values from 200,000 to 250,000. For water softening, using a much higher solids concentration, such as 20 to 40% by volume, results in G values from 250 to $400\,sec^{-1}$ and GT values from 300,000 to 400,000 (AWWA, 1969).

COAGULATION AND FLOCCULATION IN WASTEWATER TREATMENT

All of the previously discussed coagulants have been used in coagulating municipal wastewaters, and some have been employed in coagulating industrial wastewaters with sizable suspended solids concentrations. Municipal wastewaters, because of their relatively high turbidities and suspended solids, usually coagulate more readily than surface waters, although the chemical dosages required are much higher. For untreated or raw municipal wastewaters, coagulant dosages in the magnitude of $300\,mg/\ell$ or more may be required.

Aluminum sulfate, in addition to coagulating colloidal and suspended solids, removes an appreciable amount of the phosphorus from the wastewater. The chemistry of phosphorus removal by alum is not completely understood; however, polyphosphates and organic phosphorus are probably removed by complex reactions and by enmeshment in and sorption by the floc. For purposes of simplicity, it is assumed that the phosphorus remaining after the initial coagulation is orthophosphate, and the removal is represented by the simplified reaction (EPA, 1973a)

$$Al_2(SO_4)_3 + 2PO_4^{-3} \rightarrow \underline{2AlPO_4} \downarrow + 3SO_4^{-2} \qquad (8.26)$$

The precipitate formed is considered to be enmeshed in the floc. The removal is dependent on the pH, and optimum removal occurs between pH 5.5 to 6.5.

Iron salts, in addition to coagulating colloidal matter and suspended solids, also remove a substantial amount of the phosphorus from municipal wastewaters. The polyphosphates and organic phosphates are removed in a manner similar to the removal by alum — that is, complex reactions and enmeshment in or sorption by the floc. The ferric ion will react with orthophosphate to produce a precipitate, ferric phosphate, $FePO_4$, which will be removed by the floc. The ferrous ion also will remove orthophosphate as a precipitate; however, the chemistry is much more involved than removal by the ferric ion. The removal of phosphorus by iron salts is dependent on the pH, and optimum removal is between pH 4.5 to 8.

Lime is frequently used as a coagulant for municipal wastewaters or effluents, and in addition to removing colloidal and suspended solids, it results in excellent phosphate removal. The precipitation reactions for lime are identical to the softening reactions by lime, which are

$$Ca(OH)_2 + Ca(HCO_3)_2 \rightarrow \underline{2CaCO_3} \downarrow + 2H_2O \qquad (8.27)$$

$$2Ca(OH)_2 + Mg(HCO_3)_2 \rightarrow \underline{2CaCO_3} \downarrow + \underline{Mg(OH)_2} \downarrow + 2H_2O \qquad (8.28)$$

To precipitate calcium carbonate requires a pH of 9.5 or more, and to precipitate magnesium hydroxide requires a pH greater than 10.8. A high pH is beneficial since the amount of phosphate ion removal increases as the pH increases. The simplified reaction for phosphorus precipitation by lime is (EPA, 1973a)

$$5Ca^{+2} + 4OH^{-1} + 3HPO_4^{-2} \rightarrow \underline{Ca_5(OH)(PO_4)_3} \downarrow + 3H_2O \qquad (8.29)$$

The solubility of the precipitate formed, calcium hydroxyapatite, decreases as the pH increases; at pH 9.0 or higher the maximum removal occurs. Another benefit of the high pH involved in phosphate precipitation is that subsequent treatment by air stripping will remove ammonia, with maximum removal occurring at pH 10.8 or greater. In lime coagulation, the calcium carbonate precipitate formed is granular, and the removal of colloidal and suspended solids is due to enmeshment in the precipitate. Conversely, magnesium hydroxide is a gelatinous precipitate that probably removes colloidal and suspended solids by enmeshment in the precipitate and sorption by the precipitate. After lime coagulation, the wastewater must be stabilized to lower the pH and to precipitate the excess lime. If stabilization is not accomplished, excessive encrustations of calcium carbonate will form on the media used in subsequent treatment by multimedia filtration. Usually stabilization is achieved by using carbon dioxide, and the stabilized wastewater will have a pH of 7.0 to 8.5. If lime coagulation precedes biological treatment, the pH must be lowered to 9.5 or lower to prevent inhibition of the biological processes.

The coagulant aids most commonly used in wastewater coagulation are polyelectrolytes, turbidity addition by means of recycled chemically precipitated sludge, and lime addition. Polyelectrolyte coagulant aids, when required, are usually added after rapid mixing. Since the aluminum and iron salts used in coagulation are acidic, relatively large coagulant dosages may require lime addition to avoid an unwanted pH drop. The coagulant solution and the milk of lime slurry must be added to the mixing basins by separate feed lines.

In the alum and iron salt coagulation of municipal wastewaters and effluents, the rapid-mixing basins usually have detention times from 1 to 2 min to disperse the chemicals adequately and initiate the coagulation process (EPA, 1973b). Although municipal wastewaters and effluents usually coagulate more readily than surface waters, longer rapid-mixing times are usually required because of the high suspended solids concentrations and large coagulant dosages. The velocity gradients for rapid mixing are usually about $300 \, sec^{-1}$ (EPA, 1975) and, because of the nature of the organic solids, are generally lower than those encountered in water treatment. Overmixing may rupture the existing wastewater solids into smaller particles, which require larger coagulant dosages and longer flocculation detention times.

Since flocculation of wastewaters and effluents usually occurs with ease, the detention times and GT values required for flocculation are generally less than those used in water treatment. For alum and iron salt coagulation, the flocculation time is usually from 15 to 30 min, typical G values are from 20 to 75 sec^{-1}, and GT values range from 10,000 to 100,000 (Metcalf & Eddy, Inc., 1979). For mechanical flocculation, the detention time is usually about 15 min (EPA, 1975). For lime coagulation, the detention times for rapid mixing are usually from 1 to 2 min (EPA, 1975). The precipitates from lime coagulation, $CaCO_3$ and $Mg(OH)_2$, benefit very little from long flocculation times, and a detention time of 5 to 10 min may be adequate. The precipitates formed have high strengths, and the G values for flocculation are usually 100 sec^{-1} or more (EPA, 1975). For the case of relatively large plants, paddle-wheel flocculation with horizontal shafts is generally used, and the peripheral speed of the paddles is usually less than 2.0 fps. If paddle-wheel flocculation is used, it is usually preferable to use tapered flocculation. For smaller plants, pneumatic mixing and flocculation have been successfully used, particularly for lime coagulation.

Solids-contact units of the slurry-recirculation type have been used successfully in some plants for coagulating municipal wastewaters and effluents. The slurry-recirculation type units are not as sensitive to varying flowrates and varying loading as the sludge-blanket filtration type. Since solids-contact units require more skill in operation than conventional mixing, flocculation, and settling, this should be considered in process selection.

In the coagulation of industrial wastewaters, a rapid-mixing time of 0.5 to 6 min and a flocculation time of 20 to 30 min have been reported (Eckenfelder, 1966, 1970, 1989). Coagulant aids, such as polyelectrolytes, are usually added after rapid mixing. It should be understood that, because of the varying nature of industrial wastes, laboratory and frequently pilot plant studies are required to determine the most effective coagulant, the coagulant aid (if necessary), the chemical dosages, the optimum rapid-mixing and flocculation times, and the G and GT values.

Chemical coagulation and precipitation have been successfully used on industrial wastewaters to remove certain heavy metal ions. The heavy metal ions of arsenic, barium, cadmium, chromium, copper, lead, mercury, nickel, silver, and zinc have been removed by chemical coagulation and precipitation (Eckenfelder, 1989, 1992). Eckenfelder also reports the successful removal of fluoride, iron, manganese, and selenium by chemical coagulation and precipitation.

REFERENCES

American Society of Civil Engineers (ASCE) and American Water Works Association (AWWA). 1990. *Water Treatment Plant Design*. 2nd ed. New York: ASCE.

American Water Works Association (AWWA). 1991. *Mixing in Coagulation and Flocculation*. Denver, Colo.: AWWA.

American Water Works Association (AWWA). 1969.

Water Treatment Plant Design. New York: AWWA.

American Water Works Association (AWWA). 1971. *Water Quality and Treatment*. New York: McGraw-Hill.

American Water Works Association (AWWA). 1990. *Water Quality and Treatment*. 4th ed. New York: McGraw-Hill.

Barnes, D.; Bliss, P. J.; Gould, B. W.; and Vallentine, H. R. 1981. *Water and Wastewater Engineering Systems*. London: Pitman Books Limited.

Black, A. P.; Buswell, A. M.; Eidsness, F. A.; and Black, A. L. 1957. Review of the Jar Test. *Jour. AWWA* 39, no. 11:1414.

Black, A. P., and Harris, R. J. 1969. New Dimensions for the Old Jar Test. *Water & Wastes Engineering*, Dec.:49.

Camp, T. R. 1955. Flocculation and Flocculation Basins. *Trans. ASCE* 120:1.

Camp, T. R. 1968. Floc Volume Concentration. *Jour. AWWA* 60, no. 6:656.

Camp, T. R. 1952. "Water Treatment" in *The Handbook of Applied Hydraulics*, ed. C. V. Davis. New York: McGraw-Hill.

Cohen, J. M. 1957. Improved Jar Test Procedure. *Jour. AWWA* 49, no. 11:1425.

Conway, R. A., and Ross, R. D. 1974. *Handbook of Industrial Waste Disposal*. New York: Van Nostrand Reinhold.

Culp, G. L., and Culp, R. L. 1974. *New Concepts in Water Purification*. New York: Van Nostrand Reinhold.

Culp, R. L., and Culp, G. L. 1978. *Handbook of Advanced Wastewater Treatment*. 2nd ed. New York: Van Nostrand Reinhold.

Degremont, Inc. 1973. *Water Treatment Handbook*. 5th ed. Rueil-Malmaison, France.

Eckenfelder, W. W., Jr. 1966. *Industrial Water Pollution Control*. New York: McGraw-Hill.

Eckenfelder, W. W., Jr. 1989. *Industrial Water Pollution Control*. 2nd ed. New York: McGraw-Hill.

Eckenfelder, W. W., Jr. 1980. *Principles of Water Quality Management*. Boston: CBI Publishing.

Eckenfelder, W. W., Jr.; Patoczka, J.; and Watkin, A. T. 1992. "Wastewater Treatment" in *Environmental Engineering in the Process Plant*, ed. N. P. Chopey. *Chemical Engineering*. New York: McGraw-Hill.

Eckenfelder, W. W., Jr. 1970. *Water Quality Engineering for Practicing Engineers*. New York: Barnes & Noble.

Environmental Protection Agency (EPA). 1973a. *Phosphorus Removal*. EPA Process Design Manual, Washington, D.C.

Environmental Protection Agency (EPA). 1973b.

Physical-Chemical Wastewater Treatment Plant Design. EPA Technology Transfer Seminar Publication, Washington, D.C.

Environmental Protection Agency (EPA). 1975. *Suspended Solids Removal*. EPA Process Design Manual, Washington, D.C.

Environmental Protection Agency (EPA). 1974. *Upgrading Existing Wastewater Treatment Plants*. EPA Process Design Manual, Washington, D.C.

Great Lakes–Upper Mississippi Board of State Sanitary Engineers. 1978. *Recommended Standards for Sewage Works*. Ten state standards, Albany, N.Y.

Hudson, H. E., Jr., and Wolfner, J. P. 1967. Design of Mixing and Flocculation Basins. *Jour. AWWA* 59, no. 10:1257.

McCabe, W. L., and Smith, J. C. 1976. *Unit Operations of Chemical Engineering*. 3rd ed. New York: McGraw-Hill.

McCabe, W. L.; Smith, J. C.; and Harriott, P. 1993. *Unit Operations of Chemical Engineering*. 5th ed. New York: McGraw-Hill.

McGhee, T. J. 1991. *Water Supply and Sewage*. 6th ed. New York: McGraw-Hill.

Metcalf & Eddy, Inc. 1979. *Wastewater Engineering: Treatment, Disposal and Reuse*. 2nd ed. New York: McGraw-Hill.

Metcalf & Eddy, Inc. 1991. *Wastewater Engineering: Treatment, Disposal and Reuse*. 3rd ed. New York: McGraw-Hill.

Montgomery, J. M., Inc. 1985. *Water Treatment: Principles and Design*. New York: Wiley.

Oldshue, J. Y. 1983. *Fluid Mixing Technology*. *Chemical Engineering*. New York: McGraw-Hill.

O'Melia, C. R. 1970. "Coagulation in Water and Wastewater Treatment" in *Advances in Water Quality Improvement by Physical and Chemical Processes*, ed. E. F. Gloyna and W. W. Eckenfelder, Jr. Austin, Tex.: University of Texas Press.

Priesing, C. P. 1962. A Theory of Coagulation Useful for Design. *Ind. and Eng. Chem.* 54, no. 8:38, and 54, no. 9:54.

Rich, L. G. 1963. *Unit Processes of Sanitary Engineering*. New York: Wiley.

Rouse, H. 1950. "Fundamental Principles of Flow" in *Engineering Hydraulics*, ed. H. Rouse. New York: Wiley.

Rushton, J. H. 1952. Mixing of Liquids in Chemical Processing. *Ind. and Eng. Chem.* 44, no. 12:2931.

Rushton, J. H.; Bissell, E. S.; Hesse, H. C.; and Everett, H. J. 1947. Designing and Utilization of Internal Fittings for Mixing Vessels. *Chem. Engr. Progr.* 43, no. 12:649.

Rushton, J. H.; Costich, E. W.; and Everett, H. J.

1950. Power Characteristics of Mixing Impellers, Parts I and II. *Chem. Eng. Progr.*, 46, no. 8:395; and 46, no. 9:467.

Rushton, J. H., and Mahoney, L. H. 1954. *Mixing Power and Pumpage Capacity*. Annual Meeting of AIME, 15 February 1954, New York.

Rushton, J. H., and Oldshue, J. Y. 1953. Mixing — Present Theory and Practice, Parts I and II. *Chem. Eng. Progr.* 49, no. 4:161; and 49, no. 5:267.

Salvato, J. A. 1992. *Environmental Engineering and Sanitation*. 4th ed. New York: Wiley.

Sanks, R. L. 1978. *Water Treatment Plant Design*. Ann Arbor, Mich.: Ann Arbor Science Publishers.

Sawyer, C. N., and McCarty, P. L. 1978. *Chemistry for Environmental Engineers*. 3rd ed. New York: McGraw-Hill.

South Lake Tahoe Public Utility District. 1971. *Advanced Wastewater Treatment as Practiced at South Tahoe*. Tech. Report for the EPA, Projet 17010 ELQ (QPRD 52-01-67), August 1971.

Steel, E. W., and McGhee, T. J. 1979. *Water Supply and Sewerage*. 4th ed. New York: McGraw-Hill.

Stumm, W., and Morgan, J. J. 1962. Chemical Aspects of Coagulation. *Jour. AWWA* 54, no. 8:971.

Stumm, W., and O'Melia, C. R. 1968. Stoichiometry of Coagulation. *Jour. AWWA* 60, no. 5:514.

Sundstrom, D. W., and Klei, H. E. 1979. *Wastewater Treatment*. Englewood Cliffs, N.J.: Prentice-Hall.

Tebbutt, T. H. Y. 1992. *Principles of Water Quality Control*. 4th ed. Oxford, England: Pergamon Press.

Water Environment Federation (WEF). 1991. *Design of Municipal Wastewater Treatment Plants*. Vols. I and II. WEF Manual of Practice No. 8, Washington, D.C.

Water Pollution Control Federation (WPCF). 1977. WPCF Manual of Practice No. 8, Washington, D.C.

Weber, W. J., Jr. 1972. *Physiochemical Processes for Water Quality Control*. New York: Wiley-Interscience.

PROBLEMS

8.1 A rapid-mixing basin is to be designed for a water coagulation plant, and the design flow for the basin is 4.0 MGD. The basin is to be square with a depth equal to 1.25 times the width. The velocity gradient is to be $900 \, sec^{-1}$ (at 50°F), and the detention time is 30 sec. Determine:

a. The basin dimensions if increments of 1 in. are used.

b. The input horsepower required.

c. The impeller speed if a vane-disc impeller with six flat blades is employed and the tank is baffled. The impeller diameter is to be 50% of the basin width.

8.2 A rapid-mixing basin is to be designed for a water coagulation plant, and the design flow for the basin is $15,140 \, m^3/d$. The basin is to be square with a depth equal to 1.25 times the width. The velocity gradient is to be $900 \, s^{-1}$ (at 10°C), and the detention time is 30 s. Determine:

a. The basin dimensions.

b. The input power required.

c. The impeller speed if a vane-disc impeller with six flat blades is employed and the tank is baffled. The impeller is to be 50% of the basin width.

8.3 A flocculation basin is to be designed for a water coagulation plant, and the design flow for the basin is 13.0 MGD. The basin is to be a cross-flow horizontal-shaft, paddle-wheel type with a mean velocity gradient of $26.7 \, sec^{-1}$ (at 50°F), a detention time of 45 min, and a GT value from 50,000 to 100,000. Tapered flocculation is to be provided, and three compartments of equal depth in series are to be used, as shown in Figure 8.19(b). The compartments are to be separated by slotted, redwood baffle fences, and the basin floor is level. The G values are to be 50, 20, and $10 \, sec^{-1}$. The flocculation basin is to have a width of 90 ft-0 in. to adjoin the settling basin. The paddle wheels are to have blades with a 6-in. width and a length of 10 ft. The outside blades should clear the floor by 1.0 ft and be 1.0 ft below the water surface. There are to be six blades per paddle wheel, and the blades should have a clear spacing of 12 in. Adjacent paddle wheels should have a clear spacing of 30 to 36 in. between blades. The wall clearance is 12 to 18 in. Determine:

a. The basin dimensions if increments of 1 in. are used.

b. The paddle-wheel design.

c. The power to be imparted to the water in each

compartment and the total power required for the basin.

d. The range in rotational speed for each compartment if 1:4 variable-speed drives are employed.

8.4 A flocculation basin is to be designed for a water coagulation plant, and the design flow is 49,200 m³/d. The basin is to be a cross-flow horizontal-shaft, paddle-wheel type with a mean velocity gradient of $26.7 s^{-1}$ (at 10°C), a detention time of 45 min, and a GT value from 50,000 to 100,000. Tapered flocculation is to be provided, and three compartments of equal depth in series are to be used, as shown in Figure 8.19(b). The compartments are to be separated by slotted, redwood baffle fences, and the basin floor is level. The G values are to be 50, 20, and $10 s^{-1}$. The flocculation basin is to have a width of 27.43 m to adjoin the settling basin. The paddle wheels are to have blades with a 150-mm width and a length of 3.05 m. The outside blades should clear the floor by 305 mm and be 305 mm below the water surface. There are to be six blades per paddle wheel, and the blades should have a clear spacing of 305 mm. Adjacent paddle wheels should have a clear spacing of 760 to 915 mm between blades. The wall clearance is 305 to 460 mm. Determine:

a. The basin dimensions.

b. The paddle-wheel design.

c. The power to be imparted to the water in each compartment and the total power required for the basin.

d. The range in rotational speed for each compartment if 1:4 variable-speed drives are employed.

8.5 A pneumatic flocculation basin is to be designed for a tertiary treatment plant having a flow of 5.0 MGD. The plant is to employ high-pH lime coagulation, and pertinent data for the flocculation basin are as follows: detention time = 5 min, G = $150 sec^{-1}$ (at 50°F), length = 2 times width, depth = 9 ft-10 in., diffuser depth = 9 ft-0 in., and air flow = 4 cfm per diffuser. Determine:

a. The basin dimensions if 1-in. increments are used.

b. The total air flow in ft³/min.

c. The number of diffusers.

8.6 A pneumatic flocculation basin is to be designed for a tertiary treatment plant having a flow of 19,000 m³/d. The plant is to employ high-pH lime coagulation, and pertinent data for the flocculation

basin are as follows: detention time = 5 min, G = $150 s^{-1}$ (at 10°C), length = 2 times width, depth = 3.0 m, diffuser depth = 2.75 m, and air flow = 6.80 m³/h per diffuser. Determine:

a. The basin dimensions.

b. The total air flow in m³/h.

c. The number of diffusers.

8.7 An impeller-powered flocculation basin is to be designed for a tertiary treatment plant having a flow of 25 MGD. The plant is to employ alum coagulation, and pertinent data for the flocculation basin are as follows: detention time = 20 min, G = $35 sec^{-1}$ (at 50°F), GT = 10,000 to 100,000, width = 1.25 times depth, length = twice width, no baffling, number of impellers = 2, number of blades per impeller = 6 pitched at 45°, impeller diameter = 30% of basin width, K_L = 70.0, and K_T = 1.65. Determine:

a. Basin dimensions if 1-in. increments are used.

b. Impeller diameter.

c. Speed of impellers in rpm.

8.8 An impeller-powered flocculation basin is to be designed for a tertiary treatment plant for a flow of 94,600 m³/d. The plant is to employ alum coagulation, and pertinent data for the flocculation basin are as follows: detention time = 20 min, G = $35 s^{-1}$ (at 10°C), GT = 10,000 to 100,000, width = 1.25 times depth, length = twice width, no baffling, number of impellers = 2, number of blades per impeller = 6 pitched at 45°, impeller diameter = 30% of basin width, K_L = 70.0, and K_T = 1.65. Determine:

a. Basin dimensions.

b. Impeller diameter.

c. Speed of impellers in rpm.

8.9 A lime-soda softening plant treats a flow of 150 MGD, and the water has 86 mg/ℓ Ca^{+2}, 35 mg/ℓ Mg^{+2}, 299 mg/ℓ HCO_3^-, and 6 mg/ℓ CO_2. The commerical grade of quicklime has a purity of 85%, and the soda ash has a purity of 95%. Determine:

a. The pounds of quicklime and soda ash required per million gallons.

b. The tons of quicklime and soda ash required per month if a month is considered to be 30 days.

8.10 A lime-soda softening plant treats a flow of 568,000 m³/d, and the water has 86 mg/ℓ Ca^{+2}, 35 mg/ℓ Mg^{+2}, 299 mg/ℓ HCO_3^-, and 6 mg/ℓ CO_2. The commerical grade of quicklime has a purity of 85%, and the soda ash is 95% pure. Determine:

a. The kilograms of quicklime and soda ash required per million liters.

b. The tonnes of quicklime and soda ash required per month if a month is considered to be 30 days.

8.11   A lime-soda softening plant treats a flow of 150 MGD, and the water has a hardness of 225 mg/ℓ as $CaCO_3$, an alkalinity of 178 mg/ℓ as $CaCO_3$, a Mg^{+2} ion content of 39 mg/ℓ, and 4 mg/ℓ of CO_2. The commerical grade of quicklime has a purity of 85%, and the soda ash has a purity of 95%. Determine:

a. The pounds of quicklime and soda ash required per million gallons.

b. The tons of quicklime and soda ash required per month if a month is considered to be 30 days.

8.12   A lime-soda softening plant treats a flow of 568,000 m³/d, and the water has a hardness of 225 mg/ℓ as $CaCO_3$, an alkalinity of 178 mg/ℓ as $CaCO_3$, a Mg^{+2} ion content of 39 mg/ℓ, and 4 mg/ℓ of CO_2. The commerical grade of quicklime has a purity of 85%, and the soda ash has a purity of 95%. Determine:

a. The kilograms of quicklime and soda ash required per million liters.

b. The tonnes of quicklime and soda ash required per month if a month is considered as 30 days.

9

SEDIMENTATION

Sedimentation is a solid-liquid separation utilizing gravitational settling to remove suspended solids. It is commonly used in water treatment, wastewater treatment, and advanced wastewater treatment. In water treatment its main applications are

1. Plain settling of surface waters prior to treatment by a rapid sand filtration plant.
2. Settling of coagulated and flocculated waters prior to rapid sand filtration.
3. Settling of coagulated and flocculated waters in a lime-soda type softening plant.
4. Settling of treated waters in an iron or manganese removal plant.

In wastewater treatment its main uses are

1. Grit or sand and silt removal.
2. Suspended solids removal in primary clarifiers.
3. Biological floc removal in activated sludge final clarifiers.
4. Humus removal in trickling filter final clarifiers.

In advanced wastewater treatment and tertiary treatment, its main purpose is the removal of chemically coagulated floc prior to filtration.

Sedimentation is one of the earliest unit operations used in water or wastewater treatment. The principles of sedimentation are the same for basins used in either water or wastewater treatment; the equipment and operational methods are also similar.

Sedimentation basins are usually constructed of reinforced concrete and may be circular, square, or rectangular in plan view. Circular tanks may be from 15 to 300 ft (4.57 to 91.43 m) in diameter and are usually from 6 to 16 ft (1.83 to 4.88 m) deep. The most common sizes are from 35 to 150 ft (10.67 to 45.72 m) in diameter, and depths are usually 10 to 14 ft (3.05 to 4.27 m). Standard-size tanks in the United States have diameters with 5-ft (1.52-m) intervals in order to accommodate commercially built sludge rake mechanisms. Square tanks have widths from 35 to 200 ft (10.67 to 60.96 m) and depths from 6 to 19 ft (1.83 to 5.79 m). Standard-size square tanks have widths with 5-ft (1.52-m) intervals. The freeboard for circular or square tanks is from 1 to 2.5 ft (0.30 to 0.76 m). Tanks that are not standard size may be furnished with specially built sludge rake mechanisms. Also, collectors for tanks with depths greater than those stated can be obtained by special order. Rectangular tanks usually have three types of sludge rake mechanisms: (1) sprocket and chain-driven rakes, (2) rakes supported from a traveling bridge, and (3) tandem scrapers built for square basins. Rectangular tanks with sprocket and chain drives have widths from 5 to 20 ft (1.52 to 6.10 m), lengths up to about 250 ft (76.2 m), and depths greater than 6 ft (1.83 m). Widths up to 80 to 100 ft (24.38 to 30.48 m) can be achieved by using four or five multiple bays with individual cleaning mechanisms. Rectangular tanks with traveling bridges that support the sludge rakes have widths from about 10 to 120 ft (3.05 to 65.57 m) and lengths from 40 to 300 ft (12.19 to 91.44 m). Traveling bridges have rapid sludge removal, and the rakes may be removed for inspection or repair without draining the basin. Rectangular tanks using two square tank sludge rake mechanisms in tandem give a settling tank with a 2 : 1 length-to-width ratio. Tanks as large as 200 ft by 400 ft (60.96 to 121.91 m) have been built in this manner, and this type of tank construction is particularly well suited for large water treatment plants. Rectangular tanks can use common wall construction and also occupy less land space than circular clarifiers of equal volume. Figure 9.1 shows a rectangular settling basin at a water treatment plant. Figures 9.2 and 9.3 show circular settling basins at wastewater treatment plants.

Coe and Clevenger (1916) presented a classification for the types of settling that may occur; this was later refined by Camp (1946) and Fitch (1956). This classification divides settling into four general types or classes that are based on the concentration of the particles and the ability of the particles to interact. A discussion of the various types of settling is presented in the following text.

TYPE I SETTLING

Type I settling, or **free settling**, is the settling of discrete, nonflocculent particles in a dilute suspension. The particles settle as separate units, and there is no apparent flocculation or interaction between the particles. Examples of type I settling are the plain sedimentation of surface waters and the settling of sand particles in grit chambers.

In type I settling, a particle will accelerate until the drag force, F_D, equals the impelling force, F_I; then settling occurs at a constant velocity, V_s. The impelling force, F_I, is

$$F_I = (\rho_s - \rho)gV \qquad \textbf{(9.1)}$$

where

F_I = impelling force

ρ_s = mass density of the particle

ρ = mass density of liquid

V = volume of particle

g = acceleration due to gravity

(a) *Two sludge rake mechanisms designed for square basins are used in tandem to form a rectangular basin with a length to width ratio of 2 : 1. The tops of the two center columns with drives are in foreground and background. View is toward basin inlet.*

(b) *These sludge rake mechanisms are in a basin identical to that shown in the top illustration. The rakes are for square basins and have hinged rakes for scraping the basin corners. View is toward basin outlet.*

FIGURE 9.1 *A Rectangular Settling Basin at a Water Treatment Plant*

FIGURE 9.2 *A Circular Primary Clarifier at an Activated Sludge Wastewater Treatment Plant*

FIGURE 9.3 *A Circular Final Clarifier Being Repaired at an Activated Sludge Plant. Note surface skimmer, bottom sludge rakes, sludge hopper, and hydrostatic blowout plugs in bottom slab.*

The drag force is given by Newton's law:

$$F_D = C_D A_c \rho \left(\frac{V_s^2}{2} \right) \qquad (9.2)$$

where

F_D = drag force

C_D = coefficient of drag, which is a function of the Reynolds number, N_{Re}

A_c = area in cross section at right angle to the velocity

ρ = mass density of liquid

V_s = settling velocity

Combining Eqs. (9.1) and (9.2) gives

$$(\rho_s - \rho)gV = C_D A_c \rho \left(\frac{V_s^2}{2} \right)$$

or

$$V_s = \sqrt{ \frac{2g}{C_D} \left(\frac{\rho_s - \rho}{\rho} \right) \frac{V}{A_c} } \qquad (9.3)$$

For spheres of diameter d, the volume, V, is

$$V = \left(\frac{\pi}{6} \right) d^3$$

The cross-sectional area, A_c, is

$$A_c = \left(\frac{\pi}{4} \right) d^2$$

From the preceding two equations, it follows that for spheres,

$$\frac{V}{A_c} = \left(\frac{\pi}{6} \right) d^3 \left(\frac{4}{\pi} \right) \frac{1}{d^2} = \frac{2}{3} d$$

Substituting this expression for V/A_c into Eq. (9.3) gives the following equations:

$$V_s = \sqrt{ \frac{4g}{3C_D} \left(\frac{\rho_s - \rho}{\rho} \right) d } \qquad (9.4a)$$

or

$$V_s = \sqrt{ \frac{4g}{3C_D} (S_s - 1) d } \qquad (9.4b)$$

where S_s = specific gravity of the particles.

The numerical value of the drag coefficient depends on whether the flow regime around the particle is laminar or turbulent. Figure 9.4 shows the drag coefficient for various shapes as a function of the Reynolds number, N_{Re}. The Reynolds number for the spheres is defined as

$$N_{Re} = \frac{V_s d}{v} = \frac{V_s d \rho}{\mu} \tag{9.5}$$

where

v = kinematic viscosity

μ = dynamic viscosity

ρ = density, $\rho = \gamma/g_c$

In Figure 9.4 there are three distinct regions: I, laminar flow; II, transition flow; and III, turbulent flow. For region I, laminar flow, N_{Re} is less than 1 and the viscous forces are more important than inertia forces. The relationship for the drag coefficient for spheres is

$$C_D = \frac{24}{N_{Re}} \tag{9.6}$$

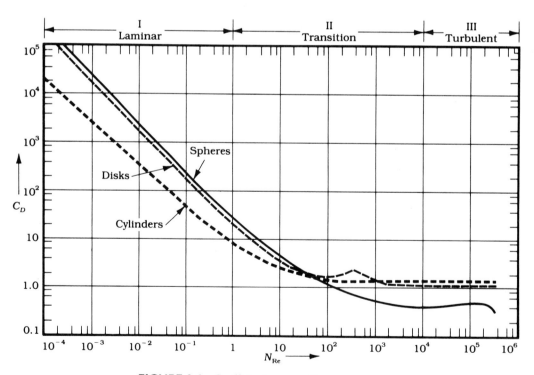

FIGURE 9.4 *Coefficients versus N_{Re}*

Adapted from "Sedimentation and the Design of Settling Tanks" by T. R. Camp. In *Transactions of the American Society of Civil Engineers* 111 (1952):895. Reprinted by permission.

For region II, transition flow, N_{Re} equals 1 to 10^4 and viscous and inertia forces are of equal importance. The relationship for the drag coefficient for spheres is (Fair *et al.*, 1968)

$$C_D = \frac{24}{N_{Re}} + \frac{3}{\sqrt{N_{Re}}} + 0.34 \tag{9.7}$$

For region III, turbulent flow, N_{Re} is greater than 10^4 and inertia forces are the most important. The drag coefficient for spheres is

$$C_D = 0.4 \tag{9.8}$$

For laminar flow (region I), Eq. (9.5) may be combined with Eq. (9.6) to eliminate N_{Re}, giving

$$C_D = \frac{24v}{V_s d}$$

Substituting this equation for C_D into Eq. (9.4b) gives the following expression:

$$V_s = \sqrt{\left(\frac{4g}{3}\right)\left(\frac{V_s d}{24v}\right)(S_s - 1)d} \tag{9.9}$$

Squaring Eq. (9.9) and rearranging gives Stokes' law:

$$V_s = \frac{g}{18v}(S_s - 1)d^2 \tag{9.10}$$

Or, since $v = \mu/\rho$, substitution into Eq. (9.10) yields

$$V_s = \frac{g}{18\mu}(\rho_s - \rho)d^2 \tag{9.11}$$

which is another form of Stokes' law. Much of the settling of dilute suspensions in water and wastewater treatment follows Stokes' law.

For transition flow (region II), the determination of the settling velocity is a trial and error solution using Eq. (9.4a) or (9.4b) and Eq. (9.7).

For turbulent flow (region III), substituting Eq. (9.8) into Eq. (9.4b) gives

$$V_s = \sqrt{3.3g(S_s - 1)d} \tag{9.12}$$

The settling velocity of sand in grit removal chambers can be determined using Eq. (9.12).

The solution for the settling velocity in regions I, II, and III can also be done using the graph in Figure (9.5) (Camp, 1952). This graph gives a direct solution for the settling velocity, V_s, if the diameter, specific gravity, and temperature are known.

The ideal basin theory by Camp (1946) assumes the following:

1. The settling is type I settling — in other words, discrete particles.

2. There is an even distribution of the flow entering the basin.

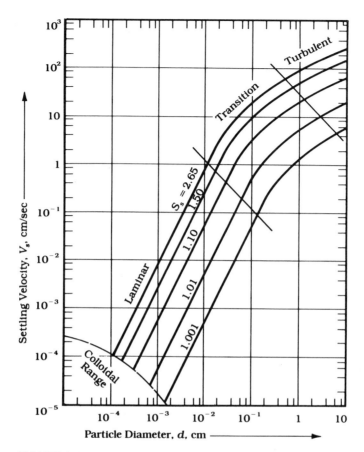

FIGURE 9.5 *Type I Settling of Spheres in Water at 10°C*
Adapted from "Water Treatment" by T. R. Camp in *Handbook of
Applied Hydraulics*, 2nd ed. Edited by C. V. Davis. Copyright ©
1952 by McGraw-Hill., Inc. Reprinted by permission.

3. There is an even distribution of the flow leaving the basin.

4. There are three zones in the basin: (1) the entrance zone, (2) the
   outlet zone, and (3) the sludge zone.

5. There is a uniform distribution of particles throughout the depth of
   the entrance zone.

6. Particles that enter the sludge zone remain there, and particles that
   enter the outlet zone are removed.

Figure 9.6 shows an ideal rectangular settling basin of a length, L, a
width, W, and a depth, H. V_0 is the settling velocity of the smallest particle
size that is 100% removed. When a particle of this size enters the basin at
the water surface, point 1, it has a trajectory as shown and intercepts the
sludge zone at point 2, which is at the downstream end. The detention time,
t, is equal to the depth, H, divided by the settling velocity, V_0, or

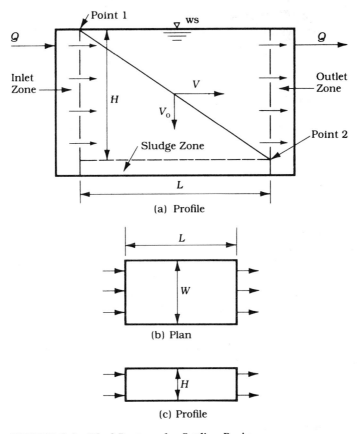

FIGURE 9.6 *Ideal Rectangular Settling Basin*

$$t = \frac{H}{V_0} \qquad (9.13)$$

The detention time, t, is also equal to the length, L, divided by the horizontal velocity, V, or

$$t = \frac{L}{V} \qquad (9.14)$$

The horizontal velocity, V, is equal to the flowrate, Q, divided by the cross-sectional area, HW, or

$$V = \frac{Q}{HW} \qquad (9.15)$$

Combining Eqs. (9.14) and (9.15) to eliminate V gives

$$t = \frac{LWH}{Q} \qquad (9.16)$$

Since LWH equals the basin volume, V, the detention time, t, is equal to the basin volume, V, divided by the flow rate, Q, or

$$t = \frac{V}{Q} \tag{9.17}$$

Equating Eqs. (9.16) and (9.13) gives

$$\frac{LWH}{Q} = \frac{H}{V_0} \tag{9.18}$$

Rearranging yields

$$V_0 = \frac{Q}{LW} \tag{9.19}$$

or

$$V_0 = \frac{Q}{A_p} = \text{overflow rate or surface loading, gal/day-ft}^2 \ (\text{m}^3/\text{d-m}^2) \tag{9.20}$$

where A_p is the plan area of the basin. Equation (9.20) shows that the overflow rate is equivalent to the settling velocity of the smallest particle size that is 100% removed.

The previous fundamentals also apply to an ideal circular settling basin, shown in Figure 9.7. The horizontal velocity, V, is given by

$$V = \frac{Q}{2\pi r H} \tag{9.21}$$

From inspection of Figure 9.7,

$$\frac{dh}{dr} = \frac{V_0}{V} \tag{9.22}$$

Substituting Eq. (9.21) into Eq. (9.22) gives

$$\frac{dh}{dr} = \frac{2\pi r H V_0}{Q} \tag{9.23}$$

Rearranging Eq. (9.23) and setting the integration limits yields

$$\int_0^H dh = \frac{2\pi H V_0}{Q} \int_{r_1}^{r_0} r\,dr \tag{9.24}$$

Integrating gives

$$H = \frac{2\pi H V_0}{Q} \left[\frac{r^2}{2} \right]_{r_1}^{r_0} \tag{9.25}$$

or

$$H = \frac{\pi H V_0}{Q}(r_0^2 - r_1^2) = \frac{H A_p V_0}{Q} \tag{9.26}$$

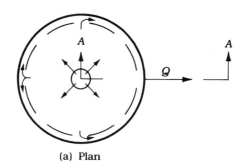

(a) Plan

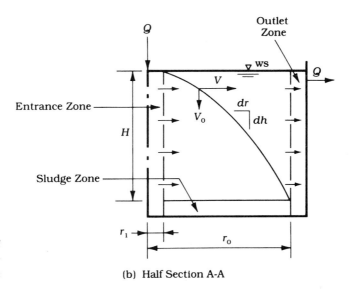

FIGURE 9.7 *Ideal*
Circular Settling
Basin

(b) Half Section A-A

where A_p = plan area of the basin. Cancelling the H terms in Eq. (9.26) and rearranging yields

$$V_0 = \frac{Q}{A_p} = \text{overflow rate or surface loading, gal/day-ft}^2 \ (\text{m}^3/\text{d-m}^2) \quad \textbf{(9.27)}$$

Equation (9.27) is identical to Eq. (9.20) for the rectangular basin.

The depth of the ideal rectangular or circular basin is given by

$$H = V_0 t \qquad \textbf{(9.28)}$$

where V_0 is the overflow rate or surface loading rate expressed as a velocity. It can be shown that an overflow rate or surface loading rate of 100 gal/day-ft^2 is equal to a settling velocity of 0.555 ft/hr (100 m^3/d-m^2 is equal to a settling velocity of 4.67 m/h). Also, a settling velocity of 1 cm/sec is equal to an overflow rate or surface loading rate of 21,200 gal/day-ft^2 (864 m^3/d-m^2). With these conversion values, settling velocities at any overflow rate or surface loading rate can be determined by proportionality.

Inspection of Figures 9.8 (Camp, 1946) and 9.9 shows that all particles with a settling velocity, V_1, greater than V_0 will be 100% removed since their trajectory intercepts the sludge zone. For particles with a settling velocity, V_2, less than V_0, the fraction removed, R_2, is equal to H_2/H or V_2/V_0. Thus,

$$R_2 = \frac{H_2}{H} = \frac{V_2}{V_0} \tag{9.29}$$

A large variation in particle size will exist in a typical suspension of particles. Thus, one must evaluate the entire range of settling velocities in determining the overall removal for a given design settling velocity or overflow rate. This requires experimental analyses usually employing the use of a settling column. In a batch settling column, samples are withdrawn at various times and various depths, and the solids concentrations are determined. An analysis of the data by an appropriate procedure will yield a settling velocity curve as shown in Figure 9.10. The fraction of the total particles removed for a design velocity, V_0, will be (Camp, 1946)

$$\text{Fraction removed} = (1 - F_0) + \frac{1}{V_0} \int_0^{F_0} V dF \tag{9.30}$$

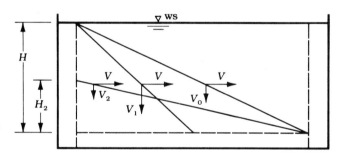

FIGURE 9.8 *Profile through an Ideal Rectangular Basin*

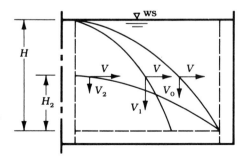

FIGURE 9.9 *Half Section through an Ideal Circular Basin*

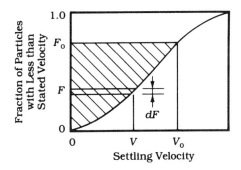

FIGURE 9.10
Type I Settling Curve

where

$$1 - F_0 \quad = \text{fraction of particles with velocity } V \text{ greater than } V_0$$

$$\frac{1}{V_0}\int_0^{F_0} V\,dF = \text{fraction of particles with velocity } V \text{ less than } V_0$$

To summarize the ideal settling basin theory, the removal of suspended solids is a function of the overflow rate or design settling velocity, V_0, the detention time, t, and the depth, H. Although the ideal settling basin analysis is theoretical, it does give a rational method for the design of sedimentation tanks. An evaluation of vast amounts of operating and research data using the theoretical parameters gives a range of overflow rates and detention times that, in most cases, has been found to be satisfactory for municipal waters and wastewaters.

TYPE II SETTLING

Type II settling is the settling of flocculent particles in a dilute suspension. The particles flocculate during settling; thus they increase in size and settle at a faster velocity. Examples of type II settling are the primary settling of wastewaters and the settling of chemically coagulated waters and wastewaters.

Batch settling tests are usually required to evaluate the settling characteristics of a flocculent suspension. A schematic drawing of a batch settling column is shown in Figure 9.11. The column should be at least 5 to 8 inches (130 to 205 mm) in diameter to minimize side-wall effects, and the height should be at least equal to the depth of the proposed settling tank. Sampling ports are provided at equal intervals in height.

The suspension must be mixed thoroughly and poured rapidly into the column in order to insure that a uniform distribution of the particles occurs throughout the height of the column. To be representative, the test must take place under quiescent conditions, and the temperature should not vary more than 1°C throughout the column height in order to avoid convection currents. Samples are removed through the ports at periodic time intervals, and the suspended solids concentrations are determined. The percent removal is calculated for each sample from the initial suspended solids

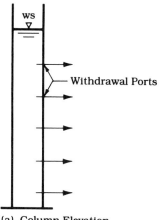

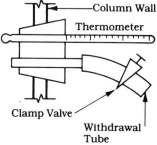

FIGURE 9.11
Batch Settling Column Details for Type II Settling

(a) Column Elevation

(b) Withdrawal Port Detail

concentration and the concentration of the sample. The percent removal is plotted on a graph as a number versus time and depth of collection for the sample. Interpolations are made between the plotted points, and curves of equal percent removal (R_A, R_B, and so on) are drawn, as shown in Figure 9.12.

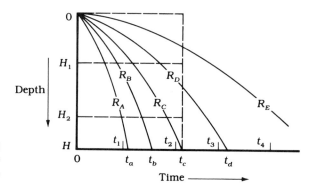

FIGURE 9.12
Settling Diagram for Type II Settling

The overflow rates, V_0, are determined for the various settling times (t_a, t_b, and so on) where the R curves intercept the x-axis. For example, for the curve R_C, the overflow rate is

$$V_0 = \frac{H}{t_c} \times \text{proper conversions} \tag{9.31}$$

where H is the height of the column and t_c is the intercept of the R_C curve and the x-axis. The fractions of solids removed, R_T, for the times t_a, t_b, and so on are then determined. For example, for time t_c, the fraction removed, R_T, would be

$$R_T = R_C + \frac{H_2}{H}(R_D - R_C) + \frac{H_1}{H}(R_E - R_D) \tag{9.32}$$

where H_2 represents the height that the particles of $(R_D - R_C)$ size would settle during t_c. These would intercept the sludge zone in a basin, as in Figure 9.13. By using the various times, t_a, t_b, and so on, the various overflow rates, V_0, and the various fractions removed, R_T, a graph of the overflow rates versus fractions removed can be constructed. Also, a graph of the fractions removed versus detention times can be made. In applying the curves to design a tank, scale-up factors of 0.65 for the overflow rate and 1.75 for detention time are used to compensate for the side-wall effects of the settling column (Eckenfelder, 1980).

EXAMPLE 9.1   *Primary Clarifier*

A primary clarifier is to be designed to treat an industrial wastewater having 320 mg/ℓ suspended solids and a flow of 2.0 MGD. A batch settling test was performed using an 8-in.-diameter column that was 10 ft long and had withdrawal ports every 2.0 ft. The reduced data giving the percent removals are as shown in Table 9.1.

Determine:

1. The design detention time and design overflow rate if 65% of the suspended solids are to be removed.

2. The diameter of the tank.

3. The design diameter if equipment is available in 5-ft increments of tank diameter.

4. The depth of the tank.

FIGURE 9.13
Profile for Type II Settling

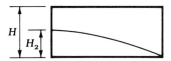

TABLE 9.1 *Percent Suspended Solids Removal at Given Depths*

TIME (min)	2 ft	4 ft	6 ft	8 ft	10 ft
0	0	0	0	0	0
10	28	18	18	12	[a]
20	48	39	25	27	[a]
30	68	50	34	31	[a]
45	70	56	53	41	[a]
60	85	66	59	53	[a]
90	88	82	73	62	[a]

[a] Data showed an increase in solids concentration.

SOLUTION   A plot of the percent removals at the various depths and times is shown in Figure 9.14. Interpolations have been made to locate the 20, 30, 40, 50, 60, and 70% removal curves, and the curves have been drawn on the plot. The 20% curve intersects the *x*-axis at 16 min; thus the overflow rate at that time is

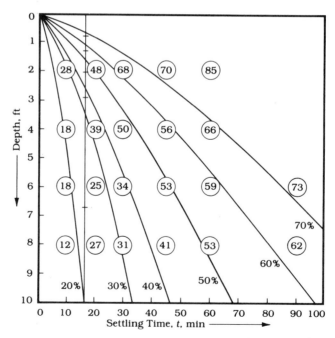

FIGURE 9.14 *Graph Showing Suspended Solids Removal (as a Percent) at Various Depths and Settling Times, for Example 9.1*

$$V_0 = \frac{10\,\text{ft}}{16\,\text{min}} \left| \frac{1440\,\text{min}}{\text{day}} \right| \frac{7.48\,\text{gal}}{\text{ft}^3}$$

$$= 6730\,\text{gal/day-ft}^2$$

The detention time in hours is 16/60 or 0.27 hr. The point midway between the 20 and 30% curves at 16 min is located as shown and is at a depth of 6.7 ft. In a like manner, the points midway between the 30 and 40, 40 and 50, 50 and 60, and 60 and 70% curves are located, and the respective depths are 2.9, 2.0, 1.3, and 0.8 ft. These values give a total fraction removed, R_T, at 16 min (0.27 hr) of

$$R_T = 20 + (6.7/10)(30 - 20) + (2.9/10)(40 - 30) + (2.0/10)(50 - 40)$$
$$+ (1.3/10)(60 - 50) + (0.8/10)(70 - 60)$$
$$= 33.7\%$$

Similarly, the overflow rates, detention times, and total fractions removed are computed for the 30, 40, 50, and 60% curves; and a summary of the reduced data is shown in Table 9.2.

A plot of the fraction removed (R_T) versus detention time (t) is shown in Figure 9.15. Also, a plot of the fraction removed (R_T) versus overflow rate (V_0) is shown in Figure 9.16. For 65% removal the detention time is 1.22 hr; thus the design detention time is (1.22)1.75 = $\boxed{2.14\,\text{hr.}}$ For 65% removal the overflow rate is 1420 gal/day-ft^2; thus the design overflow rate is $(1420)(0.65) = \boxed{923\,\text{gal/day-ft}^2}$. The required area is

$$A = \frac{2,000,000\,\text{gal}}{\text{day}} \left| \frac{\text{day-ft}^2}{923\,\text{gal}} = 2167\,\text{ft}^2 \right.$$

Thus the diameter, D, is

$$D = \left[\frac{4}{\pi}(2167) \right]^{1/2} = \boxed{52.5\,\text{ft}}$$

The design diameter is $\boxed{55\,\text{ft}}$ for standard-size sludge rakes.

TABLE 9.2 *Reduced Data for 20, 30, 40, 50, and 60% Curves*

TIME t (hr)	OVERFLOW RATE V_0 (gal/day-ft^2)	FRACTION REMOVED R_T (%)
0.27	6730	33.7
0.55	3260	48.7
0.77	2340	56.7
1.13	1590	63.8
1.60	1120	68.6

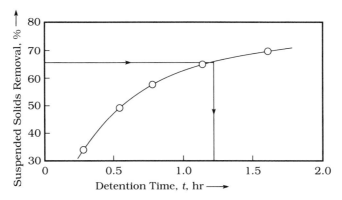

FIGURE 9.15 *Suspended Solids Removal versus Detention Time, for Example 9.1*

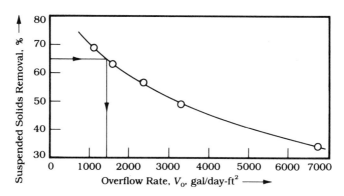

FIGURE 9.16 *Suspended Solids Removal versus Overflow Rate, for Example 9.1*

The required depth, H, is

$$H = \frac{2{,}000{,}000\,\text{gal}}{24\,\text{hr}} \left| \frac{2.14\,\text{hr}}{} \right| \frac{\text{ft}^3}{7.48\,\text{gal}} \left| \frac{4}{\pi} \right| \frac{}{(55\,\text{ft})^2}$$

$$= \boxed{10.03\,\text{ft}}$$

EXAMPLE 9.1 SI *Primary Clarifier*

A primary clarifier is to be designed to treat an industrial wastewater having 320 mg/ℓ suspended solids and a flow of 7570 m³/d. A batch settling test was performed using a 205-mm-diameter column that was 3.05 m long and had withdrawal ports every 0.61 m. The reduced data giving the percent removals are shown in Table 9.3.

Determine:

1. The design detention time and design surface loading rate if 65% of the suspended solids are to be removed.
2. The diameter and depth of the tank.

SOLUTION A plot of the percent removals at the various depths and times is shown in Figure 9.17. Interpolations have been made to locate the 20, 30, 40, 50, 60, and 70% removal curves. The 20% curve intersects the *x*-axis at 16 min; thus the surface loading at that time is

$$V_0 = \frac{3.05 \text{ m}}{16 \text{ min}} \left| \frac{1440 \text{ min}}{d} \right| \frac{\text{m}^2}{\text{m}^2}$$

$$= 275 \text{ m}^3/\text{d-m}^2$$

The detention time in hours is 16/60 or 0.27 h. The point midway between the 20 and 30% curves at 16 min is located as shown and is at a depth of 2.04 m. In a like manner, the points midway between the 30 and 40, 40 and 50, 50 and 60, and 60 and 70% curves are located, and the respective depths are 0.88, 0.61, 0.40, and 0.24 m. These values give the total fraction removed (R_T) at 16 min (0.27 h) as

$$R_T = 20 + (2.04/3.05)(30 - 20) + (0.88/3.05)(40 - 30)$$
$$+ (0.61/3.05)(50 - 40) + (0.40/3.05)(60 - 50) + (0.24/3.05)(70 - 60)$$
$$= 33.7\%$$

Similarly, the surface loading rates, detention times, and total fractions removed are computed for the 30, 40, 50, and 60% curves, and a summary of the reduced data is shown in Table 9.4.

A plot of the fraction removed (R_T) versus detention time (*t*) is shown in Figure 9.18. Also, a plot of the fraction removed (R_T) versus surface

TABLE 9.3 *Percent Suspended Solids Removal at Given Depths*

TIME (min)	0.61 m	1.22 m	1.83 m	2.44 m	3.05 m
0	0	0	0	0	0
10	28	18	18	12	[a]
20	48	39	25	27	[a]
30	68	50	34	31	[a]
45	70	56	53	41	[a]
60	85	66	59	53	[a]
90	88	82	73	62	[a]

[a] Data showed an increase in solids concentration.

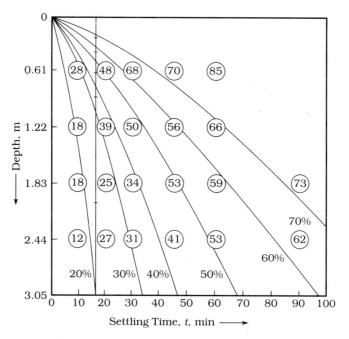

FIGURE 9.17 *Graph Showing Suspended Solids Removal (as a Percent) at Various Depths and Settling Times, for Example 9.1 SI*

TABLE 9.4 *Reduced Data for 20, 30, 40, 50, and 60% Curves*

TIME t (h)	SURFACE LOADING V_0 (m³/d-m²)	FRACTION REMOVED R_T (%)
0.27	275	33.7
0.55	133	48.7
0.77	95.3	56.7
1.13	64.8	63.8
1.60	45.6	68.6

loading rate (V_0) is shown in Figure 9.19. For 65% removal the detention time is 1.22 h; thus the design detention time is $(1.22)1.75 = \boxed{2.14\,\text{h.}}$ For 65% removal the surface loading is $58.0\,\text{m}^3/\text{d-m}^2$; thus the design surface loading is $(58.0)0.65 = \boxed{37.7\,\text{m}^3/\text{d-m}^2.}$ The required area is

$$A = \frac{7570\,\text{m}^3}{\text{d}} \left| \frac{\text{d-m}^2}{37.7\,\text{m}^3} = 201\,\text{m}^2 \right.$$

Thus the diameter, D, is

$$D = \left[\frac{4}{\pi}(201) \right]^{1/2} = \boxed{16.0\,\text{m}}$$

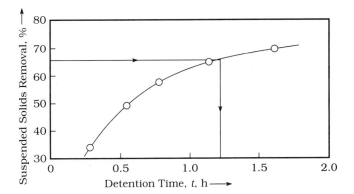

FIGURE 9.18 *Suspended Solids Removal versus Detention Time, for Example 9.1 SI*

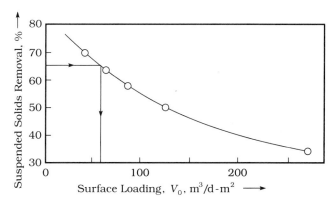

FIGURE 9.19 *Suspended Solids Removal versus Surface Loading for Example 9.1 SI*

The required depth, H, is

$$H = \frac{7570\,\text{m}^3}{24\,\text{h}} \left| \frac{2.14\,\text{h}}{} \right| \frac{}{201\,\text{m}^2} = \boxed{3.36\,\text{m}}$$

TYPE III AND TYPE IV SETTLING

Type III settling, or **zone** or **hindered settling**, is the settling of an intermediate concentration of particles in which the particles are so close together that interparticle forces hinder the settling of neighboring particles. The particles remain in a fixed position relative to each other, and all settle at a constant velocity. As a result, the mass of particles settle as a zone. At the top of the settling mass, there will be a distinct solid-liquid interface between the settling particle mass and the clarified liquid. An example of type III settling is the settling that occurs in the intermediate depths in a final clarifier for the activated sludge process. Type IV settling, or **compression settling**, is the settling of particles that are of such a high concentration that the particles touch each other and settling can occur only by compression of the compacting mass. An example of type IV settling is the compression settling that occurs in the lower depths of a final clarifier for the activated sludge process. Both discrete and flocculent particles may settle by zone or compression settling; however, flocculent particles are the most common type encountered.

The settling of a flocculent suspension of activated sludge placed in a graduated cylinder is as shown in Figure 9.20(a). At first, time $t = 0$ and the particles have zone settling (zs). They have the same relative position with respect to each other. The concentration of particles is so great that they interfere with the velocity fields of each other, and the rate of settling is a function of the solids concentration. At time $t = t_1$, the sludge mass has settled until a clear water zone exists above the sludge. Below the region of zone or hindered settling, the concentration of the particles has become so great that many of the particles have made physical contact with each other. This is transition settling (TS) from zone settling to compression settling (CS). Below the transition zone is the compression settling zone, where all of the particles are in contact with each other and compression has begun. At time

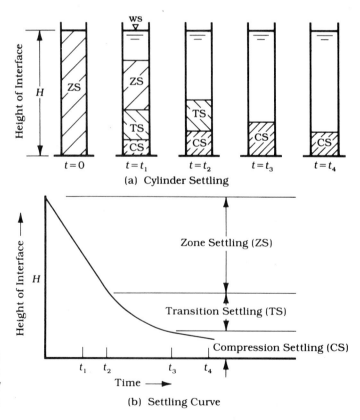

(a) Cylinder Settling

(b) Settling Curve

FIGURE 9.20
Settling of a Concentrated Suspension

$t = t_2$, the zone settling region has disappeared and all particles are undergoing transition or compression settling. At time $t = t_3$, the transition zone has disappeared and all the particles are in a state of compression settling. At time $t = t_4$, the compression settling is almost complete. Figure 9.20(b) shows the settling curve for the sludge–water interface in the batch settling test.

Figure 9.21(a) shows a cross section of a circular final clarifier for the activated sludge process and illustrates the classes of settling which occur. The clear water zone is usually about 5 to 6 ft (1.5 to 1.8 m) deep, and the total depth for zone or hindered, transition, and compression settling is usually about 5 to 7 ft (1.5 to 2.1 m).

Batch settling tests as previously described can be used in the laboratory to obtain the parameters needed for the design of an activated sludge final clarifier. In final clarifiers, both clarification of the liquid wastewater effluent and thickening of the solids must be accomplished. In the design of final clarifiers, the flowrate, Q, is usually taken as the average daily flow. During peak hours of the day, however, the flowrate may be considerably more and could be up to about five times the average daily flow for municipal wastewaters. This has caused many final clarifiers to have appreciable solids spill over the effluent weirs during the peak of the day. Therefore, the plan area for the peak hour of the day should be checked. In many cases, the peak flow condition controls the design area for the final clarifier for an activated sludge process.

An approach to the design of final clarifiers for the activated sludge processes and for sludge thickeners is based on the solids flux concept (Dick, 1970). The solids flux is the rate of solids thickening per unit area in plan view — in other words, the lb/hr-ft^2 (kg/h-m^2). As the solids settle in clarifiers and thickeners, they must be thickened from the initial concentration, C_0, to the underflow concentration, C_u, at the bottom of the tank. As the solids move downward, at some level in the tank a limiting solids

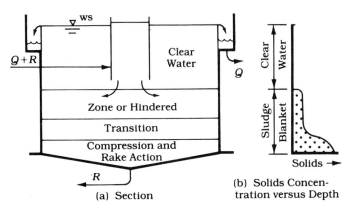

FIGURE 9.21 *Settling in a Final Clarifier for the Activated Sludge Process*

flux, G_L, occurs. This flux must not be exceeded, or the solids will build up and spill over with the effluent from the tank. The movement of the solids downward occurs by hindered settling and also by the bulk flow downward due to the underflow. The data required for the flux design method are determined from batch settling tests. Numerous concentrations of the sludge are allowed to settle to obtain the hindered settling velocities. The hindered or zone settling velocities, V_0, are measured using a slowly stirred graduated cylinder. A plot is made of the hindered settling velocity, V_0, versus the solids concentration, C, as shown in Figure 9.22. The solids flux is computed at numerous concentrations, since it is obtained by multiplying the velocity by the solids concentration. The resulting curve of flux versus concentration is shown in Figure 9.23.

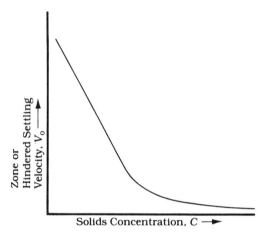

FIGURE 9.22 *Zone or Hindered Settling Velocity versus Solids Concentration*

FIGURE 9.23
Solids Flux versus Solids Concentration

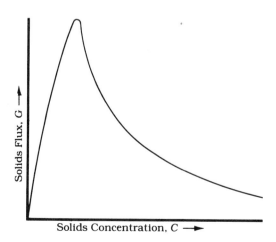

At any level in the settling tank, the movement of solids by settling is

$$G_s = C_t V_t \tag{9.33}$$

where

G_s = solids flux by gravity

C_t = solids concentration

V_t = hindered settling velocity

The movement of the solids due to bulk flow is given by

$$G_b = C_t V_b \tag{9.34}$$

where

G_b = bulk flux

V_b = bulk velocity

The total solids flux for gravity settling and bulk movement is therefore

$$G_t = G_s + G_b = C_t V_t + C_t V_b \tag{9.35}$$

where G_t = total flux.

The bulk velocity is given by

$$V_b = \frac{Q_u}{A} \tag{9.36}$$

where

Q_u = flowrate of the underflow

A = plan area of the tank

The mass rate of solids settling — that is, the weight of the solids settling per unit time — is

$$M_t = Q_0 C_0 = Q_u C_u \tag{9.37}$$

where

M_t = rate of solids settling

Q_0 = influent flowrate to the tank

C_0 = influent solids concentration

The limiting cross-sectional area, A, required is given by

$$A = \frac{M_t}{G_L} = \frac{Q_0 C_0}{G_L} \tag{9.38}$$

where G_L = limiting flux.

Rearranging Eq. (9.37) gives $Q_u = M_t/C_u$, and combining this with Eq. (9.36) and Eq. (9.38) gives

$$V_b = \frac{Q_u}{A} = \frac{M_t}{C_u A} = \frac{G_L}{C_u} \qquad (9.39)$$

These relationships are shown in Figure 9.24. Selecting an underflow concentration, C_u, and drawing a tangent to the flux curve gives the y-axis intercept as G_L, the limiting flux value. The slope of the tangent is equal to V_b, the bulk velocity. The value of the gravity flux is G_s, whereas the value of the bulk flux is $G_L - G_s$. These concepts are illustrated in Example 9.2 with batch settling data.

EXAMPLE 9.2   *Final Clarifier*

Batch settling tests have been performed using an acclimated activated sludge to give the data in Table 9.5.

The design mixed liquor flow to the final clarifier is 2530 GPM (gallons per minute), the MLSS is 2500 mg/ℓ, and the underflow concentration is 12,000 mg/ℓ. Determine the diameter of the final clarifier and the design diameter if equipment is available in 5-ft increments of tank diameter.

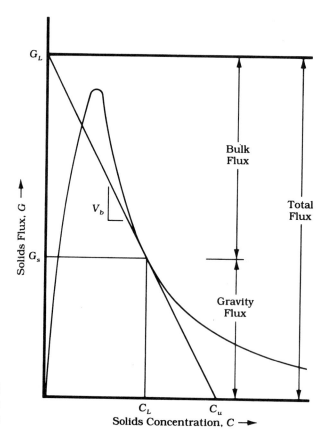

FIGURE 9.24
Solids Flux versus
Solids Concentration

TABLE 9.5 *Concentrations, Settling Velocities, and Solids Flux for Various Tests*

TEST no.	C (mg/ℓ)	V (ft/hr)	G = CV (lb/hr-ft²)
1	12,460	0.409	0.318
2	9,930	0.817	0.506
3	7,450	1.53	0.712
4	5,220	3.28	1.07
5	3,140	9.65	1.89
6	1,580	13.7	1.35

SOLUTION The settling curve showing the settling velocity versus solids concentration is shown in Figure 9.25. The flux curve showing the solids flux versus solids concentration is shown in Figure 9.26. A tangent to the curve drawn from $C_u = 12,000\,\text{mg/ℓ}$ gives a G_L value of $1.80\,\text{lb/hr-ft}^2$. Using a scale-up factor of 1.5 gives $G_L = 1.80/1.5$ or $1.20\,\text{lb/hr-ft}^2$. The rate at which the solids settle, M_t, is equal to $Q_0 C_0$, or $M_t = (2530\,\text{gal/min})(60\,\text{min/hr})(8.34\,\text{lb/gal})$ $(2500/10^6)$ or $3165\,\text{lb/hr}$. From Eq. (9.38) the area required is M_t/G_L, or $A = (3165\,\text{lb/hr})/(1.20\,\text{lb/hr-ft}^2)$ or $2638\,\text{ft}^2$. The required diameter is given by

$$D = \left[\frac{4}{\pi}(2638\,\text{ft}^2)\right]^{1/2}$$

$$= \boxed{58.0\,\text{ft}}$$

The design diameter is $\boxed{60\,\text{ft}}$ for standard-size sludge rakes.

EXAMPLE 9.2 SI *Final Clarifier*

Batch settling tests have been performed using an acclimated activated sludge to give the data in Table 9.6.

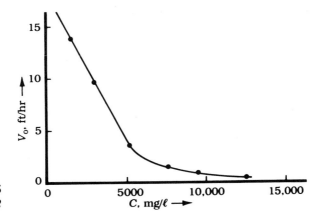

FIGURE 9.25
Example 9.2

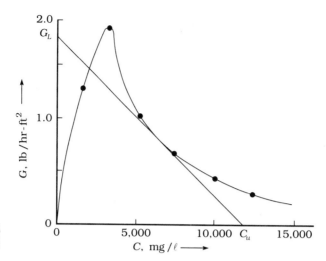

FIGURE 9.26
Example 9.2

TABLE 9.6 *Concentrations, Settling Velocities, and Solids Flux for Various Tests*

TEST	C (mg/ℓ)	V (m/h)	G = CV (kg/h-m²)
1	12,460	0.125	1.56
2	9,930	0.249	2.47
3	7,450	0.465	3.46
4	5,220	1.00	5.22
5	3,140	2.94	9.24
6	1,580	4.18	6.60

The design mixed liquor flow to the final clarifier is 160 ℓ/s, the MLSS is 2500 mg/ℓ, and the underflow concentration is 12,000 mg/ℓ. Determine the diameter of the final clarifier.

SOLUTION

The settling curve showing the settling velocity versus solids concentration is shown in Figure 9.27. From the previous data, the flux curve showing the solids flux versus solids concentration is shown in Figure 9.28. A tangent to the curve drawn from $C_u = 12,000$ mg/ℓ gives a G_L value of 8.90 kg/h-m². Using a scale-up factor of 1.5 gives $G_L = 8.90/1.5$ or 5.93 kg/h-m². The rate at which the solids settle, M_t, is equal to Q_0C_0, or $M_t = (160 ℓ/s)(60 s/min)/(60 min/h)(2.50 g/ℓ)(kg/1000 g) = 1440$ kg/h. From Eq. (9.38) the area required is M_t/G_L, or $A = (1440$ kg/h$)($h-m²/5.93 kg$) = 242.8$ m². The required diameter is given by

$$D = \left[\frac{4}{\pi}(242.8 \text{ m}^2)\right]^{1/2}$$

$$= \boxed{17.6 \text{ m}}$$

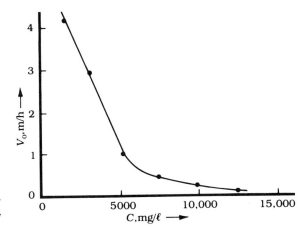

FIGURE 9.27
Example 9.2 SI

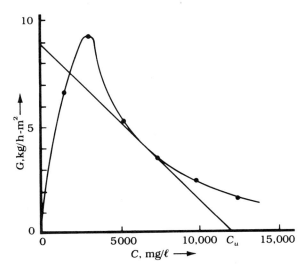

FIGURE 9.28
Example 9.2 SI

ACTUAL SEDI-MENTATION BASINS

Actual settling basins are rectangular, square, or circular in plan area. A single rectangular basin will cost more than a circular basin of the same size; however, if numerous tanks are required, the rectangular units can be constructed with common walls and be the most economical. Rectangular basins have a disadvantage if they have sprocket and chain drives for the sludge rakes, because these will have more wear than the rotary type scraper mechanisms used for circular settling tanks.

Figure 9.29 shows a schematic of the inlet to a rectangular tank if the unit is adjacent to a flocculation basin. This occurs frequently in the chemical coagulation of waters and wastewaters. The flocculation basin will be the same width as the settling tank but is usually not as deep. The two basins are separated by a wood baffle fence or a concrete wall with numerous ports. The inlet water will enter uniformly across the basin. This inlet

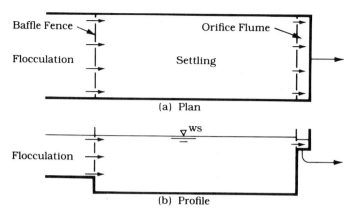

FIGURE 9.29 *Inlet and Outlet Details for a Rectangular Settling Tank with Orifice Flume Outlet Preceded by Flocculation*

arrangement closely approaches that of an ideal rectangular tank; the only difference is that the inlet zone does not extend down to the full depth of the settling tank but extends down to the depth of the flocculator. If the rectangular basin does not adjoin a flocculator, as shown in Figure 9.30, the inlet water is distributed uniformly across the basin by a flume with ports into the tank, and the inlet zone does not extend down the full depth of the tank as depicted in an ideal tank. However, a baffle in front of the flume will disperse the water downward to give a deeper inlet zone. Figure 9.30 also shows one type of outlet for a rectangular basin. A weir is used which spills into the effluent flume and extends across the entire width of the basin. If, however, the water is a chemically coagulated water, a weir should be

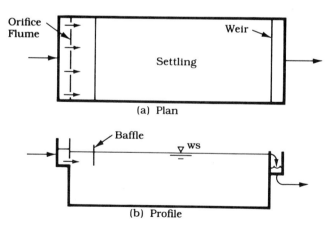

FIGURE 9.30 *Inlet and Outlet Details for a Rectangular Settling Tank with Orifice Flume Inlet and Weir Channel Outlet*

avoided because the turbulence will break up much of the fine floc and result in poor filter performance. For chemically treated waters it is best to have an orifice flume across the basin width, as in Figure 9.29. An orifice flume does not have a high degree of turbulence and will not break up fine floc. In either of these outlet arrangements, the flow regime is conservative, because the outlet zone does not extend down the full depth as depicted in the ideal tank. Figure 9.31 shows the details of a rectangular settling tank of the type used in wastewater treatment. It has mechanical collection of the sludge and also surface skimming. In water treatment, surface skimmers are not required. Large rectangular tanks, such as those used in water treatment plants having a capacity above about 2 MGD (7.6 MLD), frequently have length to width ratios of 2:1. Each tank uses two rotary scrapers that are designed for square basins.

In circular tanks, the flow enters either the center of the tank (center feed) or the periphery of the tank (side feed). Figure 9.32 shows the inlets to center-feed tanks. If the tank is less than about 30 ft (9.14 m) in diameter, the inlet pipe will enter through the wall and discharge into the baffle well,

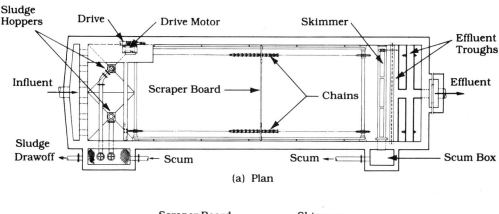

(a) Plan

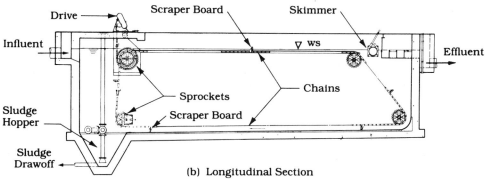

(b) Longitudinal Section

FIGURE 9.31 *Rectangular Settling Tank*

Courtesy of Walker Process Equipment, Division of McNish Corporation.

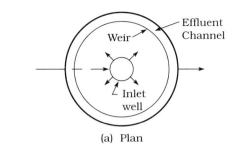

(a) Plan

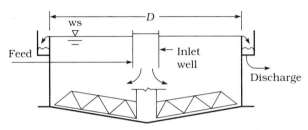

(b) Section, *D* < 30 to 35 ft (9.14 to 10.7m)

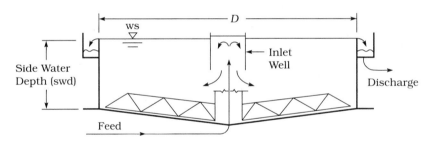

(c) Section, *D* > 30 to 35 ft (9.14 to 10.7m)

FIGURE 9.32 *Inlet and Outlet Details for Circular Tanks (Center Feed)*

as shown in Figure 9.32(b). Then, the flow enters in a downward direction. If the tank is greater than about 30 ft (9.14 m) in diameter, the inlet pipe will run underneath the tank and discharge vertically in the center at the baffle well, as shown in Figure 9.32(c). The depth of a circular clarifier is considered to be the depth at the side of the tank, as shown in Figure 9.32(c), and is referred to as the side water depth (swd). This depth is used for determining tank volume and detention time. The outlets for both tanks consist of a weir channel around the periphery giving a uniform flow removal. The depth of the outlet zone is not as great as for an ideal basin; therefore, it is conservative. Figures 9.33 and 9.34 show the details of the center-feed circular clarifiers that are used in wastewater treatment. They have both mechanical

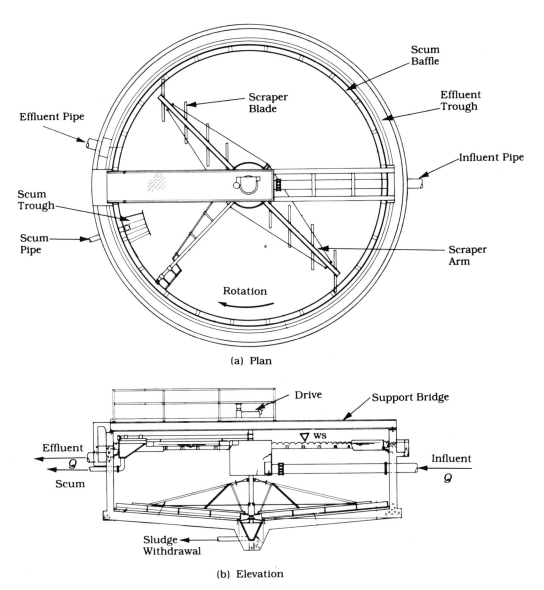

(a) Plan

(b) Elevation

FIGURE 9.33 *Circular Settling Tank (Center Feed by Pipe through Wall)*

Courtesy of Infilco Degremont, Inc.

sludge rakes and surface skimming. The unit in Figure 9.33 has the influent pipe entering through the clarifier wall and extending to the baffle well. Figure 9.34 shows a unit where the influent pipe runs under the clarifier, rises at the center, and discharges into the inlet well. Circular tanks used in water treatment are similar to those used in wastewater treatment except that surface skimmers are not required. The bottom of a circular tank slopes

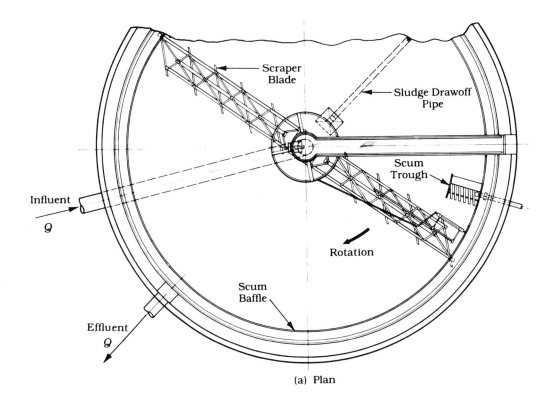

(a) Plan

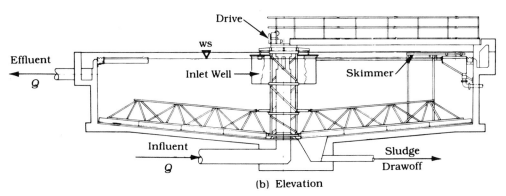

(b) Elevation

FIGURE 9.34 *Circular Settling Tank (Center Feed by Pipe under Tank Bottom)*

Courtesy of Infilco Degremont, Inc.

to the center at a slope which is usually 1:12; thus it forms a flat inverted cone. In design, the volume of the cone is not considered in the design volume, which is taken as being the plan area times the depth of the water at the side of the tank. The sludge is usually collected in a hopper near the center of the tank.

Figure 9.35 shows the inlet details for a periphery-feed tank. As the flow enters, it is deflected so that it moves around the periphery in an orifice channel, as shown in Figure 9.35(a). From the channel the flow discharges through the orifices into the clarifier, as shown in Figure 9.35(b). Sometimes, instead of an orifice channel, there will simply be a skirt surrounding the inside of the tank, and the liquid flows out under the skirt as shown in Figure 9.35(c). Peripheral entry does not give as uniform a flow as the previously discussed tanks. The outlet consists of a weir channel in the center of the basin, and since the outlet zone does not extend the full depth of the tank, it is conservative.

Actual settling basins are affected by dead spaces in the basins, eddy currents, wind currents, and thermal currents. In the ideal settling basin, all of the fluid elements pass through the basin at a time equal to the theoretical detention time, t, which is equal to V/Q. Actual basins, however, have most of the fluid elements passing through at a time shorter than the theoretical detention time, although some fluid elements take longer. Dead spaces and eddy currents have rotational flow and do very little sedimentation, since the

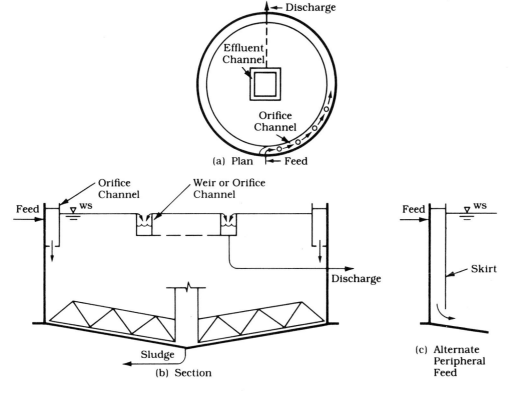

FIGURE 9.35 *Inlet and Outlet Details for a Circular Tank (Peripheral Feed)*

inflow to and the outflow from these spaces is very small. As a result, the net volume available for settling is reduced and the mean flow-through time for the fluid elements is decreased. Also, wind and thermal currents create flows that pass directly from the inlet to the outlet of the basin, which decreases the mean flow-through time. The magnitude of the effects of dead spaces, thermal currents, and so on, and the hydraulic characteristics of a basin, may be measured by using tracer studies. A slug of tracer is added to the influent and the tracer concentration is observed at the outlet, as shown in Figure 9.36. If there are dead spaces, the following relationship occurs (Camp, 1946, 1952):

$$\frac{\text{Mean } t}{\text{Theoretical } t} < 1 \qquad (9.40)$$

If there are no dead spaces, the relationship is

$$\frac{\text{Mean } t}{\text{Theoretical } t} = 1 \qquad (9.41)$$

If short circuiting is occurring, the time relationship is

$$\frac{\text{Median } t}{\text{Mean } t} < 1 \qquad (9.42)$$

If there is no short circuiting,

$$\text{Mean } t = \text{Median } t \qquad (9.43)$$

If a basin is unstable, the time-concentration plot cannot be reproduced in a series of tracer tests. Consequently, erratic basin performance can be expected.

Figure 9.37 shows the results of tracer studies on three types of settling

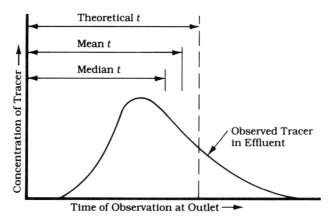

FIGURE 9.36 *Settling Basin Characteristics as Shown by Tracer Studies*

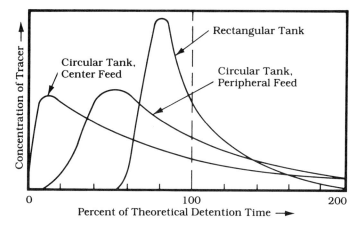

FIGURE 9.37 *Tracer Studies on Circular and Rectangular Settling Tanks*

basins. It can be seen that the rectangular basin approached the ideal more closely than the circular type. For the circular type, the peripheral feed had better performance than the center-feed tank.

In water and wastewater engineering, most of the suspensions are flocculant to a certain degree. Flocculant particles of the same initial size and density as discrete particles will intercept the sludge zone in a shorter time because of flocculation and more rapid settling. Therefore, if the ideal settling basin theory is applied to slightly flocculant particles, it will be conservative. Although there are differences between the ideal basin and actual basins, the ideal settling basin theory gives the most rational approach to design and reveals that the most important design parameters are (1) the overflow rate or design settling velocity, (2) detention time, and (3) depth.

SEDIMENTATION IN WATER TREATMENT PLANTS

In water treatment, sedimentation of both untreated waters (plain sedimentation) and chemically coagulated waters is practiced. If a water has a high turbidity due to silt, plain sedimentation may be used to reduce the turbidity. Plain sedimentation is frequently used for waters having consistent turbidities greater than $1000 \, mg/\ell$. Some rivers, such as the Mississippi, may have infrequent turbidities as high as $40,000 \, mg/\ell$. When plain sedimentation is used, the detention time may be as much as 30 days, and as a result of the extremely large volume, these basins are usually earthen and are constructed using dikes. In most cases, a water to be settled has been coagulated by the addition of chemicals such as those employed in rapid sand filtration plants and lime-soda softening plants.

The settling characteristics of the floc or precipitate depend upon the characteristics of the water, the coagulant used, and the degree of flocculation. The only method to determine accurately the settling velocities and the required overflow rates and detention times is to perform experimental settling tests. Generally, overflow rates of 500 to 1000 gal/day-ft^2 (20.4 to 40.8 m^3/d-m^2) and detention times of 2 to 8 hr (4 to 6 hr being most common) are used for waters coagulated with alum or iron salts in rapid sand filtration plants, and weir or orifice channel loadings usually range from 12,000 to 22,000 gal/day-ft (149 to 273 m^3/d-m). Since alum produces a light floc, overflow rates are usually from 500 to 800 gal/day-ft^2 (20.4 to 32.6 m^3/d-m^2), and weir or orifice channel loadings are from 12,000 to 18,000 gal/day-ft (149 to 224 m^3/d-m). Iron salts produce a more dense floc, and overflow rates are usually from 700 to 1000 gal/day-ft^2 (28.6 to 40.8 m^3/d-m^2), and weir or orifice channel loadings are from 16,000 to 22,000 gal/day-ft (199 to 273 m^3/d-m). The overflow rates, detention times, and weir or orifice channel loadings are based on the average daily flow. In lime-soda softening plants, the overflow rates are usually 700 to 1500 gal/day-ft^2 (28.6 to 61.2 m^3/d-m^2), the detention times are 4 to 8 hr, and the weir or orifice channel loadings are from 22,000 to 26,000 gal/day-ft (273 to 323 m^3/d-m).

EXAMPLE 9.3   *Clarifier for Water Treatment*

A rectangular clarification basin is to be designed for a rapid sand filtration plant. The flow is 8 MGD, the overflow rate or surface loading is 600 gal/day-ft^2, and the detention time is 6 hr. Two sludge scraper mechanisms for square tanks are to be used in tandem to give a rectangular tank with a length to width ratio of 2:1. Determine the design dimensions of the basin if sludge rakes are available in 5-ft increments of length and the design depth.

SOLUTION   The plan area required = $(8.0 \times 10^6 \text{ gal/day})(\text{day-ft}^2/600 \text{ gal})$ or 13,333 ft^2. Since the length, L, is twice the width, W, $(2W)(W) = 13,333$ ft^2. From this, $W = 81.65$ ft; thus, the next standard size is 85 ft. Therefore, the plan design dimensions of the basin are

$$\boxed{\text{Width} = 85 \text{ ft, Length} = 170 \text{ ft}}$$

The actual overflow rate = $(8.0 \times 10^6 \text{ gal/day}) \div (85 \text{ ft})(170 \text{ ft})$ or 554 gal/day-ft^2. Since the depth, H, is equal to the settling rate times the detention time, $H = (554 \text{ gal/day-ft}^2)(\text{ft}^3/7.48 \text{ gal})(\text{day}/24 \text{ hr})(6 \text{ hr})$ or 18.52 ft. Thus,

$$\boxed{\text{Depth} = 18.52 \text{ ft.}}$$

EXAMPLE 9.3 SI   *Clarifier for Water Treatment*

A rectangular clarification basin is to be designed for a rapid sand filtration plant. The flow is 30,300 m^3/day, the overflow rate or surface loading is 24.4 m^3/d-m^2, and the detention time is 6 h. Two sludge scraper mechanisms for square tanks are to be used in tandem to give a rectangular tank with a length to width ratio of 2:1. Determine the dimensions of the basin.

SOLUTION The plan area required $= (30,000\,m^3/d)(d\text{-}m^2/24.4\,m^3)$ or $1242\,m^2$. Since the length, L, is twice the width, W, $(2W)(W) = 1242\,m^2$. From this, $W = 24.9\,m$; thus $L = 2(24.9) = 49.8\,m$. Therefore, the plan dimensions are

$$\boxed{\text{Width} = 24.9\,m, \text{Length} = 49.8\,m}$$

Since the depth, H, is equal to the settling rate times the detention time, $H = (24.4\,m^3/d\text{-}m^2)(d/24\,h)(6\,h) = 6.10\,m$. Thus,

$$\boxed{\text{Depth} = 6.10\,m}$$

SEDIMEN-TATION IN WASTEWATER TREATMENT PLANTS

In conventional wastewater treatment plants, primary sedimentation is used to remove as much settleable solids as possible from raw wastewaters. Secondary settling in activated sludge plants is employed to remove the MLSS and in trickling filter plants to remove any growths that may slough off the filters. As a result, good secondary settling produces a high-quality effluent low in suspended solids. In advanced or tertiary wastewater treatment plants, sedimentation is used for coagulated wastewaters to remove flocculated suspended solids and/or chemical precipitates.

Primary Sedimentation

Recommended criteria for primary clarifiers treating municipal wastewaters are listed in Table 9.7.

The detention times based on the average daily flow are usually from about 45 min to 2 hr; however, the depths and overflow rates listed in Table 9.7 should control in design. Multiple tanks should be used when the flow exceeds 1.0 MGD (3.8 MLD). For plants having a capacity less than 1.0 MGD (3.8 MLD) peak weir loadings should not exceed 20,000 gal/day-ft

TABLE 9.7 *Overflow Rates and Depths for Primary Clarifiers*

| TYPE OF TREATMENT | OVERFLOW RATE gal/day-ft^2 (m^3/d-m^2) | | DEPTH ft (m) |
	Average	Peak	
Primary Settling Followed by Secondary Treatment	800–1200 (32.6–48.9)	2000–3000 (81.5–122)	10–12 (3.0–3.7)
Primary Settling with Waste Activated Sludge	600–800 (24.5–32.6)	1200–1500 (48.9–61.1)	12–15 (3.7–4.6)

From EPA, *Suspended Solids Removal*, EPA Process Design Manual, January 1975.

(248 m³/d-m). For plants having a capacity greater than 1.0 MGD (3.8 MLD), peak loadings should not exceed 30,000 gal/day-ft (373 m³/d-m). A surface skimmer and a baffle are necessary for primary clarifiers to remove scum from the water surface. Although individual tank performance varies, an estimate of the BOD₅ and suspended solids removal can be made using the detention time and overflow rate based on the average daily flow and Figures 9.38 and 9.39.

Some regulatory agencies specify the minimum depth required for wastewater clarifiers. Other agencies specify the minimum detention time and minimum overflow rate or surface loading rate, and the engineer must determine the minimum depth. The overflow rate or surface loading rate must be converted to a settling rate, and the depth is equal to the settling rate times the detention time.

EXAMPLE 9.4   *Primary Clarifier*

A primary clarifier for a municipal wastewater treatment plant is to be designed for an average flow of 2.0 MGD. The state's regulatory agency criteria for primary clarifiers are as follows: peak overflow rate = 2200 gal/day-ft², average overflow rate = 900 gal/day-ft², minimum side water depth = 10 ft, and peak weir loading = 30,000 gal/day-ft. The ratio of the peak hourly flow to the average hourly flow = 2.75. Determine:

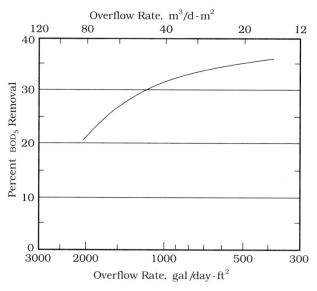

FIGURE 9.38 *Percent BOD₅ Removal versus Overflow Rate in Performance of Primary Clarifiers Treating Municipal Wastewaters*

Adapted from *Recommended Standards for Sewage Works* by the Upper Mississippi and Great Lakes Boards of Public Health Engineers, 1978.

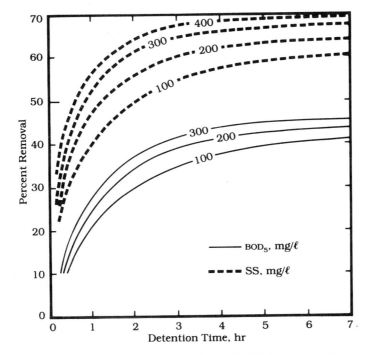

FIGURE 9.39 *Percent* BOD₅ **and Suspended Solids Removal versus Detention Time in Performance of Primary Clarifiers Treating Municipal Wastewaters**

Adapted from *Water Supply and Sewage*, 4th ed., by E. W. Steel. Copyright © 1960 by McGraw-Hill, Inc. Reprinted by permission.

1. The clarifier diameter.
2. The design diameter if equipment is available in 5-ft increments in diameter.
3. The side water depth.
4. The peak weir loading if peripheral weirs are used. Is it acceptable?

SOLUTION The required area based on average flow = (2,000,000 gal/day) (day-ft²/900 gal) = 2222 ft². The required area based on the peak flow = (2,000,000 gal/day)(2.75)(day-ft²/2200 gal) = 2500 ft². The peak flow controls. Thus 2500 ft² = ($\pi/4$)D^2 or D = .56.4 ft.

Therefore, the clarifier diameter is $\boxed{56.4\,\text{ft}}$ and the design diameter is $\boxed{60\,\text{ft}}$ for standard-size equipment. The side water depth is $\boxed{10\,\text{ft.}}$

The length of the peripheral weir = (π)(60 ft) or 189 ft. Thus the peak weir loading = (2,000,000 gal/day)(2.75)/(189 ft) = $\boxed{29{,}100\,\text{gal/day-ft.}}$ Since this is less than the allowable of 30,000 gal/day-ft, it is $\boxed{\text{acceptable.}}$ *Note:*

If the weir loading exceeded the allowable, the diameter of the clarifier would have to be increased until the allowable loading was obtained. In this case weir loading, not overflow or surface loading rate, would determine the diameter of the clarifier.

EXAMPLE 9.4 SI   *Primary Clarifier*

A primary clarifier for a municipal wastewater treatment plant is to be designed for an average flow of $7570 \, m^3/d$. The regulatory agency criteria for primary clarifiers are as follows: peak overflow rate = $89.6 \, m^3/d\text{-}m^2$, average overflow rate = $36.7 \, m^3/d\text{-}m^2$, minimum side water depth = $3.0 \, m$, and peak weir loading = $389 \, m^3/d\text{-}m$. The ratio of the peak hourly flow to the average hourly flow is 2.75. Determine:

1. The diameter of the clarifier.

2. The depth of the clarifier.

3. The peak weir loading if peripheral weirs are used. Is it acceptable?

SOLUTION   The required area based on average flow = $(7570 \, m^3/d) \, (d\text{-}m^2/36.7 \, m^3)$ = $206 \, m^2$. The required area based on the peak flow = $(7570 \, m^3/d) \, (2.75)$ $(d\text{-}m^2/89.6 \, m^3)$ = $232 \, m^2$. The peak flow controls. Thus $232 \, m^2 = (\pi/4)D^2$ or $D = \boxed{17.2 \, m.}$ The depth of the clarifier = $\boxed{3.0 \, m.}$ The length of the peripheral weir = $(\pi)(17.2 \, m)$ = $54.0 \, m$. Thus the peak weir loading = $(7570 \, m^3/d) \, (2.75)/(54.0 \, m)$ = $\boxed{386 \, m^3/d\text{-}m.}$ Since this is less than the allowable of $389 \, m^3/d\text{-}m$, it is $\boxed{acceptable.}$ *Note*: If the weir loading exceeded the allowable, the diameter of the clarifier would have to be increased until the allowable loading was obtained. In this case weir loading, not overflow or surface loading rate, would determine the diameter of the clarifier.

Secondary Sedimentation   Recommended criteria for secondary clarifiers treating municipal wastewaters are shown in Table 9.8. Table 9.9 gives recommended depths for final clarifiers for the activated sludge process. Figure 9.40 shows the details of a final clarifier for the activated sludge process.

    The detention time based on the average daily flow is usually from about 1.0 to 2.5 hr; however, the depths, overflow rates, and solids loadings listed in Table 9.8 should control in design. Multiple tanks should be used when the plant capacity exceeds 1.0 MGD (3.8 MLD), and peak weir loadings similar to those used for primary clarifiers should not be exceeded. Final clarifiers should be provided with skimmers and baffles to remove any floating materials.

    Final clarifiers for the activated sludge process frequently use suction

TABLE 9.8 *Overflow Rates, Solids Loadings, and Depths for Secondary Clarifiers*

TYPE OF TREATMENT	OVERFLOW RATE gal/day-ft² (m³/d-m²)		SOLIDS LOADING lb/day-ft² (kg/d-m²)		DEPTH ft (m)
	Average	Peak	Average	Peak	
Activated Sludge (except Extended Aeration)	400–800 (16.3–32.6)	1000–2000 (40.8–81.6)	20–30 (98–147)	50 (244)	12–15 (3.7–4.6)
Activated Sludge, Extended Aeration	200–400 (8.15–16.3)	800 (32.6)	20–30 (98–147)	50 (244)	12–15 (3.7–4.6)
Activated Sludge, Pure Oxygen	400–800 (16.3–32.6)	1000–2000 (40.8–81.6)	25–35 (122–171)	50 (244)	12–15 (3.7–4.6)
Trickling Filters	400–600 (16.3–24.5)	1000–2000 (40.8–81.6)	—	—	10–12 (3.0–3.7)

From EPA, *Suspended Solids Removal*, EPA Process Design Manual, January 1975.

TABLE 9.9 *Suggested Depths for Final Clarifiers for the Activated Sludge Process*

TANK DIAMETER ft (m)		SIDE WATER DEPTH ft (m)
Up to 40 ft	(Up to 12.2 m)	11 (3.35)
40–70	(12.2–21.3)	12 (3.65)
70–100	(21.3–30.5)	13 (3.96)
100–400	(30.5–42.7)	14 (4.27)
>140	(>42.7)	15 (4.57)

Adapted from *Wastewater Treatment Plant Design*, American Society of Civil Engineers, Manuals and Reports on Engineering Practice No. 36. Copyright © 1977 by the American Society of Civil Engineers. Reprinted by permission.

removal of the settled sludge. In these units, the scraper blades are mounted in pairs, each pair forming a vee (V) in plan view. As the sludge rakes move, the sludge is collected in each vee and removed by a suction pipe mounted above each pair of blades. The sludge is discharged from the suction pipes into a collection well in the center of the unit. The sludge is removed from the collection well by gravity flow in the recycled sludge pipe. The water surface of the sludge in the collection well is slightly lower than the water surface in the clarifier, and this elevation difference is the head

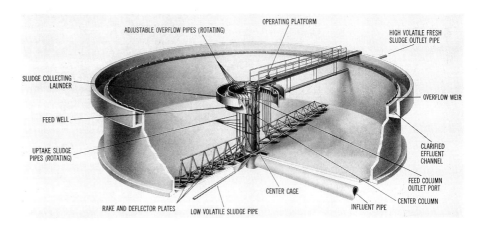

OPERATING PLATFORM
ADJUSTABLE OVERFLOW PIPES (ROTATING)
HIGH VOLATILE FRESH SLUDGE OUTLET PIPE
SLUDGE COLLECTING LAUNDER
OVERFLOW WEIR
FEED WELL
CLARIFIED EFFLUENT CHANNEL
UPTAKE SLUDGE PIPES (ROTATING)
FEED COLUMN OUTLET PORT
CENTER COLUMN
RAKE AND DEFLECTOR PLATES   LOW VOLATILE SLUDGE PIPE   CENTER CAGE   INFLUENT PIPE

FIGURE 9.40 *Final Clarifier for the Activated Sludge Process. Sludge removal is by suction withdrawal.*
Courtesy of Dorr-Oliver, Inc.

required to operate the suction pipes for sludge removal. Clarifiers having suction type sludge removal usually have a sludge blanket less than about 2 ft (0.6 m) deep, and the residence time of the sludge in the clarifier is minimal.

EXAMPLE 9.5   *Final Clarifier*

A final clarifier is to be designed for an activated sludge treatment plant serving a municipality. The state's regulatory agency criteria for final clarifiers used for activated sludge are as follows: peak overflow rate = 1400 gal/day-ft^2, average overflow rate = 600 gal/day-ft^2, peak solids loading = 50 lb/day-ft^2, peak weir loading = 30,000 gal/day-ft, and depth = 11 to 15 ft. The flow to the reactor basin prior to junction with the recycle line = 3.0 MGD. The maximum recycled sludge flow is 100% of the influent flow and is constant throughout the day. The MLSS is 3000 mg/ℓ, and the ratio of the peak hourly influent flow to the average hourly flow is 2.50. Determine:

1. The clarifier diameter.
2. The design diameter if equipment is available in 5-ft increments in diameter.
3. The depth of the clarifier.
4. The peak weir loading if peripheral weirs are used. Is it acceptable?

SOLUTION   The recycle is (1.0)(3.0 MGD) = 3.0 MGD. The average mixed liquor flow = 3.0 + 3.0 = 6.0 MGD. The peak mixed liquor flow = (2.50)(3.0) + 3.0 = 7.5 + 3.0 = 10.5 MGD. The area for clarification based on the average flow = (3,000,000 gal/day)(day-ft^2/600 gal) = 5000 ft^2. The area for clarification based on the peak flow = (7,500,000 gal/day)(day-ft^2/1400 gal) = 5357 ft^2.

The peak solids flow = $(10,500,000 \text{ gal/day})(8.34 \text{ lb/gal})(3000/10^6)$ = $262,710 \text{ lb/day})$. The area for solids loading = $(262,710 \text{ lb/day})(\text{day-ft}^2/50 \text{ lb})$ = 5254 ft^2. Thus the peak overflow rate controls. The diameter is given by $(\pi/4)(D^2) = 5357 \text{ ft}^2$ or the clarifier diameter = $\boxed{82.6 \text{ ft.}}$

The design diameter = $\boxed{85 \text{ ft}}$ for standard-size equipment. For an 85-ft-diameter clarifier, Table 9.9 gives a suggested depth of 13 ft, so the side water depth = $\boxed{13 \text{ ft.}}$

The peak weir loading = $(7,500,000 \text{ gal/day})/(\pi)(85.0 \text{ ft})$ or $\boxed{28,100 \text{ gal/day-ft.}}$ Since this is less than the allowable of 30,000 gal/day-ft, the design is $\boxed{\text{satisfactory.}}$

EXAMPLE 9.5 SI   *Final Clarifier*

A final clarifier is to be designed for an activated sludge treatment plant serving a municipality. The state's regulatory agency criteria for final clarifiers used for activated sludge are as follows: peak overflow rate = $57.0 \text{ m}^3/\text{d-m}^2$, average overflow rate = $24.4 \text{ m}^3/\text{d-m}^2$, peak solids loading = 244 kg/d-m^2, peak weir loading = $373 \text{ m}^3/\text{d-m}$, and depth = 3.35 to 4.57 m. The flow to the reactor basin prior to junction with the recycle line = $11,360 \text{ m}^3/\text{day}$. The maximum recycled sludge flow is 100% of the influent flow and is constant throughout the day. The MLSS = $3000 \text{ mg}/\ell$, and the ratio of the peak hourly influent flow to the average hourly flow is 2.50. Determine:

1. The diameter of the clarifier.

2. The depth of the clarifier.

3. The peak weir loading if peripheral weirs are used. Is it acceptable?

SOLUTION   The recycle is $(1.0)(11,360 \text{ m}^3/\text{d}) = 11,360 \text{ m}^3/\text{d}$. The average mixed liquor flow = $11,360 + 11,360 = 22,720 \text{ m}^3/\text{d}$. The peak mixed liquor flow = $(2.50)(11,360) + 11,360 = 28,400 + 11,360 = 39,760 \text{ m}^3/\text{d}$. The area for clarification based on the average flow = $(11,360 \text{ m}^3/\text{d})(\text{d-m}^2/24.4 \text{ m}^3)$ = 466 m^2. The area for clarification based on the peak flow = $(28,400 \text{ m}^3) \times (\text{d-m}^2/57.0 \text{ m}^3) = 498 \text{ m}^2$. The peak solids flow = $(39,760 \text{ m}^3/\text{d})(1000 \ell/\text{m}^3) \times (3000 \text{ mg}/\ell)(\text{kg}/10^6 \text{ mg}) = 119,280 \text{ kg/d}$. The area for solids loading = $(119,280 \text{ kg/d})(\text{d-m}^2/244 \text{ kg}) = 489 \text{ m}^2$. Thus the peak overflow rate controls. The diameter is given by $(\pi/4)(D^2) = 498 \text{ m}^2$ or the clarifier diameter = $\boxed{25.2 \text{ m.}}$ For a clarifier of this diameter, Table 9.9 gives a suggested depth of 3.96 m, so the side water depth = $\boxed{3.96 \text{ m.}}$ The peak weir loading = $(28,400 \text{ m}^3/\text{d})/(\pi)(25.2 \text{ m}) = \boxed{359 \text{ m}^3/\text{d-m.}}$ Since this is less than the allowable of $373 \text{ m}^3/\text{d-m}$, the design is $\boxed{\text{satisfactory.}}$

Chemical Treatment Sedimentation

The peak overflow rates used for coagulation in tertiary treatment or for coagulation of raw municipal wastewaters and secondary effluents depend mainly upon the type of coagulant employed. Recommended peak overflow rates are shown in Table 9.10.

A detention time of at least 2 hours based on the average daily flow should be provided. Alum or iron salt coagulated wastewaters should not have a weir loading greater than 10,000 to 15,000 gal/day-ft (124 to 186 m^3/d-m) based on the average daily flow. Lime coagulated wastewaters should not have an average weir loading greater than 20,000 to 30,000 gal/day-ft (248 to 372 m^3/d-m). The expected performance from chemical coagulation of wastewaters can best be determined from pilot plant studies. Chemical sludges produced from chemical coagulation may vary from 0.5 to more than 1.0% of the plant capacity and have solids concentrations from 1 to 15%, depending on the chemical used and on basin efficiency.

Lime-feed or lime-sludge drawoff lines should be glass-lined or PVC pipe to facilitate cleaning of encrustations. Also, recycling of sludge from the bottom of the clarifiers to the rapid-mix tank should be provided to assist in coagulation. For lime-settling basins the mechanical sludge rakes should be the bottom scraper type, and not the suction pick-up type, because of the dense sludge to be removed.

INCLINED-SETTLING DEVICES

Inclined settling devices include inclined-tube settlers and the Lamella separator, both of which have overflow rates much higher than conventional settling basins.

Inclined-Tube Settlers

Figure 9.41 shows the details of the inclined-tube settler. The water to be clarified passes upward through the tubes, and as settling occurs the solids are collected on the bottom of the tubes, as shown in Figure 9.41(a). The tubes are inclined at an angle of 45° to 60°, which is steep enough to cause the settled sludge to slide down the tubes. The sludge falls from the tubes to

TABLE 9.10 *Peak Overflow Rates for Various Coagulants*

COAGULANT	PEAK OVERFLOW RATE gal/day-ft² (m^3/d-m²)
Alum	500–600 (20.4–24.5)
Iron salts	700–800 (28.5–32.6)
Lime	1400–1600 (57.1–65.2)

From EPA, *Suspended Solids Removal*, EPA Process Design Manual, January 1975.

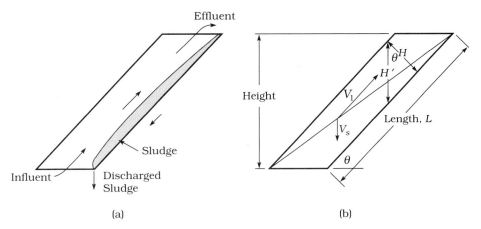

FIGURE 9.41 *Inclined-Tube Settler*

the bottom of the clarifier, where it is removed by the sludge rakes. The tube cross section may be of numerous geometric shapes; however, a square or rectangular cross section is the most common type. Inclined-tube settlers are made in modules consisting of numerous inclined tubes. Figure 9.42 shows modules installed in a circular clarifier, whereas Figure 9.43 shows modules installed in a rectangular clarifier.

The advantage of a tube settler over a conventional tank can be illustrated by using the ideal settling theory and a theoretical problem. If the flow to a conventional rectangular settling tank is 1.0 MGD, (3785 m³/d), the detention time is 2 hr, the overflow rate is 1000 gal/day-ft² (40.75 m³/d-m²), and the length to width ratio is 4:1, then the tank will be 63.25 feet (19.28 m) long, 15.81 feet (4.82 m) wide, and 11.1 feet (3.40 m) deep. The overflow rate of 1000 gal/day-ft² (40.75 m³/d-m²) corresponds to a settling velocity of 5.55 ft/hr (1.70 m/h). According to the ideal settling basin theory, a settling particle must intersect the sludge zone before it reaches the outlet end of a settling chamber in order to be removed. Thus if V_s is the settling velocity, H is the depth, V_1 is the horizontal velocity, and L is the chamber

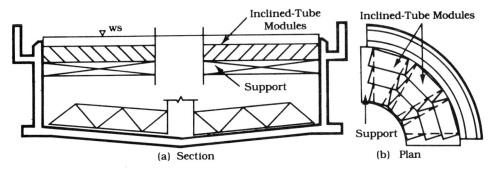

FIGURE 9.42 *Inclined-Tube Settlers in a Circular Clarifier*

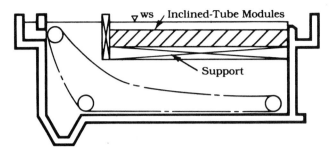

FIGURE 9.43 *Inclined-Tube Settlers in a Rectangular Clarifier*

length, the critical trajectory of a settling particle is such that the following relationship holds true:

$$\frac{V_1}{L} = \frac{V_s}{H} \tag{9.44}$$

or

$$V_1 = V_s L/H \tag{9.45}$$

Applying Eq. (9.45) to the hypothetical problem gives

$$V_1 = \left(5.55 \frac{\text{ft}}{\text{hr}}\right)\left(\frac{63.25\,\text{ft}}{11.10\,\text{ft}}\right) = 31.6\,\text{ft/hr}$$

or

$$V_1 = \left(1.70 \frac{\text{m}}{\text{h}}\right)\left(\frac{19.28\,\text{m}}{3.40\,\text{m}}\right) = 9.64\,\text{m/h}$$

Thus the horizontal velocity is 31.6 ft/hr (9.64 m/h). For an inclined-tube settler, a particle must settle through the distance H', as shown in Figure 9.41(b). Thus, for an inclined tube, Eq. (9.44) becomes

$$\frac{V_1}{L} = \frac{V_s}{H'} \tag{9.46}$$

Assume, for the hypothetical problem, that modules of tube settlers are to be placed in the rectangular tank. The modules are 3.0 ft (0.914 m) high, and the tubes are 2 in. (0.0508 m) deep and inclined at a 45° angle. The value of H' is given by

$$H' = \frac{H}{\cos \theta} \tag{9.47}$$

or

$$H' = \left(\frac{2\,\text{in.}}{\cos 45°}\right)\left(\frac{\text{ft}}{12\,\text{in.}}\right) = 0.236\,\text{ft}$$

or

$$H' = \frac{0.0508 \text{ m}}{\cos 45°} = 0.07184 \text{ m}$$

The value of L is

$$L = \text{module height}/\sin \theta \qquad \textbf{(9.48)}$$

or

$$L = 3.0 \text{ ft}/\sin 45° = 4.24 \text{ ft}$$

or

$$L = 0.914 \text{ m}/\sin 45° = 1.293 \text{ m}$$

Now, applying Eq. (9.46) to get the velocity through the tube gives

$$V_1 = \left(\frac{L}{H'}\right)V_s = \left(\frac{4.24}{0.236}\right)\left(\frac{5.55 \text{ ft}}{\text{hr}}\right) = 99.7 \text{ ft/hr}$$

or

$$V_1 = \left(\frac{L}{H'}\right)V_s = \left(\frac{1.293 \text{ m}}{0.07184 \text{ m}}\right)\left(\frac{1.70 \text{ m}}{\text{h}}\right) = 30.6 \text{ m/h}$$

Thus the velocity through the inclined-tube settler can be 99.7 ft/hr (30.6 m/h) and still have the same degree of solids removal as the horizontal settling unit. This is 99.7/31.6, or 3.2 (30.6/9.64 or 3.2) times as much flow as the conventional basin can accommodate. Thus the advantages of the inclined-tube settlers are readily apparent.

Usually, the overflow rates used for inclined-tube settlers are from three to six times as great as those used for conventional settling tanks. Laminar flow is necessary for efficient settling since turbulent flow would scour the settled solids. Laminar conditions are made possible by the use of tubes with small hydraulic radii.

Figures 9.42 and 9.43 show modules of inclined-tube settlers installed in a new or existing circular or rectangular clarifier. In either type clarifier, a large portion of the plan area, usually 67 to 80%, is occupied by the tube modules. New settling tanks using inclined-tube settlers will have much less area requirements than conventional settling tanks; however, one of the most common uses of inclined tubes is to increase the capacity of existing clarifiers,

In water treatment where the water temperature is above 50°F, it has been reported that effluent turbidities from 1 to 7 JTU (Jackson Turbidity Units) may be achieved, depending upon the overflow rate, if the raw water turbidity is less than 1000 JTU (Culp and Culp, 1974). When the water temperature is below 40°F, expected turbidities are from 1 to 10 JTU if the raw water turbidity is less than 1000 JTU. Since the settling velocity is dependent upon the water temperature, better results are obtained with warm waters. The usual overflow rates based upon the area covered by the tube modules are from 3600 to 6000 gal/day-ft^2 (147 to 245 m^3/d-m^2). The tube settlers are ideal for increasing the capacity of existing clarifiers.

In wastewater treatment, tube settlers have been successfully used in secondary settling for activated sludge and trickling filter plants and for settling of coagulated wastewaters. They are not well suited as primary clarifiers because biological growths develop within the tubes. In particular, they are useful in increasing the capacities of existing final clarifiers. When they are used for final clarifiers, the activated sludge mixed liquor or trickling filter effluent is discharged below the tube modules. Some sludge settles and the final effluent is clarified as it passes upward through the tube settlers. The sludge from the settlers falls from the modules to the bottom of the clarifier and is removed with the other sludge. Overflow rates as high as 4000 gal/day-ft^2 (163 m^3/d-m^2) have been used, which are about five times the overflow rates normally used for conventional settling tanks. Some installations have had microbial slime buildups inside the tube settlers. This can be minimized by installing an air grid underneath the modules and using air scouring periodically to clean off the growths.

Lamella Separator   The Lamella separator is similar to the inclined-tube settlers except that inclined plates are used to form the settling compartments, and the sludge and water flow is cocurrent instead of countercurrent. The manufacturer recommends it only for use with coagulated waters and wastewaters.

INLET AND OUTLET HYDRAULICS   To prevent short circuiting and basin instability, it is essential that the influent flow enter a sedimentation basin uniformly and also that the effluent flow leave uniformly. Figure 9.44(a) shows the plan of a typical rectangular basin with an influent orifice flume and an effluent weir channel. Figure 9.44(b) shows a cross section through the influent flume, and Figure 9.44(c) shows a cross section through the effluent channel. Figure 9.45(a) shows a profile of the influent flume, whereas Figure 9.45(b) shows a profile through the effluent channel. Although the profile shows the influent flume with circular orifices, square orifices are also used. The effluent weir may be a suppressed weir or a series of 90° V-notch weirs, as shown in Figure 9.45(b) for a portion of the channel. If V-notch weirs are used, they are usually at about 8-in. (205-mm) centers. The elevation of the water surface in the effluent box, shown in Figure 9.45(b), is set by the elevation of the water surface in the next down-line treatment unit and by the total head loss between the two water surfaces.

In the design of an orifice flume, as shown in Figures 9.44(a) and (b) and 9.45(a), good design practice requires that the discharge from the most distant orifice from the influent pipe be at least 95% of the discharge from the closest orifice. Usually, the friction and form losses in the flume are very small compared to the head losses through the orifices, which makes this criterion easy to satisfy. The discharge from an orifice is given by $Q = 0.6 \times A\sqrt{2gh}$, where Q = cfs (m^3/s), A = orifice area (ft^2 or m^2), h = head loss (ft or m), and g = acceleration due to gravity. The velocity in the

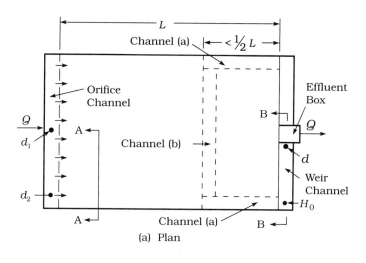

(a) Plan

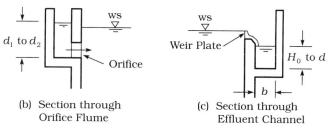

(b) Section through
Orifice Flume

(c) Section through
Effluent Channel

FIGURE 9.44 *Inlet and Outlet Details for a Rectangular Tank*

influent pipe and flume should be sufficient to maintain the suspended solids in suspension, yet not over 3 to 4.5 fps (0.9 to 1.37 m/s) to avoid unnecessary head losses. If the influent is coagulated water, low velocities should be used to avoid shearing the floc. For detailed procedures on determining head losses in influent orifice flumes, the reader is referred to the book titled *Water Treatment Plant Design* by J. M. Montgomery Inc. (1985).

In the design of a weir channel, as shown in Figures 9.44(a) and (c) and 9.45(b), the flow enters uniformly along the channel since the suppressed weir crest and the crests of the V-notch weirs are level. The discharge over a suppressed weir is given by

$$Q = CLH^{3/2} \tag{9.49}$$

where

Q = discharge, cfs (m^3/s)
C = 3.33 for USCS units and 1.84 for SI units
L = weir length, ft (m)
H = head, ft (m)

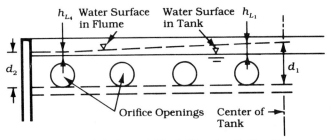

(a) Section A-A through Tank Showing Orifice Flume

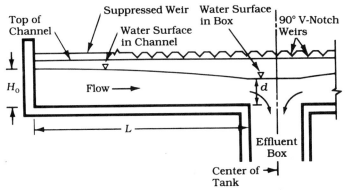

(b) Section B-B through Effluent Channel and Effluent Box

FIGURE 9.45 *Sections through Orifice Flume and Effluent Channel*

If 90° V-notch weirs are used, the discharge is given by

$$Q = CH^{5/2} \tag{9.50}$$

where

Q = discharge, cfs (m³/s)

C = 2.54 for USCS units and 1.40 for SI units

H = head, ft (m)

Usually, 90° V-notch weirs are used for settling tanks in water and wastewater treatment plants instead of suppressed weirs, because they are affected less by sustained winds blowing across the tanks. A sustained wind will cause setup in which the water surface is not level. When setup occurs, the upwind water surface drops and the downwind water surface rises, and if peripheral weirs are used for a circular tank, the discharge per unit length of channel at the upwind side is less than the discharge per unit length at the downwind side. Although both V-notch and suppressed weirs are affected by setup, the increase or decrease in discharge for V-notch weirs is significantly lower than that for suppressed weirs. Wind setup also affects weirs in rectangular tanks. Orifice channels are used mainly in water treatment plants. Effluent channels are lateral spillway channels, and in 1940, Professor T. R. Camp published his equation for their design. His equation was presented in the first edition of this textbook, and the friction term in his equation was based on Darcy's equation. At the time Camp published his equation in 1940, many equations were used for open channels, and he chose to use Darcy's because it was a fundamental equation, not an empirical one like the others. Since Manning's equation is presently the most commonly used for open channels, Camp's equation has been modified (Reynolds *et al.*, 1989) to include a friction term based on Manning's equation. The modified equation for a level lateral spillway channel is

$$H_0 = \sqrt{d^2 + \frac{2Q^2}{gb^2d} + \frac{2}{3} \cdot \frac{n^2LQ^2}{C^2b^2\bar{r}^{4/3}\bar{d}}} \qquad (9.51)$$

where

H_0 = upstream water depth, ft (m)

d = downstream water depth, ft (m)

Q = total discharge, cfs (m³/s)

g = acceleration due to gravity, ft/sec² (m/s²)

b = channel width, ft (m)

n = Manning's friction factor, 0.029 to 0.032 for concrete

L = channel length, ft (m)

C = 1.49 for USCS units and 1.0 for SI units

$\bar{r}$ = mean hydraulic radius, ft (m)

$\bar{d}$ = mean depth, ft (m)

The values of the Darcy friction factor, f, found by Camp correspond to those of the Manning friction factors, n, given here. The values of n are higher than normally found for concrete because of the turbulence in the channel created by the flow spilling laterally into the channel. For a level channel the water surface profile is parabolic, and for this case, the mean depth, $\bar{d}$, can be approximated by

$$\bar{d} = H_0 - 1/3(H_0 - d) \qquad (9.52)$$

The mean hydraulic radius, $\bar{r}$, is given by

$$\bar{r} = \frac{\bar{d}b}{2\bar{d} + b} \qquad (9.53)$$

In using Eq. (9.51), a trial calculation ignoring friction is made to determine the H_0 value without friction. Next, Eq. (9.52) can be used to get the mean depth, $\bar{d}$, and Eq. (9.53) can be used to get the mean hydraulic radius, $\bar{r}$. Eq. (9.51) can then be used to get the H_0 value for the first trial. The procedure is repeated until a constant H_0 value is obtained, which usually requires three successive trials.

Usually a freefall of 3 to 4 in. (76 to 102 mm) is allowed from the weir crest to the maximum water surface which occurs at H_0. The value of the downstream depth, d, is set by the elevation of the water surface in the effluent box. The minimum value of d is the critical depth, y_c, and this occurs when the water surface in the effluent box is low enough to cause $d = y_c$. The critical depth for a rectangular channel is given by $y_c = (q^2/g)^{1/3}$, where y_c = critical depth (ft or m), q = discharge per unit width of channel (that is, $q = Q/b$, cfs/ft or m³/s-m), and g = acceleration due to gravity. If the water surface in the effluent box is low enough to cause a freefall (that is, the water surface is below the effluent channel), the value of d is y_c since critical depth occurs just before a freefall.* If the weir loading causes the required weir length to be greater than the tank width, the channel may be extended along the sides of the basin up to a length of one-half the basin length, as shown by channel (a) in Figure 9.44(a). If more channel is still required, a self-supporting channel may be constructed across the basin, as shown by channel (b) in Figure 9.44(a). If a channel is constructed across the basin, it should have about 10- to 15-ft (3.0 to 4.6 m) minimum clearance between it and the channel across the end of the tank. Quite often at water treatment plants with basins having a 2:1 length to width ratio, to obtain an effluent channel of sufficient length the channel is placed 5 to 8 ft (1.8 to 2.4 m) from the walls and is supported on cantilever beams. If a weir channel is used, this allows flow to spill over both sides. If an orifice channel is used, this allows flow to enter both sides and also the bottom. The channel runs across the end of the basin and up both sides (not more than one-half the length) to form a U-shaped layout when viewed from above the basin. A head loss of $1.5V^2/2g$ occurs at 90° bends in a channel. If the effluent channel is an orifice channel, which is frequently the case in water treatment, it may still be designed using Eq. (9.51), and the discharge from the last orifice should be at least 95% of the discharge of the first orifice.

For a peripheral-feed circular clarifier with an influent orifice flume, the flume may be designed as described for a rectangular tank. The effluent channels may be designed by using Eq. (9.51). For a center-feed circular clarifier, the entrance loss equals the velocity head. The peripheral effluent channel may be designed by using Eq. (9.51) and assuming that the flow splits directly opposite the effluent box so that one-half the total flow enters each side of the effluent box.

EXAMPLES 9.6 AND 9.6 SI

Inlet and Outlets

A rectangular settling basin is 70 ft (21.33 m) wide by 140 ft (42.67 m) long and has a flow of 4.90 MGD or 7.58 cfs (0.2146 m³/s). The inlet flume is an orifice channel with eight orifices that are $8\frac{1}{2}$ in. (216 mm) in diameter, each with an area of 0.394 ft² (0.0366 m²). The difference in the elevations

---

* Critical depth can be assumed at the freefall.

of the water surface at the influent pipe (that is, the center of the flume) and at the last orifice in the flume is 0.02 ft (0.0061 m). This loss in head is due to the friction and form losses in the flume. The effluent weir plate consists of 90° V-notch weirs spaced at 8-in. (203-mm) centers. The effluent channel is 160 ft (48.77 m) in length and extends across the downstream end of the basin, which is 70 ft (21.33 m), and 45 ft (13.72 m) upstream along each side. The effluent channel is rectangular in cross section and is 1.75 ft (0.533 m) in width. There is a 4-in. (102-mm) freefall between the crests of the V-notch weirs and the maximum water depth. The Manning friction coefficient is 0.032, and there is a freefall at the effluent box. Determine:

1. The ratio of the flow from the last influent orifice to the flow from the influent orifice nearest to the influent pipe.

2. The head on the V-notch weirs.

3. The water depth in the effluent channel at the effluent box and the depth at the upstream end of the channel. Make the depth at the effluent box 4 in. (102 mm) greater than the critical depth to insure submerged conditions for the outlet pipe.

4. The vertical distance from the crests of the V-notch weirs to the flowline of the effluent channel.

SOLUTION For the solution using USCS units, the average flow through the orifices is 7.58 cfs/8 or 0.95 cfs per orifice. The head loss at the first orifice, h_{L_1}, is given by $0.95 = (0.6)(0.394)(2gh_{L_1})^{1/2}$ from which $h_{L_1} = 0.25$ ft. The head loss at the last orifice, h_{L_4}, is $h_{L_1} - 0.02$ or $h_{L_4} = 0.25 - 0.02 = 0.23$ ft. The flow from the last orifice is $Q_4 = (0.6)(0.394)[(2g)(0.23)]^{1/2}$, from which $Q_4 = 0.91$ cfs. The ratio of the flows through the orifices is $(Q_4/Q_1)(100\%)$ or $(0.91/0.95)(100\%)$ or $\boxed{95.8\%.}$ The number of V-notch weirs is 160 ft/0.667 ft $= 240$. The flow for each is 7.58/240 or 0.0316 cfs. The head on the weirs is given by $0.0316 = 2.54 H^{5/2}$ or $H = \boxed{0.17\,\text{ft,}}$ which is $\boxed{2.04\,\text{in.}}$ The discharge per foot width of channel is $(7.58\,\text{cfs})(1/2)(1/1.75\,\text{ft})$ or 2.166 cfs/ft. Thus the critical depth is $y_c = [(2.166)^2/32.17]^{1/3}$ or 0.53 ft. Therefore, the depth at the effluent box is $0.53 + 4/12 = \boxed{0.86\,\text{ft.}}$ Using Eq. (9.51) and ignoring the friction term gives $H_0 = 1.04$ ft. For the first trial with friction, the mean depth $\bar{d} = 1.04 - 1/3(1.04 - 0.86) = 0.9800$ ft. The mean hydraulic radius $\bar{r} = (0.9800)(1.75)/(2 \times 0.9800 + 1.75) = 0.4623$ ft. The channel length is $(45 + 70 + 45)(1/2) = 80$ ft. Substituting these and other variables into Eq. (9.51) gives $H_0 = 1.19$ ft. For the second trial with friction, the mean depth $\bar{d} = 1.19 - 1/3(1.19 - 0.86) = 1.080$ ft. The mean hydraulic radius $\bar{r} = (1.080)(1.75)/(2 \times 1.080 + 1.75) = 0.4834$ ft. Substituting these and other variables into Eq. (9.51) gives $H_0 = 1.17$ ft. For the third trial with friction, the mean depth $\bar{d} = 1.17 - 1/3(1.17 - 0.86) = 1.067$ ft. The mean hydraulic radius $\bar{r} = (1.067)(1.75)/(2 \times 1.067 + 1.75) = 0.4807$ ft. Substituting these and other variables into Eq. (9.51) gives

$$H_0 = \sqrt{(0.86)^2 + \frac{(2)(7.58/2)^2}{(32.17)(1.75)^2(0.86)} + \frac{2}{3} \cdot \frac{(0.032)^2(80)(7.58/2)^2}{(1.49)^2(1.75)^2(0.4807)^{4/3}(1.067)}}$$

From this, $H_0 = 1.17$ ft. The flow at the 90° bend in the channel is $(7.58\,\text{cfs})(1/2)(45)/(45 + 35)$ or 2.132 cfs. The approximate velocity at the bend is $(2.132\,\text{cfs})/[(1.75)(1.17 + 0.86)(1/2)] = 1.20$ fps. Thus the head loss is $(1.5)(1.20)^2/2g$ or 0.03 ft. The upstream channel depth is $1.17 + 0.03 = \boxed{1.20\,\text{ft.}}$ The difference in elevation from the weir crests to the flowline of the effluent channel is $1.20 + 4/12$ or $\boxed{1.53\,\text{ft.}}$

For the solution using SI units, the average flow through the orifices is $0.2146\,\text{m}^3/\text{s} \div 8$ or $0.0268\,\text{m}^3/\text{s}$ per orifice. The head loss at the first orifice, h_{L_1}, is given by $0.0268 = (0.6)(0.0366)(2gh_{L_1})^{1/2}$, from which $h_{L_1} = 0.0759$ m. The head loss at the last orifice, h_{L_4}, is $0.0759 - 0.0061$ or $h_{L_4} = 0.0698$ m. The flow from the last orifice is $Q_4 = (0.6)(0.0366)[(2g)(0.0698)]^{1/2}$ or $0.0257\,\text{m}^3/\text{s}$. The ratio of the flows through the orifices is $(Q_4/Q_1)(100\%)$ or $(0.0257)/(0.0268)(100\%)$ or $\boxed{95.9\,\text{percent.}}$ The number of V-notch weirs is $(13.72\,\text{m} + 21.33\,\text{m} + 13.72\,\text{m})/(0.203\,\text{m}) = 240$. The flow through each is $0.2146/240$ or $0.000894\,\text{m}^3/\text{s}$. The head on the weirs is given by $0.000894 = 1.40\,H^{5/2}$ or $H = \boxed{0.053\,\text{m.}}$ The discharge per meter width of channel is $(0.2146\,\text{m}^3/\text{s})(1/2)(1/0.533\,\text{m}) = 0.2013\,\text{m}^3/\text{s}$. Thus the critical depth is $y_c = [(0.2013)^2/9.806]^{1/3} = 0.160$ m. Therefore, the depth at the effluent box is $0.160 + 0.102 = \boxed{0.262\,\text{m.}}$ Using Eq. (9.51) and ignoring the friction term gives $H_0 = 0.317$ m. For the first trial with friction, the mean depth $\bar{d} = 0.317 - 1/3(0.317 - 0.262) = 0.2987$ m. The mean hydraulic radius $\bar{r} = (0.2987)(0.533)/(2 \times 0.2987 + 0.533) = 0.1408$ m. The channel length is $(13.72 + 21.33 + 13.72) = 24.39$ m. Substituting these and other variables into Eq. (9.51) gives $H_0 = 0.362$ m. For the second trial with friction, the mean depth $\bar{d} = 0.362 - 1/3(0.362 - 0.262) = 0.3287$ m. The mean hydraulic radius $\bar{r} = (0.3287)(0.533)/(2 \times 0.3287 + 0.533) = 0.1472$ m. Substituting these and other variables into Eq. (9.51) gives $H_0 = 0.356$ m. For the third trial with friction, the mean depth $\bar{d} = 0.356 - 1/3(0.356 - 0.262) = 0.3247$ m. The mean hydraulic radius $\bar{r} = (0.3247)(0.533)/(2 \times 0.3247 + 0.533) = 0.1464$ m. Substituting these and other variables into Eq. (9.51) gives

$$H_0 = \sqrt{(0.262)^2 + \frac{2(0.2146/2)^2}{(9.806)(0.533)^2(0.262)} + \frac{2}{3} \cdot \frac{(0.032)^2(24.39)(0.2147/2)^2}{(0.533)^2(0.1464)^{4/3}(0.3247)}}$$

From this, $H_0 = 0.357$ m. The flow at the 90° bend in the channel is $(0.2146\,\text{m}^3/\text{s})(1/2)(13.72)/(13.72 + 21.33/2) = 0.0604\,\text{m}^3/\text{s}$. The approximate velocity at the bend is $(0.0604\,\text{m}^3/\text{s})/[(0.533)(0.357 + 0.262)(1/2)] = 0.366$ m/s. Thus the head loss is $(1.5)(0.366)^2/2g$ or 0.010 m. The upstream depth is $0.357 + 0.010 = \boxed{0.367\,\text{m.}}$ The difference in elevation from the V-notch weir crests to the flowline of the channel is $0.367 + 0.102 = \boxed{0.469\,\text{m.}}$

In the hydraulic design of a water or wastewater treatment plant, the total head losses between the various treatment units are determined for the maximum expected flowrate and are used in setting the water surface elevations in the various units. Take, for example, an activated sludge plant with a pump station between the preliminary treatment units and the primary clarifier and with a river as the receiving body of water. The total head losses are determined between the water surfaces of the primary clarifier and the aeration tank, the aeration tank and the final clarifier, the final clarifier and the chlorine contact tank, and the chlorine tank and the river. Then the maximum elevation of the river water surface is determined. To determine the minimum allowable water surface elevation of the chlorine contact tank, the head loss between the tank and the river is added to the elevation of the river water surface. To determine the minimum allowable water surface elevation of the final clarifier, the head loss between the clarifier and the contact tank is added to the elevation of the water surface in the chlorine contact tank. In a similar manner, the minimum allowable water surface elevations are determined for the aeration tank and the primary clarifier. The water surface elevations in the various units must be equal to or greater than the minimum allowable elevations in order to avoid backwater from the river. Usually the water surface elevations of the various units are above the minimum allowable elevations, but the drop in elevation between the units is only sufficient to accommodate the maximum expected flow.

REFERENCES

American Society of Civil Engineers (ASCE). 1982. *Wastewater Treatment Plant Design*. ASCE Manual of Practice No. 36. New York: ASCE.

American Society of Civil Engineers (ASCE). 1990. *Water Treatment Plant Design*. 2nd ed. New York: McGraw-Hill.

American Water Works Association (AWWA). 1990. *Water Quality and Treatment*. 4th ed. New York: McGraw-Hill.

American Water Works Association (AWWA). 1969. *Water Treatment Plant Design*. New York: AWWA.

American Water Works Association (AWWA). 1971. *Water Quality and Treatment*. New York: McGraw-Hill.

Barnes, D.; Bliss, P. J.; Gould, B. W.; and Vallentine, R. H. 1981. *Water and Wastewater Engineering Systems*. London: Pitman Books Limited.

Camp, T. R. 1940. Lateral Spillway Channels. *Trans. ASCE* 105:606.

Camp, T. R. 1946. Sedimentation and the Design of Settling Tanks. *Trans. ASCE* 111:895.

Camp, T. R. 1952. "Water Treatment." In *The Handbook of Applied Hydraulics*. 2nd ed. Edited by C. V. Davis. New York: McGraw-Hill.

Coe, H. S., and Clevenger, G. H. 1916. Methods for Determining the Capacities of Slime-Settling Tanks. *Trans. Am. Inst. Mining Met. Engrs.* 55:356.

Conway, R. A., and Ross, R. D. 1980. *Handbook of Industrial Waste Disposal*. New York: Van Nostrand Reinhold.

Culp, G. L., and Conley, W. 1970. "High-Rate Sedimentation with the Tube-Clarifier Concept." In *Advances in Water Quality Improvement by Physical and Chemical Processes*. Edited by E. F. Gloyna and W. W. Eckenfelder. Austin, Tex.: University of Texas Press.

Culp, R. L., and Culp, G. L. 1978. *Handbook of Advanced Wastewater Treatment*. 2nd ed. New York: Van Nostrand Reinhold.

Culp, G. L., and Culp, R. L. 1974. *New Concepts in Water Purification*. New York: Van Nostrand Reinhold.

Degremont, Inc. 1973. *Water Treatment Handbook.* Caxton Hill, Hertford, England: Stephen Austin and Sons.

Dick, R. I. 1970. Role of Activated Sludge Final Settling Tanks. *Jour. SED* 96, SA2:423.

Eckenfelder, W. W. 1966. *Industrial Water Pollution Control.* New York: McGraw-Hill.

Eckenfelder, W. W., Jr. 1989. *Industrial Water Pollution Control.* 2nd ed. New York: McGraw-Hill.

Eckenfelder, W. W., Jr. 1980. *Principles of Water Quality Management.* Boston, Mass.: CBI Publishing.

Eckenfelder, W. W., Jr. 1970. *Water Quality Engineering for Practicing Engineers.* New York: Barnes & Noble.

Eckenfelder, W. W., Jr., and O'Connor, D. J. 1961. *Biological Waste Treatment.* London: Pergamon Press.

Environmental Protection Agency (EPA). 1973. *Phosphorus Removal.* EPA Process Design Manual, Washington, D.C.

Environmental Protection Agency (EPA). 1975. *Suspended Solids Removal.* EPA Process Design Manual, Washington, D.C.

Environmental Protection Agency (EPA). 1974. *Upgrading Existing Wastewater Treatment Plants.* EPA Process Design Manual. Washington, D.C.

Fair, G. M.; Geyer, J. C.; and Okun, D. A. 1968. *Waste and Wastewater Engineering. Vol. 2: Water Purification and Wastewater Treatment and Disposal.* New York: Wiley.

Fitch, E. B. 1956. Sedimentation Process Fundamentals. In *Biological Treatment of Sewage and Industrial Wastes.* Vol. 2, edited by J. McCabe and W. W. Eckenfelder, Jr. New York: Reinhold.

Great Lakes–Upper Mississippi River Board of State Public Health and Environmental Managers. 1990. *Recommended Standards for Wastewater Facilities.* Ten state standards. Albany, N.Y.

Great Lakes–Upper Mississippi River Board of State Public Health and Environmental Managers. 1992. *Recommended Standards for Water Works.* Ten state standards. Albany, N.Y.

Great Lakes–Upper Mississippi Board of State Sanitary Engineers. 1978. *Recommended Standards for Sewage Works.* Ten state standards. Albany, N.Y.

Hazen, A. 1904. On Sedimentation. *Trans. ASCE* 53:45.

Horan, N. J. 1990. *Biological Wastewater Treatment Systems: Theory and Operation.* Chichester, West Sussex, England: Wiley.

Ingersoll, A, C.; McKee, J. E.; and Brooks, N.

H. 1956. Fundamental Concepts of Rectangular Settling Tank. *Trans. ASCE* 121:1179.

Kynch, G. J. 1952. A Theory of Sedimentation. *Trans. Faraday Soc.* 48:161.

Lapple, C. E., and Shepherd, C. B. 1940. Calculation of Particle Trajectories. *Ind. and Eng. Chem.* 32, no. 5:605.

McCabe, W. L., and Smith, J. C. 1967. *Unit Operations of Chemical Engineering.* 2nd ed. New York: McGraw-Hill.

McGhee, T. J. 1991. *Water Supply and Sewage.* 6th ed. New York: McGraw-Hill.

Metcalf & Eddy, Inc. 1979. *Wastewater Engineering: Treatment, Disposal and Reuse.* 2nd ed. New York: McGraw-Hill.

Metcalf & Eddy, Inc. 1991. *Wastewater Engineering: Treatment, Disposal and Reuse.* 3rd ed. New York: McGraw-Hill.

Montgomery, J. M., Inc. 1985. *Water Treatment: Principles and Design.* New York: Wiley.

Novotny, V.; Imhoff, K.; Olthof, M.; and Krenkel, P. 1989. *Karl Imhoff's Handbook of Urban Drainage and Wastewater Disposal.* New York: Wiley.

O'Connor, D. J., and Eckenfelder, W. W. 1956. Evaluation of Laboratory Settling Data for Process Design. In *Biological Treatment of Sewage and Industrial Wastes.* Vol. 2, edited by J. McCabe and W. W. Eckenfelder, Jr. New York: Reinhold.

Perry, R. H., and Chilton, C. H. 1973. *Chemical Engineer's Handbook.* 5th ed. New York: McGraw-Hill.

Planz, P. 1969. Performance of (Activated Sludge) Secondary Sedimentation Basins. In *Proceedings of the Fourth International Conference.* Prague: International Association on Water Pollution Research.

Qasim, S. R. 1985. *Wastewater Treatment Plant Design.* New York: Holt, Rinehart and Winston.

Ramalho, R. S. 1977. *Introduction to Wastewater Treatment Processes.* New York: Academic Press.

Reynolds, T. D.; Payne, F.; and Myers, M. 1989. A Modification of Camp's Lateral Spillway Equation. Presented at the Fall 1989 Meeting of the Texas Section of the ASCE. Dallas, Texas.

Rich, L. G. 1971. *Unit Operations of Sanitary Engineering.* New York: Wiley.

Salvato, J. A. 1992. *Environmental Engineering and Sanitation.* 4th ed. New York: Wiley.

Sanks, R. L. 1978. *Water Treatment Plant Design.* Ann Arbor, Mich.: Ann Arbor Science Publishers.

Schroeder, E. D. 1977. *Water and Wastewater Treatment.* New York: McGraw-Hill.

Steel, E. W., and McGhee, T. J. 1979. *Water Supply and Sewerage.* New York: McGraw-Hill.

Sundstrom, D. W., and Klei, H. E. 1979. *Wastewater Treatment.* Englewood Cliffs, N.J.: Prentice-Hall.

Talmage, W. P., and Fitch, E. B. 1955. Determining Thickener Unit Areas. *Ind. and Eng. Chem.* 47, no. 1:38.

Tebbutt, T. H. Y. 1992. *Principles of Water Quality Control.* 4th ed. Oxford, England: Pergamon Press Limited.

Water Environmental Federation (WEF). 1992. *Design of Municipal Wastewater Treatment Plants.*

Vols. I and II. WEF Manual of Practice No. 8.

Water Pollution Control Federation (WPCF). 1985. *Clarifier Design.* WPCF Manual of Practice FD-8.

Water Pollution Control Federation (WPCF). 1977. *Wastewater Treatment Plant Design.* WPCF Manual of Practice No. 8.

Weber, W. J. 1972. *Physicochemical Processes for Water Quality Control.* New York: Wiley-Interscience.

Zipparro, V. J., and Hasen, H. 1993. *Davis' Handbook of Hydraulics.* 4th ed. New York: McGraw-Hill.

PROBLEMS

9.1 A circular primary clarifier is to be designed for a municipal wastewater treatment plant having an average flow of 1.5 MGD; during the maximum flow during the day, the flowrate is 2.6 times the average hourly flow. Pertinent data are overflow rate based on average daily flow = 800 gal/day-ft², and overflow rate based on peak hourly flow = 1800 gal/day-ft². The minimum depth = 10 ft. Determine:

a. The diameter of the tank.

b. The design diameter of the tank if sludge rake equipment is available in 5-ft increments in diameter.

c. The design depth of the tank.

d. The maximum depth of water over the effluent weirs if the weirs are 90° triangular type spaced at 8-in. centers.

e. The weir depth if there is 1 in. of freeboard at maximum flow.

f. The depth of water in the effluent channel at the effluent box if the channel is 1 ft-3 in. wide. Make the depth 4 in. greater than critical depth, y_c, to insure submerged conditions for the outlet pipe.

g. The depth of water, H_0, in the effluent channel on the opposite side of the tank from the effluent box if the channel is level and Manning's $n = 0.032$.

h. The depth of the effluent channel if the triangular weirs have a 4-in. freefall to the maximum water surface in the channel.

9.2 A circular primary clarifier is to be designed for a municipal wastewater treatment plant having an average flow of 5680 m³/d; during the maximum flow during the day, the flowrate is 2.6 times the average hourly flow. Pertinent data are: overflow rate based on the average daily flow = 32.6 m³/d-m²

and overflow rate based on peak flow = 73.3 m³/d-m². The minimum depth = 3.05 m. Determine:

a. The diameter of the tank.

b. The design depth of the tank.

c. The maximum depth of water over the effluent weirs if the weirs are 90° triangular weirs spaced at 203-mm centers.

d. The weir depth if there is 25 mm of freeboard at maximum flow.

e. The depth of water in the effluent channel at the effluent box if the channel is 0.381 m wide. Make the depth 102 mm greater than critical depth, y_c, to insure submerged conditions for the outlet pipe.

f. The depth of water, H_0, in the effluent channel on the opposite side of the tank from the effluent box if the channel is level and Manning's $n = 0.032$.

g. The depth of the effluent channel if the triangular weirs have a 102-mm freefall to the maximum water surface in the channel.

9.3 A batch settling test has been performed on an industrial wastewater having an initial suspended solids of 597 mg/ℓ to develop criteria for the design of a primary clarifier. The test column was 5 in. (125 mm) in diameter and 8 ft (2.44 m) high, and sampling ports were located 2, 4, 6, and 8 ft (0.61, 1.22, 1.83, and 2.44 m) from the water surface in the column. The suspended solids remaining after the various sampling times are given in Table 9.11. The wastewater flow is 2.5 MGD (9460 m³/d). Determine:

a. The design overflow rate and detention time if 65% percent of the suspended solids are to be removed. Use scale-up factors of 0.65 for the overflow rate and 1.75 for the detention time.

TABLE 9.11

DEPTH ft (m)	TIME (min)				
	10	20	30	45	60
2 (0.61)	394	352	243	182	148
4 (1.22)	460	406	337	295	216
6 (1.83)	512	429	376	318	306
8 (2.44)	1018	1142	1208	1315	1405

b. The diameter if a circular clarifier is used.

c. The design diameter of the clarifier if equipment is available in 5-ft increments of tank diameter.

d. The depth of the tank.

9.4 Rework Problem 9.3 using SI units. The data are given in Table 9.11. Determine:

a. The design overflow rate and detention time if 65% of the suspended solids are to be removed. Use scale-up factors of 0.65 for the overflow rate and 1.75 for the detention time.

b. The diameter if a circular clarifier is used.

c. The depth of the tank.

9.5 Assume that the primary clarifier in Problem 9.1 is for an activated sludge plant. Pertinent data are: the head loss between the water surfaces of the effluent box and the reactor basin is 0.65 ft, and the elevation of the water surface in the reactor basin is 320.87. Determine the elevation of the water surface in the effluent box and the water surface elevation in the primary clarifier at the maximum flowrate.

9.6 Assume that the primary clarifier in Problem 9.2 is for an activated sludge plant. Pertinent data are: the head loss between the water surfaces of the effluent box and the reactor basin is 0.183 m, and the elevation of the water surface in the reactor basin is 97.80 m. Determine the elevation of the water surface in the effluent box and the water surface elevation in the primary clarifier at the maximum flowrate.

9.7 An activated sludge plant has a pump station between the preliminary treatment facilities and the primary clarifier. The flowsheet consists of a primary clarifier, reactor basin, final clarifier, and chlorine contact tank. The head losses in feet are as follows: head on primary weirs = 0.12, freefall to $H_0 = 0.33$, $H_0 = 1.14$, $d = 0.84$, head loss between effluent box and reactor basin = 0.36, head loss between reactor basin and final clarifier = 0.39; head on final clarifier weirs = 0.12, freefall to $H_0 = 0.33$, $H_0 = 1.14$, $d = 0.84$, head loss from final clarifier effluent box to chlorine contact tank = 0.32; head on weir in chlorine contact tank = 0.16, freefall to $H_0 = 0.33$, $H_0 = 1.08$, $d = 0.76$, and head loss from contact tank effluent box to river water surface = 0.82. The maximum recorded river water surface elevation = 320.85. Determine the total head loss through the plant and the minimum allowable water surface elevations for the primary clarifier, the reactor basin, the final clarifier, and the chlorine contact tank.

9.8 An activated sludge plant has a pump station between the preliminary treatment facilities and the primary clarifier. The flowsheet consists of a primary clarifier, reactor basin, final clarifier, and chlorine contact tank. The head losses in meters are as follows: head on primary weirs = 0.037, freefall to $H_0 = 0.102$, $H_0 = 0.347$, $d = 0.256$, head loss between effluent box and reactor basin = 0.110, head loss between reactor basin and final clarifier = 0.119, head on final clarifier weirs = 0.037, freefall to $H_0 = 0.102$, $H_0 = 0.347$, $d = 0.256$, head loss from final clarifier effluent box to chlorine contact tank = 0.098, head on weirs in chlorine contact tank = 0.049, freefall to $H_0 = 0.102$, $H_0 = 0.392$, $d = 0.232$, and head loss from contact tank effluent box to the river water surface = 0.250. The maximum recorded river water surface elevation = 97.79. Determine the total head loss through the plant and the minimum water surface elevations for the primary clarifier, the reactor basin, the final clarifier, and the chlorine contact tank.

9.9 Ferrous sulfate is used as a coagulant and lime as a coagulant aid at a water treatment plant. The specific gravity of the floc and adhered suspended material is 1.005 at a temperature of 10°C. Determine:

a. The overflow rate in gal/day-ft^2 if the flow is 7.0 MGD and the clarifier is 60 ft by 120 ft.

b. The settling rate in cm/sec for the overflow rate in part (a).

c. The diameter in cm of the smallest floc particle that is of a size that is 100% removed if it is assumed that the particles are spherical in shape. (First determine the approximate particle size from Figure 9.5, then determine the final size from the settling equations.)

9.10 Ferrous sulfate is used as a coagulant and lime as a coagulant aid at a water treatment plant. The specific gravity of the floc and adhered suspended material is 1.005 at a temperature of 10°C. Determine:

a. The overflow rate in m^3/d-m^2 if the flow is 26,500 m^3/d and the clarifier is 18.29 m by 36.57 m.

b. The settling rate in cm/s for the overflow rate in part (a).

c. The diameter in cm of the smallest floc particle that is of a size that is 100% removed if it is assumed that the particles are spherical in shape. (First determine the approximate particle size from Figure 9.5, then determine the final size from the settling equations.)

9.11 When alum is used as a coagulant at a water treatment plant, the specific gravity of the floc and adhered suspended material is 1.002 at a temperature of 10°C. Determine using USCS or SI units:

a. The settling velocity in cm/sec if the overflow rate is 800 gal/day-ft^2 (32.6 m^3/d-m^2).

b. The diameter in cm of the smallest floc particle that is of a size that is 100% removed if it is assumed that the particles are spherical in shape. (First determine the approximate particle size from Figure 9.5, then determine the final size from the settling equations.)

9.12 When municipal wastewater undergoes primary settling, the specific gravity of the settled particles is about 1.001 at 10°C. Determine using USCS or SI units:

a. The settling rate in cm/sec for a peak overflow rate of 3000 gal/day-ft^2 (122 m^3/d-m^2).

b. The diameter in cm of the smallest particle that is 100% removed if it is assumed that the particles are spherical in shape. (First determine the approximate particle size from Figure 9.5, then determine the final size from the settling equations.)

9.13 When municipal activated sludge undergoes final settling, the specific gravity of the settled particles is about 1.005 at 10°C. Determine using USCS or SI units:

a. The settling rate in cm/sec for a peak overflow rate of 2000 gal/day-ft^2 (81.5 m^3/d-m^2).

b. The diameter in cm of the smallest particle that is 100% removed if it is assumed that the particles are spherical in shape. (First determine the approximate particle size from Figure 9.5, then determine the final size from the settling equations.)

9.14 A primary clarifier is to be designed for a municipal wastewater treatment plant. The flow is 1.10 MGD and the peak flow during the day is 3.10 times the average hourly flow. Pertinent data are: peak overflow rate = 1800 gal/day-ft^2, maximum peak weir loading = 30,000 gal/day-ft, and minimum tank depth = 7 ft. Determine:

a. The clarifier diameter.

b. The design diameter if equipment is available in 5-ft increments in diameter.

c. The side water depth.

d. The peak weir loading. Is it acceptable?

9.15 A primary clarifier is to be designed for a municipal wastewater treatment plant. The flow is 4160 m^3/d, and the peak flow during the day is 3.10 times the average hourly flow. Pertinent data are: peak overflow rate = 73.3 m^3/d-m^2, maximum peak weir loading = 373 m^3/d-m, and minimum tank depth = 2.13 m. Determine:

a. The clarifier diameter.

b. The side water depth.

c. The peak weir loading. Is it acceptable?

9.16 A primary clarifier for a municipal wastewater treatment plant is to be designed for an average flow of 1.8 MGD (6810 m^3/d). The state's regulatory agency criteria for primary clarifiers are as follows: peak overflow rate = 2200 gal/day-ft^2 (89.6 m^3/d-m^2), average overflow rate = 900 gal/day-ft^2 (36.7 m^3/d-m^2), depth = 10 ft (3.05 m) for diameters up to 40 ft (12.19 m), 11 ft (3.35 m) for diameters 40 to 140 ft (12.19 to 42.67 m), and 12 ft (3.66 m) for diameters greater than 140 ft (42.67 m), and peak weir loading = 30,000 gal/day-ft (373 m^3/d-m). The ratio of the peak hourly flow to the average hourly flow is 2.80. Using USCS units, determine:

a. The clarifier diameter.

b. The design diameter if sludge rake equipment is available in 5-ft increments in diameter.

c. The side water depth.

d. The peak weir loading if peripheral weirs are used. Is it acceptable?

e. The percent BOD$_5$ removal to be expected on the basis of the average overflow rate.

f. The percent suspended solids removal on the basis of the average detention time if the influent suspended solids are 240 mg/ℓ.

9.17 Rework the primary clarifier design in Problem 9.16 using SI units. Determine:

a. The clarifier diameter.

b. The side water depth.

c. The peak weir loading if peripheral weirs are used. Is it acceptable?

d. The percent BOD$_5$ removal to be expected on the basis of the average overflow rate.

e. The percent suspended solids removal on the basis of the average detention time if the influent suspended solids are 240 mg/ℓ.

9.18 A final clarifier is to be designed for an activated sludge plant serving a municipality. The state's regulatory agency criteria for final clarifiers used for activated sludge are as follows: peak overflow rate = 1200 gal/day-ft^2 (48.9 m^3/d-m^2), average overflow rate = 500 gal/day-ft^2 (20.4 m^3/d-m^2), peak solids loading = 50 lb/day-ft^2 (244 kg/d-m^2), peak weir loading = 30,000 gal/day-ft (373 m^3/d-m), and depth = 11 to 15 ft (3.35 to 4.57 m). The average flow to the reactor basin prior to the junction with the recycle sludge is 1.8 MGD (6810 m^3/d). The maximum recycled sludge flow is 100% of the average influent flow and is constant throughout the day. The design MLSS is 3000 mg/ℓ, and the ratio of the peak hourly flow to the average hourly flow is 2.80. Using USCS units, determine:

a. The clarifier diameter.

b. The design diameter if sludge rake equipment is available in 5-ft increments in diameter.

c. The side water depth if Table 9.9 is used to determine depth.

d. The peak weir loading if peripheral weirs are used. Is it acceptable?

9.19 Rework the final clarifier design in Problem 9.18 using SI units. Determine:

a. The clarifier diameter.

b. The clarifier depth if Table 9.9 is used to determine depth.

c. The peak weir loading. Is it acceptable?

9.20 A rectangular clarification basin is to be designed for a rapid sand filtration plant using ferrous sulfate as a coagulant and lime as a coagulant aid. The flow is 10.0 MGD (37,850 m^3/d), the overflow rate is 700 gal/day-ft^2 (28.5 m^3/d-m^2), the detention time is 5 hr, and the weir loading is 22,000 gal/day-ft (273 m^3/d-m). Two sludge rake mechanisms for square basins are to be used in tandem to give a length to width ratio of 2:1. The effluent channel is to have weirs on both sides of the channel, and the channel is to be supported by cantilever beams extending from the walls, as shown in Figure 9.46(a). The center of the channel is 5 ft (1.52 m) from the walls. The channel will extend across the end of the basin, as shown in Figure 9.46(b), and will extend up along the sides of the basin to a distance x. Using USCS units, determine:

a. The plan dimensions of the basin.

b. The design plan dimensions of the basin if sludge rake equipment is available in 5-ft increments in length.

c. The design depth of the basin.

d. The length of the effluent channel.

e. The distance, x, that the channel will extend up along the sides of the basin. *Note:* The length of the channel used in computing the distance, x, is the centerline length.

f. Does the channel extend up more than one-half the length of the basin?

9.21 Rework Problem 9.20 using SI units. Determine:

a. The plan dimensions of the basin.

b. The depth of the basin.

c. The length of the effluent weir channel.

d. The distance, x, that the channel will extend up along the sides of the basin.

e. Does the channel extend up more than one-half the length of the basin?

9.22 An existing rectangular clarifier at a water treatment plant is 18 ft (5.49 m) wide and 72 ft (21.94 m) long. It was designed for an overflow rate of 600 gal/day-ft^2 (24.4 m^3/d-m^2) and is currently operating at the design capacity. Inclined-tube settlers are to be added and will occupy 80% of the plan area and will have a design overflow rate of 5000 gal/day-ft^2 (204 m^3/d-m^2). Determine the present capacity in gal/day (m^3/d) and the increased capacity in gal/day (m^3/d).

9.23 Rework Problem 9.22 using SI units. Determine the present capacity in m^3/d and the increased capacity in m^3/d.

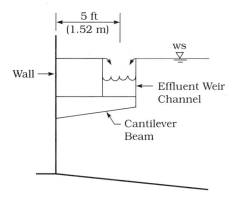

(a) Section A-A through Wall

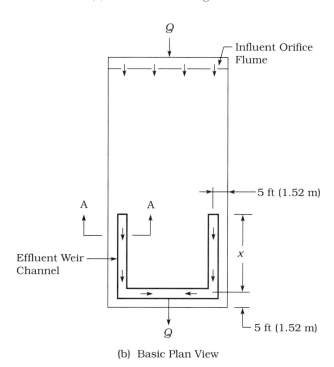

(b) Basic Plan View

FIGURE 9.46 *Rectangular Settling Basin and Details for Problems 9.20 and 9.21*

9.24 A final clarifier is to be designed for an activated sludge plant treating an industrial wastewater. Sludge settling studies using an acclimated sludge have been made giving the settling velocity as a function of solids concentration, and the results are shown in Table 9.12. The design mixed liquor flow to the final clarifier is 2640 gallons per minute (9990

liters per minute), the MLSS is 2500 mg/ℓ, and the underflow concentration is 10,000 mg/ℓ. A scale-up factor of 1.5 is to be used for the solids flux value. Using USCS units, determine:
a. A plot of settling velocity, ft/hr, on the y-axis versus solids concentration, mg/ℓ, on the x-axis.
b. The solids flux, lb/hr-ft², for each test, and plot

TABLE 9.12 *Concentrations and Settling Velocities for the Acclimated Activated Sludge for Problems 9.24 and 9.25*

TEST	C, mg/ℓ	V ft/hr	(m/h)
1	12,100	0.109	(0.0332)
2	11,200	0.162	(0.0494)
3	8,100	0.391	(0.119)
4	4,900	1.77	(0.539)
5	2,950	6.02	(1.84)
6	2,490	7.97	(2.43)
7	1,620	11.0	(3.36)

a curve of solids flux, lb/hr-ft^2, on the *y*-axis versus solids concentration, mg/ℓ, on the *x*-axis.

c. Determine the design solids flux, $G_{L(\text{Design})}$, in lb/hr-ft^2, using a scale-up factor of 1.5. *Note*: $G_{L(\text{Design})} = G_L/1.5$.

d. Determine the diameter of the final clarifier.

e. Determine the design diameter of the final clarifier if sludge rake equipment is available in 5-ft. increments in diameter.

9.25 Rework Problem 9.24 using SI units. Determine:

a. A plot of settling velocity, m/h, on the *y*-axis versus solids concentration, mg/ℓ, on the *x*-axis.

b. The solids flux, kg/h-m^2, for each test, and plot a curve of solids flux, kg/h-m^2, on the *y*-axis versus solids concentration, mg/ℓ, on the *x*-axis.

c. Determine the design solids flux, $G_{L(\text{Design})}$, in kg/h-m^2, using a scale-up factor of 1.5. *Note*: $G_{L(\text{Design})} = G_L/1.5$.

d. Determine the diameter of the final clarifier.

9.26 An activated sludge treatment plant treats a wastewater with a peak flow of 4.09 ft^3/sec (0.116 m^3/s). The plant layout and profile are shown in Figure 9.47. The process lines P-1, P-2, P-3, and P-4 are designed for gravity flow and are to have a maximum velocity of 4.5 ft/sec (1.37 m/s) to avoid excessive head losses. Process line P-2 will have a maximum flow of 4.09 ft^3/sec (0.116 m^3/s) and is 35 ft long (10.7 m). Line P-2 discharges into the reactor basin. The primary clarifier effluent channel is 1 ft-3 in. (0.381 m) wide and has a rectangular cross section. The downstream depth in the effluent channel, *d*, is 4 in. (102 mm) greater than critical depth, y_c, to insure submerged outlet conditions.

The primary clarifier V-notch peripheral weirs are 90° triangular weirs at 8-in. (203-mm) centers. There is a 4-in. (102-mm) freefall below the V-notch weir crests at the upstream end where the depth, H_0, occurs. Determine using USCS units:

a. The theoretical diameter for process line P-2.

b. The design diameter for process line P-2 if standard cast iron pipe sizes from 10 to 20 in. are 10, 12, 14, 16, 18, and 20 in.

c. The head on the 90° V-notch weirs.

d. The values of *d* and H_0 for the effluent chanel if $n = 0.032$.

e. The head losses for process line P-2. *Note*: The head losses include the entrance loss ($0.5 V^2/2g$), the friction loss for 35 ft of cast iron pipe using a Hazen Williams coefficient of 100, and the exit loss ($V^2/2g$). For USCS units, the Hazen Williams equation is $V = 1.318 C_{\text{HW}} R^{0.63} S^{0.54}$, where V = velocity in feet per second, C_{HW} = Hazen Williams friction coefficient, R = hydraulic radius in feet = $D/4$ for a circular pipe with a diameter D, and S = slope of the energy gradient.

f. The drop in water surface elevation, ΔZ_1, between the water surface in the primary clarifier and the water surface in the effluent box. *Note*: The drop in elevation between the water surface in one tank and the water surface in a downstream box or tank can be determined by writing Bernoulli's equation between the two points, which is $V_1^2/2g + p_1/\gamma + Z_1 = V_2^2/2g + p_2/\gamma + Z_2 + \Sigma h_L$. Since V_1 and V_2 are essentially zero, and since, if relative pressures are used, p_1/γ and p_2/γ are zero, Bernoulli's equation becomes $Z_1 - Z_2 = \Delta Z = \Sigma h_L$. For the drop in water surface between the primary clarifier and its outlet box, the Σh_L equals the sum of the head

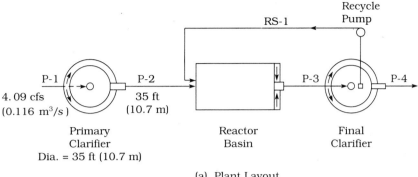

(a) Plant Layout

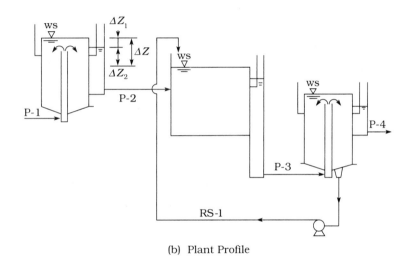

(b) Plant Profile

FIGURE 9.47 *Plant Layout and Profile for Problems 9.26 and 9.27*

losses, which are the head on the V-notch weirs, the freefall, and $H_0 - d$.

g. The drop in water surface between the water surface in the effluent outlet box and the water surface in the reactor basin, ΔZ_2.

h. The total drop in water surface elevation, $\Delta Z_1 + \Delta Z_2$, between the water surface in the primary clarifier and the water surface in the reactor basin.

9.27 Rework Problem 9.26 using SI units. In part (b), the standard pipe sizes are 250, 300, 350, 400, 450, and 500 mm. In part (e), the Hazen Williams equation for SI units is $V = 0.8464 C_{HW} R^{0.63} S^{0.54}$, where V = velocity in meters per second, C_{HW} = Hazen Williams friction coefficient, R = hydraulic radius in meters = $D/4$ for a circular pipe with a diameter D, and S = slope of the energy gradient.

10

FILTRATION

Filtration is a solid-liquid separation in which the liquid passes through a porous medium or other porous material to remove as much fine suspended solids as possible. It is used in water treatment to filter chemically coagulated and settled waters to produce a high-quality drinking water. In wastewater treatment it is used to filter (1) untreated secondary effluents, (2) chemically treated secondary effluents, and (3) chemically treated raw wastewaters. In all three of the uses in wastewater treatment, the objective is to produce a high-quality effluent.

Filters may be classified according to the types of media used as follows:

1. Single-medium filters: These have one type of medium, usually sand or crushed anthracite coal.
2. Dual-media filters: These have two types of media, usually crushed anthracite and sand.
3. Multimedia filters: These have three types of media, usually crushed anthracite, sand, and garnet.

In water treatment all three types are used; however, the dual- and multimedia filters are becoming increasingly popular. In advanced and tertiary wastewater treatment, nearly all the filters are dual- or multimedia types.

The principles of filtration are the same for filters used in either water or wastewater treatment, and the filter structures, equipment, accessories, and method of operation are similar for both types of service.

SINGLE-MEDIUM FILTERS

Rapid sand filters used in water treatment practice are usually of the gravity type and are commonly housed in open concrete basins. Figure 10.1 is a photograph of a row of filters at a rapid sand filtration plant. Figure 10.2 shows a perspective of three gravity filters and also a cutaway section showing the filter sand, the underlying gravel, and the underdrain system. Figure 10.3 shows a schematic section that gives more details such as sand depth, gravel depth, and the underdrain system. Figure 10.4 shows the filter layout, and Figure 10.5 shows the filter piping and underdrain system with all of the control valves and the rate of flow controllers. Figure 10.6 shows the pipe gallery at a rapid sand filtration plant with a row of filters on each side of the gallery. Although open gravity filters are the most common, pressure filters, shown in Figures 10.7 through 10.9, are also used.

FIGURE 10.1
Row of Rapid Sand Filters at a Water Treatment Plant

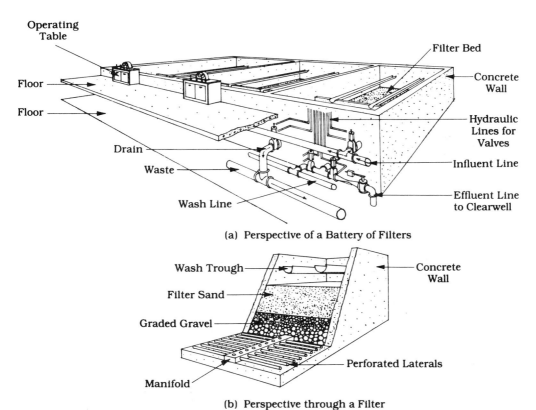

(a) Perspective of a Battery of Filters

(b) Perspective through a Filter

FIGURE 10.2 *Gravity Filters and Accessories*
Courtesy of the National Lime Association.

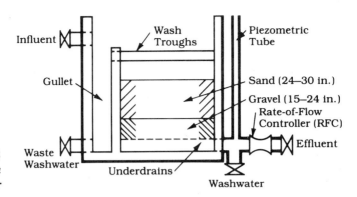

FIGURE 10.3
Section through a Rapid Sand Filter

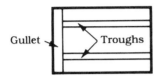

FIGURE 10.4
Wash System Layout

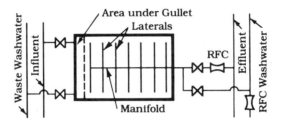

FIGURE 10.5
Filter Piping Layout

FIGURE 10.6 *Pipe Gallery at a Modern Rapid Sand Filtration Plant. The gallery serves a row of filters on both the right-hand and left-hand sides. The large pipe in the center with branches going out to filters on both sides is the washwater line. The large pipes coming out and then dropping down vertically are the waste washwater lines. The controllers and pipes in the right and left foreground are for the filtered effluent water.*

FIGURE 10.7
Profile through a Horizontal Pressure Filter

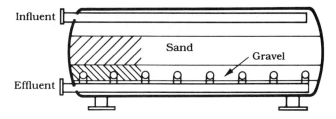

FIGURE 10.8 *Cross Section through a Horizontal Pressure Filter*

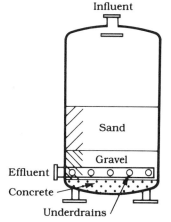

FIGURE 10.9 *Cross Section through a Vertical Pressure Filter*

Figure 10.10 shows pressure filters at a tertiary wastewater treatment plant. Frequently, crushed anthracite coal is employed instead of quartz sand. The sand bed is usually 24 to 30 in. in depth (610 to 760 mm), and the underlaying gravel is usually 15 to 24 in. thick (380 to 610 mm).

Figure 10.11 shows a schematic section through a filter during the filtration cycle. Approximately 3 to 4 ft (0.91 to 1.2 m) of water is above the sand, and the water passes downward through the media into the underdrain system. From there it flows through the rate of flow controller which controls the rate at which the water is filtered. During the filter run, the valve

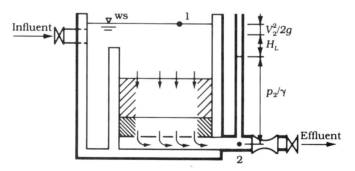

FIGURE 10.10 *Pressure Filters at a Tertiary Wastewater Treatment Plant*

FIGURE 10.11 *Schematic Section Showing a Filter during Filtration*

positions, shown in Figure 10.5, are as follows: (1) Influent and effluent valves are open and (2) washwater and washwater waste valves are closed. The action of the sand in removing finely suspended floc smaller than the pore openings consists mainly of adhesion, flocculation, sedimentation, and straining. As the water moves downward through the pore spaces, some of the fine suspended floc collides with the sand surfaces and adheres to the sand particles. As the water passes through pore constrictions, some of the fine floc is brought together, flocculation occurs, and the enlarged floc settles on the top of the sand particles immediately below the constrictions. Also, the buildup of floc that has been removed in the filter creates a straining action, and some of the incoming floc is removed by straining. During a filter run, the accumulated floc causes the pore spaces to become smaller, the velocities to increase, and some of the removed floc to be carried deeper within the filter bed. Straining may also occur at the surface of the filter if large particles of floc are strained and form a compressible cake that assists in filtering smaller particles. In summary, the removal of suspended solids is by surface removal at the top of the bed and depth removal within the filter bed itself (Baumann and Oulman, 1970). For water treatment, depth removal is usually the most important in rapid sand filtration.

When a clean filter is put into operation, the floc accumulation is in the top layers of the sand; however, as the time of operation increases, the floc accumulation extends deeper into the filter bed. The accumulated floc causes an increase in the hydraulic head loss. The magnitude of the head loss, H_L, is illustrated by writing Bernoulli's energy equation between Point 1 on the water surface in Figure 10.11 and Point 2 at the center of the effluent line. The resulting equation is

$$\frac{V_1^2}{2g} + \frac{p_1}{\gamma} + Z_1 = \frac{V_2^2}{2g} + \frac{p_2}{\gamma} + Z_2 + H_L \qquad (10.1)$$

where

V_1, V_2 = respective velocities

p_1, p_2 = respective pressures

Z_1, Z_2 = respective elevation heads

γ = specific weight of water

g = acceleration due to gravity

H_L = head loss in feet

If the relative pressure is used, $p_1 = 0$. Also, $V_1 = 0$ and the datum may be selected so that $Z_2 = 0$. Incorporating these values in Eq. (10.1) and rearranging gives

$$\frac{p_2}{\gamma} = Z_1 - \frac{V_2^2}{2g} - H_L \qquad (10.2)$$

Since it is common to have 4 ft (1.22 mm) of water over the sand, 2 ft-6 in. (0.76 m) of sand, 1 ft-6 in. (0.49 m) of gravel, and a 1-ft (0.30-m) depth for the underdrains, this gives a Z_1 value of 4.0 + 2.5 + 1.5 + 0.5 or 8.5 ft (1.22 + 0.76 + 0.49 + 0.15 or 2.62 m). Also, pipes are usually designed for about 4 fps (1.22 mps); therefore, Eq. (10.2) yields an expression for the gauge pressure head at Point 2, which is

$$\frac{p_2}{\gamma} = 8.5 \, \text{ft} - \frac{(4)^2}{2g} - H_L = 8.25 \, \text{ft} - H_L \tag{10.3a}$$

or

$$\frac{p_2}{\gamma} = 2.62 \, \text{m} - \frac{(1.22)^2}{2g} - H_L = 2.54 \, \text{m} - H_L \tag{10.3b}$$

When a clean filter is put on line, the head loss, H_L, is about 0.5 to 1.5 ft (0.15 to 0.46 m), depending upon the filtration rate, but as the filter run progresses, the head loss increases. From Eq. (10.3a) and (10.3b), it can be seen that when the head loss is 8.25 ft (2.54 m), the relative pressure at Point 2 is zero. Any further increase in head loss creates a negative pressure, which is undesirable. In practice, when the head loss reaches 6 to 8 ft (1.8 to 2.4 m), the filter is backwashed. The amount of washwater required is from 1 to 5% of the water filtered, a typical value being 2 to 3%.

The shape of the curve showing the head loss as a function of filtrate volume for a particular filter is dependent upon the type of filter action, as depicted by Figures 10.12, 10.13, and 10.14 (Baumann and Oulman, 1970). If the filter action is by surface removal of compressible solids, the head-loss curve will be exponential, as shown in Figure 10.12. This type of curve is associated with fine grained media and low filtration rates. The filter action by microscreens and diatomaceous earth filters is surface removal. If the filter action is depth removal of flocculated suspended solids, the head-loss

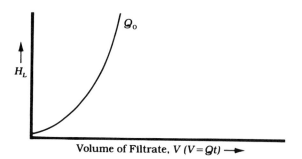

FIGURE 10.12 *Head-Loss Curve for Surface Removal of Compressible Solids*

Adapted from "Sand and Diatomite Filtration Practices" by E. R. Baumann and C. S. Oulman. In *Water Quality Improvement by Physical and Chemical Processes.* ed. E. F. Gloyna and W. W. Eckenfelder, Jr. Copyright © 1970 by the University of Texas Press. Reprinted by permission.

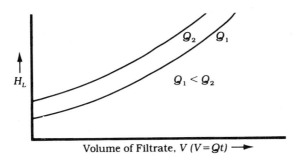

FIGURE 10.13 *Head-Loss Curve for Depth Removal of Flocculent Solids*

Adapted from "Sand and Diatomite Filtration Practices" by E. R. Baumann and C. S. Oulman. In *Water Quality Improvement by Physical and Chemical Processes*. ed. E. F. Gloyna and W. W. Eckenfelder, Jr. Copyright © 1970 by the University of Texas Press. Reprinted by permission.

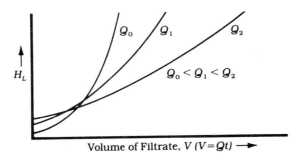

FIGURE 10.14 *Head-Loss Curves for Combined Surface and Depth Removal of Flocculent Solids*

Adapted from "Sand and Diatomite Filtration Practices" by E. R. Baumann and C. S. Oulman. In *Water Quality Improvement by Physical and Chemical Processes*. ed. E. F. Gloyna and W. W. Eckenfelder, Jr. Copyright © 1970 by the University of Texas Press. Reprinted by permission.

curve will be rather flat, as shown in Figure 10.13. This type of curve is encountered with deep granular filters at relatively high filtration rates. If the filter action is by surface removal and depth removal of flocculated suspended solids, the head-loss curves will be as shown in Figure 10.14. At low filtration rates, surface removal is predominant and the curve is similar to that in Figure 10.12. At higher filtration rates, the solids penetrate deep within the filter. The principal action is depth removal, and the head-loss curves are similar to those in Figure 10.13. In rapid sand filtration at the usual filtration rates, depth removal is usually the main filter action and a flat head-loss curve results.

Surface removal will result when the feed water contains large floc and high turbidity. The top pore spaces will rapidly become clogged, resulting in relatively short filter runs. Depth removal will result when the feed water contains small floc and low turbidity, resulting in deeper penetration within the filter bed and relatively long filter runs.

Filter sands are characterized by the **effective size** and the **uniformity coefficient**. The effective size is equal to the sieve size in millimeters that will pass 10% (by weight) of the sand. The uniformity coefficient is equal to the sieve size passing 60% of the sand divided by the size passing 10%. Most rapid sand filters have sands with an effective size of 0.35 to 0.50 mm; however, some have sand with an effective size as high as 0.70 mm. The uniformity coefficient, which is a measure of gradation, is generally not less than 1.3 or more than 1.7.

The gravel serves to support the sand bed and is usually placed in several layers. The total depth may be from 6 to 24 in. (150 to 610 mm); however, 18 in. (460 mm) is typical. The size of the top layer of gravel depends upon the sand size, whereas the size of the bottom gravel depends upon the type of underdrain system. Usually five layers are used, and the gravel grades from less than $\frac{1}{16}$ in. (1.6 mm) at the top to 1 to 2 in. (25 to 50 mm) at the bottom.

The underdrain system serves to collect the filtered water from the bed during the filtration cycle. During the washing cycle, it serves to distribute the backwash water. The rate of flow of the backwash governs the hydraulic design of the filter since it is several times greater than the filtration rate. Underdrain systems are of basically two types: (1) a manifold with perforated lateral pipes or (2) a false bottom. In the manifold and perforated pipe system, the perforations are directed downward so that the high velocity of the backwash water is dissipated by the filter bottom and the surrounding gravel. The false bottom consists of a perforated bottom with a waterway underneath that removes the filtered water and permits the backwash to enter the filter bed.

The standard rate of filtration has been considered to be 2 gal/min-ft^2 (1.36 ℓ/s-m^2) of filter bed area since this is a common rate at which the first rapid sand filters were operated. Present coagulation and sedimentation practice allows the use of higher filtration rates. Frequently, plants are rated at 2 gal/min-ft^2 (1.36 ℓ/s-m^2), but provisions are made for operation at rates up to 5 gal/min-ft^2 (3.40 ℓ/s-m^2). Although most filters are operated at a constant filtration rate, a declining rate of filtration is sometimes employed. In this type of operation, the rate of filtration is decreased as the filter run progresses and the degree of clogging increases. Frequently, this results in longer filtration runs and better effluent quality. It is limited to medium to large plants, because the filters must be staggered in the degree of clogging to permit a constant rate in the total water production from the plant.

Gravity filters may employ a single filter, as shown in Figure 10.15, or a double filter within each concrete basin, as shown in Figure 10.16. The single filter arrangement is the most popular, and the length to width ratio

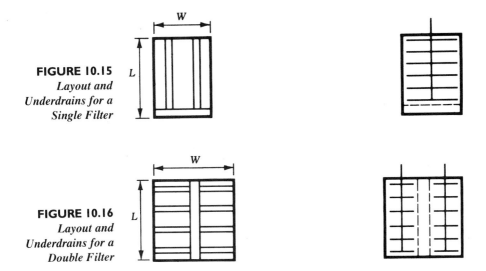

FIGURE 10.15
Layout and Underdrains for a Single Filter

FIGURE 10.16
Layout and Underdrains for a Double Filter

varies from $1:1.5$ to $1:2$. The double filter is built almost square with a length to width ratio of $1:1$. The largest filters that have been built are about $2100\,\text{ft}^2$ ($195\,\text{m}^2$) in plan area.

If the filter operation is optimum, the maximum allowable head loss, H_a, occurs simultaneously with the maximum allowable effluent turbidity, C_a, as shown in Figure 10.17. In many cases, this does not occur and the termination of the filter run is controlled by whichever occurs first — that is, H_a or C_a. The length of the filter run will depend upon the quality of the feed water, and filter runs may range from less than a day to several days. Backwashing removes the floc that has accumulated upon and within the filter bed. In modern filtration practice, a surface wash or air scour system is considered essential for high filter performance. To wash a filter, the influent valve, as shown in Figure 10.5, is closed, and when the water has filtered down below the wash troughs, the effluent valve is closed. The waste

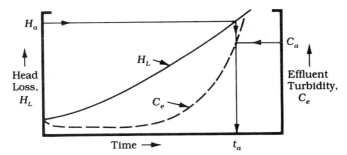

FIGURE 10.17 *Head Loss and Effluent Turbidity versus Filter Run Time for Optimum Performance*

washwater valve is opened and the surface wash is started at a rate of about 0.5 gal/min-ft^2 (0.34 ℓ/s-m^2). After about 1 minute of surface washing, the backwash flow is initiated by gradually opening the washwater influent valve, and the bed is allowed to expand to the desired height, as shown in Figure 10.18(a). The backwash flow should be from 15 to 20 gal/min-ft^2 (10.2 to 13.6 ℓ/s-m^2), and the bed expansion should be from 20 to 50% to suspend the bottom sand grains. The optimum backwash flow will depend upon the washwater temperature, because a cold washwater will expand the bed more than a warm one. The backwashing is continued until the waste washwater appears relatively clear, and the surface wash is terminated at 1 to 2 minutes prior to the end of the backwash. The surface wash system is misnamed

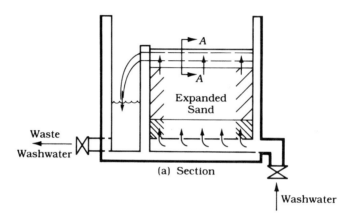

(a) Section

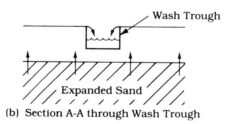

(b) Section A-A through Wash Trough

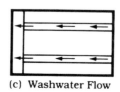

(c) Washwater Flow

FIGURE 10.18 *Schematic Showing Filter during Backwashing*

because it not only washes the filter surface prior to backwashing but also scours the expanded bed during the backwash; thus it really should be called an auxiliary scour system. Usually 3 to 10 minutes of backwash flow is required for complete washing, and the total off-line time will be up to 20 minutes. After backwashing, the initial water filtered should be wasted until the effluent turbidity is acceptable. The washwater may be supplied by a pump which pumps directly from the clearwell or by an elevated storage tank at the plant site. Usually, the volume of washwater is about 1 to 5% of the water filtered, 2 to 3% being typical.

Figure 10.19 shows the backwashing of a rapid sand filter at a lime-soda softening plant. The picture was made at the beginning of the filter backwash, and as a result, the uprising washwater is quite turbid. Figure 10.20 shows a filter console table at a rapid sand filtration plant with all the flow and valve position gauges and the valve control levers. Figure 10.21 shows the filter floor at a rapid sand filtration plant. The floor serves two rows of filters; one row is to the left outside of the building and one row is to the right outside of the building.

Hydraulics of Filtration

The head loss through a clean bed of porous media having a relatively uniform diameter, as given by the Carman-Kozeny equation, may be developed starting with the Darcy-Weisbach equation, which is

$$h_L = f \frac{LV^2}{D_c 2g} \qquad (10.4)$$

FIGURE 10.19 *Backwashing of Rapid Sand Filters at a Lime-Soda Softening Water Treatment Plant*

FIGURE 10.20 *Filter Console Table at a Water Treatment Plant. Gauges include rate of filtration, head-loss indicator, washwater flow rate, and gauges for the pneumatic control system for all valves. The levers are for controlling all valves.*

FIGURE 10.21 *Filter Floor at a Water Treatment Plant. Note consoles for all filters.*

where

h_L = frictional head loss

f = dimensionless friction factor

L = conduit length

D_c = conduit diameter

V = mean conduit velocity

g = acceleration due to gravity

The flow channels in a porous bed are irregular; thus the diameter, D_c, may be replaced by the term $4r$, where r is the hydraulic radius for a conduit diameter. If D is the bed depth, substituting this and $D_c = 4r$ into Eq. (10.4) gives

$$h_L = f\frac{DV^2}{8rg}$$ (10.5)

If there are n particles in the bed and the particle volume is v_p, the total volume of particles is nv_p. If the porosity is ε, the total bed volume is $nv_p/(1 - \varepsilon)$. The total channel volume is the void space or $\varepsilon nv_p/(1 - \varepsilon)$. If the wetted surface is considered to be the total surface of the particles, it is ns_p, where s_p is the surface area per particle. The hydraulic radius, r, is the total channel volume divided by the wetted surface, or

$$r = \left(\frac{\varepsilon}{1 - \varepsilon}\right)\frac{v_p}{s_p}$$ (10.6)

For spherical particles,

$$\frac{v_p}{s_p} = \frac{\pi d^3/6}{\pi d^2} = \frac{d}{6}$$ (10.7)

For irregularly shaped particles,

$$\frac{v_p}{s_p} = \phi\frac{d}{6}$$ (10.8)

where ϕ is the shape factor. Shape factors are 1 for spheres, 0.73 for crushed coal and angular sand, 0.82 for rounded sand, and 0.75 for average sand (Carman, 1937). The approach velocity, V_a, is equal to the flow, Q, divided by the filter surface, A. Thus the velocity through the pore spaces is

$$V = \frac{V_a}{\varepsilon}$$ (10.9)

Substituting Eqs. (10.6), (10.8), and (10.9) into Eq. (10.5) gives

$$h_L = f'\frac{D}{\phi d}\frac{1 - \varepsilon}{\varepsilon^3}\frac{V_a^2}{g}$$ (10.10)

which is the Carman-Kozeny relationship, where f' = dimensionless friction factor (Carman, 1937; Kozeny, 1927). The friction factor f' is given by (Ergun, 1952)

$$f' = 150\left(\frac{1 - \varepsilon}{N_{\text{Re}}}\right) + 1.75 \qquad (10.11)$$

The Reynolds number, N_{Re}, is defined as

$$N_{\text{Re}} = \frac{\phi d V_a}{v} = \frac{\phi \rho d V_a}{\mu} \qquad (10.12)$$

where ρ = mass density, μ = dynamic viscosity, and v = kinematic viscosity. The head loss through a clean bed of porous media having a relatively uniform diameter is also given by the Rose equation (Rose, 1945), which is

$$h_L = \frac{1.067}{\phi} \frac{C_D}{g} D \frac{V_a^2}{\varepsilon^4} \frac{1}{d} \qquad (10.13)$$

where C_D is the coefficient of drag. The coefficient of drag for $N_{\text{Re}} < 1$ is given by

$$C_D = \frac{24}{N_{\text{Re}}} \qquad (10.14)$$

For $N_{\text{Re}} > 1$ but $< 10^4$, the drag coefficient is (Fair *et al.*, 1968)

$$C_D = \frac{24}{N_{\text{Re}}} + \frac{3}{\sqrt{N_{\text{Re}}}} + 0.34 \qquad (10.15)$$

For beds with varying particle size, the Rose equation is

$$h_L = \frac{1.067}{\phi} \frac{C_D}{g} D \frac{V_a^2}{\varepsilon^4} \sum \frac{x}{d} \qquad (10.16)$$

where x is the weight fractions for particle sizes, d. For stratified beds with uniform porosity, the Rose equation is

$$h_L = \frac{1.067}{\phi} \frac{D}{g} \frac{V_a^2}{\varepsilon^4} \sum \frac{C_D x}{d} \qquad (10.17)$$

In both equations, the summation terms may be obtained from computations using sieve analyses.

The head loss through the underdrain system is usually negligible compared to the head loss through the bed. Although the Carman-Kozeny and Rose equations are limited to clean filter beds, they also illustrate the relationship between the head loss and the degree of clogging. As a filter bed clogs, the effective porosity, ε, decreases, which results in an increase in head loss, h_L.

EXAMPLE 10.1 *Head Loss through a Filter Bed*

A rapid sand filter has a sand bed 24 in. in depth. Pertinent data are: specific gravity of the sand = 2.65, shape factor (ϕ) = 0.82, porosity (ε) = 0.45, filtration rate = 2.5 gpm/ft², and operating temperature = 50°F (10°C). The sieve analysis of the sand is shown in columns (1) and (2) of Table 10.1. Determine the head loss for the clean filter bed using the Rose equation for a stratified bed.

SOLUTION The reduced data are given in Table 10.1.

The mean diameter, d, column (3), is the average of the sieve size openings. The Reynolds number $N_{\text{Re}} = \phi d V_a / v$. The velocity, V_a, and the viscosity, v, are

$$V_a = (2.5 \, \text{gal/min})(\text{ft}^3/7.48 \, \text{gal})(\text{min}/60 \, \text{sec})$$
$$= 0.00557 \, \text{ft/sec}$$
$$v = 1.3101 \, \text{centistokes at } 50°F$$
$$= (1.3101)(1.075 \times 10^{-5})$$
$$= 1.4084 \times 10^{-5} \, \text{ft}^2/\text{sec}$$

The Reynolds number for sieve size 14–20 is

$$N_{\text{Re}} = (0.82)(0.003283 \, \text{ft})(0.00557 \, \text{ft/sec}) \div (0.000014084 \, \text{ft}^2/\text{sec})$$
$$= 1.065$$

The N_{Re} values for the other sieve sizes are computed the same way, and these N_{Re} values are shown in column (4). The drag coefficient $C_D = 24/N_{\text{Re}}$ and the C_D values are shown in column (5). For the first sieve size, $N_{\text{Re}} = 1.065$, but since it is approximatley 1, the C_D value may still be

TABLE 10.1 *Stratifiled Data from Sieve Analysis*

(1) SIEVE SIZE	(2) WEIGHT RETAINED (%)	(3) d (ft)	(4) N_{Re}	(5) C_D	(6) $C_D x/d$ (ft⁻¹)
14–20	0.87	0.003283	1.065	22.54	60
20–28	8.63	0.002333	0.757	31.70	1173
28–32	26.30	0.001779	0.577	41.59	6148
32–35	30.10	0.001500	0.486	49.38	9909
35–42	20.64	0.001258	0.408	58.82	9651
42–48	7.09	0.001058	0.343	69.97	4689
48–60	3.19	0.000888	0.288	83.33	2993
60–65	2.16	0.000746	0.242	99.17	2871
65–100	1.02	0.000583	0.189	126.98	2222

$$\Sigma C_D x/d = 39{,}716 \, \text{ft}^{-1}$$

computed using $C_D = 24/N_{Re}$. The $C_D x/d$ values are shown in column (6) and the $\Sigma C_D x/d = 39{,}716\,\text{ft}^{-1}$. The Rose equation is

$$h_L = \frac{1.067}{\phi} \frac{D}{g} \frac{V_a^2}{\varepsilon^4} \sum \frac{C_D x}{d}$$

$$h_L = \frac{1.067}{0.82} \left| \frac{2.0\,\text{ft}}{} \right| \frac{\text{sec}^2}{32.17\,\text{ft}} \left| \frac{(0.00557\,\text{ft})^2}{\text{sec}^2} \right| \frac{}{(0.45)^4} \left| \frac{39{,}716}{\text{ft}} \right.$$

$$= \boxed{2.43\,\text{ft}}$$

EXAMPLE 10.1 SI *Head Loss through a Filter Bed*

A rapid sand filter has a sand bed 0.610 m in depth. Pertinent data are: specific gravity of the sand = 2.65, shape factor (ϕ) = 0.82, porosity (ε) = 0.45, filtration rate = 1.70 ℓ/s-m², and operating temperature = 10°C. The sieve analysis of the sand is shown in columns (1) and (2) of Table 10.2. Determine the head loss for a clean filter bed using the Rose equation for a stratified bed.

SOLUTION The reduced data are given in Table 10.2. The mean diameter, d, column (3), is the average of the sieve size openings. The Reynolds number $N_{Re} = \phi d V_a/v$. The velocity, V_a, and the viscosity, v, are

$$V_a = (1.70\,\ell/\text{s-m}^2)(\text{m}^3/1000\,\ell) = 0.00170\,\text{m/s}$$

$$v = 1.3101 \text{ centistokes at } 10°\text{C}$$
$$= (1.3101)(1.0 \times 10^{-6})$$
$$= 1.3101 \times 10^{-6}\,\text{m}^2/\text{s}$$

TABLE 10.2 *Stratified Data from Sieve Analysis*

(1) SIEVE SIZE	(2) WEIGHT RETAINED (%)	(3) d (m)	(4) N_{Re}	(5) C_D	(6) $C_D x/d$ (m⁻¹)
14–20	0.87	0.0010006	1.065	22.54	196
20–28	8.63	0.0007111	0.757	31.70	3,847
28–32	26.30	0.0005422	0.577	41.59	20,174
32–35	30.10	0.0004572	0.487	49.28	32,444
35–42	20.64	0.0003834	0.408	58.82	31,665
42–48	7.09	0.0003225	0.343	69.97	15,383
48–60	3.19	0.0002707	0.288	83.33	9,820
60–65	2.16	0.0002274	0.242	99.17	9,420
65–100	1.02	0.0001777	0.189	126.98	7,289

$$\Sigma C_D x/d = 130{,}238\,\text{m}^{-1}$$

The Reynolds number for sieve size 14–20 is

$$N_{Re} = (0.82)(0.0010006\,\text{m})(0.00170\,\text{m/s}) \div (0.000001310\,\text{m}^2/\text{s})$$
$$= 1.065$$

The N_{Re} values for the other sieve sizes are computed the same way, and these N_{Re} values are shown in column (4). The drag coefficient $C_D = 24/N_{Re}$ and the C_D values are shown in column (5). The C_Dx/d values are shown in column (6), and the $\Sigma C_Dx/d = 130{,}238\,\text{m}^{-1}$. The Rose equation is

$$h_L = \frac{1.067}{\phi}\frac{D}{g}\frac{V_a^2}{\varepsilon^4}\sum\frac{C_Dx}{d}$$

$$h_L = \frac{1.067}{0.82}\left|\frac{0.610\,\text{m}}{}\right|\left|\frac{\text{s}^2}{9.806\,\text{m}}\right|\left|\frac{(0.00170\,\text{m})^2}{\text{s}^2}\right|\left|\frac{1}{(0.45)^4}\right|\left|\frac{130{,}238}{\text{m}}\right|$$

$$= \boxed{0.743\,\text{m}}$$

Hydraulics of Expanded Beds

The hydraulics of expanded beds may be analyzed for both uniform and stratified beds. For a uniform bed of depth D, backwashing will expand the bed to expanded depth D_e. During backwashing, the frictional resistance of the particles equals the head loss of the liquid expanding the bed (Fair *et al.*, 1968; Fair and Hatch, 1933). Thus,

$$h_L\rho g = (\rho_s - \rho)(1 - \varepsilon_e)(D_e)g \qquad (10.18)$$

where ρ_s = mass density of the particles and ε_e = porosity of the expanded bed. Canceling g and rearranging Eq. (10.18) gives

$$h_L = \left(\frac{\rho_s - \rho}{\rho}\right)(1 - \varepsilon_e)(D_e) \qquad (10.19)$$

The value of ε_e can be determined from

$$\varepsilon_e = \left(\frac{V_b}{V_s}\right)^{0.22} \qquad (10.20)$$

where V_b = upflow velocity of the backwash water and V_s = settling velocity of the particles. Consequently, a bed of uniform particles will expand when

$$V_b = V_s\,\varepsilon_e^{4.5} \qquad (10.21)$$

The volume of the sand in an unexpanded bed will equal the volume of sand in an expanded bed or, stated mathematically,

$$(1 - \varepsilon)AD = (1 - \varepsilon_e)AD_e \qquad (10.22)$$

where A = bed area. Rearranging gives

$$D_e = \left(\frac{1 - \varepsilon}{1 - \varepsilon_e}\right)D \qquad (10.23)$$

Substituting Eq. (10.20) for ε_e in Eq. (10.23) gives

$$D_e = \left[\frac{1 - \varepsilon}{1 - (V_b/V_s)^{0.22}}\right]D \tag{10.24}$$

For stratified beds, the smaller particles in the upper layers expand first. Once V_b is sufficient to fluidize the largest particles, the entire bed will be expanded. The expansion of the bed is represented by a modification of Eq. (10.23);

$$D_e = (1 - \varepsilon)D \sum \frac{x}{1 - \varepsilon_e} \tag{10.25}$$

where x is the weight fraction of particles with an expanded porosity, ε_e.

EXAMPLE 10.2 *Filter Backwashing*

A rapid sand filter having the same sand analysis as in Example 10.1 is to be backwashed. Determine:

1. The backwash velocity required to expand the bed.
2. The backwash flow required to expand the bed.
3. The head loss at the beginning of the backwash.
4. The depth of the expanded sand bed.

SOLUTION The reduced data are given in Table 10.3.

The backwash velocity to expand the bed requires the settling velocity of the largest particles. The settling velocity, V_s, is given by

$$V_s = \left[\frac{4g}{3C_D}(S_s - 1)d\right]^{1/2}$$

TABLE 10.3 *Reduced Data from Sieve Analysis*

(1) SIEVE SIZE	(2) WEIGHT RETAINED (%)	(3) d (ft)	(4) V_s (ft/sec)	(5) ε_e	(6) $\dfrac{x}{1 - \varepsilon_e}$
14–20	0.87	0.003283	0.498	0.454	0.016
20–28	8.63	0.002333	0.334	0.494	0.171
28–32	26.30	0.001779	0.245	0.529	0.558
32–35	30.10	0.001500	0.202	0.552	0.672
35–42	20.64	0.001258	0.164	0.578	0.489
42–48	7.09	0.001058	0.136	0.602	0.178
48–60	3.19	0.000888	0.111	0.630	0.086
60–65	2.16	0.000746	0.091	0.657	0.063
65–100	1.02	0.000583	0.068	0.701	0.034

$$\sum \frac{x}{1 - \varepsilon_e} = 2.267$$

The drag coefficient, C_D, for the transition range that applies to this problem is given by

$$C_D = \frac{24}{N_{\text{Re}}} + \frac{3}{\sqrt{N_{\text{Re}}}} + 0.34$$

The N_{Re} value is

$$N_{\text{Re}} = \phi \frac{d V_s}{v}$$

For the first sieve size, $d = (0.003283\,\text{ft})(30.48\,\text{cm/ft})$ or $0.10\,\text{cm}$. From Figure 9.5, a particle $0.10\,\text{cm}$ in diameter and having a specific gravity of 2.65 has a settling velocity, V_s, of about $14\,\text{cm/sec}$. The viscosity, v, is 1.3101 centistokes $= 1.3101 \times 10^{-2}\,\text{cm}^2/\text{sec}$. The approximate N_{Re} is

$$N_{\text{Re}} = \frac{0.82 \,\left|\, 0.10 \,\right|\, 14 \,\left|\, 10^2 \right.}{\left.\,\right|\, \left.\,\right|\, 1.3101 \,\left|\, \right.} = 87.6$$

$$C_D = \frac{24}{87.6} + \frac{3}{\sqrt{87.6}} + 0.34 = 0.935$$

$$V_s = \left[\frac{4}{3} \,\left|\, \frac{32.17\,\text{ft}}{\text{sec}^2} \,\right|\, \frac{}{0.935} \,\left|\, 2.65 - 1 \,\right|\, 0.003283\,\text{ft} \,\right]^{1/2}$$

$$= 0.498\,\text{ft/sec}$$

$$V_b = V_s \varepsilon^{4.5}$$

$$= (0.498\,\text{ft/sec})(0.45)^{4.5}$$

$$= \boxed{0.0137\,\text{ft/sec} = 0.822\,\text{ft/min}}$$

$$\text{Rate} = (0.822\,\text{ft/min})(7.48\,\text{gal/ft}^3)$$

$$= \boxed{6.15\,\text{gpm/ft}^2}$$

To determine the head loss at the beginning of the backwash, $1 - \varepsilon$ and D are substituted for $1 - \varepsilon_e$ and D_e, respectively, in Eq. (10.19) to give

$$h_L = \left(\frac{\rho_s - \rho}{\rho} \right)(1 - \varepsilon)(D)$$

$$= (S_s - 1)(1 - \varepsilon)(D)$$

$$= (2.65 - 1)(1 - 0.45)(2\,\text{ft})$$

$$= \boxed{1.82\,\text{ft}}$$

The depth of the expanded bed requires the expanded porosities, which are given by

$$\varepsilon_e = \left(\frac{V_b}{V_s} \right)^{0.22}$$

$$= \left(\frac{0.0137}{V_s} \right)^{0.22}$$

The settling velocity, V_s, for the first sieve size is 0.498 ft/sec from the previous calculations. In a like manner, the settling velocities for all sieve sizes were determined and are shown in column (4). The value of ε_e for the first sieve size is

$$\varepsilon_e = \left(\frac{0.0137}{0.498}\right)^{0.22}$$
$$= 0.454$$

In a like manner, the ε_e values for all sieve sizes were determined and are shown in column (5). Dividing the weight fraction retained on the first sieve (0.0087) by $(1 - 0.454)$ gives $x/(1 - \varepsilon_e)$ of 0.016. In a like manner, all values of $x/(1 - \varepsilon_e)$ were determined and are shown in column (6). The Σ value is 2.267. The expanded bed depth for the stratified bed is

$$D_e = (1 - \varepsilon)D \sum \frac{x}{1 - \varepsilon_e}$$
$$= (1 - 0.45)(2\,\text{ft})(2.267)$$
$$= \boxed{2.49\,\text{ft}}$$

Note: The backwash rate in this problem is the minimum required to expand the bed. Actually, 15 to 20 gpm/ft^2 is used to expand a bed for proper agitation during cleansing.

EXAMPLE 10.2 SI   *Filter Backwashing*

A rapid sand filter having the same sand analysis as in Example 10.1 SI is to be backwashed. Determine:

1. The backwash velocity required to expand the bed.
2. The backwash flow required to expand the bed.
3. The head loss at the beginning of the backwash.
4. The depth of the expanded sand bed.

SOLUTION   The reduced data are given in Table 10.4. The backwash velocity to expand the bed requires the settling velocity of the largest particles. The settling velocity, V_s, is given by

$$V_s = \left[\frac{4g}{3C_D}(S_s - 1)d\right]^{1/2}$$

The drag coefficient, C_D, for the transition range that applies to this problem is given by

$$C_D = \frac{24}{N_{\text{Re}}} + \frac{3}{\sqrt{N_{\text{Re}}}} + 0.34$$

TABLE 10.4 *Reduced Data from Sieve Analysis*

(1) SIEVE SIZE	(2) WEIGHT RETAINED (%)	(3) d (m)	(4) V_s (m/s)	(5) ε_e	(6) $\dfrac{x}{1-\varepsilon_e}$
14–20	0.87	0.0010006	0.152	0.454	0.016
20–28	8.63	0.0007111	0.102	0.495	0.171
28–32	26.30	0.0005422	0.0747	0.530	0.560
32–35	30.10	0.0004572	0.0616	0.553	0.673
35–42	20.64	0.0003834	0.0500	0.579	0.490
42–48	7.09	0.0003225	0.0415	0.604	0.179
48–60	3.19	0.0002707	0.0338	0.631	0.086
60–65	2.16	0.0002274	0.0277	0.660	0.064
65–100	1.02	0.0001777	0.0207	0.703	0.034

$$\sum \frac{x}{1-\varepsilon_e} = 2.273$$

The N_{Re} value is

$$N_{Re} = \phi \frac{dV_s}{\nu}$$

For the first sieve size, $d = 0.0010006\,\text{m}$ or $0.10\,\text{cm}$. From Figure 9.5, a particle $0.10\,\text{cm}$ in diameter and having a specific gravity of 2.65 has a settling velocity, V_s, of about $14\,\text{cm/s}$. The viscosity, ν, is 1.3101 centistokes or $1.3101 \times 10^{-2}\,\text{cm}^2/\text{s}$. The approximate N_{Re} is

$$N_{Re} = \frac{0.82 \mid 0.10 \mid 14 \mid 10^2}{\mid 1.3101 \mid} = 87.6$$

$$C_D = \frac{24}{87.6} + \frac{3}{\sqrt{87.6}} + 0.34 = 0.935$$

$$V_s = \left[\frac{4}{3} \mid \frac{9.806\,\text{m}}{\text{s}^2} \mid \frac{1}{0.935} \mid 2.65 - 1 \mid 0.0010006\,\text{m}\right]^{1/2}$$

$$= 0.152\,\text{m/s}$$

$$V_b = V_s \varepsilon^{4.5}$$
$$= (0.152\,\text{m/s})(0.45)^{4.5}$$
$$= \boxed{0.00418\,\text{m/s}}$$

$$\text{Rate} = (0.00418\,\text{m/s})(1000\,\ell/\text{m}^3) = \boxed{4.18\,\ell/\text{s-m}^2}$$

To determine the head loss at the beginning of the backwash, $1 - \varepsilon$ and D are substituted for $1 - \varepsilon_e$ and D_e, respectively, in Eq. (10.19) to give

$$h_L = \left(\frac{\rho_s - \rho}{\rho}\right)(1 - \varepsilon)(D)$$

$$= (S_s - 1)(1 - \varepsilon)(D)$$

$$= (2.65 - 1)(1 - 0.45)(0.610\,\text{m})$$

$$= \boxed{0.554\,\text{m}}$$

The depth of the expanded bed requires the expanded porosity, which is given by

$$\varepsilon_e = \left(\frac{V_b}{V_s}\right)^{0.22}$$

$$= \left(\frac{0.00418}{V_s}\right)^{0.22}$$

The settling velocity, V_s, for the first sieve size is $0.152\,\text{m/s}$ from the previous calculations. In a like manner, the settling velocities for all sieve sizes were determined and are shown in column (4). The value of ε_e for the first sieve size is

$$\varepsilon_e = \left(\frac{0.00418}{0.152}\right)^{0.22}$$

$$= 0.454$$

In a like manner, the ε_e values for all sieve sizes were determined and are shown in column (5). Dividing the weight fraction retained on the first sieve (0.0087) by $(1 - 0.454)$ gives $x/(1 - \varepsilon_e)$ of 0.016. In a like manner, all values of $x/(1 - \varepsilon_e)$ were determined and are shown in column (6). The Σ value is 2.273. The expanded bed depth for a stratified bed is

$$D_e = (1 - \varepsilon)D \sum \frac{x}{1 - \varepsilon_e}$$

$$= (1 - 0.454)(0.610\,\text{m})(2.273)$$

$$= \boxed{0.757\,\text{m}}$$

Note: The backwash rate in this problem is the minimum required to expand the bed. Actually, 10.2 to 13.6 ℓ/s-m^2 is used to expand a bed for proper agitation during cleansing.

Operational Problems   The major operating problems encountered in the use of rapid sand filtration are mud accumulations or mudballs, bed shrinkage, and air binding. Mudball formation is a condition which may occur when the filter feed contains a muddy floc and the filter is not adequately backwashed. The muddy floc will accumulate on the surface of the sand bed, forming a muddy mat that will penetrate any cracks in the top of the sand. If a surface wash is not used, some of the mud may be pressed together to form small muddy balls during the backwash. With subsequent cycles of filtration and backwashing, these balls enlarge and become caked with sand and may eventually settle to the gravel layer. They interfere with uniform filtration and cause inadequate backwashings. Mudball formation may be minimized by the use of a surface wash that breaks up any muddy mat formation.

Bed shrinkage may occur if the sand grains become covered with a soft slime coating. This causes the bed to compact as the filter run progresses and results in cracks in the bed surface and along the side walls of the filter. These cracks are undesirable because they may allow improperly filtered water to pass through the bed, and fine muddy floc may accumulate in them to start mudball formation. Slime coatings on the filter sand may be minimized by the use of a surface wash system.

Air binding is caused by the release of air gases dissolved in the water, such as nitrogen and oxygen, thus creating air bubbles in the sand bed. Air binding usually results when a filter is operated under a negative head, and it may interfere with the rate of filtration. Also, at the beginning of the backwash, the violent agitation due to the rising air bubbles may cause a loss of sand. The principal method of control is the avoidance of negative head or pressures.

MULTIMEDIA FILTERS

These filters, which have more than one medium, may be open gravity filters, as shown in Figure 10.2, or pressure filters, as shown in Figures 10.7 and 10.9. In water treatment, they have become more popular in recent years. In advanced and tertiary waste treatment, they are the main type of filters that have been used successfully. Dual-media filter beds usually employ anthracite and sand; however, other materials have been used, such as activated carbon and sand. Multimedia filter beds generally use anthracite, sand, and garnet. However, other materials have been used, such as activated carbon, sand, and garnet. Also, dual- and multimedia filters using ion exchange resins as one of the media have been tried. In some of these filters, the media may have additional characteristics other than removing particulates. For example, activated carbon removes dissolved organic substances.

The main advantages of multimedia filters compared to single-medium filters are longer filtration runs, higher filtration rates, and the ability to filter a water with higher turbidity and suspended solids. The advantages of the multimedia filters are due to (1) the media particle size, (2) the different specific gravities of the media, and (3) the media gradation. These result in a filter with a larger percent of the pore volume being available for solids storage. In the single-medium filter, the pore volume available for solids storage is in the top portion of the bed, whereas in the multimedia filter, the available pore volume is extended deep within the filter bed. Because of the deep penetration of accumulated floc, these filters are frequently referred to as "deep bed filters." The single-medium filters are rarely used in wastewater or advanced wastewater treatment because of short filter runs. As a result of the large pore volume available for floc storage, the multimedia filters can be used in advanced or tertiary wastewater treatment and still have a reasonable filter run.

Dual-Media Filters

The dual-media filter, consisting of a layer of coarse anthracite coal above a layer of fine sand, is one technique for increasing the pore volume of a filter. Figure 10.22(b) shows the grain and pore size in a dual-media filter, and Figure 10.22(a) shows these characteristics for a single-medium filter. It can be seen from the pore size profile that the available pore volume of the dual-media filter will be greater than the single-medium filter. The available pore volume, however, will not be as large as the total pore volume because of the fine to coarse gradation within each layer. Ideally, the available pore volume would be maximum at the top of the filter and gradually decrease to a minimum at the bottom of the filter.

Usually, a dual-media filter consists of an 18- to 24-in. (457- to 610-mm) layer of crushed anthracite coal overlaying a 6- to 12-in. (152- to 305-mm) layer of sand. Coal has a specific gravity 1.2 to 1.6, and sand has a specific gravity of 2.65. During the first backwash, the sand layer remains below the coal as a result of its higher specific gravity and its grain size relative to the coal particles. After the first backwash, there will not be a distinct interface between the two layers, but instead there will be a blended region of both coal particles and sand grains.

The size and characteristics of the anthracite and sand media and the thickness of the layers depend upon whether the filter is to be used for water or wastewater treatment. Filtration rates may vary from 2 to 10 gal/min-ft^2 (1.36 to 6.79 ℓ/s-m^2); however, a rate ranging from 3 to 6 gal/min-ft^2 (2.04 to 4.08 ℓ/s-m^2) is common.

Mixed-Media Filters

The ideal filter has a pore size and gradation as shown in Figure 10.22(c). The pore size is greatest at the top of the bed and gradually decreases to a minimum at the bottom. The available pore volume, like the pore size, is maximum at the top of the bed and decreases to a minimum at the bottom. The media have a gradation which is from coarse at the top to fine at the bottom. The ideal filter may be approached by using a dual-media filter of crushed anthracite coal above sand and placing a third, very dense medium below the sand. This allows the third medium to be very fine and still remain in the lower depths during backwashing. The resulting filter is referred to as a mixed-media filter since there is some intermixing between the layers during backwashing. Garnet, which has a specific gravity of about 4.2, has been found to be ideal as the third medium. Ilmenite, having a specific gravity of about 4.5, is also used but to a lesser extent. The anthracite, sand, and garnet or ilmenite are properly sized to allow some intermixing of the media during backwash. After backwashing there will be no distinct interface between the media layers. The filter bed will approach the ideal, as shown in Figure 10.22(c), which has a gradual decrease in pore size with increasing depth.

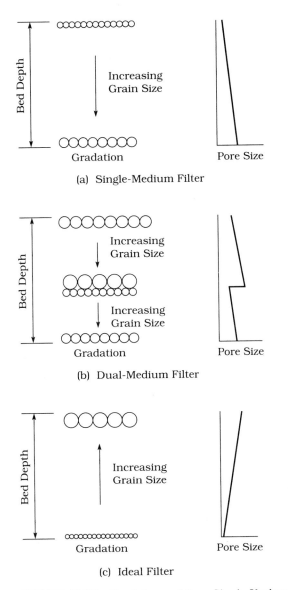

(a) Single-Medium Filter

(b) Dual-Medium Filter

(c) Ideal Filter

FIGURE 10.22 *Gradation and Pore Size in Various Filters*

Since the pore size decreases from the top to the bottom of the filter, the filter will have a large available pore volume extending throughout the depth of the filter bed. About 3 in. (76 mm) of coarse garnet or ilmenite is placed under the third layer to prevent fine particles from entering the underlaying gravel.

The size and characteristics of the media and the thickness of the layers placed in the filter depend upon the type of service for the filter — that is, whether it is for water or wastewater treatment. The filtration rates that have been used in water treatment or advanced waste treatment are from 2 to 12 gal/min-ft^2 (1.36 to 8.15 ℓ/s-m^2); however, a rate ranging from 3 to 6 gal/min-ft^2 (2.04 to 4.08 ℓ/s-m^2) is common.

FILTER LAYOUT, APPURTENANCES, AND DETAILS

Usually gravity filters are constructed of poured-in-place reinforced concrete. The minimum number of filters required is two; however, four are preferable. The filters are placed side by side in a row, and the pipe gallery, which contains all the necessary piping, valves, and so on, runs parallel to the filter row. In cold climates the filters are usually enclosed in a building; however, they may be in the open if a perforated pipe furnishing compressed air bubbles is placed around the perimeter of each filter immediately above the filter bed. The slight agitation by the air bubbles prevents freezing. The pipe gallery and operating floor are always enclosed to protect personnel and equipment. A dehumidified pipe gallery will reduce maintenance on controls, valves, and other equipment. Pressure filters are usually cylindrical in shape and are prefabricated of steel with a maximum diameter of 10 to 12 ft (3.05 to 3.66 m) and a maximum length of 60 ft (18.29 m). They should be equipped with a sight glass for observation of the bed during backwashing and should have an access manhole for necessary maintenance. Hydraulic removal of the filter media should be provided.

The two basic types of control systems are the manual and the fully automatic. The use of accurate and reliable equipment for measuring the effluent turbidity not only gives a continuous record of filter performance but greatly assists in filter operation. Fully automatic systems which have a programmer activated by the effluent turbidity or head loss are available. Once the effluent turbidity or head loss reaches a preset level, the programmer takes the filter off-line, backwashes it, and places it back on-line.

In most filters presently built, the control of the flow is by a rate of flow controller. The rate of flow controller maintains a uniform flowrate with a constant water depth over the filter bed. Uniform flow is maintained by varying the head loss between the filter bed surface and the downstream side of the rate of flow controller. The controller usually consists of a venturi with a variable-opening diaphragm or butterfly valve on the downstream side. The valve is activated by the difference in pressure between the upstream side and the throat of the venturi. Another method for flow control without the use of a rate of flow controller consists of a weir in the inlet to the filter and a weir in the effluent channel discharging into the clearwell. The weir downstream of the filter keeps a minimum water depth over the filter bed. The weir in the inlet maintains a constant rate of flow to the filter, and its flow rate is independent of the depth of water over the filter bed since the weir crest is above the water surface. As the filter run progresses, the depth of water over the filter bed increases because of the increase in head loss. The major disadvantage of this type of flow control is that the filter walls must be from 5 to 6 feet (1.52 to 1.83 m) deeper than required when a rate of flow controller is used (Baumann and Oulman,

1970). If pressure filters are used, another way to control the filtration rate is by the use of pumps that pump at a relatively constant rate to the filters.

Another auxiliary scour system, termed **air scouring**, is available in lieu of the surface wash. To wash with this system, the water is filtered down to about 6 in. (150 mm) above the bed. Air is applied to the underdrain system at 2 to 5 cfm/ft^2 (0.61 to 1.52 m^3/min-m^2) for 3 to 10 minutes, and then the backwash is initiated at 2 to 5 gpm/ft^2 (1.36 to 3.40 ℓ/s-m^2). Once the water level is about 1 ft (0.3 m) below the washwater troughs, the air is stopped and the backwash is operated at the normal rate for the usual period of time (Culp and Culp, 1978).

Washwater troughs are usually spaced at a clear distance of 5 to 6 ft (1.52 to 1.83 m) from each other. They serve to remove the backwash, and, since the distance from the top of all troughs to the underdrains is a constant value, they aid in maintaining a uniform backwash. The trough bottoms should be at least 6 in. (150 mm) above the expanded bed during backwashing. Troughs are made of materials such as reinforced concrete, fiberglass, and enameled steel. Troughs are frequently of precast reinforced concrete and have a rectangular cross section, as shown in Figure 10.23(a). Figure 10.23(b) shows the washwater overflowing into a trough. The total discharge is twice that of a suppressed weir or

$$Q = 2CLH^{3/2} \qquad (10.26)$$

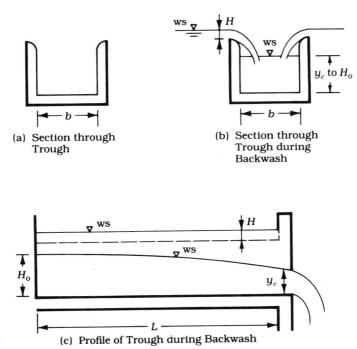

FIGURE 10.23
Wash Trough Details

(a) Section through Trough

(b) Section through Trough during Backwash

(c) Profile of Trough during Backwash

where

Q = total discharge, cfs (m³/s)

C = 3.33 for USCS units and 1.86 for SI units

L = trough length, ft (m)

H = head on weir, ft (m)

Figure 10.23(c) shows the profile of a trough during backwashing. Shortly before the freefall into the gullet, the water is at the critical depth, y_c^*. For a rectangular channel the critical depth is

$$y_c = \sqrt[3]{\frac{q^2}{g}}$$ (10.27)

where

y_c = critical depth, ft (m)

q = discharge per ft (m) of width, $q = Q/b$, cfs/ft (m³/s-m)

g = acceleration by gravity

Using the modified Camp equation for a lateral spillway channel (Reynolds et al., 1989), the upstream depth, H_0, for a level, rectangular channel with a freefall is

$$H_0 = \sqrt{y_c^2 + \frac{2Q^2}{gb^2 y_c} + \frac{2}{3} \cdot \frac{n^2 L Q^2}{C^2 b^2 \bar{r}^{4/3} \bar{d}}}$$ (10.28)

where

H_0 = upstream depth, ft (m)

y_c = downstream depth at the freefall, ft (m)

Q = total discharge, cfs (m³/s)

g = acceleration due to gravity, ft/sec² (m/s²)

b = channel width, ft (m)

n = Manning's friction factor, 0.029 to 0.032 for concrete

L = channel length, ft (m)

C = 1.49 for USCS units and 1.0 for SI units

$\bar{r}$ = mean hydraulic radius, ft (m)

$\bar{d}$ = mean depth, ft (m)

For a level channel the water surface profile is parabolic, and for this case, the mean depth, $\bar{d}$, can be approximated by

$$\bar{d} = H_0 - \tfrac{1}{3}(H_0 - y_c)$$ (10.29)

---

* Critical depth can be assumed at the freefall.

The mean hydraulic radius, $\bar{r}$, is given by

$$\bar{r} = \frac{\bar{d}b}{2\bar{d} + b} \qquad (10.30)$$

In using Eq. (10.28), a trial calculation ignoring friction is made to determine the H_0 value without friction. Next, Eq. (10.29) can be used to get the mean depth, $\bar{d}$, and Eq. (10.30) can then be used to get the mean hydraulic radius, $\bar{r}$. Next Eq. (10.28) can be used to get the H_0 value for the first trial. The procedure is repeated until a constant H_0 value is obtained, which usually requires three successive trials. The friction loss will increase the value of H_0 without friction by 6 to 16% of the water surface drawdown. A distance of at least 3 in. (75 mm) should be provided from the maximum depth to the top of the trough.

EXAMPLE 10.3 *Wash Trough Design*

A rapid sand filter has washwater troughs that are 16 ft long and spaced 6.20 ft from center to center. The precast concrete troughs are rectangular in cross section with an inside width of 18 in. The backwash flow is 15 gpm/ft². It is desired to determine the water depth over the trough sides, H, the depth at the freefall, y_c, the upstream water depth, H_0, and the trough depth if the top of the trough is at least 3 in. above the maximum water surface in the trough. The Manning's friction factor $n = 0.032$.

SOLUTION The total flow, Q, is $(16\,\text{ft})(6.20\,\text{ft})(15\,\text{gpm/ft}^2)$ or 1488 gpm, which is 3.32 cfs. Rearranging Eq. (10.26) for USCS units gives

$$H = \left[\frac{Q}{2(3.33)L} \right]^{2/3} = [(3.32)/(2)(3.33)(16.0)]^{2/3}$$

$$= \boxed{0.099\,\text{ft or } 1.19\,\text{in.}}$$

The value of q is $3.32/1.5$ or 2.21 cfs/ft. The depth of the water at the freefall is given by

$$y_c = \sqrt[3]{\frac{q^2}{g}} = \sqrt[3]{\frac{(2.21)^2}{32.17}}$$

$$= \boxed{0.53\,\text{ft or } 6.40\,\text{in.}}$$

The upstream water depth without friction is given by

$$H_0 = \sqrt{y_c^2 + \frac{2Q^2}{gb^2 y_c}} = \sqrt{(0.53)^2 + \frac{(2)(3.32)^2}{(32.17)(1.5)^2(0.53)}}$$

$$= 0.925\,\text{ft}$$

For the first trial, the mean depth $\bar{d} = 0.925 - \frac{1}{3}(0.925 - 0.53) = 0.793\,\text{ft}$. The mean hydraulic radius $\bar{r} = (0.793)(1.5)/(0.793 \times 2 + 1.5) = 0.385\,\text{ft}$. Substituting these and other variables into the lateral spillway equation gives

$H_0 = 0.982$ ft. For the second trial, the mean depth $\bar{d} = 0.982 - \frac{1}{3}(0.982 - 0.53) = 0.831$ ft. The mean hydraulic radius $\bar{r} = (0.831)(1.5)/(0.831 \times 2 + 1.5) = 0.394$ ft. Substituting these and other variables into the lateral spillway equation gives $H_0 = 0.977$ ft. For the third trial, the mean depth $\bar{d} = 0.977 - \frac{1}{3}(0.977 - 0.53) = 0.828$ ft. The mean hydraulic radius $\bar{r} = (0.828)(1.5)/(0.828 \times 2 + 1.5) = 0.394$ ft. Substituting these and other variables into the lateral spillway equation gives

$$H_0 = \sqrt{(0.53)^2 + \frac{(2)(3.32)^2}{(32.17)(1.5)^2(0.53)} + \frac{2}{3} \cdot \frac{(0.032)^2(16)(3.32)^2}{(1.49)^2(1.5)^2(0.394)^{4/3}(0.828)}}$$

$$= \boxed{0.978 \text{ ft or } 11.74 \text{ in.}}$$

If the troughs are available in 3-in. increments in depth, the trough depth should be

$$\text{Depth} = 11.74 \text{ in.} + 3 \text{ in.} = 14.74 \text{ in. or } \boxed{15 \text{ in.}}$$

EXAMPLE 10.3 SI   *Wash Trough Design*

A rapid sand filter has washwater troughs that are 5.0 m long and spaced at 1.9 m from center to center. The precast concrete troughs are rectangular in cross section with an inside width of 0.45 m. The backwash flow is 10 ℓ/s-m². It is desired to determine the water depth over the trough sides, H, the depth at the freefall, y_c, the upstream water depth, H_0, and the trough depth if the top of the trough is at least 7.5 cm above the maximum water surface in the trough.

The Manning's friction factor $n = 0.032$.

SOLUTION   The total flow, Q, is $(5.0 \text{ m})(1.9 \text{ m})(10 \text{ ℓ/s-m}^2) = 95$ ℓ/s, or $Q = (95 \text{ ℓ/s})(\text{m}^3/1000 \text{ ℓ}) = 0.095$ m³/s. The discharge over a suppressed weir is given by $Q = 1.86LH^{3/2}$, where $Q = \text{m}^3/\text{s}$, $L = \text{m}$, and $H = \text{m}$. Rearranging this equation for two weirs gives

$$H = \left[\frac{Q}{2(1.86)L}\right]^{2/3} = [(0.095)/(2)(1.86)(5.0)]^{2/3}$$

$$= \boxed{0.030 \text{ m or } 3.0 \text{ cm}}$$

The value of q is $(0.095 \text{ m}^3/\text{s})/0.45 \text{ m} = 0.211$ m³/s-m. The water depth at the freefall is

$$y_c = \sqrt[3]{\frac{q^2}{g}} = \sqrt[3]{\frac{(0.211)^2}{9.806}}$$

$$= \boxed{0.166 \text{ m}}$$

The upstream water depth without friction is given by

$$H_0 = \sqrt{y_c^2 + \frac{2Q^2}{gb^2 y_c}}$$

$$= \sqrt{(0.166)^2 + \frac{(2)(0.095)^2}{(9.806)(0.45)^2(0.166)}}$$

$$= \boxed{0.287\,\text{m}}$$

For the first trial, the mean depth, $\bar{d} = 0.287 - \frac{1}{3}(0.287 - 0.166) = 0.247\,\text{m}$. The mean hydraulic radius $\bar{r} = (0.247)(0.45)/(0.247 \times 2 + 0.45) = 0.118\,\text{m}$. Substituting these and other variables into the lateral spillway equation gives $H_0 = 0.305\,\text{m}$. For the second trial, the mean depth $\bar{d} = 0.305 - \frac{1}{3}(0.305 - 0.166) = 0.259\,\text{m}$. The mean hydraulic radius $\bar{r} = (0.259)(0.45)/(0.259 \times 2 + 0.45) = 0.120\,\text{m}$. Substituting these and other variables into the lateral spillway equation gives $H_0 = 0.304\,\text{m}$. For the third trial, the mean depth $\bar{d} = 0.304 - \frac{1}{3}(0.304 - 0.166) = 0.258\,\text{m}$. The mean hydraulic radius $\bar{r} = (0.258)(0.45)/(0.258 \times 2 + 0.45) = 0.120\,\text{m}$. Substituting these and other variables into the lateral spillway equation gives

$$H_0 = \sqrt{(0.166)^2 + \frac{(2)(0.095)^2}{(9.806)(0.45)^2(0.166)} + \frac{2}{3} \cdot \frac{(0.032)^2(5.0)(0.095)^2}{(1.0)^2(0.45)^2(0.120)^{4/3}(0.258)}}$$

$$= \boxed{0.304\,\text{m}}$$

The trough depth should be

$$\text{Depth} = 0.304 + 0.075 = \boxed{0.379\,\text{m}}$$

FILTRATION IN WATER TREATMENT

The slow sand filter, which was developed during the mid-1800s, was the first type of filter used for water treatment. Plain sedimentation of the water prior to filtration was usually provided. These filters were single-medium filters having an effective sand size of about 0.2 to 0.4 mm and were operated at filtration rates of 0.05 to 0.15 gal/min-ft^2 (0.034 to 0.11 ℓ/s-m^2). The filters were cleaned manually, usually every four to six weeks, by scraping off the top layers of clogged sand and cleaning the sand with a scouring device. Because of the large land area requirements and the manual labor involved, the slow sand filter was replaced by the rapid sand filter. Although there are some slow sand filter plants in the United States, the majority of filtration plants are rapid sand filtration plants. In the United States, some very clear surface waters are treated only by disinfection and then sent to the distribution system. However, since the cyst of the protozoan *Giardia lamblia* is very resistant to disinfection, these cities are being encouraged to add filtration. Since the waters are very clear, slow sand filtration is feasible because very long filter runs occur, and many of these cities are adding slow sand filtration.

The rapid sand filter is always preceded by chemical coagulation, flocculation, and sedimentation. The first filters, which operated at about 2 gal/min-ft^2 (1.36 ℓ/s-m^2), consisted of a quartz sand bed overlaying a gravel layer. The turbidity removals ranged from 90 to 98% if the feed water turbidity was between 5 to 10 JTU. Although the standard rate of filtration is generally considered to be 2 gal/min-ft^2 (1.37 ℓ/s-m^2), most rapid sand filters are operated at 3 to 5 gal/min-ft^2 (2.04 to 3.40 ℓ/s-m^2) and have coarse sand beds. The primary filter action in rapid sand filtration is usually depth removal. As shown in Table 10.5, the sand beds are usually 24 to 30 in. (610 to 760 mm) thick and have an effective size of 0.35 to 0.70 mm and a uniformity coefficient less than 1.7.

Calcium carbonate encrustations on the sand grains may occur when lime-soda softening is employed. These enlarge the sand grains, which is undesirable. Their formation may be controlled by lowering the pH by carbonation prior to filtration to precipitate excess lime and stabilize the water.

Since the development of the dual-media and mixed-media filters, most new plants have adopted these types of filters. The primary filter action is depth removal. The characteristics of the dual- and mixed-media filters used in water treatment are shown in Tables 10.6 and 10.7. The principal advantages of these filters over sand filters are higher filtration rates and longer filter runs because of the increased volume for floc storage within the

TABLE 10.5 *Single-Medium Filter Characteristics for Water Treatment*

	VALUE	
CHARACTERISTIC	**Range**	**Typical**
Sand medium:		
Depth		
in.	24–30	27
(mm)	(610–760)	(685)
Effective size, mm	0.35–0.70	0.60
Uniformity coefficient	<1.7	<1.7
Anthracite medium:		
Depth		
in.	24–30	27
(mm)	(610–760)	(685)
Effective size, mm	0.70–0.75	0.75
Uniformity coefficient	<1.75	<1.75
Filtration rate:		
gpm/ft^2	2–5	4
(ℓ/s-m^2)	(1.36–3.40)	(2.72)

TABLE 10.6 *Dual-Media Filter Characteristics for Water Treatment*

CHARACTERISTIC	VALUE	
	Range	Typical
Anthracite:		
Depth		
in.	18–24	24
(mm)	(460–610)	(610)
Effective size, mm	0.9–1.1	1.0
Uniformity coefficient	1.6–1.8	1.7
Sand:		
Depth		
in.	6–8	6
(mm)	(150–205)	(150)
Effective size, mm	0.45–0.55	0.5
Uniformity coefficient	1.5–1.7	1.6
Filtration rate:		
gpm/ft^2	3–8	5
(ℓ/s-m^2)	(2.04–5.44)	(3.40)

TABLE 10.7 *Mixed-Media Filter Characteristics for Water Treatment*

CHARACTERISTIC	VALUE	
	Range	Typical
Anthracite:		
Depth		
in.	16.5–21	18
(mm)	(420–530)	(460)
Effective size, mm	0.95–1.0	1.0
Uniformity coefficient	1.55–1.75	<1.75
Sand:		
Depth		
in.	6–9	9
(mm)	(150–230)	(230)
Effective size, mm	0.45–0.55	0.50
Uniformity coefficient	1.5–1.65	1.60
Garnet:		
Depth		
in.	3–4.5	3
(mm)	(75–115)	(75)
Effective size, mm	0.20–0.35	0.20
Uniformity coefficient	1.6–2.0	<1.6
Filtration rate:		
gpm/ft^2	4–10	6
(ℓ/s-m^2)	(2.72–6.80)	(4.08)

filter. As a result, less backwash water is required per unit volume of filtrate produced.

Direct filtration, which is sometimes used, is rapid sand filtration without sedimentation, and in this process, the flowsheet consists of mixing, flocculation, and rapid sand filtration. The filter removes all of the floc formed; consequently, this process is limited to waters that are extremely clear (less than 10 turbidity units) throughout the year. The criteria for rapid sand filters used for direct filtration is the same as those for ordinary rapid sand filters.

FILTRATION IN WASTEWATER TREATMENT

Filtration in advanced waste treatment may be employed for the following purposes: (1) filtration of secondary effluents, (2) filtration of chemically treated secondary effluents, and (3) filtration of chemically treated primary or raw wastewaters. With the exception of some new filtration techniques and the intermittent sand filter, the filters used in advanced or tertiary wastewater treatment are usually dual-media or mixed-media filters. Characteristics of the dual-media filters are shown in Table 10.8; characteristics of the mixed-media filters are shown in Table 10.9. The principal difference between filters used in water treatment and filters used in wastewater treatment is the size of the media. For wastewater treatment, the

TABLE 10.8 *Dual-Media Filter Characteristics for Advanced or Tertiary Wastewater Treatment*

	VALUE	
CHARACTERISTIC	**Range**	**Typical**
Anthracite:		
Depth		
in.	12–24	18
(mm)	(305–610)	(460)
Effective size, mm	0.8–2.0	1.2
Uniformity coefficient	1.3–1.8	1.6
Sand:		
Depth		
in.	6–12	12
(mm)	(150–305)	(305)
Effective size, mm	0.4–0.8	0.55
Uniformity coefficient	1.2–1.6	1.5
Filtration rate:		
gpm/ft^2	2–10	5
(ℓ/s-m^2)	(1.36–6.79)	(3.40)

Adapted from *Wastewater Engineering, Treatment, Disposal and Reuse*, 2nd ed., by Metcalf & Eddy, Inc. Copyright © 1979 by McGraw-Hill, Inc. Reprinted by permission.

TABLE 10.9 *Multimedia or Mixed-Media Filter Characteristics for Advanced or Tertiary Wastewater Treatment*

CHARACTERISTIC	VALUE	
	Range	Typical
Anthracite:		
Depth		
in.	8–20	16
(mm)	(205–510)	(405)
Effective size, mm	1.0–2.0	1.4
Uniformity coefficient	1.4–1.8	1.5
Sand:		
Depth		
in.	8–16	10
(mm)	(205–405)	(255)
Effective size, mm	0.4–0.8	0.5
Uniformity coefficient	1.3–1.8	1.6
Garnet:		
Depth		
in.	2–6	4
(mm)	(50–150)	(100)
Effective size, mm	0.2–0.6	0.3
Uniformity coefficient	1.5–1.8	1.6
Filtration rate:		
gpm/ft^2	2–10	5
(ℓ/s-m^2)	(1.36–6.79)	(3.40)

Adapted from *Wastewater Engineering: Treatment, Disposal and Reuse*, 2nd ed., by Metcalf & Eddy, Inc. Copyright © 1979 by McGraw-Hill, Inc. Reprinted by permission.

granules must be larger so that the filter will have the desired flowrate capacity and the required storage volume for the accumulated floc.

Of the numerous variables that affect filter performance in advanced wastewater treatment, two of the most important are the floc strength (the ability to withstand shear forces) and the concentration of the suspended solids. Biological flocs are usually more resistant to shear forces than chemical flocs, particularly flocs from alum and iron salt coagulation. In filtering untreated secondary effluents, the primary filter action is surface removal, and as a result, excessive head losses usually terminate the filter runs. The deterioration in the quality of the filtrate rarely determines the end of a filter run. Chemical flocs from alum and iron salt coagulation tend to penetrate deep into a filter bed; thus the main filtering action is depth removal. The termination of the filter run is usually due to filtrate quality deterioration, and breakthrough usually occurs at relatively low head losses, such as 3 to 6 feet (0.9 to 1.8 m) (Tchobanoglous and Eliassen, 1970). Polymer filter aids may be added to the feed to a filter to strengthen alum or iron salt flocs prior to filtration, thus allowing higher filtration rates than usually employed and also longer filter runs. In coagulation with lime, the calcium carbonate precipitate is relatively strong and tends to be removed on the filter surface; thus surface removal is important in this case. The calcium carbonate may form a dense mat which must be broken up by the surface wash prior to backwashing.

In order to prevent microbial slime buildup on the filter media, an auxiliary scour system, either the surface wash or air scour type, is necessary when filtering wastewaters. Calcium carbonate encrustations on the filter media may occur when high pH lime coagulation is employed. Stabilization is necessary to prevent operational problems.

There is a general relationship between the filter influent suspended solids concentration, the filtration rate, and the filter run. For example, if the influent suspended solids are $20 \, mg/\ell$, the filtration rate is $4 \, gal/min\text{-}ft^2$ ($2.72 \, \ell/s\text{-}m^2$) and the filter run is 72 hr, then the filter run at $6 \, gal/min\text{-}ft^2$ ($4.08 \, \ell/s\text{-}m^2$) would be approximately $72 \, hr \times 4/6 = 48 \, hr$ ($72 \, h \times 2.72/4.08 = 48 \, h$). If the suspended solids were reduced to $10 \, mg/\ell$ and the filtration rate were $4 \, gal/min\text{-}ft^2$ ($2.72 \, \ell/s\text{-}m^2$), the filter run would be approximately $72 \, hr \times 20/10 = 144 \, hr$ ($72 \, h \times 20/10 = 144 \, h$).

Filtration of Secondary Effluents

A review of the data from seven tertiary treatment plants (EPA, 1975; Zenz *et al.*, 1972), both pilot and full scale, shows an average suspended solids removal of 66.2%, an average filter run of 15.6 h, an average filtration rate of $3.7 \, gpm/ft^2$ ($2.51 \, \ell/s\text{-}m^2$), and an average suspended solids concentration of $18.3 \, mg/\ell$ in the feed to the filters. Two of the plants had dual-media filters, whereas the remaining five had mixed-media filters. In general, the mixed-media filters gave better performance in terms of both suspended solids removal and duration of the filter runs.

The filtrability of the suspended solids in untreated secondary effluents increases with an increase in the mean cell residence time and the hydraulic detention time in activated sludge plants. Apparently the floc strength is increased with an increase in the mean cell residence time and the hydraulic detention time. The expected effluent quality indicated by Culp and Culp

TABLE 10.10 *Expected Effluent Suspended Solids versus Type of Secondary Treatment*

EFFLUENT	EFFLUENT SUSPENDED SOLIDS (mg/ℓ)
Extended aeration	1–5
Conventional activated sludge	3–10
Contact stabilization	6–15
Two-stage trickling filter	6–15
High-rate trickling filter	10–20

(1978) for untreated secondary effluents filtered by multimedia filtration is shown in Table 10.10.

Intermittent sand filters have been used to give a combined physical-biological treatment for effluents from oxidation ponds or lagoons (Marshall and Middlebrooks, 1974; Reynolds *et al.*, 1976). The intermittent operation results in both aerobic digestion and dewatering of the filtered solids, thus reducing the required maintenance. Although the land requirements are quite large, the intermittent sand filters are a feasible method for treating effluents from small installations.

Filtration of Chemically Coagulated Effluents

A review of the data from four tertiary treatment plants (EPA, 1975; South Tahoe Public Utility District, 1971), both pilot and full scale, shows an average suspended solids removal of 74.2%, an average filter run of 33.7 hr, an average filtration rate of 3.0 gpm/ft^2 (2.04 ℓ/s-m^2), and an average suspended solids of 9.3 mg/ℓ in the feed to the filters. Lime clarification was practiced at three of the plants and alum clarification at one. Three of the plants had dual-media filters, whereas one had a mixed-media filter. The mixed-media filter gave better performance in terms of suspended solids removal, although the filter runs from both types of filters were about equal. The mixed-media filter, however, was filtering at an average rate of 3.4 gpm/ft^2 (2.31 ℓ/s-m^2), compared to 2.8 gpm/ft^2 (1.90 ℓ/s-m^2) for the dual-media filters.

Filtration of Chemically Treated Primary or Raw Wastewaters

A review of the data from four large pilot plants using physical-chemical treatments of primary or raw wastewater (Bishop *et al.*, 1971 & 1972; EPA, 1975; Villers *et al.*, 1971) showed an average suspended solids removal of 73.0% at an average filtration rate of 3.3 gpm/ft^2 (2.24 ℓ/s-m^2). All four plants employed lime clarification and dual-media filters either 24 or 36 in. (610 to 914 mm) deep. The feed at two of the plants had an average suspended solids concentration of 122 mg/ℓ, whereas the feed at the other two plants had an average of 131 mg/ℓ. The plants with the lower feed suspended solids had an average filter run of 31 hr, whereas the two plants having the higher feed suspended solids had an average filter run of 24 hr.

UPFLOW FILTRATION

Another technique to obtain the ideal filter in which the direction of filtration is from coarse to fine media is the upflow filter, shown in Figure 10.24. This filter is a single-medium filter using sand, and once hydraulic gradation has occurred by backwashing, the water being filtered is passed upward through the bed. Fluidization of the bed will occur if the head loss is sufficient to expand the medium, and as a result, floc breakthrough will occur. Fluidization may be avoided by making the bed extremely deep or by placing a restraining grid on the top of the bed, as shown in Figure 10.24. The bar spacing must be large enough to permit bed expansion during backwashing, and the arching action of the sand grains allows such spacing. In the cleaning cycle, the bed is agitated with the air scour system to break

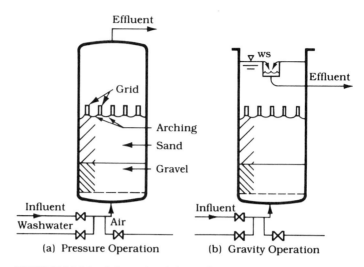

FIGURE 10.24   *Schematic of the Upflow Filter*

the arching action and loosen deposits. Once this has occurred, the backwash water is started at a sufficient rate to expand the bed further, resulting in cleansing. Once the backwashing is complete, the washwater is discontinued and the sand is allowed to settle to its original position.

The use of the upflow filter for drinking waters is limited because of a possible floc breakthrough during filtration. However, it has been used for industrial water treatment and wastewater treatment. It may be used as a pressure or a gravity filter.

Typical design parameters for upflow sand filters used in wastewater treatment are as follows (EPA, 1975): (1) filtration rate from 2 to 3 gal/min-ft^2 (1.36 to 2.04 ℓ/s-m^2), (2) terminal head loss from 6 to 20 ft (1.8 to 6.1 m), and (3) bed depth of 60 in. (1.52 m) of 2- to 3-mm sand and 4 in. (100 mm) of 10- to 15-mm sand.

Upflow filters treating secondary effluents having an average suspended solids content of 17 mg/ℓ have shown an average solids removal of 64.6% (EPA, 1975; WPCF, 1977). The filtration rate was from 2 to 5 gal/min-ft^2 (1.36 to 3.40 ℓ/s-m^2), with an average of 4.4 gal/min-ft^2 (2.99 ℓ/s-m^2). Filter runs varied from 7 to 150 hr.

MISCELLANEOUS FILTERS

Other filters such as the diatomaceous earth filter and microscreens are used for special purposes. The diatomaceous earth filter is a pressure filter employing filter elements coated with a precoat of diatomaceous earth. These filters are used for small-scale operations such as swimming pool filtration. The microscreen consists of a rotating cylindrical drum covered with a wire mesh filter cloth. The water being filtered passes from the inside to the outside of the drum, and the filtered deposits are washed away in a discharge trough by water jets. Microscreens have been used to a limited extent in tertiary treatment; however, difficulties have been encountered in processing a water with a fluctuating flow rate and a fluctuating suspended solids concentration. They have been successfully used to remove algal growths from drinking waters prior to rapid sand filtration.

REFERENCES

American Society of Civil Engineers (ASCE). 1982. *Wastewater Treatment Plant Design.* ASCE Manual of Practice No. 36. New York: ASCE.

American Society of Civil Engineers (ASCE). 1990. *Water Treatment Plant Design.* 2nd ed. New York: McGraw-Hill.

American Water Works Association (AWWA). 1969. *Water Treatment Plant Design.* New York: AWWA.

American Water Works Association (AWWA). 1990. *Water Quality and Treatment.* 4th ed. New York: McGraw-Hill.

American Water Works Association (AWWA). 1971. *Water Quality and Treatment.* New York: McGraw-Hill.

Barnes, D.; Bliss, P. J.; Gould, B. W.; and Vallentine, H. R. 1981. *Water and Wastewater Engineering Systems.* London: Pitman Books Limited.

Baumann, E. R., and Oulman, C. S. 1970. "Sand and Diatomite Filtration Practices." In *Water Quality Improvement by Physical and Chemical Processes,* edited by E. F. Gloyna and W. W. Eckenfelder, Jr. Austin, Tex.: University of Texas Press.

Bishop, D. F.; O'Farrell, T. P.; and Stamberg, J. B. 1971. Advanced Waste Treatment Systems at the Environmental Protection Agency–District of Columbia Pilot Plant. Paper presented at the 68th national meeting of the American Institute of Chemical Engineers (AICh.E.), Houston, Tex.

Bishop, D. F.; O'Farrell, T. P.; and Stamberg, J. B. 1972. Physical-Chemical Treatment of Municipal Wastewater. *Jour. WPCF* 44, no. 3:361.

Bishop, D. F.; O'Farrell, T. P.; and Stamberg, J. B. 1970. Physical-Chemical Treatment of Municipal Wastewaters. Paper presented before the 43rd annual meeting, Water Pollution Control Federation, Boston, Mass.

Camp, T. R. 1940. Lateral Spillway Channels. *Trans. ASCE* 105:606.

Camp, T. R. 1964. Theory of Water Filtration. *Jour. SED ASCE* 90, SA4:1.

Camp, T. R. 1952. "Water Treatment." In *The Handbook of Applied Hydraulics,* edited by

C. V. Davis. New York: McGraw-Hill.

Carman, P. C. 1937. Fluid Flow through Granular Beds. *Trans. Inst. Chem. Engrs. (London)* 15:150.

Cleasby, J. L. 1972. "Filtration." In *Physicochemical Processes for Water Quality Control,* edited by W. J. Weber, Jr. New York: Wiley.

Conway, R. A., and Ross, R. D. 1980. *Handbook of Industrial Waste Disposal.* New York: Van Nostrand Reinhold.

Culp, G., and Conley, W. 1970. "High Rate Filtration with the Mixed-Media Concept." In *Water Quality Improvement by Physical and Chemical Processes,* edited by E. F. Gloyna and W. W. Eckenfelder. Jr. Austin, Tex.: University of Texas Press.

Culp, G. L., and Culp, R. L. 1974. *New Concepts in Water Purification.* New York: Van Nostrand Reinhold.

Culp, R. L., and Culp, G. L. 1978. *Handbook of Advanced Wastewater Treatment.* 2nd ed. New York: Van Nostrand Reinhold.

Degremont, Inc. 1973. *Water Treatment Handbook.* Caxton Hill, Hertford, England: Stephen Austin and Sons.

Eckenfelder, W. W., Jr. 1989. *Industrial Water Pollution Control.* 2nd ed. New York: McGraw-Hill.

Eckenfelder, W. W., Jr. 1980. *Principles of Water Quality Management.* Boston: CBI Publishing.

Environmental Protection Agency (EPA). 1973. *Phosphorus Removal.* EPA Process Design Manual. Washington, D.C.

Environmental Protection Agency (EPA). 1975. *Suspended Solids Removal.* EPA Process Design Manual. Washington, D.C.

Environmental Protection Agency (EPA). 1974. *Upgrading Existing Wastewater Treatment Plants.* EPA Process Design Manual. Washington, D.C.

Ergun, S. 1952. *Chem. Engr. Progress* 48:89.

Fair, G. M.; Geyer, J. C.; and Okun, D. A. 1968. *Water and Wastewater Engineering.* Vol. 2. *Water Purification and Wastewater Treatment and Disposal.* New York: Wiley.

Fair, G. M., and Hatch, L. P. 1933. Fundamental

Factors Governing the Streamline Flow of Water through Sand. *Jour. AWWA* 25:1551.

Great Lakes–Upper Mississippi River Board of State Public Health and Environmental Managers. 1990. *Recommended Standards for Wastewater Facilities.* Ten state standards. Albany, N.Y.

Great Lakes–Upper Mississippi River Board of State Public Health and Environmental Managers. 1992. *Recommended Standards for Water Works.* Ten state standards. Albany, N.Y.

Great Lakes–Upper Mississippi Board of State Sanitary Engineers. 1978. *Recommended Standards for Sewage Works.* Ten state standards. Albany. N.Y.

Horan, N. J. 1990. *Biological Wastewater Treatment Systems: Theory and Operation.* Chichester, West Sussex, England: Wiley.

Hoover, C. P. 1946. *Water Supply and Treatment.* National Lime Association, Washington, D.C.

Kozeny, G. 1927. *Sitzber. Akad. Wiss. Wein, Math. –Naturw. Kl., Abt. IIa* 136.

Marshall, G. R., and Middlebrooks, E. J. 1974. Intermittent Sand Filtration to Upgrade Existing Wastewater Treatment Facilities. Utah Water Research Laboratory, PRJEW 115-2, Utah State University, Logan, Utah.

McCabe, W. L., and Smith, J. C. 1967. *Unit Operations of Chemical Engineering.* 2nd ed. New York: McGraw-Hill.

McGhee, T. J. 1991. *Water Supply and Sewage.* 6th ed. New York: McGraw-Hill.

Metcalf & Eddy, Inc. 1979. *Wastewater Engineering: Treatment, Disposal and Reuse.* 2nd ed. New York: McGraw-Hill.

Metcalf & Eddy, Inc. 1991. *Wastewater Engineering: Treatment, Disposal and Reuse.* 2nd ed. New York: McGraw-Hill.

Montgomery, J. M., Inc. 1985. *Water Treatment: Principles and Design.* New York: Wiley.

Neptune MicroFloc Inc. 1975. *Mixed Media.* Bulletin no. KL 4206, Corvallis, Oreg.

Novotny, V.; Imhoff, K.; Olthof, M.; and Krenkel, P. 1989. *Karl Imhoff's Handbook of Urban Drainage and Wastewater Disposal.* New York: Wiley.

O'Melia, C. R., and Stumm, W. 1967. Theory of Water Filtration. *Jour. AWWA* 59, no. 11:1393.

Perry, R. H., and Chilton, C. H. 1973. *Chemical Engineer's Handbook.* 5th ed. New York: McGraw-Hill.

Qasim, S. R. 1985. *Wastewater Treatment Plant Design.* New York: Holt, Rinehart and Winston.

Reynolds, J. H.; Harris, S. E.; Hill, D. W.; Filip, D. S.; and Middlebrooks, E. J. 1976. "Inter-mittent Sand Filtration for Upgrading Waste Stabilization Ponds." In *Ponds as a Wastewater Treatment Alternative,* edited by E. F. Gloyna, J. F. Malina, Jr., and E. M. Davis. Austin, Tex.: University of Texas Press.

Reynolds, T. D.; Payne, F.; and Myers, M. 1989. A Modification of Camp's Lateral Spillway Equation. Presented at the Fall 1989 Meeting of the Texas Section of the ASCE. Dallas, Texas.

Rich, L. G. 1971. *Unit Operations of Sanitary Engineering.* New York: Wiley.

Rose, H. E. 1945. *Proc. Inst. Mech. Engrs. (London)* 153:141, 154; also, 160:493 (1949).

Salvato, J. A. 1992. *Environmental Engineering and Sanitation.* 4th ed. New York: Wiley.

Sanks, R. L. 1978. *Water Treatment Plant Design.* Ann Arbor, Mich.: Ann Arbor Science Publishers.

Schroeder, E. D. 1977. *Water and Wastewater Treatment.* New York: McGraw-Hill.

South Tahoe Public Utility District. 1971. *Advanced Wastewater Treatment as Practiced at South Tahoe.* Tech. Report for the EPA, Project 17010 ELQ (WPRD 52-01-67).

Steel, E. W., and McGhee, T. J. 1979. *Water Supply and Sewerage.* New York: McGraw-Hill.

Sundstrom, D. W., and Klei, H. E. 1979. *Wastewater Treatment.* Englewood Cliffs, N.J.: Prentice-Hall.

Tchobanoglous, G., and Eliassen, R. 1970. Filtration of Treated Sewage Effluent. *Jour. SED ASCE* 96, no. SA2:243.

Tebbutt, T. H. Y. 1992. *Principles of Water Quality Control.* 4th ed. Oxford, England: Pergamon Press Limited.

Villers, R. V.; Berg, E. L.; Brunner, C. N.; and Masse, A. N. 1971. Municipal Wastewater Treatment by Physical and Chemical Methods. *Water and Sewage Works.* Reference no. 1971, p. R-62.

Water Environmental Federation (WEF). 1992. *Design of Municipal Wastewater Treatment Plants.* Vols. I and II. WEF Manual of Practice No. 8.

Water Pollution Control Federation (WPCF). 1977. *Wastewater Treatment Plant Design.* WPCF Manual of Practice No. 8. Washington, D.C.

Weber, W. J., Jr. 1972. *Physicochemical Processes for Water Quality Control.* New York: Wiley-Interscience.

Zenz, D. R; Lue-Hing, C.; and Obayashi, A. 1972. *Preliminary Report on Hanover Park Bay Project,* EPA Grant no. WPRD 92-01-68 (R-2).

Zipparro, V. J., and Hasen, H. 1993. *Davis' Handbook of Hydraulics.* 4th ed. New York: McGraw-Hill.

PROBLEMS

10.1 A rapid sand filter has a sand bed 30 in. in depth. Pertinent data are: specific gravity of the sand = 2.65, shape factor (ϕ) = 0.75, porosity (ε) = 0.41, filtration rate = 2.25 gpm/ft^2, and operating temperature = 50°F (10°C). The sieve analysis of the sand is given in Table 10.11. Determine the head loss for the clean filter bed in a stratified condition using the Rose equation.

10.2 A rapid sand filter has a sand bed 760 mm in depth. Pertinent data are: specific gravity of the sand = 2.65, shape factor (ϕ) = 0.75, porosity (ε) = 0.41, filtration rate = 1.53 ℓ/s-m^2, and operating temperature = 10°C. The sieve analysis of the sand is given in Table 10.11. Determine the head loss for the clean filter bed in a stratified condition using the Rose equation.

10.3 A rapid sand filter having the same sand analysis as in Problem 10.1 is to be backwashed. Determine:
a. The backwash velocity required to expand the bed, ft/sec.
b. The backwash flow in gpm/ft^2 required to expand the bed.
c. The head loss at the beginning of the backwash, ft.
d. The depth of the expanded bed, ft.
e. The depth of the expanded bed if the backwash rate is 15 gpm/ft^2.

10.4 A rapid sand filter having the same sand analysis as in Problem 10.1 is to be backwashed. Determine:

a. The backwash velocity required to expand the bed, m/s.
b. The backwash flow in ℓ/s-m^2 required to expand the bed.
c. The head loss at the beginning of the backwash, m.
d. The depth of the expanded bed, m.
e. The depth of the expanded bed if the backwash rate is 10.2 ℓ/s-m^2, m.

10.5 A trimedia filter has an 18-in. crushed anthracite coal layer (specific gravity = 1.20), a 9-in. sand layer (specific gravity = 2.65), and a 3-in. garnet layer (specific gravity = 4.20). The average sizes of particles in the three layers are 1.5 mm, 0.80 mm, and 0.30 mm, respectively. The ϕ values are 0.9 and the porosities of the respective layers are 0.40, 0.45, and 0.50. The filtration rate is 5 gpm/ft^2 and the water temperature is 50°F (10°C). Determine the head loss for the clean filter. *Note:* Treat each layer as a bed of uniform particles using the Rose equation, Eq. (10.13). Then the head loss through the bed is equal to the sum of the losses for the three layers.

10.6 A trimedia filter has a 560-mm crushed anthracite coal layer (specific gravity = 1.20), a 230-mm sand layer (specific gravity = 2.65), and a 75-mm garnet layer (specific gravity = 4.20). The average sizes of the particles in the three layers are 1.5 mm, 0.80 mm, and 0.30 mm, respectively. The ϕ values are 0.9 and the porosities of the respective layers are 0.40, 0.45, and 0.50. The filtration rate is

TABLE 10.11

SIEVE SIZE	WEIGHT RETAINED %
14–20	0.44
20–28	14.33
28–32	43.22
32–35	27.07
35–42	9.76
42–48	4.22
48–60	0.54
60–65	0.29
65–100	0.13

3.40 ℓ/s-m^2 and the water temperature is 10°C. Determine the head loss for the clean filter. *Note:* Treat each layer as a bed of uniform particles using the Rose equation, Eq. (10.13). Then the head loss through the bed is equal to the sum of the losses for the three layers.

10.7 A single-medium rapid sand filter has a maximum grain size of 0.22 cm and a porosity (ε) of 0.40. The specific gravity of the sand is 2.65. The bed will be fluidized during backwashing when the superficial velocity of the backwash water, V_b, is equal to or greater than $\varepsilon^{4.5}V_s$, where V_s is the settling velocity of the largest sand particle. Determine the minimum backwash rate in gpm/ft^2 to fluidize the bed. *Note:* To obtain sufficient bed expansion, the actual backwash rate is usually several magnitudes greater than the minimum rate.

10.8 A single-medium rapid sand filter has a maximum grain size of 0.22 cm and a porosity (ε) of 0.40. The specific gravity of the sand is 2.65. The bed will be fluidized during backwashing when the superficial velocity of the backwash water, V_b, is equal to or greater than $\varepsilon^{4.5}V_s$, where V_s is the settling velocity of the largest sand particle. Determine the minimum backwash rate in ℓ/s-m^2 to fluidize the bed. *Note:* To obtain sufficient bed expansion, the actual backwash rate is usually several magnitudes greater than the minimum rate.

10.9 A gravity-type mixed-media filter has wash-water troughs that are level, are 18 ft long and 1.25 ft wide, and are spaced 6 ft apart center to center. The wash rate is 15 gpm/ft^2. The Manning's n value is 0.032. Determine:

a. The depth of water above the trough crest if the trough is level.

b. The depth of the water in the trough where the water spills over into the gullet. For a freefall from a rectangular open channel, the water depth is equal to the critical depth, y_c. This is given by $y_c = (q^2/g)^{1/3}$, where q = the discharge per unit width (cfs/ft) and g = the acceleration due to gravity.

c. The upstream water depth in the wash trough, H_0.

d. The trough depth if the trough crest is 3 in.

above the maximum water depth in the trough, H_0. Make the trough depth an increment of 1 in.

10.10 A gravity-type mixed-media filter has wash-water troughs that are level, are 5.50 m long and 0.381 m wide, and are spaced 1.83 m apart center to center. The wash rate is 10.2 ℓ/s-m^2. The Manning's n value is 0.032. Determine:

a. The depth of water above the trough crest if the trough is level.

b. The depth of water in the trough where the water spills over into the gullet. For a freefall from a rectangular open channel, the water depth is equal to the critical depth, y_c. This is given by $y_c = (q^2/g)^{1/3}$, where q = the discharge per unit width (m^3/s-m) and g = the acceleration due to gravity.

c. The upstream water depth in the wash trough, H_0.

d. The trough depth if the trough crest is 75 mm above the maximum water depth in the channel, H_0.

10.11 A mixed-media filter at a water treatment plant is treating a clarified water having a turbidity of 5 JTU. The filtration rate has been 3 gpm/ft^2, which results in a 3-day filter run. The filter runs are terminated when the head loss reaches 7 ft. Determine:

a. The expected filter run if the filtration rate is 5 gpm/ft^2.

b. The expected filter run if the turbidity is 10 JTU and the filtration rate is 3 gpm/ft^2.

c. The expected filter run if the turbidity is 8 JTU and the filtration rate is 4 gpm/ft^2.

10.12 A mixed-media filter at a water treatment plant is treating a clarified water having a turbidity of 5 JTU. The filtration rate has been 2.04 ℓ/s-m^2, which results in a 3-day filter run. The filter runs are terminated when the head loss reaches 2.1 m. Determine:

a. The expected filter run if the filtration rate is 3.40 ℓ/s-m^2.

b. The expected filter run if the turbidity is 10 JTU and the filtration rate is 2.04 ℓ/s-m^2.

c. The expected filter run if the turbidity is 8 JTU and the filtration rate is 2.72 ℓ/s-m^2.

11

AMMONIA REMOVAL

Usually it is desirable to have a low ammonia concentration in a final effluent because it is toxic to fish at concentrations greater than about $3\,mg/\ell$ and also because it will be bio-oxidized by nitrifying microorganisms to form nitrates using molecular oxygen from the receiving water body. Nitrates are undesirable because they are nutrients that stimulate algal and aquatic growths. Ammonia removal may be accomplished by physical, chemical, or biological means.

PHYSICAL OPERATIONS

Induced-draft stripping towers and spray ponds are physical operations used to remove ammonia from wastewaters. In order for the ammonia to be stripped from a wastewater it must be in the dissolved gas form (NH_3), which requires a pH of about 10.8 or greater.

Induced-Draft Stripping Towers

Figure 11.1 shows a cross section through the stripping tower used at the Orange County Water District, Santa Ana, California, and Figure 11.2 shows a schematic drawing of this type of tower.

The type of tower shown in Figures 11.1 and 11.2 is a countercurrent system since the solvent gas (air) flows upward through the packing while the solvent liquid (water) flows downward. The tower consists of a fan, a packing to give intimate air and water contact, a tray or other distribution system to irrigate the top of the packing uniformly, a grid to support the packing and distribute the incoming air, a drift eliminator to collect liquid droplets in the exit air, and the tower structure. In the induced-draft tower, the fan at the top of the tower draws the air from the tower; thus the tower operates at a pressure slightly less than the outside pressure. Generally, a two-speed fan is used to give flexibility in operation. The pressure drop in the tower depends primarily upon the air and water flowrates, the viscosities, and the packing design. The open packing used in ammonia stripping towers usually has a pressure drop of 0.1 to 0.2 in. per foot depth (8.3 to 16.7 mm per meter).

The original packing in the ammonia stripping tower at South Tahoe was a wood-grid type using $\frac{3}{8}$ in. by $1\frac{1}{2}$-in. (9.5 mm by 38 mm) redwood slats having a clear spacing of about $1\frac{1}{2}$ in. (38 mm). Over a period of time, calcium carbonate encrustations developed, and since there was no access to the packing for cleaning, it eventually clogged. To avoid the encrustation problem, a plastic type packing was designed and is in use in the ammonia stripping tower installation at the advanced wastewater treatment plant at

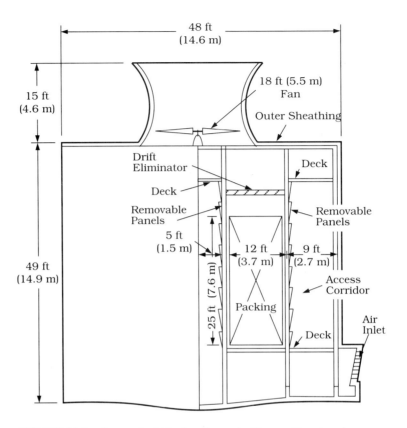

FIGURE 11.1   *Ammonia Stripping Tower in Orange County, California*

the Orange County Water District, Santa Ana, California. The grid type of packing consists of $\frac{1}{2}$-in.-diameter (13-mm) PVC pipe placed horizontally at 3-in. (76-mm) center to center spacing and placed vertically at 2-in. (51-mm) center to center spacing. The packing is placed in alternate layers at right angles. Pilot studies at the Orange County Water District showed that encrustations could be washed off by high-pressure hoses, and access corridors in the tower were incorporated to allow washing of the plastic packing while the packing is in place. Also, the packing was constructed in modular frames that could be removed from the tower for cleaning if necessary.

In order to derive equations for the design of ammonia stripping towers, it is necessary to understand the equilibrium relationships for a gas-liquid system, the fundamentals of gas mass transfer, and the material balances for a stripping tower. The equilibrium relationship for ammonia is given by Henry's law. One form of this law is

$$p_A = mX \tag{11.1}$$

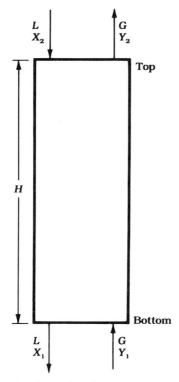

H = Tower height
L = Wastewater mass flowrate
G = Air mass flowrate
X_1, X_2 = Ammonia concentrations in the wastewater (as mass ratios)
Y_1, Y_2 = Ammonia concentrations in the air (as mass ratios)

FIGURE 11.2 *Schematic Drawing of an Ammonia Stripping Tower*

where

p_A = partial pressure of the ammonia in the air mixture in contact with the solution at equilibrium

m = constant

X = concentration of the ammonia in the solution at equilibrium, expressed as a mole or mass ratio

Henry's law is valid for a limited range in solute gas concentrations, since it assumes a straight-line relationship between the partial pressure and the concentration. The equilibrium partial pressures for ammonia gas dissolved in water at various temperatures are given in Table 11.1.

TABLE 11.1 *Ammonia Partial Pressure versus Temperature*

TEMPERATURE (°C)	p_A (mm Hg)	X (gm NH_3/10^6 gm H_2O)
0	0.0112	50
10	0.0189	50
20	0.0300	50
25	0.0370	50
30	0.0479	50
40	0.0770	50
50	0.1110	50

Taken from J.H. Perry, *Chemical Engineers Handbook*, 4th ed. (New York: McGraw-Hill, 1969).

The equilibrium ammonia concentration in an air mixture as expressed by mass or weight ratios is related to its partial pressure by

$$Y^* = \left(\frac{p_A}{P_t}\right)\left(\frac{M_A}{M_{Air}}\right) \tag{11.2}$$

where

Y^* = mass or weight ratio of the ammonia

P_t = total pressure of the atmosphere

p_A = partial pressure of the ammonia

M_A = molecular weight of the ammonia, 17 gm/gm mole

M_{Air} = molecular weight of air, 29 gm/gm mole

When the ammonia solubility, its equilibrium partial pressure, and the atmospheric pressure are known, it is possible to determine the coordinates for an equilibrium curve that shows the equilibrium concentrations of ammonia in the air and in the water.

In ammonia stripping it is sometimes useful to know the percent ammonia in the solution that is in the form of ammonia gas. The ammonia gas is in equilibrium with the ammonium ion as given by

$$NH_3 + H_2O \rightleftharpoons NH_4^+ + OH^- \tag{11.3}$$

As the pH is increased, the equilibrium shifts more to the left. The percent ammonia in the gas form at 25°C is as follows (Metcalf & Eddy, 1979):

$$NH_3 \text{ (percent)} = \frac{100}{1 + 1.75 \times 10^{+9}[H^+]} \tag{11.4}$$

where H^+ = hydrogen ion concentration. At 25°C and pH 10.8, 97.3% of the ammonia will be in the form of ammonia gas molecules dissolved in water. Since the partial pressure of ammonia in air is essentially zero, ammonia stripping will occur at a neutral pH range; however, the operation would have an extremely poor efficiency since most of the ammonia is in the form of the ammonium ion. Raising the pH to about 10.8 causes the major portion of the ammonia to be in the form of ammonia gas molecules; thus stripping readily occurs and the stripping towers will have high efficiencies.

EXAMPLE 11.1 AND 11.1 SI

Equilibrium Curve for Ammonia Stripping

An ammonia stripping tower is to be designed to operate at 750-mm atmospheric pressure and an average air and water temperture of 20°C (68°F). Determine the concentrations expressed as mass ratios for the equilibrium curve.

SOLUTION

The equilibrium mass of ammonia in the air is given by Eq. (11.2), which is

$$Y^* = (p_A/P_t)(M_A/M_{\text{Air}}) = (p_A/750)(17/29)$$

From the previous data, $p_A = 0.030 \text{ mm Hg}$ for 50 parts ammonia per 10^6 parts water; thus, $Y^* = (0.030/750)(17/29) = 2.34 \times 10^{-5}$ part ammonia per part air. The coordinates for the equilibrium curve are $\boxed{X_1 = 0, \ Y_1 = 0}$ and $\boxed{X_2 = 5 \times 10^{-5}, \ Y_2 = 2.34 \times 10^{-5}.}$

To determine the theoretical air required for an ammonia stripping tower, consider the schematic drawing shown in Figure 11.2. At steady state, a materials balance on the ammonia gives [Input] = [Output] since there is no chemical reaction. Thus the materials balance is

$$LX_2 + GY_1 = LX_1 + GY_2 \tag{11.5}$$

which may be simplified to

$$L(X_2 - X_1) = G(Y_2 - Y_1) \tag{11.6}$$

If it is assumed that there is no ammonia in the air entering ($Y_1 = 0$) and, for engineering purposes, the ammonia concentration in the effluent is negligible ($X_1 \approx 0$), then the water-to-air ratio is given by a simplification of Eq. (11.6):

$$\frac{L}{G} = \frac{Y_2}{X_2} \tag{11.7}$$

It can be seen from Eq. (11.7) that the theoretical water-to-air flow ratio is equal to the slope of the equilibrium curve for the temperature and elevation of the tower under consideration. The reciprocal of the slope,

$$\frac{G}{L} = \frac{1}{\text{slope}} = \frac{1}{(Y_2)/(X_2)} \tag{11.8}$$

gives the air-to-water flow ratio for the theoretical air flow required. Usually, the design air flow is 1.50 to 1.75 times the theoretical.

EXAMPLE 11.2   *Ammonia Stripping Tower Design: Packed Tower*

An induced-draft packed countercurrent ammonia stripping tower is to be designed for an advanced wastewater treatment plant using the equilibrium curve data from Example 11.1. Pertinent data for the design are as follows: design flow = 2.0 MGD, hydraulic load = 2.00 gpm/ft^2, design air flow = 1.75 times the theoretical, average water and air temperature = 20°C (68°F), atmospheric pressure = 750 mm Hg, and the wastewater pH = 10.8. Determine:

1. Theoretical air required, lb/hr-ft^2
2. Design air required, lb/hr-ft^2
3. Design air required, cfm/ft^2 and ft^3/gal

SOLUTION   The slope of the equilibrium curve is Y_2/X_2 or $(23.4 \times 10^{-6}$ lb NH$_3$/lb air$)$/$(50 \times 10^{-6}$ lb NH$_3$/lb water$)$ or 0.468 lb water/lb air. Thus the air-to-water ratio, G/L, is $1/0.468 = 2.14$ lb air/lb water. The mass flowrate of the water, L, is $(2.00$ gal/min-ft$^2)(8.34$ lb/gal$)(60$ min/hr$) = 1001$ lb water/hr-ft^2. Therefore, the theoretical air flow required, G, is $(2.14$ lb air/lb water$)(1001$ lb water/hr-ft$^2)$, or

$$G(\text{theoretical}) = \boxed{2140 \text{ lb air/hr-ft}^2}$$

The design air flow is

$$G(\text{design}) = (2140)1.75$$
$$= \boxed{3750 \text{ lb air/hr-ft}^2}$$

For 20°C (68°F or 528°R) and 750 mm Hg atmospheric pressure, the design air flow, Q_G, is $(3750$ lb air/hr-ft$^2)(1$ lb-mole/29 lb$)(359$ ft^3/1 lb-mole$)(528°R/492°R)(760$ mm/750 mm$)(hr/60$ min$)$ or

$$Q_G(\text{design}) = \boxed{841 \text{ cfm/ft}^2}$$
$$\text{Air per gallon} = 841/2.00 = \boxed{421 \text{ ft}^3/\text{gal}}$$

EXAMPLE 11.2 SI   *Ammonia Stripping Tower Design: Packed Tower*

An induced-draft packed countercurrent ammonia stripping tower is to be designed for an advanced wastewater treatment plant using the equilibrium curve data from Example 11.2. Pertinent data for the design are as follows: design flow = 7600 m^3/d, hydraulic load = 4.90 m^3/h-m^2, design air flow = 1.75 times the theoretical, average water and air temperature = 20°C, atmospheric pressure = 750 mm Hg, and the wastewater pH = 10.8. Determine:

1. Theoretical air required, kg/h-m^2
2. Design air required, kg/h-m^2
3. Design air required, m^3/h-m^2 and m^3/ℓ

SOLUTION The slope of the equilibrium curve is Y_2/X_2 or $(23.4 \times 10^{-6} \, \text{kg NH}_3/\text{kg air})/(50 \times 10^{-6} \, \text{kg NH}_3/\text{kg water})$ or 0.468 kg water/kg air. Thus the air-to-water ratio, G/L, is $1/0.468 = 2.14$ kg air/kg water. The mass flowrate of the water, L, is $(4.90 \, \text{m}^3/\text{h-m}^2)(1000 \, \ell/\text{m}^3)(\text{kg}/\ell) = 4900$ kg water/h-m^2. Therefore, the theoretical air flow required, G, is $(2.14 \, \text{kg air/kg water})(4900 \, \text{kg water/h-m}^2)$ or

$$G(\text{theoretical}) = \boxed{10{,}500 \, \text{kg air/h-m}^2}$$

The design air flow is

$$G(\text{design}) = (10{,}500)1.75$$
$$= \boxed{18{,}400 \, \text{kg air/h-m}^2}$$

For 20°C (293°K) and 750 mm Hg atmospheric pressure, the design air flow, Q_G, is $(18{,}400 \, \text{kg air/h-m}^2)(1 \, \text{g-mole}/29 \, \text{g})(1000 \, \text{g/kg})(22.4 \, \ell/\text{g-mole})(\text{m}^3/1000 \, \ell)$ (760 mm/750 mm)(293°K/273°K) or

$$Q_G \, (\text{design}) = \boxed{15{,}500 \, \text{m}^3/\text{h-m}^2}$$

$$\text{Air per liter} = (15{,}500 \, \text{m}^3/\text{h-m}^2)/(4.90 \, \text{m}^3/\text{h-m}^2)(1000 \, \ell/\text{m}^3)$$
$$= \boxed{3.16 \, \text{m}^3/\ell}$$

The temperatures of the wastewater and the air (since it indirectly affects the wastewater temperature) are important in ammonia stripping tower operation because of two effects. First, if the wet bulb temperature drops to 32°F (0°C), the tower will probably be affected by ice formation within the tower. The winter operational data from South Tahoe have shown the average daily exit water temperature from the tower (38.6°F or 3.67°C) to be only 0.8°F or 0.44°C below the average daily ambient air temperature (39.4°F or 4.11°C). Consequently, the average daily exit water temperature can be expected to be near the average daily ambient air temperature. Second, because the solubility of ammonia increases with a decrease in temperature, more air is required for stripping as the temperature decreases. Figure 11.3 shows the ammonia equilibrium curves for various temperatures at a total atmospheric pressure of 760 mm. Figure 11.4 shows the theoretical air flow required versus temperature. At 5°C the air required

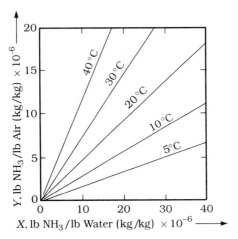

FIGURE 11.3 *Equilibrium Curves for Ammonia in Air and Water at an Atmospheric Pressure of 760 mm Hg*

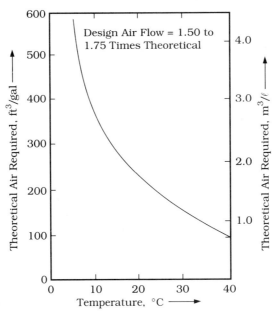

FIGURE 11.4 *Theoretical Air Required for Ammonia Stripping at Various Temperatures at 760 mm Hg, pH 11*

is about 3.5 times that at 30°C. Thus tower operations are limited to warm operating conditions. In cold climates, it may be feasible to use a stripping tower during warm months and, during cold months, to use an alternative method for removal, such as spray ponds.

Design considerations for ammonia stripping towers are as follows: unit hydraulic loading = 1 to 3 gpm/ft^2 (40.7 to 122 ℓ/min-m^2), 1 to 2 gpm/ft^2 (40.7 to 81 ℓ/min-m^2) being typical; design air flow = 1.5 to 1.75 times the theoretical air flow; maximum air pressure drop for the entire tower < 2 to 3 in. (51 to 76 mm) of water; fan motor speed = 1 or 2 speed; fan tip speed = 9000 to 12,000 ft/min (2740 to 3660 m/min); packing depth = 20 to 25 ft (6.1 to 7.6 m); packing spacing = 2 to 4 in. (51 to 102 mm) horizontal and vertical, wood or plastic packing; and influent water pH = 10.8 to 11.5. The packing employed usually consists of an expanded grid made of PVC plastic pipe or redwood slats. Also, a polypropylene grid has been used.

Ponds   Gonzales and Culp (1973) have used high-pH spray ponds for ammonia removal at South Tahoe. Some of the factors influencing the performance of a spray pond are water temperature, air temperature, initial and final solute gas concentration in the water, solute gas concentration in the air, air movement, pH, detention time, depth of pond, area of pond, aeration time (spray height), droplet size, spray nozzle spacing, layout pattern of the spray nozzles, and the number of water turnovers that are sprayed. At South Tahoe, high-pH spray ponds have been used as an alternative method for ammonia removal in cold weather. The sprays employed have large nozzles and produce coarse sprays with a height of about 15 ft (4.6 m); the pond depth is about 2.0 ft (0.61 m). It has been found that a 20-hr detention time and a water turnover of 5.2 cycles results in 52% removal of ammonia. Although the spray ponds do not give as good an ammonia removal as a stripping tower, pond operation at temperatures less than freezing is possible. Thus high-pH spray ponds are a feasible method for ammonia removal at low climatic temperatures.

CHEMICAL PROCESSES   Breakpoint chlorination and ion exchange (using clinoptilolite) are two unit processes that may be used for ammonia removal.

Breakpoint Chlorination   Chlorination by chlorine gas or hypochlorite salts will oxidize ammonia to form intermediate chloramines and, finally, to form nitrogen gas and hydrochloric acid. The reaction of chlorine gas with water to produce hypochlorous acid (HOCl) is

$$\text{Cl}_2 + \text{H}_2\text{O} \rightarrow \text{HCl} + \text{HOCl} \underset{\text{pH}<7}{\overset{\text{pH}>8}{\rightleftharpoons}} \text{H}^+ + \text{OCl}^- \tag{11.9}$$

Hypochlorous acid will react with ammonia in a wastewater to produce monochloramine (NH$_2$Cl), dichloramine (NHCl$_2$), and trichloramine (NCl$_3$). These reactions are dependent upon the pH, temperature, reaction time, and initial chlorine-to-ammonia ratio. Monochloramine and dichloramine are formed in the pH range of 4.5 to 8.5. Above pH 8.5, the predominant form of the chloramines is monochloramine, whereas below pH 4.5, the predominant form is trichloramine. If the amount of chlorine added is greater than that required for the breakpoint and the pH is about 7 to 8, the intermediate-formed monochloramine is oxidized to yield nitrogen gas that is liberated from the system. The series of steps required to oxidize the ammonia are as follows (Cassel *et al.*, 1971):

$$Cl_2 + H_2O \rightarrow HOCl + HCl \tag{11.10}$$

$$NH_4^+ + HOCl \rightarrow NH_2Cl + H_2O + H^+ \tag{11.11}$$

$$2NH_2Cl + HOCl \rightarrow N_2 \uparrow + 3HCl + H_2O \tag{11.12}$$

The overall reaction obtained from Eqs. (11.10), (11.11), and (11.12) is

$$3Cl_2 + 2NH_4^+ \rightarrow N_2 \uparrow + 6HCl + 2H^+ \tag{11.13}$$

The stoichiometric amount of chlorine required is 3 moles of chlorine per 2 moles of ammonia or 7.5 lb of chlorine per pound ammonia nitrogen (kg/kg). Studies have shown that for raw municipal wastewaters, secondary effluents, and lime-clarified and filtered secondary effluents, the chlorine required per pound (kg) of ammonia nitrogen is 10:1, 9:1, and 8:1, respectively. The amount of chlorine required is independent of temperature from 40° to 100°F (4.4° to 37.8°C), and 95 to 99% of the ammonia will be oxidized to nitrogen gas. There may be some undesirable side reactions that produce dichloramine, trichloramine (which is very offensive in odor), and nitrate ions. If, however, the pH is maintained in the range from about 7 to 8, the side reactions are minimized. If the alkalinity is insufficient to maintain this pH range, a base such as NaOH must be added to neutralize the hydrochloric acid formed. The reactions given by Eqs. (11.10) through (11.13) occur rapidly and, under proper conditions, are completed within several minutes. The disadvantages of breakpoint chlorination are that the dissolved solids are increased and that the process is about twice as expensive as ammonia stripping. Also, because of the adverse health implications of trihalomethanes that are formed during chlorination, the use of breakpoint chlorination for ammonia removal may decline in the future.

Ion Exchange Cation exchange using a natural zeolite called clinoptilolite has been employed to remove ammonia (Mercer *et al.*, 1970). In this process the secondary effluent is treated by multimedia filtration, carbon adsorption, and then ion exchange using a fixed-bed column containing clinoptilolite. Upon exhaustion, the clinoptilolite column is regenerated by the use of a brine solution of about 2% NaCl, which is renovated by electrolysis. The spent regenerant contains mainly NH_4^+, Ca^{+2}, Mg^{+2}, Na^+, and Cl^- ions. Prior to electrolysis, Na_2CO_3 and NaOH are added to precipitate as much $CaCO_3$ and $Mg(OH)_2$ as possible. The regenerant passes through electrolytic cells that produce Cl_2 at the anode and H_2 at the cathode. The chlorine reacts with the ammonia by breakpoint chlorination to produce N_2 gas, and

N_2 and H_2 gases are removed as off-gases. After the electrolytic treatment, the regenerant has been renovated so that it may be reused.

BIOLOGICAL PROCESSES

Conceptually, ammonia removal (or, in a more general sense, nitrogen removal) that employs biological processes requires the nitrogen to be in either the ammonia/ammonium form or the nitrate form prior to ultimate nitrogen *removal* from the waste stream. Ultimate nitrogen removal by the physical and chemical means discussed previously in this chapter requires a waste stream entering the physical or chemical process that is rich in ammonia. Hence biological processes must be operated in a non-nitrifying mode preceding the nitrogen removal process. Ultimate removal of nitrogen by biological means requires a waste stream entering the nitrogen removal process that is rich in nitrate. Biological processes must therefore be operated in a nitrifying mode preceding the nitrogen removal process. Nitrogen conversion to the form appropriate for the nitrogen removal process that has been selected is managed by controlling several biochemical reactions.

Aerobic biochemical action converts carbonaceous and organic nitrogen matter to ammonia, NH_3, or to the ammonium ion, NH_4^+. The form of the nitrogen is pH dependent. Continued aerobic biochemical action — nitrification — converts the ammonia/ammonium to nitrite, NO_2^-, and finally to nitrate, NO_3^-. Biochemical action under conditions devoid of oxygen — denitrification — converts nitrate and nitrite to free nitrogen gas, N_2.

Nitrification is principally accomplished by two aerobic autotrophic bacterial genera, *Nitrosomonas* and *Nitrobacter*. *Nitrosomonas* converts the ammonium ion to the nitrite ion, whereas *Nitrobacter* converts the nitrite ion to the nitrate ion. The biochemical equation for nitrification proposed by McCarty (1970) is

$$NH_4^+ + 1.682O_2 + 0.182CO_2 + 0.0455HCO_3^- \rightarrow$$
$$0.0455\,C_5H_7NO_2 + 0.955NO_3^- + 0.909H_2O + 1.909H^+ \quad \textbf{(11.14)}$$

From Eq. (11.14) it can be seen that the oxygen required for nitrification is appreciable.

Denitrification is biochemical conversion of the nitrate ion to form nitrogen gas. The process is accomplished by several genera of facultative heterotrophic bacteria. The denitrifiers include *Pseudomonas*, *Micrococcus*, *Archromobacter*, and *Bacillus*. In the absence of oxygen, these organisms use nitrogen in either nitrite or nitrate as the final electron acceptor in the process termed nitrate dissimilation. The term *anoxic*, rather than *anaerobic*, has been applied to this process to distinguish the oxygen-devoid environmental conditions from the metabolic pathway implied by use of the term *anaerobic* (EPA, 1975). However, *anaerobic* will be used in this discussion to describe the environmental conditions without implying reference to the metabolic pathway. Managing nitrogen conversion for ultimate nitrogen removal usually requires a supplemental carbon source, such as methanol, which is added to promote the denitrification reaction. The biochemical equation proposed by McCarty *et al.* (1969) for denitrification is

$$NO_3^- + 1.08CH_3OH + H^+ \rightarrow$$
$$0.065C_5H_7NO_2 + 0.47N_2 + 0.76CO_2 + 2.44H_2O \quad \textbf{(11.15)}$$

Nitrogen *removal* by the chemical and physical processes discussed previously in this chapter, such as ammonia stripping, requires the nitrogen to be in the non-nitrified, NH_3 or NH_4^+ form. The single-stage, activated sludge process, as shown in Figure 11.5, may be operated to accomplish carbonaceous BOD conversion with minimal nitrification, thereby retaining the nitrogen in the ammonium form. Biological processes other than activated sludge may be used preceding the ammonia removal process. Examples include the suspended-growth-media oxidation ditch and fixed-media biological processes such as trickling filters and rotating biological contactors. Operation of the biological process used is controlled such that carbonaceous BOD conversion proceeds to the level necessary, but nitrification, Eq. (11.14), is not allowed to proceed. This condition may be achieved by limiting the mean cell residence time in the suspended-media processes, or its equivalence in the fixed-media processes, to relatively low values as discussed in Chapters 15 and 17. The biological process is followed by the selected chemical or physical process for ammonia removal prior to discharge of the final effluent.

Nitrogen *removal* by biological processes requires the nitrogen to be in the nitrified, NO_3^+, form. The single-stage, activated sludge process, as shown in Figure 11.5, may be operated to accomplish both carbonaceous BOD conversion and nitrification. An alternative activated sludge process in which the carbonaceous BOD conversion and nitrification occur in separate stages is shown in Figure 11.6. The second-stage reactor may also be a plug-flow reactor. Although the activated sludge process is most often used to produce highly nitrified conditions, other suspended-media (*e.g.*, oxidation ditch) or fixed-media (*e.g.*, trickling filter or rotating biological contactor) processes also may be used. Operation of the biological process used is

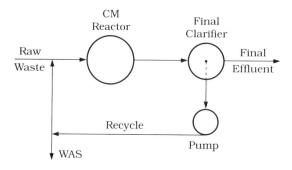

FIGURE 11.5
Single-Stage Activated Sludge Flow

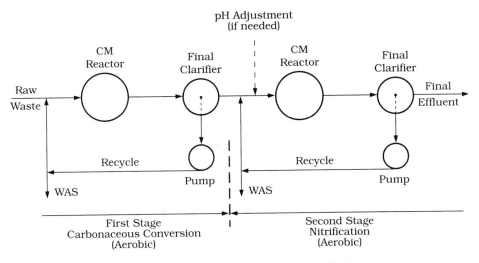

FIGURE 11.6 *Two-Stage Biological Carbonaceous Conversion-Nitrification*

controlled to achieve not only the desired degree of carbonaceous BOD conversion but also a high degree of nitrification.

Three-Stage Nitrification-Denitrification Process

The three-stage nitrification-denitrification process, as shown in Figure 11.7, results in the ultimate removal of ammonia in addition to organic, nitrite, and nitrate nitrogen. Also, phosphorus may be removed by adding a coagulant prior to the final clarifier for the first and third stages.

In the first stage, the principal biochemical action is the conversion of carbonaceous and organic nitrogen matter to terminate the nitrogen in the form of ammonia. Usually, a completely mixed reactor is used. If

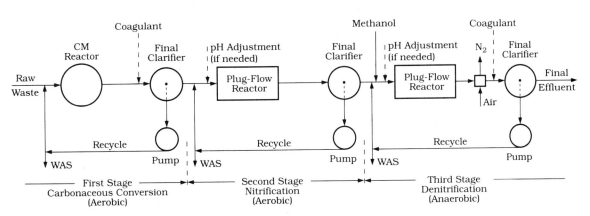

FIGURE 11.7 *Three-Stage Biological Nitrification-Denitrification*

phosphorus removal is desired, a coagulant may be added prior to the first-stage final clarifier. The first stage will have a BOD$_5$ of about 50 mg/ℓ.

In the second stage, the main biochemical action is the conversion of ammonia nitrogen to the nitrate ion — nitrification, Eq. (11.14). A plug-flow reactor is usually used.

In the third stage, the main biochemical action is the anaerobic conversion of the nitrate ion to nitrogen gas — denitrification, Eq. (11.15). In the plug-flow anaerobic mixing reactor, the nitrate ion is converted to nitrogen gas, and no oxygen will be present in the reactor. Odors should not occur because the anaerobic bacteria prefer the nitrate ion (NO_3^{-1}) to the sulfate ion (SO_4^{-2}) as a source of combined oxygen. After the mixture leaves the reactor, it passes through a nitrogen stripping tank with a detention time of 5 to 10 min. Compressed air is released at the bottom to strip out the nitrogen gas. After stripping, the mixture passes to the final clarifier, where the anaerobic sludge is separated and the effluent leaves. If a nitrogen stripping tank is not provided, floating sludge will occur in the final clarifier. If phosphorus removal is desired, a coagulant may be added prior to the third-stage final clarifier. In both the second and third stages, plug flow is approached by using long rectangular reactors with a large length-to-width ratio or rectangular reactors with baffles.

The optimum pH, the detention time based on the influent flow (that is $\theta_i = V/Q$), and the mean cell residence time (θ_c) for the three stages are shown in Table 11.2. The following is a partial summary of design parameter values (EPA, 1975) for the kinetic design of the denitrification stage of suspended-growth systems employing methanol as the supplementary carbon source. The amount of methanol required is generally 3.0 lb of methanol per pound of NO_3^--N removed (kg/kg). Alkalinity, as $CaCO_3$, is produced at a rate of 3.0 mg per milligram of nitrogen reduced. The half-saturation constant for nitrate ranges from 0.08 mg NO_3^--N per liter for a no-cellular-recycle system to 0.16 mg/ℓ for a system with cellular recycle to the denitrification stage. The denitrification cellular growth, or yield, coefficient ranges from 0.6 to 1.20 lb vss/lb NO_3^--N removed (kg/kg), and the decay coefficient is 0.04 per day. Peak denitrification rates are temperature dependent (values given are lb NO_3^--N removed or oxidized per lb MLVSS per

TABLE 11.2 *Operational Parameters for the Three-Stage Biological Nitrification-Denitrification Process*

	pH	θ_i (hrs)	θ_c (days)
1st Stage	6.5–8.0	<3	<5
2nd Stage	7.8–9.0	<3	10–25
3rd Stage	6.5–7.5	<2	<5

Taken from Environmental Protection Agency, Advanced Waste Treatment and Water Reuse Symposium, sessions 1–5. Dallas, Texas, 1971.

day or kg/kg per day): at 10°C, reported values range from 0.05 to 0.10; at 15°C, from 0.05 to 0.17; at 20°C, from 0.15 to 0.43; and at 25°C, from 0.20 to 0.48. Example 11.3 illustrates the use of these kinetic design parameters and equations from the EPA's *Process Design Manual for Nitrogen Control* (1975). The pertinent equations give the required minimum mean cell residence time (MCRT) as

$$\frac{1}{\theta_c^{\min}} = Y\bar{q} - k_e$$

where

$\theta_c^{\min}$ = miminum MCRT for denitrification, days

Y = gross yield, mg VSS/mg NO_3^--N removed

$\bar{q}$ = peak nitrate removal rate, mg NO_3^--N removed/mg VSS/day

k_e = decay coefficient, day^{-1}

The design mean cell residence time to achieve the desired final effluent nitrate concentration is given by

$$\frac{1}{\theta_c} = \frac{Y\bar{q}(D_0 - D_1)}{(D_0 - D_1) + K_D \ln\dfrac{D_0}{D_1}} - k_e \tag{11.17}$$

where

θ_c = design mean cell residence time, days

K_D = half-saturation constant, mg NO_3^--N/ℓ

D_0 = influent NO_3^--N, mg/ℓ

D_1 = effluent NO_3^--N, mg/ℓ

The design nitrate removal rate, q, is given by a rearrangement of the following equation:

$$\frac{1}{\theta_c} = Yq - k_e \tag{11.18}$$

where θ_c, Y, and k_e are as previously defined, and

q = design nitrate removal rate, mg NO_3^--N removed/mg VSS/day

The hydraulic detention time, θ, is given by rearrangement of the following equation:

$$q = \frac{D_0 - D_1}{X_1 \theta} \tag{11.19}$$

where q, D_0, and D_1 are as previously defined, and

θ = hydraulic detention time, days

X_1 = volatile suspended solids in denitrification reactor, mg/ℓ

The mass of sludge leaving the system per day is given by rearrangement of the following equation:

$$\theta_c = \frac{\text{mass MLVSS in reactor}}{\text{mass MLVSS leaving system/day}} \qquad (11.20)$$

where θ_c is as previously defined. The mass of sludge to be wasted per day, X_w, is given by

$$X_w = \text{mass MLVSS leaving system/day} - $$
$$\text{mass MLVSS leaving in final effluent/day} \qquad (11.21)$$

where X_w is mass MLVSS wasted/day.

EXAMPLE 11.3
AND 11.3 SI

Denitrification Stage Design: Plug Flow

The denitrification stage (third stage) is to be designed for the three-stage nitrification-denitrification treatment process shown in Figure 11.7. Pertinent data for the design are as follows: design flow = 10 MGD (38 MLD), temperature = 20°C, cellular recycle will be used, NO_3^--N from the second stage = 28 mg/ℓ, maximum allowable NO_3^--N in the final effluent = 1.0 mg/ℓ, maximum allowable suspended solids in the final effluent = 5.0 mg/ℓ, and volatile suspended solids are 80% of suspended solids. Use median values for the kinetic coefficients needed.

SOLUTION

Determine the required minimum mean cell residence time from Eq. (11.16), which is

$$\frac{1}{\theta_c^{\min}} = Y\bar{q} - k_e$$

Using a yield coefficient = 0.9 mg MLVSS/mg NO_3^--N removed, a peak nitrate removal rate = 0.29 mg NO_3^--N oxidized/mg MLVSS-day, and the decay coefficient = 0.04 day^{-1} yields

$$\frac{1}{\theta_c^{\min}} = \left(0.9\,\frac{\text{mg MLVSS}}{\text{mg }NO_3^-\text{-N}}\right)\left(0.29\,\frac{\text{mg }NO_3^-\text{-N}}{\text{mg MLVSS day}}\right) - 0.04\,\frac{1}{\text{day}}$$
$$= 0.221 \text{ day}^{-1}$$
$$\theta_c^{\min} = \boxed{4.52 \text{ days}}$$

Determine the design mean cell residence time required to achieve the desired final effluent nitrate concentration from Eq. (11.17), which is

$$\frac{1}{\theta_c} = \frac{Y\bar{q}(D_0 - D_1)}{(D_0 - D_1) + K_D \ln\dfrac{D_0}{D_1}} - k_e$$

Using a half-saturation constant = 0.16 mg NO_3^--N/ℓ in addition to the kinetic constant values used previously, and pertinent design values,

$$\frac{1}{\theta_c} = \frac{\left(0.9 \dfrac{\text{mg MLVSS}}{\text{mg NO}_3^--\text{N}}\right)\left(0.29 \dfrac{\text{mg NO}_3^-0\text{-N}}{\text{mg MLVSS day}}\right)\left(28 \dfrac{\text{mg NO}_3^--\text{N}}{\ell} - 1 \dfrac{\text{mg NO}_3^--\text{N}}{\ell}\right)}{\left(28 \dfrac{\text{mg NO}_3^--\text{N}}{\ell} - 1 \dfrac{\text{mg NO}_3^--\text{N}}{\ell}\right) + \left(0.16 \dfrac{\text{mg NO}_3^--\text{N}}{\ell}\right)\ln\left(\dfrac{28 \dfrac{\text{mg NO}_3^--\text{N}}{\ell}}{1 \dfrac{\text{mg NO}_3^--\text{N}}{\ell}}\right)} - 0.04 \frac{1}{\text{day}}$$

$$= 0.216 \, \text{day}^{-1}$$

$$\theta_c = \boxed{4.63 \, \text{days}}$$

Determine the design nitrate removal rate from Eq. (11.18), which is

$$\frac{1}{\theta_c} = Yq - k_e$$

$$\frac{1}{4.63 \, \text{days}} = \left(0.9 \frac{\text{mg MLVSS}}{\text{mg NO}_3^--\text{N}}\right)\left(q \frac{\text{mg NO}_3^--\text{N}}{\text{mg MLVSS day}}\right) - 0.04 \frac{1}{\text{day}}$$

$$q = \boxed{0.284 \frac{\text{mg NO}_3^--\text{N removed}}{\text{mg MLVSS day}}}$$

Determine the hydraulic detention time at design flow. MLSS in the denitrification reactor $= 3000 \, \text{mg}/\ell$ with a volatile solids content of 80%.

$$\text{MLVSS}, \; X_1 = (3000 \, \text{mg}/\ell)(0.8) = 2400 \, \text{mg}/\ell$$

Rearranging Eq. (11.19) yields

$$\theta = \frac{\left(28 \dfrac{\text{mg NO}_3^--\text{N}}{\ell} - 1 \dfrac{\text{mg NO}_3^--\text{N}}{\ell}\right)}{\left(0.284 \dfrac{\text{mg NO}_3^--\text{N}}{\text{mg MLVSS day}}\right)\left(2400 \dfrac{\text{mg MLVSS}}{\ell}\right)}$$

$$= \boxed{0.04 \, \text{day} = 0.96 \, \text{hr}}$$

For the USCS solution, the sludge wasting, X_w, required is as follows:

$$\text{MLVSS in reactor} = XV = X\theta Q$$

$$\text{Mass MLVSS} = \left(2400 \frac{\text{lb MLVSS}}{10^6 \, \text{lb mixed liquor}}\right)(0.04 \, \text{day})\left(10 \times 10^6 \frac{\text{gal}}{\text{day}}\right)$$

$$\times \left(8.34 \frac{\text{lb}}{\text{gal}}\right) = 8006 \, \text{lb MLVSS}$$

From Eq. (11.20), which is

$$\theta_c = \frac{\text{mass MLVSS}}{\text{mass MLVSS leaving system/day}}$$

$$\text{Total MLVSS leaving system/day} = \frac{8006 \, \text{lb MLVSS}}{4.63 \, \text{days}} = 1729 \frac{\text{lb MLVSS}}{\text{day}}$$

$$\text{Mass MLVSS in final effluent} = \left(10 \times 10^6 \frac{\text{gal}}{\text{day}}\right)\left(5\frac{\text{lb MLSS}}{10^6 \text{ lb mixed liquor}}\right)$$
$$\times \left(0.8\frac{\text{lb MLVSS}}{\text{lb MLSS}}\right)\left(8.34\frac{\text{lb}}{\text{gal}}\right)$$

$$\text{Mass MLVSS leaving in final effluent} = 334\frac{\text{lb MLVSS}}{\text{day}}$$

By difference, the approximate MLVSS mass to be wasted per day is

$$X_w = 1729\frac{\text{lb MLVSS}}{\text{day}} - 334\frac{\text{lb MLVSS}}{\text{day}} = \boxed{1395\frac{\text{lb MLVSS}}{\text{day}}}$$

The methanol requirement using 3.0 lb methanol per lb NO_3^--N removed is

$$\left(3.0\frac{\text{lb methanol}}{\text{lb } NO_3^- \text{-N removed}}\right)\left(27\frac{\text{lb } NO_3^- \text{-N removed}}{10^6 \text{ lb waste}}\right)\left(10 \times 10^6 \frac{\text{gal waste}}{\text{day}}\right)$$
$$\times \left(8.34\frac{\text{lb waste}}{\text{gal waste}}\right) = \boxed{6760 \text{ lb methanol/day}}$$

For the SI solution, the sludge wasting required is as follows:

$$\text{Mass MLVSS} = \left(2400\frac{\text{mg MLVSS}}{\ell \text{ mixed liquor}}\right)(0.04 \text{ d})\left(38 \times 10^6\frac{\ell}{\text{d}}\right)\left(1\frac{\text{kg}}{10^6 \text{ mg}}\right)$$
$$= 3648 \text{ kg MLVSS}$$

$$\text{Total MLVSS leaving system/day} = \frac{3648 \text{ kg MLVSS}}{4.63 \text{ d}} = 788\frac{\text{kg MLVSS}}{\text{d}}$$

$$\text{Mass MLVSS in final effluent} = \left(38 \times 10^6\frac{\ell}{\text{d}}\right)\left(5\frac{\text{mg MLSS}}{\ell \text{ mixed liquor}}\right)$$
$$\times \left(0.8\frac{\text{mg MLVSS}}{\text{mg MLSS}}\right)\left(\frac{1 \text{ kg}}{10^6 \text{ mg}}\right)$$

$$\text{Mass MLVSS leaving in final effluent} = 152\frac{\text{kg MLVSS}}{\text{d}}$$

By difference, the approximate MLVSS mass to be wasted per day is

$$X_w = 3648\frac{\text{kg MLVSS}}{\text{d}} - 152\frac{\text{kg MLVSS}}{\text{d}} = \boxed{3496\frac{\text{kg MLVSS}}{\text{d}}}$$

The methanol requirement using 3.0 kg methanol per kg NO_3^--N removed is

$$\left(3.0\frac{\text{kg methanol}}{\text{kg } NO_3^- \text{-N removed}}\right)\left(27\frac{\text{mg } NO_3^- \text{-N removed}}{\ell \text{ waste}}\right)\left(38 \times 10^6 \frac{\ell \text{ waste}}{\text{d}}\right)$$
$$\times \left(\frac{1 \text{ kg}}{10^6 \text{ mg}}\right) = \boxed{3080 \text{ kg methanol/d}}$$

The use of a three-stage nitrification-denitrification system employing chemical precipitation of phosphorus in the third-stage final clarifier has resulted in 86.3% COD, 95.5% suspended solids, 96.2% organic nitrogen, 97.3% ammonia nitrogen, and 88.1% phosphorus removals. The effluent contained 0.4 mg/ℓ organic, 0.3 mg/ℓ ammonia, 0.3 mg/ℓ nitrite, and 0.9 mg/ℓ nitrate nitrogen; the total phosphorus content was 1.5 mg/ℓ (EPA, 1971).

Four-Stage Bardenpho Process

The four-stage Bardenpho process, as shown in Figure 11.8, results in the ultimate removal of ammonia in addition to organic, nitrite, and nitrate nitrogen. Rather than requiring methanol as a carbon source, denitrification is accomplished using carbon in the raw waste and carbon from endogenous respiration. Although operational costs are reduced because an additional carbon source is not required, capital costs are increased because larger reactors are required (EPA, 1975).

In the first stage, the principal biochemical action is the conversion of nitrate (recycled from the second stage) to nitrogen gas — that is, denitrification. The raw waste provides carbon used in the denitrification. In the second stage, the principal biochemical actions are carbonaceous BOD oxidation and nitrification of ammonia from the first stage to nitrate. The remaining nitrate from the combined first and second stages enters the third stage. In the third stage, the principal biochemical action is denitrification of the remaining nitrate. In the fourth stage, the principal biochemical action is

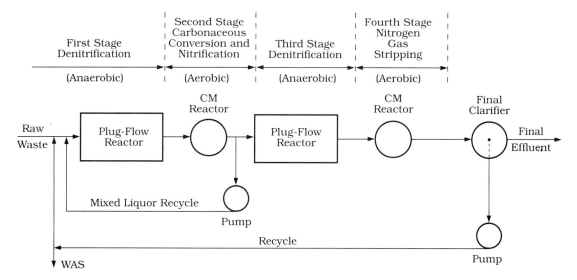

FIGURE 11.8 *Four-Stage Bardenpho Process*

the nitrification of remaining ammonia from the third stage. Physical mixing in the fourth stage serves to strip nitrogen gas produced in preceding stages (Metcalf & Eddy, 1991; EPA, 1975).

Five-Stage Bardenpho Process

The five-stage Bardenpho process, as shown in Figure 11.9, results in the ultimate removal of ammonia in addition to organic, nitrite, and nitrate nitrogen, and phosphorus removal. As shown in Figure 11.9, an anaerobic first stage is added to the four-stage Bardenpho process. The alternating anaerobic and aerobic conditions to which the microorganisms are subjected result in increased uptake of orthophosphate, polyphosphate, and organic phosphorus. The phosphorus is used by the bacteria for cell maintenance, synthesis, and energy transport and is also stored for later use. The increased level of phosphorus in the cell mass is removed with wasted sludge. The various stages of this modification of the four-stage Bardenpho process serve the same functions as previously discussed. In this modification, the fifth, aerobic stage also serves to reduce the amount of phosphorus in the clarified waste leaving the final clarifier (Metcalf & Eddy, 1991). Phosphorus removal by the five-stage Bardenpho process is discussed in more detail in Chapter 23.

Oxidation (Stabilization) Ponds

Although not considered to be a major nitrogen removal process, oxidation ponds (stabilization ponds) can effect removal of nitrogen compounds. Partial biological conversion of organic nitrogen to ammonia/ammonium, partial removal of ammonia/ammonium by stripping, and partial removal of nitrate by nitrification-denitrification occurring in oxidation ponds has been reported (NDC, 1978). Nitrogen and phosphorus are used by algae within the

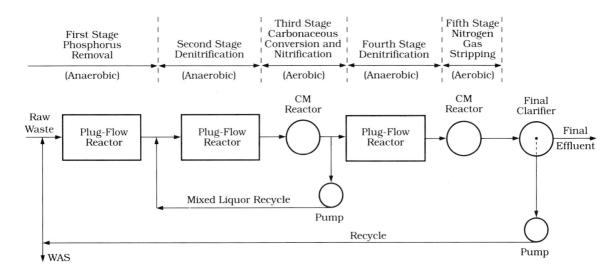

FIGURE 11.9 *Five-Stage Bardenpho Process*

pond, and nutrient removal from the pond effluent is accomplished by periodic removal of algae from the pond.

Ammonia stripping, breakpoint chlorination, ion exchange, and the biological nitrification-denitrification processes are the main methods of ultimate nitrogen removal. Specific design criteria for these processes are presented in detail in an EPA (1975) publication.

REFERENCES

AWARE, Inc. 1974. *Process Design Techniques for Industrial Waste Treatment*, edited by C. E. Adams and W. W. Eckenfelder, Jr. Nashville, Tenn.: Enviro Press.

Barnes, D., and Bliss, P. J. 1983. *Biological Control of Nitrogen in Wastewater Treatment*. New York: E. & F. N. Spon.

Barth, E. F.; Brenner, R. C.; and Lewis, R. F. 1968. Chemical-Biological Control of Nitrogen and Phosphorus in Wastewater Effluents. *Jour. WPCF* 40:2040.

Cassel, A. F.; Pressley, T. A.; Schuk, W. W.; and Bishop, D. F. 1971. Physical-Chemical Nitrogen Removal from Municipal Wastewaters. Paper presented at the 68th national meeting of the American Institute of Chemical Engineers (AlCh.E.), Houston, Tex.

Chilton, T. H., and Colburn, A. P. 1935. Distillation and Adsorption in Packed Columns. *Ind. and Eng. Chem.* 27, no. 3:255.

Colburn, A. P. 1939. The Simplified Calculation of Diffusional Processes, General Consideration of Two-Film Resistances. *Trans. AICh.E.* 35:211.

Conway, R. A., and Ross, R. D. 1980. *Handbook of Industrial Waste Disposal*. New York: Van Nostrand Reinhold.

Culp, G. L., and Hamann, C. L. 1974. Advanced Waste Treatment Process Selection; Parts 1, 2, and 3. *Public Works*. March, April, and May.

Culp, R. L., and Culp, G. L. 1978. *Handbook of Advanced Wastewaster Treatment*. 2nd ed. New York: Van Nostrand Reinhold.

Eckenfelder, W. W., Jr. 1980. *Principles of Water Quality Management*. Boston: CBI Publishing.

Eliassen, R., and Tchobanoglous, G. 1969. Removal of Nitrogen and Phosphorus from Waste Water. *Environmental Science and Technology* 3, no. 6.

Environmental Protection Agency (EPA). 1971. Advanced Waste Treatment and Water Reuse Symposium, sessions 1–5, Dallas, Tex.

Environmental Protection Agency (EPA). 1975. *Nitrogen Control*. EPA Process Design Manual, Washington, D.C.

Environmental Protection Agency (EPA). 1993. *Nitrogen Control*. EPA Manual, Washington, D.C.

Environmental Protection Agency (EPA). 1974. Physical-Chemical Nitrogen Removal. EPA Technology Transfer Seminar Publication, Washington, D.C.

Gonzales, J. G., and Culp, R. L. 1973. New Developments in Ammonia Stripping. *Public Works*. May and June.

Lewis, W. K., and Whitman, W. G. 1924. Principles of Gas Absorption. *Ind. and Eng. Chem.* 16, no. 12:1215.

McCabe, W. L., and Smith, J. C. 1967. *Unit Operations of Chemical Engineering*. New York: McGraw-Hill.

McCarty, P. L. 1970. Biological Processes for Nitrogen Removal: Theory and Application. *Proceedings Twelfth Sanitary Engineering Conference*. Urbana, Ill.: University of Illinois.

McCarty, P. L.; Beck, L.; and St. Amant, P. 1969. Biological Denitrification of Wastewaters by Addition of Organic Materials. *Proceedings of the 24th Annual Purdue Industrial Waste Conference*. Part 2.

Mercer, B. W.; Ames, L. L.; Touhill, C. J.; Van Slyke, W. J.; and Dean, R. B. 1970. Ammonia Removal from Secondary Effluents by Selective Ion Exchange. *Jour. WPCF* 42, no. 2. part 2: R95.

Metcalf & Eddy, Inc. 1979. *Wastewater Engineering: Treatment, Disposal and Reuse*. 2nd ed. New York: McGraw-Hill.

Metcalf & Eddy, Inc. 1991. *Wastewater Engineering: Treatment, Disposal and Reuse*. 3rd ed. New York: McGraw-Hill.

Mulbarger, M. C. 1971. Nitrification and Denitrification in Activated Sludge. *Jour. WPCF* 43, no. 10:2059.

Noyes Data Corporation (NDC) 1978. *Nitrogen Control and Phosphorus Removal in Sewage Treatment.* Edited by D. J. DeRenzo. Park Ridge, N.J.: NDC.

O'Farrell, T. P.; Frauson, F. P.; Cassel, A. F.; and Bishop, D. F. 1972. Nitrogen Removal by Ammonia Stripping. *Jour. WPCF* 44, no. 8:1527.

Perry, J. H. 1969. *Chemical Engineers Handbook.* 4th ed. New York: McGraw-Hill.

Pressley, T. A.; Bishop, D. F.; Pinto, A. P.; and Cassel, A. F. 1973. *Ammonia-Nitrogen Removal by Breakpoint Chlorination.* Report prepared for the Environmental Protection Agency, contract no. 14-12-818.

Ramalho, R. S. 1977. *Introduction to Wastewater Treatment Processes.* New York: Academic Press.

Reynolds, T. D., and Westervelt, R. 1969. Ion Exchange as a Tertiary Treatment. Paper presented at the American Society of Civil Engineers annual meeting and national meeting on Water Resources Engineering. New Orleans, La.

Schroeder, E. D. 1977. *Water and Wastewater Treatment.* New York: McGraw-Hill.

South Tahoe Public Utility District. 1971. *Advanced Wastewater Treatment as Practiced at South Tahoe.* Tech. Report for the EPA. Project 17010 ELQ (WPRD 52-01-67).

Steel, E. W., and McGhee, T. J. 1979. *Water Supply and Sewerage.* 6th ed. New York: McGraw-Hill.

Sundstrom, D. W., and Klei, H. E. 1979. *Wastewater Treatment.* Englewood Cliffs, N.J.: Prentice-Hall.

Wanielista, P. E., and Eckenfelder, W. W., Jr., eds. 1979. *Advances in Water and Wastewater Treatment: Biological Nutrient Removal.* Ann Arbor, Mich.: Ann Arbor Science.

Water Environment Federation (WEF) and American Society of Civil Engineers (ASCE) 1991. *Design of Municipal Wastewater Treatment Plants.* WEF Manual of Practice No. 8, ASCE Manual and Report on Engineering Practice No. 76. Brattleboro, Vt.: Book Press.

Water Pollution Control Federation (WPCF) 1983. *Nutrient Control.* Manual of Practice FD-7, Facilities Design. Washington, D.C.

Water Pollution Control Federation (WPCF). 1977. *Wastewater Treatment Plant Design.* WPCF Manual of Practice No. 8. Washington, D.C.

Weber, W. J., Jr. 1972. *Physicochemical Processes for Water Quality Control.* New York: Wiley-Interscience.

Wild, H. E.; Sawyer, C. N.; and McMahon, T. C. 1971. Factors Affecting Nitrification Kinetics. *Jour. WPCF* 43. no. 9:1845.

PROBLEMS

11.1 An induced-draft packed countercurrent ammonia stripping tower is to be designed for a tertiary treatment plant for a municipal wastewater. Pertinent data are flow = 3.0 MGD, operating pressure = 710 mm, minimum operating water and air temperature = 15°C, design hydraulic loading = 2 gpm/ft^2, design air flow = 1.75 times the theoretical, and wastewater pH = 10.9. Determine:

a. Theoretical air required, lb/hr-ft^2.

b. Design air required, lb/hr-ft^2.

c. Design air required, cfm/ft^2 and ft^3/gal.

d. Plan dimensions if tower is square.

11.2 An induced-draft packed countercurrent ammonia stripping tower is to be designed for a tertiary treatment plant for a municipal wastewater. Pertinent data are flow = 11,400 m^3/d, minimum operating water and air temperature = 15°C, design hydraulic loading = 4.85 m^3/h-m^2, design air flow =

1.75 times the theoretical, and wastewater pH = 10.9. Operating pressure = 710 mm. Determine:

a. Theoretical air required, kg/h-m^2.

b. Design air required, kg/h-m^2.

c. Design air required, m^3/h-m^2 and m^3 of air/ℓ of wastewater.

d. Plan dimensions if tower is square.

11.3 A tertiary treatment plant for a municipal wastewater treatment is to use breakpoint chlorination for ammonia removal. Pertinent data are flow = 2.0 MGD and ammonia concentration = 28 mg/ℓ as NH$_3$-N. Determine:

a. The pounds of chlorine theoretically required per month.

b. The pounds of chlorine actually required per month.

11.4 A tertiary treatment plant for a municipal wastewater treatment is to use breakpoint chlorination for ammonia removal. Pertinent data are flow = 8.0 MLD and ammonia concentration = 28 mg/ℓ as NH_3-N. Determine:

a. The kilograms of chlorine theoretically required per month.

b. The kilograms of chlorine actually required per month.

11.5 A plug-flow denitrification stage is to be designed for a three-stage nitrification-denitrification treatment process. Pertinent data for the design are as follows: design flow = 5.0 MGD, temperature = 25°C, cellular recycle will be used, NO_3^--N from the second stage = 25 mg/ℓ, maximum allowable NO_3^--N in the final effluent = 0.5 mg/ℓ, maximum allowable suspended solids in the final effluent = 5 mg/ℓ, and volatile suspended solids are 80% of suspended solids. Use median values for the kinetic coefficients needed. Determine:

a. Minimum solids retention time for denitrification.

b. Design solids retention time.

c. Design nitrate removal rate, lb NO_3^--N/lb MLVSS/day.

d. Hydraulic detention time at the design flow.

e. Sludge wasting required, lb MLVSS/day.

f. Methanol required, lb methanol/day.

11.6 A plug-flow denitrification stage is to be designed for a three-stage nitrification-denitrification treatment process. Pertinent data for the design are as follows: design flow = 19,000 m³/d, temperature = 25°C, cellular recycle will be used, NO_3^--N from the second stage = 25 mg/ℓ, maximum allowable NO_3^--N in the final effluent = 0.5 mg/ℓ, maximum allowable suspended solids in the final effluent = 5 mg/ℓ, and volatile suspended solids are 80% of suspended solids. Use median values for the kinetic coefficients needed. Determine:

a. Minimum solids retention time for denitrification.

b. Design solids retention time.

c. Design nitrate removal rate, kg NO_3^--N/kg MLVSS/d.

d. Hydraulic detention time at the design flow.

e. Sludge wasting required, kg MLVSS/d.

f. Methanol required, kg methanol/d.

11.7 A municipal wastewater activated sludge treatment facility requires upgrading to meet effluent nitrogen discharge criteria. The existing treatment process has demonstrated excellent flexiblity and can be operated to produce either a non-nitrified or a highly nitrified effluent by adjusting the mean cell residence time. Ammonia stripping and denitrification for nitrogen control are being considered. Explain, including in your discussion appropriate chemical equations and relative lengths for mean cell residence time, how the existing treatment facility should be generally operated to support each of the nitrogen control processes under consideration.

12

ADSORPTION

Adsorption consists of using the capacity of an adsorbent to remove certain substances from a solution. Activated carbon is an adsorbent that is widely used in water treatment, advanced wastewater treatment, and the treatment of certain organic industrial wastewaters, because it adsorbs a wide variety of organic compounds and its use is economically feasible. In water treatment it is used to remove compounds that cause objectionable taste, odor, or color. In advanced wastewater treatment it is used to adsorb organic compounds, and in industrial wastewater treatment it is mainly used to adsorb toxic organic compounds. It is generally used in granular form in batch, column (both fixed bed and countercurrent bed), or fluidized-bed operations, fixed-bed columns being the most common. Occasionally, activated carbon is used in powdered form and is not recovered for regeneration; however, such application is usually limited to water treatment where the amounts of carbon used are not appreciable. Adsorbents other than activated carbon are used to a lesser extent in environmental engineering.

ADSORPTION

Adsorption is the collection of a substance onto the surface of the adsorbent solids, whereas **absorption** is the penetration of the collected substance into the solid. Since both of these frequently occur simultaneously, some choose to call the phenomena **sorption**. Although both adsorption and absorption occur in sorption by activated carbon and other solids, the unit operation is usually referred to as adsorption.

Adsorption may be classified as (1) **physical adsorption** or (2) **chemical adsorption**. Physical adsorption is primarily due to van der Waals forces and is a reversible occurrence. When the molecular forces of attraction between the solute and the adsorbent are greater than the forces of attraction between the solute and the solvent, the solute will be adsorbed onto the adsorbent surface. An example of physical adsorption is the adsorption by activated carbon. Activated carbon has numerous capillaries within the carbon particles, and the surface available for adsorption includes the surfaces of the pores in addition to the external surface of the particles. Actually, the pore surface area greatly exceeds the surface area of the particles, and most of the adsorption occurs on the pore surfaces. For activated carbon the ratio of the total surface area to the mass is extremely large. In chemical adsorption, a chemical reaction occurs between the solid and the adsorbed solute, and the reaction is usually irreversible. Chemical adsorption is rarely used in environmental engineering; however, physical adsorption is widely used.

Activated carbon is made from numerous materials such as wood, sawdust, fruit pits and coconut shells, coal, lignite, and petroleum base residues. The manufacture essentially consists of carbonization of the solids followed by activation using hot air or steam.

When activated carbon particles are placed in a solution containing an organic solute and the slurry is agitated or mixed to give adequate contact, the adsorption of the solute occurs. The solute concentration will decrease from an initial concentration, C_0, to an equilibrium value, C_e, if the contact time is sufficient during the slurry test. Usually, equilibrium occurs within about 1 to 4 hr. By employing a series of slurry tests, it is usually possible to obtain a relationship between the equilibrium concentration (C_e) and the amount of organic substance adsorbed (x) per unit mass of activated carbon (m).

The Freundlich isotherm, which is an empirical formulation, frequently will represent the adsorption equilibrium over a limited range in solute concentration. One form of the equation is

$$\frac{x}{m} = X = KC_e^{1/n} \qquad (12.1)$$

where

x = mass of solute adsorbed

m = mass of adsorbent

X = mass ratio of the solid phase — that is, the mass of adsorbed solute per mass of adsorbent

C_e = equilibrium concentration of solute, mass/volume

K, n = experimental constants

Inspection of Eq. (12.1) shows that the adsorption isotherm should plot a straight line on log-log graph paper if the x-axis represents the solid-phase concentration, x/m, and the y-axis represents the liquid-phase concentration, C_e. The slope of the line will be $1/n$, and once the value of $1/n$ is known, the value of K may be determined. One of the most important aspects of the Freundlich isotherm in relation to the feasibility of using carbon adsorption is the numerical value of n and the value of x/m when $C_e = C_0$. The n value is the same regardless of the units used for the equilibrium concentration. The constant, K, however, does fluctuate with different units employed for the equilibrium concentrations. The larger the n value and the x/m value (when $C_e = C_0$), the more economically feasible is the use of carbon adsorption. Usually, the x/m value is from 0.2 to 0.8 gm COD (chemical oxygen demand) per gram of carbon.

Another isotherm, which frequently will represent adsorption equilibrium, is the Langmuir isotherm, which is

$$\frac{x}{m} = X = \frac{aKC_e}{1 + KC_e} \qquad (12.2)$$

where

a = mass of adsorbed solute required to saturate completely a unit mass of adsorbent

K = experimental constant

The Langmuir isotherm was derived assuming that (1) there is a limited area available for adsorption, (2) the adsorbed solute material on the surface is only one molecule in thickness, and (3) the adsorption is reversible and an equilibrium condition is achieved. When the adsorbent is placed in a solution, the solute is adsorbed; desorption also occurs, but the rate of adsorption is greater than the rate of desorption. Finally, an equilibrium condition is attained where the rate of adsorption is equal to the rate of desorption. The formulation, like the Freundlich isotherm, is valid only over a limited range in solute concentrations.

The rate of adsorption is limited by one of the various mass transport mechanisms involved. These are (1) the movement of the solute from the bulk solution to the liquid film or boundary layer surrounding the adsorbent solid, (2) the diffusion of the solute through the liquid film, (3) the diffusion of the solute inward through the capillaries or pores within the adsorbent solid, and (4) the adsorption of the solute onto the capillary walls or surfaces. Step 2 is frequently termed **film diffusion**, and step 3 is usually referred to as **pore diffusion**. Weber (1972) has found that in a stirred batch operation or in a continuous-flow operation operated at the velocities used in water or wastewater treatment, the rate of adsorption is usually limited by film diffusion (step 2) or, in some cases, by pore diffusion (step 3).

EXAMPLE 12.1   *Batch-Type Adsorption Tests*

A phenolic wastewater that has a phenol concentration of $0.400 \, \text{gm}/\ell$ as TOC (total organic carbon) is to be treated by granular activated carbon. Batch tests (slurry-type) have been performed in the laboratory to obtain the relative adsorption. Determine the Freundlich isotherm constants.

SOLUTION   The data from the batch tests are shown in columns (1), (2), and (3) of Table 12.1.

The reduced data are shown in columns (4) and (5). A plot of the data is shown in Figure 12.1 and the slope of the line is equal to 0.23. Therefore, $n = 1/\text{slope} = 1/0.23 = \boxed{4.35.}$ Taking the logarithm of both sides of Eq. (12.1) gives $\log X = \log K + 1/n \log C_e$. When $C_e = 1.0$, $X = K$, and thus from an extension of the graph, $K = \boxed{0.195.}$ The isotherm is therefore

$$X = x/m = \boxed{0.195 \, C_e^{1/4.35}}$$

TABLE 12.1 *Reduced Data from Batch or Slurry-Type Tests*

(1) m (gm/ℓ)	(2) C_0 (gm/ℓ)	(3) C_e (gm/ℓ)	(4) $\Delta C, x$ (gm/ℓ)	(5) $X = x/m$ (gm/gm)
0.52	0.400	0.322	0.078	0.150
2.32	0.400	0.117	0.283	0.122
3.46	0.400	0.051	0.349	0.101
3.84	0.400	0.039	0.361	0.094
4.50	0.400	0.023	0.377	0.084
5.40	0.400	0.012	0.388	0.072
6.67	0.400	0.0061	0.3939	0.059
7.60	0.400	0.0042	0.3958	0.052
8.82	0.400	0.0011	0.3989	0.045

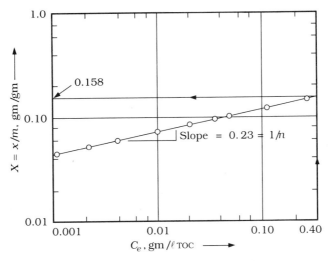

FIGURE 12.1 *Graph for Determining Freundlich Isotherm Constants, Example 12.1*

The maximum value of x/m is 0.158 gm TOC/gm carbon, which occurs when $C_e = C_0 = 0.400$ gm/ℓ TOC.

COLUMN CONTACTING TECHNIQUE AND EQUIPMENT

Adsorption using a granular adsorbent may be accomplished by batch, column, or fluidized-bed operations. The usual contacting systems are fixed beds or countercurrent moving beds because of their lower labor costs and high utilization of the adsorption capacity of the adsorbent. The fixed beds may employ downflow or upflow of the liquid; however, downflow is more popular since the granular adsorbent bed may also serve as a filter for suspended solids in addition to effecting the adsorption of organic substances. The countercurrent moving beds employ upflow of the liquid and downflow of the adsorbent solid since the adsorbent can be moved by the force of gravity. Both fixed beds and moving beds may use gravity or pressure liquid flow. The maximum available diameter of steel cylinders is about 12 ft (3.7 m), and pressurized flow is usually limited to these types of vessels. If the contactor is to be larger than about 12 ft (3.7 m) in diameter, it is usually built on the site, is of concrete, and generally employs gravity flow.

Figure 12.2 shows a typical fixed-bed carbon column facility used in wastewater treatment employing a single column with downflow of the liquid. The column is similar to a pressure filter and has an inlet distributor,

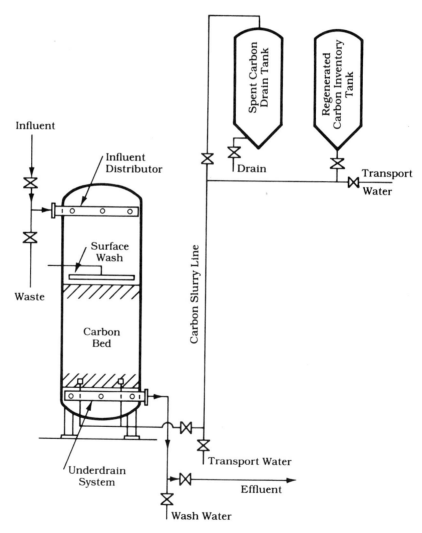

FIGURE 12.2
Fixed-Bed Adsorption System Using a Single Carbon Column

an underdrain system, and a surface wash. During the adsorption cycle the influent flow enters through the inlet distributor at the top of the column, and the wastewater flows downward through the bed and leaves through the underdrain system. The unit hydraulic flowrate employed is usually from 2 to 5 gpm/ft^2 (1.4 to 3.4 ℓ/s-m^2). When the head loss becomes excessive as a result of the accumulated suspended solids, the column is taken off-line and backwashed. Backwashing consists of operating the surface scour for a short time and then using the backwash water to expand the bed by 10 to 50%. This will require a backwash of 10 to 20 gpm/ft^2 (6.8 to 13.6 ℓ/s-m^2), depending upon the carbon particle size. To remove carbon for regeneration, the column is taken off-line and the bed is expanded by 20 to 35% using the backwash at a rate of 8 to 18 gpm/ft^2 (5.4 to 12.2 ℓ/s-m^2), depending upon the carbon particle size. The carbon-water slurry is removed from the bottom of the bed by opening the bottom valve and the transport water valve, thus sending the carbon-water slurry to the spent carbon drain tanks. After draining, the spent carbon is regenerated using a furnace, which is usually a multiple-hearth type, and the regenerated carbon is stored in the regenerated carbon inventory tank. In Figure 12.2, the regenerated carbon inventory tank is shown smaller than the carbon in the column for illustrative purposes; however, it actually holds enough carbon to fill an empty column. An empty column is refilled with carbon by using a carbon-water slurry transported from the regenerated carbon inventory tank. The slurry enters the bottom of the column, and, once filled to the proper level, the column is placed back on-line. If the wastewater flow is continuous and no storage for the flow is provided, two columns are required since a column must be taken off-line to regenerate its carbon. Two columns are provided in nearly all cases, and they are usually used in series.

Figure 12.3 shows a typical countercurrent moving-bed carbon column facility used in wastewater treatment employing upflow of the water. Two or more columns are usually provided and are always operated in series. The influent wastewater flow enters the bottom of the first column by means of a manifold system employing screens that uniformly distribute the flow across the bottom; the liquid flows upward through the column. The unit hydraulic flow rate employed is usually from 2 to 10 gpm/ft^2 (1.4 to 6.8 ℓ/s-m^2). The effluent is collected by a screen and manifold system at the top of the column and flows to the bottom manifold of the second column. The carbon flow is not continuous but instead is pulsewise. Spent carbon is removed and fresh carbon added when 5 to 10% of the total carbon in the column is spent. The wastewater flow does not have to be discontinued when carbon is withdrawn and more carbon added. To remove spent carbon, the inlet carbon valve at the top of the column is opened. If the effluent flow rises in the carbon filling chamber, some of the influent wastewater flow should be bypassed to the second column to lower the hydraulic gradient below the filling chamber. The jet water is started at the bottom of the column, and the outlet carbon valve and transport water valve are opened. Because of the jet action and the weight of the carbon, the carbon-water slurry will flow

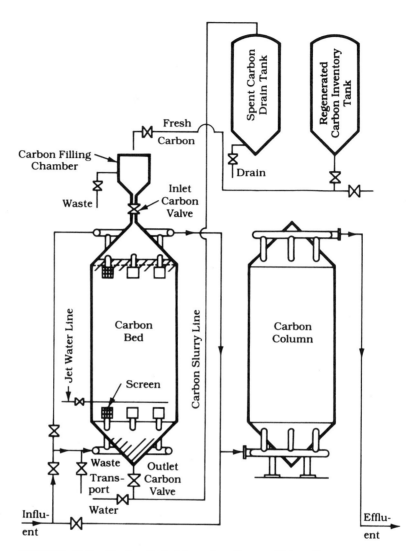

FIGURE 12.3 *Moving-Bed Adsorption System Using Two Carbon Columns*

from the column to the spent carbon drain tank, where the water is removed as drainage. Once sufficient carbon has been withdrawn, the outlet carbon valve, the transport water valve, and the jet water valve are closed. To add fresh carbon, the fresh carbon-water slurry is allowed to flow into the carbon filling chamber. The fresh carbon flows downward into the column, and the transport water from the filling chamber flows to waste. After fresh carbon is added, the inlet carbon valve is closed, and if any wastewater has been bypassed to the second column, the bypass flow is discontinued. To wash the column to remove accumulated suspended solids that have been filtered consists of reversing the flow through the column. The influent is temporarily sent to the second column, although some of the influent is temporarily sent downward through the first column to backwash it. The backwash is for about 10 min at a rate of 10 to 12 gpm/ft^2 (6.8 to 8.1 ℓ/s-m^2), and the waste washwater leaves through the bottom manifold. After backwashing, the column is placed back on-line and is again operated in series with the second column. Although the type of column shown in Figure 12.3 is usually operated as a countercurrent moving bed, it may also be employed as a fixed bed.

Figure 12.4 shows typical equipment required to regenerate spent granular activated carbon. It consists of the spent carbon drain tank, a multiple-hearth furnace (usually six hearths), a quench tank, and the carbon wash and inventory tank. The furnace is heated to 1500 to 1700°F (816 to

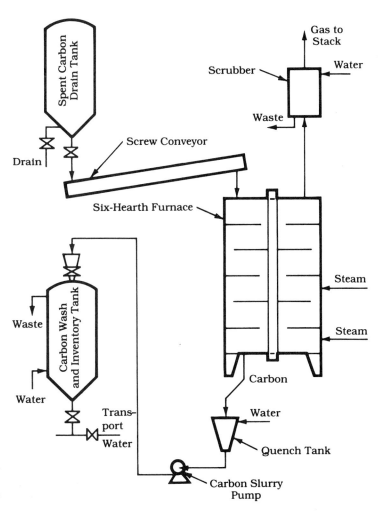

FIGURE 12.4
Typical Carbon Regenerating Facility

927°C), and then the spent carbon is fed by a screw conveyer to the multiple-hearth furnace. As the carbon passes through the first two hearths, it is dried and the adsorbed organic matter is baked. By the time the carbon has reached the third hearth the adsorbed organic materials have been ignited, thus evolving as gases and leaving a carbon residue in the pores of the activated carbon. The activating step consists of oxidizing the remaining carbon residue by the combined effects of heat, steam produced by evaporating water, oxygen and carbon dioxide present, and additional steam added. The regenerated carbon is discharged from the furnace bottom into the quench tank, where cold water is introduced to cool the carbon. The activated carbon slurry is then pumped to the carbon wash and inventory tank. The upward-flowing wash water in the tank removes the fines that have been produced by the breakdown of larger carbon particles during regeneration. The regenerated carbon is stored in the carbon inventory tank and is sent to the carbon columns as a slurry.

FIXED-BED ADSORPTION COLUMNS

A fixed-bed adsorption column is shown in Figure 12.5, and a typical breakthrough curve for this type of column is shown in Figure 12.6. The breakthrough curve for a column shows the solute concentration in the effluent on the y-axis versus the effluent throughput volume on the x-axis. The length of the column in which adsorption occurs is termed the **sorption tone**, Z_s. It is in this zone that the solute is transferred from the liquid to the solid phase — that is, mass transfer of the solute occurs. Frequently, the sorption zone is termed the **mass transfer zone**. Above this zone the solute in the liquid phase is in equilibrium with that sorbed on the solid phase, and the solute concentration is C_0. Above the sorption zone the equilibrium

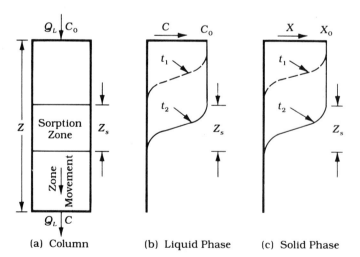

(a) Column   (b) Liquid Phase   (c) Solid Phase

FIGURE 12.5 *A Fixed-Bed Column and Its Liquid- and Solid-Phase Concentrations*

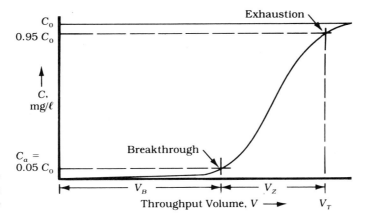

FIGURE 12.6
Typical Breakthrough
Curve

solid-phase concentration is X_0 (q_0). The value of X_0 or q_0 is equal to the x/m value from slurry test data when $C_e = C_0$ (see Example 12.1). The mass transfer fronts for the liquid and solid phases are shown at two different times, t_1 and t_2, in Figure 12.5. Both fronts have the same geometry at t_2 as they had at t_1; however, they have moved downward. The liquid flow in Figure 12.5 is shown moving downward through the column; however, if the flow is upward through the column, the discussed fundamentals remain the same. As the sorption zone passes down the column, the concentration of the solute in the effluent is theoretically zero. Actually, a slight amount of leakage or bypassing of the organic solute occurs, but it is negligible for illustrating the fundamentals. Once the sorption zone reaches the bottom of the column, the effluent solute concentration becomes a finite value and breakthrough begins, as shown in Figure 12.6. As the sorption zone disappears, the effluent solute concentration increases to C_0 and the column is exhausted, resulting in a breakthrough curve as shown in Figure 12.6. In Figure 12.6 the allowable breakthrough concentration, C_a, is considered to be 0.05 C_0. The allowable breakthrough concentration does not necessarily have to be this value, although 5% is commonly used. The exhaustion is considered to occur at $C = 0.95\,C_0$. For the case of a symmetrical breakthrough curve (which is the usual case), it has been shown that the length of the sorption zone, Z_s, is related to the column height, Z, and the throughput volumes, V_B and V_T, by (Michaels, 1952)

$$Z_s = Z\left[\frac{V_Z}{V_T - 0.5V_Z}\right] \qquad (12.3)$$

where $V_Z = V_T - V_B$. The area above the breakthrough curve represents the mass or amount of solute adsorbed by the column and is equal to $\int(C_0 - C)dV$ from $V = 0$ to $V =$ the allowable throughput volume under consideration. At the allowable breakthrough volume, V_B, the area above the breakthrough curve is equal to the mass or amount of solute adsorbed by the column. At complete exhaustion, $C = C_0$ and the area above the breakthrough curve is equal to the maximum amount of solute adsorbed. At complete exhaustion, the entire adsorption column is in equilibrium with the influent and effluent flows. Also, the solute concentration in the influent is equal to the concentration in the effluent.

It is not possible to design a column accurately without a test column breakthrough curve for the liquid of interest and the adsorbent solid to be used. In the subsequent text, a scale-up and kinetic approach to design adsorption columns is presented. In both of the approaches a breakthrough curve from a test column, either laboratory or pilot scale, is required, and the column should be as large as possible to minimize side-wall effects. Neither of the design procedures requires the adsorption to be represented by an isotherm such as the Freundlich equation. A disadvantage of the scale-up and kinetic design approaches is that they do not take into account the effect of the unit hydraulic flowrate. This is incorporated in a third design approach using mass transfer fundamentals that has been developed and published by the coauthor (Reynolds and Pence, 1980).

Scale-Up Approach This method was developed by Fornwalt and Hutchins (1966) for the design of carbon adsorption columns. The principal experimental information required is a breakthrough curve from a test column, either laboratory or pilot scale, that has been operated at the same liquid flowrate in terms of bed volumes per unit time, Q_b, as the design column. Since the contact time, T_c, is equal to ε/Q_b where ε is the pore fraction, the design column will have the same contact time as the test column. Since the contact times are the same, it is assumed that the volume of liquid treated per unit mass of adsorbent, $\hat{V}_B$, for a given breakthrough in the test column is the same as for the design column. Before a breakthrough test can be performed, it is necessary to select a satisfactory liquid flowrate, Q_b, in bed volumes per unit time. This may be estimated from calculations using such information as the required breakthrough volume, the solute concentration, the maximum solid-phase concentration, and other pertinent data. Usually, the value of Q_b is from 0.2 to 3.0 bed volumes per hour.

The bed volume of the design column is given by

$$\text{Bed volume } (BV) = \frac{Q}{Q_b} \tag{12.4}$$

where Q is the design liquid flowrate. The mass or weight of the adsorbent, M, for the design column is determined from

$$M = (BV)(\rho_s) \tag{12.5}$$

where ρ_s is the adsorbent bulk density. From the breakthrough curve for the laboratory- or pilot-scale column, the breakthrough volume, V_B, is determined for the allowable effluent solute concentration, C_a. The volume of liquid treated per unit mass of adsorbent, $\hat{V}_B$, is then determined by

$$\hat{V}_B = \frac{V_B}{M} \tag{12.6}$$

where M is the mass of the adsorbent in the test column. The mass of adsorbent exhausted per hour, M_t, for the design column is computed from

$$M_t = \frac{Q}{\hat{V}_B} \qquad (12.7)$$

where Q is the design liquid flowrate. The breakthrough time, T, is

$$T = \frac{M}{M_t} \qquad (12.8)$$

where M is the mass of adsorbent in the design column. The calculated breakthrough volume, V_B, for the allowable breakthrough concentration, C_a, for the design column is

$$V_B = QT \qquad (12.9)$$

If the calculated breakthrough time, T, from Eq. (12.8) or the calculated breakthrough volume, V_B, from Eq. (12.9) is not acceptable, another liquid flowrate, Q_b, to give the required time or volume should be determined from the available breakthrough data. The breakthrough test on the laboratory- or pilot-scale column should be repeated using the new Q_b value. Then the computations using Eqs. (12.4) through (12.9) should be repeated. Usually, one breakthrough test is adequate, although in some cases two tests must be done. The main advantages of this design procedure are its simplicity and the relatively few experimental data required.

EXAMPLE 12.2   *Fixed-Bed Column Design by the Scale-Up Approach*

A phenolic wastewater having a TOC of 200 mg/ℓ is to be treated by a fixed-bed granular carbon adsorption column for a wastewater flow of 40,000 gal/day, and the allowable effluent concentration, C_a, is 10 mg/ℓ as TOC. A breakthrough curve, shown in Figure 12.7, has been obtained from an experimental pilot column operated at 1.67 BV/hr. Other data concerning the pilot column are as follows: inside diameter = 3.75 in., length = 41 in., mass of carbon = 2980 gm (6.564 lb), liquid flowrate = 12.39 ℓ/hr, unit liquid flowrate = 0.71 gpm/ft^2, and packed carbon density = 25 lb/ft^3 (401 gm/ℓ). Using the scale-up approach for design, determine:

1. The design bed volume, ft^3.
2. The design mass of carbon required, lb.
3. The breakthrough time, T, hr.
4. The breakthrough volume, V_B, gal.

SOLUTION   The design bed volume is

$$BV = \frac{Q}{Q_b} = \frac{40{,}000\,\text{gal}}{24\,\text{hr}} \left| \frac{\text{hr}}{1.67BV} \right| \frac{\text{ft}^3}{7.48\,\text{gal}}$$

$$= \boxed{133\,\text{ft}^3}$$

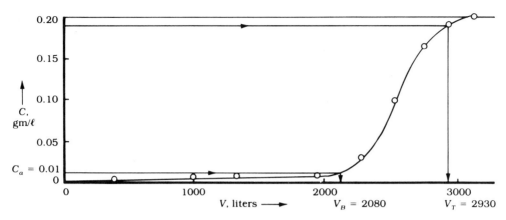

FIGURE 12.7 *Breakthrough Curve for Examples 12.2 and 12.2 SI*

The mass of carbon required is

$$M = (BV)(\rho_s) = (133\,\text{ft}^3)(25\,\text{lb/ft}^3) = \boxed{3330\,\text{lb}}$$

From the breakthrough curve, shown in Figure 12.7, the volume treated at the allowable breakthrough, C_a, of $10\,\text{mg}/\ell$ TOC is $2080\,\ell$ or $549.5\,\text{gal}$. The solution treated per pound of carbon, $\hat{V}_B$, is $(549.5\,\text{gal}/6.564\,\text{lb})$ or $83.71\,\text{gal/lb}$. The number of pounds of carbon exhausted per hour, M_t, is equal to the flow rate divided by the volume treated per pound of carbon or $(40{,}000\,\text{gal}/24\,\text{hr})(\text{lb}/83.71\,\text{gal})$, which is equal to $19.91\,\text{lb/hr}$. Thus the breakthrough time, T, is

$$T = \frac{M}{M_t} = (3330\,\text{lb})(\text{hr}/19.91\,\text{lb})$$

$$= \boxed{167\,\text{hr or 6.96 days}}$$

The breakthrough volume, V_B, for the design column is

$$V_B = QT = (40{,}000\,\text{gal}/24\,\text{hr})(167\,\text{hr})$$

$$= \boxed{278{,}000\,\text{gal}}$$

EXAMPLE 12.2 SI   *Fixed-Bed Column Design by the Scale-Up Approach*

A phenolic wastewater having a TOC of $200\,\text{mg}/\ell$ is to be treated by a fixed-bed granular carbon adsorption column for a wastewater flow of $150\,\text{m}^3/\text{d}$, and the allowable effluent concentration, C_a, is $10\,\text{mg}/\ell$ as TOC. A breakthrough curve, shown in Figure 12.7, has been obtained from an experimental pilot column operated at $1.67\,BV/\text{h}$. Other data concerning the pilot column are as follows: inside diameter = 9.50 cm, length = 1.04 m, mass of carbon = 2.98 kg, liquid flowrate = 12.39 ℓ/h, unit liquid flowrate = 0.486 ℓ/s-m², and packed carbon density = 400 kg/m³. Using the scale-up approach for design, determine:

1. The design bed volume, m³.
2. The design mass of carbon required, kg.
3. The breakthrough time, T, in hours and days.
4. The breakthrough volume, V_B, m³.

SOLUTION The design bed volume is

$$BV = \frac{Q}{Q_b} = \frac{150 \text{ m}^3}{24 \text{ hr}} \left| \frac{\text{h}}{1.67 \, BV} \right. = \boxed{3.74 \text{ m}^3}$$

The mass of carbon required is

$$M = (BV)(\rho_s) = (3.74 \text{ m}^3)(400 \text{ kg/m}^3) = \boxed{1500 \text{ kg}}$$

From the breakthrough curve, as shown in Figure 12.7, the volume treated at the allowable breakthrough, C_a, of 10 mg/ℓ TOC is 2080 ℓ. The solution treated per kilogram carbon, $\hat{V}_B$, is (2080 ℓ/2.98 kg) or 698.0 ℓ/kg. The number of kilograms of carbon exhausted per hour, M_t, is equal to the flowrate divided by the volume treated per kilogram of carbon or (150 m³/24 h)(kg/698.0 ℓ) (1000 ℓ/m³) = 8.954 kg/h. The breakthrough time, T, is

$$T = \frac{M}{M_t} = (1500 \text{ kg})(\text{h}/8.954 \text{ kg}) = \boxed{168 \text{ h or 7 d}}$$

The breakthrough volume, V_B, for the design column is

$$V_B = QT = (150 \text{ m}^3/\text{d})(7 \text{ d}) = \boxed{1050 \text{ m}^3}$$

If the design procedure does not give the required breakthrough but gives a breakthrough time or volume near the required breakthrough, the design column volume can be computed from a rearrangement of Eq. (12.6), which is $M = V_B/\hat{V}_B$. The term $\hat{V}_B$ is the volume treated per unit weight of carbon in the breakthrough test. If, however, the design procedure gives a breakthrough time or volume that differs appreciably from that required, another test must be run. Using the data from the first test, it is possible to estimate the flowrate, Q_b, in BV/hr to be used in the second test.

Kinetic Approach This method utilizes a kinetic equation based on the derivation by Thomas (1948). The kinetic equation may also be derived from an extension of the Bohart and Adams (1920) equation (Loebenstein, 1975). The principal experimental information required is a breakthrough curve from a test column, either laboratory or pilot scale.

The expression by Thomas for an adsorption column is as follows:

$$\frac{C}{C_0} \cong \frac{1}{1 + e^{\frac{k_1}{Q}(q_0 M - C_0 V)}} \qquad \textbf{(12.10)}$$

where

C = effluent solute concentration

C_0 = influent solute concentration

k_1 = rate constant

q_0 = maximum solid-phase concentration of the sorbed solute — for example, gm per gm*

M = mass of the adsorbent — for example, gm*

V = throughput volume — for example, liters

Q = flow rate — for example, liters per hour

Assuming the left side equals the right side, cross multiplying gives

$$1 + e^{\frac{k_1}{Q}(q_0 M - C_0 V)} = \frac{C_0}{C} \qquad \textbf{(12.11)}$$

Rearranging and taking the natural logarithms of both sides yield the design equation

$$\ln\left(\frac{C_0}{C} - 1\right) = \frac{k_1 q_0 M}{Q} - \frac{k_1 C_0 V}{Q} \qquad \textbf{(12.12)}$$

From Eq. (12.12), it can be seen that this is a straight-line equation of the type $y = b + mx$. The terms are $y = \ln(C_0/C - 1)$, $x = V$, $m = k_1 C_0/Q$, and $b = k_1 q_0 M/Q$.

The laboratory- or pilot-scale column used to obtain the breakthrough curve for the kinetic design approach should be operated at approximately the same flowrate in terms of bed volumes per hour as the design column. One advantage of the kinetic approach is that the breakthrough volume, V, may be selected in the design of a column.

EXAMPLE 12.3   *Fixed-Bed Column Design by the Kinetic Approach*

A phenolic wastewater having a TOC of 200 mg/ℓ is to be treated by a fixed-bed granular carbon adsorption column for a wastewater flow of 40,000 gal/day, and the allowable effluent concentration, C_a, is 10 mg/ℓ as TOC. A breakthrough curve has been obtained from an experimental pilot column operated at 1.67 BV/hr. This is the same pilot column that was used in Example 12.2 and the resulting breakthrough curve is shown in Figure 12.7. Other data concerning the pilot column are as follows: inside diameter =

---

*The solid-phase concentration could be expressed as lb per lb, the mass could be lb, and the flowrate could be gal per hr, and so on.

3.75 in., length = 41 in., mass of carbon = 2980 gm (6.564 lb), liquid flowrate = 12.39 ℓ/hr, unit liquid flowrate = 0.71 gpm/ft^2, and the packed carbon density = 25 lb/ft^3 (401 gm/ℓ). The design column is to have a unit liquid flowrate of 3 gpm/ft^2, and the allowable breakthrough volume is 280,000 gal. (This approximates the breakthrough volume in Example 12.2 and is used so that a comparison can be made between the design approach in Example 12.2 and the approach in this example.) Using the kinetic approach for design, determine:

1. The design reaction constant, k_1, gal/hr-lb.
2. The design maximum solid-phase concentration, q_0, lb/lb.
3. The pounds of carbon required for the design column.
4. The diameter and height of the design column if the diameters are in 6 in. increments.
5. The pounds of carbon required per 1000 gal of waste treated.

SOLUTION The data from the breakthrough test are given in columns (1) and (2) of Table 12.2. The reduced data for the portion of the curve where break-through occurs are given in columns (3), (4), and (5). The plot of $\ln(C_0/C - 1)$ versus V is shown in Figure 12.8. The slope = $k_1 C_0/Q$ or $(\ln 8.8 \times 10^6 - \ln 1.0)/(0 - 2470)\ell = 6.474 \times 10^{-3}\,\ell^{-1}$. The value of $k_1 = (\text{slope})(Q/C_0)$ or $k_1 = (6.474 \times 10^{-3}/\ell)(12.39\,\ell/\text{hr})(\ell/0.200\,\text{gm})(\text{gal}/3.785\,\ell)(454\,\text{gm/lb})$ = $\boxed{48.1\,\text{gal/hr-lb.}}$ The y-axis intercept, b, equals $\ln 8.80 \times 10^6$ or 15.99. Since $b = k_1 q_0 M/Q$, rearranging gives $q_0 = bQ/k_1 M = (15.99)(12.39\,\ell/\text{hr})$ (hr-lb/48.1 gal) $\times$ (gal/3.785 ℓ)(1/6.564 lb) = $\boxed{0.166\,\text{lb/lb.}}$ The mass of carbon, M, may be computed from

$$\ln\left(\frac{C_0}{C} - 1\right) = \frac{k_1 q_0 M}{Q} - \frac{k_1 C_0 V}{Q}$$

TABLE 12.2 *Reduced Data from Breakthrough Test*

(1) V (liters)	(2) C (gm/ℓ)	(3) C/C_0	(4) C_0/C	(5) $C_0/C - 1$
378	0.009			
984	0.011			
1324	0.008			
1930	0.009	0.045	22.1	21.1
2272	0.030	0.150	6.67	5.67
2520	0.100	0.500	2.00	1.00
2740	0.165	0.825	1.21	0.21
2930	0.193	0.965	1.036	0.036
3126	0.200	1.000	1.00	0.00

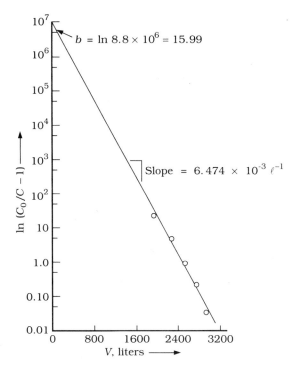

FIGURE 12.8
Plot for Examples 12.3 and 12.3 SI

$$\ln\left(\frac{0.200}{0.010} - 1\right) = \frac{48.1\,\text{gal}}{\text{hr-lb}} \left| \frac{0.166\,\text{lb}}{\text{lb}} \right| \frac{\text{day}}{40,000\,\text{gal}}$$

$$\times \frac{24\,\text{hr}}{\text{day}} \left| M - \frac{48.1\,\text{gal}}{\text{hr-lb}} \right| \frac{0.200\,\text{gm}}{\ell} \left| \frac{3.785\,\ell}{\text{gal}} \right.$$

$$\times \frac{\text{lb}}{454\,\text{gm}} \left| \frac{\text{day}}{40,000\,\text{gal}} \right| \frac{280,000\,\text{gal}}{} \left| \frac{24\,\text{hr}}{\text{day}} \right.$$

From this, $M = \boxed{3430\,\text{lb.}}$ The column diameter is $D = [(40,000\,\text{gal/day}) \times (\text{day}/1440\,\text{min})(\text{min-ft}^2/3.0\,\text{gal})(4/\pi)]^{1/2} = 3.43\,\text{ft}$ or $\boxed{3\,\text{ft-6 in.}}$ Volume $= (3430\,\text{lb})(1\,\text{ft}^3/25\,\text{lb}) = 137.2\,\text{ft}^3$. The carbon height is $Z = (137.2\,\text{ft}^3)(4/\pi) \times (1/3.5\,\text{ft})^2 = \boxed{14.3\,\text{ft.}}$ The number of pounds of carbon required per 1000 gal is $(3430\,\text{lb}/280,000\,\text{gal})(1000\,\text{gal}/T\,\text{gal}) = \boxed{12.3\,\text{lb}/1000\,\text{gal.}}$ The carbon required by the kinetic approach is 3430 lb and by the scale-up approach is 3330 lb.

EXAMPLE 12.3 SI *Fixed-Bed Column Design by the Kinetic Approach*

A phenolic wastewater having a TOC of 200 mg/ℓ is to be treated by a fixed-bed granular carbon adsorption column for a wastewater flow of 150 m³/d, and the allowable effluent concentration, C_a, is 10 mg/ℓ as TOC. A breakthrough curve has been obtained from an experimental pilot column operated at $1.67\,BV/h$. This is the same pilot column that was used in Example 12.2, and the resulting breakthrough curve is shown in Figure 12.7. Other data concerning the pilot column are as follows: inside diameter $= 9.5\,cm$, length $= 1.04\,m$, mass of carbon $= 2.98\,kg$, liquid flowrate $= 12.39\,\ell/h$, unit liquid flowrate $= 0.486\,\ell/s\text{-}m^2$, and the packed carbon density $= 400\,kg/m^3$. The design column is to have a unit liquid flowrate of $2.04\,\ell/s\text{-}m^2$, and the allowable breakthrough volume is $1060\,m^3$. (This approximates the breakthrough volume in Example 12.2 SI and is used so that a comparison can be made between the design approach in Example 12.2 SI and the approach in this example.) Using the kinetic approach for design, determine:

1. The design reaction constant, k_1, $\ell/s\text{-}kg$.
2. The design maximum solid-phase concentration, q_0, kg/kg.
3. The carbon required for the design column, kg.
4. The diameter and height of the design column, m.
5. The kilograms of carbon required per cubic meter of waste treated.

SOLUTION The data from the breakthrough test are given in columns (1) and (2) of Table 12.2. The reduced data for the portion of the curve where breakthrough occurs are given in columns (3), (4), and (5). The plot of $\ln(C_0/C - 1)$ versus V is shown in Figure 12.8. The slope $= k_1C_0/Q$ or $(\ln 8.8 \times 10^6 - \ln 1.0)/(0 - 2470)\ell = 6.474 \times 10^{-3}\,\ell^{-1}$. The value of $k_1 = (\text{slope})(Q/C_0)$ or $k_1 = (6.474 \times 10^{-3}/\ell)(12.39\,\ell/h)(h/3600\,s)(\ell/0.200\,gm)(1000\,gm/kg) = \boxed{0.111\,\ell/s\text{-}kg.}$ The y-axis intercept, b, equals $\ln 8.8 \times 10^6$ or 15.99. Since $b = k_1q_0M/Q$, rearranging gives $q_0 = bQ/k_1M = (15.99)(12.39\,\ell/h) \times (s\text{-}kg/0.111\,\ell)(h/3600\,s)(1/2.98\,kg) = \boxed{0.166\,kg/kg.}$ The mass of carbon, M, may be computed from Eq. (12.12), which is

$$\ln\left(\frac{C_0}{C} - 1\right) = \frac{k_1q_0M}{Q} - \frac{k_1C_0V}{Q}$$

$$\ln\left(\frac{0.200}{0.010} - 1\right) = \left(\frac{0.111\,\ell}{s\text{-}kg}\right)\left(\frac{0.166\,kg}{kg}\right)\left(\frac{d}{150\,m^3}\right)\left(\frac{86,400\,s}{d}\right)\left(\frac{m^3}{1000\,\ell}\right)(M\,kg)$$

$$- \left(\frac{0.111\,\ell}{s\text{-}kg}\right)\left(\frac{0.200\,gm}{\ell}\right)\left(\frac{d}{150\,m^3}\right)(1060\,m^3)\left(\frac{86,400\,s}{d}\right)\left(\frac{1\,kg}{1000\,gm}\right)$$

From this $M = \boxed{1550\,kg.}$ The column diameter is $D = [(150\,m^3/d) \times (d/86,400\,s)(s\text{-}m^2/2.04\,\ell)(1000\,\ell/m^3)(4/\pi)]^{1/2} = \boxed{1.04\,m.}$ Volume $= (1550\,kg)(1\,m^3/400\,kg) = 3.9\,m^3$. The carbon height is $Z = (3.9\,m^3) \times (4/\pi)(1/1.04\,m)^2 = \boxed{4.6\,m.}$ The carbon required is $(1550\,kg/1060\,m^3) = \boxed{1.46\,kg/m^3.}$ The carbon required by the kinetic approach is $1550\,kg$ and by the scale-up approach is $1500\,kg$.

**Geometric
Considerations**

For a fixed-bed column it is desirable to have a large height-to-diameter ratio, Z/D, because the percent utilization of the maximum capacity of the adsorbent increases with this ratio. Usually, Z/D is from $3:1$ to $5:1$, and the unit liquid loading is from 4 to 10 gpm/ft^2 (2.7 to 6.8 ℓ/s-m^2). If head space is limited, several short columns in series may be used to simulate a taller column.

**Fixed-Bed
Columns in Series**

In most cases, it is advantageous to use two or more smaller fixed-bed columns operated in series rather than one large column containing the same amount of adsorbent. Fixed-bed columns are used in series when the headroom is limited, and several small columns are used to simulate one large column. Also, fixed-bed columns in series are used when the increased utilization of the maximum adsorption capacity of the adsorbent will pay for the additional capital and operating costs of two or more smaller columns in lieu of one large column.

The operation of two columns in series may be illustrated by considering two granular activated carbon columns designated A and B, A being the lead column. Once the sorption zone has passed from column A to column B, the carbon in column A is completely exhausted and the column is taken off-line, leaving column B on-line. After the carbon in column A is removed and replaced with fresh carbon, the column is placed back on-line but downstream from column B. Once the sorption zone has passed from column B to column A, column B is taken off-line and recharged with fresh carbon in the same manner as column A. This type of operation allows the carbon to become completely exhausted prior to regeneration. Thus the carbon required per 1000 gal (1000 ℓ) treated is less than that for a single, large fixed-bed column containing the same amount of carbon as the two smaller columns.

**MOVING-BED
COUNTER-
CURRENT
ADSORPTION
COLUMNS**

The moving-bed adsorption column (pulsed-bed) is a steady-state countercurrent operation since the adsorbent solid is moving downward through the column while the liquid is flowing upward. It is a common method of operation; approximately one-half the carbon columns used in advanced and tertiary wastewater treatment are of this type. A material balance on a moving-bed column at steady state gives

$$Q_L(C_1 - C_2) = L_S(X_1 - X_2) \tag{12.13}$$

where Q_L is the liquid flowrate, L_S is the adsorbent flowrate, C_1 and C_2 are the liquid-phase solute concentrations at the bottom and top of the column, and X_1 and X_2 are the solid-phase concentrations at the bottom and top of the column. The minimum theoretical height of a countercurrent adsorption column is equal to the height of the adsorption zone, Z_s. A column of the minimum theoretical height would require extreme care in operation to maintain the entire mass transfer front within the column. Therefore, in practice, the height of a countercurrent column is usually equal to the design height of a fixed-bed column for the same design situation. Once a column has been designed, V_T may be determined from the previous equations, and the height of the adsorption zone may be estimated from Eq. (12.3).

EXAMPLE 12.4 AND 12.4 SI *Moving-Bed Countercurrent Adsorption Column Design*

The phenolic wastewater in Example 12.3 is to be treated by a moving-bed countercurrent granular carbon adsorption column, and the allowable effluent concentration, C_a, is $10\,mg/\ell$ as TOC. Determine the carbon required in lb/1000 gal and in kg/m^3.

SOLUTION Equation (12.13) may be rearranged to give

$$L_s/Q_L = (C_1 - C_2)/(X_1 - X_2)$$

Here $C_1 = 0.200\,gm/\ell$, $C_2 = 0.010\,gm/\ell$, $X_1 = 0.166\,gm/gm$ (kg/kg), and $X_2 = 0\,gm/gm$ (kg/kg).

The USCS solution is

$$L_s/Q_L = (0.200 - 0.010)(gm/\ell)/(0.166 - 0)(gm/gm) = 1.145\,gm/\ell$$

The carbon per 1000 gal is $(1.145\,gm/\ell)(lb/454\,gm)(3.785\,\ell/gal)(1000\,gal/T\,gal) = \boxed{9.55\,lb/1000\,gal.}$

The SI solution is

$$L_s/Q_L = (0.200 - 0.010)(gm/\ell)/(0.166 - 0)(kg/kg) = 1.145\,gm/\ell$$

The carbon required is $(1.145\,gm/\ell)(kg/1000\,gm)(1000\,\ell/m^3) = \boxed{1.15\,kg/m^3.}$

The countercurrent system is more efficient than a single fixed-bed column: the carbon required in Example 12.3 was $12.3\,lb/1000\,gal$ and in Example 12.3 SI was $1.46\,kg/m^3$.

EXAMPLE 12.5 *Sorption Zone Height*

The adsorption column from Example 12.3 contained 3430 lb carbon and was 3.50 ft in diameter and 14.3 ft in height. The flow was 40,000 gal/day, and the allowable breakthrough was $C_a = 5\%\ C_0$ or $0.010\,gm/\ell$. Determine the height of the adsorption zone, Z_s, using the scale-up approach and the kinetic approach.

SOLUTION From the breakthrough curve, Figure 12.7, $V_B = 2080\,\ell$ at $C = 5\%\ C_0$ and $V_T = 2930\,\ell$ at $C = 95\%\ C_0$. The test column contained 6.564 lb carbon; thus the volume treated per pound of carbon is $\hat{V}_B = (2080\,\ell)(gal/3.785\,\ell)(1/6.564\,lb)$ or 83.72 gal/lb, and $\hat{V}_T = (2930\,\ell)(gal/3.785\,\ell)(1/6.564\,lb)$ or 117.93 gal/lb. Thus V_B for the design column $= (83.72\,gal/lb)(3430\,lb) = 287,160\,gal$, and $V_T = (117.93\,gal/lb)(3430\,lb) = 404,500\,gal$. The value $V_Z = 404,500 - 287,160 = 117,340\,gal$. The sorption zone is now obtained from Eq. (12.3):

$$Z_s = 14.3 \, \text{ft}\left[\frac{117{,}340}{404{,}500 - 0.5(117{,}340)}\right]$$

$$= \boxed{4.85 \, \text{ft}}$$

From Example 12.3, $V_B = 280{,}000$ gal at $C = 5\%$ C_0. Using the equation in Example 12.3 to determine V_T gives $\ln(1/0.95 - 1) = (48.1 \, \text{gal/hr-lb})(0.166 \, \text{lb/lb})(\text{day}/40{,}000 \, \text{gal}) \times (24 \, \text{hr/day})(3430 \, \text{lb}) - (48.1 \, \text{gal/hr-lb})(0.200 \, \text{gm}/\ell)(3.785 \, \ell/\text{gal})(\text{lb}/454 \, \text{gm})(\text{day}/40{,}000 \, \text{gal})(24 \, \text{hr/day}) \, V_T$. From the equation, $V_T = 402{,}666$ gal. The value $V_Z = 402{,}666 - 280{,}000 = 122{,}666$ gal. The sorption zone height is now obtained from Eq. (12.3):

$$Z_s = 14.3 \, \text{ft}\left[\frac{122{,}666}{402{,}666 - 0.5(122{,}666)}\right]$$

$$= \boxed{5.14 \, \text{ft}}$$

EXAMPLE 12.5 SI *Sorption Zone Height*

The adsorption column from Example 12.3 SI contained 1550 kg carbon and was 1.04 m in diameter and 4.6 m in height. The flow was 150 m^3/d, and the allowable breakthrough was $C_a = 5\%$ C_0 or 0.010 gm/ℓ. Determine the height of the adsorption zone, Z_s, using the scale-up approach and the kinetic approach.

SOLUTION From the breakthrough curve, Figure 12.7, $V_B = 2080 \, \ell$ at $C = 5\%$ C_0, and $V_T = 2930 \, \ell$ at $C = 95\%$ C_0. The test column contained 2.98 kg carbon; thus the volume treated per kilogram of carbon is $\hat{V}_B = (2080 \, \ell)(1/2.98 \, \text{kg})$ or $698 \, \ell/\text{kg}$, and $\hat{V}_T = (2930 \, \ell)(1/2.98 \, \text{kg})$ or $983 \, \ell/\text{kg}$. Thus V_B for the design column $= (698 \, \ell/\text{kg})(1550 \, \text{kg})(\text{m}^3/1000 \, \ell) = 1082 \, \text{m}^3$, and $V_T = (983 \, \ell/\text{kg})(1550 \, \text{kg})(\text{m}^3/1000 \, \ell) = 1524 \, \text{m}^3$. The value $V_Z = 1524 - 1082 = 442 \, \text{m}^3$. The sorption zone is now obtained from Eq. (12.3):

$$Z_s = 4.6 \, \text{m}\left[\frac{442}{1524 - 0.5(442)}\right]$$

$$= \boxed{1.56 \, \text{m}}$$

From Example 12.3 SI, $V_B = 1060 \, \text{m}^3$ at $C = 5\%$ C_0. Using the equation in Example 12.3 SI to determine V_T gives $\ln(1/0.95 - 1) = (0.111 \, \ell/\text{s-kg})(0.166 \, \text{kg/kg})(\text{d}/150 \, \text{m}^3)(1550 \, \text{kg})(\text{m}^3/1000 \, \ell)(86{,}400 \, \text{s/d}) - (0.111 \, \ell/\text{s-kg})(0.200 \, \text{gm}/\ell)(\text{d}/150 \, \text{m}^3)(V_T \, \text{m}^3)(86{,}400 \, \text{s/d})(\text{kg}/1000 \, \ell)$. From the equation, $V_T = 1517 \, \text{m}^3$. The value $V_Z = 1517 - 1060 = 457 \, \text{m}^3$. The sorption zone height is now obtained from Eq. (12.3):

$$Z_s = 4.6 \, \text{m}\left[\frac{457}{1517 - 0.5(457)}\right]$$

$$= \boxed{1.63 \, \text{m}}$$

FLUIDIZED BEDS

Occasionally fluidized beds are used in water treatment and in the advanced treatment of wastewaters. The fluidized bed consists of a bed of granular adsorbent solids. The liquid flows upward through the bed in the vertical direction. The upward liquid velocity is sufficient to suspend the adsorbent solids so that the solids do not have constant interparticle contact. At the top of the solids there is a distinct interface between the solids and the effluent liquid. The principal advantage of the fluidized bed is that liquids with appreciable suspended solids content may be given adsorption treatment without clogging the bed, since the suspended solids pass through the bed and leave with the effluent. Usually fluidized beds are operated in a continuous countercurrent fashion. The ratio of the adsorbent required per given amount of liquid, L_s/Q_L, is the same as for a countercurrent moving-bed adsorption column. For further reading, the text by Weber (1972) is recommended.

TEST COLUMNS

The larger the laboratory or pilot column, the smaller will be the scale-up effects. A laboratory column should be at least 1 in. (2.54 cm) in diameter (ID) and the height should be at least 24 in. (61 cm). Carbon manufacturers have numerous good publications on laboratory techniques. It has been found that side-wall effects (the channeling of flow down the inside of a test column) are of significant magnitude if the unit liquid flowrate is appreciable. The maximum unit liquid flowrates to prevent appreciable side-wall effects are as follows: $1\frac{3}{8}$ in. (3.5 cm) ID, 0.50 gpm/ft^2 (0.34 ℓ/s-m^2); $2\frac{3}{4}$ in. (7 cm) ID, 1.00 gpm/ft^2 (0.68 ℓ/s-m^2); and $3\frac{3}{4}$ in. (9.5 cm) ID, 1.50 gpm/ft^2 (1.02 ℓ/s-m^2).

DESIGN CONSIDERA-TIONS

Carbon contactors may employ gravity-flow or pressurized-flow contactors. For pressurized contactors, which are usually fixed-bed columns, the carbon beds may be operated in a downflow or upflow fashion; however, the downflow fashion is the most common. For downflow fixed-bed columns, the carbon size is usually 8×30 mesh, whereas for upflow fixed-bed columns, the carbon size is usually 12×40 mesh. Carbon beds have depths from 10 to 40 ft (3.05 to 12.2 m), 15 to 20 ft (4.6 to 6.1 m) being the most common. Column height must be sufficient to allow for 50% bed expansion for the backwashing of downflow fixed beds or for 15% bed expansion for the backwashing of upflow fixed beds. For fixed-bed columns, it is desirable to have a large carbon bed depth-to-diameter ratio, Z/D, because the percent utilization of the adsorbent increases with this ratio. Usually Z/D is from 1.5:1 to 4:1. Prefabricated steel cylinders may be obtained in diameters up to 12 ft (3.7 m) in 3-in. (7.6-cm) intervals; however, 6-in. (15.2-cm) intervals are more common, and diameters range from 2 ft-6 in. (0.76 m) to 12 ft. (3.7 m). The operating pressure drop through a carbon bed depends mainly upon the unit hydraulic loading and the carbon size. Usually it is less than 1.0 psi per foot (22.6 kPa per meter) of bed depth. For downflow or upflow fixed-bed columns, the filtration rate is from 2 to 5 gpm/ft^2 (1.4 to 3.4 ℓ/s-m^2), whereas for pulsed-bed columns, the filtration rate is from 2 to 10 gpm/ft^2 (1.4 to 6.8 ℓ/s-m^2). Backwash rates for downflow fixed beds using

8×30 mesh carbon are from 10 to 20 gpm/ft^2 (6.8 to 13.6 ℓ/s-m^2), which gives bed expansions of 10 to 50%. Backwashing of fixed-bed upflow columns using 12×40 mesh carbon consists merely of increasing the upflow rate to 10 to 12 gpm/ft^2 (6.8 to 8.1 ℓ/s-m^2), which gives a bed expansion of 10 to 15%. With upflow columns, a bed expansion of 10% is adequate for washing particulates from the bed. The backwash for downflow or upflow fixed-bed columns should last from 10 to 15 min. Backwashing of pulsed-bed columns consists of reversing the flow at a backwash rate of 10 to 12 gpm/ft^2 (6.8 to 8.1 ℓ/s-m^2) for about 10 to 15 min. Since the backwash is downward, expansion of the bed does not occur.

Gravity-flow carbon beds usually are the downflow type using 8×30 mesh carbon, and the filtration rates are from 2 to 4 gpm/ft^2 (1.4 to 2.7 ℓ/s-m^2). Most gravity-flow beds are open beds constructed using reinforced concrete, and a bed depth of at least 10 ft (3.05 m) is used. Backwash rates are from 15 to 20 gpm/ft^2 (10.2 to 13.6 ℓ/s-m^2), which will expand the bed depth from about 25 to 50%. Backwashing should last for 10 to 15 min.

Expanded or fluidized beds are usually constructed using reinforced concrete and are open beds with a depth of at least 10 ft (3.05 m). The filtration rates are from 6 to 10 gpm/ft^2 (4.1 to 6.8 ℓ/s-m^2), and the carbon size is usually 12×40 mesh. Bed expansion is about 10%, which requires an upflow of 6 gpm/ft^2 (4.1 ℓ/s-m^2) for 12×40 mesh carbon or 10 gpm/ft^2 (6.8 ℓ/s-m^2) for 8×30 mesh carbon. Backwashing is not required since a 10% bed expansion allows the particulates to pass through the bed.

In order to have filter runs with a suitable duration, the feed flow to the carbon contactors, with the exception of the fluidized bed, should have less than 5 mg/ℓ suspended solids. For the usual flow schemes, this places the carbon contactors as one of the last treatment processes.

Because of the wide variety of industrial and municipal wastewaters treated by activated carbon, the numerous organic compounds present, and the different characteristics of the various activated carbons, it is usual practice to report results in terms of the most significant parameters. Usually, these are the unit hydraulic loading in gpm/ft^2 (ℓ/s-m^2), the empty-bed contact time, the pounds of COD removed per pound of carbon (kg/kg), and the effluent COD. For municipal wastewaters treated by activated carbon in tertiary treatment or physical-chemical treatment, the unit hydraulic loadings are in the ranges that have been previously discussed for the various contacting methods. The empty-bed contact time is from 30 to 45 min, and the organic removal is from 0.2 to 0.8 lb COD per pound of carbon (kg/kg). The effluent COD will be from 0.5 to 15 mg/ℓ.

REFERENCES

Ademoroti, C. M. A. 1986. Water Purification by Fluidized Bed Technique. *Water Research* 20: 1105–1109.

Bohart, G. S., and Adams, E. Q. 1920. Some Aspects of the Behavior of Charcoal with Respect to Chlorine. *Jour. ACS* 42:523.

Burleson, N. K.; Eckenfelder, W. W., Jr.; and Malina, J. F., Jr. 1968. Tertiary Treatment of Secondary Industrial Effluents by Activated Carbon. *Proceedings of the 23rd Annual Purdue Industrial Waste Conference*. Part 1.

Conway, R. A., and Ross, R. D. 1980. *Handbook of Industrial Waste Disposal*. New York: Van Nostrand Reinhold.

Cookston, J. T., Jr. 1970. Design of Activated Carbon Adsorption Beds. *Jour. WPCF* 42, no. 12:2124.

Culp, R. L., and Culp, G. L. 1978. *Handbook of Advanced Wastewater Treatment*. 2nd ed. New York: Van Nostrand Reinhold

Eckenfelder, W. W., Jr. 1980. *Principles of Water Quality Management*. Boston: CBI Publishing.

Environmental Protection Agency (EPA). 1973. *Carbon Adsorption*. EPA Process Design Manual. Washington, D.C.

Fornwalt, J. J., and Hutchins, R. A. 1966. Purifying Liquids with Activated Carbon. *Chem. Eng.* (11 April): 1979; (9 May):155.

Keinath, T. M., and Weber, W. J., Jr. 1968. A Predictive Model for the Design of Fluid-Bed Adsorbers. *Jour. WPCF* 40, no. 5:741.

Loebenstein, W. V. 1975. Comparison of Column Decolorization Experiments with Theory. *Proceedings of the Fifth Technical Session on Bone Char*.

Lukchis, G. M. 1973. Adsorption Systems, Part I, II, and III. *Chem. Eng.* (11 June):111; (9 July):83; (6 August):83.

Metcalf & Eddy, Inc. 1979. *Wastewater Engineering: Treatment, Disposal and Reuse*. 2nd ed. New York: McGraw-Hill.

Metcalf & Eddy, Inc. 1991. *Wastewater Engineering: Treatment, Disposal and Reuse*. 3rd ed. New York: McGraw-Hill.

Meyers, A. L., and Belfort, G., eds. 1984. *Fundamentals of Adsorption*. *Proceedings of the Engineering Foundation Conference*, Schloss Elmau, Bavaria, West Germany, May 6–11, 1983. New York: AIChE. New York: Engineering Foundation.

Michaels, A. S. 1952. Simplified Method of Interpreting Kinetics Data in Fixed-Bed Ion Exchange. *Ind. and Eng. Chem.* 44:1922–1930.

Montgomery, J. M., Inc. 1985. *Water Treatment Principles and Design*. New York: Wiley.

Pittsburg Activated Carbon Division. 1966. The Laboratory Evaluation of Granular Activated Carbons for Liquid Phase Applications. Calgon Corp., Pittsburgh, Pa.

Ramalho, R. S. 1977. *Introduction to Wastewater Treatment Processes*. New York: Academic Press.

Reynolds, T. D., and Pence, R. F. 1980. Design of an Activated Carbon System for Wood Preserving Wastes. *Proceedings of the 35th Annual Purdue Industrial Waste Conference*. West Lafayette, Ind.

Ruthven, D. M. 1984. *Principles of Adsorption and Adsorption Processes*. New York: Wiley.

Sanks, R. L. 1978. *Water Treatment Plant Design*. Ann Arbor, Mich.: Ann Arbor Science Publishers.

Schroeder, E. D. 1977. *Water and Wastewater Treatment*. New York: McGraw-Hill.

Sundstrom, D. W., and Klei, H. E. 1979. *Wastewater Treatment*. Englewood Cliffs, N.J.: Prentice-Hall.

Suzuki, M. 1990. *Adsorption Engineering*. Tokyo: Kodansha Ltd. New York: Elsevier Science Publishing.

Thomas, H. C. 1948. Chromatography: A Problem in Kinetics. *Annals of the New York Academy of Science* 49:161.

Water Environment Federation (WEF) and American Society of Civil Engineers (ASCE). 1991. *Design of Municipal Wastewater Treatment Plants*. WEF Manual of Practice No. 8. ASCE Manual and Report on Engineering Practice No. 76. Brattleboro, Vt.: Book Press.

Water Pollution Control Federation (WPCF). 1977. *Wastewater Treatment Plant Design*. Manual of Practice No. 8. Washington, D.C.

Weber, W. J., Jr. 1972. *Physicochemical Processes for Water Quality Control*. New York: Wiley-Interscience.

Weber, W. J., Jr., and Ying, W. 1979. Bio-physicochemical Adsorption Model Systems for Wastewater Treatment. *Jour. WPCF* 51, no. 11: 2661.

Witco Chemical Company. Column Evaluation Techniques. *Tech. Bull.* 5–6. New York.

PROBLEMS

12.1 A pilot column breakthrough test has been performed using the phenolic wastewater in Example 12.1. Pertinent design data are inside diameter = 3.75 in., length = 41 in., mass of carbon = 2980 gm (6.564 lb), liquid flowrate = 17.42 ℓ/hr, unit liquid flowrate = 1.00 gpm/ft^2, and packed carbon density

= 25 lb/ft^3 (401 gm/ℓ). The breakthrough data are given in Table 12.3.

Determine:

a. The liquid flowrate in bed volumes per hour and the volume of liquid treated per unit mass of carbon — in other words, the gal/lb at an allowable breakthrough of 35 mg/ℓ TOC.

b. The kinetic constants k_1 in gal/hr-lb and q_0 in lb/lb.

12.2 A pilot column breakthrough test has been performed using the phenolic wastewater in Example 12.1. Pertinent design data are inside diameter = 0.095 m, length = 1.04 m, mass of carbon = 2.98 kg,

liquid flowrate = 17.42 ℓ/hr, unit liquid flowrate = 0.679 ℓ/s-m^2, and packed carbon density = 401 gm/ℓ. The breakthrough data are given in Table 12.4.

Determine:

a. The liquid flowrate in bed volumes per hour and the volume of liquid treated per unit mass of carbon — in other words, the ℓ/kg at an allowable breakthrough of 35 mg/ℓ TOC.

b. The kinetic constants k_1 in ℓ/s-kg and q_0 in kg/kg.

12.3 The phenolic wastewater in Problem 12.1 is to be treated by a fixed-bed granular carbon adsorption column for a wastewater flow of

TABLE 12.3

THROUGHPUT VOLUME, V (gal)	EFFLUENT TOC, C (mg/ℓ)
4	12
18	16
42	24
72	16
100	16
180	20
255	28
292	32
321	103
340	211
372	350
409	400

TABLE 12.4

THROUGHPUT VOLUME, V (ℓ)	EFFLUENT TOC, C (mg/ℓ)
15	12
69	16
159	24
273	16
379	16
681	20
965	28
1105	32
1215	103
1287	211
1408	350
1548	400

60,000 gal/day. The allowable breakthrough concentration, C_a, is 35 mg/ℓ as TOC. The design column will have a unit liquid flowrate of 3.5 gpm/ft^2. The design procedure is to be the scale-up approach using the values from Problem 12.1. Determine the column dimensions if the diameters are in 6 in. increments.

12.4 The phenolic wastewater in Problem 12.2 is to be treated by a fixed-bed granular carbon adsorption column for a wastewater flow of 227,100 ℓ/d. The allowable breakthrough concentration, C_a, is 35 mg/ℓ as TOC. The design column will have a unit liquid flowrate of 2.38 ℓ/s-m^2. The design procedure is to be the scale-up approach; use the values from Problem 12.2. Determine the column dimensions.

12.5 The phenolic wastewater in Problem 12.1 is to be treated by a fixed-bed granular carbon adsorption column for a wastewater flow of 60,000 gal/day. The allowable breakthrough concentration, C_a, is 35 mg/ℓ as TOC. The design column will have a unit liquid flowrate of 3.5 gpm/ft^2, and the on-line time is 7 days. The design procedure is to be the kinetic approach using the k_1 and q_0 values from Problem 12.1. Determine:
a. The column dimensions if the diameters are in 6 in. increments.
b. The number of columns in series to be used if the available headroom is 18 ft.

12.6 The phenolic wastewater in Problem 12.2 is to be treated by a fixed-bed granular carbon adsorption column for a wastewater flow of 227,100 ℓ/d. The allowable breakthrough concentration, C_a, is 35 mg/ℓ as TOC. The design column will have a unit liquid flowrate of 2.38 ℓ/s-m^2, and the on-line time is 7 days. The design procedure is to be the kinetic approach; use the k_1 and q_0 values from Problem 12.2. Determine:
a. The column dimensions.
b. The number of columns in series to be used if the available headroom is 5.5 m.

12.7 The phenolic wastewater in Problem 12.1 is to be treated by a countercurrent moving-bed column (pulsed-bed), and the allowable breakthrough concentration, C_a, is 35 mg/ℓ as TOC. The wastewater flow = 60,000 gal/d. Determine the pounds of carbon required per day.

12.8 The phenolic wastewater in Problem 12.2 is to be treated by a countercurrent moving-bed column (pulsed-bed), and the allowable break-through concentration, C_a, is 35 mg/ℓ as TOC. The wastewater flow = 227,100 ℓ/d. Determine the kilograms of carbon required per day.

12.9 A tertiary treatment plant treats a municipal secondary effluent and is to employ fixed-bed carbon adsorption columns. Parallel treatment using two rows of two columns in series (in other words, a total of four columns) will be used. The flowsheet consists of lime coagulation, settling, ammonia stripping, recarbonation, settling, multimedia filtration, carbon adsorption, and chlorination. From batch-type slurry tests it has been found that 0.42 lb of COD/lb carbon is adsorbed when $C_e = C_0$. Pertinent data are as follows: flow = 1.5 MGD, COD in feed to carbon columns = 20 mg/ℓ, contact time function based on an empty carbon bed = 30 min, and unit liquid flowrate = 6.50 gpm/ft^2. Each column is kept on-line until the entire mass of carbon in the column is completely exhausted. Determine:
a. The volume of each column, ft^3.
b. The diameter and height of each column, ft if the diameters are in 6 in. increments.
c. The mass of carbon in each column if the packed density is 25 lb/ft^3.
d. The pounds of carbon exhausted per day if the COD removed is assumed to be 98%.
e. The on-line time for each of the columns.

12.10 A tertiary treatment plant treats a municipal secondary effluent and is to employ fixed-bed carbon adsorption columns. Parallel treatment using two rows of two columns in series (in other words, a total of four columns) will be used. The flowsheet consists of lime coagulation, settling, ammonia stripping, recarbonation, settling, multimedia filtration, carbon adsorption, and chlorination. From batch-type slurry tests it has been found that 0.42 gm COD/gm carbon is adsorbed when $C_e = C_0$. Pertinent data are as follows: flow = 5700 m^3/d, COD in feed to carbon columns = 20 mg/ℓ, contact time function based on an empty carbon bed = 30 min, and unit liquid flowrate = 4.4 ℓ/s-m^2. Each column is kept on-line until the entire mass of carbon in the column is completely exhausted. Determine:
a. The volume of each column, m^3.
b. The diameter and height of each column, m.
c. The mass of carbon in each column if the packed density is 400 kg/m^3.
d. The kilograms of carbon exhausted per day if the COD removed is assumed to be 98%.
e. The on-line time for each of the columns.

13

ION EXCHANGE

The ion exchange process consists of a chemical reaction between ions in a liquid phase and ions in a solid phase. Certain ions in the solution are preferentially sorbed by the ion exchanger solid, and because electroneutrality must be maintained, the exchanger solid releases replacement ions back into the solution. For instance, in the softening of water by the ion exchange process, the calcium and magnesium ions are removed from the solution, and the exchanger solid releases sodium ions to replace the removed calcium and magnesium ions. The reactions are stoichiometric and reversible and obey the law of mass action.

The first commercially used ion exchange materials were naturally occurring porous sands that were commonly called zeolites. These minerals have a deficit of positive atoms within their crystalline structure, and as a result, they have a net negative charge that is balanced by exchangeable cations held within the pore capillaries. Zeolites were the first ion exchangers used to soften waters; however, they have been almost completely replaced in recent years by synthetic organic exchange resins that have a much higher ion exchange capacity. Synthetic cation exchange resins are polymeric materials that have reactive groups, such as the sulfonic, phenolic, and carboxylic, that are ionizable and may be charged with exchangeable cations. Also, synthetic anion exchange resins are available that have ionizable groups, such as the quaternary ammonium or amine groups, that may be charged with exchangeable anions. Thus synthetic resins are available that have both cation and anion exchange capabilites. Certain minerals, such as montmorillonite clays, have appreciable cation exchange capacity and are used for special ion exchange applications.

Ion exchange is used extensively in both water and wastewater treatment. Some of the common applications are water softening, demineralization, desalting, ammonia removal, treatment of heavy metal wastewaters, and treatment of some radioactive wastes. (1) One of the largest uses of ion exchange in environmental engineering is the softening of water by the exchange of sodium ions for calcium and magnesium ions. Since removal of all of the hardness is undesirable for a domestic water supply, a portion of the flow may bypass the exchangers to give a blended water of the desired hardness. This process is termed split-flow softening. (2) Ion exchange is used for the removal of all cations and anions from a water. In total demineralization, the cationic resins are charged with the hydrogen ion and the anionic resins are charged with the hydroxyl ion. The cationic resins exchange hydrogen ions for cations, and the anionic resins exchange hydroxyl ions for anions. Thus the treated water has only hydrogen and hydroxyl ions, which makes it essentially pure water. Industries using high-pressure boilers require demineralized water as boiler water. In addition, there are other industries that require demineralized water. (3) Ion exchange may be used for partial demineralization of wastewaters in tertiary treatment and of brackish waters for water supplies. Several ion exchange techniques may be employed, one of which is split-flow demineralization. (4) The natural zeolite, clinoptilolite, may be used to remove ammonia in advanced waste treatment plants in lieu of or in addition to other methods of ammonia removal. (5) Ion exchangers may be used to remove heavy metallic ions from certain wastewaters. The heavy metallic ions are thus concentrated in the spent regenerate. An example is the treatment of wastewaters from a metal plating industry that contain zinc, cadmium, copper, nickel, and chromium. (6) Clays and other minerals possessing ion exchange capacities are used to treat low- or moderate-level radioactive wastes to remove preferentially such heavy metallic radionuclides as Cs^{137}.

A number of texts and references are available on the theory and application of the ion exchange process and laboratory evaluation procedures (Betz, 1962; Dorfner, 1971; Dow, 1964; Helfferich, 1962; Kitchener, 1961; Kunin, 1960).

THEORY

The softening of a hard water by an exchanger solid, either a zeolite or a synthetic resin, may be represented by the following reactions:

$$Ca^{+2} + 2Na \cdot Ex \rightleftarrows Ca \cdot Ex_2 + 2Na^+ \tag{13.1}$$

$$Mg^{+2} + 2Na \cdot Ex \rightleftarrows Mg \cdot Ex_2 + 2Na^+ \tag{13.2}$$

where Ex represents the exchanger solid. As shown by the reactions, a hard water may be softened by exchanging Na^+ from the exchanger solid for the Ca^{+2} and Mg^{+2} in the solution. After the solid is saturated with the Ca^{+2} and Mg^{+2}, it may be regenerated by a strong salt solution since the reactions are reversible. The regeneration reactions are

$$Ca \cdot Ex_2 + 2Na^+ \xrightarrow[\longleftarrow]{\text{strong brine}} 2Na \cdot Ex + Ca^{+2} \tag{13.3}$$

$$Mg \cdot Ex_2 + 2Na^+ \xrightarrow[\longleftarrow]{\text{strong brine}} 2Na \cdot Ex + Mg^{+2} \tag{13.4}$$

After regeneration, the exchanger solid is washed to remove the remaining brine and then is placed back on-line to soften more water.

In the demineralization of water, the water is first passed through cation exchange resins charged with the hydrogen ion. The cation removal may be represented by the reaction

$$M^{+x} + xH \cdot Re \rightleftarrows M \cdot Re_x + xH^+ \tag{13.5}$$

where M^+ represents the cationic species present and x is the valence number. After passing through the cation exchange resins, the water passes through anion exchange resins charged with the hydroxyl ion. The anion removal may be represented by the reaction

$$A^{-z} + z\mathrm{Re} \cdot \mathrm{OH} \rightleftharpoons \mathrm{Re}_z \cdot A + z\mathrm{OH}^- \tag{13.6}$$

where A^- represents the anion species present and z is the valence number. After the resins become exhausted, the cation exchange resins are regenerated using a strong mineral acid such as sulfuric or hydrochloric acid. In a similar manner, the anion exchange resins are regenerated using a strong base such as sodium hydroxide.

Exchange resins are usually bead- or granular-shaped, having a size of about 0.1 to 1.0 mm. The ion exchange ability of resins may be classified as "strong" or "weak" according to the characteristics of the exchange capability (Dorfner, 1971). The strong acid cation resins have strong reactive sites such as the sulfonic group ($-SO_3H$), and the resins readily remove all cations. Conversely, the weak acid cation exchange resins have weak reactive sites such as the carboxylic group ($-COOH$), and these resins readily remove cations from the weaker bases such as Ca^{+2} and Mg^{+2} but have limited ability to remove cations from strong bases such as Na^+ and K^+. The strong base anion exchange resins have reactive sites such as the quaternary ammonium group, and these resins readily remove all anions. The weak base anion exchange resins have reactive sites such as the amine group, and these resins remove mainly anions from strong acids such as SO_4^{-2}, Cl^-, and NO_3^-, with limited removal for HCO_3^-, CO_3^{-2}, or SiO_4^{-4}.

Ion exchange resins or zeolites have a limited number of exchange sites available, and the total solid-phase concentration, $\hat{q}_0$, is termed the **ion exchange capacity**. For cation exchange resins, it is usually between 200 to 500 meq per 100 gms. Since a cation exchanger must remain electrically neutral during the exchange reaction, all of the exchange sites must be occupied by sufficient cations to balance the negative charge of the exchanger. Thus, for a system involving Ca^{+2}, Mg^{+2}, and Na^+, the sum of the solid-phase concentrations of these ions must, at any time, be equal to the cation exchange capacity, $\hat{q}_0$. Electroneutrality applies to anion exchangers as well as to cation exchangers.

Since ion exchange is a chemical reaction, the law of mass action may be applied. The generalized equation for cation exchange by a resin may be represented by

$$M_1^+ + \mathrm{Re} \cdot M_2 \rightleftharpoons M_2^+ + \mathrm{Re} \cdot M_1 \tag{13.7}$$

where M_1^+, M_2^+ are cations of different species and Re is the resin. The mass action constant is

$$K_{M_2^+}^{M_1^+} = \frac{[\mathrm{Re} \cdot M_1][M_2^+]}{[\mathrm{Re} \cdot M_2][M_1^+]} \tag{13.8}$$

$$= \left[\frac{M_1}{M_2}\right]_{\mathrm{solid}} \times \left[\frac{M_2}{M_1}\right]_{\mathrm{solution}} \tag{13.9}$$

where

$$K\frac{M_1^+}{M_2^+} = \text{mass action constant or selectivity coefficient}$$

In Eqs. (13.8) and (13.9) the bracketed terms represent equilibrium concentrations expressed in an appropriate concentration unit. The magnitude of K represents the relative preference for ion exchange. Thus it is the relative preference of the resin to sorb cation M_1^+ as compared to cation M_2^+.

The greater the selectivity coefficient, K, the greater is the preference for the ion by the exchanger. An ion exchanger tends to prefer (1) ions of higher valence, (2) ions with a small solvated volume, (3) ions with greater ability to polarize, (4) ions that react strongly with the ion exchange sites of the exchanger solid, and (5) ions that participate least with other ions to form complexes. For the usual cation exchangers, the preference series for the most common cations is as follows (Helfferich, 1962):

$$Ba^{+2} > Pb^{+2} > Sr^{+2} > Ca^{+2} > Ni^{+2} > Cd^{+2} > Cu^{+2}$$
$$> Co^{+2} > Zn^{+2} > Mg^{+2} > Ag^+ > Cs^+ > K^+$$
$$> NH_4^+ > Na^+ > H^+$$

This series is for strong acid resins — that is, those having strong reactive sites such as the sulfonic group ($-SO_3H$). Weak acid resins — that is, those having weak reactive sites such as the carboxylic group ($-COOH$) — will have the H^{+1} position to the left of that shown here, depending upon the strength of the reactive site. For very weak sites, the H^{+1} may fall to the left as far as Ag^{+1}. For the usual anion exchangers the preference series for the most common anions is as follows (Helfferich, 1962):

$$SO_4^{-2} > I^- > NO_3^- > CrO_4^{-2} > Br^- > Cl^- > OH^-$$

This series is for strong base resins — that is, those having strong reactive sites such as the quaternary ammonium group. For weak base resins — those having weak reactive sites such as the secondary or tertiary amine group — the OH^- will fall farther to the left, depending upon the strength of the reactive group. It should be understood that the previous series are to be used as guides and that exceptions occur. Since equilibria are involved, the most suitable way to determine the uptake by a certain exchanger solid for a particular ion in a solution is to perform sorption tests. These may be done by (1) using a column to obtain the breakthrough curve for the particular ion of interest or (2) using batch-type slurry tests to obtain the uptake of the particular ion.

The rate of ion exchange depends upon the rates of the various transport mechanisms involved and the rate of the exchange reaction itself. These mechanisms are as follows: (1) movement of the ions from the bulk solution to the film or boundary layer surrounding the exchanger solid, (2) diffusion of the ions through the film to the solid surface, (3) diffusion of the ions inward through the pores of the solid to the exchange sites, (4) exchange of the ions by the reaction, (5) diffusion of the exchanged ions outward through the pores to the solid surface, (6) diffusion of the exchanged ions through the liquid film or boundary layer surrounding the solid, and (7) movement of the exchanged ions into the bulk solution. Weber (1972) has shown that in a stirred batch process or in a continuous-flow process operated at the velocities used in water and waste treatment systems, the rate of exchange is usually controlled by step 2 or, in some cases, by step 3. Step 2, the diffusion of ions through the film or boundary layer, is frequently referred to as **film diffusion**; step 3, the diffusion of ions through the pores, is termed **pore diffusion**.

CONTACTING TECHNIQUES AND EQUIPMENT

All of the operating techniques ordinarily used for adsorption are also used for ion exchange. Thus there are batch, column, and fluidized-bed processes. Column processes may consist of fixed-bed columns in which the beds are stationary or of moving-bed columns in which both the liquid and the ion exchanger solid move countercurrently. Also, fluidized beds may be operated countercurrently. Of all the contacting systems, the fixed-bed columns are the most common, and the second most popular are the counter-current moving-bed columns. The popularity of these two contacting techniques is due mainly to their reduced labor costs. Frequently in ion exchange applications, there may be several columns or stages in series.

Figure 13.1 shows a schematic drawing of typical equipment required for a fixed-bed column used for water softening with a sodium-charged exchange resin. The equipment consists of the exchange column, a salt storage tank, a regeneration solution tank, and all required appurtenances.

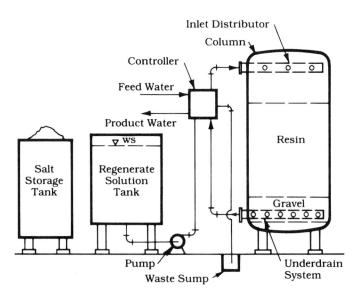

FIGURE 13.1 *Typical Fixed-Bed Column Equipment for Softening Using a Sodium-Charged Resin*

The exchange column is usually a steel cylindrical tank with an inlet distribution system and an underdrain collection system. The resins inside the vessel are supported on a graded gravel layer. The controller controls the flow to and from the column and contains the valves necessary to change the system from the softening cycle to the regeneration cycle and vice versa.

During the softening cycle, the water enters the top of the bed and flows downward at a constant flowrate of about 1 to 8 gpm/ft^2 (0.68 to 5.4 ℓ/s-m^2). Once the allowable breakthrough of hardness occurs in the effluent, the controller is activated so that backwashing is accomplished to remove any suspended material that may have accumulated by filtering action during the softening cycle. After backwashing, the salt solution, which is usually 5 to 10% salt, is passed through the exchanger bed at a controlled rate to regenerate the resins. Once regeneration is completed, a slow rinse flow is passed through the bed to displace the remaining regenerate solution to waste. This is followed by a fast, short rinse to remove the last traces of the regenerate solution from the bed. Once the fast rinse is completed, the column is placed back on-line to continue the softening process.

DESIGN PROCEDURES

The breakthrough curves for an ion exchange column and an adsorption column are similar, and the contacting techniques are almost identical. Consequently, the same procedures used for the design of adsorption columns may be used for ion exchange columns. The scale-up approach by Fornwalt and Hutchins (1966) and the kinetic approach by Thomas (1948) also may be applied to the design of ion exchange columns by the same procedures presented in Chapter 12. A laboratory- or pilot-scale breakthrough curve is required for both procedures. The breakthrough curve for a column shows the solute or ion concentration in the effluent on the y-axis versus the effluent throughput volume on the x-axis. The area above the breakthrough curve represents the amount of solute or ions taken up by the column and is $\int (C_0 - C)dV$ from $V = 0$ to $V =$ the allowable throughput volume under consideration. At the allowable breakthrough volume, V_B, the area above the breakthrough curve is equal to the amount of ions removed by the column. At complete exhaustion, $C = C_0$ and the area above the breakthrough curve is equal to the maximum amount of ions removed by the column. At complete exhaustion, the entire exchange column is in equilibrium with the influent and effluent flows. Also, the ion concentration in the influent is equal to the ion concentration in the effluent.

Another design approach is to determine the meqs or equivalent weights of the ions removed by a test column using the breakthrough curve. The throughput volume under consideration should be the allowable breakthrough volume, V_B, at the allowable breakthrough concentration, C_a. The ratio of the amount of ions removed per unit mass of exchanger is computed. Then, using this ratio, the mass of exchanger required is calculated from the allowable breakthrough volume for the design column and the concentration of the polyvalent metallic ions to be removed from the liquid flow. For this method to be valid, the flow rate used for the test column in terms of bed volumes per hour must be similar to the flow rate of the design column. A design procedure based on mass transfer fundamentals and the unit transfer concept is also available (Michaels, 1952).

SOFTENING AND DEMIN- ERALIZATION

Softening may be achieved by using a strong acid cation exchanger on the sodium cycle (that is, charged with Na^+), as depicted in Figure 13.2. The Ca^{+2}, Mg^{+2}, and other divalent or polyvalent metallic ions are sorbed by the exchanger solid, and Na^+ ions are released into the solution. Since electroneutrality must be maintained, one meq of divalent or polyvalent cations removed causes one meq of sodium ions to be released. Once breakthrough occurs, the bed is regenerated using a strong brine (NaCl) solution.

Demineralization where silica reduction is not required may be accomplished by the flowsheet shown in Figure 13.3. This consists of a strong acid cation exchanger on the hydrogen cycle, a weak base anion exchanger on the hydroxyl cycle, and a carbon dioxide stripping or degasification unit. Once the water passes through the cation exchanger, the cations have been removed and exchanged for hydrogen ions, resulting in an acidic water. The bicarbonates and carbonates present are converted to carbonic acid because of the low pH. The water then passes through a weak base anion exchanger, and the anions, with the exception of silicates, are removed. The water then passes through a carbon dioxide stripper (degasifier), where excess carbon dioxide is removed. The product water will contain some silicate ions since weak base anion exchangers have limited removal for these. If the carbon dioxide content is not objectionable or the feed water has a low alkalinity, carbon dioxide stripping is not required. Regeneration is achieved using a solution of a strong mineral acid, such as sulfuric or hydrochloric for the cation exchanger and a sodium hydroxide solution for the anion exchanger.

Demineralization where silica removal is required may be accomplished

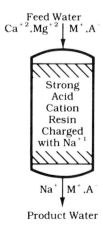

FIGURE 13.2 *Ion Exchange Softening*

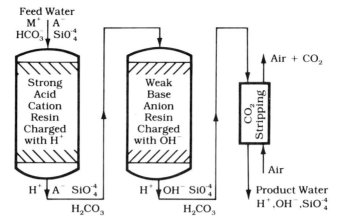

FIGURE 13.3 *Ion Exchange Demineralization*

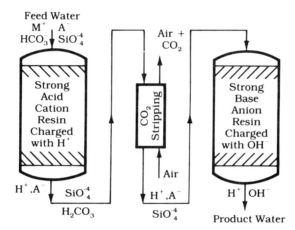

FIGURE 13.4 *Ion Exchange Demineralization*

by the flowsheet shown in Figure 13.4. This consists of a strong acid cation exchanger on the hydrogen cycle, a carbon dioxide stripper or degasifier, and a strong base anion exchanger on the hydroxyl cycle. Once the water passes through the cation exchanger, it goes through the carbon dioxide stripper where excess carbon dioxide is removed. The water then goes through the strong base anion exchanger, which removes all anions, including silicates. If the feed water has a low alkalinity, the carbon dioxide stripper may be omitted. The product water will be essentially pure since the cations and anions have been removed, leaving only the hydrogen and hydroxyl ions, which form water.

Demineralization may also be accomplished by a mixed-resin bed consisting of strong acid cation exchange resins mixed with strong base anion exchange resins, as depicted in Figure 13.5 (a). The processes shown in Figures 13.3 and 13.4 give up to 99% dissolved solids removal; however, a mixed-resin bed will give greater removal. Once breakthrough occurs in the

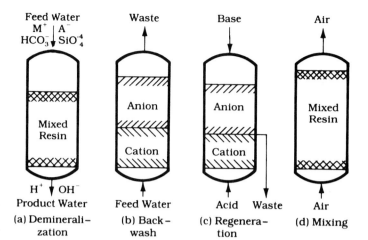

FIGURE 13.5
Mixed-Resin
Demineralization

(a) Demineralization
(b) Backwash
(c) Regeneration
(d) Mixing

mixed bed, it is backwashed to separate the resins. Since the anion resins are less dense than the cation resins, they will move to the top of the bed, as shown in Figure 13.5 (b). Regeneration is accomplished using a strong mineral acid solution and a strong base solution, as illustrated in Figure 13.5 (c). After regeneration, air is introduced into the bottom of the bed to remix the resins, as shown in Figure 13.5 (d). Frequently, mixed-resin beds are used to polish waters demineralized by the processes shown in Figures 13.3 and 13.4.

Other demineralization processes employing more than two exchange columns may be used and are reported in the literature (Betz, 1962; Dorfner, 1971; Helfferich, 1962; Kitchener, 1961; Kunin, 1960; Nordell, 1961; Perry and Chilton, 1973; Weber, 1972).

AMMONIA REMOVAL

Ion exchange for ammonia removal was introduced in Chapter 11. The naturally occurring zeolite, clinoptilolite, appears to be the ion exchange medium of choice for removal of ammonium ion, NH_4^+, in wastewater treatment. Ion exchange using clinoptilolite for ammonia (ammonium ion) removal in wastewater treatment is discussed in several texts and references (Metcalf & Eddy, 1991; EPA, 1975; WPCF, 1983). Clinoptilolite exhibits an ion capture series that is particularly suitable for ammonium ion removal in wastewater treatment. The ion capture series is (WPCF, 1983)

$$K^+ > NH_4^+ > Na^+ > Ca^{+2} > Fe^{+3} > Mg^{+2}$$

Municipal wastewaters typically contain sufficiently small potassium ion concentrations relative to the ammonium ion concentrations such that the ammonium ion readily exchanges in clinoptilolite. The performance of ion exchange beds of 20×50-mesh clinoptilolite varies little between flows of 7.5 to 15 BV/hr. An operational pH range of 4 to 8 results in optimum exchange capacity. Removals of the ammonium ion range from 0.25 to

0.32 meq NH_4^+/gm clinoptilolite for exchange bed depths of 3 ft (0.91 m) to 6 ft (1.83 m). Typical bed depths are 4 to 5 ft (1.2 to 1.5 m). Backwash rates range from 6 to 8 gpm/ft^2 (4.1 to 5.4 ℓ/s-m^2).

Regeneration methods used for the clinoptilolite include high-pH regeneration using either lime, $Ca(OH)_2$, or sodium hydroxide, NaOH; and neutral-pH regeneration using sodium chloride, NaCl. A major disadvantage of the high-pH regeneration method is the precipitation of magnesium hydroxide, $Mg(OH)_2$, and calcium carbonate, $CaCO_3$, which can foul the exchange media, thereby reducing the exchange capacity. Disadvantages of the neutral-pH regeneration method include an increased bed-volume requirement and increased regeneration time. The primary advantage of the high-pH method is the reduced number of bed-volumes required for regeneration. The primary advantage of neutral-pH regeneration is reduced scaling. The methods used for regenerate recovery are gas stripping, which includes air stripping, steam stripping, or closed gas stream stripping followed by gas adsorption; and breakpoint chlorination by electrolytic generation of chlorine. In general, the use of ion exchange for ammonia removal in wastewater treatment is advantageous when both stringent effluent criteria must be met and difficulties are encountered in achieving nitrification. Because the exchange capacity of clinoptilolite can be accurately predicted, the need for pilot tests is minimized. Detailed design data and criteria are contained in *Nutrient Control* (WPCF, 1983) and *Nitrogen Control* (EPA, 1975).

EXAMPLE 13.1 *Ion Exchange in Waste Treatment*

An industrial wastewater with 107 mg/ℓ of Cu^{+2} (3.37 meq/ℓ) is to be treated by an exchange column. The allowable effluent concentration, C_a, is 5% C_0. A breakthrough curve, shown in Figure 13.6, has been obtained from an

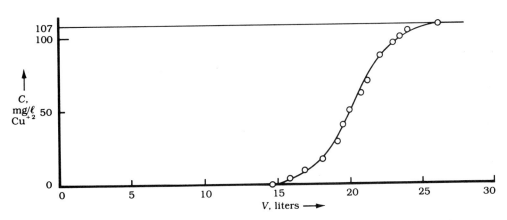

FIGURE 13.6 *Breakthrough Curve for Examples 13.1 and 13.1 SI*

experimental laboratory column on the sodium cycle. Data concerning the column are as follows: inside diameter = 0.5 in., length = 18 in., mass of resin = 41.50 gm on a moist basis (23.24 gm on a dry basis), moisture = 44%, bulk density of resin = 44.69 lb/ft^3 on a moist basis, and liquid flow-rate = 1.0428 ℓ/day. The design column will have a flowrate of 100,000 gal/day, the allowable breakthrough time is 7 days of flow, and the resin depth is approximately twice the column diameter. Using the kinetic approach to column design, determine:

1. The pounds of resin required.
2. The diameter and depth if the diameter is in 6 in. increments.
3. The height of the sorption zone.

SOLUTION In this problem, the design equation to be used is Eq. (13.12). The data for the breakthrough test are given in columns (1) and (2) of Table 13.1.

The plot of $\ln(C_0/C - 1)$ versus V is shown in Figure 13.7. The slope = $k_1 C_0/Q$ or 0.7583 ℓ^{-1}. The value of k_1 = (slope)(Q/C_0) or k_1 = $(0.7583/\ell)$ $(1.0428\,\ell/\text{day})(\ell/3.37\,\text{meq})(1000\,\text{meq/eq})(\text{gal}/3.785\,\ell)$ = 61.994 gal/day-eq. The y-axis intercept, b, equals $\ln 4.540 \times 10^6$ or 15.33. Since $b = k_1 q_0 M/Q$, rearranging gives $q_0 = bQ/k_1 M$ = (15.33)(1.0428 ℓ/day)(day-eq/61.994 gal) × $(\ell/23.24\,\text{gm})(454\,\text{gm/lb})(\text{gal}/3.785\,\ell)$ = 1.3309 eq/lb. The mass of resin may be computed from

$$\ln\left(\frac{C_0}{C} - 1\right) = \frac{k_1 q_0 M}{Q} - \frac{k_1 C_0 V}{Q}$$

or

TABLE 13.1 *Reduced Data from Breakthrough Test*

(1) V (liters)	(2) C (mg/ℓ)	(3) C (meq/ℓ)	(4) C/C_0	(5) C_0/C	(6) $C_0/C - 1$
15.9	4.45	0.14	0.041	24.29	23.29
16.9	9.85	0.31	0.091	10.97	9.97
18.1	17.16	0.54	0.159	6.30	5.30
19.1	27.56	0.88	0.259	3.86	2.86
19.5	40.03	1.26	0.371	2.70	1.70
20.0	49.56	1.56	0.459	2.18	1.18
20.7	62.90	1.98	0.582	1.72	0.72
21.2	68.89	2.20	0.647	1.55	0.55
22.0	86.41	2.72	0.800	1.25	0.25
22.9	94.03	2.96	0.871	1.15	0.15
23.4	98.17	3.09	0.917	1.09	0.09
24.0	102.93	3.24	0.961	1.05	0.05
26.0	107.00	3.37	1.000	1.01	0.01

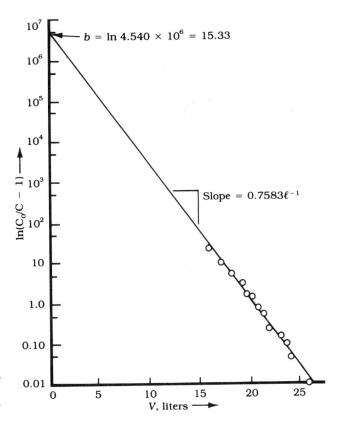

FIGURE 13.7 *Plot for Examples 13.1 and 13.1 SI*

$$\ln\left(\frac{C_0}{0.05\,C_0} - 1\right) = \frac{61.994\,\text{gal}}{\text{day-eq}} \left|\; \frac{1.3309\,\text{eq}}{\text{lb}}\right.$$

$$\times \frac{\text{day}}{100{,}000\,\text{gal}} \left|\; M\; - \;\frac{61.994\,\text{gal}}{\text{day-eq}}\right|\frac{3.37\,\text{meq}}{\ell}$$

$$\times \frac{\text{eq}}{1000\,\text{meq}} \left|\; \frac{3.785\,\ell}{\text{gal}}\right|\frac{\text{day}}{100{,}000\,\text{gal}} \left|\; 700{,}000\,\text{gal}\right.$$

From this, $M = \boxed{10{,}300\,\text{lb dry weight.}}$ Ft^3 = $(10{,}300\,\text{lb})(1/0.56)(\text{ft}^3/44.69\,\text{lb})$ = 411.6 ft^3. Since $(\pi/4)(D^2)(2D)$ = 411.6 ft^3, D = 6.40 ft. Use $D = \boxed{6\,\text{ft-6 in.}}$ and Z = $(411.6\,\text{ft}^3)(4/\pi)(1/6.50\,\text{ft})^2 = \boxed{12.4\,\text{ft.}}$ At 5% C_0 breakthrough, the breakthrough volume = 700,000 gal. Using 95% C_0 and the previous equation gives a throughput volume of 1,444,760 gal. Thus V_T = 1,444,760 gal, V_B = 700,000 gal, and V_z = 1,444,760 – 700,000 = 744,760 gal. Substituting into the equation $Z_s = Z[V_z/(V_T - 0.5V_z)]$ gives

$$Z_s = 12.4\,\text{ft}\left[\frac{744{,}760}{1{,}444{,}760 - 0.5(744{,}760)}\right]$$

$$= \boxed{8.61\,\text{ft}}$$

EXAMPLE 13.1 SI   *Ion Exchange in Waste Treatment*

An industrial wastewater with $107 \, mg/\ell$ of Cu^{+2} $(3.37 \, meq/\ell)$ is to be treated by an exchange column. The allowable effluent concentration, C_a, is 5% C_0. A breakthrough curve, shown in Figure 13.6, has been obtained from an experimental laboratory column on the sodium cycle. Data concerning the column are as follows: inside diameter = 1.3 cm, length = 45.7 cm, mass of resin = 41.50 gm on a moist basis (23.24 gm on a dry basis), moisture = 44%, bulk density of resin = $716.5 \, kg/m^3$ on a moist basis, and liquid flowrate = $1.0428 \, \ell/d$. The design column flowrate will be $378,500 \, \ell/d$, the allowable breakthrough time is 7 days of flow, and the resin depth is approximately twice the column diameter. Using the kinetic approach to column design, determine:

1. The kilograms of resin required.
2. The diameter and depth.
3. The height of the sorption zone.

SOLUTION   In this problem, the design equation to be used is Eq. (13.12). The data for the breakthrough test are given in columns (1) and (2) of Table 13.1.

The plot of $\ln(C_0/C - 1)$ versus V is shown in Figure 13.7. The slope = $k_1 C_0/Q$ or $0.7583 \, \ell^{-1}$. The value of k_1 = (slope)(Q/C_0) or k_1 = $(0.7583/\ell)$ $(1.0428 \, \ell/d)(\ell/3.37 \, meq)(1000 \, meq/eq)$ = $234.6 \, \ell/d\text{-eq}$. The y-axis intercept, b, equals $\ln 4.540 \times 10^6$ or 15.33. Since $b = k_1 q_0 M/Q$, rearranging gives $q_0 = bQ/k_1 M$ = $(15.33)(1.0428 \, \ell/d)(d\text{-eq}/234.6 \, \ell)(\ell/23.24 \, gm) \times (1000 \, gm/kg)$ = $2.932 \, eq/kg$. The mass of resin may be computed from

$$\ln\left(\frac{C_0}{C} - 1\right) = \frac{k_1 q_0 M}{Q} - \frac{k_1 C_0 V}{Q}$$

or

$$\ln\left(\frac{C_0}{0.05 C_0} - 1\right) = \left(\frac{234.6 \, \ell}{d\text{-eq}}\right)\left(\frac{2.932 \, eq}{kg}\right)\left(\frac{d}{378,500 \, \ell}\right)\left(\frac{M \, kg}{}\right)$$

$$- \left(\frac{234.6 \, \ell}{d\text{-eq}}\right)\left(\frac{3.37 \, meq}{\ell}\right)\left(\frac{eq}{1000 \, meq}\right)\left(\frac{d}{378,500 \, \ell}\right)\left(\frac{7 \, d \times 378,500 \, \ell}{d}\right)$$

From this, $M = \boxed{4670 \, kg \text{ dry weight.}}$ Resin volume = $(4670 \, kg)(1/0.56) \times (m^3/716.5 \, kg)$ = $11.6 \, m^3$. Since $(\pi/4)(D^2)(2D)$ = $11.6 \, m^3$, $\boxed{D = 1.95 \, m.}$ Z = $(11.6 \, m^3)(4/\pi)(1/1.95 \, m)^2$ = $\boxed{3.88 \, m.}$ At 5% C_0 break-through, the breakthrough volume = $2.65 \times 10^6 \, \ell$. Using 95% C_0 and the previous equation gives a throughput volume of $5.47 \times 10^6 \, \ell$. Thus $V_T = 5.47 \times 10^6 \, \ell$, $V_B = 2.65 \times 10^6 \, \ell$, and $V_Z = 5.47 \times 10^6 - 2.65 \times 10^6 = 2.82 \times 10^6 \, \ell$. Substituting into the equation $Z_S = Z[V_Z/(V_T - 0.5 V_Z)]$ gives

$$Z_S = 3.88 \, m\left[\frac{2.82 \times 10^6}{5.47 \times 10^6 - 0.5(2.82 \times 10^6)}\right] = \boxed{2.69 \, m}$$

EXAMPLE 13.2 *Ion Removal*

For the test column and breakthrough curve given in Example 13.1, determine the meq of Cu^{+2} ion removed per 100 gm resin on a dry weight basis at the allowable breakthrough volume, V_B, for $C_a = 0.05C_0$. Also, determine the meq of Cu^{+2} ion removed per 100 gm resin on a dry weight basis at complete exhaustion. The dry weight of resin used was 23.24 gm.

SOLUTION

The meq weight of Cu^{+2} is 63.54/2 or 31.77 mg. The area above the breakthrough curve from V = zero to V = the volume under consideration is equal to the ion content removed by the exchanger. Figure 13.8 shows the breakthrough curve and the area above the breakthrough curve out to V_B, the allowable breakthrough volume for $C_a = 0.05C_0$. The area, A_1, is 1735 mg or 1735/31.77, which is equal to 54.61 meq. From the allowable breakthrough volume, V_B, to the volume at exhaustion, the area, A_2, is 437 mg or 437/31.77, which is equal to 13.76 meq. At exhaustion, the total area is 54.61 + 13.76 or 68.37 meq. The copper removed at $C_a = 0.05C_0$ is 54.61 meq/23.24 gm or $\boxed{235 \text{ meq}/100 \text{ gm dry weight.}}$ The copper removed at exhaustion is 68.37 meq/23.24 gm or $\boxed{294 \text{ meq}/100 \text{ gm dry weight.}}$

EXAMPLE 13.3 *Ion Exchange Softening*

A well water is to be softened by split-flow ion exchange with the exchanger on the sodium cycle. The flow is 50 gpm, the hardness = 225 mg/ℓ as $CaCO_3$, the desired hardness is 50 mg/ℓ as $CaCO_3$, and the moisture content of the resin is 45%. A test column has been used in the laboratory to obtain a breakthrough curve. The computed hardness removed by the resin at the allowable breakthrough concentration $C_a = 0.05C_0$ was 282 meq/100 gm

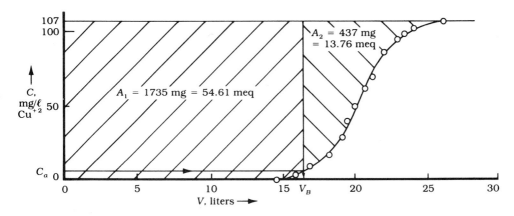

FIGURE 13.8 *Breakthrough Curve for Example 13.2*

resin on a dry weight basis. Determine the pounds of resin required if the allowable breakthrough is 7 days.

SOLUTION   A materials balance on the flow downstream from the exchanger is $Q_e(0) + (Q_b)(225) = (Q_e + Q_b)50$, where Q_e = flow through the exchanger and Q_b = flow bypassing the exchanger. From the balance, $Q_b = 0.28570Q_e$. Since $Q_b + Q_e = 50$ gpm, $0.2857Q_e + Q_e = 50$ gpm or $Q_e = 38.89$ gpm. The hardness of the well water in meq/ℓ is (225 mg/ℓ)(meq/50 mg) or 4.50 meq/ℓ. The hardness removed in 7 days is (38.89 gal/min)(1440 min/day)(7 day) (3.785 ℓ/gal)(4.50 meq/ℓ) = 6,676,930 meq. Thus the amount of resin = (6,676,930 meq)(100 gm/282 meq)(lb/454 gm) = $\boxed{5220 \text{ lb dry weight.}}$ On a moist basis, this is (5220 lb)[1/(1 − 0.45)] = $\boxed{9490 \text{ lb moist weight.}}$

EXAMPLE 13.3 SI   *Ion Exchange Softening*

A well water is to be softened by split-flow ion exchange with the exchanger on the sodium cycle. The flow is 3.2 ℓ/s, the hardness is 225 mg/ℓ as $CaCO_3$, the desired hardness is 50 mg/ℓ as $CaCO_3$, and the moisture content of the resin is 45%. A test column has been used in the laboratory to obtain a breakthrough curve. The computed hardness removed by the resin at the allowable breakthrough concentration $C_a = 0.05C_0$ was 282 meq/100 gm resin on a dry weight basis. Determine the kilograms of resin required if the allowable breakthrough is 7 days.

SOLUTION   A materials balance on the flow downstream from the exchanger is $Q_e(0) + Q_b(225) = (Q_e + Q_b)50$, where Q_e = flow through the exchanger and Q_b = flow bypassing the exchanger. From the balance, $Q_b = 0.2857Q_e$. Since $Q_b + Q_e = 3.2$ ℓ/s, $0.2857Q_e + Q_e = 3.2$ ℓ/s or $Q_e = 2.49$ ℓ/s. The hardness of the well water in meq/ℓ is (225 mg/ℓ)(meq/50 mg) or 4.50 meq/ℓ. The hardness removed in 7 days is (2.49 ℓ/s)(60 s/min)(1440 min/d) × (7 d) (4.50 meq/ℓ) = 6,776,784 meq. Thus the amount of resin = (6,776,784 meq)(100 g/282 meq)(kg/1000 g) = $\boxed{2400 \text{ kg dry weight.}}$ On a moist basis, this is (2400 kg)[1/(1 − 0.45)] = $\boxed{4360 \text{ kg moist weight.}}$

DESIGN CONSIDER-ATIONS   Zeolite or resin beds used for softening usually have an exchanger bed 2.0 to 8.5 ft (0.61 to 2.59 m) in depth and operate at 1 to 8 gpm/ft^2 (0.68 to 5.4 ℓ/s-m^2). Since ion exchangers are regenerated without removing the exchanger solid from the vessel, the height-to-diameter ratio is not as critical as for carbon adsorption columns. Usually, it is from 1.5:1 to 3:1. The column height should be sufficient to allow for expansion of the bed during the backwash. On backwashing, zeolites expand to about 25% of the bed depth, whereas synthetic resins expand to about 75 to 100% of the bed depth. The maximum column height is usually 12 ft (3.66 m). If a column height greater than 12 ft (3.66 m) is required, two columns in series can be used. Prefabricated steel cylinders may be obtained in diameters up to 12 ft (3.66 m) in 3-in. (7.6-cm) intervals; but 6-in. (15.2-cm) intervals are more common; diameters range from 2 ft-6 in. to 12 ft (0.76 to 3.66 m). At an allowable breakthrough of 5% C_0, usually 65 to 85% of the ion exchange capacity will be used for removing hardness. Synthetic cation exchange resins with strong acid exchange sites have ion exchange capacities from about 350 to 520 meq/100 gm dry resin. Moist densities are from 43.0 lb/ft^3 (689 kg/m^3) to 54.0 lb/ft^3 (866 kg/m^3), and the moisture content may be from 40 to 60%. Dry resins may swell up to 155% of their original volume upon becoming moist. The ion exchange capacity, density, and moisture content depend upon the particular resin under consideration. Regeneration of synthetic resins on the sodium cycle is done with a 5 to 25% brine solution. Usually a 5 to 10% solution is used. In regeneration, the brine solution is passed through the bed in either a downward or an upward direction. The unit liquid flow rate is 1 to 2 gpm/ft^2 (0.68 to 1.36 ℓ/s-m^2). After regeneration, a slow rinse flow is passed through the bed in the direction of the softening flow to rinse the brine from the bed. After the slow rinse, a short, fast rinse is used to flush any remaining brine from the bed. Usually 30 to 150 gal (4000 to 20,000 ℓ) of rinse water are required per ft^3 (m^3) resin. Salt requirements are from 5 to 20 lb salt per ft^3 (80 to 320 kg/m^3) resin bed, 5 to 10 lb per ft^3 (80 to 160 kg/m^3) being typical.

Strong acid cation exchangers on the hydrogen cycle are regenerated using a H_2SO_4 or HCl solution. If H_2SO_4 is used, a 2% solution should first be passed through the exchanger, followed by a 10% solution. The lower concentration used first is to avoid precipitation of $CaSO_4$ in the exchanger bed. The H_2SO_4 required is from 6 to 12 lb per ft^3 (96 to 192 kg/m^3) resin. If HCl is used, a 15% solution should be employed, and the HCl required is from 5 to 10 lb per ft^3 (80 to 160 kg/m^3) resin. Strong base exchangers on the hydroxyl cycle are regenerated with a 2 to 10% NaOH solution. Regeneration requires from 3 to 6 lb NaOH per ft^3 (48 to 96 kg/m^3) resin.

Test columns should be regenerated in the same manner as a design column in order for the breakthrough curve to be representative. That is, the resin should be regenerated with the regenerate ion, and then a test performed to complete exhaustion. Next the resin should be regenerated, and the test should be rerun to obtain the design breakthrough curve.

REFERENCES

American Water Works Association (AWWA) Research Foundation. 1973. *Desalting Techniques for Water Supply Quality Improvement*. Report for the Office of Saline Water, Department of the Interior.

Betz Laboratories. 1962. *Betz Handbook of Industrial Water Conditioning*. 6th ed. Trevose, Pa.: Betz Laboratories, Inc.

Conway, R. A., and Ross, R. D. 1980. *Handbook of Industrial Waste Disposal*. New York: Van Nostrand Reinhold.

Culp, R. L., and Culp, G. L. 1978. *Handook of*

Advanced Wastewater Treatment. 2nd ed. New York: Van Nostrand Reinhold.

Dorfner, K. 1971. *Ion Exchangers, Properties and Applications.* 3rd ed. Ann Arbor, Mich.: Ann Arbor Science Publishers.

Dow Chemical Co. 1964. DOWEX: *Ion Exchange.* Midland, Mich.: Dow Chemical Co.

Eckenfelder, W. W., Jr. 1980. *Principles of Water Quality Management.* Boston: CBI Publishing.

Environmental Protection Agency (EPA). 1971. Advanced Waste Treatment and Water Reuse Symposium. Sessions 1–5, Dallas, Tex.

Environmental Protection Agency (EPA). 1975. *Nitrogen Control.* EPA Process Design Manual. Washington, D.C.

Fornwalt, H. J., and Hutchins, R. A. 1966. Purifying Liquids with Activated Carbon. *Chem. Eng.* (11 April):179; (9 May):155.

Helfferich, F. 1962. *Ion Exchange.* New York: McGraw-Hill.

Kitchener, J. A. 1961. *Ion Exchange Resins.* London: Methuen & Co.; and New York: Wiley.

Kunin, R. 1960. *Elements of Ion Exchange.* New York: Reinhold.

Metcalf & Eddy, Inc. 1979. *Wastewater Engineering: Treatment, Disposal and Reuse.* 2nd ed. New York: McGraw-Hill.

Metcalf & Eddy, Inc. 1991. *Wastewater Engineering: Treatment, Disposal and Reuse.* 3rd ed. New York: McGraw-Hill.

Michaels, A. S. 1952. Simplified Method of Interpreting Kinetic Data in Fixed-Bed Ion Exchange. *Ind. and Eng. Chem.* 44, no. 8:1922–1930.

Montgomery, J. M., Inc. 1985. *Water Treatment Principles and Design.* New York: Wiley.

Nordell, E. 1961. *Water Treatment for Industrial and Other Uses.* 2nd ed. New York: Reinhold.

Noyes Data Corporation (NDC). 1978. *Nitrogen Control and Phosphorus Removal in Sewage Treatment.* Edited by D. J. DeRenzo. Park Ridge, N.J.: NDC.

Perry, R. H., and Chilton, C. H. 1973. *Chemical Engineers' Handbook.* 5th ed. New York: McGraw-Hill.

Ramalho, R. S. 1977. *Introduction to Wastewater Treatment Processes.* New York: Academic Press.

Reynolds, T. D., and Westervelt, R. 1969. Ion Exchange as a Tertiary Treatment. Paper presented at the American Society of Civil Engineers annual meeting and national meeting on Water Resources Engineering, New Orleans, La.

Rich, L. G. 1963. *Unit Processes of Sanitary Engineering.* New York: Wiley.

Sanks, R. L. 1978. *Water Treatment Plant Design.* Ann Arbor, Mich.: Ann Arbor Science Publishers.

Schroeder, E. D. 1977. *Water and Wastewater Treatment.* New York: McGraw-Hill.

Sundstrom, D. W., and Klei, H. E. 1979. *Wastewater Treatment.* Englewood Cliffs, N.J.: Prentice-Hall.

Thomas, H. C. 1948. Chromatography: A Problem in Kinetics. *Annals of the New York Academy of Science* 49:161.

Water Pollution Control Federation (WPCF). 1983. *Nutrient Control.* Manual of Practice FD-7, Facilities Design. Washington, D.C.

Weber, W. J., Jr. 1972. *Physicochemical Processes for Water Quality Control.* New York: Wiley-Interscience.

PROBLEMS

13.1 A water is to be softened for an industrial water supply. The Ca^{+2} content is $107 \, mg/\ell$, and the Mg^{+2} content is $18 \, mg/\ell$. The allowable breakthrough, C_a, is 5% C_0, where C_0 is the hardness of the untreated water. A pilot column 4 in. in diameter and containing 3 ft of resin has been operated at a flowrate of 0.59 gal/hr to obtain breakthrough data. The following throughput volumes in gallons at the respective C_e/C_0 values (where C_e is the effluent hardness) were obtained: 620, 0.03; 768, 0.07; 820, 0.16; 860, 0.24; 895, 0.36; 922, 0.47; 940, 0.52; 980, 0.66; 1008, 0.74; 1066, 0.87; and 1220, 0.99. The resin has a specific weight of 44 lb/ft³, and

the design flowrate is 150,000 gal/day. Using the scale-up design approach, determine:

a. The exchange column volume and pounds of resin.

b. The rate at which the resins are expended per hour.

c. The design column life in hours and days.

d. The diameter and height if the resin height is about twice the diameter and diameters are in 6 in. increments up to a maximum of 12 ft.

13.2 A water is to be softened for an industrial water supply. The Ca^{+2} content is $107 \, mg/\ell$ and the Mg^{+2} content is $18 \, mg/\ell$. The allowable breakthrough, C_a, is 5% C_0, where C_0 is the hardness of the untreated water. A pilot column 10.2 cm in diameter and containing 91.4 cm of resin has been operated at a flowrate of 2.23 ℓ/h to obtain breakthrough data. The following throughput volumes in liters at the respective C_e/C_0 values (where C_e is the effluent hardness) were obtained: 2347, 0.03; 2907, 0.07; 3104, 0.16; 3255, 0.24; 3388, 0.36; 3490, 0.47; 3558, 0.52; 3709, 0.66; 3815, 0.74; 4035, 0.87; and 4618, 0.99. The resin has a specific weight of 705.4 kg/m^3, and the design flowrate is 568 m^3/d. Using the scale-up design approach, determine:

a. The exchange column volume and kilograms of resin.

b. The rate at which the resins are expended per hour.

c. The design column life in hours and days.

d. The diameter and height if the resin height is twice the diameter.

13.3 A countercurrent ion exchange column removes copper ions (Cu^{+2}) from a wastewater stream. The copper ion concentration is 350 mg/ℓ and there is 99.9% removal. The wastewater has a flow upward at 4.5 gpm/ft^2 and the ion exchange resin flows downward. The ion exchange capacity of the resin is 4.50 meq/gm and the resin flow leaving the column is saturated with copper ions (that is, essentially all the exchange sites are occupied by copper ions). Determine:

a. The resin mass flow rate, lb/hr-ft^2.

b. The gallons of wastewater treated by 1 lb of resin.

c. The diameter of the column in feet if the flowrate is 100 gpm and diameters are in 6 in. increments.

13.4 A countercurrent ion exchange column removes copper ions (Cu^{+2}) from a wastewater stream. The copper ion concentration is 350 mg/ℓ and there is 99.9% removal. The wastewater has a flow upward at 3.0 ℓ/s-m^2 and the ion exchange resin flows downward. The ion exchange capacity of the resin is 4.50 meq/gm and the resin flow leaving the column is saturated with copper ions (that is, essentially all the exchange sites are occupied by copper ions). Determine:

a. The resin mass flow rate, kg/h-m^2.

b. The liters of wastewater treated by 1 kg of resin.

c. The diameter of the column in meters if the flow rate is 6.3 ℓ/s.

13.5 An industrial wastewater containing 205 mg/ℓ Cd^{+2} is to be treated by an ion exchange column,

and the allowable effluent concentration is 5% C_0. A breakthrough curve has been obtained using a small laboratory column on the sodium cycle. Data concerning the column are as follows: inside diameter = 0.5 in., mass of resin = 45.28 gm on a moist basis, moisture content = 48%, bulk density of the resin = 738 g/ℓ on a moist basis, and liquid flowrate = 1.120 ℓ/day. The following throughput volumes in liters at the respective effluent cadmium concentrations in mg/ℓ were obtained: 5.0, 0.0; 10.0, 0.0; 11.0, 14.06; 11.5, 26.43; 12.0, 42.17; 12.5, 68.60; 13.0, 104.59; 13.5, 141.70; 14.0, 163.07; 15.0, 196.81; 15.5, 202.43; 16.0, 205.00; and 16.5, 205.00. The design column will have a flowrate of 60,000 gal/day, the allowable breakthrough time is 7 days of flow, and the resin depth is approximately twice the column diameter. Using the kinetic approach to design, determine:

a. The pounds of resin required.

b. The column diameter and depth if diameters are in 6 in. increments.

c. The height of the adsorption zone.

13.6 An industrial wastewater containing 205 mg/ℓ Cd^{+2} is to be treated by an ion exchange column, and the allowable effluent concentration is 5% C_0. A breakthrough curve has been obtained using a small laboratory column on the sodium cycle. Data concerning the column are as follows: inside diameter = 1.3 cm, mass of resin = 45.28 gm on a moist basis, moisture content = 48%, bulk density of the resin = 738 gm/ℓ on a moist basis, and liquid flowrate = 1.120 ℓ/d. The following throughput volumes in liters at the respective effluent cadmium concentrations in mg/ℓ were obtained: 5.0, 0.0; 10.0; 0.0; 11.0, 14.07; 11.5, 26.43; 12.0, 42.17; 12.5, 68.60; 13.0, 104.59; 13.5, 141.70; 14.0, 163.07; 15.0, 196.81; 15.5, 202.43; 16.0, 205.00; and 16.5, 205.00. The design column will have a flowrate of 2.63 ℓ/s, the allowable breakthrough time is 7 days of flow, and the resin depth is approximately twice the column diameter. Using the kinetic approach to design, determine:

a. The kilograms of resin required.

b. The column diameter and depth.

c. The height of the adsorption zone.

13.7 The wastewater from a metal plating industry is to be treated by ion exchange on the sodium cycle. Two columns in series are to be used, and the lead column will reach 100% exhaustion before it is taken off-line. The flow is 82,000 gal/day and the concentrations of the ions to be removed are 15 mg/ℓ zinc (Zn^{+2}), 20 mg/ℓ cadmium (Cd^{+2}), 25 mg/ℓ copper (Cu^{+2}), and 20 mg/ℓ nickel (Ni^{+2}). The cation exchange resin to be used has a capacity of 375 meq/100 gm on a moist basis and 42% moisture. If the allowable breakthrough time is 7 days, determine the pounds of dry resin required and also the pounds of moist resin required.

13.8 The wastewater from a metal plating industry is to be treated by ion exchangers on the sodium cycle. Two columns in series are to be used, and the lead column will reach 100 percent exhaustion before it is taken off-line. The flow is 310 m³/d and the concentrations of the ions to be removed are 15 mg/ℓ zinc (Zn^{+2}), 20 mg/ℓ cadmium (Cd^{+2}), 25 mg/ℓ copper (Cu^{+2}), and 20 mg/ℓ nickel (Ni^{+2}). The cation exchange resin to be used has a capacity of 375 meq/100 gm on a moist basis and 42% moisture. If the allowable breakthrough time is 7 days, determine the kilograms of dry resin

required and also the kilograms of moist resin required.

13.9 A well water is to be softened by split-flow ion exchange with the exchange resin on the sodium cycle. The flow is 40 gpm, the hardness is 195 mg/ℓ as $CaCO_3$, the desired hardness is 50 mg/ℓ as $CaCO_3$, the resin removes 296 meq/100 gm on a dry basis at an allowable breakthrough of 5% C_0, and the moisture content is 45%. The allowable breakthrough time is 7 days. Determine the pounds of resin required on both a moist and a dry basis.

13.10 A well water is to be softened by split-flow ion exchange with the exchange resin on the sodium cycle. The flow is 2.53 ℓ/s, the hardness is 195 mg/ℓ as $CaCO_3$, the desired hardness is 50 mg/ℓ as $CaCO_3$, the resin removes 296 meq/100 gm on a dry basis at an allowable breakthrough of 5% C_0, and the moisture content is 45%. The allowable breakthrough time is 7 days. Determine the kilograms of resin required on both a moist and a dry basis.

14

MEMBRANE PROCESSES

In the membrane processes, separation of a substance from a solution containing numerous substances is possible by the use of a selectively permeable membrane. The solution containing the components is separated from the solvent liquid by the membrane, which must be differently permeable to the components. The main membrane processes are (1) dialysis, (2) electrodialysis, and (3) reverse osmosis. Figure 14.1 shows the effective separation ranges of the membrane processes and allows a comparison to other separation techniques. Each membrane process requires a driving force to cause mass transfer of the solute. The driving force is a difference in concentration for dialysis, in electric potential for electrodialysis, and in pressure for reverse osmosis. The main difficulty with the membrane processes is that the rate of mass transfer per unit area of the membrane (that is, mass flux) is relatively small.

DIALYSIS

Although dialysis is not used to an appreciable extent in environmental engineering, a knowledge of dialysis is necessary to understand the principles involved in electrodialysis and reverse osmosis.

Theory

Dialysis consists of separating solutes of different ionic or molecular size in a solution by means of a selectively permeable membrane. The driving force for dialysis is the difference in the solute concentration across the membrane. In a batch dialysis cell, the solution to be dialyzed is separated from the solvent solution by the semipermeable membrane. The smaller ions or molecules pass through the membrane, whereas the larger ions and molecules do not pass through because of the relative size of the membrane pore openings. The passage of the smaller solutes is from the solution side to the solvent side because this is in the direction of a drop in concentration of the solutes.

The mass transfer of solute passing through the membrane in a batch dialysis cell at a given time is

$$M = KA\Delta C \tag{14.1}$$

where

M = mass transferred per unit time — for example, gm/hr

K = mass transfer coefficient — for example, $gm/(hr\text{-}cm^2)(gm/cm^3)$

A = membrane area — for example, cm^2

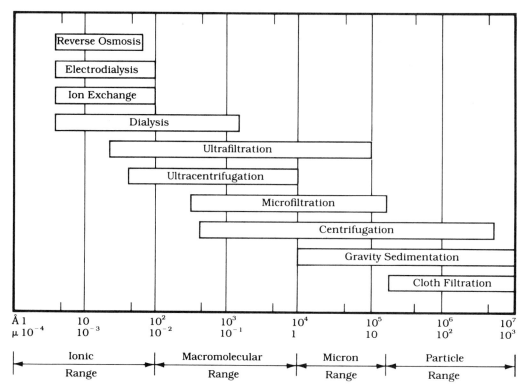

FIGURE 14.1 *Effective Ranges of Some Separation Techniques*

Adapted from "Membrane Separation Processes" by R. E. Lacey in *Chemical Engineering*, 4 September 1972. Copyright © 1972 by McGraw-Hill, Inc. Reprinted by permission.

ΔC = difference in concentration of the solute passing through the membrane — for example, gm/cm^3

Since a batch dialysis cell represents an unsteady-state condition, the term ΔC decreases with an increase in time. In continuous-flow dialysis cells, the flow of solvent and solution is countercurrent, and in application, numerous cells are pressed together in a stack and connected in parallel. The purpose of using a stack is to make the membrane area/stack volume as high as possible, thus resulting in a compact unit.

Application   Sodium hydroxide has been recovered from textile mill wastewaters by a continuous-flow dialysis stack (Nemerow and Steel, 1955). The flowrate was from 420 to 475 gal per day (1590 to 1800 ℓ/d), and the recovery of sodium hydroxide was from 87.3 to 94.6%. Although good results were obtained, dialysis is limited to small flows because the mass transfer coefficient, K, is relatively small.

**ELECTRO-
DIALYSIS**

In environmental engineering, electrodialysis is presently used and has potential for more uses. In this process, the driving force for mass transfer is an electromotive force.

Theory

If dialysis is to be used to separate inorganic electrolytes from a solution, the presence of an electromotive force across the selectively permeable membrane will result in an increased rate of ion transfer. In this manner, the salt concentration of the treated solution is decreased. Since electrodialysis demineralizes, it has been used to produce fresh water from brackish water and seawater. Also, it has been used to demineralize effluents in tertiary treatment.

An electrodialysis stack consisting of three cells is shown in Figure 14.2. When a direct current is applied to the electrodes, all positively charged ions (cations) tend to migrate toward the cathode. Also, all negatively charged ions (anions) tend to migrate toward the anode. The cations can pass through the cation-permeable membranes (designated C in Figure 14.2), but they are obstructed from passing through the anion-permeable membranes (designated A). In a similar manner, the anions can pass through the anion-permeable membranes but are obstructed from passing through the cation-permeable membranes. As shown in Figure 14.2, alternate compartments are formed in which the ionic concentration is greater or less than the concentration in the feed solution. As a result, the flows from the stack consist of the product water, which has a low electrolyte concentration, and the brine solution, which has a high electrolyte concentration. The cells in

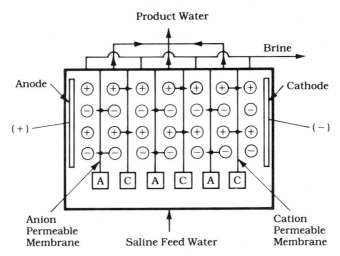

FIGURE 14.2 *Schematic of a Continuous-Flow Electrodialysis Stack*

Adapted from "Membrane Separation Processes" by R. E. Lacey in *Chemical Engineering*, 4 September 1972, Copyright © 1972 by McGraw-Hill, Inc. Reprinted by permission.

the stack are connected for parallel flow. Gases are frequently formed at the electrodes, such as hydrogen at the cathode and oxygen and chlorine at the anode.

An electrodialysis membrane is a porous, sheetlike, structural matrix made of synthetic ion exchange resin. The matrix of a cation-permeable membrane has a negative charge because of the ionization of its cation exchange sites. Exchangeable cations within the pore spaces balance the negative charge of the matrix. When a current flows, cations enter the pores and pass through the membrane since the electrical migratory forces acting upon the cations are greater than the attractive forces between the cations and the cation-permeable membrane. Since the matrix is negatively charged, it repels the negatively charged anions. Anion-permeable membranes are made in a similar manner and allow anions to pass through but repel cations. Because of the nature of the membranes, some consider electro-dialysis a specialized application of ion exchange.

The current required for an electrodialysis stack may be calculated using Faraday's laws of electrolysis. One Faraday (F) of electricity (96,500 ampere-seconds or coulombs) will cause one gram equivalent weight of a substance to migrate from one electrode to another. The number of gram equivalent weights removed per unit time in an electrolytic cell is

$$\text{Number of gram equivalent weights/unit time} = QNE_r \qquad \textbf{(14.2)}$$

where

Q = solution flowrate, liters/sec

N = normality of the solution — that is, the number of gram equivalent weights/liter

E_r = electrolyte removal as a fraction

Since one Faraday (96,500 ampere-seconds) is required per gram equivalent weight removed and a cell has a given current effciency, the current required for a single cell can be derived from Eq. (14.2) as follows:

$$I = \frac{FQNE_r}{E_c} \qquad \textbf{(14.3)}$$

where

I = current in amperes

F = Faraday's constant (96,500 ampere-seconds per gram equivalent weight removed)

E_c = current efficiency as a fraction

Since the same electric current is passed through all the cells in a stack, it is effectively used n times, where n is the number of cells. Thus the current for the stack is given by a modification of Eq. (14.3), which is

$$I = \frac{FQNE_r}{nE_c} \tag{14.4}$$

where n = number of cells.

The electrodialysis stacks used in desalting usually have from 100 to 250 cells (200 to 500 membranes). The current efficiency (E_c) for a particular electrodialysis stack and feed water must be determined experimentally. Usually, the current efficiency as a fraction is 0.90 or more. The electrolyte removal as a fraction (E_r) is usually from 0.25 to 0.50.

The capacity of an electrodialysis cell to pass an electric current is related to the current density and to the normality of the feed water. The current density is defined as the current divided by the membrane area and is usually expressed as ma/cm^2. The normality of the feed water is, by definition, the number of gram equivalent weights per liter of solution. The parameter relating these variables is the current density/normality ratio. The ratio may vary from 400 to 700 when the current density is expressed as ma/cm^2. If the current density/normality ratio is too high, it will cause regions of low ionic concentrations near the membranes, resulting in polarization. This is to be avoided since high electrical resistance occurs, thus causing high electrical consumption.

The resistance of an electrodialysis stack treating a particular feed water must be determined experimentally. Once the electrical resistance, R, and the current flow, I, are known, the voltage required, E, is given by Ohm's law, which is $E = RI$, where E = volts, R = ohms, and I = amperes. The power required is given by $P = EI = RI^2$, where P = power in watts.

EXAMPLE 14.1 AND 14.1 SI

Electrodialysis

An electrodialysis stack having 200 cells is to be used to partially demineralize 90,000 gal per day (341,000 ℓ/d or 3.943 ℓ/s) of advanced treated wastewater so that it can be used by an industry. The salt content is 4000 mg/ℓ and the cation or anion content is 0.066 gram equivalent weights per liter. Pilot-scale studies using a multicellular stack have been made. It was found that the current efficiency, E_c, was 90%, the efficiency of salt removal, E_r, was 50%, the resistance was 4.5 ohms, and the current density/normality ratio was 400. The stack is to have 200 cells. Determine:

1. The current, I, required.
2. The area of the membranes.
3. The power required.

SOLUTION

The current, I, is given by Eq. (14.4):

$$I = \frac{FQNE_r}{nE_c}$$

For the USCS solution,

$$I = \frac{96,500 \,\text{amp-sec}}{\text{gm-eq-wt}} \left| \frac{90,000 \,\text{gal}}{\text{day}} \right| \frac{0.066 \,\text{gm-eq-wt}}{\ell}$$

$$\times \frac{3.785 \,\ell}{\text{gal}} \left| \frac{\text{day}}{86,400 \,\text{sec}} \right| 0.50 \left| \frac{}{200} \right| 0.90$$

$$= \boxed{69.8 \,\text{amps}}$$

For the SI solution,

$$I = \left(\frac{96,500 \,\text{a-s}}{\text{gm-eq-wt}}\right)\left(\frac{3.943 \,\ell}{\text{s}}\right)\left(\frac{0.066 \,\text{gm-eq-wt}}{\ell}\right)\left(\frac{0.5}{1}\right)\left(\frac{1}{200}\right)\left(\frac{1}{0.90}\right)$$

$$= \boxed{69.8 \,\text{amps}}$$

Current density/normality ratio = 400. Since normality equals the number of gram equivalent weights per liter, normality = 0.066. The current density is therefore equal to (400)(normality) = (400)(0.066) = 26.4 ma/cm^2. Thus the membrane area is

$$\text{Area} = (69.8 \,\text{amps})(\text{cm}^2/26.4 \,\text{ma})(1000 \,\text{ma/amp})$$

$$= \boxed{2640 \,\text{cm}^2}$$

If the membranes are square in shape, the size is 51.4 cm × 51.4 cm. The power required is

$$\text{Power} = RI^2 = (4.5 \,\text{ohms})(69.8 \,\text{amp})^2$$

$$= \boxed{21,900 \,\text{watts}}$$

Application   The technical feasibility of demineralizing sea water and brackish waters by electrodialysis has been demonstrated by pilot installations and, in the case of brackish waters, by full-scale plants (AWWA, 1973; Browning, 1970; Collier and Fulton, 1967). Electrodialysis stacks used for demineralization usually consist of several hundred cells (two membranes per cell). From Eq. (14.4), it can be seen that the electrical energy requirements are directly proportional to the amount of salt removed. Consequently, the electrical costs are governed by both the dissolved solids content of the feed water and the desired dissolved solids content of the product water. Also, energy consumption increases with desposition of scale upon the membranes, and the amount of scale is related to the various dissolved salts in the feed water. Because of the scaling problem and high electrical consumption, electro-dialysis is not well suited to the demineralization of sea water. However, electrodialysis is adaptable to the demineralization of brackish waters, and numerous desalination plants using electrodialysis have been constructed throughout the world for brackish water desalting. Studies have shown that electrodialysis is particularly adaptable for the deionization of brackish waters of $5000\,mg/\ell$ dissolved solids or less to produce a product water with dissolved solids of about $500\,mg/\ell$. Membrane replacement costs and electrical power costs are about 40% of the total cost.

Pilot plant studies have shown that electrodialysis is a technically feasible method of demineralizing secondary effluents; however, scaling problems and organic fouling of the membranes will occur (Brunner, 1967; Culp, 1975; EPA, 1971; HEW, 1962, 1965; FWQA, 1970). It has been found that 25 to 50% of the dissolved salts can be removed in a single pass through an electrodialysis stack. In order to obtain greater salt removal, several electrodialysis stacks may be placed in series. Scaling and organic fouling may be reduced by proper pretreatments such as coagulation, settling, filtration, and activated carbon adsorption. Scaling may also be reduced by the addition of a small amount of acid to the feed stream, whereas organic fouling may be reduced by cleaning the membranes with an enzyme detergent solution. Scaling and membrane fouling must be controlled in order for electrodialysis to be economically feasible. Since reverse osmosis desalts at a comparable cost and gives other pollutant removals, it has a greater potential in tertiary treatment than electrodialysis.

REVERSE OSMOSIS

In environmental engineering, reverse osmosis is presently used and has potential for more uses. For this process, the driving force for mass transfer is a hydrostatic pressure difference.

Theory

Reverse osmosis consists of separating a solvent, such as water, from a saline solution by the use of a semipermeable membrane and a hydrostatic pressure. Consider the batch dialysis cell shown in Figure 14.3 (a). The solvent flow is from the solvent side through the semipermeable membrane to the saline side because this is in the direction of a drop in solvent concentration. This transfer of solvent water through a semipermeable

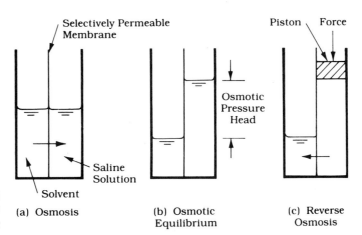

FIGURE 14.3
Osmosis and Reverse Osmosis

(a) Osmosis

(b) Osmotic Equilibrium

(c) Reverse Osmosis

membrane from the solvent side to the saline side is referred to as **osmosis**. Eventually the system will reach equilibrium, as shown in Figure 14.3 (b), where the hydrostatic pressure is referred to as the **osmotic pressure**. If a force is applied to a piston to produce a pressure greater than the osmotic pressure, as shown in Figure 14.3 (c), there will be a transfer of the solvent in the reverse direction. This mass transfer of a solvent using a semipermeable membrane and a hydrostatic pressure is referred to as **reverse osmosis**. Reverse osmosis is a useful separation method since it permits the passage of water and rejects the passage of molecules other than water and of ions. The reverse osmosis cell in Figure 14.3 (c) is a batch operation, whereas the cell in Figure 14.4 is a continuous-flow operation.

Reverse osmosis is similar to ultrafiltration and microfiltration because all three techniques utilize semipermeable membranes and hydrostatic pressures to force the solvent through the membranes. In ultrafiltration and microfiltration, the separation is due mainly to a filtering action, not a reverse osmotic action; also, the size of the substances that are rejected by the ultrafiltration and microfiltration membranes is larger, as shown in Figure 14.1.

The osmotic pressure of solutions of electrolytes may be determined by

$$\pi = \phi v \frac{n}{V} RT \tag{14.5}$$

where

π = osmotic pressure

ϕ = osmotic coefficient

v = number of ions formed from one molecule of electrolyte

n = number of moles of electrolyte

V = volume of solvent

R = universal gas constant

T = absolute temperature

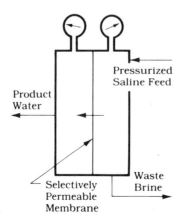

FIGURE 14.4
A Continuous-Flow Reverse Osmosis Unit

The osmotic coefficient, ϕ, depends upon the nature of the substance and on its concentration. The osmotic pressure for sea water, which has 35,000 mg/ℓ dissolved solids, is 397 psi (2740 kPa) at 25°C (77°F). Since a salt solution of 35,000 mg/ℓ dissolved solids has an osmotic pressure of 397 psi (2740 kPa), it may be assumed for practical purposes that an increase of 1000 mg/ℓ salt concentration results in an increase of 11.3 psi (78 kPa) in osmotic pressure.

The schematic diagram of a continuous-flow reverse osmosis unit is shown in Figure 14.4. The saline feed is pressurized so that the differential pressure between the two compartments is greater than the difference in osmotic pressure. Although the transfer of solvent will begin when the pressure difference exceeds the osmotic pressure difference, the rate of solvent mass transfer increases as the pressure difference increases. In practice, the pressures used for the feed stream are from 250 to 800 psi (1720 to 5520 kPa). The design pressure depends mainly upon the osmotic pressure differential between the feed and product solution, the characteristics of the membrane, and the temperature.

The main design parameters for a reverse osmosis unit are the production per unit area of membrane and the product water quality (Kaup, 1973). The production is measured by the flux of water through the membrane — for example, gal/day-ft^2 (ℓ/d-m^2). The flux of a membrane depends on the membrane characteristics (that is, thickness and porosity) and the system conditions (that is, temperature, differential pressure across the membrane, salt concentration, and flow velocity of the water through the membrane). In practice, the water flux is simply related to the pressures by (Kaup, 1973)

$$F_w = K(\Delta p - \Delta \pi) \tag{14.6}$$

where

F_w = water flux — that is, gal/day-ft^2 (ℓ/d-m^2)

K = mass transfer coefficient for a unit area of membrane, gal/day-ft^2-psi (ℓ/d-m^2-kPa)

Δp = pressure difference between the feed and product water

$\Delta \pi$ = osmotic pressure difference between the feed and product water or the feed osmotic pressure minus the product water osmotic pressure

The membrane flux value furnished by a manufacturer is usually for 25°C (77°F). As temperature varies, the diffusivity and viscosity vary also, and this in turn causes the flux to vary. Membrane area corrections (A_T/A_{25}) due to the respective temperatures are as follows: 10°C, 1.58; 15°C, 1.34; 20°C, 1.15; 25°C, 1.00; and 30°C, 0.84. The term A_T/A_{25} is the ratio of the areas required for temperatures of T (°C) and 25°C (Kaup, 1973).

The flux value will gradually decrease during the lifetime of a membrane because of a slow densification of the membrane structure, which results in a decrease in the pore passages. This gradual flux reduction occurs in all membranes, and it is permanent. The membrane must be replaced when the flux has reached the minimum acceptable value. Manufacturers can provide initial flux values and the expected rate of flux reduction for various operating pressures. The relationship of flux to duration will plot a straight line on log-log paper. Usually, the life span of a membrane is from two months to two years.

The most common membrane used is made of cellulose acetate. These membranes are relatively tight; that is, they have low water permeability and can reject over 99% of the salts. However, the water flux is very low (about 10 gal/day-ft^2 or 400 ℓ/d-m^2). The main types of mounting hardware employed in reverse osmosis equipment modules are classified as: (1) tubular, (2) hollow fiber, and (3) spiral wound. Figures 14.5 and 14.6 show schematic drawings of the tubular and the hollow-fiber types of mountings. In the tubular mounting, the feed water enters the inside of the tubular membrane, which is encased by a porous support tube that gives the required tensile strength, as shown in Figure 14.5 (a). The portion of the water that passes through the membrane leaves as product water, while the remaining water leaves the end of the tube as brine. In the hollow-fiber mounting, a portion of the feed water passes from the outside of the tube to the inside, as shown in Figure 14.6 (a) and leaves as the product water, while the remaining water leaves as brine. In the spiral-wound mounting, a porous hollow tube is spirally wrapped with a porous sheet for the feed flow and

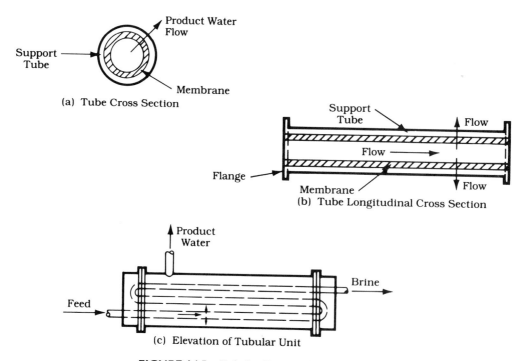

FIGURE 14.5 *Tubular Reverse Osmosis Equipment*

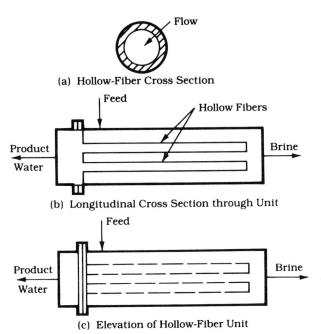

FIGURE 14.6
Hollow-Fiber Reverse Osmosis Equipment

(a) Hollow-Fiber Cross Section

(b) Longitudinal Cross Section through Unit

(c) Elevation of Hollow-Fiber Unit

with a membrane sheet and a porous sheet for the product water flow to give a spiral sandwich-type wrapping. The spiral module is encased in a pressure vessel, and the feed flow through the porous sheet is in an axial direction to the porous tube. As the feed flow passes through the porous sheet, a portion of the flow passes through the membrane into the porous sheet for the product water. From there the product water flows spirally to the porous center tube and is discharged from the conduit in the tube. The brine is discharged from the downstream end of the porous sheet for the feed flow. Some of the characteristics of the various mountings are given in Table 14.1 (SI values: Table 14.2).

TABLE 14.1 *Characteristics of Different Reverse Osmosis Modules, USCS Values*

MODULE TYPE	SURFACE AREA (ft² MEMBRANE AREA PER ft³ OF EQUIPMENT)	WATER FLUX (gal/day-ft²)	WATER PRODUCED (gal/day PER ft³ OF EQUIPMENT)
Tubular	20	32	640
Hollow fiber (cellulose acetate)	2500	10	25,000
Hollow fiber (nylon)	5400	1	5,400
Spiral wound	250	32	8,000

Environmental Protection Agency, Advanced Waste Treatment and Water Reuse Symposium. Sessions 1–5, Dallas, Tex., 1971.

TABLE 14.2 *Characteristics of Different Reverse Osmosis Modules, SI Values*

MODULE TYPE	SURFACE AREA (m² MEMBRANE AREA PER m³ OF EQUIPMENT)	WATER FLUX (ℓ/d-m²)	WATER PRODUCED (ℓ/s PER m³ OF EQUIPMENT)
Tubular	66	1,300	1
Hollow fiber (cellulose acetate)	8,200	400	39
Hollow fiber (nylon)	17,700	41	8
Spiral wound	820	1,300	12

Adapted from Environmental Protection Agency, Advanced Waste Treatment and Water Reuse Symposium. Sessions 1–5, Dallas, Tex., 1971.

From these data, it appears that hollow fibers are best; however, in practice their hydraulic inadequacies are serious.

EXAMPLE 14.2 *Reverse Osmosis*

A reverse osmosis unit is to demineralize 200,000 gal of tertiary treated effluent per day. Pertinent data are as follows: mass transfer coefficient = 0.0350 gal/(day-ft²)(psi) at 25°C, pressure difference between the feed and product water = 350 psi, osmotic pressure difference between the feed and product water = 45 psi, lowest operating temperature = 10°C, and $A_{10°}$ = 1.58 $A_{25°}$. Determine the membrane area required.

SOLUTION The water flux is given by Eq. (14.6):

$$F_w = [0.0350 \text{ gal/(day-ft}^2)(\text{psi})](350 \text{ psi} - 45 \text{ psi})$$
$$= 10.675 \text{ gal/(day-ft}^2) \text{ at } 25°C$$

The area at 10°C is given by

$$A = (200,000 \text{ gal/day})[(\text{day-ft}^2)/10.675 \text{ gal}](1.58)$$
$$= \boxed{29,600 \text{ ft}^2}$$

EXAMPLE 14.2 SI *Reverse Osmosis*

A reverse osmosis unit is to demineralize 760,000 ℓ/d of tertiary treated effluent. Pertinent data are as follows: mass transfer coefficient = 0.2068 ℓ/(d-m²)(kPa) at 25°C, pressure difference between the feed and product water = 2400 kPa, osmotic pressure difference between the feed and product water = 310 kPa, lowest operating temperature = 10°C, and $A_{10°}$ = 1.58 $A_{25°}$. Determine the membrane area required.

SOLUTION The water flux is given by Eq. (14.6):

$$F_w = [0.2068 \, \ell/(\text{d-m}^2)(\text{kPa})](2400 \, \text{kPa} - 310 \, \text{kPa})$$
$$= 432.21 \, \ell/(\text{d-m}^2) \text{ at } 25°\text{C}$$

The area at 10°C is given by

$$A = (760,000 \, \ell/\text{d})[(\text{d-m}^2)/(432.21 \, \ell)](1.58)$$
$$= \boxed{2780 \, \text{m}^2}$$

Application The technical feasibility of demineralizing sea water and brackish waters by reverse osmosis has been demonstrated by pilot plant installations (AWWA, 1973; Browning, 1970; Collier and Fulton, 1967).

Pilot plant work in the United States has yielded a product water of 500 mg/ℓ dissolved solids from sea water and a product water of 250 mg/ℓ from a brackish water with 4500 mg/ℓ dissolved solids. Operating pressures up to 1500 psi (10,300 kPa) for sea water and up to 750 psi (5200 kPa) for the brackish water were used in the pilot plant work. From pilot plant work it was found that the membranes have a short life and membrane replacement would represent about half the cost of desalting seawater. Because of the energy requirements for reverse osmosis and the short membrane life, reverse osmosis is best suited to the demineralization of brackish waters, not sea water. It has been shown that reverse osmosis is particularly well suited to desalting a feed water having a dissolved solids content from 3000 to 10,000 mg/ℓ. Membrane replacement expense is the largest single cost item, and this amounts to about 32% of the total cost.

Reverse osmosis field units are used by the U.S. armed forces (the Army and the Marine Corps) for water purification (U.S. Army, 1987; Starks, 1994; Thomas, 1994). Three models of the field units, which are termed Reverse Osmosis Water Purification Units or ROWPUs, can produce potable water at rates of 10, 50, and 100 gpm (38, 190, and 380 ℓ/m), respectively. The ROWPUs employ coagulation and multimedia filtration for initial solids removal. Coagulation is accomplished by addition of a cationic polymer. Filtration is accomplished in two stages: (1) filtration through a medium consisting of gravel, coarse garnet, fine garnet, silica sand, and anthracite; and (2) filtration through cartridge filters composed of spiral-type filter tube elements. After filtration, water is treated by reverse osmosis with maximum operating pressures of 960 psi (6620 kPa) for salt water and 500 psi (3448 kPa) for fresh water. The ROWPUs have been used to treat a salt water containing 35,000 mg/ℓ total dissolved solids and produce water containing approximately 1000 mg/ℓ total dissolved solids.

Pilot plant studies have shown that reverse osmosis is a technically feasible method for demineralizing secondary effluents; however, some organic fouling of the membranes will occur. The modified cellulose acetate membranes have been found to be the most applicable. These membranes will reject from 90 to 99% of the salts and 90% of the organic materials. Organic fouling of the membranes is usually not serious enough to require prior treatments such as activated carbon adsorption. Periodic cleaning of the membranes with an enzyme detergent solution usually controls the fouling. Reverse osmosis has a large potential for wastewater treatment since a high degree of removal for all contaminants can be accomplished by the unit operation. Pilot plants have shown that typical removals from secondary effluents treated by reverse osmosis units operated at 450 psi (3100 kPa) and 8 gal/day-ft^2 (2.8 ℓ/d-m^2) are as follows: total organic carbon, 90%; total dissolved solids, 93%; phosphate, 94%; organic nitrogen, 86%; ammonia nitrogen, 85%; and nitrate nitrogen, 65% (Culp, 1975).

REFERENCES

AWWA Water Desalting and Reuse Committee. 1989. Committee Report: Membrane Desalting Technologies. *Jour. AWWA* 81, no. 11:30.

American Water Works Association (AWWA) Research Foundation. 1973. *Desalting Techniques for Water Supply Quality Improvement*. Report for Office of Saline Water, U.S. Department of the Interior.

Baier, J. W.; Lykins, B. W., Jr.; Fronk, C. A.; and Kramer, S. J. 1987. Using Reverse Osmosis to Remove Agricultural Chemicals from Groundwater. *Jour. AWWA* 79, no. 8:55.

Browning, J. E. 1970. Zeroing in on Desalting. *Chem. Eng.* (23 March):64.

Brunner, C. A. 1967. Pilot-Plant Experiences in Demineralization of Secondary Effluents Using Electrodialysis. *Jour. WPCF* 39, no. 10, part 2:R1.

Castellan, G. W. 1964. *Physical Chemistry*. Reading, Mass.: Addison-Wesley.

Collier, E. P., and Fulton, J. F. 1967. Water Desalination. Department of Energy, Mines and Resources, Ottawa, Canada.

Conlon, W. J., and McClellan, S. A. 1989. Membrane Softening: A Treatment Process Comes of Age. *Jour. AWWA* 81, no. 11:47.

Conway, R. A., and Ross, R. D. 1980. *Handbook of Industrial Waste Disposal*. New York: Van Nostrand Reinhold.

Culp, G. L. 1975. *Treatment Processes for Wastewater Reclamation for Groundwater Recharge*. Preliminary Report for the State of California, Department of Water Resources.

Culp G. L., and Hamann, C. L. 1974. Advanced Waste Treatment Process Selection: Parts 1, 2 and 3. *Public Works* (March, April, and May).

Culp, R. L., and Culp, G. L. 1978. *Handbook of Advanced Wastewater Treatment*. 2nd ed. New York: Van Nostrand Reinhold.

Eckenfelder, W. W., Jr. 1980. *Principles of Water Quality Management*. Boston: CBI Publishing.

El-Rehaili, A. M. 1991. Reverse Osmosis Applications in Saudi Arabia. *Jour. AWWA* 83, no. 6:72.

Environmental Protection Agency (EPA). 1971. Advanced Waste Treatment and Water Reuse Symposium. Sessions 1–5, Dallas, Tex.

Halper, B. M., and Olie, J. 1972. 1.25 MGD Electrodialysis Plant in Israel. *Jour. AWWA* 64, no. 11:735.

Katz, W. E. 1961. Preliminary Evaluation of Electric Membrane Processes for Chemical Processing Applications. In *Separation Processes in Practice*, edited by R. F. Chapman. New York: Reinhold.

Kaup, E. C. 1973. Design Factors in Reverse Osmosis. *Chem. Eng.* (2 April):48.

King, J. C. 1971. *Separation Processes*. New York: McGraw-Hill.

Lacy, R. E. 1972. Membrane Separation Processes. *Chem. Eng.* (4 September): 56.

Lo, T., and Sudak, R. G. 1992. Removing Color from a Groundwater Source. *Jour. AWWA* 83, no. 1:79.

Mason, E. A., and Juda, W. 1959. Applications of Ion Exchange Membranes in Electrodialysis. *Chem. Eng. Progr. Symposium Series* 55, no. 24:155.

Mason, E. A., and Kirkham, T. A. 1959. Design of Electrodialysis Equipment. *Chem. Eng. Progr. Symposium Series* 55, no. 24:173.

Metcalf & Eddy, Inc. 1979. *Wastewater Engineering:*

Treatment, Disposal and Reuse. 2nd ed. New York: McGraw-Hill.

Metcalf & Eddy, Inc. 1991. *Wastewater Engineering: Treatment, Disposal and Reuse*. 3rd ed. New York: McGraw-Hill.

Montgomery, J. M., Inc. 1985. *Water Treatment Principles and Design*. New York: Wiley.

Nemerow, N. C., and Steel, W. R. 1955. *Proceedings of the 11th Annual Purdue Industrial Waste Conference*.

Noyes Data Corporation (NDC). 1978. *Nitrogen Control and Phosphorus Removal in Sewage Treatment*. Edited by D. J. DeRenzo. Park Ridge, N.J.: NDC.

Perry, R. H., and Chilton, C. H. 1973. *Chemical Engineers Handbook*. 5th ed. New York: McGraw-Hill.

Ramalho, R. S. 1977. *Introduction to Wastewater Treatment Processes*. New York: Academic Press.

Rich, L. G. 1963. *Unit Processes of Sanitary Engineering*. New York: Wiley.

Robinson, R. A., and Stokes, R. H. 1959. *Electrolyte Solutions*. London: Butterworth.

Sanks, R. L. 1978. *Water Treatment Plant Design*. Ann Arbor, Mich.: Ann Arbor Science Publishers.

Schroeder, E. D. 1977. *Water and Wastewater Treatment*. New York: McGraw-Hill.

Sourirajan, S. 1970. *Reverse Osmosis*. New York: Academic Press.

Sourirajan, S., and Matsuura, T., eds. 1985. *Reverse Osmosis and Ultrafiltration*. American Chemical Society. Washington, D.C.

Starks, M. L. 1994. ROWPUs on the Beach. *Quartermaster Professional Bulletin*, Spring:20. U.S. Army Quartermaster Center and School. Fort Lee, Va.

Sundstrom, D. W., and Klei, H. E. 1979. *Wastewater Treatment*. Englewood Cliffs, N.J.: Prentice-Hall.

Thomas, D. M. 1994. ROWPU Operations: The Cold Facts. *Quartermaster Professional Bulletin*, Spring:13. U.S. Army Quartermaster Center and School. Fort Lee, Va.

U.S. Army, and U.S. Marine Corps. 1987. *Operator's Manual, Water Purification Unit, Reverse Osmosis, TM 5-4610-215-10*. Headquarters, Department of the Army; and Headquarters, U.S. Marine Corps. Washington, D.C.

U.S. Department of Health, Education and Welfare (HEW). 1962. *Advanced Waste Treatment Research-1, Summary Report, June 1960 to Dec. 1961*.

U.S. Department of Health, Education and Welfare (HEW). 1965. *Advanced Waste Treatment Research, Summary Report, Jan. 1962 to June 1964*, AWTR-14.

U.S. Department of Interior (FWQA). 1970. Advanced Waste Treatment Seminar. Sessions 1–4, San Francisco, Calif.

Water Pollution Control Federation (WPCF). 1983. *Nutrient Control*. Manual of Practice FD-7, Facilities Design. Washington, D.C.

Weber, W. J., Jr. 1972. *Physicochemical Processes for Water Quality Control*. New York: Wiley-Interscience.

PROBLEMS

14.1 An electrodialysis stack is to be used to partially demineralize 100,000 gal/day of advanced treated wastewater so that it can be used for reuse by an industry. The membranes are 30 × 30 in., and the salt concentration is 4500 mg/ℓ. The cations and anions are 0.0742 eq/ℓ. The product water must not have more than 2250 mg/ℓ salt content. Pertinent data are as follows: stack resistance = 4.5 ohms, current efficiency = 90%, ratio of maximum current density to normality = 400 (with current density as ma/cm^2), power costs = 2$\frac{1}{2}$ ¢/kw-hr, brine density = 67 lb/ft^3, and brine = 12% salt. Determine:

a. The removal efficiency, number of membranes, power consumption (watts), and power costs per 1000 gal.

b. The product water and brine flows, gal/day.

14.2 An electrodialysis stack is to be used to partially demineralize 380 m^3/d of advanced treated wastewater so that it can be used for reuse by an industry. The membranes are 76 × 76 cm, and the salt concentration is 4500 mg/ℓ. The cations and anions are 0.0742 eq/ℓ. The product water must not have more than 2250 mg/ℓ salt content. Pertinent data are as follows: stack resistance = 4.5 ohms, current efficiency = 90%, ratio of maximum current density to normality = 400 (with current density as ma/cm^2), power costs = 2$\frac{1}{2}$ ¢/kwh, brine density = 1074 kg/m^3, and brine = 12% salt. Determine:

a. The removal efficiency, number of membranes, power consumption (watts), and power costs per 1000 ℓ.

b. The product water and brine flows, m^3/d.

14.3 A reverse osmosis unit is to demineralize 100,000 gal of tertiary treated effluent per day. Pertinent data are as follows: mass transfer coefficient = 0.030 gal/(day-ft^2)(psi), pressure difference between the feed and product water = 380 psi, osmotic pressure difference between the feed and product water = 45 psi, lowest operating temperature = 10°C, and membrane area per unit volume of equipment = 2500 ft^2/ft^3. Determine:

a. The membrane area required.

b. The space required for the equipment, ft^3.

14.4 A reverse osmosis unit is to demineralize 760 m^3/d of tertiary treated effluent. Pertinent data are as follows: mass transfer coefficient = 0.207 ℓ/(d-m^2)(kPa), pressure difference between the feed and product water = 2400 kPa, osmotic pressure difference between the feed and product water = 310 kPa, lowest operating temperature = 10°C, and membrane area per unit volume of equipment = 2500 m^2/m^3. Determine:

a. The membrane area required.

b. The space required for the equipment, m^3.

15

ACTIVATED SLUDGE

The **activated sludge** process utilizes a fluidized, mixed growth of micro-organisms under aerobic conditions to use the organic materials in the wastewater as substrates, thus removing them by microbial respiration and synthesis. The main units of the system, as shown in Figure 15.1 (a), consist of a biological reactor basin with its oxygen supply (the aeration tank), a solid-liquid separator (the final clarifier), and the recycle sludge pumps.

The feed wastewater flow, Q, mixes with the recycled activated sludge flow, R, immediately prior to entering the biological reactor or immediately after entering. The activated sludge-wastewater mixture is termed the **mixed liquor**, and the **mixed liquor suspended solids** (MLSS) usually range from 2000 to 4000 mg/ℓ by dry weight. Upon entering the reactor, the activated sludge rapidly adsorbs the suspended organic solids in the wastewater, this period lasting from about 20 to 45 minutes. After adsorption, the adsorbed organic solids are solubilized and oxidized by biological oxidation as the mixed liquor moves through the aeration tank. The soluble organic substances, on the other hand, are usually sorbed (that is, both adsorbed and absorbed) at the greatest rate at the upstream end of the tank. The rate of sorption gradually decreases as the mixed liquor passes through the tank. The sorbed soluble organic materials are oxidized by biological oxidation with a reaction time usually less than that required for the adsorbed suspended organic substances. The oxygen supply for the aeration tank is usually furnished by diffused compressed air, as shown in Figure 15.1 (b), or by mechanical surface aeration as shown in Figure 15.1 (c); however, pure oxygen has been used in some instances. Aeration by diffused compressed air or mechanical means has a dual purpose, because it must supply the required oxygen for the aerobic bio-oxidation and provide sufficient mixing for adequate contact between the activated sludge and the organic substances in the wastewater. When pure oxygen is used as the oxygen source, it must be supplemented with mechanical mixing devices to furnish the required mixing.

At the downstream end of the reactor, the adsorbed and absorbed organic materials have been bio-oxidized and the mixed liquor flows to the solid-liquid separator (the final clarifier). There, the active biological solids settle to the bottom, as shown in Figure 15.1 (d), and the separated treated wastewater spills over the peripheral weirs into the effluent channels. Usually, effluent disinfection is required for municipal effluents prior to discharge into the receiving body of water; however, many industrial effluents do not require disinfection because there are no pathogens present.

411

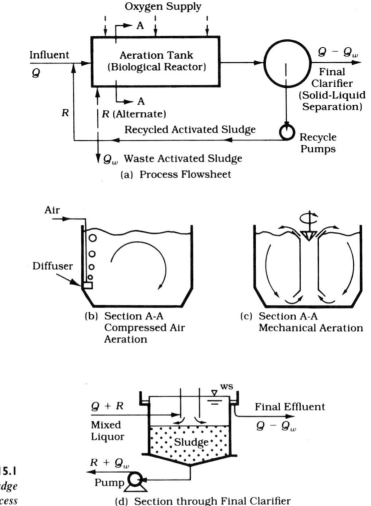

FIGURE 15.1
Activated Sludge Process

The activated sludge from the bottom of the final clarifier is pumped by recycle pumps, and the recycled activated sludge flow mixes with the incoming feed wastewater flow as previously described and as shown in Figure 15.1 (a). Figure 15.2 shows a rectangular reactor basin or aeration tank with diffused compressed air, and Figure 15.3 shows a square reactor basin or aeration tank (completely mixed) with mechanical aeration.

The desired mixed liquor suspended solids (MLSS) concentration in the reactor basin is maintained at a constant level by recycling a fixed amount of the settled activated sludge from the solid-liquid separator. The recycle ratio, R/Q, is dependent upon the desired MLSS concentration and the concentration of the settled activated sludge in the recycle flow as measured by the sludge density index (SDI). In this control test, a volume of mixed

FIGURE 15.2 *Rectangular Reactor Basin with Diffused Compressed Air*

liquor is taken from the downstream end of the reactor basin, and 1 liter of the mixture is placed in a 1-liter graduate cylinder. At the same time, the suspended solids concentration in the mixed liquor is determined. After the sludge has settled in the cylinder for 30 min, the volume occupied by the settled sludge is read from the graduations. The concentration of the settled sludge in mg/ℓ is then computed; it represents the sludge density index (SDI). This test approximates the settling that occurs in the final clarifier. If the sludge density index is 10,000 mg/ℓ and the desired MLSS is 2500 mg/ℓ, a material balance for the activated sludge at the junction of the influent and recycle lines is

$$Q(0) + R(10{,}000) = (Q + R)(2500)$$

and, from this,

$$R/Q = 2500/7500 = \tfrac{1}{3} \quad \text{or} \quad 33.3\%$$

Thus the recycled sludge flowrate, R, must be 33.3% of the incoming flowrate, Q, in order to maintain the desired mixed liquor suspended solids at a constant concentration.

FIGURE 15.3   *Square Reactor Basin (Completely Mixed) with Mechanical Aeration*

The reciprocal of the sludge density index, after appropriate unit conversions are made, is the sludge volume index (SVI). This is the volume in mℓ occupied by 1 gram of settled activated sludge, and it is a measure of the settling characteristics of the sludge. In a properly operating diffused air activated sludge treatment plant, the SVI is usually from 50 to 150 mℓ/gm.

Because in biological oxidation the substrate is used for respiration and for synthesis of new microbial cells, the net cell production (the waste activated sludge) must be removed from the system, as shown in Figure 15.1 (a), in order to maintain a constant mixed liquor suspended solids concentration in the reactor. Usually, the waste activated sludge flowrate amounts to 1 to 6 percent of the incoming feed wastewater flowrate.

The term *activated* stems from the sorptive properties of the biological solids. At the downstream end of the reactor, the activated sludge is in a low substrate environment and has utilized its sorbed organic substances; as a result, it has a relatively high sorption capacity for suspended and dissolved organic material. In the upstream region of the reactor, the sludge will have used most of its sorption capacity and will not be reactivated until it has biologically oxidized the sorbed organic material.

A simplified biochemical equation for the utilization of organic matter as a substrate for respiration and cell synthesis in the activated sludge process is

$$\text{Organic matter} + O_2 \xrightarrow[\text{microbes}]{\text{Aerobic}} \text{New cells} + \text{Energy for cells} + CO_2 + H_2O + \text{Other end products}$$

(15.1)

Some of the other end products are NH_4^+, NO_2^-, NO_3^-, and PO_4^{-3}. The empirical equation that has usually been found to represent activated sludge is $C_5H_7O_2N$ (McKinney, 1962), which has a molecular weight of 113. The net mass of cells produced daily represents the mass of cells that must be disposed of daily as waste activated sludge; it is equal to the total cell mass synthesized minus the cell mass that is endogenously decayed. The most common organic materials in municipal wastewaters not containing appreciable amounts of industrial wastes are carbohydrates, fats, proteins, urea, soaps, detergents, and their degradation or breakdown products. Carbohydrates and fats contain the elements carbon, hydrogen, and oxygen; proteins, in addition to these elements, contain nitrogen, sulfur, and phosphorus. Urea consists of carbon, hydrogen, oxygen, and nitrogen. Soaps consist mainly of carbon, hydrogen, and oxygen; detergents, in addition to these elements, contain phosphorus. Also there may be trace amounts of other substances, such as pesticides, herbicides, and other agricultural chemicals that enter by infiltration. The organic substances and compounds present in industrial wastewaters vary greatly from one type of industry to another. Also, there may be variation within an industry, such as the petrochemical industry. The organic materials remaining after biological treatment (that is, the nonbiodegradable fraction) include very slowly degraded and bioresistant substances.

In the design of an activated sludge process, the engineer must determine: (1) The volume of the reactor basin or basins (V in ft^3 or m^3), the number of reactor basins, and the geometric dimensions of each basin. Small plants have only one basin, whereas large plants have two or more. (2) The sludge production per day (X_w) as mass dry weight solids per day (*i.e.*, lb/day or kg/day). This is required for the design of the sludge digestion system. (3) The oxygen required per day (O_r) as mass oxygen per day (*i.e.*, lb/day or kg/day). This is required for the design of the aeration system. (4) The final clarifier design. Small plants have only one final clarifier, whereas large plants have two or more. The final clarifier serves as the solid-liquid separator to recover the biomass and to clarify the effluent before it leaves the plant. The reactor basin volume is usually determined from (1) kinetic relationships, (2) space loading relationships, or (3) empirical relationships, although using kinetic relationships is the most recent trend.

MICRO-ORGANISMS

The mixed culture (a growth of two or more species) present in activated sludge is a dynamic system; the number of species and the particular species and their populations depend upon the specific wastewater being treated and the environmental conditions in the reactor-clarifier system. The microorganisms include bacteria (both single and multicellular), protozoa, fungi, rotifers, and sometimes nematodes. The principal organisms involved in the bio-oxidation of organic substances in wastewaters are the single-celled bacteria.

Most of the single-celled bacteria used in wastewater treatment are soil microorganisms, and very few are of enteric origin. The majority of the bacterial species, with the exception of some groups such as the nitrifying bacteria, are saprophytic heterotrophs because they require nonliving, preformed organic materials as substrates. The nitrifying bacteria, which convert the ammonium ion to nitrite and the nitrite ion to nitrate, are autotrophs because they use carbon dioxide as their carbon source instead of preformed organic substances. The microbes present consist of both aerobes and facultative anaerobes since they use free molecular oxygen in their respiration process. The principal bacterial genera found in activated sludge when treating municipal wastewaters are: *Achromobacter*, *Arthrobacter*, *Cytophaga*, *Flavobacterium*, *Alkaligenes*, *Pseudomonas*, *Vibrio*, *Aeromonas*, *Bacillus*, *Zoogloea*, *Nitrosomonas*, and *Nitrobacter* (Lighthart and Loew, 1972; McKinney, 1962a). The cells of the first four genera are the most numerous (Lighthart and Loew, 1972). In addition to the nitrifying bacteria, other specialized groups, such as some sulfur and iron bacteria, are present, although in small numbers. The enteric bacteria rapidly die off in the reactor basin because they cannot compete with the other microbes in the existing environment. Although some individual microbial cells are present in activated sludge, the majority are present as zoogloeal biomass particles, which consist of mixed species of cells embedded in masses of polysaccharide gums from the slime layers of living and lysed bacterial cells. The zoogloeal particles, frequently termed **floc**, are desirable because they have appreciable sorptive properties and settle quickly, as shown in Figure 2.6 (a). The filamentous organisms, such as the bacterium *Sphaerotilis natans* and most fungi, are usually not numerous and are not desired because, in large numbers, they can create a sludge with poor settling characteristics, as shown in Figure 2.6 (b). The protozoa do not use the organic substances in the wastewater themselves but instead feed on the bacterial population. The protozoa found include the stalked and free-swimming Ciliata and, to a lesser extent, the Suctoria. Rotifers are usually not numerous but are found in activated sludges that have undergone an appreciable aeration time. They feed on activated sludge fragments that are too large for protozoa. Nematodes are not numerous; however, they use organic materials not readily oxidized by other microorganisms.

The active biological solids in activated sludge are frequently assumed to be represented by the mixed liquor suspended solids (MLSS); however, the mixed liquor volatile suspended solids (MLVSS) are a more accurate representation.

The **growth phases** of microorganisms are primarily related to the number of viable cells present and the amount of substrate or limiting nutrient present, in addition to other environmental factors. Many of the concepts involved in the continuous growth of microbes, as in the activated sludge process, are illustrated by growth relationships of batch cultures.

Consider an aerated laboratory vessel containing a substrate and inoculated with an acclimated mixed culture of microbes (one developed to use the particular substrate). Figure 15.4 shows the cells and substrate concentrations versus time after inoculation for an initial substrate to microbe ratio of 680/495 or 1.374 mg COD/mg activated sludge. The rate of substrate conversion at any time, t, is equal to the slope of a tangent to the substrate curve at that time. A lag phase, based on the number of cells, exists because the maximum substrate utilization rate does not occur at time zero, but instead is at about 1.5 hours. The log growth phase ends at about 2.5 hours, because the inflection point on the cell curve is at that time and, also, the rate of substrate utilization begins to decrease. The declining growth phase exists from about 2.5 to 12 hours, because at 12 hours the cell mass concentration has reached a maximum value. Beyond that time the endogenous decay rate predominates, the endogenous phase exists, and the mass of microorganisms has begun to decrease.

Figure 15.5 shows the cell and substrate concentrations versus time for the same substrate but with a much larger inoculum. The initial substrate to

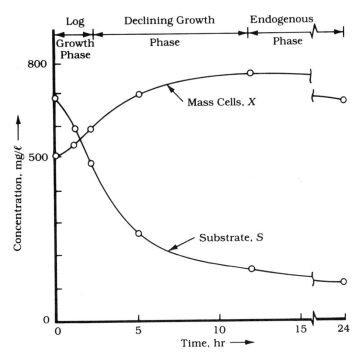

FIGURE 15.4 *Batch Activated Sludge Data: Initial S/X Ratio = 680/495 = 1.374 mg COD/mg cells*

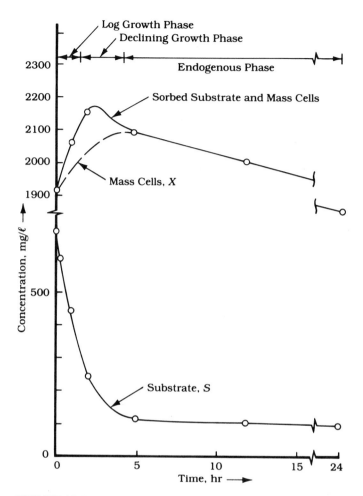

FIGURE 15.5 *Batch Activated Sludge Data: Initial S/X Ratio = 680/1920 = 0.354 mg COD/mg cells*

microorganism ratio is 680/1920 or 0.354 mg COD/mg activated sludge. Since the maximum substrate utilization rate is at time zero, there is no lag phase; this is due to the extremely large inoculum. The log growth phase ends at approximately 1.5 hours, and it can be seen from the graph that the cell growth rate of the microorganisms starts decreasing at that time. The declining growth phase exists from about 1.5 to 4.5 hours, and after 4.5 hours, the endogenous phase exists and the cell mass decreases with respect to time. Also depicted in the graph is the mass of substrate sorbed by the activated sludge from time zero to about 5 hours.

If the biological reactor in Figure 15.1 is a long, narrow tank, a quantity of the MLSS will pass through the reactor with a travel time equal to the theoretical detention time. Curves similar to those shown in Figure 15.5 will represent the cell-substrate relationship for the biological reactor. The maximum substrate removal is at the upstream end of the reactor, and the rate is equal to the slope of a tangent to the substrate curve at time zero. If the reactor has a reaction time greater than about 4.5 hours, the growth phases that will be present are the log growth, declining growth, and endogenous phases. The log growth phase will exist in the upstream reach in the reactor, the declining growth phase will be in the middle reach, and the endogenous phase will be present in the downstream reach. Although endogenous degradation is the principal means by which changes in cell mass occur in the endogenous phase, it is continuous through all phases; the endogenous decay rate is merely concealed by the larger log growth and declining growth rates.

BIOCHEMICAL KINETICS

In an earlier chapter, titled Biological Concepts, it was shown that the rate of substrate utilization by microbes can be represented by the Michaelis-Menten equation or by the specific rate of substrate utilization equation. If a batch activated sludge reactor is inoculated and the values of the cell mass and substrate mass concentrations versus time of reaction are obtained, the data may be analyzed by either of the two approaches to give the kinetic constants required for design. The inoculum must come from an acclimated parent culture developed in a continuous-flow reactor in order for meaningful data to be obtained.

The specific rate of substrate utilization equation is based on two limiting cases of the Michaelis-Menten equation. In the first case, the substrate concentration is relatively large and the Michaelis-Menten equation is reduced to a pseudo–zero-order reaction. In the second case, the substrate concentration is relatively small and the Michaelis-Menten equation is reduced to a pseudo–first-order reaction. If the specific rate of substrate utilization is a pseudo–zero-order reaction, the rate equation is (Eckenfelder, 1966, 1989)

$$-\frac{1}{X}\frac{dS}{dt} = K \tag{15.2}$$

where

$(1/X)(dS/dt)$ = specific rate of substrate utilization, mass/(mass microbes) × (time)

dS/dt = rate of substrate utilization, mass/(volume)(time)

K = rate constant, time^{-1}

The negative sign indicates that the substrate is decreasing with respect to time. Rearranging Eq. (15.2) for integration gives

$$\int_{S_0}^{S_t} dS = -K\bar{X}\int_0^t dt \tag{15.3}$$

where

K = rate constant, time^{-1}

$\bar{X}$ = average cell mass concentration during the biochemical reaction — that is, $\bar{X} = (X_0 + X_t)\frac{1}{2}$, where X_0 and X_t are the cell mass concentrations at the respective times $t = 0$ and $t = t$, mass/volume

S_t = substrate concentration at time t, mass/volume

S_0 = substrate concentration at time $t = 0$, mass/volume

Integration of Eq. (15.3) yields

$$S_t - S_0 = -K\bar{X}t \tag{15.4}$$

or

$$S_t = S_0 - K\bar{X}t \tag{15.5}$$

Equation (15.5) is of the form $y = mx + b$; thus plotting S_t on the y-axis versus $\bar{X}t$ on the x-axis on arithmetical paper will result in a straight line if the reaction is pseudo-zero order, and the slope will equal $-K$.

If the specific rate of substrate utilization is a pseudo–first-order reaction, the rate equation is (Eckenfelder, 1966, 1989)

$$-\frac{1}{X}\frac{dS}{dt} = KS \tag{15.6}$$

where K is the rate constant, volume/(mass microbes)(time). Rearranging for integration gives

$$\int_{S_0}^{S_t}\frac{dS}{S} = -K\bar{X}\int_0^t dt \tag{15.7}$$

Integration results in

$$\ln S\bigg]_{S_0}^{S_t} = -K\bar{X}t \tag{15.8}$$

Equation (15.8) may be simplified to

$$\ln S_t = \ln S_0 - K\bar{X}t \tag{15.9}$$

Equation (15.9) is of the form $y = mx + b$; thus plotting S_t on the y-axis versus $\bar{X}t$ on the x-axis on semilog paper will result in a straight line if the reaction is pseudo-first order, and the line slope will equal $-K$.

Since in wastewaters there are numerous substrates present, the rate constant, K, for the kinetic equations and the constants for the Michaelis-Menten equation represent overall average values.

The parameter used for the substrate in the previous kinetic equations must be in terms of biodegradable material since the substrate represents the food substance or substances for the microbes. It could be the 5-day biochemical oxygen demand (BOD$_5$), the biodegradable fraction of the chemical oxygen demand (COD), the biodegradable fraction of the total organic carbon (TOC), or the biodegradable fraction of any other organic parameter used.

The value of the rate constant, K, for the kinetic equations depends primarily on the specific wastewater of interest, because the species of organic substrates will vary for different wastewaters and, in particular, for industrial wastewaters and municipal wastewaters containing significant amounts of industrial wastes. Because only certain microbial species in an acclimated sludge can bio-oxidize a particular organic compound and because each species has its own particular rate of utilization, it follows that the overall rate constant, K, will vary for different types of wastewaters. Table 15.1 gives some typical wastewaters along with the reaction order and rate constant based on the biodegradable TOC by acclimated activated sludges. It can be seen that the variation in K is widespread and that even for a certain type of industry, such as the petrochemical industry, there is a significant variation in the rate constant. The petrochemical wastewaters having the lower K values are from plants manufacturing organic compounds relatively difficult to bio-oxidize, such as insecticides, herbicides, fungicides, organic compounds for production of hard plastics, and certain organic solvents. The petrochemical wastewaters having the higher K values are from plants producing compounds more easily biodegraded, such as un-saturated hydrocarbons for synthetic rubber and flexible plastic manufacture. It can be noted that the municipal wastewater has substrates more easily degraded than any of the industrial wastewaters; however, this is not always true. Based on limited data, the rate constant, K, for municipal wastewaters

TABLE 15.1 *Reaction Orders and Rate Constants for Some Selected Wastewaters*

TYPE OF WASTEWATER	PSEUDOREACTION ORDER	REACTION RATE CONSTANT K^a ℓ/(gm MLVSS)(hr) AT 25°C
Pulp and paper mill	First	0.375
Pulp and paper mill	First	0.528
Chemical manufacture	First	0.479
Chemical manufacture	First	0.601
Oil refinery	First	0.504
Oil refinery	First	0.660
Petrochemical manufacture	First	0.592
Petrochemical manufacture	First	0.686
Petrochemical manufacture	First	0.713
Petrochemical manufacture	First	0.911
Petrochemical manufacture	First	1.221
Petrochemical manufacture	First	1.333
Municipal (domestic)	First	1.717

[a] Based on biodegradable total organic carbon (TOC).

ranges from 0.10 to 1.25 ℓ/(gm MLSS)(hr) using total BOD$_5$ as a measure of the organic content — that is, both soluble and insoluble BOD$_5$. Because of the wide range in K values, care should be exercised in arbitrarily assuming a K value for design when no bench-scale or pilot-scale studies have been done. In the absence of such studies, a K value in the range of 0.10 to 0.40 ℓ/(gm MLSS)(hr) is recommended.

The rate constant, K, may be determined from batch reactor studies, as shown in Example 15.1, or from three or more continuous-flow, completely mixed activated sludge reactors, as shown later in this chapter. If possible, pilot plant studies should be used since they simulate field conditions better than laboratory studies. Figures 15.6 (a) and (b) show details of a module of four batch reactors both empty and in operation. Figures 15.7 (a) and (b) show details of a continuous-flow reactor both empty and in operation.

EXAMPLE 15.1
AND 15.1 SI

Biochemical Kinetics

A kinetic study of a soluble organic wastewater has been done in the laboratory using a batch reactor inoculated from a parent acclimated culture developed in a continuous-flow activated sludge reactor. The COD and MLSS concentrations for the various reaction times were as shown in Table 15.2.

At 24 hours, the BOD$_5$ = 4.2 mg/ℓ, the BOD$_5$ = 0.35 BOD$_u$ (ultimate first-stage BOD), and the BOD$_u$ is equal to the degradable COD. The MLVSS = 88% of the MLSS. Determine the reaction order and the reaction rate constant in terms of MLSS and MLVSS.

(a)   (b)

(a)   *Module of Four Batch Activated Sludge Reactors Empty*

(b)   *Module of Four Batch Activated Sludge Reactors in Operation*

FIGURE 15.6   *A Module of Four Batch Reactors Both Empty and in Operation*

(a)

(b)

(a) *Continuous-Flow Completely Mixed Activated Sludge Reactor Empty. Baffle between mixing and settling chambers may be lowered or raised to adjust recycle.*

(b) *Continuous-Flow Completely Mixed Activated Sludge Reactor in Operation. Effluent leaves by vertical tube (not visible) passing through bench top.*

FIGURE 15.7 *A Continuous-Flow Completely Mixed Reactor Both Empty and in Operation*

TABLE 15.2 *Kinetic Data from Batch Test*

REACTION TIME (hr)	COD (mg/ℓ)	MLSS (mg/ℓ)
0	680	1910
1	440	2180
2	240	2210
3	165	2190
4	128	2130
5	115	2090
24	102	1860

SOLUTION The nondegradable COD is equal to the COD at 24 hours minus the BOD$_u$ or $102 - 4.2/0.35 = 102 - 12 = 90 \, \text{mg}/\ell$. Assuming a pseudo–first-order reaction, the data are shown in Table 15.3 for a time up to 5 hours. The data point at 24 hours is not used in the reduced data in Table 15.3, because the COD is so low that it will not give a representative point.

TABLE 15.3   *Reduced Data from Batch Test*

TIME, t (hr)	DEGRADABLE COD (mg/ℓ)	X MLSS (mg/ℓ)	$\bar{X}$ MLSS (mg/ℓ)	$\bar{X}t$ (mg/ℓ)(hr)
0	590	1910		
1	350	2180	2045	2,045
2	150	2210	2060	4,120
3	75	2190	2050	6,150
4	38	2130	2020	8,080
5	25	2090	2000	10,000

The plot of $\ln S_t$ versus $\bar{X}t$ is shown in Figure 15.8. Since it is a straight line, the reaction is pseudo-first order and from the slope, K (MLSS) $= 0.316\,\ell/$(gm MLSS-hr). K (MLVSS) $= K$ (MLSS)$/0.88$; therefore, K (MLVSS) $= 0.316/0.88$ or K (MLVSS) $= 0.359\,\ell/$(gm MLVSS-hr).

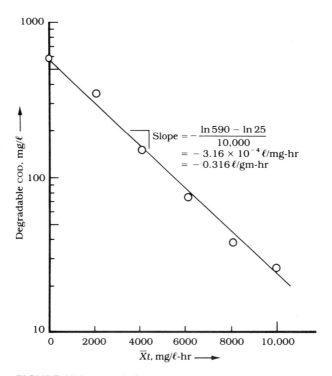

FIGURE 15.8   *Graph for Example 15.1 and 15.1 SI*

**FOOD-TO-
MICROBE
RATIO (F/M)
AND MEAN
CELL
RESIDENCE
TIME (θ_c)**

The instantaneous food-to-microorganism ratio is equal to the specific rate of substrate utilization, $(1/X)(dS/dt)$, and integration of this equation yields

$$\frac{F}{M} = \frac{\Delta S}{\bar{X}\Delta t} \tag{15.10}$$

where $\Delta S/\Delta t$ is the rate of substrate removal, mass/time. The units for the F/M ratio are mass substrate/(mass microbes)(time) — that is, lb BOD$_5$/(lb MLVSS)(day)(kg/kg-day). The mean cell residence time, θ_c, is

$$\theta_c = \frac{\bar{X}}{X_w} \tag{15.11}$$

where the active biological solids in the reactor and in the waste activated sludge flow are $\bar{X}$ and X_w, respectively. The units for the mean cell residence time are days, and, frequently, the mean cell residence time is referred to as the sludge age. Both the food-to-microorganism ratio and the mean cell residence time are used to characterize the performance of an activated sludge process. For example, a high food-to-microorganism ratio and a low mean cell residence time usually produces filamentous growths that have poor settling characteristics. On the other hand, a low food-to-microorganism ratio and a large mean cell residence time can cause the biological solids to undergo excessive endogenous degradation and cell dispersion. For municipal wastewaters, the mean cell residence time must be at least 3 to 4 days to attain proper settling in the final clarifier. If significant nitrification is desired for a municipal wastewater, the mean cell residence time must be at least 10 days, as shown in Figure 15.9, and the dissolved oxygen (DO) at least 2.0 mg/ℓ.

The relationship between the mean cell residence time, θ_c, and the food-to-microbe ratio, F/M, can be derived by starting with the equation for cell production, which is

$$\frac{\Delta X}{\Delta t} = Y\frac{\Delta S}{\Delta t} - k_e\bar{X} \tag{15.12}$$

where

$\Delta X/\Delta t$ = rate of cell production, mass/time

Y = cell yield coefficient, mass cells created/mass substrate removed

k_e = endogenous decay coefficient, mass cells/(total mass cells) × (time)

$\bar{X}$ = average cell concentration, mass

Dividing by $\bar{X}$ gives

$$\frac{\Delta X/\Delta t}{\bar{X}} = Y\frac{\Delta S/\Delta t}{\bar{X}} - k_e \tag{15.13}$$

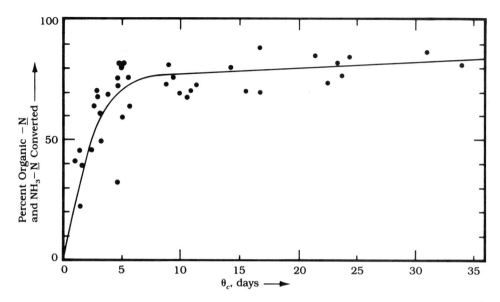

FIGURE 15.9   *Nitrification versus Mean Cell Residence Time*
Adapted from "Nitrification in Wastewater Treatment Plants (Activated Sludge)" by A. C. Petrasek. Dissertation, Civil Engineering Department, Texas A&M University, 1975. Reprinted by permission.

The mean cell residence time, θ_c, is the average time a cell remains in the system; thus,

$$\theta_c = \frac{\bar{X}}{\Delta X/\Delta t} \tag{15.14}$$

The food-to-microbe ratio, F/M, is the rate of substrate removal per unit weight of the cells, or

$$\frac{F}{M} = \frac{\Delta S/\Delta t}{\bar{X}} \tag{15.15}$$

Substituting Eqs. (15.14) and (15.15) into Eq. (15.13) gives

$$\frac{1}{\theta_c} = Y\frac{F}{M} - k_e \tag{15.16}$$

The F/M ratio is also given by Eq. (15.10), which is another form of Eq. (15.15). Substituting Eq. (15.10) into Eq. (15.16) gives

$$\frac{1}{\theta_c} = Y\frac{\Delta S}{\bar{X}\Delta t} - k_e \tag{15.17}$$

In this expression, $F/M = \Delta S/(\bar{X}\Delta t)$. In many cases, Eqs. (15.16) and (15.17) are approximations because the coefficients Y and k_e frequently have been found to decrease as θ_c increases (WPCF, 1977). An example using Eq. (15.17) is as follows: If the influent to an aeration tank has a BOD$_5$ = 150 mg/ℓ, the effluent BOD$_5$ = 10 mg/ℓ, the MLSS = 2500 mg/ℓ, Y = 0.7 lb-MLSS/lb BOD$_5$ (kg/kg), k_e = 0.05 day^{-1}, and the aeration time (Δt) = 6 hr, then $\theta_c^{-1} = (0.7)(150 - 10)/(2500)(6/24) - 0.05$ or $\theta_c = 9.4$ days.

PLUG-FLOW AND DISPERSED PLUG-FLOW REACTORS

Figure 15.10 shows a schematic profile through a plug-flow or dispersed plug-flow reactor basin, along with its final clarifier and recycle pumps. In the plug-flow reactor there is negligible diffusion along the flow path through the reactor, whereas in the dispersed plug-flow reactor there is significant

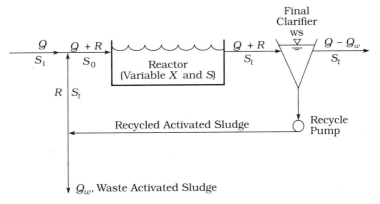

FIGURE 15.10 *Plug-Flow or Dispersed Plug-Flow Activated Sludge Reactor*

diffusion along the flow path through the reactor. The conventional activated sludge process and the tapered aeration activated sludge process both use plug-flow and dispersed plug-flow reactor basins.

The **conventional** activated sludge process, as shown in Figure 15.11,

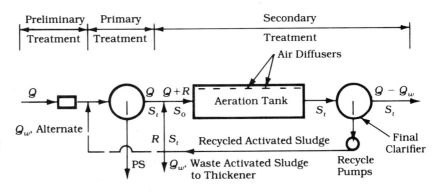

FIGURE 15.11 *Conventional Activated Sludge Process*

uses a rectangular aeration tank as the reactor basin; the food-to-microbe ratio, F/M, is from 0.2 to 0.4 lb BOD_5/lb MLSS-day (kg BOD_5/kg MLSS-day). The space loading is usually from 20 to 40 lb BOD_5/day-1000 ft³ of basin volume (0.3 to 0.6 kg BOD_5/day-m³). Other design and operational parameters for the process are shown in Table 15.4.

The wastewater is preliminary treated to remove screened materials and grit, such as sand and silt; then the wastewater is primary treated to remove as much settleable suspended solids as possible. The primary sludge must be treated by some method such as anaerobic digestion to bio-oxidize the organic solids and thus stabilize the primary sludge solids. The clarified wastewater, Q, is mixed with the return activated sludge flow, R, to inoculate the wastewater with active biological solids, and the mixed liquor, $Q + R$, enters the biological reactor. As the mixed liquor passes through the reactor, the activated sludge solids sorb the organic matter, both the insoluble and the soluble portions, and bio-oxidize the materials to produce carbon dioxide, water, and other end products and to synthesize new cells. As the bio-oxidation proceeds in the reactor, endogenous degradation is occurring simultaneously. Aerobic conditions are maintained by compressed air or mechanical aeration devices, spaced uniformly along the length of the reactor, and the minimum operating dissolved oxygen level is 2.0 mg/ℓ (Great Lakes Board, 1990).

The mixed liquor passes from the reactor to the final clarifier (solid-liquid separator), where the activated sludge solids settle by gravity. The effluent spills over the clarifier weirs and is usually disinfected by chlorine prior to discharge. The settled activated sludge is pumped by recycle pumps, and the major portion, R, of the recycled sludge is mixed with the feed wastewater to inoculate the flow, Q, to the reactor. The recycle ratio, R/Q, is from 25 to 100%, depending on the desired MLSS and the sludge density index (SDI).

A portion of the recycled activated sludge, Q_w, is wasted to remove the net mass synthesized in the system. In plants having a capacity less than several million gallons a day (million liters a day), the waste activated sludge, Q_w, is usually mixed with the feed to the primary clarifier, as shown in Figure 15.11. It is subsequently settled and removed with the primary sludge, PS. In this manner, the sludge is thickened and a large portion of its water is removed prior to sludge treatment. In large plants, the waste activated sludge flow, Q_w, is usually thickened by separate sludge thickeners. The primary sludge and waste activated sludge solids must be oxidized or stabilized, usually by anaerobic or aerobic digestion, prior to disposal. Usually the waste activated sludge flow is from about 1 to 6% of Q by volumetric flowrate.

The **tapered aeration** activated sludge process, shown in Figure 15.12, is only a modification of the conventional process. It is identical with the conventional process except that the aeration devices are spaced along the length of the reactor in accordance to the oxygen demand. At the upstream end of the reactor, as shown in Figure 15.13, the oxygen demand is high

TABLE 15.4 *Design and Operational Parameters for Activated Sludge Treatment of Municipal Wastewaters*

| TYPE OF PROCESS | MEAN CELL RESIDENCE TIME, θ_c, days | FOOD-TO-MICROBE RATIO | SPACE LOADING | | HYDRAULIC RETENTION TIME IN AERATION BASIN θ, hr | MIXED-LIQUOR SUSPENDED SOLIDS (MLSS), mg/ℓ | RECYCLE RATIO, R/Q | FLOW REGIME[a] | BOD REMOVAL EFFICIENCY, % |
			lb BOD$_5$ / day·1000 ft^3	kg BOD$_5$ / day·m^3					
Conventional	5–15	0.2–0.4	20–40	0.3–0.6	4–8	1500–3000	0.25–1.0	PF, DPF	85–95
Tapered aeration	5–15	0.2–0.4	20–40	0.3–0.6	4–8	1500–3000	0.25–1.0	PF, DPF	85–95
Completely mixed	5–30	0.1–0.6	50–120	0.8–2.0	3–6	2500–4000	0.25–1.5	CM	85–95
Step aeration	5–15	0.2–0.4	40–60	0.6–1.0	3–5	2000–3500	0.25–0.75	PF, DPF	85–95
Modified aeration	0.2–0.5	1.5–5.0	75–150	1.2–2.4	1.5–3	200–500	0.05–0.15	PF, DPF	60–75
Contact stabilization	5–15	0.2–0.6	60–75	1.0–1.2			0.50–1.5		
Contact basin					0.5–1.0	1000–3000		PF, DPF	80–90
Stabilization basin					3–6	4000–10,000		PF, DPF	
High-rate aeration	5–10	0.4–1.5	100–1000	1.6–16	2–4	4000–10,000	1.0–5.0	CM	75–90
Extended aeration	20–30	0.05–0.15	10–25	0.16–0.4	18–36	3000–6000	0.75–1.50	PF, DPF	75–95
Pure oxygen	8–20	0.25–1.0	100–200	1.6–3.2	1–3	3000–8000	0.25–0.5	CM	85–95

[a] PF = plug flow, DPF = dispersed plug flow, CM = completely mixed.
Adapted from *Wastewater Engineering: Treatment, Disposal and Reuse* by Metcalf & Eddy, Inc., 3rd ed. Copyright © 1991 by McGraw-Hill, Inc.; and from *Design of Municipal Wastewater Treatment Plants*, Vol. 1, WEF Manual of Practice No. 8 and ASCE Manual and Report on Engineering Practice No. 76. Copyright © 1991 by Water Environment Federation and American Society of Civil Engineers. Reprinted by permission.

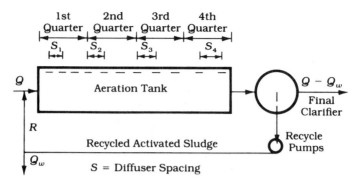

FIGURE 15.12   *Tapered Aeration Activated Sludge Process*

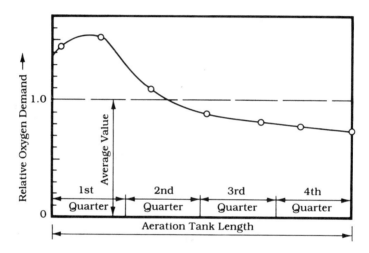

FIGURE 15.13   *Relative Oxygen Demand for the Conventional or Tapered Aeration Activated Sludge Process (Municipal Wastewater)*

because the BOD_5 is the maximum in the reactor. As the mixed liquor passes through the reactor, the BOD_5 gradually declines; thus, the oxygen demand also gradually declines. Both the BOD_5 and the oxygen demand become minimum at the downstream end. The oxygen required for each quarter for treating municipal wastewater with a reaction time less than 8 hr is shown in Table 15.5.

The flow regime in biological reactors used in the conventional or tapered aeration process is dispersed plug flow; however, it may approach plug flow in some cases. In a plug-flow reactor, as previously mentioned, there is negligible induced mixing between elements of the fluid along the axial direction of flow. Actually, in the conventional or tapered aeration

TABLE 15.5 *Oxygen Demand versus Reactor Length for Municipal Wastewaters*

QUARTER LENGTH OF REACTOR	QUARTER OXYGEN DEMAND/ TOTAL OXYGEN DEMAND FOR THE ENTIRE REACTOR (%)
1st	35
2nd	26
3rd	20
4th	19

reactor, there is some longitudinal or axial mixing, but in long, narrow tanks, the flow regime may approach plug flow. The longitudinal mixing is characterized by the dispersion number $d = D/vL$, where D is the dispersion coefficient, v is the axial velocity, and L is the tank length. For rectangular activated sludge reactor basins, the dispersion number, d, is from 0 to 0.2 for plug-flow basins and is from 0.2 to 4.0 for dispersed plug-flow reactor basins (Metcalf & Eddy, Inc., 1979).

The performance of a plug-flow biological reactor exhibiting a pseudo–first-order reaction is the same as for a batch reactor, as given by Eq. (15.9), which may be rearranged to yield (Eckenfelder, 1989)

$$\frac{S_t}{S_0} = e^{-K\bar{X}\theta} \qquad (15.18)$$

where the detention time, θ, for the plug-flow reactor is equal to the reaction time, t, for the batch reactor. The reaction or detention time, θ, for a plug-flow reactor may be determined from Eq. (15.18). For a dispersed plug-flow reactor, the detention time, θ, in the previous plug-flow equation must be increased to maintain the same S_t value. The ratio of the detention times of a dispersed plug-flow and plug-flow reactor is the same as the ratio of their volumes. Thus the defention time ratio may be determined from Figure 15.14, and then the detention time required for a dispersed plug-flow reactor may be determined. The volume of the reactor, V, for either plug flow or dispersed plug flow is given by

$$V = (Q + R)\theta \qquad (15.19)$$

where V is the reactor volume.

For the conventional or tapered aeration activated sludge process, the aeration time (based on the flow, Q) is usually from 4 to 8 hr, the mean cell residence time is from 5 to 15 days, the recycle ratio is usually from 25 to 100%, and the MLSS concentration is usually from 1500 to 3000 mg/ℓ (WEF, 1992; Metcalf & Eddy, Inc., 1991). The performance is from 85 to 95% BOD$_5$ removal and from 85 to 95% suspended solids removal. Either process will give good nitrification if the mean cell residence time is greater than 10 days and the DO is adequate.

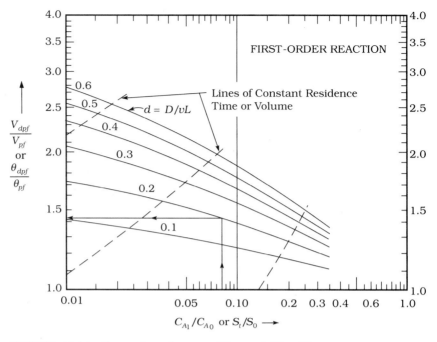

FIGURE 15.14 *Comparison between Dispersed Plug-Flow and Ideal Plug-Flow Reactors for a First-Order Reaction of the Type, A → Products*

Adapted with permission from "Backmixing in the Design of Chemical Reactors" by O. Levenspiel and K. B. Dischoff in *Industrial and Engineering Chemistry* 51, no. 12 (December 1959):1431; and from "Reaction Rate Constant May Modify the Effects of Backmixing" by O. Levenspiel and K. B. Bischoff in *Industrial and Engineering Chemistry* 53, no. 4 (April 1961):313. Copyright 1959, 1961 American Chemistry Society.

EXAMPLE 15.2 AND 15.2 SI

Plug-Flow and Dispersed Plug-Flow Reactor Basins

A municipal wastewater having an influent BOD_5 of 240 mg/ℓ is to be treated by the tapered aeration activated sludge process. The flow diagram is as shown in Figure 15.11; the primary clarifier removes 33% of the influent BOD_5. The effluent must have a BOD_5 of 10 mg/ℓ or less. The design MLSS is 2500 mg/ℓ, the MLVSS is 75% of the MLSS, and the SDI is 10,000 mg/ℓ. The reaction is pseudo–first order with a reaction constant of 0.218 ℓ/(gm MLVSS-hr). The influent flow is 5.0 MGD (18.9 MLD). Determine:

1. The reaction or detention time if the flow regime in the reactor is plug flow, and the reactor basin volume.

2. The reaction or detention time if the flow in the reactor is dispersed plug flow with a dispersion number, $d = D/vL$, of 0.20, and the reactor basin volume.

SOLUTION The BOD$_5$ in the primary effluent is $(240 \text{ mg}/\ell)(1 - 0.33) = 161 \text{ mg}/\ell$. A materials balance on the biological solids at the junction of the primary effluent and the return sludge line is

$$(Q)(0) + (R)(10{,}000) = (Q + R)(2500)$$

The recycle ratio, R/Q, from this equation is 33.3%. A material balance on the BOD$_5$ at the junction is

$$(Q)(161) + (R)(10) = (Q + R)(S_0)$$

or

$$(Q)(161) + (0.333\,Q)(10) = (Q + 0.333\,Q)(S_0)$$

The BOD$_5$ in the mixed liquor, S_0, from this equation is $123 \text{ mg}/\ell$. The performance equation for a plug-flow reactor is

$$\frac{S_t}{S_0} = e^{-K\bar{X}\theta}$$

or

$$\frac{10}{123} = e^{-(0.218)(2.50)(0.75)\theta}$$

Therefore,

$$\boxed{\theta = 6.14 \text{ hours for a plug-flow reactor}}$$

For the solution in USCS units, the plug-flow reactor basin volume is

$$V = \left(\frac{5 \times 10^6 \text{ gal}}{24 \text{ hr}}\right)\left(\frac{\text{ft}^3}{7.48 \text{ gal}}\right)(1 + 0.333)(6.14 \text{ hr}) = \boxed{228{,}000 \text{ ft}^3}$$

For the solution in SI units, the plug-flow reactor basin volume is

$$V = \left(\frac{18{,}900{,}000 \, \ell}{24 \text{ h}}\right)\left(\frac{\text{m}^3}{1000 \, \ell}\right)(1 + 0.333)\,(6.14 \text{ h}) = \boxed{6450 \text{ m}^3}$$

The fraction of BOD$_5$ remaining is

$$\frac{S_t}{S_0} = \frac{10}{123} = 0.0813$$

As shown in Figure 15.14, the ratio of the dispersed plug-flow volume to the plug-flow volume, $V_{\text{dpf}}/V_{\text{pf}}$, is 1.42 for a dispersion number of 0.20 and a fraction remaining equal to 0.0813. Since the volume ratio is the same as the detention time ratio,

$$\theta = (6.14 \text{ hours})(1.42)$$

or

$$\boxed{\theta = 8.72 \text{ hours for a dispersed plug-flow reactor}}$$

For the solution in USCS units, the dispersed plug-flow reactor basin volume is

$$V = \left(\frac{5 \times 10^6 \, \text{gal}}{24 \, \text{hr}}\right)\left(\frac{\text{ft}^3}{7.48 \, \text{gal}}\right)(1 + 0.333)(8.72 \, \text{hr}) = \boxed{324{,}000 \, \text{ft}^3}$$

For the solution in SI units, the dispersed plug-flow reactor basin volume is

$$V = \left(\frac{18{,}900{,}000 \, \ell}{24 \, \text{h}}\right)\left(\frac{\text{m}^3}{1000 \, \ell}\right)(1 + 0.333)(8.72 \, \text{h}) = \boxed{9150 \, \text{m}^3}$$

The design of a dispersed plug-flow reactor basin is a trial-and-error process. First, a dispersion number, $d = D/vL$, is assumed and the reactor detention time and volume are determined. The basin geometry is selected and then the actual dispersion number, $d = D/vL$, is determined. If the actual dispersion number differs significantly from the assumed dispersion number, then another basin geometry is selected and a second trial is made. Successive trials are made until the actual dispersion number is essentially equal to the assumed dispersion number, and once this occurs, the basin is properly designed. In rare cases, a proper basin geometry cannot be obtained because of physical restraints, and for these cases, a new dispersion number must be assumed and the reactor detention time and volume are determined. Then successive trials are made until a basin geometry is obtained which has a dispersion number essentially equal to the assumed number. Once this occurs, the basin is properly designed. The terms in the dispersion number, $d = D/vL$, are D = dispersion coefficient (ft²/hr or m²/h), v = axial velocity through the reactor basin (ft/hr or m/h), and L = reactor basin length (ft or m). The dispersion coefficient, D, for spiral-flow rectangular activated sludge reactor basins using compressed air has been found to be related to the air flowrate, Q_a, and to the basin width, W (Murphy and Boyko, 1970) and is given by

$$D = CW^2 Q_a^{0.346} \tag{15.20}$$

where

D = dispersion coefficient, ft²/hr (m²/h)
C = 3.118 for USCS and SI units
W = basin width, ft (m)
Q_a = standard air flow, scfm/1000 ft³ (m³/min-1000 m³)

Usually the air flowrate, Q_a, is from 20 to 30 standard cubic feet per minute per 1000 ft³ of basin volume (20–30 m³/min-1000 m³). In compressed air design, standard conditions are 20°C and sea level (1 atmosphere).

EXAMPLE 15.3   *Determining the Dispersion Number for a Trial Reactor Basin*

A dispersed plug-flow reactor basin is to be designed, and the first trial has resulted in a reactor basin that is 15 ft deep, 30 ft wide, and 200 ft long. The total flow to the basin is 2.8 MGD, and the air flowrate, Q_a, is 25 scfm/1000 ft³. Determine the dispersion number, $d = D/vL$.

SOLUTION The dispersion coefficient is given by

$$D = 3.118W^2Q_a^{0.346}$$

or

$$D = 3.118(30)^2(25^{0.346}) = 8547\,\text{ft}^2/\text{hr}$$

The axial velocity, v, $= (2.8 \times 10^6\,\text{gal}/24\,\text{hr})(\text{ft}^3/7.48\,\text{gal}) \div (15\,\text{ft} \times 30\,\text{ft}) = 34.7\,\text{ft}/\text{hr}$. The dispersion number, $d = D/vL$, is given by

$$d = (8547\,\text{ft}^2/\text{hr}) \div (34.7\,\text{ft}/\text{hr} \times 200\,\text{ft}) = \boxed{1.23}$$

EXAMPLE 15.3 SI *Determining the Dispersion Number for a Trial Reactor Basin*

A dispersed plug-flow reactor basin is to be designed, and the first trial has resulted in a reactor basin that is 4.57 m deep, 9.14 m wide, and 61 m long. The total flow to the basin is 10,600,000 ℓ/day, and the air flowrate, Q_a, is 25 m^3/min-1000 m^3. Determine the dispersion number, $d = D/vL$.

SOLUTION The dispersion coefficient is given by

$$D = 3.118W^2Q_a^{0.346}$$

or

$$D = 3.118(9.14)^2(25^{0.346}) = 793\,\text{m}^2/\text{h}$$

The axial velocity, v, $= (10,600,000\,\ell/24\,\text{h})(\text{m}^3/1000\,\ell) \div (4.57\,\text{m} \times 9.14\,\text{m}) = 10.57\,\text{m}/\text{h}$. The dispersion number, $d = D/vL$, is given by

$$d = (793\,\text{m}^2/\text{h}) \div (10.57\,\text{m/h} \times 61\,\text{m}) = \boxed{1.23}$$

COMPLETELY MIXED REACTORS

Figure 15.15 shows a schematic profile through a completely mixed reactor basin, along with its final clarifier and recycle pumps. In the completely mixed reactor, there is sufficient stirring for perfect mixing to occur, which assumes that the fluid entering the reactor is instantaneously mixed into the fluid already present and that the reactor contents are uniform throughout the tank volume. If the stirring is adequate and the fluid is not too viscous, such as wastewaters, this assumption of perfect mixing is closely approached in practice. In completely mixed activated sludge reactor basins, the dispersion number, $d = D/vL$, varies from 4.0 to ∞ (Metcalf & Eddy, Inc., 1979).

The completely mixed activated sludge process, as shown in Figures 15.16 and 15.17, generally uses circular or square aeration tanks as the reactor basins; however, slightly rectangular tanks have also been employed. The food-to-microbe ratio, F/M, is generally from 0.1 to 0.6 lb BOD₅/lb MLSS-day (kg BOD₅/kg MLSS-day). The space loading is usually from 50 to 120

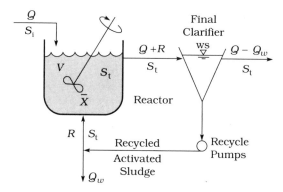

FIGURE 15.15   *Completely Mixed Activated Sludge Reactor*

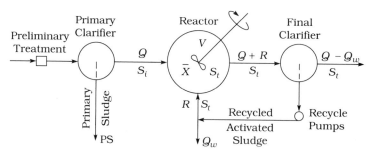

FIGURE 15.16   *Completely Mixed Activated Sludge Process*

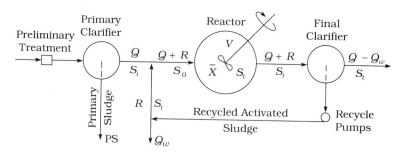

FIGURE 15.17   *Completely Mixed Activated Sludge Process*

lb-BOD_5/day-1000 ft^3 (0.8 to 2.0 kg BOD_5/day-m^3). Other design and operating parameters for the process are shown in Table 15.4.

After preliminary and primary treatment, as shown in Figure 15.16, the clarified wastewater flow, Q, enters the reactor and is rapidly dispersed throughout the reactor volume. The recycled activated sludge flow, R, is discharged directly into the reactor in the vicinity of the aerators so that it is quickly mixed throughout the reactor volume. Since the recycled sludge flow, R, contains the active biological solids, its introduction into the reactor inoculates the reactor with an active microbial mass. An alternative method of introducing the recycled sludge flow, as shown in Figure 15.17, is to mix the recycled sludge flow, R, with the clarified wastewater flow, Q, to inoculate the incoming wastewater with the active biological solids. The mixed-liquor flow, $Q + R$, then enters the reactor, where it is quickly dispersed throughout the reactor volume.

Introducing the recycled sludge flow directly to the reactor, as shown in Figure 15.16, assists in minimizing the effect of a toxic slug of material in the feed wastewater. In this respect, it is preferable to discharge the recycled sludge directly to the reactor instead of mixing it with the influent feed flow. The contents in the reactor are uniform throughout the volume and have the same characteristics as the reactor effluent. In the reactor, the active biological solids sorb the organic matter, both insoluble and soluble fractions, and bio-oxidize these materials to produce aerobic end products and synthesize new microbial cells. As the bio-oxidation proceeds, endogenous degradation is also occurring. Aerobic conditions are usually maintained by mechanical aeration devices, and the minimum operating dissolved oxygen level is $2.0 \, mg/\ell$ (Great Lakes–Upper Mississippi River Board of State Public Health and Environmental Managers, 1990). The solid-liquid separation in the final clarifier, the recycled sludge system, and the method of disposal of the waste activated sludge are carried out in the same manner as in the conventional and tapered aeration activated sludge processes.

The completely mixed activated sludge process is unique because the contents throughout the reactor volume have the same characteristics as the mixed-liquor flow leaving the reactor. Thus the soluble substrate concentration in the reactor is the same as in the discharged mixed-liquor flow and in the final effluent. Also, the total substrate concentration in the reactor, both soluble and insoluble fractions, has essentially the same concentration as in the flow to the final clarifier and in the final effluent.

The completely mixed activated sludge process has several advantages over the conventional and tapered aeration processes, as well as other activated sludge modifications, because it has (1) maximum equalization in the oxygen uptake rate, (2) maximum dampening of slug or shock loads because they are quickly dispersed throughout the reactor volume, (3) maximum neutralization of the carbon dioxide produced during the aerobic bio-oxidation, (4) maximum reduction in the toxicity of a slug of toxic substance because it is quickly mixed throughout the reactor volume (consequently, it will be of lower concentration than at the upstream end of a conventional or tapered aeration activated sludge reactor), (5) relatively uniform environmental conditions for the active biological mass, and (6) greater flexibility than the other activated sludge processes. The principal disadvantage of the completely mixed process is that the reactor volume for a given organic removal for a soluble organic wastewater must be larger than the volume for a conventional process or most of the other process modifications. It is particularly applicable for treating industrial wastewaters having a high organic content. If a 20- to 36-hour reaction time is used, the rate of oxygen demand in mg/ℓ-hour will be relatively low, and the large endogeneous decay of the sludge minimizes the amount of waste activated sludge to be disposed.

Numerous researchers have developed design equations for the completely mixed activated sludge process (Stack and Conway, 1959; Busch, 1961; McKinney, 1962b; Rich, 1963; Pipes *et al.*, 1964; Reynolds and Yang, 1966; Smith and Paulson, 1966; McCarty and Lawrence, 1970; Eckenfelder, 1970; Eckenfelder *et al.*, 1975), and they have usually been based on kinetic equations, growth equations, and material balances on the substrate or biological solids or a combination of these. In 1966 the coauthor presented a model based on growth relationships and material balances on the substrate and the biological cell mass (Reynolds and Yang, 1966). All these approaches have shown similar agreement.

For a kinetic derivation, consider the flowsheet of a completely mixed reactor shown in Figure 15.16. Although the subsequent kinetic derivation is based on the flowsheet shown in Figure 15.16, it is also applicable to the flowsheet in Figure 15.17. It is assumed that the reaction is pseudo–first order and, thus, the rate of substrate utilization is

$$-\frac{1}{X}\left(\frac{dS_t}{dt}\right) = KS_t \qquad (15.21)$$

and the material balance on the substrate is given by

$$[\text{Accumulation}] = [\text{Input}] - \begin{bmatrix} \text{Decrease} \\ \text{due to} \\ \text{reaction} \end{bmatrix} - [\text{Output}] \qquad (15.22)$$

Thus,

$$dS_t \cdot V = QS_i dt + RS_t dt - V[dS_t]_{\text{Growth}} - (Q + R)S_t dt \qquad (15.23)$$

From Eq. (15.21), $[dS_t]_{\text{Growth}} = K\bar{X}S_t dt$. Substituting this in Eq. (15.23) and simplifying gives

$$dS_t \cdot V = QS_i dt - VK\bar{X}S_t dt - QS_t dt \qquad (15.24)$$

Dividing by Vdt produces

$$\frac{dS_t}{dt} = \frac{Q}{V}S_i - K\bar{X}S_t - \frac{Q}{V}S_t \qquad (15.25)$$

Defining θ as V/Q and substituting yields

$$\frac{dS_t}{dt} = \frac{S_i}{\theta} - K\bar{X}S_t - \frac{S_t}{\theta} \qquad (15.26)$$

The term dS_t/dt represents the change in S_t from one time to another successive time. If $S_t = 10\,\text{mg}/\ell$ on one day and $10\,\text{mg}/\ell$ on the next successive day, then $dS_t/dt = (10 - 10)(\text{mg}/\ell)/(1 \text{ day}) = 0$. Thus, for steady state, $dS_t/dt = 0$. Therefore, for steady state, Eq. (15.26) becomes

$$0 = \frac{S_i - S_t}{\theta} - K\bar{X}S_t \tag{15.27}$$

Rearranging gives the design equation

$$\theta = \frac{S_i - S_t}{K\bar{X}S_t} \tag{15.28}$$

The volume for a completely mixed reactor is given by

$$V = Q\theta \tag{15.29}$$

Eq. (15.28) may be rearranged to (Eckenfelder, 1989)

$$\frac{S_i - S_t}{\bar{X}\theta} = KS_t \tag{15.30}$$

Thus, if $(S_i - S_t)/\bar{X}\theta$ is plotted on the y-axis and S_t on the x-axis, the data will plot a straight line with a slope $= K$ if the reaction is pseudo–first order.

In the application of the previous equations to the flowsheet shown in Figure 15.17, it would be imagined that the recycled activated sludge flow, R, does not join the incoming wastewater flow, Q, but instead is discharged directly into the reactor basin, as shown in Figure 15.16. The detention time, θ, would be determined from Eq. (15.28), and the reactor volume, V, from Eq. (15.29). Thus the flowsheet in Figure 15.17 is solved in the same fashion as the flowsheet in Figure 15.16.

For the completely mixed activated sludge process, the reaction or aeration time (based on the flow, Q) is usually from 4 to 36 hours (3 to 6 hours for municipal wastes), the mean cell residence time is from 5 to 30 days, the recycle ratio is from 25 to 150%, and the MLSS concentration is from about 2500 to 4000 mg/ℓ (Metcalf & Eddy, Inc., 1991). Since the process is very flexible, it is possible to design a completely mixed system to operate as a high-rate aeration process at one extreme, or, at the other extreme, it may be designed to operate as an extended aeration process. The performance in terms of BOD$_5$ removal and suspended solids removal will depend upon the particular design; however, BOD$_5$ and suspended solids removals are usually from 85 to 95%. For municipal wastewaters, nitrification will occur if the mean cell residence time is at least 10 days and the DO is adequate.

EXAMPLE 15.4 AND 15.4 SI

Completely Mixed Reactor

A completely mixed activated sludge reactor is to be designed to treat a petrochemical industrial wastewater having a negligible suspended solids content and a COD of 960 mg/ℓ. The design MLSS is 3000 mg/ℓ. The effluent COD must be less than 120 mg/ℓ. Pilot-scale studies have been conducted, and it was found that the biochemical reaction is pseudo–first order. The rate constant based on MLVSS is 0.548 ℓ/(gm)(hr) at 18°C. The MLVSS is 70% of the MLSS, and the nonbiodegradable COD is 95 mg/ℓ. The flowsheet is the same as shown in Figure 15.16. The influent flow is 5.0 MGD (18.9 MLD). Determine the reaction time, θ, and the reactor volume, V.

SOLUTION   The reaction time is given by the equation

$$\theta = \frac{S_i - S_t}{K\bar{X}S_t}$$

where S_i and S_t are expressed as biodegradable COD. Thus $S_i = 960 - 95 = 865$ mg/ℓ and $S_t = 120 - 95 = 25$ mg/ℓ. The value $\bar{X}$ is $(3.00)0.70 = 2.10$ gm volatile suspended solids per liter. Thus the reaction time is

$$\theta = \frac{865 - 25}{(0.548)(2.10)(25)}$$

from which

$$\boxed{\theta = 29.2 \text{ hours}}$$

For the solution in USCS units, the reactor volume is

$$V = \left(\frac{5 \times 10^6 \text{ gal}}{24 \text{ hr}}\right)\left(\frac{\text{ft}^3}{7.48 \text{ gal}}\right)(29.2 \text{ hr}) = \boxed{813,000 \text{ ft}^3}$$

For the solution in SI units, the reactor volume is

$$V = \left(\frac{18,900,000 \ \ell}{24 \text{ h}}\right)\left(\frac{\text{m}^3}{1000 \ \ell}\right)(29.2 \text{ h}) = \boxed{23,000 \text{ m}^3}$$

Note: If the flowsheet were as shown in Figure 15.17, it would be imagined that the recycled activated sludge flow, R, did not join with the influent wastewater flow, Q, but instead discharged directly into the reactor basin. The problem would be solved in the same manner as above, and the answers would be the same.

OTHER ACTIVATED SLUDGE PROCESS MODIFICA-TIONS

Several other modifications of the activated sludge process have been made, and each was developed to attain a particular operational or design objective. With the exception of the modified aeration process, all subsequent activated sludge processes will achieve nitrification if the mean cell residence time is sufficient and the environmental conditions are favorable, particularly the dissolved oxygen level. Operating and design parameters for all of these modifications are shown in Table 15.4.

Step Aeration

The step aeration process, shown in Figure 15.18, has a food-to-microorganism ratio, F/M, of 0.2 to 0.4 lb BOD$_5$/lb MLSS-day (kg/kg-day). The space loading is from 40 to 60 lb BOD$_5$/day-1000 ft^3 (0.6 to 1.0 kg BOD$_5$/day-m^3). This

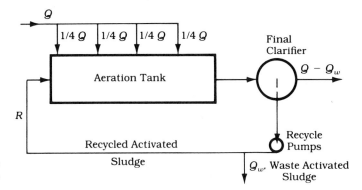

FIGURE 15.18 *Step Aeration Process*

particular activated sludge process modification was developed to even out the oxygen demand of the mixed liquor throughout the length of the reactor. This is accomplished by making each of the step loadings to the aeration tank equal to one-fourth the total wastewater inflow. Although the oxygen demand is greater at the upstream end of the reactor than at the downstream end, it has been evened out considerably compared to the conventional activated sludge process. The aeration tank is a dispersed plug-flow reactor with step inputs of the feed flow.

In the step aeration process, the reaction or aeration time is normally 3 to 5 hours (based on the flow, Q), the mean cell residence time is 5 to 15 days, the recycle ratio is from 25 to 75%, and the MLSS concentration is from about 2000 to 3500 mg/ℓ (WEF, 1991; Metcalf & Eddy, 1991). The performance is usually from 85 to 95% BOD$_5$ and suspended solids removal.

Modified Aeration

The modified aeration process operates with a food-to-microbe ratio, F/M, of 1.5 to 5 lb BOD$_5$/lb MLSS-day (kg/kg-day) and was designed to provide a lower degree of treatment than the other activated sludge processes. The space loading is usually 75 to 150 lb BOD$_5$/day-1000 ft^3 (1.2 to 2.4 kg/day-m^3). Usually, the aeration time is from 1.5 to 3 hours (based on the flow, Q) and the BOD$_5$ removal is from 60 to 75%. The mean cell residence time is from 0.2 to 0.5 day, the recycle ratio is from 5 to 15%, and the MLSS concentration is from 200 to 500 mg/ℓ (WEF, 1991; Metcalf & Eddy, Inc., 1991). Since the mean cell residence time is so small, the suspended solids concentration in the effluent is high in relation to the other activated sludge processes. With municipal wastewaters virtually no nitrification occurs.

Contact Stabilization or Biosorption

The sorption (that is, both adsorption and absorption) characteristics of a wastewater with soluble organic materials and a wastewater with mainly suspended and colloidal organic materials are illustrated in Figure 15.19. Particulate organic solids exhibit rapid sorption by activated sludge flocs, whereas soluble organic materials are sorbed more slowly. Municipal wastewaters usually have about 80 to 85% of the organic content in the form of particulate matter, and an aeration time less than 1 hour usually removes

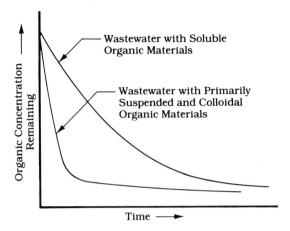

FIGURE 15.19 *Sorption Characteristics of Soluble and Particulate Organic Materials*

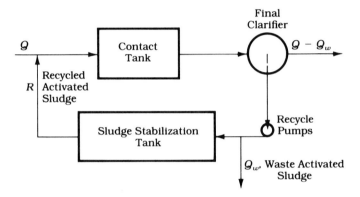

FIGURE 15.20 *The Contact Stabilization or Biosorption Process*

from 80 to 90% of the BOD$_5$. The contact stabilization or biosorption process, shown in Figure 15.20, was designed to provide two reactors, one for the sorption of organic materials and one for bio-oxidation of the sorbed materials. Usually, this type of plant does not have a primary clarifier system. The food-to-microbe ratio, F/M, is from 0.2 to 0.6 lb BOD$_5$/lb MLSS-day (kg/kg-day). The space loading is usually from 60 to 75 lb/day-1000 ft^3 (1.0 to 1.2 kg BOD$_5$/day-m^3).

In the contact tank, which has a contact time of 30 to 60 minutes (based on Q), the active biological solids sorb the suspended organic matter and much of the dissolved organic substances, and the active biological solids are then separated from the treated wastewater in the final clarifier. The separated solids flow or recycled sludge flow is from about 50 to 150% of the incoming wastewater flow rate to the plant. The biological solids are aerated in a sludge bio-oxidation or stabilization tank for a 3- to 6-hour reaction time (based on R), and in the reactor, the sorbed organic materials are bio-oxidized to yield end products and new microbial cells. Separating the sorption process from the bio-oxidation process by providing two reactors requires a total aeration volume of only 50 to 60% of that for a conventional plant. The contact tank is usually 30 to 35% of total tank volume at a plant.

The contact stabilization process was developed for municipal waste-waters that have an appreciable amount of the organic matter in the form of particulate solids that are readily sorbed. For other wastewaters, bench-scale or pilot-scale studies should be made to determine the feasibility of the process for that particular wastewater, because in many cases, sufficient sorption of the organic substances in the wastewater does not occur in the short time provided in the contact tank. This may preclude the use of this type of activated sludge process. The flow regime in the contact tank is usually dispersed plug flow, and in the bio-oxidation tank it is dispersed plug flow.

In the contact stabilization process, the mean cell residence time is from 5 to 15 days, the MLSS concentration in the contact tank is from 1000 to 3000 mg/ℓ, and the MLSS concentration in the bio-oxidation tank is about 4000 to 10,000 mg/ℓ (WEF, 1991; Metcalf & Eddy, Inc., 1991). The performance is usually from 80 to 90% BOD$_5$ and suspended solids removal.

High-Rate Aeration This process uses high MLSS concentrations combined with high space loadings. This combination allows high food-to-microbe ratios, long mean cell residence times, and relatively short hydraulic detention times. Because of the high MLSS concentrations, adequate mixing is important. The MLSS concentration is from 4000 to 10,000 mg/ℓ, and the space loading is from 100 to 1000 lb BOD$_5$/lb MLSS-day (1.6 to 16 kg/day-m^3). The food-to-microbe ratio, F/M, is from 0.4 to 1.5 lb BOD$_5$/lb MLSS-day (kg/kg-day), and the mean cell residence time is from 5 to 10 days. The hydraulic detention time (based on Q) is from 2 to 4 hours, and the recycled activated sludge ratio is from 100% to 500% (WEF, 1991; Metcalf & Eddy, Inc., 1991). The performance is usually from 75% to 90% BOD$_5$ and suspended solids removal.

Extended Aeration The extended aeration process, shown in Figures 15.21 and 15.22, has a food-to-microorganism ratio, F/M, from 0.05 to 0.15 lb BOD$_5$/lb MLSS-day and was developed to minimize waste activated sludge production by providing a large endogenous decay of the sludge mass. The space loading is usually from 10 to 25 lb BOD$_5$/day-1000 ft^3 (0.16 to 0.40 kg BOD$_5$/day-m^3). The process is designed so that the mass of cells synthesized per day equals the mass of cells endogenously degraded per day. Thus there is theoretically no net production of cell mass. The wastewater usually receives only preliminary treatment prior to entering the reactor, because most of these plants do not have a primary clarifier system. The reaction time is from about 18 to 36 hours (based on Q), and the reactors are either square, round, or rectangular in plan view or a racetrack design. The racetrack design, commonly called

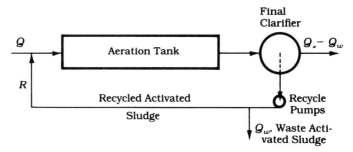

FIGURE 15.21 *Extended Aeration Process (Rectangular Tank Layout)*

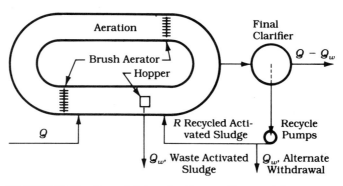

FIGURE 15.22 *Extended Aeration Process (Racetrack Layout, Commonly Called an Oxidation Ditch)*

an **oxidation ditch**, uses brush-type aerators that have a rotating axle with radiating steel bristles or some other axle-type aeration device, as shown in Figure 15.23 and Figure 15.24 (a) and (b).

To make the net sludge production equal zero, it is necessary that the synthesized degradable cell mass equal the endogenously degraded cell mass. Some of the cell mass synthesized, such as peptidoglycans or mucopeptides used in the microbial cell walls, is nonbiodegradable in the reaction times employed. Zero net cell production is given by

$$X_w = Y_b S_r - k_e f \bar{X} = 0 \tag{15.31}$$

where Y_b is the biodegradable yield coefficient, S_r is the substrate removed per day, and f is the fraction of degradable solids in the system, which is usually 0.7 to 0.8. The total mass of cells in the process, $\bar{X}$, is much larger than that in a conventional plant having the same capacity. Rearranging Eq. 15.31 gives (Eckenfelder, 1989)

$$\bar{X} = \frac{Y_b S_r}{k_e f} \tag{15.32}$$

FIGURE 15.23 *A Large Oxidation Ditch Facility*
Courtesy Lakeside Equipment Corporation.

This equation gives the total cell mass that is required in the system, and knowing the design MLSS value, it is possible to determine the reactor volume required. Another design procedure is to assume a reaction time and a recycle ratio. Then, after X and the reactor volume have been computed, the minimum required MLSS concentration may be determined. Although there is no net cell production, there is an accumulation of lysed cell fragments that are very slowly degraded. This necessitates the periodic wasting of some of the sludge. The fraction of viable cells in the mixed liquor is so low that usually the waste sludge may be dewatered without requiring digestion.

The circular or square reactors have a completely mixed flow regime, whereas the rectangular reactors are dispersed plug flow and the racetrack type (oxidation ditch) approaches plug flow. In the oxidation ditch, the velocity must be from 0.8 to 1.2 fps (0.24 to 0.37 mps) to maintain the solids in suspension. Two sets of rotors should be provided, and each should have the capability of furnishing the required oxygen demand with one rotor out of service. The ditch should be lined with reinforced concrete. Provisions to vary the rotor depth of submergence should be provided. In the extended aeration process, the mean cell residence time is from 20 to 30 days, the recycle ratio is from 75 to 150%, and the MLSS concentration is from about 3000 to 6000 mg/ℓ (WEF, 1991; Metcalf & Eddy, Inc., 1991). Performance in terms of BOD$_5$ removal is from 75 to 95%. Because of the long aeration time and mean cell residence time, considerable amounts of cell fragments usually will be present in the final effluent. Consequently, the effluent suspended solids are relatively high compared to the other modifications of the activated

(a) *Oxidation Ditch with Two Rotating Brush Aerators. The ditch is concrete lined to prevent erosion.*

(b) *Rotating Brush Aerators for an Oxidation Ditch. The aerators furnish the required oxygen and impart sufficient velocity to the mixed liquor to maintain the solids in suspension.*

FIGURE 15.24 *Oxidation Ditch Wastewater Treatment Plant*

sludge process. The extended aeration process is very applicable for small installations, such as small communities and isolated institutions. The reactor basin volume for an extended aeration plant is given by the same equation as for a plug-flow or dispersed plug-flow reactor, which is $V = (Q + R)\theta$.

EXAMPLE 15.5   *Oxidation Ditch*

An oxidation ditch is to be designed for a community of 6000 persons. The flow is 100 gal/cap-day, and the influent BOD$_5$ is 225 mg/ℓ. The plant has 90% BOD$_5$ removal, the yield coefficient Y_b is 0.65 lb MLVSS per lb BOD$_5$ oxidized, the endogenous coefficient is 0.06 day^{-1}, the biodegradable fraction of the total solids is 0.8, and the MLVSS is 50% of MLSS. Determine:

1. The reactor volume if the reaction time is 1 day and the expected recycle ratio is 1.00.

2. The operating MLSS concentration.

3. The layout dimensions if the ditch has a trapezoidal cross section and the distance across the racetrack layout is 150 ft from ditch centerline to centerline. The design cross section is 15 ft wide at the base, has 2:1 side slopes, and has a 6-ft water depth.

SOLUTION   The flow is (6000 persons)(100 gal/cap-day) or 600,000 gal per day. Since $V = (Q + R)\theta$, the volume is

$$V = \frac{600,000 \text{ gal}}{\text{day}} \left| \frac{1 + 1}{} \right| \frac{1 \text{ day}}{} \left| \frac{\text{ft}^3}{7.48 \text{ gal}} \right.$$

$$= \boxed{160,000 \text{ ft}^3}$$

The BOD$_5$ removed per day is

$$S_r = \frac{600,000 \text{ gal}}{\text{day}} \left| \frac{8.34 \text{ lb}}{\text{gal}} \right| \frac{225}{10^6} \left| 0.90 \right.$$

$$= 1013 \text{ lb BOD}_5/\text{day}$$

The volatile solids are

$$\bar{X} = \frac{Y_b S_r}{k_e f} = \frac{0.65 \text{ lb MLVSS}}{\text{lb BOD}_5} \left| \frac{1013 \text{ lb BOD}_5}{\text{day}} \right| \frac{\text{day}}{0.06} \left| \frac{1}{0.8} \right.$$

$$= 13,718 \text{ lb MLVSS}$$

Operating MLSS is

$$\bar{X} = \frac{13,718 \text{ lb MLVSS}}{160,000 \text{ ft}^3} \left| \frac{1.0 \text{ lb MLSS}}{0.50 \text{ lb MLVSS}} \right| \frac{1 \text{ ft}^3}{62.4 \text{ lb}} \left| \frac{10^6}{} \right.$$

$$= \boxed{2750 \text{ ppm} = 2750 \text{ mg}/\ell}$$

The cross-sectional area of the ditch is (15 ft + 39 ft)(1/2)(6 ft) or 162 ft^2. The ditch length, L, is $L = (160,000 \text{ ft}^3)(1/162 \text{ ft}^2) = \boxed{987.7 \text{ ft.}}$ The length of the curved part of the racetrack is (π)150 ft or 471.2 ft. The straight length of the ditch, L_s, is $L_s = (987.7 \text{ ft} - 471.2 \text{ ft})(1/2) = \boxed{258 \text{ ft.}}$

EXAMPLE 15.5 SI   *Oxidation Ditch*

An oxidation ditch is to be designed for a community of 6000 persons. The flow is 380 ℓ/cap-d, and the influent BOD$_5$ is 225 mg/ℓ. The plant has 90% BOD$_5$ removal, the yield coefficient Y_b is 0.65 gm MLVSS per gm BOD$_5$ oxidized, the endogenous coefficient is 0.06 d^{-1}, the biodegradable fraction of the total solids is 0.8, and the MLVSS is 50% of the MLSS. Determine:

1. The reactor volume if the reaction time is 1 d and the expected recycle ratio is 1.00.

2. The operating MLSS concentration.

3. The layout dimensions if the ditch has a trapezoidal cross section and the distance across the racetrack layout is 45 m from ditch centerline to centerline. The design cross section is 4.5 m wide at the base, has 2 : 1 side slopes, and has a 1.8-m water depth.

SOLUTION   The flow is (6000 persons)(380 ℓ/cap-day)(m^3/1000 ℓ) = 2280 m^3/day. Since $V = (Q + R)\theta$, the volume is

$$V = \frac{2280 \, \text{m}^3}{\text{d}} \, \bigg| \, \frac{1 + 1}{} \, \bigg| \, \frac{1 \, \text{d}}{}$$

$$= \boxed{4560 \, \text{m}^3}$$

The BOD$_5$ removed per day is

$$S_r = \frac{2280 \, \text{m}^3}{\text{d}} \, \bigg| \, \frac{1000 \, \ell}{\text{m}^3} \, \bigg| \, \frac{225 \, \text{mg}}{\ell} \, \bigg| \, \frac{\text{kg}}{10^6 \, \text{mg}} \, \bigg| \, \frac{0.9}{}$$

$$= 461.7 \, \text{kg}$$

The volatile solids are

$$\bar{X} = \frac{Y_b S_r}{k_e f} = \frac{0.65 \, \text{gm MLVSS}}{\text{gm BOD}_5} \, \bigg| \, \frac{461.7 \, \text{kg BOD}_5}{\text{d}} \, \bigg| \, \frac{\text{d}}{0.06} \, \bigg| \, \frac{1}{0.8}$$

$$= 6252 \, \text{kg MLVSS}$$

Operating MLSS is

$$\bar{X} = \frac{6252 \, \text{kg MLVSS}}{4560 \, \text{m}^3} \, \bigg| \, \frac{1.0 \, \text{kg MLSS}}{0.5 \, \text{kg MLVSS}} \, \bigg| \, \frac{10^6 \, \text{mg}}{\text{kg}} \, \bigg| \, \frac{\text{m}^3}{1000 \, \ell}$$

$$= \boxed{2740 \, \text{mg}/\ell}$$

The cross-sectional area of the ditch is (4.5 m + 11.7 m)(1/2)(1.8 m) = 14.58 m^2. The ditch length, L, is L = (4560 m^3)(1/14.58 m^2) = $\boxed{312.8 \, \text{m}}$. The length of the curved part of the racetrack is (π)45 m or 141.4 m. The straight length of the ditch, L_s, is L_s = (312.8 m − 141.4 m)(1/2) = $\boxed{85.7 \, \text{m}}$.

PURE OXYGEN ACTIVATED SLUDGE

The pure oxygen activated sludge process uses covered aeration tanks under a slight pressure and pure oxygen as the oxygen supply instead of air, as shown in Figure 15.25. The tanks are compartmented to give a series of completely mixed biological reactors, and rotary devices, such as turbines, are employed for mixing. Usually, at least 3 stages are provided, as shown in Figure 15.26. The partial pressure of the oxygen in the atmosphere above

FIGURE 15.25
Reactor Basins at a Pure Oxygen Activated Sludge Plant

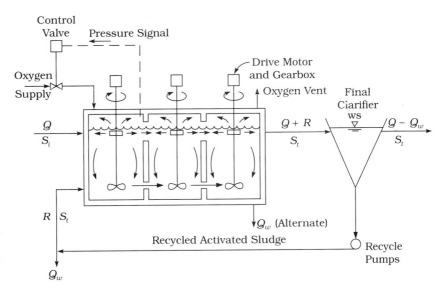

FIGURE 15.26
Pure Oxygen Activated Sludge Process

the mixed liquor is about 0.8 atm; thus the oxygen saturation concentration in the mixed liquor is about 4 times the concentration normally encountered. In pure oxygen activated sludge systems, the food-to-microbe ratio, F/M, is from 0.25 to 1.0 lb BOD_5/day-lb MLSS (kg/kg-day), the aeration time (based on Q) is 1 to 3 hours, the mean cell residence time is 8 to 20 days, the recycle ratio is from 25 to 50%, the MLSS concentration is 3000 to 8000 mg/ℓ, and the operating dissolved oxygen level is 4 to 10 mg/ℓ. The space loading is usually 100 to 200 lb BOD_5/day-1000 ft^3 (1.6 to 3.2 kg BOD_5/day-m^3). The advantages reported for pure oxygen systems are reduced reaction times, decreased amounts of waste activated sludge, increased sludge settling characteristics, and reduced land requirements.

TEMPERATURE EFFECTS

The biochemical reactions employed in microbial metabolism for substrate utilization and endogenous degradation are enzyme-catalyzed reactions. Reactions catalyzed by enzymes are temperature dependent, and in general, the reaction rate or reaction velocity approximately doubles for each 10°C rise in temperature up to a temperature at which the enzymes are denatured. In the activated sludge process, both the reaction rate constant for substrate utilization, K, and the rate constant for endogenous degradation, k_e, are temperature dependent.

The relationship between the reaction rate constant for substrate utilization and the temperature is

$$K_2 = K_1 \cdot \theta^{(T_2 - T_1)} \tag{15.33}$$

where

K_1, K_2 = reaction rate constants at the respective temperatures, T_1 and T_2, °C

θ = temperature correction coefficient

T_1 = temperature of the mixed liquor, °C, for K_1

T_2 = temperature of the mixed liquor, °C, for K_2

Usually θ is from 1.03 to 1.09 (Eckenfelder, 1989). The variation of θ is due to the difference between various activated sludge systems, and this is dependent upon the specific wastewater of interest. Also, it is indicated that the θ value is dependent upon the food-to-microorganism ratio, F/M, with high F/M ratios resulting in high θ values, and low F/M ratios resulting in low θ values.

The effect of temperature upon the endogenous degradation rate constant, k_e, may be estimated from a relationship similar to Eq. (15.33), which is

$$k_{e_2} = k_{e_1} \cdot \theta^{(T_2 - T_1)} \tag{15.34}$$

where

k_{e_1}, k_{e_2} = rate constants for endogenous degradation at the respective temperatures, T_1 and T_2, °C

θ = temperature correction coefficient

T_1 = temperature of the mixed liquor, °C, for k_{e_1}

T_2 = temperature of the mixed liquor, °C, for k_{e_2}

The value of θ in this equation depends upon the specific activated sludge system and wastewater of interest, and the θ value may range from 1.065 to 1.085 (Wuhrmann, 1964).

Temperature affects the rate of microbial growth because it is also dependent upon enzyme-catalyzed reactions. For example, one study of a particular microbial species showed that the generation time (cell division or fission time) at 4°C was 180 minutes, whereas at 42°C it decreased to about 20 minutes (Clifton, 1957).

EXAMPLE 15.6 AND 15.6 SI | *Temperature Effects on an Activated Sludge Process*

The temperature of the mixed liquor in Example 15.4 and 15.4 SI was 18°C (64.4°F). In this problem, assume that the temperature correction coefficient θ is 1.05 for a mixed liquor temperature of 28°C (82.4°F). Determine:

1. The effluent COD.

2. The percent COD removal or conversion at 28°C.

3. The percent COD removal or conversion at 18°C.

SOLUTION | From Example 15.4 and 15.4 SI, the reaction constant at 18°C is 0.548 ℓ/(gm)(hr), the total COD in the feed flow is 960 mg/ℓ, and the nonbiodegradable COD is 95 mg/ℓ. The flowsheet for the process is shown in Figure 15.16. The reaction constant at 28°C is

$$K_2 = (0.548)1.05^{(28-18)} = 0.893 \; \ell/(gm)(hr)$$

The biodegradable COD in the feed flow is $960 - 95$ or 865 mg/ℓ. The reaction time for a completely mixed reactor is

$$\theta = \frac{S_i - S_t}{K \bar{X} S_t}$$

Substituting known values gives

$$29.2 = \frac{865 - S_t}{(0.893)(2.10)(S_t)}$$

from which $S_t = 15.5$ mg/ℓ. This is the biodegradable COD; thus, the total COD is $15.5 + 95$ or

Effluent COD = 110.5 or $\boxed{111 \text{ mg}/\ell \text{ total COD}}$

The fraction removal at 28°C is $(960 - 111) \div 960$ or 0.884. Therefore, the percent COD removal at 28°C is $\boxed{88.4\%.}$ The fraction removal at 18°C is $(960 - 120) \div 960$ or 0.875. Thus the percent COD removal at 18°C is $\boxed{87.5\%.}$

OTHER KINETIC RELATIONSHIPS

Another kinetic approach to biochemical reactions is that based on the Michaelis-Menten relationship or the Monod relationship. The fundamental expressions for the Monod approach are

$$\frac{dX}{dt} = \mu X = Y\frac{dS}{dt}$$ (15.35)

where

dX/dt = rate of cell growth, mass/(volume)(time)
μ = growth coefficient, time^{-1}

The growth coefficient is given by Monod's expression, which is

$$\mu = \mu_{max}\left(\frac{S}{K_s + S}\right)$$ (15.36)

where

μ_{max} = maximum value of the growth coefficient, time^{-1}
K_s = substrate concentration with $\mu = \frac{1}{2}\mu_{max}$

The rate of endogenous decay is given by

$$\frac{dX}{dt} = k_e X$$ (15.37)

Consider the completely mixed reactors shown in Figures 15.27 and 15.28. For the **completely mixed system without recycle**, a material balance on the

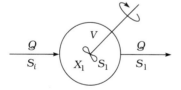

FIGURE 15.27
Completely Mixed Activated Sludge without Recycle

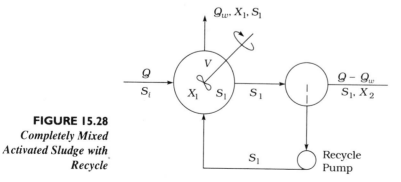

FIGURE 15.28
Completely Mixed Activated Sludge with Recycle

cells gives [Accumulation] = [Increase due to growth] − [Decrease due to endogenous decay] − [Output] or

$$dX_1 \cdot V = V\mu X_1 dt - Vk_e X_1 dt - QX_1 dt \tag{15.38}$$

Dividing by Vdt gives

$$\frac{dX_1}{dt} = \mu X_1 - k_e X_1 - \frac{Q}{V} X_1 \tag{15.39}$$

Since $dX_1/dt = 0$ for steady state, and the detention time, θ_i, based on the influent flow is $\theta_i = V/Q$, Eq. (15.39) can be rearranged to give

$$\mu = \frac{1}{\theta_i} + k_e \tag{15.40}$$

A material balance on the substrate gives [Accumulation] = [Input] − [Output] − [Decrease due to growth] or

$$dS_1 \cdot V = QS_i dt - QS_1 dt - V[dS_1]_{\text{Growth}} \tag{15.41}$$

The decrease due to growth is given by a rearrangement of Eq. (15.35), which is $[dS_1]_{\text{Growth}} = (\mu/Y)(X_1)dt$. Substituting this into Eq. (15.41) gives

$$dS_1 \cdot V = QS_i dt - QS_i dt - V\frac{\mu}{Y} X_1 dt \tag{15.42}$$

Dividing by Vdt gives an expression for unsteady-state conditions that is as follows:

$$\frac{dS_1}{dt} = \frac{Q}{V} S_i - \frac{Q}{V} S_i - \frac{\mu}{Y} X_1 \tag{15.43}$$

Since $dS_1/dt = 0$ for steady state and $\theta_i = V/Q$, Eq. (15.43) can be rearranged to give

$$\mu = \frac{Y}{X_1 \theta_i} (S_i - S_1) \tag{15.44}$$

For the nonrecycle system, $\theta_i = \theta_c$; thus, equating Eqs. (15.40) and (15.44) and rearranging gives the design equation

$$X_1 = \frac{Y(S_i - S_1)}{1 + k_e \theta_c} \tag{15.45}$$

Using Eq. (15.45) and knowing the other parameters, it is possible to determine the MLSS required, X_1. Another useful equation is given by equating Eqs. (15.40) and (15.36) and substituting $\theta_c = \theta_i$ and $S_1 = S$ to give the design equation

$$\frac{1}{\theta_c} = \mu_{\max}\left(\frac{S_1}{K_s + S_1}\right) - k_e \tag{15.46}$$

Equation (15.46) may be rearranged to give the design equation

$$S_1 = \frac{K_s(1 + k_e\theta_c)}{\theta_c(\mu_{\max} - k_e) - 1} \qquad (15.47)$$

Equations (15.45), (15.46), and (15.47) represent design equations and can be used for the design of aerated lagoons, which essentially are activated sludge systems without recycle.

To determine the values of Y and k_e, Eqs. (15.44) and (15.40) may be equated and rearranged to give

$$\frac{S_i - S_1}{X_1\theta_i} = \frac{k_e}{Y} + \frac{1}{Y}\frac{1}{\theta_i} \qquad (15.48)$$

This equation is of the form $y = b + mx$. Using a continuous-flow reactor or several continuous-flow reactors operated at several flow rates will give the data to plot $(S_i - S_1)/X_i\theta_i$ on the y-axis and $1/\theta_i$ on the x-axis. The slope of the line equals $1/Y$ and the y-intercept is k_e/Y. To determine K_s and $\mu_{\max}$, Eqs. (15.40) and (15.36) can be equated and rearranged to give the following expression relating K_s and $\mu_{\max}$ to the terms S_1, k_e, and θ_i:

$$\left(\frac{\theta_i}{1 + k_e\theta_i}\right)S_1 = \frac{K_s}{\mu_{\max}} + \frac{1}{\mu_{\max}}S_1 \qquad (15.49)$$

This equation is of the form $y = b + mx$; thus, plotting $[\theta_i/(1 + k_e\theta_i)]S_1$ values on the y-axis and S_1 values on the x-axis will give a line with a slope equal to $1/\mu_{\max}$ and a y-axis intercept of $K_s/\mu_{\max}$.

For the **completely mixed system with recycle**, shown in Figure 15.28, a materials balance on the cells gives [Accumulation] = [Increase due to growth] − [Decrease due to decay] − [Output] or

$$dX_1 \cdot V = V\mu X_1 dt - Vk_e X_1 dt - Q_w X_1 dt - (Q - Q_w)X_2 dt \quad (15.50)$$

Dividing by Vdt gives

$$\frac{dX_1}{dt} = \mu X_1 - k_e X_1 - \frac{Q_w X_1}{V} - \frac{(Q - Q_w)X_2}{V} \qquad (15.51)$$

Since $dX_1/dt = 0$ for steady state, setting Eq. (15.51) equal to zero and dividing by X_1 and rearranging gives

$$\mu = \frac{Q_w X_1 + (Q - Q_w)X_2}{VX_1} + k_e \qquad (15.52)$$

Since $\theta_c = (VX_1)/[Q_w X_1 + (Q - Q_w)X_2]$, Eq. (15.52) yields

$$\mu = \frac{1}{\theta_c} + k_e \qquad (15.53)$$

A materials balance on the substrate gives [Accumulation] = [Input] − [Output] − [Decrease due to growth] or

$$dS_1 \cdot V = QS_i dt - Q_w S_1 dt - (Q - Q_w)S_1 dt - V[dS_1]_{\text{Growth}} \quad (15.54)$$

Inserting the value $[dS_1]_{Growth} = (\mu/Y)X_1dt$ yields

$$dS_1 \cdot V = QS_idt - Q_wS_1dt - (Q - Q_w)S_1dt - V\frac{\mu}{Y}X_1dt \quad \textbf{(15.55)}$$

Dividing by Vdt and simplifying gives

$$\frac{dS_1}{dt} = \frac{Q}{V}S_i - \frac{Q}{V}S_1 - \frac{\mu}{Y}X_1 \quad \textbf{(15.56)}$$

Since $dS_1/dt = 0$ for steady state and $\theta_i = V/Q$, Eq. (15.56) can be rearranged to give

$$\mu = \frac{Y}{X_1}\left(\frac{S_i - S_1}{\theta_i}\right) \quad \textbf{(15.57)}$$

Equating Eqs. (15.57) and (15.53) and simplifying give the design equation

$$X_1 = \frac{\theta_c}{\theta_i} \cdot \frac{Y(S_i - S_1)}{1 + k_e\theta_c} \quad \textbf{(15.58)}$$

or

$$\theta_i = \frac{\theta_c}{X_1} \cdot \frac{Y(S_i - S_1)}{1 + k_e\theta_c} \quad \textbf{(15.59)}$$

Another useful equation is given by equating Eqs. (15.53) and (15.36) and substituting $S_1 = S$ to give the design equation

$$\frac{1}{\theta_c} = \mu_{max}\left(\frac{S_1}{K_s + S_1}\right) - k_e \quad \textbf{(15.60)}$$

which is identical to Eq. (15.46) for the nonrecycle system. Equation (15.60) can be rearranged to give the design equation

$$S_1 = \frac{K_s(1 + k_e\theta_c)}{\theta_c(\mu_{max} - k_e) - 1} \quad \textbf{(15.61)}$$

which is identical to Eq. (15.47) for the nonrecycle system.

In the design of an activated sludge plant by equations based on the Monod relationship, it is necessary to use one equation that employs the parameters μ_{max} and K_s. It is not valid to assume a mean cell residence time, θ_c, a Y value, and a k_e value and, knowing the other parameters, determine the reactor detention time, θ_i, from Eq. (15.59). For the usual case, the substrate concentration in the effluent, S_1, is known, and μ_{max}, K_s, Y, and k_e are determined from kinetic studies. The required mean cell residence time, θ_c, can then be determined from Eq. (15.60). Then, knowing S_i and assuming an X_1 value, one can determine the required reactor detention time, θ_i from Eq. (15.59). The required reactor basin volume is given by $V = Q\theta_i$. Both extensive laboratory studies and field studies have found typical Monod parameters for municipal wastewaters to be as shown in Table 15.6.

TABLE 15.6 *Typical Monod Growth and Kinetic Coefficients for the Activated Sludge Process Treating Municipal Wastewaters*

COEFFICIENT	BASIS	RANGE	TYPICAL
μ_{max}	day^{-1}	2–10	5
K_s	mg/ℓ BOD$_5$	25–100	60
	mg/ℓ COD	15–70	40
Y	mgVSS^a/mg BOD$_5$	0.4–0.8	0.6
k_e	day^{-1}	0.025–0.075	0.06

^{a} MLVSS is typically 70% to 80% of MLSS.
Adapted from *Wastewater Engineering: Treatment, Disposal and Reuse* by Metcalf & Eddy, Inc., 3rd ed. Copyright © 1991 by McGraw-Hill, Inc. Reprinted by permission.

EXAMPLE 15.7 AND 15.7 SI *Completely Mixed Reactor Designed by Monod Equations*

A completely mixed activated sludge process, treating a municipal wastewater, has a primary clarifier effluent BOD$_5$ of 135 mg/ℓ. The design MLSS is 4000 mg/ℓ, and the MLVSS is 75% of the MLSS. The plant permit is for an effluent BOD$_5$ of 10 mg/ℓ and 15 mg/ℓ suspended solids. The effluent suspended solids have a BOD$_5$ of 0.5 mg BOD$_5$/mg suspended solids. If $\mu_{max} = 3.0$ day^{-1}, $K_s = 60$ mg/ℓ, $Y = 0.60$ mg MLVSS/mg BOD$_5$, $k_e = 0.06$ day^{-1}, and the influent flow is 5.0 MGD (18.9 MLD), determine the required reactor detention time, θ_i, and the reactor basin volume, V.

SOLUTION The soluble effluent BOD$_5$ is 10 mg/ℓ − (0.5)(15 mg/ℓ) or 2.5 mg/ℓ. The required mean cell residence time, θ_c, is given by

$$\frac{1}{\theta_c} = \mu_{max}\left(\frac{S_1}{K_s + S_1}\right) - k_e$$

Thus,

$$\frac{1}{\theta_c} = (3.0 \text{ day}^{-1})\left(\frac{2.5}{60 + 2.5}\right) - 0.06 \text{ day}^{-1}$$

or

$$\theta_c = 16.7 \text{ days}$$

The hydraulic detention time, θ_i, is given by

$$\theta_i = \frac{\theta_c}{X_1} \cdot \frac{Y(S_i - S_1)}{1 + k_e\theta_c}$$

Since Y is in terms of MLVSS, and X_1 must be in the same terms, $X_1 = (4000 \text{ mg/}\ell)(0.75) = 3000 \text{ mg/}\ell$. Thus,

$$\theta_i = \frac{16.7}{3000} \cdot \frac{0.6(135 - 2.5)}{1 + (0.06)(16.7)}$$

$$= 0.2211 \text{ days}$$

$$= \boxed{5.31 \text{ hr}}$$

For the solution in USCS units, the reactor basin volume is given by

$$V = Q\theta_i$$

or

$$V = \left(\frac{5,000,000\,\text{gal}}{24\,\text{hr}}\right)\left(\frac{\text{ft}^3}{7.48\,\text{gal}}\right)(5.31\,\text{hr}) = \boxed{148,000\,\text{ft}^3}$$

For the solution in SI units, the reactor basin volume is given by

$$V = \left(\frac{18,900,000\,\ell}{24\,\text{h}}\right)\left(\frac{\text{m}^3}{1000\,\ell}\right)(5.31\,\text{h}) = \boxed{4180\,\text{m}^3}$$

Note: In this problem S_1 was expressed in terms of soluble BOD$_5$, which will give the most accurate solution; however, the total BOD$_5$ can be used instead if the soluble fraction is not known.

To determine the values of Y and k_e, Eqs. (15.57) and (15.53) can be equated and simplified to give

$$\frac{S_i - S_1}{X_1\theta_i} = \frac{k_e}{Y} + \frac{1}{Y}\frac{1}{\theta_c} \tag{15.62}$$

This equation is of the form $y = b + mx$. Using one or more continuous-flow reactors operated at different θ_c values will give the data to plot $(S_i - S_1)/X_1\theta_i$ on the y-axis and $1/\theta_c$ on the x-axis. The slope of the resulting straight line is $1/Y$, and its y-axis intercept is k_e/Y. To determine K_s and $\mu_{\max}$, Eqs. (15.53) and (15.36) can be equated, S_1 substituted for S, and the equation rearranged to give

$$\left(\frac{\theta_c}{1 + k_e\theta_c}\right)S_1 = \frac{K_s}{\mu_{\max}} + \frac{1}{\mu_{\max}}S_1 \tag{15.63}$$

This equation is of the form $y = b + mx$. Using one or more continuous-flow reactors operated at several θ_c values will give the data to plot $[\theta_c/(1 + k_e\theta_c)]S_1$ on the y-axis and S_1 on the x-axis. The slope of the resulting straight line is $1/\mu_{\max}$ and the y-axis intercept is $K_s/\mu_{\max}$.

For the case of a **plug-flow reactor**, the performance equation that can be derived using the Monod relationship is

$$\frac{1}{\theta_c} = \frac{\mu_{\max}(S_i - S_1)}{(S_i - S_1) + CK_s} - k_e \tag{15.64}$$

where $C = (1 + \alpha)\ln[(\alpha S_1 + S_i)/(1 + \alpha)S_1]$ and $\alpha = R/Q$. In these expressions, S_i is the substrate concentration in the influent flow prior to mixing with the recycle. The limit of C as $\alpha \to 0$ is equal to $\ln(S_i/S_1)$. If $\alpha < 1$, this approximation may be made, and Eq. (15.64) reduces to

$$\frac{1}{\theta_c} = \frac{\mu_{\max}(S_i - S_1)}{(S_i - S_1) + K_s\ln(S_i/S_1)} - k_e \tag{15.65}$$

The relationship for the cell concentration is

$$X_1 = \frac{\theta_c}{\theta_i} \cdot \frac{Y(S_i - S_1)}{1 + k_e\theta_c} \tag{15.66}$$

or

$$\theta_i = \frac{\theta_c}{X_1} \cdot \frac{Y(S_i - S_1)}{1 + k_e\theta_c} \tag{15.67}$$

From the previous relationships the hydraulic detention time, θ_i, may be determined for a given set of conditions. The volume for a plug-flow reactor, V_{pf}, may be determined from $V_{pf} = Q\theta_i$.

For the case of the **dispersed plug-flow reactor**, the previous equations may be used to get the time required, θ_{pf}, for a plug-flow reactor. Then this volume can be used to obtain the time, θ_{dpf}, for a dispersed plug-flow reactor using Figure 15.14. This method was illustrated in Example 15.2 and 15.2 SI.

The plug-flow equations, Eqs. (15.63) and (15.64), are rarely used in design because they have not been well verified by field studies. When substituted into the plug-flow equations, typical parameters for municipal wastewaters result in an extremely small detention time.

Some investigators (McCarty and Lawrence, 1970) in their research work have used a modification of the Monod and the Michaelis-Menten equations as a basis for their derivations. The modified equation in the form usually used is

$$\frac{dS}{dt} = kX\left(\frac{S}{K_s + S}\right)$$

where

$\quad k\ $ = maximum rate constant for substrate utilization

$\quad K_s$ = substrate concentration when the rate of substrate utilization is half the maximum rate

The design equations that have been derived are similar to Eqs. (15.45), (15.46), (15.47), (15.58), (15.60), (15.61), (15.64), (15.65), and (15.66) except that the value Yk is equal to μ_{max} in these equations.

Inspection of the previous derivations shows that Eqs. (15.46) and (15.47) for the nonrecycle reactor are identical to Eqs. (15.60) and (15.61) for the recycle reactor. For both systems there will be a minimum mean cell residence time, θ_w, at which washout of the cells from the reactors will occur. The washout mean cell residence time, θ_w, can be determined from Eqs. (15.46) and (15.60) by substituting S_i for S_1, because this will occur for a washout condition.

SPACE OR VOLUMETRIC LOADING RELATIONSHIPS

The design of the reactor basin volume by space or volumetric loading rates is the simplest and oldest method of determining the reactor basin volume. It is mainly limited to small municipal wastewater treatment plants with negligible amounts of industrial wastewaters present in the municipal wastewater. The space or volumetric loading is the lb BOD$_5$/day-1000 ft^3 (kg BOD$_5$/day-m^3) or reactor basin volume. The reactor basin volume is equal to the lb (kg) BOD$_5$ applied per day to the reactor basin, divided by the space or volumetric loading. A disadvantage to this method of design is that it ignores the concentration of the MLVSS or MLSS. A reactor with an MLSS of 4000 mg/ℓ will have a much better performance than one with an MLSS of 2000 mg/ℓ; however, space or volumetric loading design ignores this.

EXAMPLE 15.8 AND 15.8 SI

Space or Volumetric Loading Design

An activated sludge treatment plant is to be designed for a city of 5000 persons, and primary clarification is to be used. The BOD$_5$ contribution is 0.17 lb BOD$_5$/cap-day (0.077 kg/cap-d). Other pertinent data are: the primary clarifier removes 33% of the influent BOD$_5$, and the space or volumetric loading allowed by the state regulatory agency is 40 lb/day-1000 ft^3 (0.64 kg/d-m^3). Determine the reactor basin volume.

SOLUTION

The BOD$_5$ applied to the reactor per day for the solution in USCS units is

$$\text{BOD}_5 \text{ applied/day} = (5000 \text{ persons})(0.17 \text{ lb BOD}_5/\text{cap-day})(1 - 0.33)$$
$$= 570 \text{ lb BOD}_5/\text{day}$$

The reactor basin volume for the solution in USCS units is

$$V = (570 \text{ lb BOD}_5/\text{day})(1000 \text{ ft}^3\text{-day}/40 \text{ lb BOD}_5) = \boxed{14{,}300 \text{ ft}^3}$$

The BOD$_5$ applied to the reactor per day for the solution in SI units is

$$\text{BOD}_5 \text{ applied/d} = (5000 \text{ persons})(0.077 \text{ kg/cap-d})(1 - 0.33)$$
$$= 258 \text{ kg BOD}_5/\text{d}$$

The reactor basin volume for the solution in SI units is

$$V = (258 \text{ kg BOD}_5/\text{d})(\text{m}^3\text{-d}/0.64 \text{ kg}) = \boxed{403 \text{ m}^3}$$

In the design by kinetic relationships, the space or volumetric loading should be determined as a check on the design calculations.

EMPIRICAL RELATIONSHIPS

Several empirical relationships have been developed to give the performance of activated sludge plants when used for treating municipal wastewaters. One of the most widely used is the expression by Fair and Thomas (1950), commonly termed the **National Research Council (NRC) equation**, which is,

$$E_s = \frac{100}{1 + 0.03\left(\dfrac{y_0}{W\theta_i}\right)^{0.42}} \tag{15.68}$$

where

E_s = efficiency of the secondary stage — that is, the aeration tank and the final clarifier, percent

y_0 = pounds (kilograms) of BOD$_5$ applied to the aeration tank per day

W = thousands of pounds (kilograms) of activated sludge in the aeration tank

θ_i = aeration time based on the influent flow, hours

In the derivation of this expression, the aeration time, θ_i, is assumed to be V/Q. Another convenient form of Eq. (15.68) is

$$W\theta_i = \frac{y_0}{4200}\left(\frac{E_s}{100 - E_s}\right)^{2.38} \tag{15.69}$$

At the time the NRC equation was developed, the only types of activated sludge plants that existed were the conventional and tapered aeration processes. Consequently, the NRC equations apply only to these processes.

EXAMPLE 15.9   *Activated Sludge Performance by the NRC Equation*

A tapered aeration activated sludge plant is to treat the wastewater from a population of 150,000 persons, and the effluent BOD$_5$ is to be 20 mg/ℓ. Other pertinent data are: flow = 100 gal/cap-day, BOD$_5$ of the primary clarifier effluent = 158 mg/ℓ, and MLSS = 2500 mg/ℓ. Determine the required aeration time in hours.

SOLUTION   The efficiency of the secondary stage is $[(158 - 20)/(158)] \times 100 = 87.34\%$. The value of y_0 is $(150,000 \text{ persons})(100 \text{ gal/cap-day})(8.34 \text{ lb/gal})(158/10^6) = 19,766$ lb BOD$_5$ per day. The value of $W\theta_i$ is given by Eq. (15.69) or

$$W\theta_i = \left(\frac{19,766}{4200}\right)\left(\frac{87.34}{100 - 87.34}\right)^{2.38}$$

$$= 466.6 \tag{15.70}$$

Since $V = Q\theta_i$ and $Q = (150,000 \text{ persons})(100 \text{ gal/cap-day}) = 15 \text{ MGD}$,

$$W = \left(\frac{15,000,000 \text{ gal}}{24 \text{ hr}}\right)\left(\frac{8.34 \text{ lb}}{\text{gal}}\right)\left(\frac{2500}{10^6}\right)\left(\frac{\theta_i}{1000}\right)$$

$$= 13.03\,\theta_i \tag{15.71}$$

Combining Eqs. (15.70) and (15.71) gives $\theta_i = \boxed{5.98 \text{ hours.}}$

EXAMPLE 15.9 SI   *Activated Sludge Performance by the NRC Equation*

A tapered aeration activated sludge plant is to treat the wastewater from a population of 150,000 persons, and the effluent BOD$_5$ is to be 20 mg/ℓ. Other pertinent data are: flow = 380 ℓ/cap-d, BOD$_5$ of the primary clarifier effluent = 158 mg/ℓ, and MLSS = 2500 mg/ℓ. Determine the required aeration time in hours.

SOLUTION   The efficiency of the seconary stage is $(158 - 20)/(158) = 87.34\%$. The value of y_0 is (150,000 persons)(380 ℓ/cap-d)(158 mg/ℓ)(gm/1000 mg)(kg/1000 gm) = 9006 kg BOD$_5$ per day. The value of $W\theta_i$ is given by Eq. (15.69) or

$$W\theta_i = \left(\frac{9006}{4200}\right)\left(\frac{87.34}{100 - 87.34}\right)^{2.38} = 212.6 \tag{15.72}$$

Since $V = Q\theta_i$ and $Q = (150,000 \text{ persons})(380 \text{ } \ell/\text{cap-d}) = 57,000,000 \text{ } \ell/\text{d}$,

$$W = \left(\frac{57,000,000 \text{ } \ell}{24 \text{ h}}\right)\left(\frac{2500 \text{ mg}}{\ell}\right)\left(\frac{gm}{1000 \text{ mg}}\right)\left(\frac{kg}{1000 \text{ gm}}\right)\left(\frac{\theta_i}{1000}\right) = 5.938 \, \theta_i \tag{15.73}$$

Combining Eqs. (15.72) and (15.73) gives $\theta_i = \boxed{5.98 \text{ h.}}$

REACTOR BASINS

Usually, plug-flow or dispersed plug-flow reactor basins are rectangular tanks constructed of reinforced concrete and employ diffused compressed air. Rectangular basins can use common wall construction when multiple basins are provided. The walls are vertical and usually have 1.25 to 2 ft (0.38 to 0.61 m) freeboard, as shown in Figure 15.29. The mixed liquor depth is usually 10 to 15 ft (3.05 to 4.57 m), and the tank width is from 1.5 to 2.0 times the depth. The length:width ratio is 5:1 or more, and the tank lengths may be up to 500 ft (152 m) or more. The air diffusers are usually mounted from 1.5 to 2.5 ft (0.46 to 0.76 m) above the tank bottom. Fillets are provided at the bottom of the walls, and the tops of the walls flare out at 45° to promote a spiral roll to the mixed liquor. Multiple basins should be provided when the flow exceeds 1.0 MGD (4 MLD). It is desirable to have a diffused air system that allows the diffusers to be raised above the mixed liquor surface for cleaning. If mechanical surface aerators are used, the tank width must be great enough to contain the spray and mist from the units; also, high water velocities must be avoided to minimize splashing against the basin walls. Three types of mechanical aerators have been successfully used in rectangular tanks; these are the slow-speed vertical turbine or radial-flow turbine, the submerged turbine with a sparger ring, and the horizontal rotating brush aerator. The slow-speed vertical turbine has a gear reduction box to obtain low rotational speeds (30 to 60 rpm), and the turbine is barely submerged below the still water surface. As the turbine rotates, the flow moves radially outward from the unit. In rectangular basins, these units

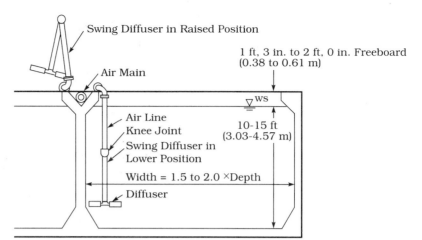

FIGURE 15.29
*Section through an
Aeration Tank*

usually have a draft tube to obtain satisfactory mixing. The submerged turbine has a slow-speed turbine mounted slightly above the tank bottom with a sparger ring beneath the turbine. The sparger ring allows compressed air to be released below the turbine and results in a relatively high oxygen transfer. When turbines are used in rectangular tanks, they usually are located halfway between the walls. The basin width is usually larger than the width when compressed air is used. The rotating brush aerator, which is also used in oxidation ditches, consists of a horizontal axle with radiating steel bristles. As the axle rotates, the bristles spray the water outward, giving mixing and also entraining air. In rectangular tanks, the rotating axles are mounted along the same wall.

Usually, completely mixed reactor basins for small installations are square or circular concrete tanks with vertical walls and may use diffused compressed air or mechnical aeration. For large basins, square or rectangular tanks with a length:width ratio of 3:1 or less are used. The sides of square or rectangular basins may be vertical reinforced concrete walls or sloped embankments having a reinforced concrete liner on the slope to avoid erosion. The floor of a basin is usually a reinforced concrete slab. Sometimes, gunite or asphalt liners are used instead of reinforced concrete because they are less expensive. Because of their relatively large areas, widths, and lengths, completely mixed basins usually employ mechanical aeration, although compressed air has successfully been used. Widths and lengths of several hundred feet are common. In particular, the slow-speed vertical turbine type aerator mounted on piers has been very popular. Sometimes these units are float mounted for flexibility. For slow-speed vertical turbines, tank depths are from 6 to 30 ft (1.8 to 9.1 m); with 12 to 15 ft (3.66 to 4.57 m) being most common. Tanks from 15 to 30 ft (4.57 to 9.14 m) deep usually require draft tubes or a second turbine mounted at about mid-depth to obtain adequate mixing. Aeration basins are usually designed so that the area of influence around each aerator is approximately a square.

Biochemical equations are useful in illustrating the principles of respiration and synthesis involved in bio-oxidation. A compound or substance is oxidized if (1) it is united with oxygen, (2) hydrogen atoms are removed, or (3) electrons are removed. Conversely, a compound or substance is reduced if (1) oxygen is removed from the substance, (2) hydrogen atoms are added, or (3) electrons are added. If the empirical equation for the substrate and the yield coefficient for the cells (that is, mass of cells produced per mass of substrate used) are known, it is possible to develop respiration and synthesis equations and also an overall biochemical equation representing the bio-oxidation (Hoover and Porges, 1952). Since biological oxidation involves oxidation-reduction reactions, these are utilized in the development of equations for biochemical reactions. For chemoorganotrophs, the organic substrate is oxidized and molecular oxygen is reduced; thus, the steps necessary to develop the respiration equation require an oxidation reaction for the substrate and a reduction reaction for the molecular oxygen.

If the empirical equation for the organic portion of a wastewater is $C_6H_{14}O_2N$ and the nitrogen end product is the ammonium ion, the development of the respiration equation for a slightly basic condition is as follows.

Step 1. Balance all the key elements except oxygen and hydrogen to form the end products of CO_2 and NH_4^+.

$$C_6H_{14}O_2N \rightarrow 6CO_2 + NH_4^+$$
$$O_2 \rightarrow 2H_2O$$

Step 2. Balance the oxygen with the oxygen of water.

$$C_6H_{14}O_2N + 10H_2O \rightarrow 6CO_2 + NH_4^+$$
$$O_2 \rightarrow 2H_2O$$

Step 3. Balance the hydrogen with the hydrogen ion.

$$C_6H_{14}O_2N + 10H_2O \rightarrow 6CO_2 + NH_4^+ + 30H^+$$
$$4H^+ + O_2 \rightarrow 2H_2O$$

Step 4. Balance the charges with electrons.

$$C_6H_{14}O_2N + 10H_2O \rightarrow 6CO_2 + NH_4^+ + 30H^+ + 31e^-$$
$$4H^+ + 4e^- + O_2 \rightarrow 2H_2O$$

Step 5. Add the two reactions to eliminate the electrons.

$$C_6H_{14}O_2N + 10H_2O \rightarrow 6CO_2 + NH_4^+ + 30H^+ + 31e^-$$
$$31H^+ + 31e^- + 7.75O_2 \rightarrow 15.5H_2O$$

$$\overline{C_6H_{14}O_2N + 7.75O_2 + H^+ \rightarrow 6CO_2 + NH_4^+ + 5.50H_2O} \tag{15.74}$$

Equation (15.74) represents the respiration in a slightly acidic solution.

Step 6. For a slightly basic solution, add hydroxyl ions to unite with the hydrogens to form water. Adding one hydroxyl ion to both sides of Eq. (15.74) yields

$$C_6H_{14}O_2N + 7.75O_2 \rightarrow 6CO_2 + NH_4^+ + 4.50H_2O + OH^- \quad \textbf{(15.75)}$$

Thus Eq. (15.75) represents the respiration equation in a slightly basic solution.

If the empirical equation for the activated sludge produced is $C_5H_7O_2N$, the synthesis equation for a slightly basic solution is developed in a manner similar to the respiration equation and is as follows:

Step 1. Balance all the key elements except oxygen and hydrogen. Since the nitrogen in the respiration equation terminated as the ammonium ion, use the ammonium ion in the balance.

$$5C_6H_{14}O_2N + NH_4^+ \rightarrow 6C_5H_7O_2N$$
$$O_2 \rightarrow 2H_2O$$

Step 2. Balance the oxygen with the oxygen of water.

$$5C_6H_{14}O_2N + NH_4^+ + 2H_2O \rightarrow 6C_5H_7O_2N$$
$$O_2 \rightarrow 2H_2O$$

Step 3. Balance the hydrogen with the hydrogen ion.

$$5C_6H_{14}O_2N + NH_4^+ + 2H_2O \rightarrow 6C_5H_7O_2N + 36H^+$$
$$4H^+ + O_2 \rightarrow 2H_2O$$

Step 4. Balance the charges with electrons.

$$5C_6H_{14}O_2N + NH_4^+ + 2H_2O \rightarrow 6C_5H_7O_2N + 36H^+ + 35e^-$$
$$4H^+ + 4e^- + O_2 \rightarrow 2H_2O$$

Step 5. Add the two reactions to eliminate the electrons.

$$5C_6H_{14}O_2N + NH_4^+ + 2H_2O \rightarrow 6C_5H_7O_2N + 36H^+ + 35e^-$$
$$35H^+ + 35e^- + 8.75O_2 \rightarrow 17.5H_2O$$
$$\overline{\phantom{5C_6H_{14}O_2N + NH_4^+ + 8.75O_2 \rightarrow 6C_5H_7O_2N + H^+ + 15.5H_2O}} \quad \textbf{(15.76)}$$
$$5C_6H_{14}O_2N + NH_4^+ + 8.75O_2 \rightarrow 6C_5H_7O_2N + H^+ + 15.5H_2O$$

Equation (15.76) represents the synthesis equation for a slightly acidic solution.

Step 6. For a slightly basic solution, add hydroxyl ions to each side to unite with the hydrogen ion to form water. Adding one hydroxyl to each side yields

$$5C_6H_{14}O_2N + NH_4^+ + 8.75O_2 + OH^- \rightarrow 6C_5H_7O_2N + 16.5H_2O \quad \textbf{(15.77)}$$

Equation (15.77) represents the synthesis equation for a slightly basic solution. In some cases, no oxygen will be required in the synthesis equation.

To combine the respiration equation with the synthesis equation to give the overall biochemical reaction requires the yield coefficient, Y, which is defined as the mass of cells produced per unit mass of substrate utilized. The theoretical oxygen demand (TOD) may be determined from the respiration equation, and since the pseudomolecular weight of the substrate is 132, the

TOD is $(7.75)(32)/132$ or $1.879\,\text{mg}$ oxygen per mg substrate. If the yield coefficient, Y, is assumed to be $0.40\,\text{mg}$ cells per mg TOD, the yield coefficient in terms of milligrams of substrate is

$$Y = (0.40\,\text{mg cells/mg TOD})(1.879\,\text{mg TOD/mg substrate})$$

$$= 0.752\,\text{mg cells/mg substrate}$$

Assuming $100\,\text{mg}$ of cells are produced, the substrate used is

$$\text{Substrate} = (100\,\text{mg cells})(1\,\text{mg substrate}/0.752\,\text{mg cells})$$

$$= 132.98\,\text{mg}$$

Letting x be the substrate required for respiration and y the substrate required for synthesis gives

$$x + y = 132.98\,\text{mg}$$

The pseudomolecular weight for the cells is 113; thus, from Eq. (15.77),

$$\frac{y}{5(132)} = \frac{100}{6(113)} \quad \text{or } y = 97.35\,\text{mg for synthesis}$$

Thus $x = 132.98 - 97.35$ or $x = 35.63\,\text{mg}$ of substrate used for respiration. The fraction of substrate used for respiration is $35.63/132.98$ or 0.268, and the fraction of substrate used for synthesis is $97.35/132.98$ or 0.732. To combine the respiration equation, Eq. (15.75), with the synthesis equation, Eq. (15.77), requires multiplying Eq. (15.75) by 0.268 and Eq. (15.77) by $0.732(\frac{1}{5})$. This yields the following equations, which may be combined:

$$0.268C_6H_{14}O_2N + 2.077O_2 \rightarrow 1.608CO_2 + 0.268NH_4^+$$
$$+ 1.206H_2O + 0.268OH^-$$

$$0.732C_6H_{14}O_2N + 0.146NH_4^+ + 1.281O_2$$
$$+ 0.146OH^- \rightarrow 0.878C_5H_7O_2N + 2.416H_2O$$

$$C_6H_{14}O_2N + 3.358O_2 \rightarrow 0.878C_5H_7O_2N + 1.608CO_2$$
$$+ 0.122NH_4^+ + 3.622H_2O + 0.122OH^- \qquad \textbf{(15.78)}$$

Thus Eq. (15.78) represents the overall biochemical equation representing the bio-oxidation of the organic wastewater having an empirical equation of $C_6H_{14}O_2N$ for the organic fraction.

The theoretical oxygen demand of the cells (TOD) may be determined from the following equation for the oxidation:

$$C_5H_7O_2N + H^+ + 5O_2 \rightarrow 5CO_2 + NH_4^+ + 2H_2O \qquad \textbf{(15.79)}$$

Thus the TOD is $(5)(32)/113$ or $1.42\,\text{mg}$ oxygen per mg cells. If the cell yield coefficient, Y, is expressed in terms of cell TOD produced per unit substrate TOD bio-oxidized, Y may be determined using values from Eq. (15.78) as

$$Y = \frac{\text{Cell TOD}}{\text{Substrate TOD}} = \frac{(0.878)(113)(1.42)}{(132)(1.879)} = 0.567 \qquad \textbf{(15.80)}$$

The oxygen coefficient, Y', is defined as the mass of oxygen used per unit mass of substrate utilized. If the substrate is expressed as TOD, the oxygen coefficient, Y', may be determined using values from Eq. (15.78) as

$$Y' = \frac{\text{Oxygen used}}{\text{Substrate TOD}} = \frac{(3.358)(32)}{(132)(1.879)} = 0.433 \qquad (15.81)$$

Thus the sum of the yield coefficient (Y) and the oxygen coefficient (Y') is

$$Y + Y' = 0.567 + 0.433 \qquad (15.82)$$

or

$$Y + Y' = 1.00 \qquad (15.83)$$

In summary, the previous biochemical equation for the aerobic bio-oxidation of the substrate, Eq. (15.78), shows that when the cells produced are expressed as TOD and the substrate utilized is expressed as TOD, the sum of the yield coefficient, Y, and the oxygen coefficient, Y', is unity (Stack and Conway, 1959). Thus if the yield coefficient, Y, is known, then Y' may be computed. If Y is in terms of cell mass produced per unit TOD utilized, then the yield coefficient in terms of cell TOD produced is $1.42Y$. Therefore,

$$1.42Y + Y' = 1.00 \qquad (15.84)$$

where Y is the mass of cells produced per unit substrate TOD oxidized.

Frequently BOD$_5$ is used instead of the TOD; for this case, the units may be converted from BOD$_5$ by applying the BOD$_5$:TOD ratio. For example, if the BOD$_5$ is 0.68 TOD, then

$$1.42Y\left(\frac{cells}{BOD_5}\right)\left(\frac{0.68\,\text{BOD}_5}{1.00\,\text{TOD}}\right) + Y'\left(\frac{oxygen}{BOD_5}\right)\left(\frac{0.68\,\text{BOD}_5}{1.00\,\text{TOD}}\right) = 1.00 \qquad (15.85)$$

where the units in italics represent the units to be utilized in the calculations. From Eq. (15.85) it can be seen that the BOD$_5$:TOD ratio merely coverts the units to TOD. From Eq. (15.85) it follows that

$$0.966 + 0.68Y' = 1.00 \qquad (15.86)$$

where

   Y = mass of cells produced per unit BOD$_5$ oxidized — for example, mg cells/mg BOD$_5$

   Y' = mass of oxygen required per unit BOD$_5$ oxidized — for example, mg oxygen/mg BOD$_5$

Equations (15.83) and (15.84) will not hold true if: (1) the wastewater has an appreciable volatile fraction that will be stripped from the water, (2) a significant amount of organic materials exert an immediate chemical oxygen demand, and (3) a sizeable amount of substrate is stored as food material. In general, though, the previous expressions are useful in determing the oxygen requirements for substrate bio-oxidation.

As shown in the previous section on biochemical equations, the bio-oxidation of an organic substrate produces a certain number of cells as a result of synthesis. At the same time, sludge is being endogenously decayed, and because endogenous degradation occurs during all growth phases, the entire mass of cells in the biological reactor is being endogeously degraded. The bio-oxidation of the ammonium ion to the nitrate ion synthesizes a certain mass of cells; however, this cell mass is usually very small in relation to the cells synthesized from the organic substrate, and it is usually neglected in determining sludge production. A material balance on the cells in the reactor-clarifier system is

$$[\text{Accumulation}] = [\text{Input}] - \left[\begin{array}{c}\text{Output due}\\ \text{to effluent}\end{array}\right] - \left[\begin{array}{c}\text{Output due}\\ \text{to wastage}\end{array}\right]$$
$$+ \left[\begin{array}{c}\text{Increase due}\\ \text{to synthesis}\end{array}\right] - \left[\begin{array}{c}\text{Decrease due}\\ \text{to decay}\end{array}\right] \qquad (15.87)$$

Because the accumulation term is zero for steady state, the input of active biological solids in the feed wastewater is negligible, and the output of active biological solids in the effluent is negligible, the material balance becomes

$$\left[\begin{array}{c}\text{Output due}\\ \text{to wastage}\end{array}\right] = \left[\begin{array}{c}\text{Increase due}\\ \text{to synthesis}\end{array}\right] - \left[\begin{array}{c}\text{Decrease due}\\ \text{to decay}\end{array}\right] \qquad (15.88)$$

The mathematical representation for the net cell production rate in Eq. (15.88) is

$$\frac{dX}{dt} = Y\frac{dS}{dt} - k_e X \qquad (15.89)$$

where

dX/dt = rate of net cell production, mass/time

Y = yield coefficient, mass microbes produced/mass substrate utilized

dS/dt = rate of substrate removal, mass/time

k_e = endogenous degradation coefficient, mass cells/(total mass cells)(time)

X = total cell mass in the biological reactor, mass

Equation (15.89) may be rearranged to give

$$dX = YdS - k_e Xdt \qquad (15.90)$$

Preparing for integration produces

$$\int_0^{X_w} dX = Y\int_0^{S_r} dS - k_e \bar{X}\int_0^1 dt \qquad (15.91)$$

Integration yields the equation by Eckenfelder and Weston (1956),

$$X_w = YS_r - k_e\bar{X} \qquad (15.92)$$

where

X_w = waste cells produced, mass/day

S_r = substrate removed, mass/day

k_e = endogenous coefficient, mass cells/(total mass cells)(day)

$\bar{X}$ = average cell concentration in the biological reactor, mass

The nature of the wastewater determines the values of Y and k_e. They may be determined from three or more batch reactors, as shown in Examples 15.14 and 15.14 SI, or from three or more completely mixed, continuous-flow reactors, as shown in Examples 15.15 and 15.15 SI.

Eq. (15.92) can be used to determine the sludge production from activated sludge reactors designed by any method previously mentioned; however, if a reactor has been designed using the Monod equation, a simpler equation is available. For these reactors, $\theta_c = \bar{X}/X_w$, from which $X_w = \bar{X}/\theta_c = (V X)/\theta_c$.

OXYGEN REQUIREMENTS

As depicted in the previous section on biochemical equations, the bio-oxidation of a substrate requires a certain amount of oxygen for respiration and, in most cases, for synthesis. Also, oxygen is required for the endogenous degradation of the cell mass and for nitrification. Thus the total oxygen required is given by (AWARE, 1974)

$$O_r = Y'S_r + k'_e \bar{X} + O_n \qquad (15.93)$$

where

O_r = total oxygen required, mass/day

Y' = oxygen coefficient, mass oxygen/mass substrate utilized

k'_e = endogenous respiration coefficient, mass oxygen/(mass cells)(day)

O_n = oxygen required for nitrification, mass/day

Research has shown that 4.33 mg of oxygen are required to convert 1 mg of ammonia nitrogen to the nitrate ion (Eckenfelder, 1989). Equation (15.93) can be used to determine the oxygen required by an activated sludge reactor designed by any of the previously mentioned methods.

EXAMPLE 15.10 AND 15.10 SI

Sludge Production and Oxygen Requirements

An activated sludge plant treats 5.0 MGD (18.9 MLD) of municipal wastewater. The recycle ratio, R/Q, is $\frac{1}{3}$, the aeration time is 6 hours, the MLSS concentration is 2500 mg/ℓ, the SDI = 10,000 mg/ℓ and the MLVSS is 75% of the MLSS. The yield coefficient, Y, is 0.60 mg MLVSS/mg BOD$_5$ removed, and the endogenous coefficient, k_e, is 0.06 day^{-1}. The oxygen coefficient, Y', is 0.62 mg oxygen/mg BOD$_5$ removed, and the endogenous oxygen coefficient, k'_e, is 0.09 mg oxygen/mg MLVSS-day. The sum of the organic and ammonia nitrogen in the primary effluent is 25 mg/ℓ, and, excluding the nitrogen synthesized, 100% is converted to nitrate. It requires 4.33 mg of oxygen to convert 1 mg of ammonia nitrogen to the nitrate ion. The flow from the primary clarifier has a BOD$_5$ of 135 mg/ℓ and the effluent from the plant has a BOD$_5$ of 10 mg/ℓ. Determine:

1. The net sludge produced as a result of BOD$_5$ removal and endogenous decay.
2. The oxygen required.
3. The waste activated sludge flow rate, Q_w, in gallons per day (liters per day) if the sludge is wasted from the recycle line and the wet specific gravity is 1.01.
4. The mean cell residence time, θ_c, for the reactor basin in days.

SOLUTION FOR USCS UNITS   The reactor volume is

$$V = \frac{5.0 \times 10^6 \,\text{gal}}{\text{day}} \left|\frac{\text{day}}{24\,\text{hr}}\right| \frac{1 + 0.333}{} \left| 6\,\text{hr} \right.$$

$$= 1.666 \times 10^6 \,\text{gal}$$

The total mass of MLVSS in the reactor is

$$\bar{X} = \frac{1.666 \times 10^6 \,\text{gal}}{} \left|\frac{8.34\,\text{lb}}{\text{gal}}\right| \frac{2500}{10^6} \left| 0.75 \right.$$

$$= 26,052\,\text{lb}$$

The substrate removed per day is

$$S_r = \frac{135 - 10}{10^6} \left|\frac{5.0 \times 10^6 \,\text{gal}}{\text{day}}\right| \frac{8.34\,\text{lb}}{\text{gal}}$$

$$= 5213\,\text{lb of BOD}_5$$

The volatile suspended solids produced is

$$X_w = YS_r - k_e\bar{X}$$
$$= \left(\frac{0.60\,\text{lb MLVSS}}{\text{lb BOD}_5}\right)\left(\frac{5213\,\text{BOD}_5}{\text{day}}\right) - \left(\frac{0.06}{\text{day}}\right)(26,052\,\text{lb})$$
$$= \boxed{1565\,\text{lb MLVSS/day}}$$

A material balance for the nitrogen is

$$[\text{Input}] = [\text{Output}] + \begin{bmatrix}\text{Decrease due} \\ \text{to synthesis}\end{bmatrix} + \begin{bmatrix}\text{Decrease due} \\ \text{to nitrification}\end{bmatrix}$$

The input is $(5 \times 10^6 \,\text{gal/day})(25/10^6)(8.34\,\text{lb/gal})$ or 1043 lb N/day. Since the cells are represented by $C_5H_7O_2N$, the percent nitrogen is $(14/113)(100)$ or 12.39%. Thus the decrease due to synthesis is $(1565\,\text{lb MLVSS/day})$ $(0.1239\,\text{lb N/lb MLVSS})$ or 194 lb N/day. Therefore,

$$[\text{Output}] + \begin{bmatrix}\text{Decrease due} \\ \text{to nitrification}\end{bmatrix} = 1043 - 194 = 849\,\text{lb N/day}$$

The [Output] term represents the organic and ammonia nitrogen in the effluent, and since there is 100% conversion, the [Output] term is zero. Thus the [Decrease due to nitrification] = 849 lb $\underline{N}$/day. The total oxygen requirement is

$$O_r = Y'S_r + k'_e\bar{X} + O_n$$

$$= \left(\frac{0.62 \text{ lb oxygen}}{\text{lb BOD}_5}\right)\left(\frac{5213 \text{ lb BOD}_5}{\text{day}}\right) + \left(\frac{0.09 \text{ lb oxygen}}{\text{lb MLVSS-day}}\right)(26,052 \text{ lb MLVSS})$$

$$+ \left(\frac{849 \text{ lb }\underline{N}}{\text{day}}\right)\left(\frac{4.33 \text{ lb oxygen}}{\text{lb }\underline{N}}\right)$$

$$= 3232 + 2345 + 3676$$

$$= \boxed{9253 \text{ lb oxygen/day}}$$

The waste activated sludge flow is given by

$$Q_w = \frac{1565 \text{ lb MLVSS}}{\text{day}} \left|\frac{\text{lb MLSS}}{0.75 \text{ lb MLVSS}}\right| \frac{10^6}{10,000}\left|\frac{\text{gal}}{8.34 \text{ lb}}\right| 1.01$$

$$= \boxed{24,800 \text{ gal/day}}$$

The mean cell residence time, θ_c, is given by

$$\theta_c = \frac{\bar{X}}{X_w} = \frac{26,052 \text{ lb MLVSS}}{1565 \text{ lb MLVSS/day}} = \boxed{16.6 \text{ days}}$$

The reactor volume is

$$V = \frac{18,900,000 \, \ell}{\text{d}}\left|\frac{\text{day}}{24 \text{ h}}\right|\frac{\ell + 0.333}{}\left|\frac{6 \text{ h}}{}\right| = 6,298,425 \, \ell$$

The total mass of MLVSS in the reactor is

$$\bar{X} = \frac{6,298,425 \, \ell}{}\left|\frac{2500 \text{ mg}}{\ell}\right|\frac{\text{gm}}{1000 \text{ mg}}\left|\frac{\text{kg}}{1000 \text{ gm}}\right|0.75 = 11,810 \text{ kg}$$

The substrate removed per day is

$$S_r = \frac{135 - 10}{}\left|\frac{\text{mg}}{\ell}\right|\frac{18,900,000 \, \ell}{\text{d}}\left|\frac{\text{gm}}{1000 \text{ mg}}\right|\frac{\text{kg}}{1000 \text{ gm}} = 2363 \text{ kg of BOD}_5$$

The volatile suspended solids produced is

$$X_w = YS_r - k_e\bar{X}$$

$$= \left(\frac{0.60 \text{ kg MLVSS}}{\text{kg BOD}_5}\right)\left(\frac{2363 \text{ kg BOD}_5}{\text{d}}\right) - \left(\frac{0.06}{\text{d}}\right)(11,810 \text{ kg MLVSS})$$

$$= \boxed{709 \text{ kg MLVSS/d}}$$

A material balance for the nitrogen is

$$[\text{Input}] = [\text{Output}] + \begin{bmatrix}\text{Decrease due} \\ \text{to synthesis}\end{bmatrix} + \begin{bmatrix}\text{Decrease due} \\ \text{to nitrification}\end{bmatrix}$$

The input is $(18,900,000\,\ell/d)(25\,mg/\ell)(gm/1000\,mg)(kg/1000\,gm) = 473\,kg\,\underline{N}/d$. Since the cells are represented by $C_5H_7O_2N$, the percent nitrogen is $(14/113)(100)$ or 12.39%. Thus the decrease due to synthesis is $(709\,kg\,MLVSS/day)(0.1239)$ or $87.8\,kg\,\underline{N}/d$. Therefore,

$$[\text{Output}] + \begin{bmatrix} \text{Decrease due} \\ \text{to nitrification} \end{bmatrix} = 473 - 87.8 = 385.2\,kg\,\underline{N}/d$$

The [Output] term represents the organic and ammonia nitrogen in the effluent, and since there is 100% conversion, the [Output] term is zero. Thus the [Decrease due to nitrification] $= 385.2\,kg\,\underline{N}/d$. The total oxygen requirement is

$$
\begin{aligned}
O_r &= Y'S_r + k'_e\bar{X} + O_n \\
&= \left(\frac{0.62\,kg\,\text{oxygen}}{kg\,BOD_5}\right)\left(\frac{2363\,kg\,BOD_5}{d}\right) + \left(\frac{0.09\,kg\,\text{oxygen}}{kg\,MLVSS\text{-}d}\right)(11{,}810\,kg\,MLVSS) \\
&\quad + \left(\frac{385.2\,kg\,\underline{N}}{d}\right)\left(\frac{4.33\,kg\,\text{oxygen}}{kg\,\underline{N}}\right) \\
&= \boxed{4196\,kg\,\text{oxygen}/d}
\end{aligned}
$$

The waste activated sludge flow is given by

$$
\begin{aligned}
Q_w &= \frac{709\,kg\,MLVSS}{d}\left|\frac{kg\,MLSS}{0.75\,kg\,MLVSS}\right|\frac{10^6\,mg\,MLSS}{kg\,MLSS}\left|\frac{ml}{1.01\,mg}\right|\frac{\ell}{1000\,ml} \\
&= \boxed{936{,}000\,\ell/d}
\end{aligned}
$$

The mean cell residence time, θ_c, is given by

$$\theta_c = \frac{\bar{X}}{X_w} = \frac{11{,}810\,kg\,MLVSS}{709\,kg\,MLVSS/d} = \boxed{16.6\,d}$$

EXAMPLE 15.11 AND 15.11 SI

Sludge Production by Reactors Designed by the Monod Equation

Determine the sludge production for the completely mixed reactor in Examples 15.7 and 15.7 SI. In these examples, $\theta_c = 16.7$ days, $V = 148,000\,ft^3$ $(4180\,m^3)$, and $X = 3000\,mg/\ell\,MLSS$.

SOLUTION FOR USCS UNITS

$$X_w = \frac{XV}{\theta_c} = \left(\frac{148{,}000\,ft^3}{16.7\,days}\right)\left(\frac{62.4\,lb}{ft^3}\right)\left(\frac{3000}{10^6}\right) = \boxed{1659\,lb\,MLSS/day}$$

SOLUTION FOR SI UNITS

$$X_w = \left(\frac{4180\,m^3}{16.7\,d}\right)\left(\frac{1000\,\ell}{m^3}\right)\left(\frac{3000\,mg}{\ell}\right)\left(\frac{g}{1000\,mg}\right)\left(\frac{kg}{1000\,g}\right) = \boxed{751\,kg/day}$$

NITRIFICATION

Most medium- to large-size wastewater treatment plants are required to achieve good nitrification — that is, the conversion of ammonia nitrogen,

NH_3-N, to nitrate nitrogen, NO_3-N — and effluent permits usually impose a maximum NH_3-N concentration allowed. If nitrification is not achieved in the treatment plant, it will occur in the receiving stream, thus placing an additional oxygen demand on the stream. All forms of nitrogen are usually reported in terms of the nitrogen concentration and not the nitrogen ion concentration involved. For example, NH_3 has a molecular weight of 14 + 3(1) = 17. The nitrogen content is 14/17 or 0.82 or 82%. Thus 30 mg/ℓ of NH_3 would be (30)(0.82) or 24.6 mg/ℓ as nitrogen. When the chemical symbol NH_3-N, NO_2-N, or NO_3-N is used in expressing the concentration of nitrogen, it indicates that the nitrogen forms are in terms of nitrogen. Twenty milligrams per liter of NH_3-N means that it is 20 mg/ℓ as nitrogen. The microbes involved are the autotrophic genera *Nitrosomonas* and *Nitrobacter*, which carry out the reaction in two steps:

$$2NH_4^+ + 3O_2 \rightarrow 2NO_2^- + 4H^+ + 2H_2O \quad \text{(\textit{Nitrosomonas})} \quad \textbf{(15.94)}$$

$$2NO_2^- + O_2 \rightarrow 2NO_3^- \quad \text{(\textit{Nitrobacter})} \quad \textbf{(15.95)}$$

For nitrification to occur in an activated sludge reactor basin, four critieria are required.

1. The design mean cell residence time, θ_c, must be sufficient for nitrifiers. In some cases the design mean cell residence time, θ_c, for the nitrifiers is greater than the design, θ_c, required for substrate removal, such as BOD$_5$ removal.

2. The design detention time, θ, should be sufficient for nitrification to occur. In some cases, the detention time for the nitrifiers, θ, is greater than the detention time, θ, for substrate removal, such as BOD$_5$ removal, and nitrification will control.

3. The operating dissolved oxygen concentration, DO, should be equal to or greater than 2.0 mg/ℓ. Below this level, the rate of nitrification is inhibited.

4. The oxygen supplied to the basin should be sufficient for nitrification. If sufficient oxygen is not supplied, substrate removal, such as BOD$_5$ removal, will occur, but the degree of nitrification will be limited.

The cell yield for *Nitrosomonas* has been reported (Eckenfelder, 1989) as 0.05 to 0.29 mg MLVSS/mg NH_3-N and for *Nitrobacter* as 0.02 to 0.08 mg MLVSS/mg NH_3-N. A value of 0.15 mg MLVSS/mg NH_3-N is commonly used for design purposes. It is generally accepted that the specific growth rate, μ, of *Nitrobacter* is higher than the specific growth rate, μ, of *Nitrosomonas*. Therefore, there will be no accumulation of nitrite, NO_2^-, in the process, and the growth rate of *Nitrosomonas* will control the overall conversion reaction.

For effective nitrification to occur, the mean cell residence time, θ_c, must be greater than $1/\mu_{max}$ of *Nitrosomonas*. That is, $\theta_c^{Min} = 1/\mu_{max}$, and lower θ_c values result in washout of the microbes. Nitrification is affected by the mixed liquor temperature, the optimum range being 25°C to 35°C. The minimum mean cell residence time is (Eckenfelder, 1989)

$$\theta_c^{\text{Min}} = 2.13e^{0.098(15-T)} \tag{15.96}$$

The U.S. Environmental Protection Agency (EPA) recommends that a safety factor of 2.5 be used for design. Thus the design $\theta_c = 2.5\,\theta_c^{\text{Min}}$. This safety factor insures that ammonia breakthrough does not occur during diurnal peak loads.

The fraction of nitrifiers in the reactor basin can be estimated from (Eckenfelder, 1989)

$$f_n = \frac{Y_n N_r}{Y_n N_r + Y S_r} \tag{15.97}$$

where

- f_n = fraction of nitrifiers in the reactor
- Y_n = growth coefficient for the nitrifiers, mass cells/mass substrate
- N_r = nitrogen substrate removed, mg/ℓ
- Y = growth coefficient for the MLVSS, mass MLVSS/mass substrate
- S_r = carbonaceous substrate removed (*e.g.*, BOD$_5$), mg/ℓ

The nitrification rate at 20°C can be estimated as 1.04 mg NH$_3$-N oxidized per milligram of nitrifying cell mass per day. The overall rate of nitrification at 20°C can be expressed as

$$R_n = 1.04 f_n X_v \tag{15.98}$$

where

- R_n = rate of nitrification, mg/ℓ-day
- f_n = fraction of nitrifiers in the mixed liquor
- X_v = mixed liquor volatile suspended solids (MLVSS), mg/ℓ

The nitrogen oxidized is the nitrogen in the influent minus the nitrogen synthesized to cells, or

$$N_r = \text{TKN} - N_s \tag{15.99}$$

where

- N_r = nitrogen removed, mg/ℓ
- TKN = total Kjeldahl nitrogen in the wastewater, mg/ℓ
- N_s = nitrogen synthesized into nitrifying cells, mg/ℓ

The total Kjeldahl nitrogen is the total organic nitrogen, Org-N, plus the ammonia nitrogen, NH$_3$-N, or

$$\text{TKN} = \text{Organic N} + \text{NH}_3\text{-N} \tag{15.100}$$

The organic nitrogen in municipal wastewater, without any industrial wastewater source of organic nitrogen, is mainly the nitrogen in proteinaceous food materials or their degradation products plus nitrogen in urea. Nitrifying microbes are inhibited by certain inorganic and organic chemicals. A discussion of these is presented by Eckenfelder (1989).

Nitrification is temperature dependent, and the effects are shown by a modification of Eq. (15.98)

$$R_n = 1.04\theta^{T-20}f_nX_v \qquad (15.101)$$

where

$\theta = 1.03$ to 1.15

$T =$ mixed liquor temperature, °C

All microbial species have an optimum pH range in which they thrive. For the nitrifers, this is from pH 6 to 7.5.

EXAMPLE 15.12
AND 15.12 SI

Nitrification

For Examples 15.10 and 15.10 SI, the organic and ammonia nitrogen in the influent to the reactor basin was 25 mg/ℓ N. The minimum temperature of the mixed liquor is 19°C and the temperature correction coefficient, θ, is 1.09. Determine:

1. The minimum mean cell residence time, θ_c^{Min}, required for the nitrifiers.
2. The design mean cell residence time, θ_c, for the nitrifiers. Does this control the design?
3. The rate of nitrification in mg/ℓ-day.
4. The detention time required for the nitrifiers. Does this control the design?

SOLUTION

Since this problem requires numerous data and parameters from Examples 15.10 and 15.10 SI, these are summarized as follows: Primary clarifier effluent BOD₅ = 135 mg/ℓ, effluent BOD₅ = 10 mg/ℓ, MLSS = 2500 mg/ℓ, MLVSS = 75% of the MLSS, organic and ammonia nitrogen in the primary clarifier effluent = 1043 lb N/day (473 kg/d), organic and ammonia nitrogen converted = 849 lb N/day (385.2 kg/d), $\theta = 6$ hours and $\theta_c = 16.6$ days. The minimum mean cell residence time, θ_c^{Min}, for the nitrifiers is

$$\theta_c^{\text{Min}} = 2.13e^{0.098(15-T)}$$
$$= 2.13e^{0.098(15-19)} = \boxed{1.44\,\text{days}}$$

The design mean cell residence time, θ_c, is

$$\theta_c = 2.5\theta_c^{\text{Min}} = (2.5)(1.44) = \boxed{3.60\,\text{days}}$$

For substrate removal (BOD₅ removal), $\theta_c = 16.6$ days > 3.60 days, so the process will nitrify and nitrification does not control the mean cell residence time, θ_c. The fraction of the nitrogen converted $= (849\,\text{lb}\,\underline{N}/\text{day})/(1043\,\text{lb}\,\underline{N}/\text{day}) = 0.814$ or $(385.2\,\text{kg}\,\underline{N}/\text{d})/(473\,\text{kg}\,\underline{N}/\text{d}) = 0.814$. Thus the nitrogen converted, N_r, is

$$N_r = (25\,\text{mg}/\ell)(0.814) = 20.4\,\text{mg}/\ell$$

The substrate removed (BOD$_5$) is $S_r = 135 - 10 = 125\,\text{mg}/\ell$. Assuming $Y_n = 0.15\,\text{lb}\,\text{MLVSS}/\text{lb}\,\underline{N}$ (kg/kg), the fraction of the mixed liquor that consists of nitrifiers is

$$
\begin{aligned}
f_n &= \frac{Y_n N_r}{Y_n N_r + Y S_r} \\
&= \frac{(0.15)(20.4)}{(0.15)(20.4) + (0.60)(125)} = 0.0392 \text{ or } 3.92\%
\end{aligned}
$$

The rate of nitrification is

$$
\begin{aligned}
R_n &= 1.04(1.09)^{T-20} f_n X_v \\
&= 1.04(1.09)^{19-20}(0.0392)(2500)(0.75) \\
&= \boxed{70.1\,\text{mg}/\ell\text{-day}}
\end{aligned}
$$

Thus the detention time required for nitrification is

$$\theta = \left(\frac{20.4\,\text{mg}}{\ell}\right)\left(\frac{\ell\text{-day}}{70.1\,\text{mg}}\right)\left(\frac{24\,\text{hr}}{\text{day}}\right) = \boxed{6.98\,\text{hr}}$$

Since $6.98\,\text{hr} > 6.0\,\text{hr}$, the detention time for nitrification does control. *Note:* In this problem typical parameters have been used, and the hydraulic detention time for nitrification was slightly more than the detention time for substrate removal. Thus if nitrification is required, the detention time for nitrification will control.

ACTIVATED SLUDGE COEFFICIENTS

The yield coefficient, Y, and the endogenous decay or degradation constant, k_e, may be determined using a series of bench-scale batch reactors inoculated with a parent activated sludge culture developed from a continuous-flow activated sludge reactor. The acclimated parent culture from a continuous-flow reactor will be biologically active, and as a result, the growths in the series of batch reactors will exhibit little, if any, lag phase. If the acclimated parent culture is from a fill and draw reactor, a lag phase usually will occur. In general, the test for activated sludge coefficients is for a 24-hr period, because a test of this duration will give more accurate values of Y and k_e than a shorter test. The proof for the test is based on cell-substrate relationships. The net rate of cell production is given by

$$\frac{dX}{dt} = Y\frac{dS}{dt} - k_e X \tag{15.102}$$

where

dX/dt = rate of cell mass production, mass/(volume)(time)

dS/dt = rate of substrate bio-oxidation, mass/(volume)(time)

X = cell mass concentration in the reactor, mass/volume

Dividing by X, the mass concentration of cells in the reactor, gives

$$\frac{1}{X}\frac{dX}{dt} = \frac{Y}{X}\frac{dS}{dt} - k_e \qquad \textbf{(15.103)}$$

If measurable increments are used, Eq. (15.103) becomes

$$\frac{1}{\bar{X}}\frac{\Delta X}{\Delta t} = \frac{Y}{\bar{X}}\frac{\Delta S}{\Delta t} - k_e \qquad \textbf{(15.104)}$$

where

ΔX = change in cell mass concentration, mass/volume

ΔS = change in substrate mass concentration, mass/volume

Δt = time increment

$\bar{X}$ = average cell mass concentration during the time increment, mass/volume—that is, $\bar{X} = (X_0 + X_t)(\frac{1}{2})$, where X_0 and X_t are the cell mass concentrations at time $t = 0$ and $t = t$

If the test is for 1 day, Δt is unity and Eq. (15.104) reduces to

$$\frac{\Delta X}{\bar{X}} = Y\frac{\Delta S}{\bar{X}} - k_e \qquad \textbf{(15.105)}$$

Equation (15.105) is of the straight-line form, $y = mx + b$; thus, if $\Delta X/\bar{X}$ for each reactor is plotted on the y-axis and $\Delta S/\bar{X}$ is plotted on the x-axis, the data result in a straight line with a slope equal to Y and a y-axis intercept equal to k_e. The oxygen coefficient, Y', may usually be computed from the yield coefficient, Y, and Eqs. (15.83), (15.84), (15.85), or a modification of these using the appropriate unit conversions. The endogenous oxygen coefficient k_e' theoretically equals $1.42\, k_e$.

It is also possible to determine Y' and k_e' by measuring the oxygen uptake rates in each reactor throughout the test duration. The mass of oxygen used, the average cell mass, and the mass of substrate removed for each reactor may be correlated to give Y' and k_e'. This is desirable because it gives a check on the Y' and k_e' values computed from the previous theoretical relationships. Usually, the measured value of k_e' is larger than that computed from $k_e' = 1.42k_e$. The equation for determining the oxygen coefficients from batch reactor data is

$$\frac{O_r}{\bar{X}} = Y'\frac{\Delta S}{\bar{X}} + k_e' \qquad \textbf{(15.106)}$$

where O_r is the oxygen used, mass/volume. The k_e' value at a temperature other than the test temperature may be obtained by multiplying the test k_e' value by the ratio of the k_e values at the two respective temperatures.

The nature of the wastewater determines the values of Y, Y', k_e, and k_e'. Y ranges from about 0.3 to 0.8 lb MLVSS/lb BOD$_5$ removed (kg/kg) and k_e from about 0.04 and 0.25 day^{-1}. For municipal wastewaters, Y is about 0.5 to 0.7 and k_e is about 0.04 to 0.10. Y' ranges from about 0.3 to 0.8 lb oxygen/lb BOD$_5$ removed (kg/kg), and k_e' ranges from about 0.05 to 0.35 lb oxygen/lb MLVSS-day (kg/kg-d). For municipal wastewaters, Y' is about 0.5 to 0.7 and k_e' is about 0.05 to 0.15. Typical values of Y, Y', k_e, and k_e' for municipal wastewaters are shown in Table 15.7.

The values of Y, k_e, Y', and k_e' may be determined from three or more batch reactors or from three or more completely mixed, continuous-flow reactors.

EXAMPLE 15.13 AND 15.13 SI

Activated Sludge Coefficients

An acclimated parent culture of activated sludge was produced using a continuous-flow activated sludge reactor for a soluble organic wastewater. Four batch reactors were inoculated using the parent culture and aerated for 24 hours. The MLVSS was 70% of the MLSS, and the COD was approximately equal to the TOD. A summary of the data is given in Table 15.8. Determine Y, k_e, and the theoretical values of Y' and k_e'.

SOLUTION

A summary of the reduced data is given in Table 15.9, where $\bar{X}$ is MLVSS. A plot of $\Delta X/\bar{X}$ versus $\Delta S/\bar{X}$ is shown in Figure 15.30. From the slope and

TABLE 15.7 *Typical Activated Sludge Coefficients for Municipal Wastewaters*

$Y = 0.60$ mg MLVSS/mg BOD$_5$ removed
$Y' = 0.62$ mg oxygen/mg BOD$_5$ removed
$k_e = 0.06$ day^{-1}
$k_e' = 0.085$ mg oxygen/mg MLVSS-day

Note: The MLVSS is usually 70% to 80% of the MLSS.

TABLE 15.8 *Data from Batch Reactor Tests*

REACTOR NO.	X_0 MLSS (mg/ℓ)	X_0 MLVSS (mg/ℓ)	X_t MLSS (mg/ℓ)	X_t MLVSS (mg/ℓ)	S_0 COD (mg/ℓ)	S_t COD (mg/ℓ)
1	420	294	737	516	706	95
2	830	581	1115	781	706	89
3	1620	1134	1910	1337	706	77
4	3670	2569	3820	2674	706	70

TABLE 15.9   *Reduced Data from Batch Reactor Tests*

REACTOR NO.	$\bar{X}$	ΔX	ΔS	$\Delta X/\bar{X}$	$\Delta S/\bar{X}$
1	405	222	611	0.548	1.509
2	681	200	617	0.294	0.906
3	1236	203	629	0.164	0.509
4	2622	105	636	0.040	0.243

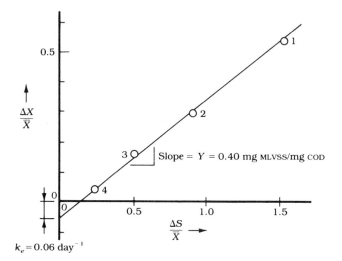

FIGURE 15.30
Graph for Examples 15.13 and 15.13 SI

y-axis intercept, $Y = 0.40\,\text{mg MLVSS/mg COD}$ and $k_e = 0.06\,\text{mg MLVSS/mg MLVSS-day.}$ The theoretical value $Y' = 1 - 1.42Y = 1 - 1.42(0.40)$ or $Y' = 0.43\,\text{mg oxygen/mg COD.}$ The theoretical value $k'_e = 1.42k_e = 1.42(0.06)$ or $k'_e = 0.09\,\text{mg oxygen/mg MLVSS-day.}$

EXAMPLE 15.14 AND 15.14 SI   *Activated Sludge Coefficients*

An acclimated parent culture of activated sludge was produced using a continuous-flow activated sludge reactor for a soluble organic wastewater. Four batch reactors were inoculated using the parent culture and aerated for 24 hours. The oxygen uptake rates in mg/ℓ-hr were measured for each reactor at periodic intervals during the 24 hours. The MLVSS was 72% of the MLSS. A summary of the data is given in Table 15.10. Determine the Y' and k'_e values.

SOLUTION   Figure 15.31 shows a plot of the oxygen uptake rate versus time for reactor no. 1, and the area under the curve, 459 mg/ℓ, represents the total oxygen

TABLE 15.10 *Data from Batch Reactor Tests*

REACTOR NO.	X_0 MLSS (mg/ℓ)	X_0 MLVSS (mg/ℓ)	X_t MLSS (mg/ℓ)	X_t MLVSS (mg/ℓ)	S_0 COD (mg/ℓ)	S_t COD (mg/ℓ)
1	817	588	1020	734	712	101
2	1636	1178	1780	1282	712	87
3	3263	2349	3342	2406	712	80
4	5161	3716	5120	3686	712	73

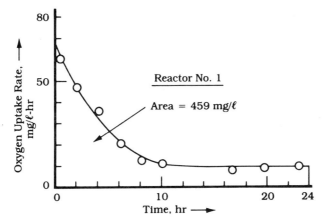

FIGURE 15.31
Graph for Examples 15.14 and 15.14 SI

TABLE 15.11 *Reduced Data from Batch Reactor Tests*

REACTOR NO.	$\bar{X}$	ΔS	O_2	$\Delta S/\bar{X}$	$O_r/\bar{X}$
1	661	611	459	0.924	0.694
2	1230	625	552	0.508	0.449
3	2378	632	780	0.266	0.328
4	3701	639	1125	0.173	0.304

used. Similar plots for the other reactors showed the oxygen used to be 552, 780, and 1125 mg/ℓ for reactors no. 2, 3, and 4, respectively. Using X as MLVSS and correlating the oxygen and other data results in the reduced data in Table 15.11.

A plot of $\Delta S/\bar{X}$ versus $O_r/\bar{X}$ is shown in Figure 15.32. From the graph, $\boxed{Y' = 0.53 \text{ mg oxygen/mg COD}}$ and $\boxed{k'_e = 0.20 \text{ mg oxygen/mg MLVSS-day.}}$

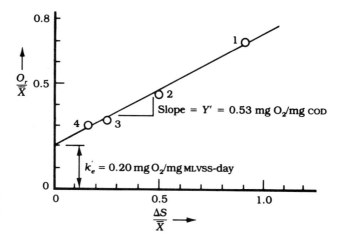

FIGURE 15.32
*Graph for Examples
15.14 and 15.14 SI*

**RATE
CONSTANTS
AND
COEFFICIENTS
FROM
CONTINUOUS-
FLOW
BIOLOGICAL
REACTORS**

It is also possible to determine the reaction rate constant, K, and the coefficients Y, k_e, Y', and k_e' from three or more continuous-flow, completely mixed lab, or pilot reactors operated at different F/M ratios or from one reactor operated at three or more F/M ratios. If possible, it is desirable to use pilot-scale reactors. These simulate field conditions better than laboratory reactors, because the quality of the wastewater stream may vary with time. It is important that the reactor or reactors have a completely mixed flow regime in order for this procedure to be valid. Usually three or more reactors are operated simultaneously to conserve time. The reactors are operated with increasing F/M values until acclimated sludges are developed and the reactors are at the desired F/M ratios. The end of the acclimation period can best be determined by measuring the organic substrate concentrations in the effluents from the reactors. Once the substrate concentration in the effluent from a reactor has reached the minimum value, the activated sludge in the reactor has become completely acclimated. After acclimation has been attained, the reactors are operated at steady state for a week or more. The influent and effluent substrate concentrations are measured daily, along with the oxygen uptake rates and the amounts of sludge that must be removed from the reactors to keep the MLSS values at the desired concentrations. The rate constant, K, can be obtained by a modification of Eq. (15.30), which is

$$\frac{S_i - S_t}{\overline{X}\theta_i} = KS_t \tag{15.107}$$

where S_i is the substrate concentration in the influent to the reactors. Since in most model reactors it is not possible to measure the recycle flow (R), $S_i = S_0$ and the reaction or detention time is given by $\theta_i = V/Q$. Plotting the $(S_i - S_i)/(\overline{X}\theta_i)$ values on the y-axis and the S_t values on the x-axis will give a straight line if the biochemical reaction is pseudo–first order. The slope of the line is equal to the rate constant, K.

The coefficients Y and k_e may be determined by a modification of Eq. (15.104), which is

$$\frac{\Delta X/\Delta t}{\bar{X}} = Y\left(\frac{S_i - S_t}{\bar{X}\theta_i}\right) - k_e \qquad (15.108)$$

Plotting the $(\Delta X/\Delta t)/\bar{X}$ values on the y-axis versus the $(S_i - S_t)/(\bar{X}\theta_i)$ values on the x-axis will give a straight line with a slope equal to Y and a y-axis intercept equal to k_e. The coefficients Y' and k'_e may be determined by a modification of Eq. (15.106), which is

$$\frac{O_r}{\bar{X}} = Y'\left(\frac{S_i - S_t}{\bar{X}\theta_i}\right) + k'_e \qquad (15.109)$$

where O_r is the oxygen used, mass/(volume)(day). Plotting the $O_r/\bar{X}$ values on the y-axis versus the $(S_i - S_t)/(\bar{X}\theta_i)$ values on the x-axis will give a straight line with a slope equal to Y' and a y-axis intercept equal to k'_e. In Eq. (15.109), the term for nitrification, O_n, has been omitted; thus the Y' value includes the oxygen required for carbonaceous removal and also nitrification. By determining the nitrogen values (all forms) in the influent and effluent flows and computing the nitrogen synthesized into cells, the nitrogen converted to NO_3^- can be determined. By using the amount of nitrogen converted to NO_3^-, the amount of oxygen required for nitrification can be computed. Subtracting the oxygen for nitrification from the O_r values gives the net oxygen required for carbonaceous removal. When this is done, the plot will give a Y' value only for carbonaceous removal. Frequently, the nitrification that occurs in a model reactor is not sufficient to require an oxygen adjustment to the O_r values, and when this occurs, the Y' value from the plot is essentially that required for carbonaceous removal alone.

**EXAMPLE 15.15
AND 15.15 SI**

Continuous-Flow Reactor Studies

The previous concepts are illustrated by the data in Table 15.12, which is from four completely mixed, continuous-flow activated sludge reactors. The acclimated cultures were developed for a soluble organic wastewater at F/M ratios of about 0.2, 0.4, 0.6, and 0.8 lb BOD$_5$/lb MLVSS-day (kg/kg-d). The

TABLE 15.12 *Data from Continuous-Flow Reactor Tests*

REACTOR NO.	$\bar{X}$ MLSS (mg/ℓ)	$\bar{X}$ MLVSS (mg/ℓ)	θ_i (hr)	S_i BOD$_5$ (mg/ℓ)	S_t BOD$_5$ (mg/ℓ)	O_2 UPTAKE (mg/ℓ-hr)	$\Delta X/\Delta t$ MLVSS/ day (mg/ℓ-day)
1	1276	919	25.3	885	61	29	309
2	1875	1350	23.7	885	47	35	340
3	2868	2065	24.4	885	31	39	313
4	5394	3884	24.9	885	12	58	136

MLVSS was 72% of the MLSS. After a week of steady-state operation, the data in Table 15.12 were obtained.

Determine K, Y, k_e, Y', and k_e'.

SOLUTION   A summary of the reduced data is given in Table 15.13, where $X =$ MLVSS.

A plot of the $(S_i - S_t)/(\overline{X}\theta_i)$ values versus the S_t values is shown in Figure 15.33. The slope from the graph gives

$$\boxed{K = 0.588\,\ell/(\text{gm MLVSS})(\text{hr})}$$

TABLE 15.13   *Reduced Data from Continuous-Flow Reactor Tests*

REACTOR NO.	$\dfrac{S_i - S_t}{\overline{X}\theta_i}$ (mg/mg-day)	$\dfrac{\Delta X/\Delta t}{\overline{X}}$ (mg/mg-day)	$\dfrac{O_r}{\overline{X}}$ (mg/mg-day)	S_t BOD$_5$ (mg/ℓ)
1	0.851	0.336	0.760	61
2	0.629	0.252	0.630	47
3	0.407	0.151	0.450	31
4	0.217	0.035	0.360	12

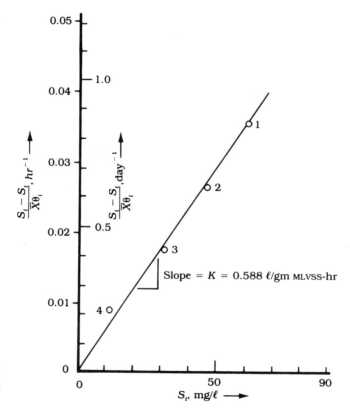

FIGURE 15.33
Graph for Examples 15.15 and 15.15 SI

Since the MLVSS is equal to 72% of the MLSS, the K based on MLSS is $(0.588)(0.72)$ or

$$K = 0.423 \, \ell/(\text{gm MLSS})(\text{hr})$$

The line in Figure 15.33 passes through the origin because the organic concentration is the BOD_5, which represents the biodegradable substrate. A plot of the $(\Delta X/\Delta t)/\bar{X}$ values versus the $(S_i - S_t)/(\bar{X}\theta_i)$ values is shown in Figure 15.34. From the graph, $Y = 0.47 \, \text{mg MLVSS/mg BOD}_5$ and $k_e = 0.05 \, \text{day}^{-1}$. A plot of the $O_r/\bar{X}$ values versus the $(S_i - S_t)/(\bar{X}\theta_i)$ values is shown in Figure 15.35. From the graph, $Y' = 0.66 \, \text{mg O}_2/\text{mg BOD}_5$ and $k_e' = 0.20 \, \text{mg O}_2/\text{MLVSS-day}$.

Note: In the previous example, the biodegradable COD or the total COD may be used instead of BOD_5. All graphs will be similar to those shown except for Figure 15.33 for the case of total COD. For this case, the straight line will not

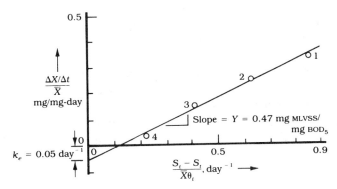

FIGURE 15.34
Graph for Examples 15.15 and 15.15 SI

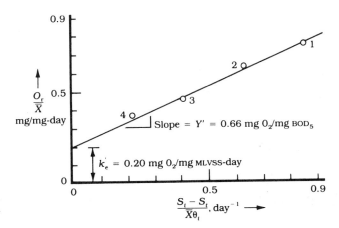

FIGURE 15.35
Graph for Examples 15.15 and 15.15 SI

pass through the origin but will intersect the x-axis at a value equal to the nondegradable COD, and the slope of the line will be equal to the reaction rate constant, K. A plot of this type is shown in Figure 15.36.

It is also possible to use three or more continuous-flow activated sludge reactors to determine the rate constants, $\mu_{\max}$ and K_{sm}, and the coefficients Y and k_e for the relationships derived from the Monod equation. For reactors without recycle, Eqs. (15.48) and (15.49) are used, and for reactors with recycle, Eqs. (15.62) and (15.63) are used. The procedure for analyzing the data was previously discussed in the text where these equations were presented.

OPERATIONAL PROBLEMS

The most common operating problem in the activated sludge process is inadequate solid-liquid separation in the final clarifier. This results in appreciable amounts of biological solids spilling over the effluent weirs and unacceptable amounts of solids leaving with the effluent. This condition, generally termed **sludge bulking**, may be caused by the occurrence of a biological sludge that has poor settling characteristics, hydraulic and solids overloading, or inadequate recycle flowrate. For diffused air activated sludge processes treating municipal wastewaters, the sludge density index is usually from about 6500 mg/ℓ (svi ≃ 150) to 20,000 mg/ℓ (svi ≃ 50). For mechanical aeration activated sludge processes treating municipal wastewaters, the sludge density index is usually about 3500 mg/ℓ (svi ≃ 300) to 5000 mg/ℓ (svi ≃ 200) (Culp, 1970; Steel and McGhee, 1979). Thus, in mechanical aeration tanks, the added agitation is sufficient to cause some shearing of the biological floc. Slow-speed turbines impart less shear forces to the water than high-speed turbines; consequently, the sludge density index is greater for the slow-speed units. If the svi values are greater than given above, a poor effluent in terms of suspended solids and BOD₅ can be expected. For municipal wastewaters, a sludge with excellent settling characteristics will be produced when the environmental conditions are favorable, the mean cell

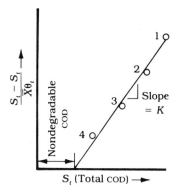

FIGURE 15.36 *Plot for Using Total COD for Determining the Reaction Constant* **K**

residence time is from 5 to 15 days, and the food-to-microbe ratio (F/M) is from 0.2 to 0.7 lb BOD₅/lb MLSS-day (kg/kg-day).

The flocculating capability of an activated sludge culture is lowest during the log growth phase (high F/M ratio), and it increases during the declining growth phase and is maximum during the beginning of the endogenous phase (low F/M ratio). The flocculating capacity is partly due to the presence of gummy slime layers or capsules surrounding the viable cells and the accumulation of this material from the attrition of the slime layers. In the log growth phase, the slime layers or capsules are the thinnest, and they reach the maximum thickness at the beginning of the endogenous phase. The slime layers are usually polysaccharide gums sometimes having proteinaceous or lipoidal material associated with them (Frobisher *et al.*, 1974). From the attrition of the slime layers comes a gelatinous agglomerating substance that is relatively bioresistant and assists in floc aggregation. The flocculating capacity of an activated sludge is also partly due to the inactivity of the microbial cells and their behavior as colloids. In the log growth phase (high F/M ratio), the cells are very active and their motility prevents their flocculation (McKinney, 1962a). In the declining growth phase and, in particular, the endogenous phase (low F/M ratio), the motility of the cells is appreciably decreased, and if the cells come in close contact because of mixing, the forces of attraction become greater than the forces of repulsion and the cells flocculate.

If the aeration time is inadequate, relatively small amounts of the polysaccharide gums will be present, thus contributing to poor settling characteristics. Also, if an excessive slug of organic material enters the reactor, the amount of agglomerating gum may not be sufficient, and as a result, poor biological flocculation will occur. In addition, an excessive slug of organic matter will cause cell activity and motility to increase, thus significantly decreasing the flocculating capacity of the biological suspension.

When filamentous microbes (such as the *Sphaerotilus* group) outgrow the desirable microorganisms, they become predominant and, as a result of their wiry nature, prevent proper sludge settling and compaction in the final clarifier. In a similar manner, excessive growths of fungi cause a poorly settling sludge. Conditions favoring filamentous growths are F/M ratios greater than about 0.8 lbs BOD₅/lb MLSS-day (kg/kg-day) for several days, wastewaters containing a high carbohydrate content, lack of nitrogen and phosphorus nutrients, and low operating dissolved oxygen concentrations. Fungi growths are also promoted by wastewaters having a high carbohydrate content and inadequate amounts of nitrogen and phosphorus nutrients.

Overaeration creates shear forces that prevent proper agglomeration of the biological floc and also shear apart floc particles. Usually, if the power level is not greater than about 0.30 hp/1000 gal (59 kW/1000 m³) or 35 scfm/1000 ft³ (35 m³/min-1000 m³), overaeration will not occur. Also, if the aeration time is too long, shearing will break apart the floc, large amounts of lysed cell fragments will be present, and poor settling will result.

Underaeration usually occurs when the oxygen level is much less than

1.0 mg/ℓ, and as a result, the inner portions of the biological particles may become anaerobic and produce gaseous end products. Also, anaerobic action causes the metabolism of carbohydrates and some other materials to be shunted, resulting in the formation of intracellular products such as glycerol, which has specific gravity less than 1. The overall effect of the entrained gaseous end products in the floc and the substances of lower specific gravity in the cells is a reduction in the specific gravity of the floc particles, which causes poor settling characteristics.

When the dissolved oxygen level in the mixed liquor is low and the sludge is retained for numerous hours in the final clarifier, anaerobic action will occur, and floating sludge masses will result that may contribute to a poor effluent. If the dissolved oxygen level in the mixed liquor is at least 2.0 mg/ℓ and the sludge retention time in the final clarifier is less than about 1 hour, this problem is usually avoided.

If any of the chemical ions, compounds, or substances given in the section in Chapter 2, Biological Concepts, titled "Environmental Factors Affecting Microbial Activity" are present in inhibitory or toxic concentrations for sufficient time, they will interfere with the microbial metabolic functions and result in a decreasing rate of substrate or BOD_5 utilization. As a result, the BOD_5 removal efficiency will decrease, and if the condition persists long enough or if the chemical concentration is great enough, killing of the microbial population will occur. When the cells die, they lyse or break apart into fragments, and this will result in a sludge having poor settling characteristics.

Excessive grease can coat the biological floc particles and cause them to become anaerobic. This coating will interfere with the sorption of organic substances, the attrition of slime layers, the agglutination of the biological floc particles, and the transfer of oxygen. If the grease content in the sludge is less than 5 to 7%, this usually will not be a problem (Steel and McGhee, 1979).

In bulking, where the cause is excessive or shock hydraulic loads, the influent flowrate reaches a point where the overflow velocity in the final clarifier exceeds the settling velocity of the mixed liquor. As a result, the biological solids accumulate in the final clarifier at a concentration near that of the mixed liquor. The sludge blanket rate of rise will be approximately equal to the difference between the overflow velocity and the sludge settling velocity (Texas Water Utilities Association, 1971). For example, if the overflow rate is 1600 gal/day-sq ft (65.2 m³/d-m²) the overflow velocity is 8.9 ft/hr (2.71 m/h), and if the mixed liquor suspended solids settling velocity is 4.5 ft/hr (1.37 m/h), the sludge blanket will rise at a rate of about 8.9 − 4.5 or 4.4 ft/hr (2.71 − 1.37 or 1.34 m/h). If the high hydraulic loading persists, the tank will become filled with activated sludge solids, and the biological solids will overflow with the effluent. When this occurs, the solids concentration in the effluent will approach that of the mixed liquor in the reactor, which is usually several thousand mg/ℓ. When the overflow velocity exceeds the sludge settling velocity, the bulking cannot be controlled by increasing the underflow or recycled sludge flowrate. The only remedy is the proper reduction in the mixed liquor flowrate to the clarifier or the addition of another clarifier.

If the solids loading to the final clarifier is excessive, the solids will not thicken and be removed as the underflow at the rate the solids enter the clarifier. As a result, the sludge blanket will rise, and if the time duration is sufficient, excessive solids will spill over the effluent weirs. When this occurs, the solids concentration in the effluent will approach that of the mixed liquor. Increasing the recycle flowrate will not eliminate the situation. The only remedy is a reduction in the mixed liquor flowrate to the final clarifier, a reduction in the mixed liquor suspended solids concentration, or the addition of another final clarifier. The design solids loading based on the average hourly flow is usually from about 0.60 to 1.20 lb/hr-ft^2 (2.93 to 5.86 kg/h-m^2), and the design solids loading based on the peak hourly flow is from about 1.25 to 2.0 lb/hr-ft^2 (6.11 to 9.77 kg/h-m^2). The larger solids loading values are for plants located in warm climates because the sludge settling rate increases with an increase in temperature.

In the operation of an activated sludge plant, it is essential to recycle the appropriate amount of activated sludge to maintain the desired MLSS concentration and also to determine daily the solids concentration in the reactor and the sludge blanket depth in the final clarifier. Normally, the top of the sludge blanket is less than half the final clarifier tank depth. If the recycle is inadequate, the top of the blanket will rise and eventually spill over the effluent weirs. This will result in an excessive suspended solids concentration in the effluent, and the solids concentration may approach or exceed that of the mixed liquor. If desired, a plant may be operated with a sludge blanket depth as small as about 2 ft (0.61 m). Final clarifiers with suction-type sludge removal usually have sludge blankets less than 2 ft (0.61 m) deep.

REFERENCES

Aiba, S.; Humphery, A. E.; and Millis, N. F. 1965. *Biochemical Engineering*. New York: Academic Press.

American Society of Civil Engineers (ASCE). 1982. *Wastewater Treatment Plant Design*. ASCE Manual of Practice No. 36. New York: ASCE.

AWARE, Inc. 1974. *Process Design Techniques for Industrial Waste Treatment*, edited by C. E. Adams, Jr. and W. W. Eckenfelder, Jr. Nashville, Tenn.: Enviro Press.

Barnes, D.; Bliss, P. J.; Gould, B. W.; and Vallentine, H. R. 1981. *Water and Wastewater Engineering Systems*. London: Pittman Books Limited.

Benefield, L. D., and Randall, C. W. 1980. *Biological Process Design for Wastewater Treatment*. Englewood Cliffs, N.J.: Prentice-Hall.

Blakebrough, N., ed. 1967. *Biochemical and Biological Engineering Science*. Vol. 1. London: Academic Press.

Boyle, W. C.; Crabtree, K.; Rohlich, G. A.; Iaccarino, E. P.; and Lightfoot, E. N. 1969. Flocculation Phenomena in Biological Systems. In *Advances in Water Quality Improvement*, edited by E. F. Gloyna and W. W. Eckenfelder, Jr. Austin, Tex.: University of Texas Press.

Brey, W. S., Jr. 1958. *Principles of Physical Chemistry*. New York: Appleton-Century-Crofts.

Busch, A. W. 1961. Treatability vs. Oxidizability of Industrial Wastes and the Formulation of Process Design Criteria. In *Proceedings of the 16th Annual Purdue Industrial Waste Conference*.

Clifton, C. E. 1957. *Introduction to Bacterial Physiology*. New York: McGraw-Hill.

Conway, R. A., and Ross, R. D. 1980. *Handbook of Industrial Waste Disposal*. New York: Van Nostrand Reinhold.

Culp, R. L. 1970. The Operation of Wastewater Treatment Plants. *Public Works Magazine* (October, November, and December).

Downing, A. L.; Painter, H. A.; and Knowles, G. 1964. Nitrification in the Activated Sludge Process. *Jour. and Proc. Inst. Sewage Purif.*, part, 2:130.

Eckenfelder, W. W., Jr. 1966. *Industrial Water Pollution Control*. New York: McGraw-Hill.

Eckenfelder, W. W., Jr. 1989. *Industrial Water Pollution Control*. 2nd ed. New York: McGraw-Hill.

Eckenfelder, W. W., Jr. 1980. *Principles of Water Quality Management*. Boston: CBI Publishing.

Eckenfelder, W. W., Jr. 1970. *Water Quality Engineering for Practicing Engineers*. New York: Barnes & Noble.

Eckenfelder, W. W., Jr.; Adams, C. E.; and Hovious, J. C. 1975. A Kinetic Model for Design of Completely Mixed Activated Sludge Treating Variable Strength Industrial Wastewaters. *Water Research* 9, no. 1:37.

Eckenfelder, W. W., Jr., and Ford, D. L. 1970. *Water Pollution Control*. Austin, Tex.: Pemberton Press.

Eckenfelder, W. W., Jr.; Irvine, R. L.; and Reynolds, T. D. 1972. Wastewater Treatment Cost Models and Required Parameters. *Technical Report to Galveston Bay Study*. Texas Water Quality Board.

Eckenfelder, W. W., Jr., and O'Connor, D. J. 1961. *Biological Waste Treatment*. London: Pergamon Press.

Eckenfelder. W. W., Jr., and Weston, R. F. 1956. Kinetics of Biological Oxidation. In *Biological Treatment of Sewage and Industrial Wastes*. Vol. 1, edited by J. McCabe and W. W. Eckenfelder, Jr. New York: Reinhold.

Fair, G. M.; Geyer, J. C.; and Okun, D. A. 1968. *Water and Wastewater Engineering*. Vol. 2. *Water Purification and Wastewater Treatment and Disposal*. New York: Wiley.

Fair, G. M., and Thomas, H. A., Jr. 1950. The Concept of Interface and Loading in Submerged, Aerobic, Biological Sewage-Treatment Systems. *Jour. and Proc. Inst. Sewage Purif.*, part 3:235.

Farquhar, G. J., and Boyle, W. C. 1971. Identification and Occurrence of Filamentous Microorganisms in Activated Sludge. *Jour. WPCF* 43, no. 4:604; and 43, no 5:779.

Ford, D. L.; Gloyna, E. F.; and Yang, Y. T. 1967. Development of Biological Treatment Data for Chemical Wastes. *Proceedings of the 22nd Annual Purdue Industrial Waste Conference*. Part 1.

Ford, D. L., and Reynolds, T. D. 1965. Aerobic Oxidation in the Thermophilic Range. *Proceedings of the 5th Industrial Water and Waste Conference*, Dallas, Tex.

Frobisher, M.; Hinsdill, R. D.; Crabtree, K. T.; and Goodheart, C. R. 1974. *Fundamentals of Microbiology*. 9th ed. Philadelphia, Pa.: Saunders.

Great Lakes–Upper Mississippi River Board of State Public Health and Environmental Managers. 1990. Recommended Standards for Wastewater Facilities. Ten state standards. Albany, N.Y.

Great Lakes–Upper Mississippi River Board of State Sanitary Engineers. 1978. Recommended Standards for Sewage Works. Ten state standards. Albany, N.Y.

Hoover, S. R., and Porges, N. 1952. Assimilation of Dairy Wastes by Activated Sludge. *Sew. and Ind. Wastes* 24, no. 3:306.

Horan, N. J. 1990. *Biological Wastewater Treatment Systems: Theory and Operation*. Chichester, West Sussex, England: Wiley.

Imhoff, K.; Muller, W. J.; and Thistlethwayte, D. K. B. 1971. *Disposal of Sewage and Other Waterborne Wastes*. 2nd ed. Ann Arbor, Mich.: Ann Arbor Science Publishers.

Levenspiel, O., and Bischoff, K. B. 1959. Backmixing in the Design of Chemical Reactors. *Ind. and Eng. Chem.* 51. no. 12:1431.

Levenspiel, O., and Bischoff, K. B. 1961. Reaction Rate Constant May Modify the Effects of Backmixing. *Ind. and Eng. Chem.* 53, no. 4:313.

Lighthart, B., and Loew, G. A. 1972. Identification Key for Bacteria Clusters from an Activated Sludge Plant. *Jour. WPCF* 44, no. 11:2078

McCarty, P. L. 1970. Biological Processes for Nitrogen Removal: Theory and Application. *Proceedings of the Twelfth Sanitary Engineering Conference*. Urbana, Ill.: University of Illinois.

McCarty, P. L., and Lawrence, A. W. 1970. Unified Basis for Biological Treatment Design and Operation. *Jour. SED* 96, no. SA3:757.

McGhee, T. J. 1991. *Water Supply and Sewage*. 6th ed. New York: McGraw-Hill.

McKinney, R. E. 1962a. *Microbiology for Sanitary Engineers*. New York: McGraw-Hill.

McKinney, R. E. 1962b. Mathematics of Complete Mixing Activated Sludge. *Jour. SED* 88, no. SA3:87.

Metcalf & Eddy, Inc. 1972. *Wastewater Engineering: Collection, Treatment and Disposal*. New York: McGraw-Hill.

Metcalf & Eddy, Inc. 1979. *Wastewater Engineering:*

Treatment, Disposal and Reuse. 2nd ed. New York: McGraw-Hill.

Metcalf & Eddy, Inc. 1991. *Wastewater Engineering: Treatment, Disposal and Reuse.* 3rd ed. New York: McGraw-Hill.

Middlebrooks, E. J., and Garland, C. F. 1968. Kinetics of Model and Field Extended-Aeration Wastewater Treatment Units. *Jour. WPCF* 40, no. 4:586.

Monod, J. 1949. The Growth of Bacterial Cultures. *Annual Review of Microbiology,* 3:371.

Novotny, V.; Imhoff, K.; Olthof, M.; and Krenkel, P. 1989. *Karl Imhoff's Handbook of Urban Drainage and Wastewater Disposal.* New York: Wiley.

Qasim, S. R. 1985. *Wastewater Treatment Plant Design.* New York: Holt, Rinehart and Winston.

Pearson, E. A. 1968. Kinetics of Biological Treatment. In *Advances in Water Quality Improvement,* edited by E. F. Gloyna and W. W. Eckenfelder, Jr. Austin, Tex.: University of Texas Press.

Petrasek, A. C. 1975. Nitrification in Wastewater Treatment Plants (Activated Sludge). Dissertation, Civil Engineering Department, Texas A&M University.

Pipes, W. O., Jr.; Grieves, R. B.; and Milbury, W. F. 1964. A Mixing Model for Activated Sludge. *Jour. WPCF* 36, no. 5:619.

Ramalho, R. S. 1977. *Introduction to Wastewater Treatment PProcesses.* New York: Academic Press.

Reynolds, T. D., and Yang, J. T. 1966. Model of the Completely Mixed Activated Sludge Process. In *Proceedings of the 21st Annual Purdue Industrial Waste Conference.* Part 2.

Rich, L. G. 1973. *Environmental Systems Engineering.* New York: McGraw-Hill.

Rich, L. G. 1963. *Unit Processes of Sanitary Engineering.* Clemson, S.C.: L. G. Rich.

Salvato, J. A. 1992. *Environmental Engineering and Sanitation.* 4th ed. New York: Wiley.

Santer, M., and Ajl, S. 1954. Metabolic Reactions of *Pasteurella pestis,* L. Terminal Oxidation. *J. Bacteriology,* 67:379.

Sawyer, C. N. 1956. Bacterial Nutrition and Synthesis. In *Biological Treatment of Sewage and Industrial Wastes.* Vol. 1, edited by J. McCabe and W. W. Eckenfelder, Jr. New York: Reinhold.

Schroeder, E. D. 1977. *Water and Wastewater Treatment.* New York: McGraw-Hill.

Servizi, J. A., and Bogan, R. H. 1964. Thermodynamic Aspects of Biological Oxidation and Synthesis. *Jour. WPCF* 38, no. 5:607.

Smith, H. S., and Paulson, W. L. 1966. Homogeneous Activated Sludge-Wastewater Treatment at Lower Cost. *Civil Engineering* (May):56.

Stack, V. T., Jr., and Conway, R. A. 1959. Design Data for Completely Mixed Activated Sludge Treatment . *Sew. and Ind. Wastes* 31:1181.

Steel, E. W., and McGhee, T. J. 1979. *Water Supply and Sewerage.* 6th ed. New York: McGraw-Hill.

Stewart, M. J. 1964. Activated Sludge Process Variations: The Complete Spectrum. Part 1: Basic Concepts: Part 2: Process Descriptions; Part 3: Effluent Quality-Process Loading Relationships. *Water and Sewage Works* 111, no. 4:153; 111, no. 5:246; 111, no. 6:295.

Sundstrom, D. W., and Klei, H. E. 1979. *Wastewater Treatment.* Englewood Cliffs, N.J.: Prentice-Hall.

Tebbutt, T. H. Y. 1992. *Principles of Water Quality Control.* 4th ed. Oxford, England: Pergamon Press Limited.

Texas Water Utilities Association. 1971. *Manual of Wastewater Operations.* 4th ed. Austin, Tex.

Water Environment Federation (WEF). 1992. *Design of Municipal Wastewater Treatment Plants,* Vol. I and II, WEF Manual of Practice No. 8.

Water Pollution Control Federation (WPCF). 1977. *Wastewater Treatment Plant Design.* WPCF Manual of Practice No. 8. Washington, D.C.

Water Resources Symposium No. 6. 1973. *Application of Commercial Oxygen to Water and Wastewater Systems,* edited by R. E. Speece and J. F. Malina, Jr. Austin, Tex.: University of Texas Press.

Weston, R. F., and Eckenfelder, W. W., Jr. 1955. Applications of Biological Treatment to Industrial Wastes: I. Kinetics and Equilibria of Oxidative Treatment. *Sew. and Ind. Wastes* 27, no. 7:802.

Wilner, B., and Clifton, C. E. 1954. Oxidative Assimilation by *Bacillus subtilis.* *J. Bacteriology* 67:571.

Wu, Y. C., and Kao, D. F. 1976. Yeast Plant Wastewater Treatment. *Jour. SED* 102, no. EE5:969.

Wuhrmann, K. 1964. Die Grundlagen Der Dimensionierung der Berlüftung bei Belebschlammanlagen Vortrag. Swiss Federal Institute of Technology, Zurich, Switzerland.

PROBLEMS

15.1 An industrial wastewater from a large milk processing plant has been treated using a continuous-flow activated sludge reactor to develop an acclimated culture of activated sludge. Using the acclimated culture, three batch reactors (each containing the same amount of wastewater) were inoculated with the same amount of activated sludge. The reactors were operated over a 24-hour duration to obtain data for kinetic evaluations. The laboratory data from the three batch tests (average values) are shown in Table 15.14.

At 24 hours, the BOD_5 = 5.2 mg/ℓ, the BOD_5 = 0.35 BOD_u, and the BOD_u = COD (biodegradable). The MLVSS values were 85.7% of the MLSS values. Determine the nonbiodegradable COD, the reaction order, and the rate constant on the basis of the biodegradable COD and both the MLVSS and the MLSS values.

Note: Do not include the 24-hour values in the plot for the reaction order and rate constant, because the COD value is too low to be representative.

15.2 A soluble organic industrial wastewater has been treated using a continuous-flow, activated sludge reactor to develop an acclimated culture of activated sludge. The desired amount of acclimated sludge was placed in a batch reactor, and the reactor was operated over a 24-hour duration to obtain data

for kinetic evaluations. The observed COD and MLSS concentrations are shown in Table 15.15.

The MLVSS = 0.85 MLSS, the BOD_5 = 7 mg/ℓ, and the BOD_5 = 0.36 BOD_u at 24 hr. The BOD_u = COD (biodegradable).

a. Determine the nondegradable COD and plot the concentration of mixed liquor suspended solids, X, and the biodegradable COD, C, versus time, t, on arithmetic paper. Is a lag phase evident?

b. Determine the reaction order and rate constant on the basis of both MLVSS and MLSS if the substrate concentration is represented by the biodegradable COD. *Note*: Do not include the 8-hour and 24-hour values in the plot for the reaction order and rate constant, because the COD values are too low to be representative.

15.3 A soluble organic industrial wastewater with a BOD_5 of 860 mg/ℓ is to be treated by a tapered aeration activated sludge plant preceded by an equalization basin, and the effluent BOD_5 is to be 50 mg/ℓ. The reaction is pseudo–first-order, the rate constant, K, is 0.118 ℓ/(gm MLVSS-hr), the MLSS concentration is 3000 mg/ℓ, the MLVSS is 75% of the MLSS, the sludge density index is 9000 mg/ℓ, and the flow is 5.0 MGD (18.9 MLD). Determine, using USCS units:

a. The reaction time and reactor basin volume for a plug-flow reactor basin.

b. The reaction time and reactor basin volume for a dispersed plug-flow reactor basin if the dispersion number, d, is 0.2.

TABLE 15.14

TIME (hr)	COD (mg/ℓ)	MLSS (mg/ℓ)
0	918	1688
0.5	821	1652
1.0	740	1716
1.5	675	1689
2.0	528	1718
2.5	428	1812
3.0	363	1875
3.5	297	2106
4.0	265	1948
5.0	220	2023
6.0	162	1982
24.0	107	1626

TABLE 15.15

TIME (hr)	MLSS X (mg/ℓ)	COD C (mg/ℓ)
0	3670	706
0.017	3672	604
0.5	3675	372
1.0	3690	276
2.0	3760	147
4.0	3860	93
8.0	3880	93
24.0	3820	89

15.4 Repeat Problem 15.3 using SI units.

15.5 Determine, using USCS units, the dispersion number, d, and the flow regime for the following:

a. A rectangular reactor basin with a flow of 7.25 MGD (27.4 MLD), $Q_a = 25$ scfm/1000 ft³ (25 m³/min-1000 m³), depth = 15 ft (4.57 m), width = 30 ft (9.14 m), length = 500 ft (154 m), and $\theta = 5.6$ hours.

b. A rectangular reactor basin with a flow of 2.53 MGD (9.58 MLD), $Q_a = 25$ scfm/1000 ft³ (25 m³/min-1000 m³), depth = 15 ft (4.57 m), width = 30 ft (9.14 m), length = 175 ft (53.3 m), and $\theta = 5.6$ hours.

15.6 Repeat Problem 15.5 using SI units.

15.7 A completely mixed activated sludge plant is to be designed for a population of 145,000 persons. A treatability study has been made using a pilot plant, and the following data have been obtained: influent flow = 95 gal/cap-day (360 ℓ/cap-d), influent BOD₅ = 205 mg/ℓ, 34% of influent BOD₅ removed by primary clarification, MLSS = 3000 mg/ℓ, MLVSS = 75% of MLSS, $K = 0.926$ ℓ/(gm MLVSS-hr), reaction is pseudo–first-order, and effluent BOD₅ = 10 mg/ℓ. Determine, using USCS units:

a. The required reaction time, θ, hours.
b. The required reactor basin volume, ft³.
c. The food-to-microbe ratio, lb/lb-day.
d. The space or volumetric loading, lb BOD₅/day-1000 ft³.

15.8 Repeat Problem 15.7 using SI units.

15.9 A tapered aeration activated sludge plant is to be designed for a population of 175,000 persons. A treatability study has been made using a pilot plant, and the following data have been obtained: influent flow = 105 gal/cap-day (397 ℓ/cap-d), influent BOD₅ = 210 mg/ℓ, 33% of influent BOD₅ removed by primary clarification, MLSS concentration = 3000 mg/ℓ, MLVSS = 75% of MLSS, sludge density index = 9000 mg/ℓ, $K = 0.249$ ℓ/(gm MLVSS-hr), reaction is pseudo–first-order, and effluent BOD₅ = 10 mg/ℓ. Determine, using USCS units:

a. The required reaction time and reactor basin volume for a plug-flow reactor basin.
b. The required reaction time and reactor basin volume for a dispersed plug-flow reactor basin with a dispersion number, d, of 0.2.
c. The food-to-microbe ratio for part (a), lb/lb-day.
d. The space or volumetric loading for part (a), lb BOD₅/day-1000 ft³.

e. The food-to-microbe ratio for part (b), lb/lb-day.
f. The space or volumetric loading for part (b), lb BOD₅/day-1000 ft³.

15.10 Repeat Problem 15.9 using SI units.

15.11 A soluble organic industrial wastewater with a COD of 2590 mg/ℓ is to be treated by a completely mixed activated sludge plant. A treatability study has been made, and the following data were obtained: design influent flow = 4.20 MGD (15.9 MLD), MLSS = 3000 mg/ℓ, MLVSS = 86.7% of the MLSS, $K = 0.236$ ℓ/(gm MLVSS-hr), reaction is pseudo–first-order, nondegradable COD = 16 mg/ℓ, and effluent COD = 150 mg/ℓ. Determine, using USCS units:

a. The required reaction time, θ, hours.
b. The required reactor basin volume, ft³.

15.12 Repeat Problem 15.11 using SI units.

15.13 A soluble organic industrial wastewater has a BOD₅ of 410 mg/ℓ and is to be treated by the activated sludge process to produce an effluent BOD₅ of 20 mg/ℓ. The flow is 2.0 MGD (15.1 MLD). The rate constant K is 0.226 ℓ/(gm MLVSS-hr). The MLSS concentration is 3000 mg/ℓ, the MLVSS is 75% of the MLSS, and the sludge density index is 9000 mg/ℓ. Determine, using USCS units:

a. The reaction time and reactor basin volume for a plug-flow reactor.
b. The reaction time and reactor basin volume for a dispersed plug-flow reactor if the dispersion number is 0.2.
c. The reaction time and reactor basin volume for a completely mixed reactor.

15.14 Repeat Problem 15.13 using SI units.

15.15 An oxidation ditch is to be designed for a community of 5000 persons. The flow is 95 gal/cap-day (360 ℓ/cap-d) and the influent BOD₅ is 200 mg/ℓ. The plant is not to have a primary clarifier. The plant is to have 90% BOD₅ removal, the yield coefficient, Y_b, is 0.55 mg MLVSS per mg of BOD₅ removed, the endogenous coefficient, k_e, is 0.05 day⁻¹, the biodegradable fraction of the biological solids is 0.8, and the MLVSS is 50% of the MLSS. Determine, using USCS units:

a. The reactor basin volume if the detention time is 1 day and the expected recycle ratio is 1.00.
b. The operating MLSS concentration.
c. The layout dimensions if the ditch has a trapezoidal cross section and the distance across the racetrack layout is 125 ft (38.1 m) from ditch cen-

terline to ditch centerline. The design cross section is 14 ft (4.27 m) at the base, has 2:1 side slopes, and has a 5-ft (1.52-m) water depth.

15.16   Repeat Problem 15.15 using SI units.

15.17   A soluble industrial wastewater has a COD of 2590 mg/ℓ and is to be treated by a completely mixed activated sludge plant. A treatability study has been made, and the following data were obtained: design flow = 5.0 MGD (18.9 MLD), MLSS = 3000 mg/ℓ, MLVSS = 86.7% of the MLSS, K = 0.236 ℓ/(gm MLVSS-hr) at 23°C, the reaction is pseudo–first-order, the nondegradable COD = 16 mg/ℓ, the minimum mixed liquor operating temperature = 19°C, the maximum mixed liquor operating temperature = 24°C, the temperature correction coefficient θ = 1.05, and the effluent COD = 180 mg/ℓ. Determine, using USCS units:

a. The required reaction time and the reactor basin volume.

b. The effluent COD when the operating mixed-liquor temperature is 24°C.

15.18   Repeat Problem 15.17 using SI units.

15.19   An activated sludge plant is to be designed for a population of 4000 persons. The influent flow is 100 gal/cap-day (380 ℓ/cap-d) and the influent BOD$_5$ is 200 mg/ℓ. Pertinent data are: Primary clarifier removes 33% of the influent BOD$_5$, and the state regulatory agency allows a space or volumetric loading of 30 lb BOD$_5$/day-1000 ft^3 (0.48 kg/d-m^3). Determine, using USCS units, the reactor basin volume.

15.20   Repeat Problem 15.19 using SI units.

15.21   A tapered aeration activated sludge plant is to be designed for a population of 65,000 persons, and the design is to be by the NRC equation. Pertinent data are: influent flow = 105 gal/cap-day (397 ℓ/cap-d), influent BOD$_5$ = 205 mg/ℓ, primary clarification removes 33% of the influent BOD$_5$, MLSS = 2000 mg/ℓ, and effluent BOD$_5$ = 20 mg/ℓ. Determine, using USCS units:

a. The required aeration time.

b. The required reactor basin volume.

15.22   Repeat Problem 15.21 using SI units.

15.23   A tapered aeration activated sludge plant is to be designed for a population of 85,000 persons. Pertinent data are: influent flow = 95 gal/cap-day (360 ℓ/cap-d), influent BOD$_5$ = 210 mg/ℓ, primary clarification removes 33% of the influent BOD$_5$,

effluent BOD$_5$ = 10 mg/ℓ, MLSS = 2500 mg/ℓ, MLVSS = 75% of the MLSS, sludge density index = 10,000 mg/ℓ, reaction time = 5.0 hr based on Q + R, Y = 0.60 mg MLVSS/mg BOD$_5$ removed, and k_e = 0.06 day^{-1}. Determine, using USCS units:

a. The reactor basin volume.

b. The waste activated sludge production per day.

c. The waste activated sludge flow, in gallons per day, if the sludge is wasted from the recycle line and the specific gravity of the wet sludge is 1.01.

d. The mean cell residence time, θ_c, for the reactor basin.

15.24   Repeat Problem 15.23 using SI units.

15.25   A tapered aeration activated sludge plant is to be designed for a population of 95,000 persons, and a treatability study has been made using a pilot plant. Pertinent data are: influent flow = 105 gal/cap-day (397 ℓ/cap-d), influent BOD$_5$ = 205 mg/ℓ, primary clarification removes 34% of the influent BOD$_5$, effluent BOD$_5$ = 10 mg/ℓ, MLSS = 2500 mg/ℓ, MLVSS = 78.2% of the MLSS, sludge density index = 10,000 mg/ℓ, K = 0.243 ℓ/(gm MLVSS-hr), reaction is pseudo–first-order, Y = 0.61 mg MLVSS/mg BOD$_5$ removed, and k_e = 0.05 day^{-1}. Determine, using USCS units:

a. The reactor basin volume for a plug-flow reactor basin.

b. The waste activated sludge produced per day.

c. The waste activated sludge flow if the sludge is wasted from the recycle line and the specific gravity of the wet sludge is 1.01.

d. The mean cell residence time, θ_c, for the reactor basin.

15.26   Repeat Problem 15.25 using SI units.

15.27   A tapered aeration activated sludge plant is to be designed for a population of 125,000 persons, and a treatability study has been made using a pilot plant. Pertinent data obtained are: influent flow = 95 gal/cap-day (360 ℓ/cap-d), influent BOD$_5$ = 210 mg/ℓ, primary clarification removes 33% of the influent BOD$_5$, effluent BOD$_5$ = 10 mg/ℓ, MLSS = 2500 mg/ℓ, MLVSS = 74.8% of the MLSS, sludge density index = 10,000 mg/ℓ, K = 0.209 ℓ/(gm MLVSS-hr), reaction is pseudo–first-order, Y = 0.60 mg MLVSS/mg BOD$_5$ removed, k_e = 0.06 day^{-1}, Y' = 0.62 mg oxygen/mg BOD$_5$ removed, k'_e = 0.085 mg oxygen/(mg MLVSS-day), organic and ammonia nitrogen in the primary clarifier effluent = 24 mg/ℓ, mixed liquor operating temperature =

22°C, temperature correction coefficient for nitrification is $\theta = 1.09$, empirical equation for cells is $C_5H_7O_2N$, and 4.33 mg of oxygen are required per milligram of nitrogen converted. Determine, using USCS units:

a. The reactor basin volume for a plug-flow reactor basin.

b. The waste activated sludge produced per day.

c. The waste activated sludge flow if the sludge is wasted from the recycle line and the wet sludge has a specific gravity of 1.01.

d. The mean cell residence time, θ_c, for the reactor basin.

e. The oxygen required per day.

f. The design mean cell residence time, θ_c, for the nitrifiers. Does this control the design?

g. The fraction of nitrifiers in the mixed liquor.

h. The rate of nitrification in mg/ℓ-day.

i. The detention time required for nitrification. Does this control the design?

j. The oxygen required per hour for the first, second, third, and fourth quarters of the reactor basin.

15.28 Repeat Problem 15.27 using SI units.

15.29 A design is needed for a tapered aeration activated sludge plant having a design or projected population of 12,000 persons. The influent BOD_5 is 200 mg/ℓ and the average annual per capita flow is 100 gal/day (380 ℓ/cap-d). The effluent BOD_5 is to be 10 mg/ℓ. The primary clarifier removes 33% of the influent BOD_5. The design is to be by the NRC equation. Determine:

a. The required BOD_5 removal as a percent for the secondary treatment.

b. The recirculation ratio, R/Q, if the MLSS concentration in the aeration tanks is to be maintained at 2500 mg/ℓ and the sludge density index is 10,000 mg/ℓ.

c. The required detention time in hours for the reactor basin and the required volume for the reactor basin.

d. The blower capacity in cfm if the air required is 1000 ft^3 per lb BOD_5 applied to the reactor basin (62.4 m^3/kg BOD_5) (not including BOD_5 in the return sludge).

e. The space loading.

15.30 Repeat Problem 15.29 using SI units.

15.31 A tapered aeration activated sludge plant is to be designed for a municipality having a population of 10,000 persons, and the BOD_5 of the effluent is to

be equal to 10 mg/ℓ. Pertinent design data are: $Y = 0.60$ mg MLVSS/mg BOD_5, $k_e = 0.06$ day^{-1}, $Y' = 0.62$ mg oxygen/mg BOD_5, $k_e^1 = 0.085$ mg oxygen/mg MLVSS-day, design MLSS = 2500 mg/ℓ, MLVSS = 0.75 MLSS, SDI = 10,000 mg/ℓ, dispersion number (D/vL) = 0.20 for the dispersed plug-flow reactor, flow = 100 gal/cap-day (380 ℓ/cap-day), BOD_5 in effluent from the primary clarifier = 135 mg/ℓ, organic and ammonia nitrogen in the effluent from the primary clarifier = 25 mg/ℓ, 4.33 mg oxygen are required per milligram of nitrogen converted, formulation for the cells in the volatile suspended solids = $C_5H_7O_2N$, plant elevation = 3000 ft (915 m) above sea level, and design temperature = 68°F (20°C). The biochemical reaction is pseudo-first-order, and the rate constant based on MLSS is 0.200 ℓ/gm-hr. Determine:

a. The required aeration time and volume of the dispersed plug-flow reactor.

b. The net sludge production per day.

c. The waste activated sludge flow per day if the specific gravity is 1.01.

d. The oxygen demand per day and per hour, and also the average oxygen uptake rate in mg/ℓ-hr.

e. The average oxygen demand for the first, second, third, and fourth quarters of the reactor basin and the oxygen uptake in mg/ℓ-hr for each quarter of the reactor basin.

f. The standard cubic feet of air required per minute (m^3/min) for each quarter of the tank length if the oxygen diffusers have an efficiency of 6% at an operating DO of 2.0 mg/ℓ.

g. The cubic feet of compressed air per pound of BOD_5 removed (m^3/kg).

Note: From the graph in Chapter 16, the oxygen content at El. 3000 ft (915 m) at 20°C is 0.01562 lb oxygen per cubic foot of air (0.250 kg/m^3)

15.32 Repeat Problem 15.31 using SI units.

15.33 A completely mixed activated sludge plant is to be designed for a soluble organic industrial waste having a flowrate of 2.0 MGD (7.57 MLD), and the total COD of the effluent during winter conditions is to be 120 mg/ℓ. Pertinent design data are: influent COD (total) = 1020 mg/ℓ, nondegradable COD = 90 mg/ℓ, $Y = 0.40$ mg MLVSS/mg COD, $k_e = 0.06$ day^{-1} (at 25°C), $Y' = 0.63$ mg oxygen/mg COD, $k_e' = 0.14$ mg oxygen/mg MLVSS-day (at 25°C), $K = 0.370$ ℓ/gm-hr based on MLVSS (at 25°C), MLVSS = 90% of MLSS, design MLSS = 3000 mg/ℓ, sludge density index (SDI) = 6000 mg/ℓ, minimum operating temperature of the mixed liquor = 18°C, maximum operating temperature of the mixed liquor = 28°C, temperature correction coefficient (θ) for the rate constant = 1.05, temperature correction coefficient (θ) for the endogenous degradation coefficient = 1.07, depth of the basin = 16 ft (4.88 m), basin geometry is square in plan view, specific gravity of waste activated sludge flow = 1.01, and four mechanical aerators are to be used. The biochemical reaction is pseudo–first-order based on the biodegradable COD. There is no nitrogen in the influent wastewater, and just enough nitrogen is added for the required cell synthesis, thus nitrification is not a consideration. For the winter operating conditions, determine:

a. The required aeration time, the reactor volume, and the reactor dimensions.
b. The net sludge production per day.
c. The waste activated sludge flow per day and as a percent of the influent flow.
d. The oxygen demand per day and per hour, and also per aerator.

Next, for summer operating conditions, determine:

e. The biodegradable COD and total COD in the effluent.
f. The net sludge production per day.
g. The waste activated sludge flow per day and as a percent of the influent flow.
h. The oxygen demand per day and per hour, and also per aerator.

Which condition (summer or winter) will control in the design of a reactor size for the maximum effluent COD? In this problem, which condition controlled the design of the waste activated sludge handling facilities? Which condition controlled the oxygen requirements?

15.34 Repeat Problem 15.33 in SI units.

15.35 A completely mixed activated sludge plant is to be designed for a population of 85,000 persons. A treatability study has been made using a pilot plant, and the following data were obtained: influent flow = 95 gal/cap-day (360 ℓ/cap-d), influent BOD$_5$ = 205 mg/ℓ, 34% of the influent BOD$_5$ was removed by primary clarification, μ_{max} = 2.8 day^{-1}, K_s = 65 mg/ℓ, Y = 0.62 mg MLVSS/mg BOD$_5$, k_e = 0.05 day^{-1}, MLSS = 4000 mg/ℓ, MLVSS = 72% of MLSS, effluent suspended solids have a BOD$_5$ of 0.5 mg BOD$_5$ per milligram of suspended solids, and plant permit for effluent is 10 mg/ℓ BOD$_5$ and 15 mg/ℓ suspended solids. Determine, using USCS units:

a. The required reaction time and reactor basin volume.
b. The waste activated sludge production per day.
c. The waste activated sludge flow per day if the sludge is wasted from the recycle line, the sludge density index is 8000 mg/ℓ, and the wet sludge has a specific gravity of 1.01.
d. The food-to-microbe ratio.
e. The space or volumetric loading.

15.36 Repeat Problem 15.35 using SI units.

15.37 Four laboratory-scale, continuous-flow, completely mixed activated sludge reactors have been operated using a wastewater from a yeast production plant. The data found (Wu and Kao, 1976) are shown in Table 15.16. Determine the parameters Y, k_e, μ_{max}, and K_s.

15.38 A completely mixed activated sludge plant is to be designed for the wastewater in Problem 15.37. The influent BOD$_5$ is the average found in the studies. The influent flow is 250,000 gal/day (946,000 ℓ/d). Determine, using USCS units:

a. The mean cell residence time, θ_c, if the effluent BOD$_5$ is 240 mg/ℓ.
b. The reaction time if the MLSS is 7500 mg/ℓ.
c. The reactor volume.
d. The waste activated sludge produced per day.

TABLE 15.16

REACTOR NO.	θ_i (hr)	X_1 MLSS (mg/ℓ)	S_i BOD$_5$ (mg/ℓ)	S_1 BOD$_5$ (mg/ℓ)	$\Delta X_1/\Delta t$ (mg/ℓ-day)
1	24	4,165	2462	546	1417
2	24	6,566	2406	305	1279
3	24	8,996	2383	224	1259
4	24	10,781	2436	172	1025

15.39 Repeat Problem 15.38 using SI units.

15.40 A dispersed plug-flow activated sludge plant is to be designed for the wastewater in Problem 15.37. The influent BOD$_5$ is the average found in the studies. The influent flow is 250,000 gal/day (946,000 ℓ/d), and the dispersion number, d, is 0.2. Determine, using USCS units:
a. The required reaction time if the MLSS concentration is 7500 mg/ℓ, the sludge density index is 12,500 mg/ℓ, and the effluent BOD$_5$ is 240 mg/ℓ.
b. The reactor basin volume.
c. The waste activated sludge produced per day.

15.41 Repeat Problem 15.40 using SI units.

15.42 For the data in Problem 15.37, determine the parameters K, Y, and k_e.

15.43 A completely mixed activated sludge plant is to be designed for the wastewater in Problem 15.37 using the parameters found in Problem 15.42. The influent BOD$_5$ is the average found in the studies. The influent flow is 250,000 gal/day (946,000 ℓ/d). Determine, using USCS units:
a. The required reaction time if the MLSS is 7500 mg/ℓ and the effluent BOD$_5$ is 240 mg/ℓ.
b. The reactor volume.
c. The waste activated sludge produced per day.

15.44 Repeat Problem 15.43 using SI units.

15.45 A dispersed plug-flow activated sludge plant is to be designed for the wastewater in Problem 15.37 using the parameters found in Problem 15.42. The influent BOD$_5$ is the average found in the studies. The influent flow is 250,000 gal/day (946,000 ℓ/d) and the dispersion number is 0.2. Determine, using USCS units:
a. The required reaction time if the MLSS concentration is 7500 mg/ℓ, the sludge density is 12,500 mg/ℓ, and the effluent BOD$_5$ is 240 mg/ℓ.
b. The reactor basin volume.

c. The waste activated sludge produced per day.

15.46 Repeat Problem 15.45 using SI units.

15.47 An organic industrial wastewater has been treated using a continuous-flow activated sludge reactor to develop an acclimated culture of activated sludge. Using the acclimated culture, four batch reactors were operated to obtain data to determine the coefficients Y, Y', k_e, and k_e'. The test was 24 hours in duration, and the laboratory data obtained are shown in Table 15.17. The MLVSS was 88% of the MLSS. Determine the coefficients Y and k_e and the theoretical coefficients Y' and k_e'.

15.48 The empirical equation for the organic fraction of an organic industrial wastewater is $C_5H_{14}O_2N$.
a. Develop the equation for the respiration of the organic material if the nitrogen is oxidized to the ammonium ion, NH_4^+, and the oxidation products for respiration are CO_2, H_2O, and NH_4^+.
b. Develop the equation for the synthesis of microbial cells if the empirical equation for the cells is $C_5H_7O_2N$.
c. Develop the overall biochemical equation for respiration and synthesis if the cell yield is 0.39 mg cells per mg TOD.
d. What is the theoretical oxygen demand per 1.0 mg organic matter?
e. What is the ratio of the ultimate first-stage BOD to the theoretical oxygen demand?
f. What is the ratio of the 5-day BOD to the theoretical oxygen demand if the 5-day BOD is 70% of the ultimate first-stage BOD?

15.49 An industrial wastewater from a sugar refining plant has sucrose, $C_{12}H_{22}O_{11}$, as essentially all of its organic content, and the nitrogen source to be added is in the form of the ammonium ion, NH_4^+.
a. Develop the equation for the respiration of sucrose.

TABLE 15.17

REACTOR NO.	X_0 MLSS (mg/ℓ)	X_t MLSS (mg/ℓ)	S_0 COD (mg/ℓ)	S_t COD (mg/ℓ)
1	500	665	680	111
2	946	1050	680	83
3	1920	1850	680	95
4	3650	3310	680	88

b. Develop the equation for the synthesis of microbial cells from sucrose if the empirical equation for the cell mass is $C_5H_7O_2N$.

c. Develop the overall biochemical equation for respiration and synthesis if the cell yield is 0.39 mg cells per mg TOD.

d. What is the theoretical oxygen demand per 1.0 mg organic matter?

e. What is the ratio of the ultimate first-stage BOD to the theoretical oxygen demand?

15.50 An activated sludge treatment plant treats a peak wastewater flow of $4.09\,\text{ft}^3/\text{sec}$ ($0.116\,\text{m}^3/\text{s}$), and the maximum recycled sludge flow is 150% of the average flow or $1.86\,\text{ft}^3/\text{sec}$ ($0.0527\,\text{m}^3/\text{s}$). The plant layout and profile are shown in Figure 15.37. The process lines P-1, P-2, P-3, and P-4 are designed for gravity flow and are to have a maximum velocity of $4.5\,\text{ft/sec}$ ($1.37\,\text{m/s}$) to avoid excessive head losses. Process line P-3 will have a maximum flow of $4.09 + 1.86 = 5.95\,\text{ft}^3/\text{sec}$ ($0.116 + 0.0527 =$

$0.1687\,\text{m}^3/\text{s}$). Line P-3 discharges into a 28-in. (710-mm) center column in the final clarifier that is 15 ft (4.57 m) high. The reactor basin is 50 ft wide (15.2 m) and has a suppressed weir at the outlet channel. The downstream depth, d, is 4 in. (102 mm) greater than critical depth to insure submerged outlet conditions. The effluent channel is 1 ft-3 in. (0.381 m) wide and has a rectangular cross section. There is a 4-in. (102-mm) freefall below the suppressed weir at the upstream end where the depth, H_0, occurs. Determine, using USCS units:

a. The theoretical diameter for process line P-3.

b. The design diameter for process line P-3 if standard cast iron pipe sizes from 10 to 20 in. are 10, 12, 14, 16, 18, and 20 in.

c. The head, H, on the suppressed weir at the outlet end of the reactor basin.

d. The values of d and H_0 for the effluent channel if $n = 0.032$.

e. The head losses for process line P-3 and the

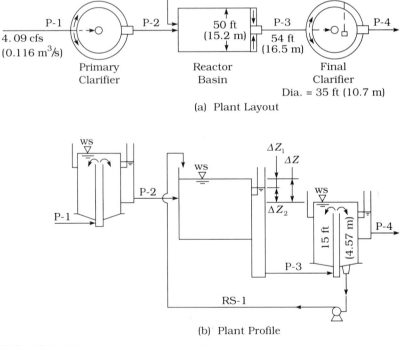

(a) Plant Layout

(b) Plant Profile

FIGURE 15.37 *Plant Layout and Profile for Problems 15.50 and 15.51*

28-in. center column in the final clarifier. *Note*: The head losses include the entrance loss, $(0.5V^2/2g)$, the friction loss for 54 ft of cast iron pipe using a Hazen Williams coefficient of 100, the loss for a sudden expansion and 90° change in direction where P-3 enters the center column $\Big($loss $= \Delta V^2/2g$, where ΔV is the vector change in velocity — that is, $\Big)$, the friction loss in the 15-ft center column using a Hazen Williams coefficient of 100, and the exit loss $(V^2/2g)$. For the USCS units, the Hazen Williams equation is $V = 1.318C_{HW}R^{0.63}S^{0.54}$, where V = velocity in ft/sec, C_{HW} = Hazen Williams friction coefficient, R = hydraulic radius in ft = $D/4$ for a circular pipe with a diameter D, and S = slope of the energy gradient.

f. The drop in water surface elevation, ΔZ_1, between the water surface in the reactor basin and that in the effluent box. *Note*: The drop in elevation between the water surface in one tank and the water surface in a downstream box or tank can be determined by writing Bernoulli's equation between the two points:

$$V_1^2/2g + p_1/\gamma + Z_1 = V_2^2/2g + p_2/\gamma + Z_2 + \Sigma h_L$$

Since V_1 and V_2 are essentially zero and, if re-lative pressures are used, p_1/γ and p_2/γ are zero, Bernoulli's equation becomes $Z_1 - Z_2 = \Delta Z = \Sigma h_L$. For the drop in water surface between the reactor basin and its outlet box, Σh_L equals the sum of the head losses, which are the head on the suppressed weir, the freefall, and $H_0 - d$.

g. The drop in water surface elevation, ΔZ_2, between the water surface in the effluent outlet box and the water surface in the final clarifier.

h. The total drop in water surface elevation, $\Delta Z_1 + \Delta Z_2$, between the water surface in the reactor basin and the water surface in the final clarifier.

Note: The suppressed weir on the outlet channel for a reactor basin should have a head of 0.10 ft (31 mm) or more in order to ensure a free nappe. If one long weir across the end of the basin gives a head less than this, then several shorter suppressed weirs should be used.

15.51 Rework Problem 15.50 using SI units. In part (b), the standard pipe sizes are 250, 300, 350, 400, 450, and 500 mm. In part (e), the Hazen Williams equation in SI units is $V = 0.8464C_{HW}R^{0.63}S^{0.54}$, where V = velocity in m/s, C_{HW} = Hazen Williams friction coefficient, R = hydraulic radius in meters = $D/4$ for a circular pipe with a diameter D, and S = slope of the energy gradient.

16

OXYGEN TRANSFER
AND MIXING

In activated sludge, aerated lagoon, and aerobic digestion processes, oxygen must be supplied to the biological solids for aerobic respiration, and mixing must be sufficient to maintain the solids in suspension. Oxygen transfer and mixing are provided by diffused compressed air, by mechanical aeration, by a combination of diffused compressed air and mechanical aeration (such as the submerged-turbine type of aerator), or by pure oxygen with mechanical agitation for mixing. This chapter covers primarily diffused compressed air aeration and mechanical aeration because these are the most common systems used. However, the principles presented also apply to pure oxygen systems.

OXYGEN TRANSFER

The transfer of a solute gas from a gas mixture into a liquid that is in contact with the mixture can be described by the two-film theory of Lewis and Whitman (1924). Figure 16.1 shows a schematic drawing of the two phases in contact with each other. The partial pressures of the solute gas in the bulk gas and at the gas interface are p_G and p_{Gi}, respectively. The concentrations of the solute gas at the liquid interface and in the bulk liquid are C_{Li} and C_L, respectively. The solute gas must diffuse through the gas film (laminar layer), pass through the interface, and then diffuse through the liquid film (laminar layer). The interface offers no resistance to the solute gas transfer. For gases that are very soluble in the liquid, the rate limiting step is the diffusion of the solute gas through the gas film. For gases that are slightly soluble in the liquid, such as oxygen in water, the rate limiting step is the diffusion of the solute gas through the liquid film. The diffusion transfer coefficient, K_L, for oxygen diffusing through the water film is given by $K_L = D/\delta_\ell$, where D is the diffusivity coefficient of oxygen in water and δ_ℓ is the film thickness. Multiplying K_L by a, the interfacial bubble area per unit volume of water, gives the overall mass transfer coefficient, $K_L a$. Since the liquid resistance is controlling, $p_G = p_{Gi}$ and $C_{Li} = C_s$, C_s being the saturation dissolved oxygen concentration in equilibrium with the oxygen partial pressure in air bubbles, p_G. The saturation dissolved oxygen concentration, C_s, is in equilibrium with the partial pressure, p_G, in accordance with Henry's law, one form of this law being $p_G = (\text{constant})(C_s)$. The driving force for mass transfer is $C_{Li} - C_L$, but since $C_s = C_{Li}$, the driving force is $C_s - C_L$. Since the rate of oxygen mass transfer is equal to the mass transfer coefficient times the driving force, the mass transfer is expressed by

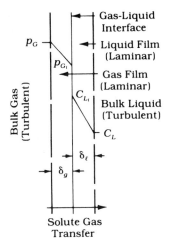

FIGURE 16.1 *Schematic Drawing Illustrating the Two-Film Theory*

$$\frac{dC}{dt} = K_L a(C_s - C_L) \tag{16.1}$$

where

dC/dt = rate of oxygen transfer, mass/(volume)(time) — for example, mg/ℓ-hr

$K_L a$ = overall liquid mass transfer coefficient, time^{-1} — for example, hour^{-1}

C_s = saturation dissolved oxygen concentration, mass/volume — for example, mg/ℓ

C_L = dissolved oxygen concentration in the liquid, mass/volume — for example, mg/ℓ

The mass transfer per unit time is given by

$$\frac{dM}{dt} = N = K_L a V(C_s - C_L) \tag{16.2}$$

where

dM/dt = rate of oxygen transfer, mass/time

N = rate of oxygen transfer, mass/time

V = liquid volume

Aeration devices for diffused compressed air may be classified as coarse or fine bubble diffusers. The efficiency of oxygen transfer depends primarily upon the design of the diffuser, the size of the bubbles produced, and the depth of submergence. The coarse bubble diffusers usually consist of a short vertical tube and/or an orifice device, as in Figure 16.2. The diffusers must be spaced far enough apart so that the upward-flowing bubble plumes do not significantly interfere with each other. The field transfer efficiency for coarse bubble diffusers is usually from 4 to 8%, a value of 6% being typical. The coarse bubble type has a lower efficiency than the fine bubble type; however, it is less susceptible to clogging. The fine bubble types, some of which are shown in Figure 16.3, are usually made of porous ceramic tubes or cylindrical metal frames covered with a wrapping such as a Dacron cloth or Saran cord. Sometimes porous ceramic plates covering an air channel cast in the tank bottom are used. The field transfer efficiency of the fine bubble type is from 8 to 12%, a value of 9% being typical. Usually, the air flow per unit for either the fine or the coarse bubble type is from 4 to 16 scfm (at 20°C and 1 atm) (0.113 to 0.453 m³/min).

For diffused aeration, Eckenfelder and Ford (1968) have presented the following formulation for the rate of oxygen mass transfer:

$$N = CG_a^{1-n}D^{0.67}(C_m - C_L) \cdot 1.02^{(T-20)} \cdot \alpha \qquad \textbf{(16.3)}$$

where

N = rate of oxygen transfer, lb/hr (kg/h)

C and n = constants

G_a = air flow, standard cubic feet per minute (at 20°C and 1 atm) (standard cubic meters per minute)

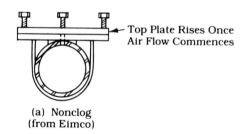

(a) Nonclog
(from Eimco)

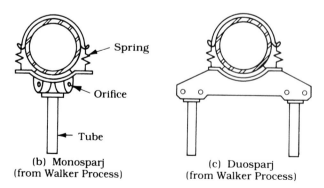

FIGURE 16.2
Coarse Bubble
Diffusers

(b) Monosparj
(from Walker Process)

(c) Duosparj
(from Walker Process)

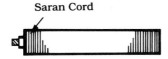

(b) Saran-Cord Wrapped
(from FMC Corp.)

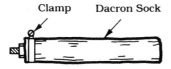

FIGURE 16.3 *Fine Bubble Diffusers*

(a) Flexofuser
(from FMC Corp.)

D	= depth to diffusers, ft (m)
C_m	= saturation dissolved oxygen in the wastewater at the mid-depth of the tank at operating conditions
T	= operating temperature, °C
a	= K_La(wastewater)/K_La(water)

It can be seen in Eq. (16.3) that the terms $CG_a^{1-n}D^{0.67} \times 1.02^{(T-20)}a$ represent K_LaV in Eq. (16.2). Usually the values of the experimental constants C and n are furnished by the manufacturer.

Since the solubility of oxygen varies with pressure, the saturation dissolved oxygen, C_m, is determined at the tank mid-depth and is given by (Oldshue, 1956)

$$C_m = C_w\left(\frac{P_r}{C} + \frac{O_e}{42}\right) \tag{16.4}$$

where

C_m = saturation dissolved oxygen concentration of wastewater at the tank mid-depth at operating conditions

C_w = saturation dissolved oxygen concentration of the wastewater at operating conditions

P_r = absolute pressure at the depth of air release, psia (kPa)

O_e = percent oxygen content in the exit air flow

C = 29.4 for USCS units and 203 for SI units

The saturation dissolved oxygen of the wastewater at the operating atmospheric pressure, C_w, is given by $C_w = \beta C_s$, where β is the ratio of the saturation dissolved oxygen for the wastewater divided by the saturation dissolved oxygen of tap water at the same temperature. Figure 16.4 shows diffused compressed air diffusers at an activated sludge plant, and Figure 16.5 shows typical air compressors at an activated sludge plant.

Mechanical aerators usually employ impellers, and, as shown in Figure 16.6, they may be float or fixed mounted. They may be classified as the high-speed axial-flow pump type, the slow-speed vertical turbine, the submerged slow-speed vertical turbine with a sparger ring, and the rotating brush

FIGURE 16.4 *Activated Sludge Aeration Tank with Swing-Type Air Diffusers*
Courtesy FMC Corporation, Materials Handling Systems Division.

FIGURE 16.5 *Air Compressors at a Diffused Compressed Air Type Activated Sludge Plant*

aerator. The pump-type aerator, as shown in Figure 16.6 (a), is a high-speed axial-flow pump with a propeller; it is used mainly for aerated lagoons. Motor sizes are from 1 to 150 hp (0.75 to 112 kW), and they have speeds from 900 to 1800 rpm, with 900 to 1200 rpm being typical. The operating water depth is from 3 to 18 ft (0.9 to 5.5 m), and draft tubes are usually required when the depth is greater than about 10 to 12 ft (3.0 to 3.7 m). Oxygen transfer occurs as the spray passes through the air and also at the impingement area, because a considerable amount of turbulence is created. These aerators are not suited for severely cold climates where freezing of the water might occur. The oxygen transfer rate is from 2.0 to 3.9 lb/hp-hr (nameplate hp) (1.22 to 2.37 kg/kW-h), with the smallest machines having the highest transfer.

The slow-speed vertical turbine, shown in Figure 16.6 (b) and (c) and in Figures 16.7 and 16.8, may be used for activated sludge, aerobic digesters, and aerated lagoons. Motor sizes are from 3 to 150 hp (2.2 to 112 kW), and turbine sizes range from about 3 to 12 ft (0.9 to 3.7 m) in diameter. The speeds are from about 30 to 60 rpm; thus they always require a gear reduction box. They may be fixed mounted, as shown in Figure 16.6 (b) and (c), or float mounted for flexibility. The operating water depth is from 3 to 30 ft (0.9 to 9.1 m), with 12 to 15 ft (3.7 to 4.6 m) being the most common. Depths from about 15 to 30 ft (4.6 to 9.1 m) require draft tubes, as shown in Figure 16.6 (b). The turbines are submerged only a few inches (cm) below the still water surface, and when they are operating, the flow goes radially outward. Air entrainment occurs in the immediate vicinity of the turbine, and also oxygen absorption occurs as a result of turbulence. The proper depth of submergence is critical for good oxygen transfer. Transfer rates are

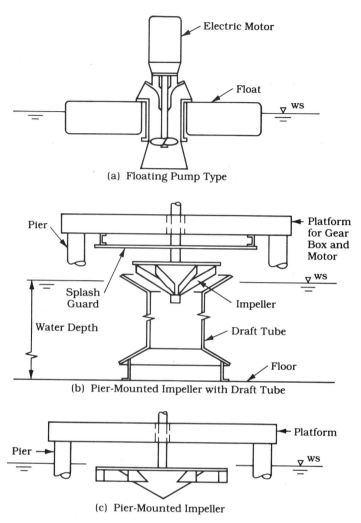

FIGURE 16.6 *Some Impellers Employed by Mechanical Aerators*

from about 2.0 to 3.9 lb/hp-hr (nameplate) (1.22 to 2.37 kg/kW-h), with the smallest machines having the greatest transfer.

For mechanical aeration, Eckenfelder and Ford (1968) have presented the following formulation for the rate of oxygen mass transfer:

$$N = N_0 \left(\frac{C_w - C_L}{9.17} \right) 1.02^{(T-20)} \alpha \tag{16.5}$$

where

N = rate of oxygen transferred at operating conditions, lb/hr or lb/(hp)(hr) (kg/h or kg/kW-h)

FIGURE 16.7 *Mechanical Surface Aerator. A drawing of the impeller and draft tube for this type of unit is shown in Figure 16.6 (b).*

FIGURE 16.8 *Mechanical Surface Aerator. A drawing of the impeller for this type of unit is shown in Figure 16.6 (c).*

N_0 = oxygen rating of the aerator, lb/hr or lb/(hp)(hr) (kg/h or kg/kW-h)

The rating of a mechanical aerator, N_0, is for 20°C and a zero dissolved oxygen concentration; thus the parenthetical term in Eq. 16.5 is the ratio of the actual driving force to the rated driving force.

The submerged slow-speed vertical turbine with a sparger ring, shown in Figure 16.9, is located about 1.5 ft (0.46 m) above the bottom of a tank; the

FIGURE 16.9 *Submerged Slow-Speed Vertical Turbine Aerator. Note the upper turbine used to provide additional mixing and the compressed air release below the lower turbine.*

sparger ring allows compressed air to be released below the turbine, thus creating oxygen transfer. Oxygen transfer is from about 2.0 to 3.0 lb/hp-hr (nameplate) (1.22 to 1.83 kg/kW-h). Usually, the turbine diameter is 0.1 to 0.2 times the tank width. In a circular tank, four baffles mounted on the wall at 90° intervals in plan view are required to minimize rotational flow. In a square tank, two baffles mounted on opposite walls are usually satisfactory. In rectangular tanks, baffles are usually not required if the ratio of length to width is 1.5:1 or more. One advantage of these units is that they can be used in deep, narrow tanks. A disadvantage is that they require a source of compressed air.

The rotating brush aerator, shown in Figure 16.10, is used mainly for oxidation ditches; however, in Europe such aerators are often used for aeration in long, narrow tanks used for the activated sludge process. The aerator consists of a long horizontal axle with radiating steel bristles partly

FIGURE 16.10 *Rotating Brush Aerator. The radiating steel bristles furnish the required agitation.*

submerged in the still water surface. As the axle rotates, the bristles spray the liquid outward and oxygen transfer occurs because of air entrainment in the immediate vicinity of the bristles, and also because of the spray and impingement area. Oxygen transfer rates are from 3.0 to 3.5 lb/hp-hr (nameplate) (1.83 to 2.13 kg/kW-h).

The value of α may be determined in the laboratory by using a vessel equipped with a diffused aeration stone and employing a constant liquid depth in the vessel and a constant air flowrate. The vessel is filled with tap water and deoxygenated by adding sodium sulfite and cobalt chloride, the cobalt serving as a catalyst. The aeration is begun, and the dissolved oxygen concentration versus time is recorded. This represents an unsteady state, and for this condition,

$$\frac{dC}{dt} = K_L a(C_s - C) \tag{16.6}$$

where C = dissolved oxygen concentration, mg/ℓ. Rearranging for integration gives

$$\int_{C_0}^{C_t} \frac{-dC}{C_s - C} = K_L a \int_0^t - dt \tag{16.7}$$

where

C_0 = dissolved oxygen concentration at time zero

C_t = dissolved oxygen concentration at time, t

Integrating and rearranging give

$$\ln(C_s - C_t) = \ln(C_s - C_0) - K_L at \qquad (16.8)$$

Equation (16.8) is of the straight-line form $y = mx + b$; thus plotting $\ln(C_s - C_t)$ versus time, t, gives a straight line with a slope equal to $K_L a$. This gives $K_L a$ for tap water. Now the vessel is emptied and refilled to the same depth with the untreated wastewater, and, using the same air flowrate, the experiment is repeated to give $K_L a$ for the untreated wastewater. Similarly, using treated wastewater, the value of $K_L a$ for treated wastewater is obtained. The value for $K_L a$ for the mixed liquor is the average of the two values. The value α is given by $\alpha = K_L a(\text{wastewater})/K_L a(\text{water})$. Here, $K_L a(\text{wastewater})$ is equal to the mean value of $K_L a(\text{untreated wastewater})$ and $K_L a$ (treated wastewater). The $K_L a$ (wastewater) is equal to $K_L a$ for the mixed liquor. In summary, $\alpha = (1/2)[K_L a(\text{untreated wastewater}) + K_L a(\text{treated wastewater})] \div [K_L a(\text{water})]$. The saturation dissolved oxygen concentration for the waste may be obtained by filling the vessel with wastewater and aerating the waste until the dissolved oxygen reaches a maximum, thus giving the saturation dissolved oxygen concentration for the wastewater. This is repeated for the treated wastewater, giving a dissolved oxygen saturation concentration for the treated waste. The value C_w is equal to the average of the two saturation values for the wastewater — that is, the values for the untreated and treated wastewater. The β value is the average C_w divided by the saturated oxygen concentration, C_s, for tap water at the same temperature. In summary, $\beta = (1/2)[C_w(\text{untreated wastewater}) + C_w(\text{treated wastewater})] \div [C_s(\text{water})]$.

In the previous aeration equations, the values of C_s and, in turn, C_w, are for the operating temperature of the mixed liquor and the operating atmospheric pressure. If the elevation of the plant is much greater than sea level, the operating pressure is less than 1 atm (760 mm Hg), and the saturated dissolved oxygen concentration is less than that given in standard solubility tables. The atmospheric pressures for various elevations are shown in Table 16.1.

The saturation dissolved oxygen concentration at an elevation above sea level is equal to the C_s value given in a standard table at 760 mm multiplied by the ratio of the atmospheric pressure at that elevation divided by 760. For instance, at an elevation of 5500 ft (1676 m), the atmospheric pressure using interpolation of the above values is 620 mm; thus the solubility of the dissolved oxygen at 20°C is $(620/760)(9.17)$ or $7.48\,\text{mg}/\ell$. If an activated sludge plant for this elevation were designed using a saturated dissolved oxygen value at 760 mm and an operating dissolved oxygen value of $2.0\,\text{mg}/\ell$, the aeration system would be underdesigned by $(9.17 - 2.0)/(7.48 - 2.0)$ minus 1 or 31%. Consequently, in some instances a correction for elevation must be made.

MIXING

Mixing has not been rationally analyzed to a great extent, and it is common to express mixing requirements in terms of power intensity or air flowrate

TABLE 16.1 *Atmospheric Pressures for Various Elevations*

ELEVATION (ft)	ATMOSPHERIC PRESSURE (mm Hg)
0	760
1000	733
2000	706
3000	680
4000	656
5000	632
6000	609
7000	586
8000	564
9000	543

NOTE: 1 m = 3.281 ft.

for a given volume. For instance, Metcalf & Eddy, Inc. (1991) gives aeration intensities required to keep activated sludge in complete suspension as 20 to 30 scfm/1000 ft^3 (20 to 30 m^3/min-1000 m^3) of tank volume. For mechanical aerators they recommend 0.75 to 1.5 hp/1000 ft^3 (19.8 to 39.5 kW/1000 m^3). For aerobic digesters, they recommend 20 to 40 scfm/1000 ft^3 (20 to 40 m^3/min-1000 m^3) of diffused air for mixing or 0.75 to 1.5 hp/1000 ft^3 (19.8 to 39.5 kW/1000 m^3) of mechanical aeration for mixing. For diffused aeration, the degree of mixing is related to the depth of the diffusers; thus the lower requirements, such as 20 scfm/1000 ft^3 (20 m^3/min-1000 m^3), are for deep tanks, and the higher requirements, such as 30 or 40 scfm/1000 ft^3 (30 or 40 m^3/min-1000 m^3), are for shallow tanks. Normally, activated sludge tanks are from about 10 to 16 ft (3.0 to 4.9 m) deep. For mechanical surface aeration, the degree of mixing is inversely related to the tank depth; thus the higher values, such as 1.5 hp/1000 ft^3 (39.5 kW/1000 m^3), are for deep tanks and the lower values, such as 0.75 hp/1000 ft^3 (19.8 kW/1000 m^3), are for shallow tanks.

COMPRESSOR REQUIREMENTS

The power required by a compressor to furnish diffused compressed air may be determined from the following equation, which gives brake (shaft) power for the adiabatic compression of a gas (Perry and Chilton, 1973).

$$P = \frac{FRT_1}{CnE}\left[\left(\frac{p_2}{p_1}\right)^n - 1\right] \tag{16.9}$$

where

P = brake or shaft power, hp (kW)

F = mass of air flow, lb/sec (kg/s)

R = 53.5 for USCS units and 0.288 for SI units

T_1 = inlet absolute temperature, °R (°K)

C = 550 for USCS units and 1.0 for SI units

n = 0.283 for air

E = efficiency of the compressor as a fraction (usually 0.70 to 0.80)

p_1 = absolute inlet pressure, lb per sq. in. absolute, psia (kilopascals, kPa)

p_2 = absolute outlet pressure, lb per sq. in. absolute, psia (kilopascals, kPa)

The mass air flow in lb/sec (kg/s) is given by the formulation

$$F = G_a \times \rho_{\text{air}} \qquad (16.10)$$

where

F = mass air flow, lb/sec (kg/s)

G_a = air flowrate, ft³/sec (m³/s)

ρ_{air} = air density, lb/ft³ (kg/m³)

The specific weight or density of air depends upon the pressure and temperature and is given by Figure 16.11. In compressor work, a standard cubic foot (cubic meter) is at 1 atm (14.7 psia), 20°C (68°F), and 36% relative humidity. The term **free air** refers to air at conditions prevailing at the compressor inlet. In using Eq. (16.9), the power requirements should be determined by the free air flow in lb/sec (kg/s) and the actual inlet

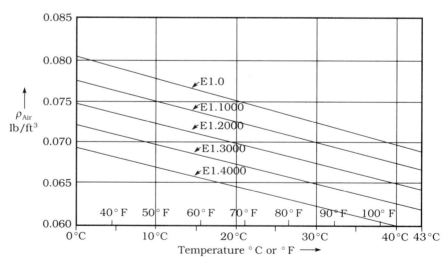

FIGURE 16.11 *Density or Specific Weight of Air (36% relative humidity)*

NOTE: 1 kg/m³ = 0.06236 lb/ft³

1 m = 3.281 ft

temperature. The two types of compressors (frequently called **blowers**) used in wastewater treatment are the centrifugal and the rotary type. The centrifugal type has a split impeller and operates on the same principle as a centrifugal pump. The rotary type is a positive displacement compressor that uses two high-speed rotors or impellers. Equation (16.9) gives the power that must be applied to the shaft of the compressor; therefore, to get the power required by the motor, the shaft power of the compressor must be divided by the efficiency of the motor. Electric motors used in compressor work have about 95% to 98% efficiency. The operating pressure is a function of the head loss through the piping and diffusers and of the depth of submergence of the diffusers. Most of the pressure is due to the depth of submergence, and operating pressures are usually 5 to 10 psi (34.5 to 69 kPa).

Figure 16.12 gives the oxygen content in air (lb oxygen/ft^3 or kg oxygen/ m^3) versus temperature, elevation, and atmospheric pressure. Note that the oxygen content at a given temperature decreases with an increase in elevation and that the oxygen content at a given elevation decreases with an increase in temperature.

EXAMPLE 16.1 *Diffused Aeration*

An activated sludge plant (tapered aeration) is located at El. 2500. The maximum oxygen demand occurs in the summer when the air temperature is 100°F and the maximum wastewater temperature is 82°F. The oxygen demand

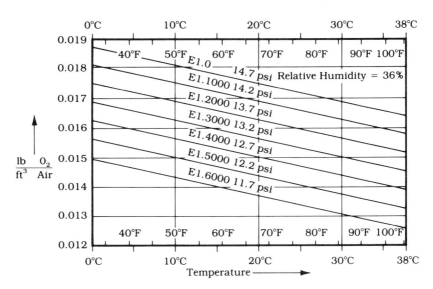

FIGURE 16.12 *Pounds of Oxygen per Cubic Foot of Air versus Elevation, Atmospheric Pressure, and Temperature*
NOTE: 1 kg/m^3 = 0.06236 lb/ft^3
      1 m   = 3.281 ft
      1 kPa   = 0.1450 psi

is 2320 lb O_2/day. The air flow per diffuser is 15 scfm (at 1 atm and 20°C), α = 0.75, β = 0.95, the operating DO = 2.0 mg/ℓ, tank depth = 15 ft, and the diffusers are 1.5 ft above the tank bottom. The oxygen demand values for the respective quarters are 35%, 26%, 20%, and 19% of the total demand. The diffuser performance is given by

$$N = 0.00170 G_a^{0.9} D^{0.67}(C_m - C_L) \cdot 1.02^{(T-20)} \cdot \alpha$$

where N = lb/hr transferred, G_a = air flow (scfm), D = diffuser depth, C_m = saturation DO at mid-depth, mg/ℓ, C_L = operating DO, mg/ℓ, T = wastewater temperature, °C, and α = $K_L a$(wastewater)/$K_L a$(tap water). Determine:

1. The oxygen transfer for the diffusers.
2. The number of diffusers.
3. The total air flow at the plant site.
4. The theoretical horsepower of the compressor if the compressor efficiency is 75%.

SOLUTION   The wastewater temperature in degrees Centigrade is $(82 - 32)5/9 = 27.8$°C. Interpolation from the oxygen solubility table in the Appendix gives C_s = 7.95 mg/ℓ at sea level. Using the atmospheric pressure table, Table 16.1, gives the pressure as 693 mm Hg. Thus C_s at El. 2500 is $(7.95)(693/760)$ = 7.25 mg/ℓ. The saturation dissolved oxygen for wastewater is given by

$$C_w = \beta C_s$$

where C_w = saturation dissolved oxygen for wastewater, β = correction coefficient, and C_s = saturation dissolved oxygen from standard tables. Thus,

$$C_w = 0.95 C_s = (0.95)(7.25) = 6.89 \text{ mg/}\ell$$

If one ignores the trace gases, the atmosphere is about 21% oxygen and 79% nitrogen. Assuming 100 moles of air to be compressed, there are about 21 moles of oxygen and 79 moles of nitrogen. Assuming 5% transfer, the oxygen in the exit gas has $(21)(1 - 0.05) = 19.95$ moles of oxygen. The percent oxygen in the exit gas is $(19.95)/(19.95 + 79)(100) = 20.16\%$. The pressure at the point of release = $(15 \text{ ft} - 1.5 \text{ ft})(\text{psi}/2.31 \text{ ft}) + (14.7 \text{ psia})(693/760) = 5.84 + 13.40 = 19.24 \text{ psia}$. The saturation dissolved oxygen at mid-depth is given by

$$C_m = C_w\left(\frac{P_r}{29.4} + \frac{O_e}{42}\right)$$

where C_m = saturation dissolved oxygen at mid-depth, C_w = saturation dissolved oxygen of the wastewater, P_r = absolute pressure at the point of release (psia), and O_e = exit air oxygen content (percent). Thus,

$$C_m = 6.89\left(\frac{19.24}{29.4} + \frac{20.16}{42}\right) = 7.82 \text{ mg/}\ell$$

Substituting values in the performance equation gives

$$N = 0.00170 \, (15)^{0.9}(13.5)^{0.67}(7.82 - 2.0) \cdot 1.02^{(27.8 - 20)} \cdot 0.75$$
$$= \boxed{0.567 \, \text{lb} \, O_2/\text{hr}}$$

The number of diffusers for the first quarter is $(2320 \, \text{lb} \, O_2/24 \, \text{hr})(0.35) \times (\text{hr}/0.567 \, \text{lb} \, O_2) = 59.7$ or 60 diffusers. In a like manner, the numbers for the other quarters are found to be 45, 35, and 33. Thus the total diffusers are $60 + 45 + 35 + 33 = \boxed{173.}$ The total air flow is $(173)(15 \, \text{scfm}) = 2595 \, \text{scfm}$. Standard conditions are 20°C and 1 atm or 14.7 psia. The atmospheric pressure at the plant site is $(693/760)(14.7) = 13.40$ psia. The 20°C temperature in °F is $(20)(9/5) + 32 = 68$°F. The absolute temperature for 68°F is °R = °F + 460 or °R = 68 + 460 = 528°R. The absolute temperature for 100°F is °R = 100 + 460 = 560°R. The general gas law is

$$\frac{P_1 V_1}{T_1} = \frac{P_2 V_2}{T_2}$$

where P_1, P_2 = absolute gas pressures, V_1, V_2 = gas volumes, and T_1, T_2 = absolute gas temperatures, °R or °K. Thus,

$$V_2 = \frac{P_1}{P_2} \cdot \frac{T_2}{T_1} \cdot V_1$$

or cfm $= (14.7 \, \text{psia}/13.40 \, \text{psia})(560°R/528°R)(2595 \, \text{cfm}) = \boxed{3020 \, \text{cfm.}}$

The power required is given by

$$\text{hp (brake or shaft)} = \frac{FRT_1}{550nE}\left[\left(\frac{p_2}{p_1}\right)^{0.283} - 1\right]$$

where F = mass flowrate of air, lb/sec, $R = 53.5$, T_1 = inlet absolute temperature, °R, p_1 = absolute inlet pressure, psia, p_2 = absolute outlet pressure, psia, $n = 0.283$ for air, and E = compressor efficiency as a fraction. Figure 16.11 shows that at El. 2500 and 100°F, $\rho_{\text{air}} = 0.064 \, \text{lb/ft}^3$, so

$$F = (3020 \, \text{ft}^3/\text{min})(0.064 \, \text{lb/ft}^3)(\text{min}/60 \, \text{sec})$$
$$= 3.22 \, \text{lb/sec}$$

The pressure at the inlet, p_1, is 13.40 psia, and the pressure at the outlet, p_2, is 19.24 psia. Inserting the various variables gives

$$\text{hp} = \left(\frac{3.22}{550}\right)\left(\frac{53.5}{0.283}\right)\left(\frac{560}{0.75}\right)\left[\left(\frac{19.24}{13.40}\right)^{0.283} - 1\right]$$
$$= \boxed{89.1 \, \text{hp}}$$

(*Note:* Many manufacturers have bulletins showing graphs of the percent oxygen transfer in tap water at a given depth. This is for 20°C, 1 atm, an operating dissolved oxygen of zero, and $\alpha = 1.0$. Using these data, the methods in this example, and Eq. (16.3), it is possible to determine the constant C for their equipment. With the constant, Eq. (16.3) can be used for design.)

EXAMPLE 16.1 SI   *Diffused Aeration*

An activated sludge plant (tapered aeration) is located at El. 762 m (El. 2500 ft). The maximum oxygen demand occurs in the summer when the air temperature is 38°C and the maximum wastewater temperature is 28°C. The oxygen demand is 1053 kg O_2/d. The air flow per diffuser is 0.425 m³/min (at 1 atm or 101.37 kPa and 20°C), $\alpha = 0.75$, $\beta = 0.95$, the operating DO = 2.0 mg/ℓ, tank depth = 4.57 m, and the diffusers are 0.45 m above the tank bottom. The diffuser performance is given by

$$N = 0.04233 G_a^{0.9} D^{0.67}(C_m - C_L) \cdot 1.02^{(T-20)} \cdot \alpha$$

where N = kg/h transferred, G_a = air flow sm³/min, C_m = saturation DO at mid-depth, mg/ℓ, C_L = operating DO, mg/ℓ, T = wastewater temperature, °C, and $\alpha = K_L a$(wastewater)/$K_L a$(tap water). Determine:

1. The oxygen transfer for the diffusers.
2. The total air flow at the plant site.
3. The theoretical compressor power, kW, if the compressor efficiency = 75%.

SOLUTION   The temperature of the wastewater is 28°C, and from the table in the Appendix, the saturation DO = 7.92 mg/ℓ at sea level. The elevation is (762 m) × (3.281 ft/m) = 2500 ft. Using the atmospheric pressure table, Table 16.1 in this chapter, gives the barometric pressure as 693 mm Hg. Thus C_s at El. 760 m is (7.92)(693/760) = 7.22 mg/ℓ. The saturation DO of the wastewater is given by

$$C_w = \beta C_s$$

where all variables have been previously defined. Thus,

$$C_w = 0.95(7.22) = 6.86 \text{ mg/ℓ}$$

If one ignores the trace gases, the atmosphere is about 21% oxygen and 79% nitrogen. Assuming 100 moles of air to be compressed, there are about 21 moles of oxygen and 79 moles of nitrogen. Assuming 5% transfer, the oxygen in the exit gas has (21)(1 − 0.05) = 19.95 moles of oxygen. The percent oxygen in the exit gas is (19.95)/(19.95 + 79)(100) = 20.16%. The pressure at the point of release (gage pressure) is (4.57 m − 0.45 m) (9810 N/m³) = 40,417 N/m² = 40.42 kPa. The absolute pressure at the point of release is 40.42 kPa + [101.37 kPa (1 atm)](693/760) = 40.42 + 92.43 = 132.85 or 133 kPa. The saturation DO at mid-depth is given by

$$C_m = C_w\left(\frac{P_r}{203} + \frac{O_e}{42}\right)$$

where all values have been previously defined. Substituting values gives

$$C_m = 6.86\left(\frac{133}{203} + \frac{20.16}{42}\right) = 7.79 \, \text{mg}/\ell$$

The depth to the diffusers = $4.57 - 0.45 = 4.12$ m. Substituting values in the performance equation gives

$$N = (0.04233)(0.425)^{0.9}(4.12)^{0.67}(7.79 - 2.0) \cdot 1.02^{28-20} \cdot 0.75$$
$$= \boxed{0.257 \, \text{kg O}_2/\text{h}}$$

The number of diffusers is $(1053 \, \text{kg/d})(\text{d}/24\,\text{h})(\text{h}/0.257\,\text{kg}) = 170.7$ or $\boxed{171.}$ The total air flow is $(171)(0.425) = 72.7 \, \text{sm}^3/\text{min}$. Standard conditions are 20°C and 1 atm. The atmospheric pressure at the plant site is $693/760 = 0.9118$ atm. The absolute temperature at 20°C = $20 + 273 = 293$°K. The absolute temperature at the plant site is $38 + 273 = 311$°K. The general gas law is

$$\frac{P_1 V_1}{T_1} = \frac{P_2 V_2}{T_2}$$

where all values have been previously defined. Thus,

$$V_2 = \frac{P_1}{P_2} \cdot \frac{T_2}{T_1} \cdot V_1$$

or $(1 \, \text{atm}/0.9118 \, \text{atm})(311°\text{K}/293°\text{K})(72.7 \, \text{m}^3/\text{min}) = \boxed{84.6 \, \text{m}^3/\text{min.}}$ The power required is given by

$$\text{kW (brake or shaft)} = \frac{FRT_1}{nE}\left[\left(\frac{p_2}{p_1}\right)^{0.283} - 1\right]$$

where all values have been previously defined. The temperature at the site is 38°C or $(38)9/5 + 32 = 100$°F. For 100°F and El. 2500, Figure 16.11 gives the air density $\rho_{\text{air}} = 0.064 \, \text{lb/ft}^3$. Thus $\rho_{\text{air}} = (0.064 \, \text{lb/ft}^3)(1 \, \text{kg/m}^3)(\text{ft}^3/0.06236 \, \text{lb}) = 1.026 \, \text{kg/m}^3$. Thus,

$$F = (84.6 \, \text{m}^3/\text{min})(1.026 \, \text{kg/m}^3)(\text{min}/60 \, \text{s})$$
$$= 1.447 \, \text{kg/s}$$

The pressure at the inlet, p_1, is $(101.37)(693/760) = 92.4 \, \text{kPa}$. The pressure at the outlet, p_2 is 133 kPa. The temperature, T_1, is 311°K. Inserting the values gives

$$\text{kW} = \left(\frac{1.447}{0.283}\right)\left(\frac{0.288}{0.75}\right)(311)\left[\left(\frac{133}{92.4}\right)^{0.283} - 1\right]$$
$$= \boxed{66.3 \, \text{kW}}$$

EXAMPLE 16.2
AND 16.2 SI

Diffused Aeration

An activated sludge plant (tapered aeration) is located at El. 2000 (610 m). The oxygen demand for the plant is 1980 lb/day (890 kg/d). The maximum air temperature is 100°F (38°C). Coarse bubble diffusers are to be used that have a field transfer efficiency of 4%. Determine, in cubic feet per minute (cubic meters per minute), the air flow required at the plant site.

SOLUTION FOR USCS UNITS

Using Figure 16.12 and interpolating gives the oxygen content as 0.0152 lb O_2/ft^3. Thus,

$$ft^3/min = \left(\frac{1980 \text{ lb } O_2}{1440 \text{ min}}\right)\left(\frac{ft^3}{0.0152 \text{ lb } O_2}\right)\left(\frac{1}{0.04}\right)$$

$$= \boxed{2260 \text{ cfm}}$$

SOLUTION FOR SI UNITS

The elevation in feet = (610 m)(3.281 m/ft) = 2000 ft. Using Figure 16.12 gives the oxygen content as 0.0152 lb O_2/ft^3. Thus kg/m^3 = (0.0152 lb/ft^3) × (ft^3/0.06236 lb)(kg/m^3) = 0.2437 kg/m^3. Accordingly,

$$m^3/min = \left(\frac{890 \text{ kg}}{1440 \text{ min}}\right)\left(\frac{m^3}{0.2437 \text{ kg}}\right)\left(\frac{1}{0.04}\right)$$

$$= \boxed{63.4 \text{ m}^3/\text{min}}$$

[*Note*: The calculations in this problem are simplified compared to those required in using Eq. (16.3); however, these calculations require an estimation of the field transfer efficiency, which is difficult to estimate. Some authorities, such as GLUMRB (1990), recommend, in absence of α, β, and C values for Eq. (16.3), a field transfer efficiency of 50% of the clean water transfer efficiency. Thus, if a clean water transfer efficiency is 8%, the field transfer efficiency is (0.50)(8) = 4%. The use of Eq. (16.3) is the recommended method for diffused aeration design. However, to simplify aeration design in the subsequent chapter on aerobic digesters, the method applied in this example is used for illustrative purposes. If an actual aerobic digester were to be designed, Eq. (16.3) would be used.]

EXAMPLES 16.3
AND 16.3 SI

Mechanical Aeration

A completely mixed activated sludge plant is located at El. 2000 ft (610 m). The oxygen demand is 2680 lb/day (1220 kg/d) during the summer when the wastewater temperature is 82°F (28°C). The α value is 0.75 and β is 0.95. Four aerators are to be used, and the manufacturer has a test certifying the transfer (N_0) as 2.2 lb O_2/hp-hr (nameplate hp) (1.34 kg/kW-h). Determine the theoretical aerator power per aerator if the operating DO is 2.0 mg/ℓ.

SOLUTION FOR USCS UNITS

$°C = (82 - 32)(5/9) = 27.8°C$. Interpolating from the table of saturation DO in the Appendix gives $C_s = 7.95\,mg/\ell$. Table 16.1 gives the barometric pressure as 706 mm Hg. Thus $C_s = (7.95)(706/760) = 7.39\,mg/\ell$. The value $C_w = (0.95)(7.39) = 7.02\,mg/\ell$. The oxygen transfer is given by

$$N = N_0\left(\frac{C_w - C_L}{9.17}\right)1.02^{(T-20)} \cdot \alpha$$

where $N = $ lb O_2 transferred per nameplate horsepower, $N_0 = $ lb O_2 transferred per nameplate hp at standard conditions (furnished by the manufacturer), $C_w = $ saturation DO of the wastewater, mg/ℓ, $C_L = $ operating DO, mg/ℓ, $T = $ wastewater temperature, °C, and $\alpha = K_La(\text{wastewater})/K_La(\text{tap water})$. Substituting gives

$$N = 2.2\left(\frac{7.02 - 2.0}{9.17}\right)1.02^{(27.8-20)} \cdot 0.75$$

$$= 1.054\,\text{lb } O_2/\text{hp-hr}$$

The theoretical nameplate horsepower is

$$\text{hp} = \left(\frac{2680\,\text{lb}}{24\,\text{hr}}\right)\left(\frac{\text{hp-hr}}{1.054\,\text{lb}}\right)\left(\frac{1}{4}\right)$$

$$= \boxed{26.5\,\text{hp (nameplate)}}$$

SOLUTION FOR SI UNITS

The DO saturation value from the table in the Appendix $= 7.92\,mg/\ell$. The elevation is $(610\,m)(3.281\,ft/m) = 2000\,ft$. Table 16.1 gives the barometric pressure as 706 mm Hg. Thus $C_S = (7.92)(706/760) = 7.36\,mg/\ell$. The value $C_w = (0.95)(7.36) = 6.99\,mg/\ell$. The oxygen transfer is given by

$$N = N_0\left(\frac{C_w - C_L}{9.17}\right)1.02^{(T-20)} \cdot \alpha$$

where all units have previously been defined. Substituting the values gives

$$N = 1.34\left(\frac{6.99 - 2.0}{9.17}\right)1.02^{28-20} \cdot 0.75$$

$$= 0.6408\,\text{kg/kW-h}$$

The theoretical power is

$$\text{kW} = \left(\frac{1220\,\text{kg}}{24\,\text{h}}\right)\left(\frac{\text{kW-h}}{0.6408\,\text{kg}}\right)\left(\frac{1}{4}\right)$$

$$= \boxed{19.8\,\text{kW}}$$

Note: It is customary to rate mechanical aerators using the rated horsepower of the electric motor as shown by the nameplate on the motor. In the absence of a test for N_0, many regulatory authorities allow the maximum to be assumed as 2.0 lb O_2/hp-hr (nameplate) (1.22 kg O_2/kW-h).

REFERENCES

American Society of Civil Engineers (ASCE). 1982. *Wastewater Treatment Plant Design.* ASCE Manual of Practice No. 36. New York: ASCE.

Dobbins, W. E. 1956. The Nature of the Oxygen Transfer Coefficient in Aeration Systems. In *Biological Treatment of Sewage and Industrial Wastes.* Vol. 1, edited by J. McCabe and W. W. Eckenfelder, Jr. New York: Reinhold.

Dreier, D. E. 1956. Theory and Development of Aeration Equipment. In *Biological Treatment of Sewage and Industrial Wastes.* Vol. 1, edited by J. McCabe and W. W. Eckenfelder, Jr. New York: Reinhold.

Eckenfelder, W. W., Jr. 1966. *Industrial Water Pollution Control.* New York: McGraw-Hill.

Eckenfelder, W. W., Jr. 1989. *Industrial Water Pollution Control.* 2nd ed. New York: McGraw-Hill.

Eckenfelder, W. W., Jr. 1980. *Principles of Water Quality Management.* Boston: CBI Publishing.

Eckenfelder, W. W., Jr., and Ford, D. L. 1966. Engineering Aspects of Surface Aeration Design. In *Proceedings of the 22nd Annual Purdue Industrial Waste Conference.*

Eckenfelder, W. W., Jr., and Ford, D. L. 1968. New Concepts in Oxygen Transfer and Aeration. In *Advances in Water Quality Improvement,* edited by E. F. Gloyna and W. W. Eckenfelder, Jr. Austin, Tex.: University of Texas Press.

Eckenfelder, W. W., Jr., and Ford, D. L. 1970. *Water Pollution Control.* Austin, Tex.: Pemberton Press.

Gaden, E. L., Jr. 1956. Aeration and Oxygen Transport in Biological Systems — Basic Considerations. In *Biological Treatment of Sewage and Industrial Wastes.* Vol. 1, edited by J. McCabe and W. W. Eckenfelder, Jr. New York: Reinhold.

Great Lakes–Upper Mississippi River Board of State Public Health and Environmental Managers. 1990. Recommended Standards for Wastewater Facilities. Ten state standards. Albany, N.Y.

Great Lakes–Upper Mississippi River Board of State Sanitary Engineers. 1978. Recommended Standards for Sewage Works. Ten state standards. Albany, N.Y.

Horan, N. J. 1990. *Biological Wastewater Treatment Systems: Theory and Operation.* Chichester, West Sussex, England: Wiley.

Lewis, W. K., and Whitman, W. G. 1924. Principles of Gas Absorption. *Ind. and Eng. Chem.* 16, no. 12:1215.

Metcalf & Eddy, Inc. 1972. *Wastewater Engineering: Collection, Treatment and Disposal.* New York: McGraw-Hill.

Metcalf & Eddy, Inc. 1991. *Wastewater Engineering: Treatment, Disposal and Reuse.* 3rd ed. New York: McGraw-Hill.

Metcalf & Eddy. Inc. 1979. *Wastewater Engineering: Treatment, Disposal and Reuse.* 2nd ed. New York: McGraw-Hill.

Novotny, V.; Imhoff, K.; Olthof, M.; and Krenkel, P. 1989. *Karl Imhoff's Handbook of Urban Drainage and Wastewater Disposal.* New York: Wiley.

Oldshue, J. Y. 1956. Aeration of Biological Systems Using Impellers. In *Biological Treatment of Sewage and Industrial Wastes.* Vol. 1, edited by J. McCabe and W. W. Eckenfelder, Jr. New York: Reinhold.

Qasim, S. R. 1985. *Wastewater Treatment Plant Design.* New York: Holt, Rinehart and Winston.

Perry, R. H., and Chilton, C. H. 1973. *Chemical Engineer's Handbook.* 5th ed. New York: McGraw-Hill.

Ramalho, R. S. 1977. *Introduction to Wastewater Treatment Processes.* New York: Academic Press.

Schroeder, E. D. 1977. *Water and Wastewater Treatment.* New York: McGraw-Hill.

Sundstrom, D. W., and Klei, H. E. 1979. *Wastewater Treatment.* Englewood Cliffs, N.J.: Prentice-Hall.

Von der Emde, W. 1968. Aeration and Developments in Europe. In *Advances in Water Quality Improvement,* edited by E. F. Gloyna and W. W. Eckenfelder, Jr. Austin, Tex.: University of Texas Press.

Water Environment Federation (WEF). 1992. *Design of Municipal Wastewater Treatment Plants.* Vol. I and II. WEF Manual of Practice No. 8.

PROBLEMS

16.1 An activated sludge plant uses compressed air and is just above sea level. Each dual tube coarse bubble diffuser, as shown in Figure 16.2 (c), has a flow of 15 scfm. Pertinent data are: $\alpha = 0.75$, $\beta = 0.95$, operating DO = 2.0 mg/ℓ, reactor basin depth = 14 ft, the diffusers are 1.5 ft above the basin bottom, and the operating mixed liquor temperature = 20°C. The performance equation for the diffusers is

$$N = 0.00175 G_a^{0.9} D^{0.67}(C_m - C_L) \cdot 1.02^{T-20} \cdot \alpha$$

Determine, using USCS units:

a. The mass of oxygen transfered per diffuser (assume 5% transfer in making calculations).

b. The field transfer efficiency as a percent.

16.2 An activated sludge plant uses compressed air and is just above sea level. Each dual tube coarse bubble diffuser, as shown in Figure 16.2 (c), has a flow of 0.425 m³/min. Pertinent data are: $\alpha = 0.75$, $\beta = 0.95$, operating DO = 2.0 mg/ℓ, reactor basin depth = 4.27 m, diffusers are 0.457 m above the basin bottom, and the operating mixed liquor temperature = 20°C. The performance equation for the diffusers is

$$N = 0.0435 G_a^{0.9} D^{0.67}(C_m - C_L) \cdot 1.02^{T-20} \cdot \alpha$$

Determine, using SI units:

a. The mass of oxygen transfered per diffuser (assume 5% transfer in making calculations).

b. The field transfer efficiency as a percent.

16.3 A manufacturer's bulletin shows a graph of the percent oxygen transfer in tap water at a 14-ft submergence depth. For 7 scfm, the graph gives a transfer efficiency of 18%. Pertinent data are: $\alpha = 1.0$, $\beta = 1.0$, operating DO = 0.0 mg/ℓ, operating temperature = 20°C, and operating pressure = 14.7 psia (sea level). Determine, using USCS units, the constant C in the following performance equation:

$$N = CG_a^{0.9} D^{0.67}(C_m - C_L) \cdot 1.02^{T-20} \cdot \alpha$$

16.4 A manufacturer's bulletin shows a graph of the percent oxygen transfer at a 4.27-m submergence depth. For an air flow of 0.198 m³/min, the graph gives a transfer of 18%. Pertinent data are: $\alpha = 1.0$, $\beta = 1.0$, operating DO = 0.0 mg/ℓ, operating temperature = 20°C, and atmospheric pressure = 101.37 kPa (sea level). Determine, using SI units, the constant C in the following performance equation:

$$N = CG_a^{0.9} D^{0.67}(C_m - C_L) \cdot 1.02^{T-20} \cdot \alpha$$

16.5 An activated sludge plant is located at El. 2500 (760 m). The oxygen demand for the plant is 2730 lb/day (1240 kg/d). The maximum air temperature is 100°F (28°C). Fine bubble diffusers are to be used that have a field transfer efficiency of 10%. Using USCS units, determine the air flow required in cubic feet per minute at the plant site.

16.6 Repeat Problem 16.5 using SI units to determine the air flow in cubic meters per minute at the plant site.

16.7 An activated sludge plant uses mechanical aerators and is located just above sea level. The manufacturer's test rating for the aerators gives $N_0 = 2.20$ lb O_2/hp-hr (nameplate) (1.34 kg/kW-h). Pertinent data are: $\alpha = 0.80$, $\beta = 0.93$, and the operating mixed-liquor temperature = 22°C. Determine, using USCS units, the oxygen transfered per unit energy expended.

16.8 Repeat Problem 16.7 using SI units.

16.9 A tapered aeration activated sludge plant treats a municipal wastewater, and the plant is located just above sea level. The plant uses compressed air with fine bubble diffusers. The diffusers are mounted in pairs, as shown in Figure 16.13, with each one of a pair being mounted opposite the other on the air line. Pertinent data are: reactor basin length = 400 ft (122 m), reactor basin depth = 15 ft (4.57 m), diffusers are 1.5 ft (0.46 m) above the basin bottom, oxygen demand = 7920 lb/day (3600 kg/day), operating mixed liquor temperature = 16°C, $\alpha = 0.75$, $\beta = 0.95$, air flow per diffuser = 7 scfm (0.199 m³/min), operating DO = 2.0 mg/ℓ, maximum air temperature = 100°F (38°C), compressor efficiency = 0.75%, and the performance equation for the diffusers is

$$N = CG_a^{0.9} D^{0.67}(C_m - C_L) \cdot 1.02^{T-20} \cdot \alpha$$

where $C = 0.00377$ for USCS units and 0.0937 for SI units. Determine, using USCS units:

a. The oxygen transfered by each diffuser (assume 9% transfer for making calculations).

b. The oxygen demand for each quarter of the reactor basin using Table 15.5.

c. The number of pairs of diffusers required for each quarter and the spacing of the pairs for each quarter.

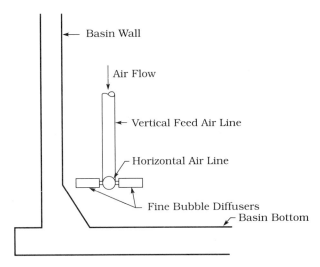

FIGURE 16.13 *Section through the Reactor Basin Showing the Fine Bubble Air Diffusers for Problems 16.9 and 16.10*

d. The total standard air flow required.

e. The actual air flow required.

f. The shaft or brake power for the compressor.

16.10 Repeat Problem 16.9 using SI units.

16.11 Repeat Problem 16.9 using USCS units for the plant if it is located at El. 2800.

16.12 Repeat Problem 16.9 using SI units for the plant if it is located at El. 853 m.

16.13 A completely mixed activated sludge plant treating an industrial wastewater has four mechanical aerators, and the plant is just above sea level. The maximum oxygen demand is 243 lb/hr (110 kg/h) and the wastewater temperature is 24°C. Pertinent data are: $\alpha = 0.72$, $\beta = 0.91$, operating DO = 2.0 mg/ℓ, and the manufacturer's test rating for the aerators (N_0) is 2.20 lb/hp-hr (nameplate) (1.34 kg/ kW-h). Determine, using USCS units:

a. The oxygen transfered in the field.

b. The power required for each aerator.

16.14 Repeat Problem 16.13 using SI units.

16.15 Repeat Problem 16.13 using USCS units for the plant if it is located at El. 2200.

16.16 Repeat Problem 16.13 using SI units for the plant if it is located at El. 670 m.

16.17 A tapered aeration activated sludge plant located at El. 2000 treats a municipal wastewater. The reactor basin is 15 ft deep and 396 ft long; the oxygen demands for the first, second, third, and fourth quarters are 65.8, 48.9, 37.6, and 35.7 lb/hr, respectively; and the diffusers are 12 in. above the basin bottom. Other pertinent data are: operating mixed liquor temperature = 23°C, $\alpha = 0.75$, $\beta = 0.95$, operating DO = 2.0 mg/ℓ, and air flow to each diffuser = 15 scfm. The performance equation for the coarse bubble dual tube diffusers, as shown in Figure 16.2 (c), is

$$N = 0.00170 G_a^{0.9} D^{0.67}(C_m - C_L) \cdot 1.02^{T-20} \cdot \alpha$$

Determine, using USCS units:

a. The rate of oxygen transfered per diffuser, mass/ hr (assume 5% transfer in making calculations).

b. The number of diffusers per quarter.

c. The spacing of the diffusers per quarter.

16.18 A tapered aeration activated sludge plant located at El. 610 m treats a municipal wastewater. The reactor basin is 4.57 m deep and 120 m long; the oxygen demands for the first, second, third, and fourth quarters are 29.9, 22.2, 17.1, and 16.2 kg/h, respectively; and the diffusers are 0.30 m above the basin bottom. Other pertinent data are: operating mixed liquor temperature = 23°C, $\alpha = 0.75$, $\beta = 0.95$, operating DO = 2.0 mg/ℓ, and the air flow to each diffuser = 0.425 m^3/min. The performance equation for the coarse bubble dual tube diffusers, as shown in Figure 16.2 (c), is

$$N = 0.0423 G_a^{0.9} D^{0.67}(C_m - C_L) \cdot 1.02^{T-20} \cdot \alpha$$

Determine, using SI units:

a. The rate of oxygen transfered per diffuser, mass/h (assume 5% transfer in making calculations).

b. The number of diffusers per quarter.

c. The spacing of diffusers per quarter.

16.19 A completely mixed activated sludge plant treating an industrial wastewater has four mechanical aerators, and the plant is at El. 1500 (457 m). The oxygen demand during the summer is 256 lb/hr (116 kg/h) and the wastewater temperature is 28°C. The oxygen demand during the winter is 184 lb/hr (83.5 kg/h) and the wastewater temperature is 18°C. The mechanical aerators to be used are rated at 2.10 lb O_2/hp-hr (1.28 kg/kW-h) and the operating DO = 2.0 mg/ℓ. Other data are α = 0.72 and β = 0.93. Determine, using USCS units:

a. The power required per aerator for the summer conditions.

b. The power required per aerator for the winter conditions.

c. Which condition controls the design?

16.20 Repeat Problem 16.19 using SI units.

16.21 A treatability study has been made on an industrial wastewater, and effluent has been obtained from the pilot plant reactor. Aeration studies were performed using an aeration vessel with a diffuser stone that used compressed air. The K_La values found for the diffuser stone and vessel were

0.388 hr^{-1} for wastewater, 0.526 hr^{-1} for effluent, and 0.554 hr^{-1} for tap water. All K_La values were at the same temperature. The saturation DO values in wastewater, effluent, and water were 8.2, 8.3, and 8.7 mg/ℓ, respectively. Determine:

a. The α value for the mixed liquor.

b. The β value for the mixed liquor.

16.22 An activated sludge plant is located at El. 1500 (457 m). The design air flow is 1620 scfm (45.9 m^3/min) and the operating pressure is 7.0 psi (48.3 kPa). Other pertinent data: maximum air temperature = 100°F (38°C), efficiency of the compressor = 75%, and the efficiency of the electric motor = 95%. Determine, using USCS units:

a. The free air flow.

b. The shaft or brake power required for the compressor.

c. The electric motor power.

16.23 Repeat Problem 16.22 using SI units.

16.24 A pump-type floating surface aerator with a 65-hp (nameplate) (48.5 kW) electric motor is to be rated at standard conditions, which are water temperature = 20°C, atmospheric pressure = 760 mm Hg, water quality equal to tap water, and the dissolved oxygen concentration = 0.0 mg/ℓ. A test has been performed in which the pump-type aerator was placed in a test tank that was 95 ft (28.95 m) in diameter and had an 8.5-ft (2.59 m) depth of water.

TABLE 16.2

TIME (min)	5 ft (1.5 m) FROM SPRAY		1 ft (0.3 m) FROM WALL	
	Point 1 Top	Point 2 Bottom	Point 3 Top	Point 4 Bottom
0	0.30	0.30	0.20	0.20
3	1.60	1.60	1.50	1.70
6	2.40	2.50	2.70	2.70
8	3.40	3.40	3.40	3.40
11	4.30	4.30	4.30	4.30
14	5.20	5.20	5.10	4.90
17	5.65	5.60	5.70	5.50
20	6.10	6.20	6.00	5.90
23	6.60	6.60	6.50	6.35
26	6.80	6.80	6.70	6.60
29	7.10	7.10	7.05	6.90

The dissolved oxygen in the water was removed by adding Na_2SO_3 and $CoCl_2$ as a catalyst. Once the dissolved oxygen had reached approximately zero and there was a negligible amount of SO_3^{-2} left, the aerator was started and the dissolved oxygen was measured at four points versus time. The DO concentrations measured are given in Table 16.2 on page 521.

Points 1 and 3 were 1.0 ft (0.3 m) below the surface, whereas points 2 and 4 were 1.0 ft (0.3 m) above the bottom. Other pertinent data are: water temperature = 28°, saturation dissolved oxygen in the water = 7.80 mg/ℓ, K_La (20°C) = $K_La(T) \div 1.02^{(T-20)}$, saturation dissolved oxygen concentration in clear tap water at 20°C = 9.17 mg/ℓ, and the line hp drawn by the electric motor = 62.0 (46.25 kW). Determine, using USCS units:

a. The pounds of oxygen transferred per hp-hr (nameplate) at 20°C and at a DO equal to zero.

b. The pounds of oxygen transferred per hp-hr (line hp) at 20°C and at a DO equal to zero.

16.25 Repeat Problem 16.24 using SI units. Express transfer as kg/kW-h (nameplate) and kg/kW-h (line).

16.26 A completely mixed activated sludge plant treats the wastewater from a refinery and petrochemical plant. The reactor basin is 120 ft × 120 ft (36.57 m × 36.57 m) and has a depth of 13 ft (3.96 m). Four 150-hp (111.9-kW) slow-speed turbines furnish the oxygen for the process. The α value is 0.9 and the β value is 0.85. A field test was made that showed the oxygen uptake rate to be 61.2 mg/ℓ-hr, and the average operating DO was 2.2 mg/ℓ. The mixed liquor temperature was 18°C, and the saturation dissolved oxygen value was 9.3 mg/ℓ. Deter-

mine, using USCS units, the oxygen transfer of the aerators in lb O_2/hp-hr.

16.27 Repeat Problem 16.26 using SI units. Express transfer as kg/kW-h.

16.28 An activated sludge plant has diffusers that have a 14-ft (4.27 m) depth of submergence. The air head loss through the diffusers is 21 in. (533 mm) of water column (the usual range is from 14 to 28 in.) (356 to 711 mm), and the air piping has a head loss of 6 in. (152 mm) of water column (the usual range is from 4 to 8 in. [102 to 203 mm] for properly designed piping). Determine the operating pressure of the compressors in psia using USCS units.

16.29 Repeat Problem 16.28 using SI units.

16.30 An activated sludge plant is located at El. 1800 (549 m). The design air flow using Eq. (16.3) is 1130 scfm (32.0 m³/min), and the operating pressure is 7.0 psi (48.3 kPa). Pertinent data are: maximum summer temperature = 100°F (38°C), efficiency of the compressors = 75%, and electric motor efficiency = 95%. Determine, using USCS units:

a. The free air flow in cfm.

b. The required motor horsepower.

16.31 Repeat Problem 16.30 using SI units. Express free air flow as m³/min and required motor power as kW.

16.32 A mechanical aerator has a test rating of 2.10 lb O_2/hp-hr (nameplate) (1.28 kg/kW-h) and has been found to have a field transfer of 0.976 lb O_2/hp-hr (0.594 kg/kW-h) when placed in a reactor basin. Pertinent data are: $\alpha = 0.72$, $\beta = 0.94$, operating temperature = 28°C, and plant El. 1500 (457 m). Determine, using USCS units, the operating DO.

16.33 Repeat Problem 16.32 using SI units.

17

TRICKLING FILTERS AND ROTARY BIOLOGICAL CONTACTORS

These processes are similar because both employ cultures of microorganisms that are attached to the surface of various media. For trickling filters the medium is stationary, whereas for rotary biocontactors the medium, in the form of disks, rotates.

TRICKLING FILTERS

As a wastewater passes through a filter bed, the microbial growths sorb an appreciable amount of the organic materials for use as food substances, mainly by aerobic bio-oxidation. Figure 17.1 is a photograph of a rotary distributor trickling filter at a municipal wastewater treatment plant. Figure 17.2 is a cutaway illustration of a rotary distributor trickling filter, showing the influent pipe, rotary distributor, filter bed, underdrain system, and effluent pipe. The final clarifier, which is immediately downstream of the filter, serves to remove microbial growths that periodically slough from the filter medium. Usually large gravel or crushed stone is used as the medium; however, synthetic plastic media or other packing are also used. In the past, the trickling filter process has been commonly used for small- to medium-sized cities, primarily because of its simplicity and dependability. For very small towns, a trickling filter plant may require an operator for only a few hours each day. Because of higher effluent standards, activated sludge processes have largely displaced trickling filters in plants presently being constructed, even for very small communities. A flowsheet for a trickling filter plant treating a municipal wastewater is shown in Figure 17.3. The wastewater receives primary treatment prior to the secondary treatment, which consists of the trickling filter and its final clarifier.

Filters have been classified as low-rate or standard-rate, intermediate-rate, high-rate, and super-rate filters according to the organic loading, the unit liquid loading, and the recycle employed. Tables 17.1 and 17.2 show the characteristics for the various classes. Most filters built in the United States are circular in plan view and use rotary distributors that rotate as a result of the jet action as the wastewater is sprayed horizontally onto the bed. Small filters, such as those 10 ft (6.1 m) or less in diameter, are driven by electric motors. **Low-rate** or **standard-rate filters**, which were developed first, have organic loadings of 200 to 800 lb BOD_5/ac-ft-day (5 to 20 lb BOD_5/1000 ft^3-day,

FIGURE 17.1 *Rotary Distributor Trickling Filter. The jet action from the nozzles drives the rotary distributor.*

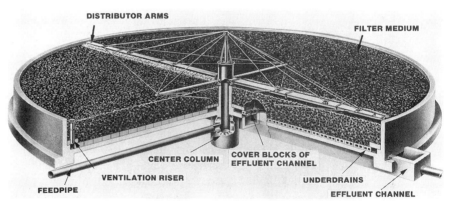

FIGURE 17.2 *Trickling Filter with a Rotary Distributor*
Courtesy of Dorr-Oliver, Inc.

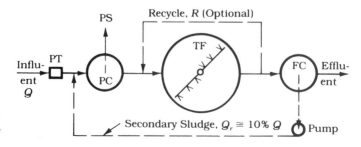

FIGURE 17.3
Flowsheet for a Trickling Filter

TABLE 17.1 *Typical Design Information for Trickling Filters*

ITEM	LOW-RATE FILTER	INTERMEDIATE-RATE FILTER	HIGH-RATE FILTER	SUPER-RATE (ROUGHING) FILTER
Hydraulic loading,				
gal/min-ft²	0.017–0.07	0.07–0.2	0.2–0.7	0.7–3.4
MG/acre-day	1–4	4–10	10–40	40–200
Organic loading,				
lb/1000 ft³-day	5–20	15–30	20–60	50–380
lb/acre-ft-day	200–800	650–1300	800–2600	2000–16,000
Depth, ft	5–10	4–8	3–6.6	14–40
Recirculation ratio	0	0–1	1–3	1–4
Filter media	Rock, slag, etc.	Rock, slag, etc.	Rock, slag, synthetic materials	Synthetic materials, redwood
Power requirements,				
hp/1000 ft³	0.07–0.15	0.07–0.30	0.22–0.38	0.38–0.76
Filter flies	Many	Intermediate	Few; larvae are washed away	Few or none
Sloughing	Intermittent	Intermittent	Continuous	Continuous
Dosing intervals	Not more than 5 min (generally intermittent)	15 to 60 sec (continuous)	Not more than 15 sec (continuous)	Continuous
Effluent	Usually fully nitrified	Partially nitrified	Nitrified at low loadings	Nitrified at low loadings

Adapted from *Wastewater Engineering: Treatment, Disposal and Reuse* by Metcalf & Eddy, Inc., 2nd. ed. Copyright © 1979 by McGraw-Hill, Inc. Reprinted by permission of McGraw-Hill, Inc.

TABLE 17.2 *Typical Design Information for Trickling Filters*

ITEM	LOW-RATE FILTER	INTERMEDIATE-RATE FILTER	HIGH-RATE FILTER	SUPER-RATE (ROUGHING) FILTER
Hydraulic loading, $m^3/m^2 \cdot d$	1–4	4–10	10–40	40–200
Organic loading, $kg/m^3 \cdot d$	0.08–0.32	0.24–0.48	0.32–1.0	0.80–6.0
Depth, m	1.5–3.0	1.25–2.5	1.0–2.0	4.5–12
Recirculation ratio	0	0–1	1–3	1–4
Filter media	Rock, slag, etc.	Rock, slag, etc.	Rock, slag, synthetic materials	Synthetic materials, redwood
Power requirements, $kW/10^3 m^3$	2–4	2–8	6–10	10–20
Filter flies	Many	Intermediate	Few; larvae are washed away	Few or none
Sloughing	Intermittent	Intermittent	Continuous	Continuous
Dosing intervals	Not more than 5 min (generally intermittent)	15 to 60 s (continuous)	Not more than 15 s (continuous)	Continuous
Effluent	Usually fully nitrified	Partially nitrified	Nitrified at low loadings	Nitrified at low loadings

Adapted from *Wastewater Engineering: Treatment, Disposal and Reuse* by Metcalf & Eddy, Inc., 2nd. ed. Copyright © 1979 by McGraw-Hill, Inc. Reprinted by permission of McGraw-Hill, Inc.

0.08 to 0.32 kg/m^3-day), unit liquid loadings of 1 to 4 MG/ac-day (0.017 to 0.07 gal/min-ft^2, 1 to 4 m^3/m^2-day), and bed depths of about 5 to 10 ft (1.5 to 3.0 m), and they employ recycle only during times, such as the night period, when the wastewater flow is inadequate to turn the rotary distributor. Low-rate trickling filter plants usually consist of one stage; thus they are single-stage. Low-rate trickling filter plants can attain high BOD$_5$ removals, such as 90% to 95%, when treating municipal wastewaters, and they can achieve better nitrification than high-rate filters; however, the volume of media required is much greater. When treating municipal wastewaters, a low-rate trickling filter plant will produce an effluent having about 12 to 25 mg/ℓ BOD$_5$ and can produce a well-nitrified effluent.

Intermediate-rate filters, which are not very common, have organic loadings of 650 to 1300 lb BOD$_5$/ac-ft-day (15 to 30 lb/1000 ft^3-day, 0.24 to 0.48 kg/m^3-day), have unit liquid loadings of 4 to 10 MG/ac-day (0.07 to 0.2 gal/min-ft^2, 4 to 10 m^3/m^2-day), and may or may not employ continuous recycle. Recycle must be employed during low-flow times in order to turn the rotary distributor. Intermediate-rate trickling filter plants may be single-stage or two-stage — that is, two filters in series. A two-stage intermediate-rate plant usually attains 85% to 90% BOD$_5$ removal. When an intermediate-rate trickling filter plant treats municipal wastewaters, the effluent will have 20 to 30 mg/ℓ BOD$_5$.

High-rate trickling filters have organic loadings of 800 to 2600 lb BOD$_5$/ac-ft-day (20 to 60 lb/1000 ft^3-day, 0.32 to 1.0 kg/m^3-day) and unit liquid loadings of 10 to 20 MG/ac-day (0.2 to 0.7 gal/min-ft^2, 10 to 40 m^3/m^2-day). Bed depths are about 3 to 6.6 ft (1 to 2 m), and they always have continuous recycle. High-rate plants can be single-stage or two-stage. A single-stage high-rate trickling filter plant can attain BOD$_5$ removals of 75% to 80%. When such a plant is used to treat municipal wastewaters, an effluent of 40 to 50 mg/ℓ BOD$_5$ is typical. A two-stage high-rate trickling filter plant can attain BOD$_5$ removals of 85% to 90%, and when municipal wastewaters are treated, an effluent of 20 to 30 mg/ℓ BOD$_5$ is typical. A high-rate trickling filter plant, even a two-stage plant, cannot produce a highly nitrified effluent when treating municipal wastewaters.

Super-rate trickling filters have organic loadings of 2000 to 16,000 lb BOD$_5$/ac-ft-day (50 to 380 lb/1000-ft^3-day, 0.80 to 6.0 kg/m^3-day) and unit liquid loadings of 40 to 200 MG/ac-day (0.7 to 3.4 gal/min-ft^2, 40 to 200 m^3/m^2-day). Bed depths are from 14 to 40 ft (4.5 to 12 m), and they always have continuous recycle. The medium is always synthetic plastic, either as modules or random packed, or redwood modules. The most common is plastic modules, which have a specific surface area (surface area per unit volume) that is from 2 to 5 times as great as that for stone media. Since the microbial growth is roughly proportional to the surface area, the super-rate filter has from 2 to 5 times the microbial growth of a rock or stone filter. An advantage of the trickling filter is that the energy costs per mass of BOD$_5$ removed are much less than for an activated sludge process. The super-rate trickling filter is used for small municipal plants or as a roughing unit ahead of an existing trickling filter or ahead of an activated sludge process. The combination of a super-rate roughing filter followed by an activated sludge process is well suited for a high-strength (that is, a high BOD_5) municipal or industrial wastewater. Small super-rate plants treating municipal wastewaters can produce a nitrified effluent if the organic loading is low. Most single-stage trickling filter plants are upgraded by adding a super-rate filter ahead of the existing trickling filter or by adding an activated sludge process downstream from the existing trickling filter. Even large two-stage high-rate trickling filter plants have been upgraded by adding an activated sludge process downstream from the trickling filters. Small single-stage or two-stage plants have been upgraded by adding a shallow polishing pond as the last treatment unit. Deep super-rate trickling filters are usually built above the ground level and are frequently called **bio-towers**. Trickling filters are usually constructed using reinforced concrete tanks and are from 10 to 250 ft (3 to 76 m) in diameter. Standard diameters are in 5-ft (1.52-m) intervals; however, any diameter may be specially made. The above-ground super-rate filter may have a reinforced concrete base and reinforced concrete walls. Sometimes the walls are plastic or aluminum sheets constructed as modules.

BIOLOGICAL FILTRATION

As wastewater passes over microbial growths, Figure 17.4, an appreciable amount of the organic material is sorbed by the growths along with molecular oxygen. Aerobic bio-oxidation occurs, and the oxidized organic and inorganic end products are released into the moving water film. The wastewater passes through a filter in a matter of minutes; however, the sorbed organic materials are retained for several hours as they undergo bio-oxidation. The microbial action is essentially biosorption, similar to that described in Chapter 15.

Most of the microbial growth is aerobic; however, the growth immediate to the medium surface is usually anaerobic. As time progresses, the thickness

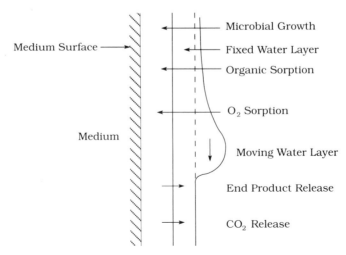

FIGURE 17.4
Biological Action of a Trickling Filter

of the microbial slime increases because of the new cell mass synthesized; finally the growth attains a thickness that causes it to slough off the medium. The amount of organic material removed per foot (meter) of bed depth is greatest at the top of the filter, and smallest at the bottom, as shown by Figure 17.5. The nitrifying microbial growths will be located in the lower depths of the filter bed. The BOD$_5$ in the effluent from a filter is the same as in the effluent from its final clarifier, as indicated in Figure 17.6 by the designation S_t. Recycling the effluent by spraying it back on the top of the filter allows the microbial growths to have successive opportunities for sorption and biological oxidation of the organic matter, thus increasing the filter efficiency.

The growths that have sloughed from the filter medium (that is, the humus) are subsequently settled in the final clarifier, and the secondary sludge is usually sent to the head of the plant, where it is resettled in the primary clarifier and thickened prior to sludge treatment.

The flow in a trickling filter passes downward in long, narrow, and irregular channels, and the flow regime approaches plug flow. Using tracers, Eckenfelder and Barnhart (1963) found some dispersion to occur.

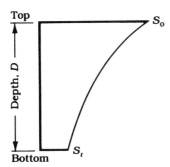

FIGURE 17.5 *Organic Concentration in the Wastewater versus Filter Depth*

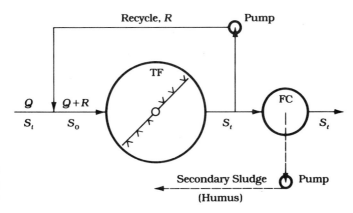

FIGURE 17.6 *Flowsheet of a Trickling Filter*

FILTER PERFORMANCE

The first performance equations for trickling filters were empirical (Fairall, 1956; National Research Council, 1946; Rankin, 1955); however, later developments have produced equations based on biochemical kinetics (Eckenfelder, 1961, 1970; Velz, 1948).

Kinetic Equations

One of the earliest performance equations based on kinetics is that by Velz (1948). He observed that the rate of organic removal per interval of depth is proportional to the remaining concentration of removable organic matter. His basic equation, which is a modification of a first-order rate equation, is $-dL/dD = kL$, where dL is an increment of remaining organic concentration, dD is an increment of depth, k is the rate constant, and L is the remaining removable organic concentration expressed as the ultimate first-stage BOD. Velz made an analogy between his rate equation and a first-order rate equation, since an increment of depth, dD, is proportional to an increment of time, dt. His integrated equation in terms of base 10 logarithms is

$$\frac{L_D}{L} = 10^{-kD} \tag{17.1}$$

where

L_D = removable ultimate first-stage BOD concentration at depth D

L = removable ultimate first-stage BOD concentration applied to the bed

k = rate constant

D = depth of bed, ft (m)

For low-rate filters treating municipal wastewaters and operated at 2 to 6 MG/ac-day (1.88 to 5.61 m³/m²-day), Velz found the k value to be 0.175 (0.574) and the removable fraction to be 90%. For high-rate filters operated at approximately 20 MG/ac-day (18.8 m³/m²-day), he found the k value to be 0.1505 (0.494) and the removable fraction to be 78.4%. Velz also developed the equation $k_2 = k_{20} \cdot 1.047^{T_2-20}$, where k_2 = rate constant at T_2°C and k_{20} = rate constant at 20°C, for correcting the rate constant for a change in temperature.

Eckenfelder (1970) has also developed a performance equation based on the specific rate of substrate removal for a pseudo–first-order reaction. It is

$$-\frac{1}{X}\frac{dS}{dt} = kS \tag{17.2}$$

where

$(1/X)(dS/dt)$ = specific rate of substrate utilization, mass/(mass microbes) × (time)

dS/dt = rate of substrate utilization, mass/(volume)(time)

k = rate constant, volume/(mass microbes)(time)

S = substrate concentration, mass/volume

Rearranging Eq. (17.2) for integration gives

$$\int_{S_0}^{S_t} \frac{dS}{S} = -k\bar{X} \int_0^t dt \qquad (17.3)$$

where

k = rate constant

$\bar{X}$ = average cell mass concentration, mass/volume

S_t = substrate concentration after the contact time, t, mass/volume

S_0 = substrate concentration applied to the filter, mass/volume

Integration of Eq. (17.3) gives

$$\frac{S_t}{S_0} = e^{-k\bar{X}t} \qquad (17.4)$$

The average cell mass concentration, $\bar{X}$, is proportional to the specific surface area of the packing, A_s — in other words, the ft^2 (m^2) of surface per bulk ft^3 (m^3) of volume. Thus,

$$\bar{X} \sim A_s \qquad (17.5)$$

where A_s is the specific surface area. The mean contact time, t, for a filter is represented by (Howland, 1957)

$$t = CD/Q_L^n \qquad (17.6)$$

where

t = mean contact time

D = depth of the bed

Q_L = unit liquid loading or surface loading

C,n = experimental constants

Substituting Eqs. (17.5) and (17.6) into Eq. (17.4) and combining the constants k and C give the performance equation by Eckenfelder (1970):

$$\frac{S_t}{S_0} = e^{-KA_s^m D/Q_L^n} \qquad (17.7)$$

where

S_t = substrate concentration in the filter effluent, mass/volume

S_0 = substrate concentration applied to the filter, mass/volume

K = rate constant

A_s = specific surface area of the packing, area/volume

D = filter depth

Q_L = unit liquid loading or surface loading

m, n = experimental constants

The value of n depends on the flow characteristics through the packing and is usually about 0.50 to 0.67. Tables 17.3 and 17.4 give specific surface areas and other characteristics for various filter media. For a particular wastewater and medium of interest, Eq. (17.7) may be simplified by combining KA_s^m to give

$$\frac{S_t}{S_0} = e^{-KD/Q_L^n} \qquad (17.8)$$

TABLE 17.3   *Specific Surface Areas for Various Trickling Filter Media*

MEDIUM	NOMINAL SIZE (in.)	SPECIFIC SURFACE AREA (ft²/ft³)
River rock		
Small	$1-2\frac{1}{2}$	16–21
Large	$4-4\frac{1}{2}$	12–15
Blast-furnace slag		
Small	2–3	16–21
Large	3–5	13–18
Plastic		
Conventional	$24 \times 24 \times 48^a$	24–30
High specific surface	$24 \times 24 \times 48^a$	30–60
Redwood	$48 \times 48 \times 20^a$	12–15

^{a} Module size.

Adapted from *Wastewater Engineering: Treatment, Disposal and Reuse* by Metcalf & Eddy, Inc. 2nd. ed. Copyright © 1979 by McGraw-Hill, Inc. Reprinted by permission of McGraw-Hill, Inc.

TABLE 17.4   *Specific Surface Areas for Various Trickling Filter Media*

MEDIUM	NOMINAL SIZE (mm)	SPECIFIC SURFACE AREA (m²/m³)
River rock		
Small	25–65	55–70
Large	100–120	40–50
Blast-furnace slag		
Small	50–80	55–70
Large	75–125	45–60
Plastic		
Conventional	$600 \times 600 \times 1200^a$	80–100
High specific surface	$600 \times 600 \times 1200^a$	100–200
Redwood	$1200 \times 1200 \times 500^a$	40–50

^{a} Module size.

Adapted from *Wastewater Engineering: Treatment, Disposal and Reuse* by Metcalf & Eddy, Inc. 2nd. ed. Copyright © 1979 by McGraw-Hill, Inc. Reprinted by permission of McGraw-Hill, Inc.

where K is the new rate constant. Germain (1966), using Eq. (17.8), reported K to be from 0.01 to 0.10 for various wastewaters, where D is in feet and Q_L is in gal/min-ft^2 for synthetic plastic packing. For municipal wastewaters treated by filters having a synthetic plastic packing (Surfpac, Dow Chemical) of 27 ft^2/ft^3 (89 m^2/m^3), he found K to be 0.088 and n to be 0.50 when the depth was in feet, the unit liquid loading Q_L was in gal/min-ft^2, and the BOD$_5$ was the total BOD$_5$ in mg/ℓ. His work was done using pilot plants. To get a design K value, it is best to use data from a pilot plant, particularly for industrial wastewaters or municipal wastewaters containing significant amounts of industrial wastewaters.

The following equation has been developed for correcting trickling filter rate constants for temperature changes (Eckenfelder, 1970):

$$K_T = K_{20} \cdot 1.035^{(T-20)} \tag{17.9}$$

where

K_T = rate constant at temperature T, °C

K_{20} = rate constant at 20°C

T = temperature, °C

One of the most common kinetic equations for filter performance when treating municipal wastewaters on stone media has been developed by Eckenfelder (1961). It is

$$\frac{S_t}{S_0} = \frac{1}{1 + C\left(\dfrac{D^{0.67}}{Q_L^{0.50}}\right)} \tag{17.10}$$

where

S_t = BOD$_5$ in the filter effluent, mg/ℓ

S_0 = BOD$_5$ in the wastewater discharged on the filter bed, mg/ℓ

C = 2.5 for USCS units and 5.358 for SI units

D = filter depth ft (m)

Q_L = unit liquid loading, MG/acre-day (m^3/day-m^2)

The values of S_t and S_0 are depicted in the flowsheet shown in Figure 17.6, and if no recycle is employed, $S_t = S_0$. Equation (17.10) may be derived from the second-order kinetic equation $(1/X)(dS/dt) = KS^2$ as follows. The integration of the second-order equation gives $S_t/S_0 = 1/(1 + S_0 K \bar{X} t)$. The mean contact time is represented by $t = CD^{0.67}/Q_L^{0.50}$. Substituting the expression for t in the integrated equation and combining the constants S_0, K, $\bar{X}$, and C give $S_t/S_0 = 1/(1 + \text{Constant } D^{0.67}/Q_L^{0.50})$. For municipal wastewaters, Eckenfelder has found the constant to be 2.5 for USCS units and 5.358 for SI units.

EXAMPLE 17.1   *Low-Rate Trickling Filter*

A city has a population of 5000 persons and is served by a low-rate trickling filter with a polishing pond downstream from the final clarifier. Pertinent data for the plant are wastewater flow = 100 gal/cap-day, BOD_5 after primary clarification = 135 mg/ℓ, BOD_5 after final clarification = 20 mg/ℓ, and filter bed depth = 4 to 6 ft. Determine a suitable diameter of the filter and its depth if distributors are available in diameters that are in increments of 5 ft.

SOLUTION   The flowsheet for a low-rate trickling filter is shown in Figure 17.6, and since the recycle, R, is used only at night, $S_i = S_0$. Assume $D = 5$ ft. Substituting $S_0 = 135$, $S_t = 20$, and $D = 5$ in Eq. (17.10) yields

$$\frac{20}{135} = \frac{1}{1 + 2.5\left(\dfrac{5^{0.67}}{Q_L^{0.50}}\right)}$$

This equation gives Q_L as 1.634 MG/ac-day. The wastewater flow is (5000 persons)(100 gal/cap-day)(1 MG/10^6 gal) or 0.50 MG/day. Thus the plan area is (0.50 MG/day)(ac-day/1.634 MG)(43,560 ft²/ac) or 13,329 ft². The diameter, D, is

$$D = \left[(13{,}329 \text{ ft}^2)\frac{4}{\pi}\right]^{1/2}$$

$$= 130.3 \text{ ft or } \boxed{135 \text{ ft for standard size}}$$

Determining the area for a 135-ft diameter gives 0.3286 acre. This results in a hydraulic load, Q_L, of 1.522 MG/ac-day. Substituting 1.522 MG/ac-day, $S_t = 20$, and $S_0 = 135$ into Eq. (17.10) gives

$$\text{Depth } (D) = \boxed{4.74 \text{ ft}}$$

EXAMPLE 17.1 SI   *Low-Rate Trickling Filter*

A city has a population of 5000 persons and is served by a low-rate trickling filter with a polishing pond downstream from the final clarifier. Pertinent data for the plant are wastewater flow = 380 ℓ/cap-d, BOD_5 after primary clarification = 135 mg/ℓ, BOD_5 after final clarification = 20 mg/ℓ, and filter bed depth = 1.5 m. Determine the diameter of the filter.

SOLUTION   The flowsheet for a low-rate trickling filter is shown in Figure 17.6, and since the recycle, R, is used only at night, $S_i = S_0$. Substituting $S_0 = 135$, $S_t = 20$, and $D = 1.5$ m in Eq. (17.10) yields

$$\frac{20}{135} = \frac{1}{1 + 5.358\left(\dfrac{1.5^{0.67}}{Q_L^{0.50}}\right)}$$

This equation gives Q_L as $1.495\,m^3/d\text{-}m^2$. The wastewater flow is $(5000$ persons$)(380\,\ell/\text{cap-d})(m^3/1000\,\ell)$ or $1900\,m^3/d$. The plan area is $(1900\,m^3/d)$ $(d\text{-}m^2/1.495\,m^3)$ or $1271\,m^2$. The diameter, D, is

$$D = \left[(1271\,m^2)\frac{4}{\pi}\right]^{1/2} = \boxed{40.2\,m}$$

EXAMPLE 17.2   *High-Rate Trickling Filter*

Assume the city in Example 17.1 is served by a high-rate trickling filter having a recycle ratio of 2 with a polishing pond downstream from the final clarifier. The BOD$_5$ after final clarification $= 30\,mg/\ell$. The depth is to be 5 to 7 ft. Determine a suitable filter diameter and depth if distributors are available in diameters that are in increments of 5 ft.

SOLUTION   The pertinent data for the plant are the same as in Example 17.1 except the effluent BOD$_5$ is $30\,mg/\ell$. A materials balance on the BOD$_5$ at the junction of the recycle flow and the primary effluent yields $(Q)135 + (2Q)30 = (1 + 2)(Q)S_0$ or $S_0 = 65\,mg/\ell$. Assume a depth of 6 ft. Substituting the values in Eq. (17.10) yields

$$\frac{30}{65} = \frac{1}{1 + 2.5\left(\dfrac{6^{0.67}}{Q_L^{0.50}}\right)}$$

This equation gives Q_L as $50.67\,MG/\text{ac-day}$. The wastewater flow, Q, is $0.50\,MGD$; thus the recycle flow, R, is $(2)0.50$ or $1.0\,MGD$. The total flow applied to the filter is $0.50 + 1.0$ or $1.5\,MGD$. The plan area is $(1.5\,MGD)(\text{ac-day}/50.67\,MG)(43,560\,ft^2/\text{ac})$ or $1290\,ft^2$. The diameter, D is

$$D = \left[(1290\,ft^2)\frac{4}{\pi}\right]^{1/2}$$

$$= 40.5\,ft \text{ or } \boxed{45\,ft \text{ for standard size}}$$

Determining the area for 45-ft diameter gives 0.0365 acre. This results in a hydraulic load, Q_L, of $41.1\,MG/\text{ac-day}$. Substituting $41.1\,MG/\text{ac-day}$, $S_t = 30$, and $S_0 = 65$ in Eq. (17.10) gives

$$\text{Depth } (D) = \boxed{5.13\,ft}$$

EXAMPLE 17.2 SI   *High-Rate Trickling Filter*

Assume the city in Example 17.1 SI is served by a high-rate trickling filter having a recycle ratio of 2 with a polishing pond downstream from the final clarifier. The BOD$_5$ after final clarification is $30\,mg/\ell$. The filter bed depth is $1.6\,m$. Determine the diameter of the filter.

SOLUTION   The pertinent data for the plant are the same as in Example 17.1 SI except that the effluent BOD_5 is $30 \, mg/\ell$. A materials balance on the BOD_5 at the junction with the recycle flow and the primary effluent yields $(Q)(135) + (2Q)(30) = (1 + 2)(Q)S_0$ or $S_0 = 65 \, mg/\ell$. Substituting the values in Eq. (17.10) yields

$$\frac{30}{65} = \frac{1}{1 + 5.358\left(\dfrac{1.6^{0.67}}{Q_L^{0.50}}\right)}$$

This equation gives Q_L as $39.59 \, m^3/d\text{-}m^2$. The wastewater flow, Q, is $1900 \, m^3/d$; thus the recycle flow, R, is $(2)1900$ or $3800 \, m^3/d$. The total flow applied to the filter is $1900 + 3800$ or $5700 \, m^3/d$. The plan area is $(5700 \, m^3/d)/(39.59 \, m^3/d\text{-}m^2)$ or $144.0 \, m^2$. The diameter, D, is

$$D = \left[(144.0 \, m^2)\frac{4}{\pi}\right]^{1/2} = \boxed{13.5 \, m}$$

EXAMPLE 17.3   *Super-Rate Trickling Filter*

A single-stage super-rate trickling filter is to be designed for a population of 10,000 persons. Pertinent data are average influent flow = $100 \, gal/cap\text{-}day$, influent BOD_5 = $200 \, mg/\ell$, primary clarifier removes 33% of the BOD_5, effluent BOD_5 = $20 \, mg/\ell$, filter depth = $20 \, ft$, and recycle ratio $R/Q = 2$. Pilot-plant studies have shown the removal constant, K, to be 0.091 at 20°C when the depth is in feet, the hydraulic loading is in gpm, and the constant $n = 0.5$. The winter wastewater temperature is 62°F (16.7°C). The synthetic plastic packing is available in modules $2 \, ft$ deep. If two filters in parallel are to be used, determine:

1. The theoretical diameter.
2. The design diameter if distributors are available in diameters that are in increments of $5 \, ft$.

SOLUTION   The performance equation is

$$\frac{S_t}{S_0} = e^{-KD/Q_L^{0.5}}$$

The temperature relationship for the removal constant K is

$$K_T = K_{20} \cdot 1.035^{(T-20)}$$

For 16.7°C, the constant K is

$$K_T = 0.091 \cdot 1.035^{(16.7-20)}$$
$$= 0.08123$$

The BOD$_5$ of the primary clarifier effluent, S_i, is $200(1 - 0.33) = 134 \, \text{mg}/\ell$. Since $R/Q = 2$, the recycle flowrate is $2Q$, where Q is the incoming flowrate. A mass balance at the junction with the recycle flow is

$$(134)(Q) + (20)(2Q) = (1 + 2)(Q)S_0$$

or

$$S_0 = 58 \, \text{mg}/\ell$$

Inserting values into the performance equation gives

$$\frac{20}{58} = e^{-(0.08123)(20)/Q_L^{0.5}}$$

which yields $Q_L = 2.33 \, \text{gpm/ft}^2$. The primary clarifier flow is $(10,000)(100) = 1,000,000 \, \text{gal/day}$ $(1.0 \, \text{MGD})$. The hydraulic load to the filters is $(1,000,000)(1 + 2) = 3,000,000 \, \text{gal/day}$. The required area of the filters is $A = (3,000,000 \, \text{gal}/1440 \, \text{min})/(\text{min-ft}^2/2.33 \, \text{gal}) = 894 \, \text{ft}^2$, or $894/2 = 447 \, \text{ft}^2$ for each filter. The theoretical diameter is

$$D = \left[\frac{4}{\pi} (447 \, \text{ft}^2) \right]^{1/2} = \boxed{23.9 \, \text{ft}}$$

The design diameter is $\boxed{25 \, \text{ft}}$ for the next standard size. Since this is a larger area than the theoretical, a check is made to determine the depth for a filter 25 ft in diameter. The area is $A = (\pi/4)(25)^2 = 491 \, \text{ft}^2$. The hydraulic load is $(3,000,000 \, \text{gal}/1440 \, \text{min})(1/2)(1/491 \, \text{ft}^2) = 2.12 \, \text{gal/min-ft}^2$. Substituting into the performance equation gives

$$\frac{20}{58} = e^{-0.08123D/(2.12)^{0.5}}$$

or $D = 19.1 \, \text{ft}$. Since the packing has modules 2 ft deep, the next smaller depth possible is $20 - 2 = 18 \, \text{ft}$. Since $19.1 > 18$, a packing depth of $\boxed{20 \, \text{ft}}$ and a filter diameter of $\boxed{25 \, \text{ft}}$ are used.

EXAMPLE 17.3 SI   *Super-Rate Trickling Filter*

A single-stage, super-rate trickling filter is to be designed for a population of 10,000 persons. Pertinent data are average influent flow $= 380 \, \ell/\text{cap-d}$, influent BOD$_5$ $= 200 \, \text{mg}/\ell$, primary clarifier removes 33% of the BOD$_5$, effluent BOD$_5$ $= 20 \, \text{mg}/\ell$, filter depth $= 6.10 \, \text{m}$, and recycle ratio $R/Q = 2$. Pilot-plant studies using a synthetic plastic packing have shown a removal constant $K = 2.26$ at 20°C when the depth is in meters, the hydraulic loading is in $\text{m}^3/\text{d-m}^2$, and the constant $n = 0.5$. The winter wastewater temperature $= 17°C$. If two filters are used in parallel, determine the diameter of the filters.

SOLUTION The performance equation is

$$\frac{S_t}{S_0} = e^{-KD/Q_L^{0.5}}$$

The temperature relationship for the removal constant K is

$$K_T = K_{20} \cdot 1.035^{(T-20)}$$

For 17°C, the constant K is

$$K_T = 2.26 \cdot 1.035^{(17-20)}$$
$$= 2.038$$

The BOD$_5$ of the primary clarifier effluent, S_i, is $200(1 - 0.33) = 134$ mg/ℓ. Since $R/Q = 2$, the recycled flowrate is $2Q$, where Q = incoming flowrate. A mass balance at the junction with the recycle flow is

$$(134)(Q) + (20)(2Q) = (1 + 2)(Q)S_0$$

or

$$S_0 = 58 \text{ mg/}\ell$$

Inserting values into Eq. (17.8) gives

$$\frac{20}{58} = e^{-(2.038)(6.10)/Q_L^{0.5}}$$

which yields $Q_L = 136$ m^3/d-m^2. The primary clarifier flow is (10,000 cap) (380 ℓ/cap-d)(m^3/1000 ℓ) = 3800 m^3/d. The hydraulic load to the filters is (3800)(1 + 2) = 11,400 m^3/d. The required area of the filters is $A = (11,400 \text{ m}^3/\text{d})(\text{m}^2\text{-d}/136 \text{ m}^3) = 83.8 \text{ m}^2$, or $83.8/2 = 41.9 \text{ m}^2$ for each filter. The required diameter is

$$D = \left[\frac{4}{\pi}(41.9 \text{ m}^2)\right]^{1/2} = \boxed{7.30 \text{ m}}$$

NRC Equations The National Research Council (NRC) made a study of the data collected from numerous wastewater treatment plants operated at military bases during World War II. Their study included low-rate, single-stage high-rate, and two-stage high-rate trickling filter plants. The NRC equations are applicable only for domestic wastewaters treated by filters with stone media. The empirical equation developed in the NRC study (1946) for a single trickling filter and its final clarifier is

$$E_s = \frac{100}{1 + C\sqrt{\dfrac{y_0}{VF}}} \tag{17.11}$$

where

E_s = filter efficiency, percent

C = 0.0085 for USCS units and 0.443 for SI units

y_0 = lb of BOD$_5$ applied to the filter per day (kg/day)

V = stone medium volume, ac-ft (m^3)

F = recycle factor, or the effective number of passes through the filter

The effective number of passes of the organic matter through the filter, F, is given by

$$F = \frac{1 + R/Q}{(1 + 0.1R/Q)^2} \qquad (17.12)$$

where R/Q is the recycle ratio. In the NRC study, it was found that the maximum feasible recycle ratio is 8. If the filter is a low-rate filter, the value of F in Eq. (17.12) is 1.

If the trickling filter is the second-stage filter of a two-stage plant, the filter must degrade materials more resistant to bio-oxidation than the first-stage filter. Because of this, the equation developed in the NRC study (1946) for a second-stage filter is

$$E_{S_2} = \frac{100}{1 + \dfrac{C}{1 - E_{S_1}} \sqrt{\dfrac{y_0'}{V_2 F_2}}} \qquad (17.13)$$

where

E_{S_2} = second-stage filter efficiency, percent

E_{S_1} = first-stage filter efficiency as a fraction

C = 0.0085 for USCS units and 0.443 for SI units

y_0' = lb (kg) of BOD$_5$ applied to the second-stage filter per day, $y_0' = y_0(1 - E_{S_1})$.

V_2 = volume of the stone medium, ac-ft (m^3)

F_2 = recycle factor, or the effective number of passes through the second-stage filter

The term $1 - E_{S_1}$ accounts for the most bioresistant materials treated by the second-stage filter.

EXAMPLE 17.4   NRC *Equations*

A two-stage high-rate trickling filter plant treats 2.0 MGD of municipal wastewater that has a BOD$_5$ of 200 mg/ℓ. The filters are of equal diameter and depth and have equal recycle ratios. Pertinent data for the plant are: BOD$_5$ removal by the primary clarifier = 33%, filter diameter = 70 ft, filter depth = 5.5 ft, and the recycle ratio = 1.0. Determine the BOD$_5$ in the final effluent.

SOLUTION   The BOD_5 applied to the first-stage filter, y_0, is $(2.0 \times 10^6 \, \text{gal/day})(8.34 \, \text{lb/gal}) \times (200/10^6)(1 - 0.33)$ or $2235 \, \text{lb/day}$. The stone volume, V, for each filter is $(70 \, \text{ft})^2(\pi/4)(5.5 \, \text{ft})(\text{ac-ft}/43,560 \, \text{ft}^3)$ or $0.4859 \, \text{ac-ft}$. The recycle factor, F, is $(1 + 1)/(1 + 0.1)^2$ or 1.65. Substituting these values into Eq. (17.11) yields

$$E_{S_1} = \frac{100}{1 + 0.0085\sqrt{\dfrac{2235}{(0.4859)(1.65)}}}$$

69.02%

This equation gives the efficiency of the first-stage filter, E_{S_1}, as 69.02%. The BOD_5 applied to the second-stage filter is therefore $(2235)(1 - 0.6902)$ or $692 \, \text{lb/day}$. Substituting the appropriate values into Eq. (17.13) yields

$$E_{S_2} = \frac{100}{1 + \dfrac{0.0085}{(1 - 0.6902)}\sqrt{\dfrac{692}{(0.4859)(1.65)}}}$$

29.37

This equation gives the efficiency of the second-stage filter, E_{S_2}, as 55.37%. The effluent BOD_5 is therefore

Effluent $\text{BOD}_5 = (200)(1 - 0.33)(1 - 0.6902)(1 - 0.5537) = \boxed{19 \, \text{mg}/\ell}$

EXAMPLE 17.4 SI   *NRC Equations*

A two-stage high-rate trickling filter plant treats $7.57 \, \text{MLD}$ of municipal wastewater. The filters are of equal diameter and depth and have equal recycle ratios. Pertinent data for the plant are BOD_5 removal by the primary clarifier $= 33\%$, filter diameter $= 21.3 \, \text{m}$, filter depth $= 1.68 \, \text{m}$, and recycle ratio $= 1.0$. Determine the BOD_5 in the final effluent.

SOLUTION   The BOD_5 applied to the first-stage filter, y_0, is $(7,570,000 \, \ell/\text{d})(200 \, \text{mg}/\ell) \times (\text{gm}/1000 \, \text{mg})(\text{kg}/1000 \, \text{gm})(1 - 0.33) = 1014 \, \text{kg/d}$. The stone volume, V, for each filter is $(21.3 \, \text{m}^2)(\pi/4)(1.68 \, \text{m}) = 598.6 \, \text{m}^3$. The recycle factor, F, is $(1 + 1)/(1 + 0.1)^2 = 1.65$. Substituting these values into Eq. (17.11) yields

$$E_{S_1} = \frac{100}{1 + 0.443\sqrt{\dfrac{1014}{(598.6)(1.65)}}}$$

This equation gives the efficiency of the first-stage filter, E_{S_1}, as 69.02%. The BOD_5 applied to the second-stage filter is therefore $(1014)(1 - 0.6902) = 314.1 \, \text{kg/d}$. Substituting the appropriate values into Eq. (17.13) yields

$$E_{S_2} = \frac{100}{1 + \dfrac{0.443}{(1 - 0.6902)}\sqrt{\dfrac{314.1}{(598.6)(1.65)}}}$$

This equation gives the efficiency of the second-stage filter, E_{S_2}, as 55.36%. The effluent BOD_5 is therefore

$$\text{Effluent BOD}_5 = (200)(1 - 0.33)(1 - 0.6902)(1 - 0.5536) = \boxed{19\,\text{mg}/\ell}$$

FLOWSHEETS FOR INTERMEDIATE- AND HIGH-RATE TRICKLING FILTER PLANTS

Figure 17.7 shows the usual flowsheets that have been employed for single-stage and two-stage intermediate- and high-rate trickling filter plants. The single-stage plant shown in Figure 17.7 (b) has the recycled flow, R, clarified prior to discharge on the filter. Also, the two-stage intermediate- and high-rate trickling filter plant shown in Figure 17.7 (d) has the recycled flows, R_1 and R_2, clarified prior to discharge on the filters. It has been found that clarification of the recycle flows has a negligible effect upon filter performance. Because clarification of the recycled flow requires a clarifier to be larger than otherwise, most plants do not use clarified recycle, and as a

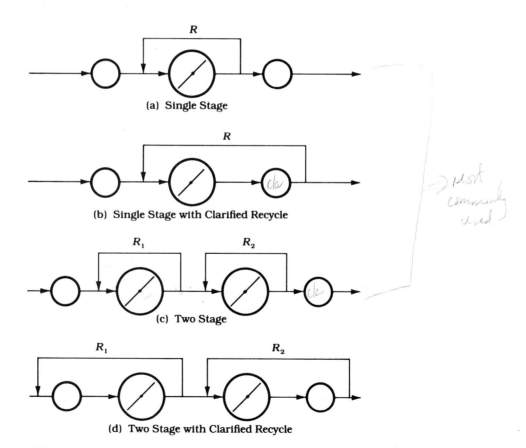

FIGURE 17.7 *Flowsheets for Intermediate- and High-Rate Trickling Filter Plants*

result, the flowsheets shown in Figures 17.7 (a) and (c) are the most commonly used. Two-stage intermediate- and high-rate trickling filter plants usually have filters of equal size, and the recycled flows, R_1 and R_2, are usually equal.

**FILTER
DETAILS**

The media used in trickling filters are usually gravel, crushed stone, or slag having a size of $2\frac{1}{2}$ to $3\frac{1}{2}$ in. (64 to 89 mm). If crushed stone is employed, the stone should be of a hard nature such as trap rock or granite. Soft stone, such as limestone, should not be used because it will gradually break apart because of the end products from microbial action and weathering. Synthetic media such as plastics are usually used when stone is not locally available. In general, the synthetic media have a larger specific area (that is, the surface area per unit bulk volume) than stone media. Surfpac is a PVC medium made in corrugated sheets similar to industrial grating. It is assembled in stacks within the filter. Actifil and Flexxirings are randomly dumped media similar to that used in packed towers for gas absorption and stripping.

Figure 17.8 (a) shows a typical underdrain block made of vitrified clay. It serves both to support the medium and also to provide small channels through which the collected effluent flows. Figure 17.8 (b) shows the plan of a typical trickling filter floor; as indicated, it slopes gradually to the effluent channel that crosses the center of the filter. The underdrain blocks are placed so that their channels align and the collected effluent flows to the effluent channel as shown in Figure 17.8 (b). Figure 17.8 (c) shows a section through the filter at right angles to the underdrain block channels. Figure 17.8 (d) shows a profile through the filter bed illustrating the collected effluent discharging from the underdrain blocks into the effluent channel. The underdrain channels and the effluent channel should be designed to flow not more than half full at the maximum hydraulic flow during the day in order to provide proper ventilation.

The most common distributor is the rotary type that is mounted on a center pier in the middle of the filter. The distributor is usually driven by the jet action of the wastewater discharging from the filter arms onto the bed. Extremely small filters, those less than about 10 ft (3.1 m) in diameter, are usually electrically driven. Also, small filters may have a single spray in the center of the filter instead of a rotary distributor. Formerly, rectangular trickling filters were used that were dosed by fixed nozzles located above the filter bed; however, frequent clogging of the nozzles has caused this type of filter to be almost totally discontinued. The hydraulic head required to drive a rotary distributor depends on the ratio of the maximum flow to the minimum flow and on whether the distributor is equipped with overflow arms or overflow jets used for high flows. If the distributor has four arms with two being used for high flows, or has double compartment arms with one compartment and its jets being used for high flows, the head required will be minimized. If the ratio of the maximum flow to the minimum flow is less than 4.0 and the distributor has overflow arms or overflow jets, the head

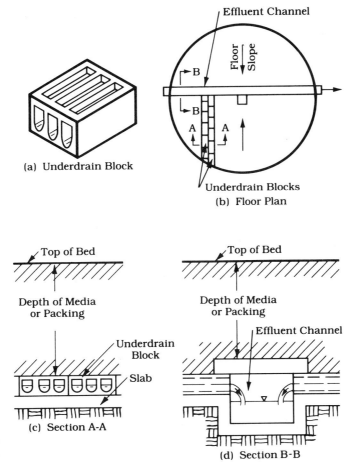

FIGURE 17.8
Trickling Filter Details

(a) Underdrain Block

(b) Floor Plan

(c) Section A-A

(d) Section B-B

required at the center of the distributor is usually about 3.50 ft (1.07 m) above the top of the filter bed.

OPERATIONAL PROBLEMS

n general, trickling filters have fewer operational problems than the activated sludge process; however, some problems may be encountered. Trickling filter humus settles rapidly in a final clarifier, and poor settling rarely results if the scraper mechanism removes the sludge as it collects at the bottom of the tank. Floating sludge masses may be encountered if anaerobic conditions occur within the settled sludge.

The most common problem with trickling filters is the presence of the *Psychoda* fly that breeds in filter beds. Ordinarily these flies remain in the immediate vicinity of the filters; however, they may be a nuisance to the operators. The most effective control is to provide for flooding the filter. If a filter is flooded and the medium allowed to stay submerged for one day of each summer month, the fly larvae will usually be killed.

Odors may be a problem when low-rate filters are employed and the wastewater is stale when it reaches the plant. However, odors are rarely a problem for high-rate trickling filters if the recycle flow is continuous throughout the day and the wastewater is not stale.

ROTARY BIOLOGICAL CONTACTORS

A single-stage rotary biological contactor, as shown in Figure 17.9, consists of a vat through which the wastewater flows and multiple plastic discs mounted on a horizontal shaft that passes through the center of the discs. The shaft is mounted at right angles to the wastewater flow, with about 40% of the total disc area being submerged. As the shaft rotates, the biological growths on the discs pass through the wastewater and sorb organic materials in addition to losing excess growths, which slough off. As the biological growths pass through the air, oxygen is adsorbed to keep the growths as aerobic as possible. The biosorption and bio-oxidation that occur are similar to those of a trickling filter. A multistage rotary biological contactor consists of two or more stages connected in series to achieve greater BOD_5 removal than occurs for a single stage. Recycle is not employed, and the growths sloughing from a single- or multistage contactor are removed by a final clarifier. The secondary sludge is mixed with the raw wastewater and resettled in the primary clarifier in order to thicken the sludge. The primary sludge, being a mixture of incoming settleable solids and secondary settlings, usually is 4.2 to 6.5% dry solids (Antonie *et al.*, 1974). Biological action is significantly reduced when the wastewater temperature drops below 55°F (13°C). In cold climates, the discs must be covered to avoid heat loss from the wastewater and to protect against freezing temperatures. In warm climates, the units do not need to be enclosed except for aesthetic purposes. In warm climates, a simple sun roof is frequently used.

FIGURE 17.9 *Rotating Biological Contactors*
Courtesy of Envirex, Inc.

The principal design parameter for rotary biological discs is the wastewater flowrate per unit surface area of the discs — that is, the hydraulic loading in gal/day-ft^2 area — that is, gpd/ft^2 (m^3/day-m^2). This is indirectly a food-to-microbe ratio since the flowrate is related to the mass of substrate per unit time and the disc area is related to the mass of microbes in the system. Usually, four stages are provided for municipal wastewaters, and the peripheral speed of the discs is about 60 ft/min (18 m/min). If nitrification is required for a municipal wastewater, five stages are usually provided.

In the treatment of municipal wastewaters with a BOD$_5$ of 200 mg/ℓ using four stages, a 20-mg/ℓ BOD$_5$ effluent can be obtained if the hydraulic loading is less than about 1.9 gpd/ft^2 (0.077 m^3/day-m^2) and the wastewater temperature is greater than 55°F (13°C). This is assuming 33% BOD$_5$ removal by the primary clarifier. The main advantages of the rotary biological contactors are the low energy requirements compared to the activated sludge process, the ability to handle shock loads, and the ability of multistage units to achieve a high degree of nitrification.

A kinetic equation for rotary biological contactors has been developed by Eckenfelder (1989) and is based on the specific rate of substrate removal for a pseudo–first-order reaction. This equation is

$$-\frac{1}{X}\frac{dS}{dt} = kS \qquad (17.14)$$

where

$(1/X)(dS/dt)$ = specific rate of substrate utilization, mass/(mass microbes) $\times$ (time)

dS/dt = rate of substrate utilization, mass/(volume)(time)

k = rate constant, volume/(mass microbes)(time)

S = substrate concentration, mass/volume

The rate of substrate utilization, dS/dt, is given by

$$\frac{dS}{dt} = Q(S_0 - S) \qquad (17.15)$$

where Q = flowrate, S_0 = substrate in the flow to the contactor, and S = substrate in the flow leaving the contactor. Substituting Eq. (17.15) into Eq. (17.14) gives

$$\frac{1}{X}Q(S_0 - S) = kS \qquad (17.16)$$

The cell mass, X, is proportional to the disc area, A; that is,

$$X \propto A \qquad (17.17)$$

Substituting $X \propto A$ in Eq. (17.16) gives

$$\frac{Q}{A}(S_0 - S) = kS \qquad (17.18)$$

The term $(Q/A)(S_0 - S)$ is equal to the rate of reaction, r. Thus,

$$\overbrace{\frac{Q}{A}(S_0 - S)}^{r} = r = kS \qquad (17.19)$$

This equation is of the form $y = mx$, so it can be graphically represented as shown in Figure 17.10. It can be seen from Figure 17.10 that there is a maximum rate that can be achieved, r_{max}, and this occurs where oxygen becomes limiting to the microbial growths. Eckenfelder (1989) presents data that verify the plot in Figure 17.10 for various industrial wastewaters. Rearranging Eq. (17.19) gives

$$\frac{Q}{A} = \frac{r_1}{S_0 - S_1} = \text{slope} \qquad (17.20)$$

where r_1 = rate of reaction for the biological contactor and S_1 = substrate concentration leaving the contactor. Thus Eq. (17.20) is for a line from S_0 to r_1, which has a slope = Q/A, as shown in Figure 17.11. Figure 17.11

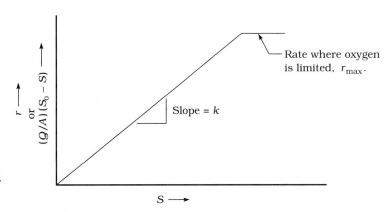

FIGURE 17.10
Plot of Equation (17.19)

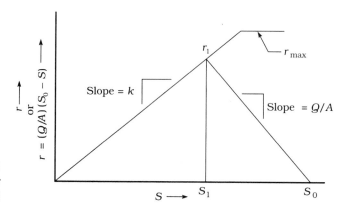

FIGURE 17.11
Plot of Equations (17.19) and (17.20)

graphically shows Eq. (17.19) and Eq. (17.20). The line from S_0 to r_1 has a slope, Q/A, equal to the hydraulic loading, and the x value at r_1 is S_1, the substrate concentration leaving the contactor. For a series of contactors, Eq. (17.20) can be generalized to give

$$\frac{Q}{A} = \frac{r_n}{S_n - S_{n-1}} \qquad (17.21)$$

which represents any stage, n, in the series of multi-stage contactors. Following the stagewise construction for four stages gives a plot as shown in Figure 17.12. Thus, by a trial-and-error solution, the hydraulic loading, Q/A, to give a desired effluent concentration can be determined. Note that this construction is similar to the construction for a series of continuously stirred tank reactors (CSTRs) as shown in Chapter 3.

An algebraic solution for a series of n rotating biological contactors can be found as follows. Dividing Eq. (17.18) by Q/A and S gives

$$\frac{S_0 - S}{S} = \frac{k}{Q/A} \qquad (17.22)$$

Thus,

$$\frac{S_0}{S} = 1 + \frac{k}{Q/A} \qquad (17.23)$$

from which

$$\frac{S}{S_0} = \frac{1}{1 + \dfrac{k}{Q/A}} \qquad (17.24)$$

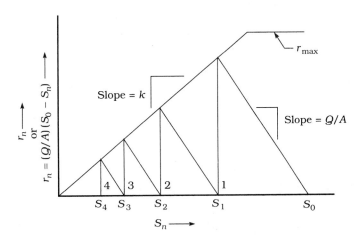

FIGURE 17.12 *Four-Stage Rotary Biological Contactor*

For the first stage,

$$S_1 = \left(\frac{1}{1 + \dfrac{k}{Q/A}}\right) S_0 \tag{17.25}$$

For the second stage,

$$S_2 = \left(\frac{1}{1 + \dfrac{k}{Q/A}}\right) S_1 \tag{17.26}$$

or

$$S_2 = \left(\frac{1}{1 + \dfrac{k}{Q/A}}\right)\left(\frac{1}{1 + \dfrac{k}{Q/A}}\right) S_0 \tag{17.27}$$

Therefore,

$$\frac{S_2}{S_0} = \left(\frac{1}{1 + \dfrac{k}{Q/A}}\right)^2 \tag{17.28}$$

Generalizing for n stages gives

$$\frac{S_n}{S_0} = \left(\frac{1}{1 + \dfrac{k}{Q/A}}\right)^n \tag{17.29}$$

where n = number of stages. Since rotating biological contactors do not have recycle, the substrate term, S_0, is equal to S_i, where S_0 = substrate concentration to the first reactor and S_i = substrate concentration in the primary clarifier effluent. Data are very limited, even for municipal waste-waters, to give a typical k value. It is best to have pilot-plant studies done for the wastewater concerned to determine the rate constant, k.

EXAMPLE 17.5
AND 17.5 SI

Rotary Biological Contactors (RBC)

A municipality with a design population of 6900 persons is to have a rotary biological contactor plant designed. Pertinent data are average flow = 100 gal/cap-day (380 ℓ/cap-d), influent BOD$_5$ = 200 mg/ℓ, primary clarifier removes 33% BOD$_5$, total effluent BOD$_5$ = 20 mg/ℓ, and number of stages = 4. If a pilot-plant study gave a k value of 1.16 gal/day-ft^2 (47.3 ℓ/d-m^2) and the algebraic method is used, determine:

1. The design hydraulic loading.
2. The BOD$_5$ after each stage.
3. The disk area per stage.
4. The total disc area.

5. The number of shafts per stage and the total shafts for the plant if a shaft has $100,000\,\text{ft}^2$ ($9289\,\text{m}^2$) of surface area.

SOLUTION The performance of rotary biological contactors is given by

$$\frac{S_n}{S_0} = \left(\frac{1}{1 + \dfrac{k}{Q/A}}\right)^n$$

The influent BOD$_5$, $S_i = S_0 = 200(1 - 0.33) = 134\,\text{mg}/\ell$, the effluent BOD$_5$, $S_4 = 20\,\text{mg}/\ell$, and $k = 1.16\,\text{gal/day-ft}^2$. Inserting these values into the performance equation gives

$$\frac{20}{134} = \left(\frac{1}{1 + \dfrac{1.16}{Q/A}}\right)^4$$

Solving this equation gives $\boxed{Q/A = 1.91\,\text{gal/day-ft}^2.}$ The BOD$_5$ after each stage can now be determined by substituting $Q/A = 1.91\,\text{gal/day-ft}^2$, $S_0 = 134\,\text{mg}/\ell$, and $k = 1.16\,\text{gal/day-ft}^2$ into the performance equation to give

$$\frac{S_n}{134} = \left(\frac{1}{1 + 1.16/1.91}\right)^n = 0.6221^n$$

or

$$S_n = (134)(0.6221)^n$$

From this equation, $\boxed{S_1 = 83\,\text{mg}/\ell, \ S_2 = 52\,\text{mg}/\ell, \ S_3 = 32\,\text{mg}/\ell,\text{ and } S_4 = 20\,\text{mg}/\ell.}$ The influent flow $= (6900\ \text{cap})(100\,\text{gal/cap-day}) = 690,000$ gal/day. The disc area per stage is

$$A = \left(\frac{690,000\,\text{gal}}{\text{day}}\right)\left(\frac{\text{day-ft}^2}{1.91\,\text{gal}}\right) = \boxed{362,000\,\text{ft}^2}$$

The total disc area is

$$A = \left(\frac{362,000\,\text{ft}^2}{\text{stage}}\right)\left(4\ \text{stages}\right) = \boxed{1,450,000\,\text{ft}^2}$$

The number of shafts per stage $= (362,000\,\text{ft}^2) \div (100,000\,\text{ft}^2/\text{shaft}) = 3.62$ or $\boxed{4}$. The total number of shafts $= (4)(4) = \boxed{16}$ for the plant.

For the solution in SI units, inserting $k = 47.3\,\ell/\text{d-m}^2$, $S_i = S_0 = 134\,\text{mg}/\ell$, and $S_4 = 20\,\text{mg}/\ell$ into the performance equation gives

$$\frac{20}{134} = \left(\frac{1}{1 + \dfrac{47.3}{Q/A}}\right)^4$$

Solving this equation gives $\boxed{Q/A = 77.7\,\ell/\text{d-m}^2.}$ The BOD$_5$ after each stage can now be determined by substituting $Q/A = 77.7\,\ell/\text{d-m}^2$, $S_0 = 134\,\text{mg}/\ell$, and $k = 47.3\,\ell/\text{d-m}^2$ into the performance equation to give

$$\frac{S_n}{134} = \left(\frac{1}{1 + 47.3/77.7}\right)^n = 0.6216^n$$

or

$$S_n = (134)(0.6216)^n$$

From this equation, $\boxed{S_1 = 83\,\text{mg}/\ell,\ S_2 = 52\,\text{mg}/\ell,\ S_3 = 32\,\text{mg}/\ell,\ \text{and } S_4 = 20\,\text{mg}/\ell.}$ The influent flow is $(6900\,\text{cap})(380\,\ell/\text{cap-d}) = 2{,}622{,}000\,\ell/\text{d}$. The disc area per stage is

$$A = \left(\frac{2{,}622{,}000\,\ell}{\text{d}}\right)\left(\frac{\text{d-ft}^2}{77.7\,\ell}\right) = \boxed{33{,}800\,\text{m}^2}$$

The total disc area is $(33{,}800)(4) = \boxed{136{,}000\,\text{m}^2.}$ The number of shafts per stage $= (33{,}800\,\text{m}^2) \div (9289\,\text{m}^2/\text{shaft}) = 3.64$ or $\boxed{4.}$ The total number of shafts $= (4)(4) = \boxed{16}$ for the plant.

EXAMPLE 17.6 AND 17.6 SI   *Rotary Biological Contactors (RBC)*

A municipality with a design population of 6900 persons is to have a rotary biological contactor plant designed. Pertinent data are average flow = 100 gal/cap-d (380 ℓ/cap-d), influent BOD$_5$ = 200 mg/ℓ, primary clarifier removes 33% of the BOD$_5$, total effluent BOD$_5$ = 20 mg/ℓ, and number of stages = 4. If a pilot-plant study gave a k value of 1.16 gal/day-ft^2 (47.3 ℓ/d-m^2) and the graphical method is used, determine:

1. The design hydraulic loading.
2. The BOD$_5$ after each stage.
3. The disc area per stage.
4. The total disc area.

SOLUTION   A plot of $Q/A(S_0 - S_n)$ versus S_n is given in Figure 17.13, and the slope of the line through the origin is k, which is 1.16 gal/day-ft^2. Slope = k = 1.16 gal/day-ft^2 = $\Delta y/\Delta x$. If $\Delta x = 50\,\text{mg}/\ell$, Δy is (1.16 gal/day-ft^2)(50 mg/ℓ) = 58 gal-mg/day-ft^2-ℓ. A line is drawn through the coordinates (0, 0) and (50, 58) to give the rate equation. The pilot-plant studies showed that at a rate of 115 gal-mg/day-ft^2-ℓ, the process was oxygen limited and was independent of $Q/A(S_0 - S_n)$. This line is also shown in the graph. For the first trial, a slope was assumed and the stagewise construction gave S_4 = 17 mg/ℓ. Since this was below 20 mg/ℓ BOD$_5$, a steeper slope was selected and

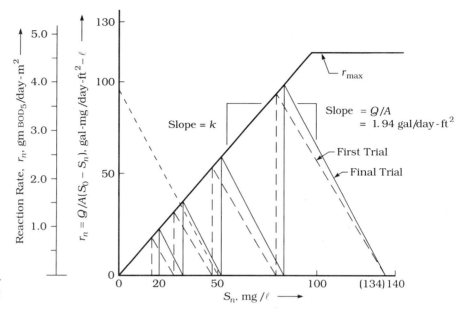

FIGURE 17.13
Graph for Example 17.6

the stagewise construction was made again. After several trials, the final trial resulted in the stagewise construction shown in Figure 17.13, which has $S_4 = 20\,\text{mg}/\ell$. By using triangles, a line parallel to the slope lines was drawn through the coordinates (50, 0) and (0, 97). The slope of this line is Q/A, and thus $Q/A = 97/50 = \boxed{1.94\,\text{gal/day-ft}^2\,.}$ This is the design hydraulic loading. The BOD$_5$ values after each stage were determined from the graph; they are $\boxed{S_1 = 83\,\text{mg}/\ell,\; S_2 = 53\,\text{mg}/\ell,\; S_3 = 33\,\text{mg}/\ell,\; \text{and}\; S_4 = 20\,\text{mg}/\ell.}$ Note that these are very close to the values determined using the algebraic method in Example 17.5.

Since the y-axis also represents the reaction rate, r, as gm BOD$_5$/day-m^2, another y-axis was drawn to represent the reaction rate in these terms. For 150 gal-mg/day-ft^2-ℓ, the r value is (150 gal-mg/day-ft^2-ℓ gal-mg/day-ft^2-ℓ) × (3.785 ℓ/gal) × (10.765 ft^2/m^2)(gm/1000 mg) = 6.11 gm BOD$_5$/day-m^2. Using this y-axis yields the r values $r_1 = 3.9\,\text{gm/day-m}^2$, $r_2 = 2.4\,\text{gm/day-m}^2$, $r_3 = 1.5\,\text{gm/day-m}^2$, and $r_4 = 0.9\,\text{gm/day-m}^2$. The disc area per stage is

$$A = (6900)\left(\frac{100\,\text{gal}}{\text{cap-day}}\right)\left(\frac{\text{day-ft}^2}{1.94\,\text{gal}}\right)$$

$$= \boxed{356{,}000\,\text{ft}^2}$$

The total disc area is

$$A = (356{,}000)(4) = \boxed{1{,}424{,}000\,\text{ft}^2}$$

For the solution in SI units, a plot of $Q/A(S_0 - S_n)$ versus S_n is given in Figure 17.14; the slope of the line through the origin is k, which is 47.3 ℓ/

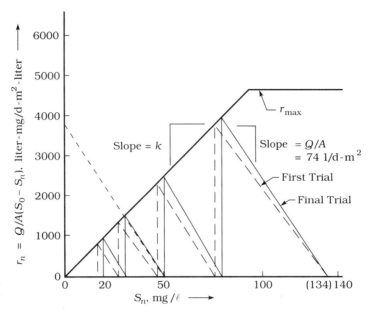

FIGURE 17.14
Graph for Example 17.6 SI

d-m^2. The slope $= k = 47.3\,\ell/\text{d-m}^2 = \Delta y/\Delta x$. If $\Delta x = 50\,\text{mg}/\ell$, Δy is $(47.3\,\ell/\text{d-m}^2)(50\,\text{mg}/\ell) = 2400\,\ell\text{-mg/d-m}^2\text{-}\ell$. A line is drawn through the coordinates $(0, 0)$ and $(50, 2400)$ to give the rate equation. The pilot-plant studies showed that at a rate of 4700 $(\ell\text{-mg/d-m}^2\text{-}\ell)$, the process was oxygen limited and was independent of $Q/A(S_0 - S_n)$. This line is also shown in the graph. For the first trial, a slope was assumed and the stagewise construction gave $S_4 = 17\,\text{mg}/\ell$. Since this was below $20\,\text{mg}/\ell$ BOD$_5$, a steeper slope was selected and the stagewise construction was made again. After several trials, the final trial resulted, as shown in Figure 17.14; it has $S_4 = 20\,\text{mg}/\ell$. By using triangles, a line parallel to the slope lines was drawn through the coordinates $(50, 0)$ and $(0, 3700)$. The slope of this line is Q/A, and therefore $Q/A = 3700/50 = \boxed{74\,\ell/\text{d-m}^2.}$ This is the design hydraulic loading. The BOD$_5$ values after each stage were determined from the graph; they are $\boxed{S_1 = 82\,\text{mg}/\ell,\ S_2 = 52\,\text{mg}/\ell,\ S_3 = 32\,\text{mg}/\ell, \text{ and } S_4 = 20\,\text{mg}/\ell.}$ Note that these are very close to the values determined by the algebraic method in Example 17.5 SI.

The disc area per stage is

$$A = (6900)\left(\frac{380\,\ell}{\text{cap-d}}\right)\left(\frac{\text{d-m}^2}{74\,\ell}\right)$$

$$= \boxed{35,500\,\text{m}^2}$$

The total disc area is

$$A = (35,500)(4) = \boxed{142,000\,\text{m}^2}$$

REFERENCES

American Society of Civil Engineers (ASCE). 1982. *Wastewater Treatment Plant Design*. ASCE Manual of Practice No. 36. New York: ASCE.

Antonie, R. L.; Kluge, D. L.; and Mielke, J. H. 1974. Evaluation of a Rotary Disk Wastewater Treatment Plant. *Jour. WPCF* 46, no. 3:498.

Balakrishnan, S.; Eckenfelder, W. W., Jr.; and Brown, C. 1969. Organics Removal by a Selected Trickling Filter Media. *Water and Wastes Eng.* 6, no. 1:A-22.

Barnes, D.; Bliss, P. J.; Gould, B. W.; and Vallentine, H. R. 1981. *Water and Wastewater Engineering Systems*. London: Pittman Books Limited.

Bio-Systems Division, Autotrol Corp. 1978. *Autotrol Wastewater Treatment Systems*. Design Manual. Milwaukee, Wis.

Conway, R. A., and Ross, R. D. 1980. *Handbook of Industrial Waste Disposal*. New York: Van Nostrand Reinhold.

Eckenfelder, W. W., Jr. 1989. *Industrial Water Pollution Control*. 2nd ed. New York: McGraw-Hill Book Co.

Eckenfelder, W. W., Jr. 1980. *Principles of Water Quality Management*. Boston: CBI Publishing.

Eckenfelder, W. W., Jr. 1961. Trickling Filter Design and Performance. *Jour. SED* 87, no. SA4, part 1:33.

Eckenfelder, W. W., Jr. 1970. *Water Quality Engineering for Practicing Engineers*. New York: Barnes & Noble.

Eckenfelder, W. W., Jr., and Barnhart, E. L. 1963. Performance of a High-Rate Trickling Filter Using Selected Media. *Jour. WPCF* 35, no. 12:1535.

Envirex, Inc. 1979. *Specific RBC Process Design Criteria*. Design Manual, Envirex, a Rexnord Division, Waukesha, Wisconsin.

EPA, 1985. *Review of Current RBC Performance and Design Procedures*, by Roy F. Weston, Inc. Technical Report, EPA Contract No. 68-03-2775 and 68-03-3019, Cincinnati, Ohio.

EPA, 1984. *Summary of Design Information on Rotating Biological Contactors*. Technical Report, 430/84-008, USEPA.

Fairall, J. M. 1956. Correlation of Trickling Filter Data. *Sew. and Ind. Wastes* 28, no. 9:1056.

Galler, W. S., and Gotaas, H. B. 1966. Optimization Analysis for Biological Filter Design. *Jour. SED* 92, no. SA1:163.

Germain, J. E. 1966. Economical Treatment of Domestic Waste by Plastic-Medium Trickling Filters. *Jour. WPCF* 38, no. 2:192.

Great Lakes–Upper Mississippi River Board of State Public Health and Environmental Managers. 1990. Recommended Standards for Wastewater Facilities. Ten state standards. Albany, N.Y.

Great Lakes–Upper Mississippi River Board of State Sanitary Engineers. 1978. Recommended Standards for Sewage Works. Ten state standards. Albany, N.Y.

Horan, N. J. 1990. *Biological Wastewater Treatment Systems: Theory and Operation*. Chichester, West Sussex, England: Wiley.

Howland, W. E. 1957. Flow Over Porous Media as in a Trickling Filter. *Proc. of the 12th Annual Purdue Ind. Wastes Conference*.

McGhee, T. J. 1991. *Water Supply and Sewage*. 6th ed. New York: McGraw-Hill.

McKinney, R. E. 1962. *Microbiology for Sanitary Engineers*. New York: McGraw-Hill.

Metcalf & Eddy, Inc. 1972. *Wastewater Engineering: Collection, Treatment, and Disposal*. New York: McGraw-Hill.

Metcalf & Eddy, Inc. 1979. *Wastewater Engineering: Treatment, Disposal and Reuse*. 2nd ed. New York: McGraw-Hill.

Metcalf & Eddy, Inc. 1991. *Wastewater Engineering: Treatment, Disposal and Reuse*. 3rd ed. New York: McGraw-Hill.

National Research Council. 1946. Trickling Filters in Sewage Treatment at Military Installations. *Sew. Works Jour.* 18, no. 5:417.

Novotny, V.; Imhoff, K. R.; Olthof, M.; and Krenkel, P. A. 1989. *Karl Imhoff's Handbook of Urban Drainage and Wastewater Disposal*. New York: Wiley.

Pano, A.; Middlebrooks, E. J.; and Reynolds, J. H. 1981. *The Kinetics of Rotating Biological Contactors Treating Domestic Wastewater*. Technical Report, UWRL/Q-81/04, Utah Water Research Laboratory, Logan, Utah.

Qasim, S. R. 1985. *Wastewater Treatment Plants*. New York: Holt, Rinehart and Winston.

Ramalho, R. S. 1977. *Introduction to Wastewater Treatment Processes*. New York: Academic Press.

Rankin, R. S. 1955. Evaluation of the Performance of Biofiltration Plants. *Trans. ASCE* 120:823.

Reck, T. 1990. *Advances in Trickling Filters*. Master of Engineering Report, Texas A&M University, College Station, Texas.

Schroeder, E. D. 1977. *Water and Wastewater Treatment*. New York: McGraw-Hill.

Schulze, K. L. 1960. Load and Efficiency of Trickling Filters. *Jour. WPCF*, 30, no. 3:245.

Stack, V. T. 1957. Theoretical Performance of

Trickling Filter Processes. *Sew. and Ind. Wastes* 29, no. 9:987.

Sundstrom, D. W., and Klei, H. E. 1979. *Wastewater Treatment*. Englewood Cliffs, N.J.: Prentice-Hall.

Velz, C. J. 1948. A Basic Law for the Performance of Biological Beds. *Sew. Works Jour.* 20, no. 4:607.

Viessman, W., and Hammer, M. J. 1993. *Water Supply and Pollution Control*. 5th ed. New York: Harper & Row.

Wadsworth, N. 1990. *Rotating Biological Contactors*. Master of Engineering Report, Texas A&M University, College Station, Texas.

Water Environment Federation (WEF). 1992. *Design of Municipal Wastewater Treatment Plants*, Vol. I and II, WEF Manual of Practice No. 8.

Water Pollution Control Federation (WPCF). 1977. *Wastewater Treatment Plant Design*. WPCF Manual of Practice no. 8, Washington, D.C.

PROBLEMS

17.1 A city of 5000 persons is to be served by a low-rate trickling filter with stone media and a polishing pond downstream from the final clarifier. Pertinent data are: wastewater flow = 95 gal/cap-day (360 ℓ/cap-d), BOD_5 after primary clarification = 135 mg/ℓ, BOD_5 after final clarification = 20 mg/ℓ, and filter bed depth = 5 to 6 ft (1.83 to 2.43 m). Determine, using USCS units and Eq. (17.10):

a. The theoretical filter diameter if the depth is assumed to be 5.5 ft.

b. The design diameter if rotary distributors are available in 5-ft increments in diameter.

c. The filter bed depth.

17.2 Repeat Problem 17.1 using SI units and Eq. (17.10) to determine the filter diameter if the depth is 1.67 m.

17.3 A city of 5000 persons is to be served by a high-rate trickling filter with stone media and a polishing pond downstream from the final clarifier. Pertinent data are: wastewater flow = 95 gal/cap-day (360 ℓ/cap-d), BOD_5 after primary clarification = 135 mg/ℓ, BOD_5 after final clarification = 30 mg/ℓ, recycle ratio $R/Q = 2$, and filter bed depth = 5 to 6 ft (1.83 to 2.43 m). Determine, using USCS units and Eq. (17.10):

a. The theoretical diameter of the filter if the depth is assumed to be 5.5 ft.

b. The design diameter of the filter if rotary distributors are available in 5-ft increments in diameter.

c. The depth of the filter.

17.4 Repeat Problem 17.3 using SI units and Eq. (17.10) to determine the diameter of the filter if the depth is 1.67 m.

17.5 A single-stage super-rate trickling filter is to be designed for a population of 8100 persons.

Pertinent data are: influent flow = 95 gal/cap-day, influent BOD_5 = 210 mg/ℓ, primary clarifier removes 33% of the influent BOD_5, effluent BOD_5 = 20 mg/ℓ, filter depth = 20 ft, and recycle ratio $R/Q = 2$. Pilot-plant studies have shown that the removal constant $K = 0.085$ at 20°C when the depth is in feet, the hydraulic loading is in gpm/ft^2, and the constant $n = 0.5$. The winter wastewater temperature = 61°F (16.1°C). The synthetic plastic packing is available in 2-ft modules. Two filters in parallel are to be used. Determine, using USCS units and Eq. (17.8):

a. The theoretical diameter, ft.

b. The design diameter if distributors are available in 5-ft increments in diameter, ft.

c. The packing depth, ft.

d. The hydraulic loading, gpm/ft^2 and MG/ac-day.

e. The organic loading in lb/day-1000 ft^3 and lb/day-ac-ft.

17.6 A single-stage super-rate trickling filter is to be designed for a population of 8100 persons. Pertinent data are: influent flow = 380 ℓ/cap-day, influent BOD_5 = 210 mg/ℓ, primary clarifier removes 33% of the influent BOD_5, effluent BOD_5 = 20 mg/ℓ, filter depth = 6.10 m, and recycle ratio $R/Q = 2$. Pilot-plant studies using the synthetic packing have shown that the removal constant $K = 2.11$ at 20°C when the depth is in meters, the hydraulic loading is in m^3/d-m^2, and the constant $n = 0.5$. The winter wastewater temperature = 16.1°C. Determine, using SI units and Eq. (17.8):

a. The filter diameter, m.

b. The hydraulic loading, m^3/d-m^2.

c. The organic loading, kg/d-m^3.

17.7 An existing two-stage high-rate trickling filter plant serves a population of 23,100 persons. Upgrading of the plant using tertiary treatment is being planned. Pertinent data are: influent $BOD_5 = 205$ mg/ℓ, influent flow = 95 gal/cap-day (360 ℓ/cap-d), BOD_5 removal by the primary clarifier = 33%, filter diameter = 75 ft (22.9 m), filter depth = 5 ft (1.52 m), and the recycle ratio $R/Q = 1$. Determine, using the NRC equations and USCS units:

a. The predicted effluent BOD_5, mg/ℓ.

b. The hydraulic loading, MG/ac-day.

17.8 Repeat Problem 17.7 using SI units. Part (a) should be in mg/ℓ and part (b) in m^3/m^2-d.

17.9 A single-stage high-rate trickling filter plant treats a municipal wastewater from a population of 10,000 persons. The plant has a polishing pond downstream from the final clarifier. Pertinent data are: wastewater flow = 100 gal/cap-day (380 ℓ/cap-day), influent $BOD_5 = 200$ mg/ℓ, BOD_5 removal by the primary clarifier = 33%, recycle ratio = 2.5, filter depth = 5 to 7 ft (1.52 to 2.13 m), and effluent BOD_5 = 25 mg/ℓ. Determine, using USCS units and the NRC equation:

a. The theoretical filter diameter if a depth of 6 ft is assumed.

b. The design diameter if rotary distributors are available in 5-ft increments in diameter.

c. The filter depth.

17.10 Repeat Problem 17.9 using SI units and the NRC equation, and determine the filter diameter if a depth of 1.83 m is assumed.

17.11 A standard-rate trickling filter plant is to treat the wastewater from a city of 6200 persons. Pertinent data are: wastewater flow = 100 gal/cap-day (380 ℓ/cap-d), influent BOD_5 = 200 mg/ℓ, BOD_5 removal by the primary clarifier = 33%, filter depth = 5 to 7 ft (1.52 to 2.13 m), and effluent BOD_5 = 20 mg/ℓ. Determine, using USCS units and the NRC equation:

a. The theoretical filter diameter if a depth of 6 ft is assumed.

b. The design diameter of the filter if rotary distributors are available in 5-ft increments in diameter.

c. The filter depth.

17.12 Repeat Problem 17.11 using SI units, and determine the filter diameter if a depth of 1.83 m is assumed.

17.13 A two-stage high-rate trickling filter plant is to treat a municipal wastewater having a flow of 2.0 MGD (7.57 MLD). Pertinent data are: influent BOD_5 = 200 mg/ℓ, BOD_5 removal by the primary clarifier = 33%, filter bed depth = 5 to 7 ft (1.53 to 2.13 m), first- and second-stage filters are equal in diameter and depth, recycle ratio = 1.0 for each filter, and final effluent BOD_5 = 20 mg/ℓ. Determine, using USCS units and the NRC equation:

a. The volume of stone required for each filter. *Note*: This is a trial-and-error solution. To shorten calculations, try an organic loading $[y_0/(V_1 + V_2)]$ of 2600 and 2700 lb BOD_5/day-ac-ft. Determine the filter volumes and then the effluent BOD_5 in mg/ℓ, and interpolate to determine the design organic loading. Then determine the effluent BOD_5 in mg/ℓ to verify the design organic loading.

b. The theoretical filter diameter if a filter bed depth of 6 ft is assumed.

c. The design filter diameter if rotary distributors are available in 5-ft increments in diameter.

d. The filter bed depth.

17.14 Repeat Problem 17.14 using SI units and the NRC equation. Determine:

a. The volume of stone required for each filter. *Note*: This is a trial-and-error solution. To shorten calculations, try an organic loading $[y_0/(V_1 + V_2)]$ of 0.957 and 0.994 kg BOD_5/d-m^3. Determine the filter volumes and then the effluent BOD_5 in mg/ℓ, and interpolate to determine the design organic loading. Then determine the effluent BOD_5 in mg/ℓ to verify the design organic loading.

b. The filter diameter if the filter bed depth is 1.83 m.

17.15 A low-rate trickling filter plant gives 95% BOD_5 removal. The primary clarifier removes 33% of the BOD_5 applied to it. The secondary sludge amounts to 0.029 lb dry solid/cap-day (0.013 kg/cap-d). If the raw wastewater has a BOD_5 of 0.17 lb/cap-day (0.077 kg/cap-d), determine the secondary sludge produced per day per pound of BOD_5 removed.

17.16 Repeat Problem 17.15 using SI units. Determine the secondary sludge produced per day per kilogram of BOD_5 removed.

17.17 A high-rate trickling filter plant gives 90% BOD_5 removal. The primary clarifier removes 33% of the BOD_5 applied to it. The secondary sludge amounts to 0.044 lb dry solid/cap-day (0.020 kg/cap-d). If the raw wastewater has a BOD_5 of 0.17 lb/cap-day (0.077 kg/cap-d), determine the secondary sludge produced per day per pound of BOD_5 removed.

17.18 Repeat Problem 17.17 using SI units. Determine the secondary sludge produced per day per kilogram of BOD_5 removed.

17.19 An industrial wastewater having a flow of 0.8 MGD and a BOD$_5$ of 320 mg/ℓ is to be treated using a high-rate trickling filter with plastic packing. The recycle ratio is 2.0, and the filter depth is to be from 18 to 20 ft. A pilot-plant study has been made, and the values K and n in Eq. (17.8) have been found to be 0.073 and 0.55, respectively. The hydraulic loading is expressed as gpm/ft^2 and the depth is in ft. The effluent BOD$_5$ is to be 20 mg/ℓ. Determine, using USCS units:
a. The theoretical filter diameter if a depth of 19 ft is assumed.
b. The design filter diameter if rotary distributors are available in 5-ft increments in diameter.
c. The filter depth.

17.20 An industrial wastewater having a flow of 3.03 MLD and a BOD$_5$ of 320 mg/ℓ is to be treated using a high-rate trickling filter with plastic packing. The recycle ratio is 2.0 and the filter depth is 5.79 m. A pilot-plant study has been made, and the values K and n in Eq. (17.8) have been found to be 2.25 and 0.55, respectively. The hydraulic loading is expressed as m^3/d-m^2 and the depth is in meters. The effluent BOD$_5$ is to be 20 mg/ℓ. Determine, using SI units, the diameter of the filter.

17.21 A municipality with a design population of 4900 persons is having a rotary biological contactor plant designed. Pertinent data are average flow = 110 gal/cap-day (416 ℓ/cap-d), influent BOD$_5$ = 205 mg/ℓ, primary clarifier removes 33% of the influent BOD$_5$, effluent BOD$_5$ = 20 mg/ℓ, and number of stages = 4. A pilot-plant study gave a k value of 1.23 gal/day-ft^2 (50.2 ℓ/d-m^2). Determine, using the algebraic method and USCS units:
a. The design hydraulic loading, gal/day-ft^2.
b. The BOD$_5$ after each stage, mg/ℓ.

c. The disc area per stage, ft^2.
d. The number of shafts per stage and the total number of shafts for the plant if a shaft has 100,000 ft^2 of surface area.

17.22 Repeat Problem 17.21 using the graphical method and USCS units.

17.23 Repeat Problem 17.21 using the algebraic method and SI units. Part (a) should be in ℓ/d-m^2, part (c) should be in m^2, and for part (d) a shaft has 9289 m^2 per shaft.

17.24 Repeat Problem 17.21 using the graphical method and SI units. Part (a) should be in ℓ/d-m^2, part (c) should be in m^2, and for part (d) a shaft has 9289 m^2 per shaft.

17.25 In Example 17.5 there were four shafts required per stage, and each shaft had 100,000 ft^2 of surface area. Determine, using USCS units:
a. The hydraulic loading per stage.
b. The BOD$_5$ in the effluent from all four stages.

17.26 In Example 17.5 SI there were four shafts required per stage, and each shaft had 9289 m^2 of surface area. Determine, using SI units:
a. The hydraulic loading per stage.
b. The BOD$_5$ in the effluent from all four stages.

17.27 In Problem 17.21 there were three shafts per stage, and each shaft had 100,000 ft^2 of surface area. Determine, using USCS units:
a. The hydraulic loading per stage.
b. The BOD$_5$ in the effluent from all four stages.

17.28 In Problem 17.23, there were three shafts per stage, and each saft had 9289 m^2 of surface area. Determine, using SI units:
a. The hydraulic loading per stage.
b. The BOD$_5$ in the effluent from all four stages.

18

STABILIZATION PONDS AND AERATED LAGOONS

The stabilization or oxidation pond, shown in Figure 18.1, consists of a quiescent diked pond in which the wastewater enters and, as a result of microbial action, the organic materials are bio-oxidized, giving CO_2, NH_3, inorganic radicals such as SO_4^{-2} and PO_4^{-3}, and new microbial cells as end products. The algal population uses the CO_2, inorganic radicals, and sunlight to produce dissolved oxygen and new algal cells as end products. Thus the microbial and algal populations have a synergistic relationship in which both groups benefit from each other. Although most stabilization ponds have effluents discharging directly into the receiving body of water, the future trend in design is to use polishing treatments such as intermittent sand beds to remove algal growths from the effluent, thus giving a better degree of treatment. Stabilization ponds are used for both municipal and industrial wastewater treatment, particularly for small municipalities and seasonal industrial wastewaters.

The aerated lagoon, shown in Figures 18.2 and 18.3, consists of a diked pond with artificial aeration, usually by floating pump-type aerators, and a downline settling tank or facultative stabilization pond that serves as the final clarifier. The biological solids developed in the aerated lagoon are removed from the effluent by the settling tank or stabilization pond. If a settling tank is used, it is usually a poured-in-place concrete clarifier with mechanical sludge rakes for continuous sludge removal. If a stabilization pond is used, the biological solids settle to the pond bottom and undergo anaerobic decomposition. The pond is drained periodically, usually every two or three years, and the digested solids are removed and disposed of by sanitary landfill. The aerated lagoon process is essentially a nonrecycle activated sludge process that usually has biological solids at a concentration of about 200 to $500 \, \text{mg}/\ell$. Aerated lagoons are widely used in industrial wastewater treatment because they are less expensive than the activated sludge process. However, they have much greater land requirements, which must be considered. Frequently, an industrial wastewater treatment facility is initially a stabilization pond system, and as the waste load increases, artificial aeration is added to convert some of the ponds into aerated lagoons.

Stabilization ponds are widely used in the southern and southwestern United States because the climate is warm and there is a high sunlight intensity. Aerated lagoons are also used in these regions because the lagoon temperature is relatively high as a result of the climate and sunlight intensity.

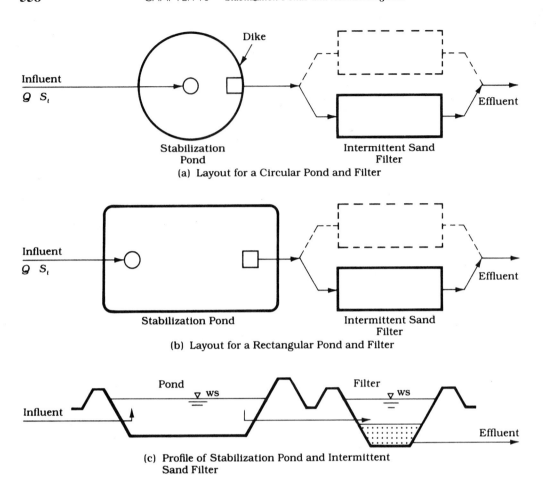

(a) Layout for a Circular Pond and Filter

(b) Layout for a Rectangular Pond and Filter

(c) Profile of Stabilization Pond and Intermittent Sand Filter

FIGURE 18.1 *Stabilization Pond System*

STABILIZATION PONDS

Stabilization ponds may be classified as aerobic, facultative, or anaerobic according to their oxygen profile. An aerobic pond has dissolved oxygen throughout its depth, whereas an anaerobic pond has no dissolved oxygen at any depth, and the facultative pond has dissolved oxygen in the upper zone of water and no dissolved oxygen in the lower zone. Most ponds are facultative since these are loaded higher than aerobic ponds, yet few odors are produced because the upper pond depth is aerobic.

The amount of oxygen present in a facultative pond depends on the organic loading and the sunlight intensity. There will be a diurnal variation in the dissolved oxygen concentration in the upper zone, and the variation will be greatest during the summer. During the night, the dissolved oxygen will be low because the microbial population and some of the algal population use dissolved oxygen. After the sun rises, photosynthesis starts to occur and the dissolved oxygen, which is an algal end product, begins

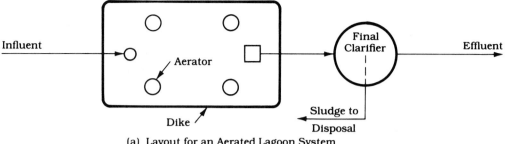

(a) Layout for an Aerated Lagoon System

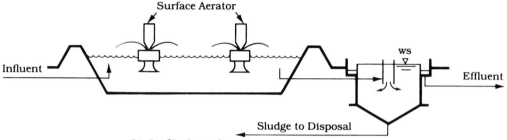

(b) Profile through an Aerated Lagoon System

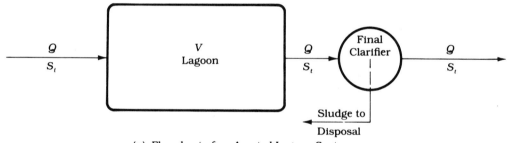

(c) Flowsheet of an Aerated Lagoon System

FIGURE 18.2 *Aerated Lagoon System*

to increase, with the maximum concentration occurring during the middle of the day. Shortly after sunset, photosynthesis ceases, dissolved oxygen production stops, and the dissolved oxygen concentration decreases because of the oxygen demand of the microbial and algal population. In the summer in the southern and southwestern United States, the dissolved oxygen may vary from about 2 to 3 mg/ℓ during the night to above 14 to 16 mg/ℓ during the day. Even for facultative ponds, anaerobic conditions may occur throughout the pond depth during overcast days. Odors, however, are usually not a problem unless windy conditions cause waves on the surface.

As the algal and microbial cells die, they lyse and the cell fragments settle to the bottom of the pond, where they undergo anaerobic decomposition. For the usual facultative pond loadings, the sludge accumulation is

FIGURE 18.3 *Large Aerated Lagoon. Each floating, pump-type aerator has two pilings to moor the aerator and prevent sinking due to leaky floats.*

very slow and may be only a fraction of an inch (several millimeters) per year.

Approximately one-third to one-half the influent organic carbon in the raw wastewater is synthesized into microbial and algal cells that leave with the pond effluent unless an effluent polishing treatment is used. Many times in the past, the BOD$_5$ reported for an effluent was for a filtered effluent sample, which was misleading because the algal and microbial growths exerted an oxygen demand in the receiving body of water. The BOD$_5$ removal based on filtered effluent samples will be from 80% to 90%, whereas the BOD$_5$ removal for unfiltered effluent samples will be from about 45% to 65%. Coliform removal on unfiltered samples is usually from 85% to 95%.

Facultative ponds may range from 3 to 8 feet (0.91 to 2.44 m) deep, with the usual water depth being from 3 to 6 feet (0.91 to 1.83 m). When future conversion to an aerobic lagoon system is anticipated, the pond should be deep to accommodate surface aerators. Ponds are usually circular or rectangular in plan view, as shown in Figures 18.1 (a) and (b), unless the terrain necessitates some other geometry. Usually, two or more ponds are provided, and the piping is arranged so that series or parallel operation is possible. For single ponds or ponds operated in parallel when treating municipal wastewaters, an organic loading of 25 to 35 lb BOD$_5$/ac-day (28.0 to 39.2 kg BOD$_5$/ha-day) is frequently used. For series operation when treating municipal wastewaters, the first pond is usually loaded as high as 75 to 80 lb BOD$_5$/ac-day (84.1 to 89.7 kg BOD$_5$/ha-day), and the downstream pond at a loading of 25 to 35 lb BOD$_5$/ac-day (28.0 to 39.2 kg BOD$_5$/ha-day).

Gloyna (1976) recommends the following equation for determining the volume of a facultative stabilization pond treating municipal wastewaters:

$$V = CQS_i[\theta^{(35-T)}]ff' \tag{18.1}$$

where

V = pond volume, ft^3 (m^3)

C = 4.7 × 10^{-3} for USCS units and 3.5 × 10^{-5} for SI units

Q = flow, gal/day (ℓ/day)

S_i = ultimate influent BOD or COD, mg/ℓ

f = algal toxicity factor; f = 1 for municipal wastewaters and many industrial wastewaters

f' = sulfide or other immediate chemical oxygen demand; f' = 1 for SO$_4^{-2}$ equivalent ion concentration of less than 500 mg/ℓ

θ = temperature coefficient

T = average water temperature for the pond during winter months, °C

The value of θ ranges from 1.036 to 1.085. The value of 1.085 is recommended because it is conservative and field data support this analysis (Gloyna, 1976). For the case where θ = 1.085 and an effective depth = 5 ft (1.52 m), the area of the pond may be computed using Eq. (18.1). The area is

$$A = CQS_i[1.085^{(35-T)}]ff' \qquad \textbf{(18.2)}$$

where

A = area of pond, acres (ha), for a depth of 5 ft + 1 ft of sludge storage (1.5 m + 0.3 m)

C = 2.148 × 10^{-2} for USCS units and 2.299 × 10^{-3} for SI units

Q = flow, MGD (MLD)

The BOD$_5$ removal efficiency can be expected to be 80% to 90% based on unfiltered influent samples and filtered effluent samples. The efficiency based on unfiltered effluent samples can be expected to vary unless a maturation pond is used as a followup unit.

The recommended minimum depth of a facultative pond is 3.28 ft (1 m). Additional depth to compensate for sludge storage is desirable. The minimum depth of about 3.28 ft (1 m) is required to control potential growth of emergent vegetation. If the depth is too great, there will be inadequate surface area to support photosynthetic action; also, deep ponds tend to stratify during hot periods. The suggested design guidelines for depth are shown in Table 18.1.

For wastewaters containing considerable amounts of biodegradable settleable solids, Gloyna (1976) recommends that an anaerobic pond precede the facultative stabilization pond. The high temperature coefficient, 1.085, indicates that pond performance is very sensitive to temperature changes. To determine the design pond loading for an industrial wastewater for which no field data exist, Gloyna (1968) and Eckenfelder and Ford (1970) have developed laboratory procedures using bench-scale stabilization ponds.

Ponds should be located so that the prevailing winds passing over

TABLE 18.1   *Design Guidelines for Depth of Stabilization Ponds*

CASE	DEPTH	RELATED CONDITIONS
1	3.28 ft (1 m)	Generally ideal condition, very uniform temperature, tropical to subtropical, minimum settleable solids
2	4.10 ft (1.25 m)	Same as Case 1 but with modest amounts of settleable solids. Surface design based on 3.28-ft depth and 0.82 ft used for reserve volume (1 m and 0.25 m)
3	4.92 ft (1.5 m)	Same as Case 2 except for significant seasonal variation in temperature and major fluctuations in daily flow. Surface design based on 3.28-ft (1-m) depth
4	6.56 ft and greater (2 m and greater)	For soluble wastewaters that are slowly biodegradable and retention is controlling

the ponds are not directed toward populated areas where possible odors would create a problem. The pond bottom and sides should be relatively impervious, and if the soil is porous, a clay liner should be used. The inlet to a pond should be designed so that the influent is distributed outward into the pond, and outlet structures should be baffled to prevent floating materials from leaving with the effluent. The earthen dikes usually have side slopes of 1:3 or less, and the inside slope is protected against wave erosion by riprap. The outside slopes should be sodded to prevent erosion due to surface runoff. Surface drainage must be excluded by dikes or ditching. At least 3 ft (0.91 m) of freeboard should be provided above the maximum water surface, and the ponds should be designed so that the water level can be varied by 6 in. (150 mm) to assist in mosquito control.

EXAMPLE 18.1 AND 18.1 SI

Stabilization Pond

A stabilization pond system is to be designed for a design population of 4800 persons, and the system is to have secondary treatment. Primary treatment consisting of a primary clarifier and an anaerobic digester is to be provided. Pertinent data are: average flow = 100 gal/cap-day (380 ℓ/cap-d), influent BOD_5 = 200 mg/ℓ, the primary clarifier removes 33% of the BOD_5, BOD_5 = 0.68 BOD_u, four ponds in parallel are to be used, length-to-width ratio = 2:1, and winter operating temperature of the pond = 45°F. The values of f and f' in Eq. (18.2) are 1. Determine, using USCS units and SI units:

1. The total area of the ponds.
2. The applied BOD_5 loading.
3. The dimensions of the ponds.

SOLUTION

For the solution in USCS units, the required area of the ponds is given by

$$A = 2.148 \times 10^{-2} Q S_i [1.085^{(35-T)}] ff'$$

$Q = (4800\,\text{cap})(100\,\text{gal/cap-day}) = 480{,}000\,\text{gal/day}$ or $0.48\,\text{MGD}$. The primary clarifier effluent has $200(1 - 0.33) = 134\,\text{mg}/\ell$ BOD$_5$. The BOD$_u$ = $134/0.68 = 197\,\text{mg}/\ell$. The temperature, T, is $(45 - 32)5/9 = 7.2°\text{C}$. Substituting into the performance equation gives

$$A = (0.02148)(0.48)(197)(1.085^{35-7.2})(1)(1) = \boxed{19.6\,\text{ac}}$$

The applied BOD$_5$ load is $(4800\,\text{cap})(100\,\text{gal/cap-day})(8.34\,\text{lb/gal})(134/10^6)$ $= 536\,\text{lb/day}$. The applied loading is $(536\,\text{lb/day})(1/19.6\,\text{ac}) = $ $\boxed{27.3\,\text{lb BOD}_5/\text{ac-day.}}$ The area of each pond is $(19.6\,\text{ac})(43{,}560\,\text{ft}^2/\text{ac})(1/4)$ $= 213{,}444\,\text{ft}^2$. Thus $(2W)(W) = 213{,}444$ or $W = \boxed{327\,\text{ft}}$ and $L = (2)(327) = $ $\boxed{654\,\text{ft.}}$

For the solution in SI units, the required area of the ponds is given by

$$A = 2.299 \times 10^{-3} Q S_i [1.085^{(35-T)}] ff'$$

$Q = (4800\,\text{cap})(380\,\ell/\text{cap-d}) = 1{,}824{,}000\,\ell/\text{d}$ or $1.824\,\text{MLD}$. The primary clarifier effluent has $200(1 - 0.33) = 134\,\text{mg}/\ell$ BOD$_5$. The BOD$_u$ = $134/0.68 = $ $197\,\text{mg}/\ell$. The temperature, T, is $(45 - 32)5/9 = 7.2°\text{C}$. Substituting into the performance equation gives

$$A = (0.002299)(1.824)(197)(1.085^{35-7.2})(1)(1) = \boxed{7.98\,\text{ha}}$$

The applied BOD$_5$ load is $(4800\,\text{cap})(380\,\ell/\text{cap-d})(134\,\text{mg}/\ell)(\text{gm}/1000\,\text{mg})$ $\times (\text{kg}/10000\,\text{gm}) = 244\,\text{kg/d}$. The applied loading is $(244\,\text{kg/d})(1/7.98\,\text{ha})$ $= \boxed{30.6\,\text{kg BOD}_5/\text{ha-d.}}$ The area of each pond is $(7.98\,\text{ha})(10{,}000\,\text{m}^2/\text{ha})$ $\times (1/4) = 19{,}950\,\text{m}^2$. Thus $(2W)(W) = 19{,}950\,\text{m}^2$ or $W = \boxed{99.9\,\text{m}}$ and $L = $ $(2)(99.9) = \boxed{200\,\text{m.}}$

AERATED LAGOONS

Aerated lagoons may be classified as aerobic or facultative lagoons according to their dissolved oxygen profile. In aerobic lagoons, the oxygen furnished is sufficient to maintain dissolved oxygen throughout the pond depth, and the mixing is sufficient to keep the biological solids in suspension. In the aerobic lagoon, final clarification is provided by a settling tank or a facultative stabilization pond. The facultative lagoon has dissolved oxygen in the upper depths of the lagoon; however, no dissolved oxygen is present in the lower depths. In the facultative lagoon, the mixing is insufficient to maintain all of the biological solids in suspension, and solids deposition occurs on the lagoon bottom. These solids undergo anaerobic decomposition and are removed at infrequent intervals, such as every few years. The facultative lagoon usually has no final clarification except that which occurs in the aerated lagoon itself; consequently, the solids content in the effluent often precludes their use. In the following discussion, the fundamentals and application of the aerobic aerated lagoon are presented.

Aerated lagoons are widely used in the treatment of biodegradable organic industrial wastewaters because they occupy less land area than stabilization ponds and have less construction and operating costs than activated sludge plants. However, they have appreciable land area requirements when compared to activated sludge plants. The cost of an aerated lagoon system is between that of a stabilization pond system and that of an activated sludge plant. The oxygen requirements for an aerobic lagoon are furnished by artificial aeration and, in particular, by mechanical surface aerators such as the floating pump-type. Occasionally, perforated pipes are laid on the lagoon bottom to disperse compressed air. Earthen dikes are used to form a lagoon and to exclude surface runoff.

The aerated lagoon is essentially an activated sludge system without recycle; thus biochemical kinetics may be used for design formulations. Because the biological solids in lagoons do not vary appreciably, the pseudo–first-order reaction representing the rate of removal is

$$-\frac{dS_t}{dt} = KS_t \tag{18.3}$$

where

dS_t/dt = rate of substrate utilization, mass/(volume) (time)

K = reaction rate constant

S_t = substrate concentration at any time, mass/volume

A material balance on the substrate is given by

$$[\text{Accumulation}] = [\text{Input}] - \begin{bmatrix} \text{Decrease} \\ \text{due to} \\ \text{reaction} \end{bmatrix} - [\text{Output}] \tag{18.4}$$

Assuming the system is completely mixed and using the designations shown in Figure 18.2 give

$$V(dS_i) = QS_i dt - V[dS_t]_{\text{Growth}} - QdS_t dt \tag{18.5}$$

From Eq. (18.3), $[dS_t]_{\text{Growth}} = KS_t dt$; thus substituting this expression in Eq. (18.5) gives

$$V(dS_t) = QS_i dt - VKS_t dt - QS_t dt \tag{18.6}$$

Dividing Eq. (18.6) by Vdt yields

$$\frac{dS_t}{dt} = \left(\frac{Q}{V}\right)S_i - KS_t - \left(\frac{Q}{V}\right)S_t \tag{18.7}$$

Since the accumulation term $(dS_t/dt) = 0$ for steady state and $(V/Q) = \theta_t$, Eq. (18.7) may be rearranged to give the following formulation (Eckenfelder, 1970; Eckenfelder and Ford, 1970):

$$\frac{S_t}{S_i} = \frac{1}{1 + K\theta_i} \tag{18.8}$$

which is the design equation for an aerated lagoon. Experimental studies can be performed using a completely mixed activated sludge unit without recycle to obtain the reaction rate constant, K. The unit should be operated at a detention time of several days until acclimation is attained and the substrate and biological solids concentrations in the effluent are measured. The flowrate is then increased, and once steady state occurs, the substrate and biological solids in the effluent are again measured. Four or five increases in the flowrate should be made to obtain sufficient data for plotting. To evaluate the data, Eq. (18.8) may be rearranged to give

$$K = \frac{S_i - S_t}{S_t \theta_i} \tag{18.9}$$

Thus, plotting $S_i - S_t$ on the y-axis versus $S_t \theta_i$ on the x-axis, as in Figure 18.4, will give a straight line with a slope of K. To evaluate the data to determine the yield coefficient, Y, and the endogenous decay constant, k_e, as shown in Figure 18.5, the following equation developed by Reynolds and Yang (1966) as well as other researchers (Metcalf & Eddy, Inc., 1979; Wu and Kao, 1976) may be used:

$$\frac{S_i - S_t}{\bar{X} \theta_i} = \frac{k_e}{Y} + \left(\frac{1}{Y}\right)\frac{1}{\theta_c} \tag{18.10}$$

where

$\bar{X}$ = average cell mass concentration during the reaction
θ_i = detention time based on the influent flow
Y = yield coefficient, mass of cells produced per mass of substrate used
k_e = endogenous decay constant, time^{-1}
θ_c = mean cell residence time, time

The mean cell residence time, θ_c, is equal to the detention time for a nonrecycle reactor. A plot of $(S_i - S_t)/(\bar{X}\theta_i)$ on the y-axis versus $1/\theta_c$ on the x-axis will yield a straight line with a slope of $1/Y$ and a y-axis intercept of k_e/Y, as shown in Figure 18.5.

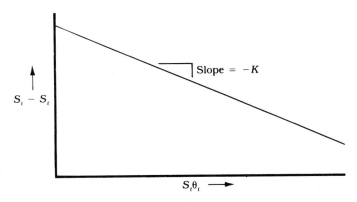

FIGURE 18.4
Graph for Determining **K**

Slope = $-K$

$S_i - S_t$

$S_t \theta_t$

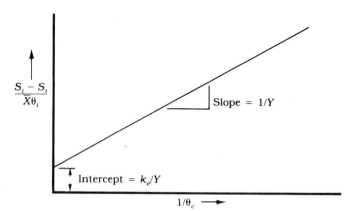

FIGURE 18.5
Graph for Determining Y and k_e

Since the biological solids concentration in an aerated lagoon is relatively low, the water temperature has a significant effect upon lagoon performance. The variation of the rate constant over a temperature range of 10°C to 30°C has been found to be represented by

$$K_{T_2} = K_{T_1} \cdot \theta^{(T_2 - T_1)} \tag{18.11}$$

where

K_{T_2} = rate constant at temperature T_2, °C

K_{T_1} = rate constant at temperature T_1, °C

θ = temperature correction coefficient, $\theta = 1.06$ to 1.10 (Eckenfelder and Ford, 1970)

The operating temperature of a pond depends on the temperature of the influent wastewater; the heat loss by convection, radiation, and evaporation; and the heat gain by solar radiation. Mancini and Barnhart (1968) have found the heat loss by evaporation and the heat gain by solar radiation to be small compared to other terms. Their findings show that the temperature of a lagoon is given by

$$T_i - T_w = \frac{fA(T_w - T_a)}{Q} \tag{18.12}$$

where

T_i = influent wastewater temperature, °F (°C)

T_w = temperature of water in the lagoon, °F (°C)

T_a = temperature of air, °F (°C)

A = lagoon area, ft² (m²)

Q = influent wastewater flow, MGD (m³/day)

f = experimental factor

For USCS units, they found the value of the factor f to be 12×10^{-6} for the central and eastern part of the United States and 20×10^{-6} for the Gulf states. For SI units, f is 0.489 for the central and eastern part of the United States and 0.815 for the Gulf states. Equation (18.12) may also be used to obtain the temperature of stabilization ponds.

The total oxygen requirements of a lagoon are given by the equation

$$O_r = Y'S_r + k'_e\bar{X} + O_n \tag{18.13}$$

where

O_r = total oxygen required, lb/day (kg/day)

Y' = oxygen coefficient, pounds of oxygen required per pound of BOD$_5$ or COD removed (kg/day)

S_r = substrate removed per day, pounds of BOD$_5$ or COD removed (kg/kg)

k'_e = endogenous oxygen coefficient, pounds of O_2 required per pound of cells decayed-day (kg/kg-day)

$\bar{X}$ = mass of biological solids in the lagoon, lb (kg)

O_n = oxygen required for nitrification, lb (kg)

The term k'_e may be assumed to equal $1.42k_e$, where k_e is the endogenous decay coefficient as time^{-1}. Frequently, the term $k'_e\bar{X}$ is very small compared to $Y'S_r$, and in such cases it may be neglected. The oxygen requirements should be at least 1.6 lb oxygen per lb BOD$_5$ (kg/kg) applied to the lagoon. The oxygen furnished by artificial aeration may be determined by equations presented in Chapter 16. The power level supplied by artificial aeration should be at least 0.15 hp/1000 ft^3 (3.95 kW/1000 m^3) to ensure maintaining the biological solids in suspension.

Since aerated lagoons are completely mixed activated sludge systems without recycle, they may also be designed by the appropriate equations presented in Chapter 15. The derivation of these kinetic equations utilized the Monod expression.

Aerated lagoons usually have a rectangular layout except where the terrain precludes the use of this geometry. When they are rectangular in layout, the length-to-width ratio is usually 2:1. At least two lagoons should be provided for flexibility, and piping should be arranged so that parallel or series operation is possible. Surface drainage must be excluded by dikes and ditching. The earthen dikes forming the lagoon usually have side slopes of 1:3 or less, and riprap or other suitable protection should be used to avoid wave erosion. The outside slopes should be sodded to avoid erosion due to rainfall runoff. The lagoon bottom and sides should be provided with a clay liner as a sealer if the soil is pervious. Inlets must distribute the influent outward into the lagoon, and outlets should be baffled to avoid floating matter from leaving the effluent. A freeboard of at least 3 ft (0.91 m) should be provided. Concrete pads must be furnished under each aerator to protect the bottom from erosion. Final clarifiers or stabilization ponds should be provided to remove biological solids from the final effluent. Lagoon depths are from 8 to 16 ft (2.43 to 4.88 m).

EXAMPLE 18.2
AND 18.2 SI

Aerated Lagoon

An aerated lagoon wastewater treatment plant is to be designed for a municipal wastewater. The design population is 5200 persons, and the aerated lagoon is to provide secondary treatment. Pertinent data are: flow = 100 gal/cap-day (380 ℓ/cap-d), influent BOD$_5$ = 200 mg/ℓ, effluent BOD$_5$ = 20 mg/ℓ, K = 2.10 day^{-1}, oxygen required = 1.6 lb oxygen per lb BOD$_5$ (kg/kg) applied to the lagoons, and the primary clarifier removes 33% of the influent BOD$_5$. Determine:

1. The required reaction time.
2. The total lagoon volume.
3. The lagoon dimensions if two lagoons in parallel are used, the length-to-width ratio is 2:1, and the depth is 10 ft (3.0 m).
4. The oxygen requirements and power requirements if the mechanical aerators furnish 1.56 lb O$_2$/hp-hr (0.949 kg/kW-h).
5. The mixing power intensity. Is it adequate?

SOLUTION

The reaction time is given by

$$K = \frac{S_i - S_t}{S_t \theta_i}$$

which can be rearranged to give

$$\theta_i = \frac{S_i - S_t}{K S_t}$$

The BOD$_5$ applied to the lagoons is $200(1 - 0.33) = 134$ mg/ℓ. Substituting the appropriate values into the performance equation gives

$$\theta_i = \frac{134 - 20}{(2.10 \text{ day}^{-1})(20)} = \boxed{2.71 \text{ days}}$$

For the solution in USCS units,

$$V = Q\theta_i$$

The flow Q = (5200 cap)(100 gal/cap-day) = 520,000 gal/day. Thus V = (520,000 gal/day)(2.71 days)(ft^3/7.48 gal) = $\boxed{188,000 \text{ ft}^3.}$ The volume per lagoon is 188,000/2 = 94,000 ft^3. Each lagoon area = 94,000 ft^3/10 ft = 9400 ft^2. Since $(2W)(W)$ = 9400 ft^2, W = $\boxed{68.6 \text{ ft,}}$ and L = (2)(68.6) = $\boxed{137 \text{ ft.}}$ The BOD$_5$ applied per day = (5200 cap)(100 gal/cap-day)(8.34 lb/gal) $(134/10^6)$ = 581 lb/day. The oxygen required = (581 lb/day)(1.6 lb O$_2$/lb) = $\boxed{930 \text{ lb/day.}}$ The power required = (930 lb/day)(day/24 hr)(hp-hr/1.56 lb) = $\boxed{24.8 \text{ hp.}}$ The power intensity = (24.8 hp)(1/188 thousand ft^3) = $\boxed{0.13 \text{ hp/1000 ft}^3.}$ Since this is less than 0.15 hp/1000 ft^3, there is not adequate mixing and more power must be added. The required power = $(0.15\,\text{hp}/1000\,\text{ft}^3)(188{,}000\,\text{ft}^3) = \boxed{28.2\,\text{hp.}}$

For the solution in SI units,

$$V = Q\theta_i$$

The flow $Q = (5200\,\text{cap})(380\,\ell/\text{cap-d})(\text{m}^3/1000\,\ell) = 1976\,\text{m}^3/\text{d}$. Thus $V = (1976\,\text{m}^3/\text{d})(2.71\,\text{d}) = \boxed{5355\,\text{m}^3.}$ The volume per lagoon is $5355/2 = 2678\,\text{m}^3$. Each lagoon area = $(2678\,\text{m}^3/3.0\,\text{m}) = 893\,\text{m}^2$. Since $(2W)(W) = 893\,\text{m}^2$, $W = \boxed{21.1\,\text{m}}$ and $L = (2)(21.1) = \boxed{42.2\,\text{m.}}$ The BOD$_5$ applied per day = $(5200\,\text{cap})(380\,\ell/\text{cap-d})(134\,\text{mg}/\ell)(\text{gm}/1000\,\text{mg})(\text{kg}/1000\,\text{gm}) = 265\,\text{kg/day}$. The oxygen required is $(265\,\text{kg/day})(1.6\,\text{kg O}_2/\text{kg}) = \boxed{424\,\text{kg/d.}}$ The power required is $(424\,\text{kg/d})(\text{d}/24\,\text{h})(\text{kW-hr}/0.949\,\text{kg}) = \boxed{18.6\,\text{kW.}}$ The power intensity = $(18.6\,\text{kW})(1/5.355\,\text{thousand}\,\text{m}^3) = \boxed{3.48\,\text{kW}/1000\,\text{m}^3.}$ Since this is less than $3.95\,\text{kW}/1000\,\text{m}^3$, the mixing is not adequate and more power must be added. The required power = $(3.95\,\text{kW}/1000\,\text{m}^3)(5355\,\text{m}^3) = \boxed{21.2\,\text{kW.}}$

REFERENCES

American Society of Civil Engineers (ASCE). 1982. *Wastewater Treatment Plant Design*. ASCE Manual of Practice No. 36. New York: ASCE.

AWARE, Inc. 1974. *Process Design Techniques for Industrial Waste Treatment*, edited by C. E. Adams, Jr. and W. W. Eckenfelder, Jr. Nashville, Tenn.: Enviro Press.

Conway, R. A., and Ross, R. D. 1980. *Handbook of Industrial Waste Disposal*. New York: Van Nostrand Reinhold.

Eckenfelder, W. W., Jr. 1989. *Industrial Water Pollution Control*. 2nd ed. New York: McGraw-Hill.

Eckenfelder, W. W., Jr. 1980. *Principles of Water Quality Management*. Boston: CBI Publishing.

Eckenfelder, W. W., Jr. 1970. *Water Quality Engineering for Practicing Engineers*. New York: Barnes & Noble.

Eckenfelder, W. W., Jr., and Ford, D. L. 1970. *Water Pollution Control*. Austin, Tex.: Pemberton Press.

Gloyna, E. F. 1968. Basis for Waste Stabilization Pond Designs. In *Advances in Water Quality Improvement*, edited by E. F. Gloyna and W. W. Eckenfelder, Jr. Austin, Tex.: University of Texas Press.

Gloyna, E. F. 1976. Facultative Waste Stabilization Pond Design. In *Ponds as a Wastewater Treatment Alternative*, edited by E. F. Gloyna, J. F. Malina, Jr., and E. M. Davis. Austin, Tex.: University of Texas Press.

Gloyna, E. F. 1965. Waste Stabilization Pond Concepts and Experiences. Lecture presented at the Tenth Summer Institute in Water Pollution Control, Manhattan College, Bronx, N.Y.

Gloyna, E. F., and Hermann, E. R. 1958. Waste Stabilization Ponds. Reprint no. 74. Bureau of Engineering Research, University of Texas, Austin, Tex.

Great Lakes–Upper Mississippi River Board of State Public Health and Environmental Managers. 1990. Recommended Standards for Wastewater Facilities. Ten state standards. Albany, N.Y.

Great Lakes–Upper Mississippi River Board of State Sanitary Engineers. 1978. Recommended Standards for Sewage Works. Ten state standards. Albany, N.Y.

Horan, N. J. 1990. *Biological Wastewater Treatment Systems: Theory and Operation*. Chichester, West Sussex, England: Wiley.

Mancini, J. L., and Barnhart, E. L. 1968. Industrial Waste Treatment in Aerated Lagoons. In *Advances in Water Quality Improvement*, edited by E. F. Gloyna and W. W. Eckenfelder, Jr. Austin, Tex.: University of Texas Press.

Metcalf & Eddy, Inc. 1972. *Wastewater Engineering: Collection, Treatment and Disposal*. New York: McGraw-Hill.

Metcalf & Eddy, Inc. 1991. *Wastewater Engineering: Treatment, Disposal and Reuse*. 3rd ed. New York: McGraw-Hill.

Metcalf & Eddy, Inc. 1979. *Wastewater Engineering, Treatment, Disposal and Reuse*. 2nd ed. New York: McGraw-Hill.

Novotny, V.; Imhoff, K. R.; Olthof, M.; and Krenkel, P. A. 1989. *Karl Imhoff's Handbook of Urban Drainage and Wastewater Disposal*. New York: Wiley.

Qasim, S. R. 1985. *Wastewater Treatment Plants*. New York: Holt, Rinehart and Winston.

Ramalho, R. S. 1977. *Introduction to Wastewater Treatment Processes*. New York: Academic Press.

Reynolds, T. D., and Yang, J. T. 1966. Model of the Completely Mixed Activated Sludge Process.

Proceedings of the 21st Annual Purdue Industrial Waste Conference. Part 2.

Sawyer, C. N. 1968. New Concepts in Aerated Lagoon Design and Operation. In *Advances in Water Quality Improvement*, edited by E. F. Gloyna and W. W. Eckenfelder, Jr. Austin, Tex.: University of Texas Press.

Schroeder, E. D. 1977. *Water and Wastewater Treatment*. New York: McGraw-Hill.

Sundstrom, D. W., and Klei, H. E. 1979. *Wastewater Treatment*. Englewood Cliffs, N.J.: Prentice-Hall.

Tebbutt, T. H. Y. 1992. *Principles of Water Quality Control*. 4th ed. Oxford, England: Pergamon Press Limited.

Texas Department of Health Resources. 1976. *Design Criteria for Sewerage Systems*. Austin, Tex.

Texas Water Resources Commission. 1993. *Design Criteria for Wastewater Treatment Facilities*. Austin, Tex.

Water Environment Federation (WEF). 1992. *Design of Municipal Wastewater Treatment Plants*, Vol. I and II, WEF Manual of Practice No. 8.

Wu, Y. C., and Kao, D. F. 1976. Yeast Plant Wastewater Treatment. *Jour. SED* 102, no. EE5:969.

PROBLEMS

18.1 A stabilization pond is to be designed for a population of 2300 persons, and the pond is to provide secondary treatment. The plant will consist of preliminary, primary, and secondary treatment. The primary sludge will be treated by an anaerobic digester. Pertinent data are: influent flow = 105 gal/cap-day (397 ℓ/cap-d), influent BOD$_5$ = 195 mg/ℓ, primary clarifier removes 33% of the BOD$_5$, BOD$_5$ = 0.68 BOD$_u$, two ponds in series are to be used, length-to-width ratio = 2:1, and pond winter operating temperature = 55°F (12.8°C). The values of f and f' are 1. Determine, using USCS units:

a. The total pond area.

b. The applied BOD$_5$ loading.

c. The dimensions of the pond.

18.2 Repeat Problem 18.1 using SI units.

18.3 An aerated lagoon wastewater treatment plant is to be designed for a municipal wastewater. The design population is 4600 persons, and the aerated lagoon is to provide secondary treatment. Pertinent data are: flow = 95 gal/cap-day (360 ℓ/cap-d), influent BOD$_5$ = 205 mg/ℓ, effluent BOD$_5$ = 20 mg/ℓ,

K = 2.18 day^{-1}, and the oxygen required is 1.6 lb oxygen per lb BOD$_5$ applied to the lagoons (kg/kg). Determine, using USCS units:

a. The required reaction time.

b. The total lagoon volume.

c. The lagoon dimensions if two lagoons in parallel are used, the length-to-width ratio is 2:1, and the depth is 10 ft.

d. The oxygen required per day and the power requirements if the mechanical aerators furnish 1.62 lb O$_2$/hp-hr.

e. The mixing power intensity. Is it adequate? If not, what power should be used?

18.4 Repeat Problem 18.3 using SI units. The depth is 3.0 m and the mechanical aerators furnish 0.986 kg O$_2$/kW-h.

18.5 A aerated lagoon wastewater treatment plant is to be designed for a municipal wastewater. The design population is 6600 persons, and the aerated lagoon is to provide secondary treatment. Pertinent data are: flow = 107 gal/cap-day (405 ℓ/cap-day), influent BOD$_5$ = 187 mg/ℓ, effluent BOD$_5$ = 20 mg/ℓ, primary clarifier removes 33% of the influent BOD_5, $K = 2.05\,day^{-1}$ at 23°C, temperature correction coefficient $\theta = 1.080$, winter operating temperature = 18°C, and the oxygen required is 1.6 lb oxygen per lb BOD_5 applied to the lagoons (kg/kg). Determine, using USCS units:

a. The required reaction time.

b. The required lagoon volume.

c. The lagoon dimensions if two lagoons in parallel are used, the length-to-width ratio is 2:1, and the lagoon depth is 10 ft.

d. The oxygen requirements and power requirements if the mechanical aerators furnish 1.65 lb O_2/hp-hr.

e. The mixing power intensity. Is it adequate? If not, what power should be used?

18.6 Repeat Problem 18.5 using SI units. The depth is 3.0 m and the mechanical aerators furnish 1.04 kg O_2/kW-h.

18.7 A facultative oxidation pond facility is to be designed for a population of 20,000 persons and is to have primary treatment prior to the ponds. The facility is to have two ponds in parallel, and the city is located in the Gulf Coast area. Pertinent data are: BOD_5 of raw wastewater = 200 mg/ℓ, 33% BOD_5 removal by primary treatment, winter wastewater temperature = 69°F (20.6°C), average January air temperature = 55°F (12.8°C), wastewater flow = 100 gal/cap-day (380 ℓ/cap-d), pond depth = 3.5 ft (1.07 m), length-to-width ratio = 2:1, and BOD_5 = 0.68 BOD_u. Determine, using USCS units:

a. The operating temperature of the ponds.

b. The total pond area.

c. The detention time, days.

d. The surface loading.

e. The dimensions of the ponds.

18.8 Repeat Problem 18.7 using SI units.

18.9 An aerobic aerated lagoon system is to be designed to treat a soluble organic industrial wastewater having a flow of 1.5 MGD (5.68 MLD). The COD of the wastewater is 1020 mg/ℓ, and the plant effluent is to have a COD not more than 40 mg/ℓ. A bench-scale laboratory study has been performed using a completely mixed reactor without recycle, and the data obtained at 23°C are shown in Table 18.2. The COD after ten days of aeration was 14 mg/ℓ and may be assumed to be the nondegradable COD. Other pertinent data are: wastewater temperature = 82°F (27.8°C), January average air temperature = 55°F (12.8°C), July average air temperature = 83°F (28.3°C), plant location is in the Gulf states, aerated lagoon depth = 8.0 ft (2.44 m), number of lagoons = 2, the lagoons are rectangular in layout, length : width = 2:1, final clarification is to be furnished by settling tanks, organic nitrogen in wastewater = 32.2 mg/ℓ, ammonia nitrogen in wastewater = 2.9 mg/ℓ, oxygen furnished by surface aerators = 1.5 lb O_2/hp-hr (0.912 kg/kW-h), BOD_5 of the untreated wastewater = 42% of the COD and 4.33 lb O_2 required per pound of nitrogen nitrified (kg/kg), and MLVSS = 88% of MLSS. Using USCS units, determine:

a. The reaction constant, K, based on biodegradable COD.

b. The yield coefficient, Y, and the endogenous decay constant, k_e.

c. The oxygen coefficient, Y', and the endogenous oxygen coefficient, k_e'.

d. The detention time in days for winter conditions.

e. The length and width of each lagoon.

f. The soluble COD in the effluent for summer conditions.

TABLE 18.2

DETENTION TIME (hr) θ	INFLUENT COD (mg/ℓ) S_i	EFFLUENT COD (mg/ℓ) S_t	BIOLOGICAL SOLIDS (mg/ℓ) X_t
11.20	1003	47	409
9.00	1003	58	401
6.00	1003	80	388
3.75	1003	162	358
3.40	1003	220	359
2.80	1003	303	324

g. The percent nitrification based on the mean cell residence time using Figure 15.9.

h. The mass of oxygen required per day for each lagoon based on summer conditions.

i. The power required for mechanical surface aerators for each lagoon.

j. The power intensity.

k. The mass of oxygen required per mass of BOD$_5$ applied.

18.10   Repeat Problem 18.9 using SI units.

18.11   A facultative oxidation pond facility is to be designed for a population of 10,000 persons and is to have primary treatment prior to the pond. The facility will have two ponds operated in parallel after primary treatment. Pertinent data are: BOD$_5$ of the raw wastewater = 200 mg/ℓ, 33% BOD$_5$ removal by the primary clarifier, wastewater flow = 100 gal/cap-day (380 ℓ/cap-d), pond depth = 3.5 ft (1.07 m), surface loading = 35 lb BOD$_5$/ac-day (39.2 kg/ha-day), and length:width = 2:1. Determine, using USCS units:

a. The total pond area.

b. The dimensions of the ponds.

c. The detention time of the ponds, days.

18.12   Repeat Problem 18.11 using SI units.

18.13   An aerated lagoon is to be designed for a pulp and paper mill wastewater. After coagulation, flocculation, and settling, the waste has a BOD$_5$ of 160 mg/ℓ. Pertinent data are: flow = 10 MGD (37.8 MLD), BOD$_5$ after treatment = 10 mg/ℓ, K = 2.4 day^{-1} at 23°C, minimum operating temperature = 55°F (12.8°C), θ = 1.080, and pond depth = 12.0 ft (3.65 m). Determine, using USCS units:

a. The detention time, days.

b. The area of the ponds.

18.14   Repeat Problem 18.13 using SI units.

18.15   A municipality of 4500 persons has a wastewater flow of 95 gal/cap-day (360 ℓ/cap-d) and the raw wastewater has a COD of 380 mg/ℓ. A wastewater stabilization pond is to treat the wastewater after preliminary treatment. The average January temperature is 47.8°F (8.8°C). Determine, using USCS units:

a. The pond area.

b. The pond depth.

18.16   Repeat Problem 18.15 using SI units.

19

ANAEROBIC DIGESTION

Anaerobic digestion, which is the most common sludge treatment, may be defined as the biological oxidation of degradable organic sludges by microbes under anaerobic conditions. Most of the microbes used in the process are facultative anaerobes; however, some, such as the methane-producing bacteria, are obligate anaerobes. The term **sludge** refers to solids that settle and are removed when a liquid with suspended solids is passed through a settling tank. Organic sludges may originate from several sources in a wastewater treatment plant, and each sludge has its own characteristics, such as solids concentration and ease of biodegradation. The various sludges from wastewater treatment plants are as follows: (1) **Raw** or **primary sludge** is sludge from the primary settling of untreated wastewater. (2) **Waste activated sludge** is excess sludge that is produced from the activated sludge process. (3) **Trickling filter secondary sludge** or **humus** is sludge that originates from the secondary settling of trickling filter effluent. (4) **Secondary sludge** is sludge from the secondary clarifier of an activated sludge or trickling filter plant. (5) **Fresh sludge** refers to untreated organic sludges. (6) **Digested sludge** is sludge that has undergone biological oxidation. (7) **Dewatered sludge** is sludge that has had the major portion of its water removed.

All organic sludges undergo considerable changes in their physical, chemical, and biological properties during anaerobic digestion. Before digestion the volatile solids will be quite high, depending on the type of sludge. Primary sludge and waste activated sludge usually have 65% to 75% volatile solids, whereas trickling filter humus has 45% to 70%. The dry solids are usually 4% to 6% for a mixture of primary and secondary sludge, and the specific gravity is about 1.01. It is difficult to separate the water from the solids, and the solids will readily decompose under anaerobic conditions. The color is usually tan. The fuel value of undigested sludges is from about 6500 to 8000 Btu/lb (15,100 to 18,600 kJ/kg) dry solid.

After digestion the volatile solids are usually reduced to 32% to 48%. The dry solids are usually from 8% to 13%, and the specific gravity is from 1.03 to 1.05. The water will readily separate from the solids, and the solids are stable and not degradable. The color will be blackish and the odor will be tarlike. The fuel value of digested sludges is from about 3500 to 4000 Btu/lb (8100 to 9300 kJ/kg) dry solid. Approximately 99.8% of the coliforms in municipal primary sludges are destroyed during digestion.

Chemically precipitated sludges, such as from an advanced wastewater treatment plant (chemical-physical treatment), usually will digest anaerobically as well as primary, waste activated sludge, and tricking filter sludges if the pH range is satisfactory.

Frequently, sludge is thickened prior to digestion because thickening reduces the daily fresh sludge volume, thus decreasing the digester size and the amount of supernatant liquor. For instance, if a fresh sludge with 4% solids content is thickened to 8%, then the thickened sludge will be half the original volume.

Anaerobic digesters are usually constructed using cylindrical reinforced concrete tanks from 12 to 45 ft (3.66 to 13.7 m) deep and from 15 to 125 ft (4.57 to 38.1 m) in diameter. The most common depth is from 20 to 35 ft (6.10 to 10.7 m). The reinforced concrete floor is shaped like an inverted cone and slopes to the center of the tank where the discharge pipe for digested sludge is located. Figure 19.1 shows a two-stage anaerobic digester system employing a fixed-cover digester for the first stage and a floating-cover digester for the second stage.

Sludge treatment and disposal is an important aspect of wastewater

FIGURE 19.1 *Two-Stage Anaerobic Digester System. The small building between the digesters houses sludge heaters, recirculation pumps, piping, valves, and other appurtenances. Mixing of the first stage is provided by gas recirculation.*
Courtesy of Walker Process Equipment, Division of McNish Corporation.

treatment since it accounts for 40% to 45% of the capital and operating costs of a wastewater treatment plant. Frequently, its importance is underestimated in planning and design.

PROCESS FUNDA- MENTALS

Anaerobic digestion employs microbes that thrive in an environment in which there is no molecular oxygen and there is a substantial amount of organic matter. The organic material is a food source for the microbes, and they convert it into oxidized materials, new cells, energy for their life processes, and some gaseous end products, such as methane and carbon dioxide. The generalized equation for anaerobic action is

$$\underset{\text{matter}}{\text{Organic}} + \underset{\text{oxygen}}{\text{Combined}} \xrightarrow[\text{microbes}]{\text{Anaerobic}} \underset{\text{cells}}{\text{New}} + \underset{\text{for cells}}{\text{Energy}} + CH_4 + CO_2 + \underset{\text{products}}{\text{Other end}}$$

(19.1)

The sources of combined oxygen include the radicals of CO_3^{-2}, SO_4^{-2}, NO_3^{-1}, and PO_4^{-3}. Some of the other end products are the gases H_2S, H_2, and N_2.

The microbial action during anaerobic digestion consists of three stages: (1) liquefaction of solids, (2) digestion of the soluble solids, and (3) gas production. Digestion is accomplished by two groups of microorganisms: (1) the organic-acid–forming heterotrophs and (2) the methane-producing heterotrophs. The organic-acid–forming heterotrophs use the complex organic substrates, such as carbohydrates, proteins, fats, oils, and their degradation products and produce organic fatty acids, primarily acetic and propionic with some butyric and valeric acids. Carbohydrates include sugars, starches, and cellulose or plant fiber. Proteins occur in all raw animal and plant food materials — in particular, animal tissue has a high protein content. Fats and oils are similar in structure; however, fats are solids at ordinary temperatures, whereas oils are liquids. Fats and oils include animal and vegetable oils and grease from animal tissue. The breakdown of carbohydrates is

$$\text{Carbohydrates} \rightarrow \underset{\text{sugars}}{\text{Simple}} \rightarrow \underset{\text{Aldehydes}}{\text{Alcohols}} \rightarrow \underset{\text{acids}}{\text{Organic}} \qquad \textbf{(19.2)}$$

The breakdown of proteins is

$$\text{Proteins} \rightarrow \text{Amino acids} \rightarrow \text{Organic acids} + NH_3 \qquad \textbf{(19.3)}$$

The breakdown of animal or vegetable fats and oils is

$$\text{Fats and Oils} \rightarrow \text{Organic acids} \qquad \textbf{(19.4)}$$

In Eqs. (19.2), (19.3), and (19.4), the final breakdown products of carbohydrates, proteins, and fats and oils are organic fatty acids frequently called **volatile acids**. Most of the organic-acid–forming bacteria are soil microorganisms and are facultative anaerobes. McKinney (1962) lists the following genera as being common in anaerobic digestion of municipal sludges: *Pseudomonas*, *Flavobacterium*, *Alcaligenes*, *Escherichia*, and *Aerobacter*. These microbes thrive in a relatively wide pH range. The methane-producing heterotrophs use the organic acids produced by the acid formers as substrates and produce methane and carbon dioxide. The methane producers grow more slowly than the acid formers and require a rather narrow pH range of about 6.7 to 7.4. The action of the methane bacteria is

$$\text{Organic acids} \rightarrow CH_4 + CO_2 \qquad (19.5)$$

The gas produced in a properly operating digester is about 55% to 75% methane, is about 25% to 45% carbon dioxide, and has trace amounts of such gases as hydrogen sulfide, hydrogen, and nitrogen. Some of the carbon dioxide from the microbial action reacts with the water present to establish the bicarbonate buffering system, which is important in the digestion process. In a properly operating digester, the organic acids are utilized as rapidly as they are produced. The bicarbonate buffering system gives flexibility in

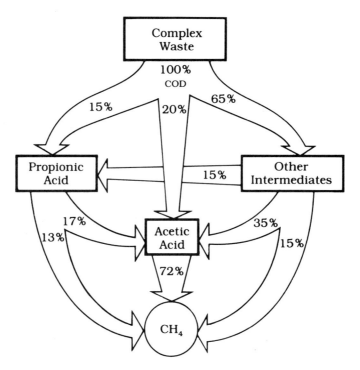

FIGURE 19.2 *Pathways for Methane Production from Complex Wastes such as Municipal Wastewater Sludges (Percentages are COD conversion.)*

Adapted from "Anaerobic Waste Treatment Fundamentals, Part 1" by P. L. McCarty in *Public Works Magazine* 95, no. 9 (September 1964):107. Reprinted by permission.

operation because, up to a certain limit, the acids may be temporarily produced faster than they are broken down, and the buffering system will maintain the pH in the proper range. If, however, excessive amounts of the acids are produced, the buffering system will be overcome and an unwanted pH drop will result that will inhibit the methane producers. The methane producers are strict anaerobes; McKinney (1962) lists the principal genera as *Methanococcus*, *Methanobacterium*, and *Methanoscarcina*. Figure 19.2 shows the pathway for methane production from organic solids found in municipal wastewaters. The figure shows that about 72% of the methane is produced from acetic acid. Figure 19.3 depicts the relationship between the bicarbonate alkalinity, the pH, and the amount of carbon dioxide in the digester gas, and also shows the range for normal operation.

Schematic diagrams of a **conventional** or **low-rate digester** and a **high-rate digester** are shown in Figure 19.4 (a) and (b). Conventional or low-rate digesters have intermittent mixing, intermittent sludge feeding, and intermittent sludge withdrawal. When mixing is not being done, the digester contents are stratified, as shown in Figure 19.4 (a). High-rate digesters have continuous mixing, as shown in Figure 19.4 (b), and continuous or intermittent sludge feeding and sludge withdrawal. If a high-rate digester is a single-stage digester, usually mixing is stopped and the contents are allowed to stratify before digested sludge and supernatant are withdrawn. Typical design

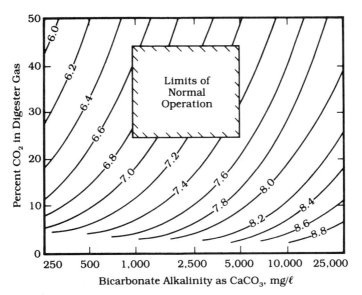

FIGURE 19.3 *Relationship between pH, Bicarbonate Concentration, and CO_2 Content near 95°F*

Adapted from "Anaerobic Waste Treatment Fundamentals, Part 2" by P. L. McCarty in *Public Works Magazine* 95, no. 10 (October 1964):123. Reprinted by permission.

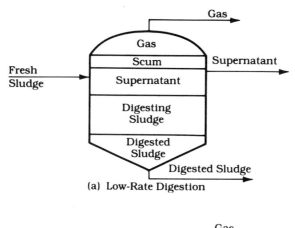

(a) Low-Rate Digestion

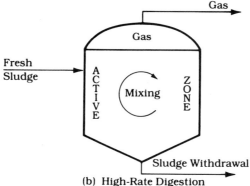

(b) High-Rate Digestion

FIGURE 19.4
*Low-Rate and
High-Rate Anaerobic
Digesters*

and operational parameters for low- and high-rate digesters are shown in Table 19.1. Most digesters, both low- and high-rate, operate in the mesophilic temperature range. Table 19.2 shows parameters for mesophilic anaerobic digestion. Low-rate digesters are semicontinuous-flow biological reactors without recycle. High-rate digesters, however, are continuous-flow biological reactors without recycle.

**LOW-RATE
DIGESTERS**

Low-rate or conventional digesters have digestion times of 30 to 60 days, organic solid loadings of 0.04 to 0.10 lb vss/ft^3-day (0.64 to 1.60 kg/m^3-day), intermittent mixing, and intermittent feeding and sludge withdrawal. They are in a stratified state except when mixing is being done.

The digestion time required to digest 90% of the degradable solids in primary sludge as a function of the digestion temperature is shown in Figure 19.5. The mesophilic range extends up to about 108°F (42.2°C), whereas the thermophilic range is above 108°F. The digestion time for mesophilic digestion decreases with an increase in temperature up to the optimum temperature of 98°F (35°C). This increase in digestion rate with an increase

TABLE 19.1 *Typical Design and Operational Parameters for Standard-Rate and High-Rate Anaerobic Digesters*

PARAMETER	LOW RATE	HIGH RATE
Digestion Time, days	30–60	10–20
Organic Solids Loading, lb vss/ft³-day (kg vss/m³-day)	0.04–0.10 (0.64–1.60)	0.15–0.40 (2.40–6.40)
Volume Criteria, ft³/capita (m³/capita)		
a. Primary Sludge	2–3 (0.0566–0.085)	$1\frac{1}{3}$–2 (0.0378–0.0566)
b. Primary Sludge and Trickling Filter Sludge	4–5 (0.113–0.142)	$2\frac{2}{3}$–$3\frac{1}{3}$ (0.0755–0.0944)
c. Primary Sludge and Waste Activated Sludge	4–6 (0.113–0.170)	$2\frac{2}{3}$–4 (0.0755–0.113)
Mixture of Primary and Secondary Sludge Feed Concentration, Percent Solids (dry basis)	2–5	4–6
Digester Underflow Concentration, Percent Solids (dry basis)	4–8	4–6

Adapted from Environmental Protection Agency, *Sludge Treatment and Disposal*, EPA Process Design Manual, 1979 Washington, D.C.

in temperature should be expected because it is a microbial process. Above 98°F the mesophilic digestion time increases. In cold weather, digesters are heated to near the optimum temperature; the heated range is usually from 85°F to 100°F (29.4°C to 37.8°C). For the thermophilic range, the optimum temperature is 130°F (54.4°C). Attempts have been made to use thermophilic digestion; however, it has not been very successful because the thermophiles are very sensitive to environmental changes.

Conventional or low-rate digesters are single-stage and may have either floating covers or fixed covers. Standard diameters for floating covers are from 15 to 125 ft (4.57 to 38.1 m) and are available in 5-ft (1.52-m) intervals. Figure 19.6 shows a section through a digester with a floating cover, along with details showing the piping required. For a floating-cover digester, when sludge is being added, recycle is done to seed the incoming fresh sludge with digesting sludge, as shown in Figure 19.7. The fresh and recycled sludge are usually added in the middle of the tank and in the gas dome of the cover. The discharge in the gas dome breaks up any scum or grease that has floated to the gas-water interface. While sludge is added, no supernatant or digested sludge is withdrawn, and fresh sludge in usually added daily. After the addition of fresh sludge into the tank, the recycle is stopped and the mixture

TABLE 19.2 *Conditions for Mesophilic Anaerobic Digestion*

Temperature	
Optimum	98° F (35°C)
Usual range of operation	85°–98° F (29.4°–35.0°C)
pH	
Optimum	7.0–7.2
Usual limits	6.7–7.4
Solids Reduction	
Volatile solids	50%–75%
Suspended solids	35%–50%
Gas Production	
Per pound (kg) of volatile solids added	6–12 ft^3 (0.37–0.75 m^3)
Per pound (kg) of volatile solids destroyed	12–18 ft^3 (0.75–1.12 m^3)
Per pound (kg) of volatile solids destroyed	1.05–1.75 lb (kg)
Per capita served (primary sludge only)	0.6–0.8 ft^3/day (0.017–0.023 m^3/day)
Per capita served (primary plus secondary sludge)	1.0–1.2 ft^3/day (0.028–0.034 m^3/day)
Gas Composition	
Methane	55%–75%
Carbon dioxide	25%–45%
Hydrogen sulfide	trace
Hydrogen	trace
Nitrogen	trace
Gas Fuel Value	530–730 Btu/ft^3 (19,700–27,200 kJ/m^3)
Volatile Acids Concentration as Acetic Acid	
Normal operation	50–250 mg/ℓ
Maximum	approx. 2000 mg/ℓ
Alkalinity Concentrations as CaCO$_3$	
Normal operation	1000–5000 mg/ℓ

stratifies, with the digesting sludge in the bottom portion of the digester and the supernatant in the top portion (see Figure 19.8). The water given up during digestion, the supernatant liquor, is withdrawn from the digester every few days. The digested sludge is usually withdrawn about every 2 weeks. If open drying beds are used and it is rainy weather, the sludge is not withdrawn but instead is kept in the digester until suitable weather occurs. The grease that rises to the top of the supernatant is kept submerged by the cover to assist in breakdown by bacterial action. The gas produced during digestion is collected in the dome at the center of the floating cover. If sludge is added over a period of time and no digested sludge or supernatant is withdrawn, the floating cover rises. In a similar manner, when digested sludge or supernatant is withdrawn, the cover will slowly drop.

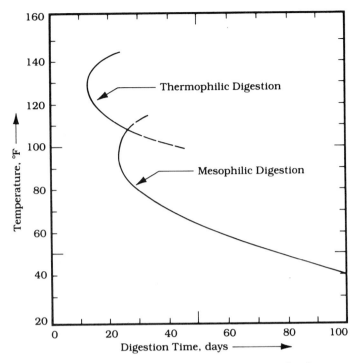

FIGURE 19.5 *Digestion Time versus Temperature for Conventional Digesters (Time required for 90% digestion of municipal wastewater sludges.)*

Adapted from *Water Purification and Wastewater Treatment and Disposal*, Vol. 2, *Water and Wastewater Engineering* by G. M. Fair, J. C. Geyer, and D. A. Okun. Copyright © 1968 by John Wiley & Sons, Inc. Reprinted by permission.

The fixed-cover digester is not as flexible in operation as the floating-cover type, because there is a limit to the amount of fresh sludge that may be added over a given period of time. Also, there is a limit to the amount of digested sludge or supernatant that may be removed at one time. When fresh sludge is added and no supernatant or digested sludge is withdrawn, the gas will be compressed and the maximum allowable pressure is about 8 in. (203 mm) of water column. When sludge or supernatant is withdrawn, the gas expands and the pressure decreases, with the minimum allowable pressure being about 3 in. (76 mm) of water column. Fixed-cover digesters have more problems with grease floating to the top of the supernatant and drying than the floating-cover type. When the grease dries, there is very little biochemical breakdown, and the grease accumulates and frequently causes operational problems, such as plugging of the supernatant outlets. Because of their operational problems, many states do not allow fixed-cover digesters to be used except when the design population is less than about 10,000 persons.

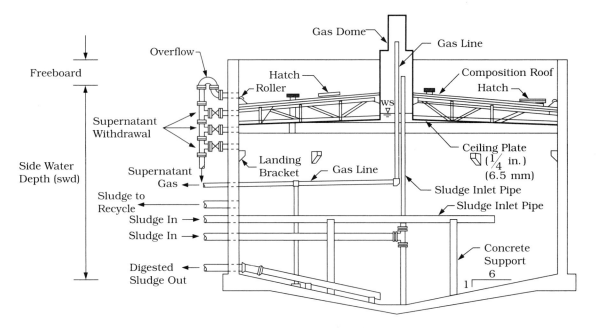

FIGURE 19.6   *Section through an Anaerobic Digester with a Floating Cover*

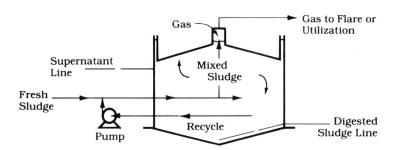

FIGURE 19.7   *Floating-Cover Digester: Sludge Being Added*

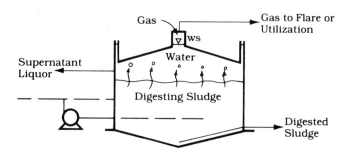

FIGURE 19.8   *Floating-Cover Digester: Sludge Digesting*

The capacity of the recirculation or recycle pump used to provide mixing should be sufficient to pump one tank volume in 30 min. Although many conventional digesters do not have any method for mixing, such digesters are subject to frequent acidification and are not considered good engineering practice. At least three supernatant withdrawal pipes should be provided along with supernatant sampling petcocks. The withdrawal pipes should be at least 2.0 ft (0.61 m) apart in elevation in order to provide selective withdrawal. The side water depth (swd) is the depth of the sludge and supernatant, if the contents are stratified, at the side of the digester. If the contents are mixed, the side water depth (swd) is the depth of the sludge mixture at the side of the digester. Usually the side water depth is from 20 to 30 ft (6.09 to 9.14 m). The freeboard is usually 2.0 ft (0.61 m) for tanks up to 50 ft (15.2 m) in diameter and is usually 2.5 ft (0.76 m) for tanks greater than 50 ft (15.2 m) in diameter.

HIGH-RATE DIGESTERS

High-rate digesters have digestion times of 10 to 20 days, organic solid loadings of 0.15 to 0.40 lb vss/day-ft^3 (2.40 to 6.40 kg/day-m^3), continuous mixing, and continuous or intermittent sludge feeding and sludge withdrawal, and the contents are in a homogeneous state. High-rate digesters usually have fixed covers.

In conventional digestion the mixing or recirculation system is usually operated only when fresh sludge solids are added, which ordinarily is less than a 1-hour period during each day. Thus, most of the time the digester contents are stratified and the digesting sludge occupies approximately the bottom half of the digester. In high-rate digestion the mixing is continuous; thus the entire digester volume is available for digesting sludge, and the mixing provides better contact between the seeded sludge and the fresh solids that have been added. This allows a high-rate digester to operate at organic loadings (that is, the pounds of volatile solids added/day-ft^3 or kg of volatile solids added/day-m^3) of several magnitudes greater than conventional digesters. Also, the detention times are much shorter than for conventional digesters. The digestion times from Figure 19.5 are for conventional or low-rate digesters; therefore, the curves are not applicable to high-rate digestion. As fresh sludge is added to a high-rate digester, the digested sludge may be displaced into a holding tank where the supernatant liquor is separated from the sludge. An alternative method for removing sludge or supernatant liquor is to stop the mixing and allow the contents to stratify. Once stratification has occurred, the withdrawals can be made. Most high-rate digesters have an additional mixing system other than a sludge recycle pump. Typical mixing systems are shown in Figure 19.9. Mixing may be provided, as shown in Figure 19.9 (a), by recycling gas to a draft tube mounted in the center of the digester. As the gas rises, it causes sludge to be lifted through the draft tube and discharged from the top of the tube. Mixing may also be provided by recycling gas, as shown in Figure 19.9 (b), and discharging it near the bottom of the digester by gas diffusers. Also,

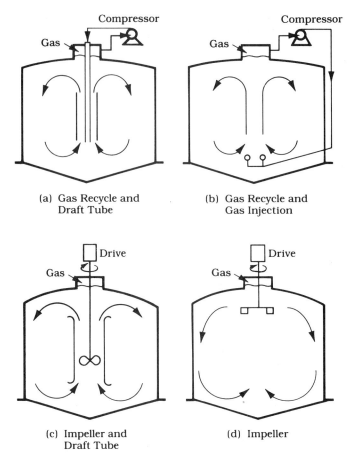

FIGURE 19.9 *Types of Mixing Systems for Anaerobic Digesters*

the gas may be discharged at about one-half the tank depth by pipes suspended from the ceiling. Mixing may be accomplished, as shown in Figure 19.9 (c), by an impeller mounted with a draft tube in the center of the digester. If the tank diameter is greater than about 50 ft (15.2 m), three or more draft tubes and turbine impellers are required. If three draft tubes and impellers are used, they are placed the same distance from the center of the digester and are located 120° apart in plan view. Mixing has also been done using turbine impellers mounted in the upper depths of a digester, as shown in Figure 19.9 (d). Mixing by gas recycle has an advantage over mechanical mixing because there are no moving mechanical parts inside the digester, where the environment is very corrosive. The high-rate digester shown in the photograph in Figure 19.10 employs an impeller and draft tube, and it corresponds to the digester schematic shown in Figure 19.9 (c).

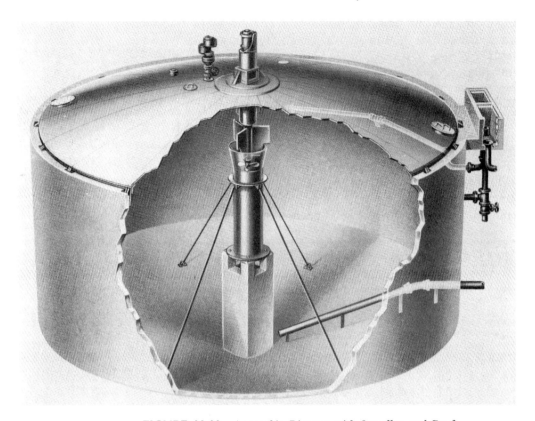

FIGURE 19.10 *Anaerobic Digester with Impeller and Draft Tube for Mixing*
Courtesy Dorr-Oliver, Inc.

TWO-STAGE DIGESTERS

A two-stage digester system such as that shown in Figures 19.1 and 19.11 is usually provided when the design population is over about 30,000 to 50,000 persons. In the first stage, the main biochemical action is liquefaction of organic solids, digestion of soluble organic materials, and gasification. In the second stage some gasification occurs; however, its main use is supernatant separation, gas storage, and digested sludge storage. The first stage is usually a high-rate digester employing a fixed cover and continuous mixing, whereas the second stage is usually a conventional digester with a floating cover and intermittent mixing. The organic loading applied to the first stage is usually several magnitudes greater than the loading applied to the second stage.

EGG-SHAPED DIGESTERS

These high-rate anaerobic digesters, shown in Figure 19.12, resemble an egg placed vertically on its end. They are widely used in Europe and Japan and are beginning to be used in the United States. They are high-rate anaerobic digesters employing an external recycle pump or other means for mixing of

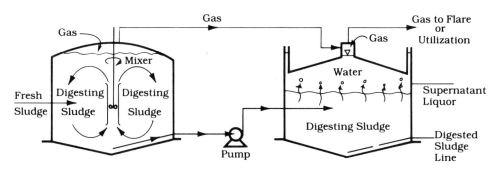

FIGURE 19.11   *Two-Stage Digestion*

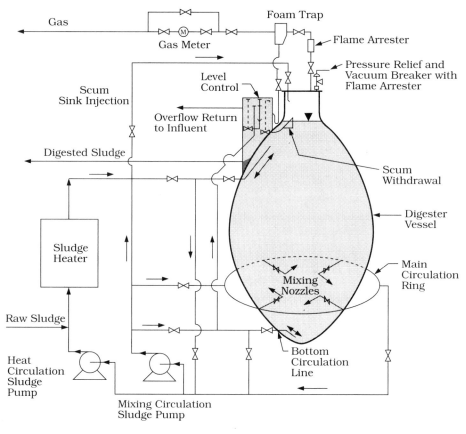

FIGURE 19.12   *Egg-Shaped High-Rate Anaerobic Digester Installation*

Courtesy of CBI Walker, Inc.

the digester contents. The tanks are constructed of either reinforced concrete or steel. Egg-shaped digesters offer several advantages over the cylindrical ones: (1) virtually no grit accumulates in the tank bottom because the conical sides are so steep that the grit is kept in suspension, (2) better mixing, (3) better control of scum at the top of the digester, and (4) smaller land requirements. The disadvantages of these digesters compared to the cylindrical ones are that (1) they are more expensive and (2) their unusual height restricts their usage, especially near residential areas. If the digester volume is less than 750,000 gal (2840 m³), mixing may be provided by an external recycle pump, as shown in Figure 19.12. For digesters with volumes greater than 750,000 gal (2840 m³), mixing is usually provided by a draft tube and impeller or a draft tube and a jet pump within the digester.

DIGESTER OPERATION

Most digesters are heated to 85°F to 100°F (29.4°C to 37.8°C) during cold weather in order to give a rapid digestion time. The digester gas produced can easily be used for heating purposes.

The optimum pH range of 7.0 to 7.2 can usually be maintained if the daily fresh sludge added is properly seeded and the sludge additions and withdrawals are not excessive. Usually, acidification will not occur if the dry solids added or withdrawn daily do not exceed 3% to 5% of the dry of solids in the digester. Acidification is characterized by a drop in pH, inhibition of the methane bacteria, a decrease in gas production, and a decrease in the methane content of the sludge gas. Also, there may be bad odors, foaming, and floating sludge. Acidification may be temporarily controlled by adding lime to raise the pH; however, a permanent solution requires changing the environmental conditions so that the methane producers are not inhibited and proper digestion can occur.

If the daily additions of fresh sludge contain inhibitory substances, such as heavy metals, they can interfere with the digestion process. When this occurs, the source of the inhibitory material must be eliminated prior to emptying the digester and restarting it.

The supernatant liquor is the water released during digestion; it may have a BOD₅ as high as 2000 mg/ℓ and a suspended solids concentration as high as 1000 mg/ℓ. Usually it is gradually fed back to the influent to the primary clarifiers.

The degree of digestion that is accomplished may be measured by the volatile solids reduction and the amount of sludge gas produced. Table 19.2 shows a summary of the general conditions for mesophilic sludge digestion and also gives the gas analyses and the amount of gas production in a properly operated anaerobic digester.

DIGESTER VOLUME

The volume of digesting sludge in a digester is a function of the volume of fresh sludge added daily, the volume of digested sludge produced daily, and the required digestion time in days. Additional volume must be provided for the supernatant liquor, gas storage, and storage of digested sludge. The volume required for gas storage is usually relatively small compared to total digester volume.

Batch digestion experiments have shown that if the supernatant is removed from a batch of digesting sludge as it is produced, the volume of the remaining digesting sludge versus digestion time is a parabolic function. For a parabolic function, the average volume is the initial volume minus two-thirds the difference between the initial and final volumes. Thus the volume of digesting sludge, V_{avg}, is given by (Fair *et al.*, 1968)

$$V_{avg} = V_1 - 2/3(V_1 - V_2) \tag{19.6}$$

where

V_{avg} = average volume of digesting sludge, ft^3/day (m^3/day)

V_1 = volume of fresh sludge added daily, ft^3/day (m^3/day)

V_2 = volume of digested sludge produced daily, ft^3/day (m^3/day)

The digestion time is a function of the tank operating temperature and can be estimated from Figure 19.5 for municipal sludges treated in conventional or low-rate digesters. The reduction in sludge volume during digestion is mainly due to the release of water from the sludge solids.

The volume of total sludge in the digester (both digesting and digested sludge) is given by

$$V_s = V_{avg} \cdot t_d + V_2 \cdot t_s \tag{19.7}$$

where

V_s = total sludge volume, ft^3 (m^3)

V_{avg} = average volume of digesting sludge, ft^3/day (m^3/day)

t_d = time required for digestion, days

V_2 = volume of digested sludge, ft^3/day (m^3/day)

t_s = time provided for sludge storage, days

The sludge volume normally occupies the bottom half of the digester, and the supernatant liquor occupies the top half of the digester, so the total digester volume, V_t, is

$$V_t = 2V_s \tag{19.8}$$

where V_t = total digester volume, ft^3 (m^3).

The digester requirements of most state health departments are based on digestion volume per capita served. These criteria are for the total volume required, which includes digesting sludge, supernatant liquor, gas, and digested sludge storage. Typical values for low- and high-rate digesters for various types of municipal sludges are shown in Table 19.1.

Digesters may also be designed on the basis of the organic loading — that is, pounds of volatile solids added per ft^3-day (kg/m^3-day) or the mean cell residence time. Typical organic solid loadings for low- and high-rate digesters are shown in Table 19.1. The mean cell residence time, θ_c, based on the solids produced per day in the digested sludge is given by

$$\theta_c = \frac{X}{\Delta X} \qquad (19.9)$$

where

θ_c = mean cell residence time, days

X = pounds (kilograms) of dry solids in the digester

ΔX = pounds (kilograms) of dry solids produced per day in the digested sludge

Frequently, the mean cell residence time is referred to as the solids retention time, θ_s.

The design of high-rate digesters is frequently done using the mean cell residence time, θ_c. For high-rate digesters, mixing is continuous; thus the system is a completely mixed, continuous-flow biological reactor without recycle. Since the number of cells in the feed to the digester is negligible compared to the cells in the digester and the digested sludge flow, the mean cell residence time, θ_c, is approximately equal to the hydraulic residence time, θ_h, and also to the solids retention time, θ_s. As the mean cell residence time decreases, a minimum time, θ_c^{Min}, will be reached at which the cells are washed from the system faster than they can multiply. Suggested values of the minimum mean cell residence time are given in Table 19.3 (McCarty, 1964).

Since the minimum cell residence time, θ_c^{Min}, is a critical condition, the design mean cell residence time, θ_c, is much longer than the minimum time. Usually, the design mean cell residence time is 2.5 times the minimum. The volume for a high-rate digester is given by

$$V = Q\theta_c = Q\theta_h \qquad (19.10)$$

where

V = total digester volume, ft^3 (m^3)

Q = fresh sludge flow, ft^3/day (m^3/day)

θ_c = design mean cell residence time, days

θ_h = design hydraulic residence time, days

TABLE 19.3 *Suggested Values of the Minimum Mean Cell Residence Time*

TEMPERATURE (°C)	θ_c^{Min} (days)
18	11
24	8
30	6
35	4
40	4

EXAMPLE 19.1   *Low-Rate Digester*

A heated, low-rate anaerobic digester is to be designed for an activated sludge plant treating the wastewater from 25,000 persons. The fresh sludge has 0.25 lb dry solids/cap-day, volatile solids are 70% of the dry solids, the dry solids are 5% of the sludge, and the wet specific gravity is 1.01. Sixty-five percent of the volatile solids are destroyed by digestion, and the fixed solids remain unchanged. The digested sludge has 7% dry solids and a wet specific gravity of 1.03. The operating temperature is 95°F and the sludge storage time is 45 days. The sludge occupies the lower half of the tank depth, and the supernatant liquor and the gas occupy the upper half of the tank depth. Determine the digester volume.

SOLUTION   The fresh sludge solids represent an Input. Fresh solids = (25,000 persons)(0.25 lb/cap-day) = 6250 lb/day. vss = (6250)(0.70) = 4375 lb/day. fss = (6250)(0.30) = 1875 lb/day. vss destroyed = (4375)(0.65) = 2844 lb/day. The digested sludge solids represent an Output. For steady state, [Output] = [Input] − [Decrease due to reaction]. Thus vss in the digested sludge = 4375 − 2844 = 1531 lb/day. fss in the digested sludge = 1875 − 0 = 1875 lb/day. The total solids in the digested sludge = 1531 + 1875 = 3406 lb/day. The fresh sludge volume = (6250 lb/day)(ft³/62.4 lb)(100 lb/5 lb)(1/1.01) or 1983 ft³/day. The digested sludge volume = (3406 lb/day)(ft³/62.4 lb)(100 lb/7 lb)(1/1.03) or 757 ft³/day. $V_{avg} = V_1 - \frac{2}{3}(V_1 - V_2) = 1983 - \frac{2}{3}(1983 - 757) = 1166$ ft³/day. From Figure 19.5 the digestion time is 23 days. Thus the total sludge volume is

$$\text{Volume of sludge} = (1166)(23) + (757)(45)$$
$$= 61,000 \, \text{ft}^3$$

Therefore, the total digester volume is

$$V_t = (61,000)(2) = \boxed{122,000 \, \text{ft}^3}$$

Note: Some sources assume that the fixed solids do not change during the digestion process, as was assumed in this problem. Other sources (Fair *et al.*, 1968) assume that 25% of the volatile solids destroyed are converted to fixed solids.

EXAMPLE 19.1 SI   *Low-Rate Digester*

A heated, low-rate anerobic digester is to be designed for an activated sludge plant treating the wastewater from 25,000 persons. The fresh sludge has 0.11 kg dry solids/cap-day, volatile solids are 70% of the dry solids, the dry solids are 5% of the sludge, and the wet specific gravity is 1.01. Sixty-five percent of the volatile solids are destroyed by digestion, and the fixed solids remain unchanged. The digested sludge has 7% dry solids and a wet specific gravity of 1.03. The operating temperature is 35°C and the sludge storage time is 45 days. The sludge occupies the lower half of the tank depth, and the supernatant liquor and the gas occupy the upper half of the tank depth. Determine the digester volume.

SOLUTION The fresh sludge solids represent an Input. Fresh solids = (25,000 persons)(0.11 kg/cap-d) = 2750 kg/d. vss = (2750)(0.70) = 1925 kg/d. fss = (2750)(0.30) = 825 kg/d. vss destroyed = (1925)(0.65) = 1251 kg/d. The digested sludge solids represent an Output. For steady state, [Output] = [Input] − [Decrease due to reaction]. Thus vss in the digested sludge = 1925 − 1251 = 674 kg/d. fss in the digested sludge = 825 − 0 = 825 kg/d. The total solids in the digested sludge = 674 + 825 = 1499 kg/d. The fresh sludge volume = (2750 kg/d)(m^3/1000 kg)(100 kg/5 kg)(1/1.01) or 54.46 m^3/d. The digested sludge volume = (1499 kg/d)(m^3/1000 kg)(100 kg/7 kg)(1/1.03) or 20.79 m^3/d. $V_{avg} = V_1 - \frac{2}{3}(V_1 - V_2) = 54.46 - \frac{2}{3}(54.46 - 20.79) = 32.01$ m^3/d. From Figure 19.5 the digestion time at 35°C (95°F) is 23 days. Thus the total sludge volume is

$$\text{Volume of sludge} = (32.01)(23) + (20.79)(45)$$
$$= 1670 \text{ m}^3$$

Therefore, the total digester volume is

$$V_t = (1670)(2) = \boxed{3340 \text{ m}^3}$$

EXAMPLE 19.2 AND 19.2 SI *High-Rate Digester*

A heated high-rate anaerobic digester is to be designed for an activated sludge plant treating the wastewater from 25,000 persons. The feed to the digester (both primary and secondary sludge) is 1980 ft^3/day (56.1 m^3/d), and the operating temperature is 95°F (35°C). Determine the digester volume using USCS units and SI units.

SOLUTION From Table 19.3, the minimum cell residence time, θ_c^{Min}, for 95°F is 4 days. The design mean cell residence time, θ_c, is 2.5 × 4 days or 10 days. For a completely mixed biological reactor without recycle, the θ_c value is equal to the hydraulic detention time, θ_h. Thus the digester volume, V, for the solution in USCS units, is

$$V = Q\theta_h = (1980 \text{ ft}^3/\text{day})(10 \text{ days}) = \boxed{19{,}800 \text{ ft}^3}$$

For the solution in SI units, the digester volume, V, is

$$V = Q\theta_h = (56.1 \text{ m}^3/\text{d})(10 \text{ d}) = \boxed{561 \text{ m}^3}$$

**MOISTURE-
WEIGHT
RELATIONSHIPS**

The specific gravity of a wet or dried sludge, s, depends upon the water content, the solids content, and the specific gravity of the dried solids, s_s. The percent water or moisture content, p_w, is given by

$$p_w = \frac{100W_w}{(W_w + W_s)} \tag{19.11}$$

where

p_w = percent water

W_w = weight of the water

W_s = weight of the dry solids

The percent solids, p_s, is given by the formulation

$$p_s = (100 - p_w) = \frac{100W_s}{W_w + W_s} \tag{19.12}$$

The specific gravity of dried sludge solids, s_s, is a function of the specific gravities of the volatile and fixed fractions, which are designated s_v and s_f, respectively. If the percent volatile material is p_v and the percent fixed material is p_f, the following may be written:

$$\frac{100}{s_s} = \frac{p_v}{s_v} + \frac{100 - p_v}{s_f} \tag{19.13}$$

Rearranging Eq. (19.13) yields

$$s_s = \frac{100s_f s_v}{100s_v + p_v(s_f - s_v)} \tag{19.14}$$

For engineering purposes, the specific gravity of the volatile fraction, s_v, may be considered as 1.0, and the fixed fraction, s_f, as 2.5. Substituting these into Eq. (19.14) and rearranging give

$$s_s = \frac{250}{100 + 1.5p_v} \tag{19.15}$$

The specific gravity of wet sludge, s, may be determined from the equation

$$s = \frac{p_w + (100 - p_w)}{p_w + (100 - p_w)/s_s} = \frac{100s_s}{p_w s_s + (100 - p_w)} \tag{19.16}$$

**SLUDGE
QUANTITIES
AND SOLIDS
CONCEN-
TRATIONS**

It has been possible, from evaluations of vast amounts of operating data and records, to make estimates of the amounts of sludge that are produced at municipal wastewater treatment plants and also to estimate the solids concentrations, volatile and fixed fractions, and other pertinent characteristics. Table 19.4 shows typical solids contributions per capita and solids contributions concentrations for various sludges from different types of treatment plants. In design, the percent solids shown in the table may be

TABLE 19.4 *Sludge Solids Contributions and Solids Contents*

TYPE OF TREATMENT AND TYPE OF SLUDGE	DRY SOLIDS (lb/cap-day)	DRY SOLIDS (kg/cap-day)	SOLIDS CONTENT (%)
Plain Sedimentation			
Fresh, wet	0.119	0.054	2.5–5.0
Digested, wet	0.075	0.034	10–15
Digested, air dried	0.075	0.034	45
Activated Sludge (Conventional)			
Excess activated sludge, wet	0.068	0.031	0.5–1.5
Primary and excess activated sludge, wet	0.187	0.085	4–5
Digested primary and excess activated sludge, wet	0.121	0.055	6–8
Digested primary and excess activated sludge, air dried	0.121	0.055	45
Thickened excess activated sludge, wet	0.068	0.031	1–2
Digested thickened excess activated sludge, wet	0.044	0.020	2–3
Trickling Filter Plant (High-rate)			
Secondary, wet	0.044	0.020	5
Primary and secondary, wet	0.163	0.074	5
Digested primary and secondary, wet	0.106	0.048	10
Digested primary and secondary, air dried	0.106	0.048	45
Trickling Filter Plant (Standard-rate)			
Secondary, wet	0.029	0.013	5–10
Primary and secondary, wet	0.148	0.067	3–6
Digested primary and secondary, wet	0.095	0.043	10
Digested primary and secondary, air dried	0.095	0.043	45
Chemically Precipitated Raw Wastewater			
Fresh wet	0.198	0.090	2–5
Digested	0.125	0.057	10

Adapted from *Sewage Treatment* by K. Imhoff and G. M. Fair. Copyright © 1956 by John Wiley & Sons, Inc. Reprinted by permission.

used without much reservation because they have been found to be relatively consistent. However, the dry solids contributions in pounds per capita per day (kg/cap-day) should be used with caution because they vary from plant to plant.

If sludge pumping records are available giving the flows of primary and secondary sludge per day and also the solids contents, these data may be used to determine the pounds (kg) of primary and secondary sludge produced per day. If these records are not available, it is necessary to use the following procedure to determine the amount of primary and secondary sludge solids. Determining the pounds (kg) of primary sludge solids produced per day requires the influent suspended solids concentration, the fraction of suspended solids removed (usually 0.60 to 0.65), and the influent flowrate of the wastewater. The suspended solids concentration should come from long-term suspended solids analyses to be representative. Determining the pounds (kg) of secondary solids from the activated sludge process requires the pounds of BOD_5 or COD removed per day (kg/day) and requires the use of the following equation (Eckenfelder and Weston, 1956):

$$\text{lb (kg) solids/day} = YS_r - k_e \bar{X} \qquad (19.17)$$

where

Y = yield coefficient, pounds of total suspended or volatile suspended solids per pound of BOD_5 or COD removed per day (kg/kg)

S_r = pounds of BOD_5 or COD removed per day (kg/day)

k_e = endogenous coefficient, day^{-1}

$\bar{X}$ = pounds of total suspended or volatile suspended solids in the aeration tank (kg)

Determining the pounds (kg) of secondary sludge solids from the trickling filter process requires the pounds (kg) of BOD_5 or COD removed per day by the secondary units and the pounds (kg) of biological solids produced per pound (kg) of BOD_5 or COD removed. Low-rate trickling filters produce from 0.25 to 0.35 lb (kg) of biological solids per lb (kg) of BOD_5 removed, whereas high-rate trickling filters produce from 0.40 to 0.50 lb (kg) biological solids per pound (kg) of BOD_5 removed. If the BOD_5/COD ratio is known, these values may be expressed in terms of COD. Determining the pounds (kg) of suspended solids produced per day by chemical coagulation of raw municipal wastewater requires the influent suspended solids concentration, the fraction of suspended solids removed (usually from 0.95 to 0.99), and the influent flowrate of the wastewater. Frequently, estimating the amount of sludge produced can be facilitated by the use of population equivalents. The population equivalents of BOD_5, suspended solids, and flow are 0.17 lb, 0.20 lb, and 100 gal per capita per day, respectively. The corresponding SI units are 0.077 kg, 0.091 kg, and 380 ℓ per capita per day. However, the increasing popularity of garbage grinders may cause these values to increase in the future.

EXAMPLE 19.3
AND 19.3 SI

Sludge Quantities

A low-rate trickling filter plant treats the wastewater from 5000 persons. Short-term analyses using composite samples of the influent wastewater show the BOD$_5$ and suspended solids concentrations to be those expected from average values for municipal wastewaters in the United States. The plant gets 95% BOD$_5$ removal, 33% BOD$_5$ removal by the primary clarifier, and 65% suspended solids removal by the primary clarifier. Assume that the primary sludge has 4% dry solids, the primary sludge specific gravity = 1.01, the secondary sludge has 5% dry solids, the specific gravity = 1.02, and the biological solids produced is 0.35 lb/lb (kg/kg) BOD$_5$ removed. Determine the primary and secondary sludge produced per day.

SOLUTION

The solution for the USCS units is as follows. The primary solids = (5000)(0.20)(0.65) = 650 lb/day. The primary sludge flow = (650 lb/day) × (100 lb/4.0 lb)(gal/8.34 lb)(1/1.01) = 1930 gal/day. Thus,

$$\text{Primary sludge} = \boxed{1930 \, \text{gal/day}}$$

The BOD$_5$ to the secondary units = (0.17)(5000)(1 − 0.33) = 569.5 lb/day. The BOD$_5$ from the plant = (0.17)(5000)(1 − 0.95) = 42.5 lb/day. The BOD$_5$ removed by the secondary units = 569.5 − 42.5 = 527 lb/day. Thus the biological solids produced per day = (527 lb/day)(0.35 lb/lb) = 184.5 lb/day. The secondary sludge flow = (184.5 lb/d)(100 lb/5 lb)(gal/8.34 lb)(1/1.02) = 434 gal/day. Thus,

$$\text{Secondary sludge} = \boxed{434 \, \text{gal/day}}$$

The solution for the SI units is as follows. The primary solids = (5000) × (0.091)(0.65) = 295.8 kg/d. The primary sludge flow = (295.8 kg) × (100 kg/4.0 kg)(ℓ/1.0 kg)(1/1.01) = 7320 ℓ/d. Thus,

$$\text{Primary sludge flow} = \boxed{7320 \, \ell/\text{d}}$$

The BOD$_5$ to the secondary units = (0.077)(5000)(1 − 0.33) = 258.0 kg/d. The BOD$_5$ from the plant = (0.077)(5000)(1 − 0.95) = 19.3 kg/d. The BOD$_5$ removed by the secondary units = 258.0 − 19.3 = 238.7 kg/d. Thus the biological solids produced per day = (238.7 kg/d)(0.35 kg/kg) = 83.55 kg/d. The secondary sludge flow = (83.55 kg/d)(100 kg/5 kg)(ℓ/1.0 kg)(1/1.02) = 1640 ℓ/d. Thus,

$$\text{Secondary sludge} = \boxed{1640 \, \ell/\text{d}}$$

DIGESTER HEAT REQUIREMENTS

In moderate and cold climates, it is necessary to heat digesters during the winter to maintain the digester temperature within the desired range. The heat furnished must be sufficient to (1) increase the temperature of the incoming fresh sludge to the temperature in the digester and (2) make up for heat losses from the digester through the walls, bottom, and cover. Usually, the heat required to raise the incoming sludge temperature to the digester temperature is larger than the heat required to make up for heat losses.

The heat required to raise the temperature of the incoming fresh sludge to the digester temperature is given by

$$Q_s = P \times \frac{100}{p_s} \times (T_d - T_s) \times \frac{1}{24} \times c_p \qquad (19.18)$$

where

Q_s = Btu/hr (J/h) required

P = pounds (kg) of fresh dry sludge solids added per day

p_s = percent dry solids in the fresh sludge

T_d = temperature in the digester, °F (°C)

T_s = temperature of the fresh sludge, °F (°C)

c_p = specific heat constant, equal to 1.0 Btu/lb-°F (4200 J/kg-°C)

The heat required to make up for losses through the top, walls, and bottom is given by

$$Q_d = CA \cdot \Delta T \qquad (19.19)$$

where

Q_d = Btu/hr (J/h) required

C = coefficient of heat flow, Btu/ft²-hr-°F (J/m²-h-°C)

A = surface area, ft² (m²)

ΔT = difference between the tank temperature and the outside material being considered, °F (°C)

The C values for concrete walls and slabs against other materials are 0.08 (1634) for dry earth, 0.30 (6128) for air, and 0.26 (5310) for wet earth. The C value for a floating cover is 0.16 (3268) (Imhoff *et al.*, 1971).

The most common method to heat a digester is to use a hot-water heater and heat exchanger outside the digester and heat the incoming and recycled sludge, as shown in Figure 19.13. The first heated digesters used hot-water heaters or boilers external to the digesters, and the hot water was circulated through coils of water pipes mounted inside the digesters. It was found, however, that sludge caked around the pipes and greatly reduced the thermal efficiency. Since digesters are rarely emptied, it was not feasible to clean the cake from the pipes. The use of an external heater, which heats the pumped sludge outside the digester as shown in Figures 19.13 and 19.14, controls the caking problem since any sludge caking will occur in the pipes in the heat exchanger. The elbows may be removed from the heat exchanger, and the cake may be rodded loose from the pipes. Then the elbows may be replaced

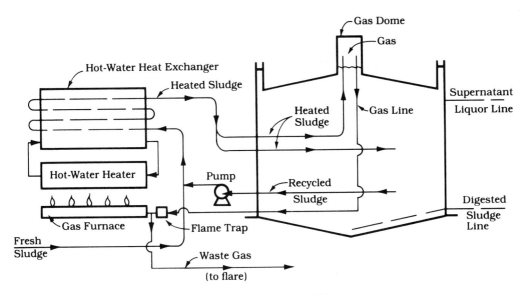

FIGURE 19.13 *Externally Heated Digester*

and heating resumed. Digesters with a mixing system consisting of recycled gas discharging in a draft tube in the center of the digester frequently are heated by internal heat exchangers mounted inside the draft tube. These heat exchangers can be removed through the access hatch in the center of the dome to clean off any caking that may have occurred.

EXAMPLE 19.4
AND 19.4 SI

Sludge Heat Requirements

An activated sludge treatment plant has an anaerobic digester and serves a population of 25,000 persons. The mixture of primary and secondary sludge amounts to 0.24 lb/cap-day (0.11 kg/cap-d), the fresh sludge has 4.5% solids on a dry basis, and the digester operating temperature is 95°F (35°C). During the coldest month, January, the sludge temperature is 55°F (12.8°C). The specific heat constant is 1.0 Btu/lb-°F (4200 J/kg-°C). Determine, using USCS units and SI units, the heat required to raise the fresh sludge temperature to that of the digester.

SOLUTION

For the solution in USCS units, the number of pounds of fresh dry sludge solids added per day is (25,000 capita)(0.24 lb/cap-day) = 6000 lb/day. The heat required, Q_s, is Q_s = (6000 lb/day)(100 lb/4.5 lb)(95°F − 55°F)(day/24 hr)(1.0 Btu/lb-°F) = 222,000 Btu/hr or

$$Q_s = \boxed{222{,}000 \text{ Btu/hr}}$$

For the solution in SI units, the number of kilograms of fresh dry sludge solids added per day is (25,000 capita)(0.11 kg/cap-d) = 2750 kg/d. The heat

FIGURE 19.14 *External Sludge Heater*
Courtesy of Walker Process Equipment, Division of McNish
Corporation.

required, Q_s, is $Q_s = (2750\,\text{kg/d})(100\,\text{kg/4.5 kg})(35°C - 12.8°C)(d/24\,h)$
$(4200\,\text{J/kg-°C}) = 237,000,000\,\text{J/h}$ or

$$Q_s = \boxed{237,000,000\,\text{J/h}}$$

EXAMPLE 19.5   *Digester Heat Losses*

The anaerobic digester for the municipality in Example 19.4 is 75 ft in
diameter, has a wall height of 30 ft, and has a floating cover. The digester is
recessed 2 ft in the earth and has a freeboard of 3 ft. For insulation, earth is
mounded around the digester up to a height of 15 ft. A wall height of 30 ft −
(2 ft + 3 ft + 15 ft), or 10 ft, is exposed to air on the outside and to digester
contents on the inside. The average monthly temperature in January is 45°F.

The digester operating temperature is 95°F, and the dry earth mounded around the digester has a temperature equal to the average of the digester operating temperature and the average monthly temperature. The earth temperature below the digester floor is 95°F, and the temperature of the earth around the wall at the 2-ft recess is 95°F. Determine the heat required to make up heat losses from the digester.

SOLUTION The area of the floating cover is $(\pi/4)(75\,\text{ft})^2 = 4418\,\text{ft}^2$. The area of the wall where air is on the outside and digester contents are on the inside is $(\pi)(75\,\text{ft})(10\,\text{ft}) = 2356\,\text{ft}^2$. The area of the wall where dry earth is on the outside and digester contents are on the inside is $(\pi)(75\,\text{ft})(15\,\text{ft}) = 3534\,\text{ft}^2$. The temperature of the dry earth mounded around the digester is (95°F + 45°F)(1/2) or 70°F. The heat required, Q_d, to make up digester heat losses is $Q_d = (4418\,\text{ft}^2)(0.16\,\text{Btu/ft}^2\text{-hr-°F})(95°F - 45°F) + (2356\,\text{ft}^2)(0.30\,\text{Btu/ft}^2\text{-hr-°F})(95°F - 45°F) + (3534\,\text{ft}^2)(0.08\,\text{Btu/ft}^2\text{-hr-°F})(95°F - 70°F) = 35,300\,\text{Btu/hr} + 35,300\,\text{Btu/hr} + 7100\,\text{Btu/hr}$ or

$$Q_d = \boxed{77,700\,\text{Btu/hr}}$$

EXAMPLE 19.5 SI *Digester Heat Losses*

The anaerobic digester for the municipality in Example 19.4 SI is 22.9 m in diameter, has a wall height of 9.14 m, and has a floating cover. The digester is recessed 0.61 m in the earth and has a freeboard of 0.91 m. For insulation, earth is mounded around the digester up to a height of 4.57 m. A wall height of 9.14 m − (0.61 m + 0.91 m + 4.57 m), or 3.05 m, is exposed to air on the outside and to digester contents on the inside. The average temperature in January is 7.2°C. The digester operating temperature is 35°C, and the dry earth mounded around the digester has a temperature equal to the average of the digester operating temperature and the average monthly temperature. The earth temperature below the digester is 35°C, and the temperature of the earth around the wall at the 0.61-m recess is 35°C. Determine the heat required to make up heat losses from the digester.

SOLUTION The area of the floating cover is $(\pi/4)(22.9\,\text{m})^2 = 411.9\,\text{m}^2$. The area of the wall where air is on the outside and digester contents are on the inside is $(\pi)(22.9\,\text{m})(3.05\,\text{m}) = 219.4\,\text{m}^2$. The area of the wall where dry earth is on the outside and digester contents are on the inside is $(\pi)(22.9\,\text{m})(4.57\,\text{m}) = 328.8\,\text{m}^2$. The temperature of the dry earth mounded around the digester is (35°C + 7.2°C)1/2 = 21.1°C. The heat required, Q_d, to make up digester heat losses is $Q_d = (411.9\,\text{m}^2)(3268\,\text{J/m}^2\text{-h-°C})(35°C - 7.2°C) + (219.4\,\text{m}^2)(6128\,\text{J/m}^2\text{-h-°C})(35°C - 7.2°C) + (328.8\,\text{m}^2)(1634\,\text{J/m}^2\text{-h-°C}) \times (35° - 21.1°C) = 82,300,000\,\text{J/h}$ or

$$Q_d = \boxed{82,300,000\,\text{J/h}}$$

DIGESTER GAS UTILIZATION

Digester gas consists of methane (about 55% to 75%), carbon dioxide (about 25% to 45%), water vapor, and trace amounts of such gases as hydrogen sulfide, hydrogen, and nitrogen. Sludge gas is explosive, and the hydrogen sulfide is a corrosive and toxic gas. The usual methods for using sludge gas are for heating purposes and for power generation. Sludge heaters will burn the gas without scrubbing; however, for use as a fuel for internal combustion engines or gas turbines, the gas must be scrubbed if the hydrogen sulfide is more than 0.015% by volume. Scrubbing will remove hydrogen sulfide and some carbon dioxide, thus increasing the fuel value in terms of Btu per ft^3 (kJ/m^3). Internal combustion engines or gas turbines may be used to drive air compressors, pumps, and generators for production of electricity. If the gas utilization facilities are extensive, the gas should be scrubbed to make it less corrosive. The amount of sludge gas produced will meet 65% to 100% of the energy requirements for a municipal wastewater treatment plant, depending on the type of secondary treatment employed. Excess sludge gas is usually burned in a flare. The gas collection and circulation system must have safety equipment such as condensate traps, pressure regulation valves, flame traps on lines to burners, engines, turbines, and flares, and a waste gas flare.

EXAMPLE 19.6 AND 19.6 SI

Sludge Gas Utilization

The sludge for the municipality in Examples 19.4 and 19.5 has 70% volatile matter; 65% of the volatile matter is destroyed during digestion. The sludge gas that is produced amounts to $14 ft^3$ per lb volatile solids destroyed ($0.874 m^3/kg$) and is 70% methane. The fuel value of methane is $963 Btu/ft^3$ ($35,850 kJ/m^3$) at standard conditions. The sludge heater that heats the incoming sludge and makes up the digester heat losses has a thermal efficiency of 60%. Determine what percent of the fuel value produced is required for the sludge heater.

SOLUTION

For the solution in USCS units, the volume of sludge gas produced per day = $(6000 lb/day)(0.70)(0.65)(14 ft^3/lb)$ = $38,220 ft^3/day$. The fuel value available is $(38,220 ft^3/day)(0.70)(963 Btu/ft^3)(day/24 hr)$ = $1,073,504 Btu/hr$. The heat required = $(222,000 Btu/hr + 77,700 Btu/hr)(1/0.60)$ = $499,500$ Btu/hr. The percent fuel value used = $(499,500/1,073,504)(100\%)$ = 46.5% or

$$\text{Fuel value used} = \boxed{46.5\%}$$

For the solution in SI units, the volume of sludge gas produced per day = $(2750 kg/d)(0.70)(0.65)(0.874 m^3/kg)$ = $1094 m^3/d$. The fuel value available is $(1094 m^3/d)(0.70)(35,850 kJ/m^3)(d/24 h)$ = $1,143,914 kJ/h$. The heat required = $(237,000,000 J/h + 82,300,000 J/h)(kJ/1000 J)(1/0.60)$ = $532,167$ kJ/h. The percent fuel value used = $(532,167/1,143,914)(100\%)$ = 46.5% or

$$\text{Fuel value used} = \boxed{46.5\%}$$

**FERTILIZER
VALUE OF
DRIED SLUDGE**

Fresh municipal sludges are not suitable as fertilizers because of hygienic and aesthetic problems. Air-dried digested sludges should not be used on crops that are eaten raw by humans, such as vegetables. Heat-dried sludges are sterile and may be used on any crops. The fertilizer value of sludges is about the same as that of barnyard manure, and in addition to their fertilizer value, sludges also serve as excellent soil conditioners.

**FLUID
PROPERTIES
OF SLUDGE**

The frictional head losses for organic municipal wastewater sludges may be estimated using hydraulic formulas, such as the Hazen-Williams equation. The head loss depends mainly upon the nature of the sludge, the type of flow (that is, laminar, transitional, or turbulent), the solids content, and the temperature. Sludges exhibit laminar, transitional, and turbulent flow at much higher Reynolds numbers than water. Laminar and transitional flow occur up to about 3.5 fps (1.07 m/s) in pipes 4 to 8 in. (100 to 200 mm) in diameter, and turbulent flow occurs above about 3.5 fps (1.07 m/s) (WPCF, 1977). Hazen-Williams coefficients for laminar and transitional flow for fresh and digested primary sludges at various solid contents are shown in Table 19.5.

Activated sludges usually have solid contents from 0.5% to 1.5%, and for these sludges, the Hazen-Williams coefficients for digested sludges may be used.

The Hazen-Williams equation for pipe flow is

$$V = CC_{HW}R^{0.63}S^{0.54} \qquad (19.20)$$

where

V = velocity, ft/sec (m/s)

C = 1.318 for USCS units and 0.8464 for SI units

C_{HW} = Hazen-Williams friction coefficient

R = hydraulic radius, ft (m)

S = slope of the energy gradient

TABLE 19.5 *Hazen-Williams Coefficients for Sludge versus Solids Content*

PERCENT SOLIDS	RAW PRIMARY[a] C_{HW}	DIGESTED C_{HW}
0	100	100
2	81	90
4	61	85
6	45	75
8.5	32	60
10	25	55

Data taken from S. G. Brisbin, "Flow of Concentrated Raw Sewage Sludges in Pipes," *Proc. SED,* ASCE 83 (1957):SA3.

EXAMPLE 19.7   *Sludge Pumping*
AND 19.7 SI

A centrifugal sludge pump with a capacity of 300 gpm (1140 ℓpm) pumps a mixture of primary and secondary sludge from the primary clarifier to the digester at a municipal treatment plant. The secondary sludge is mixed with the influent to the primary clarifier and allowed to resettle with the primary settlings in order to thicken the secondary sludge. The piping from the primary clarifier to the sludge pump and from the sludge pump to the digester is 8 in. (200 mm) in diameter. The discharge pipe in the digester runs vertically up to the gas dome of the floating cover, and the elevation difference between the water surface of the primary clarifier and the end of the vertical discharge pipe is 18.60 ft (5.67 m). The gas is under a pressure of 8 in. (200 mm) of water column. The sludge mixture has 4.5% dry solids and the specific gravity is 1.01. The length of 8 in. (200-mm) pipe is 325 ft (99.1 m), and the fittings or form losses are equivalent to 150 ft (45.7 m) of pipe. The pump is operated by a time clock, and it runs at a preset time interval once every hour. Determine the head at which the pump operates.

SOLUTION   Writing Bernoulli's equation between Point 1 at the primary clarifier water surface and Point 2 at the discharge end of the piping gives

$$\frac{V_1^2}{2g} + \frac{p_1}{\gamma} + Z_1 + H_p = \frac{V_2^2}{2g} + \frac{p_2}{\gamma} + Z_2 + 0.5\frac{V_2^2}{2g} + H_L$$

where

$$
\begin{aligned}
V_1, V_2 &= \text{velocities} \\
p_1, p_2 &= \text{pressures} \\
Z_1, Z_2 &= \text{elevation data} \\
H_p &= \text{head on pump} \\
0.5V_2^2/2g &= \text{entrance loss} \\
H_L &= \text{total head losses}
\end{aligned}
$$

For the solution in USCS units, assume a datum that makes Z_1 = El. 0.0 ft. Since $V_1 = 0$, $P_1 = 0$, and $Z_1 = 0$, Bernoulli's equation becomes

$$H_p = 1.5\frac{V_2^2}{2g} + \frac{8}{12}\text{ft} + 18.60\text{ ft} + H_L$$

$$= 1.5\frac{V_2^2}{2g} + 19.27\text{ ft} + H_L$$

Table 19.5 gives C_{HW} = 61 for 4% solids and C_{HW} = 45 for 6% solids. Interpolation for 4.5% solids gives C_{HW} = 57. The flow of 300 gpm is equal to 0.668 ft³/sec. The area for an 8-in. pipe is 0.349 ft². thus the velocity is (0.668 ft³/sec)(1/0.349 ft²) = 1.91 fps. The hydraulic radius = (8/12 ft)($\frac{1}{4}$) = 0.1667 ft. Inserting the known values into the Hazen-Williams equation gives

$$1.91 = (1.318)(57)(0.1667)^{0.63}S^{0.54}$$

which yields

$$S = 0.0090 \text{ ft/ft or } 9.00 \text{ ft/1000 ft}$$

Substituting into Bernoulli's equation gives

$$H_p = (1.5)\left[\frac{(1.91)^2}{2g}\right] + 19.27 + \left(\frac{9.00}{1000}\right)(325 + 150)$$

or

$$H_p = 0.08 + 19.27 + 4.28 = \boxed{23.63 \text{ ft}}$$

For the solution in SI units, assume a datum that makes Z_1 = El. 0.0 m. Since $V_1 = 0$, $p_1 = 0$, and $Z_1 = 0$, Bernoulli's equation becomes

$$H_p = 1.5\frac{V_2^2}{2g} + 0.200 \text{ m} + 5.67 \text{ m} + H_L$$

$$= 1.5\frac{V_2^2}{2g} + 5.87 \text{ m} + H_L$$

Table 19.5 gives C_{HW} = 61 for 4% solids and C_{HW} = 45 for 6% solids. Interpolation for 4.5% solids gives C_{HW} = 57. The flow of 1140 ℓpm is equal to 1.14 m³/min or 0.0190 m³/s. The area for a 200-mm diameter pipe is 0.03142 m²; thus the velocity is $(0.0190 \text{ m}^3/\text{s})(1/0.03142 \text{ m}^2) = 0.6047 \text{ m/s}$. The hydraulic radius is (0.200 m/4) = 0.050 m. Inserting known values into the Hazen-Williams equation gives

$$0.6047 = (0.8464)(57)(0.050)^{0.63}S^{0.54}$$

which yields

$$S = 0.00995 \text{ m/m or } 9.95 \text{ m/1000 m}$$

Substituting into Bernoulli's equation gives

$$H_p = (1.5)\left[\frac{(0.6047)^2}{2g}\right] + 5.87 + \left(\frac{9.95}{1000}\right)(99.1 + 45.7)$$

$$= 0.03 + 5.87 + 1.44 = \boxed{7.37 \text{ m}}$$

SLUDGE THICKENING

At many wastewater treatment plants, particularly large ones, the fresh sludge is thickened to increase its solids content prior to digestion. Thickening prior to digestion is becoming more common because it reduces the daily fresh sludge volume, thus decreasing the required size of the digester and also the amount of supernatant liquor to be disposed. Thickening may be accomplished by gravity thickeners, which are the most widely used, or by centrifuges. Gravity thickeners are similar to circular clarifiers; the most common type has vertical pickets mounted on the trusswork for the bottom scraper blades. The pickets extend up to about half the depth of the tank, and as they rake through the sludge they break up sludge arching and release much of the entrained water. Gravity thickeners usually thicken a sludge to about twice the original solids content, thus decreasing the volume of the fresh sludge to about half the original volume. Surface loadings are usually 600 to 800 gal/day-ft^2 (24.4 to 32.6 m^3/day-m^2) based on the supernatant flow. The allowable solids loading in lb/day-ft^2 (kg/d-m^2) depends upon the nature of the sludge. The thickening of various sludges has given the following percent solids in the thickened flow: (1) Raw primary sludges at 20 to 30 lb/day-ft^2 (97.6 to 146 kg/day-m^2) gave 8% to 10% solids. (2) Mixtures of raw primary and waste activated sludge at 6 to 10 lb/day-ft^2 (29.3 to 48.8 kg/day-m^2) gave 5% to 8% solids. (3) Mixtures of raw primary and trickling filter humus at 10 to 12 lb/day-ft^2 (48.8 to 58.6 kg/day-m^2) gave 7% to 9% solids. (4) Waste activated sludge at 5 to 6 lb/day-ft^2 (24.4 to 29.3 kg/day-m^2) gave 2.5% to 3.0% solids. (5) Trickling filter humus at 8 to 10 lb/day-ft^2 (39.1 to 48.8 kg/day-m^2) gave 7% to 9% solids (EPA, 1979).

SLUDGE DEWATERING

Anaerobically digested sludge may be dewatered by air drying, vacuum filters, centrifuges, filter presses, and sludge lagoons. Sludge air dried on sand beds may have a solids content as high as 30% to 45%, and it can usually be removed once the solids are about 25%. The solids content obtained by dewatering using mechanical means depends primarily upon the nature of the sludge, its solids content, and whether or not a conditioner is used.

Vacuum filtration of a chemically conditioned digested primary sludge at filter loadings of 5 to 8 lb/hr-ft^2 (24.4 to 39.1 kg/h-m^2) has produced a cake with 25% to 32% solids. Vacuum filtration of a chemically conditioned digested mixture of primary and secondary sludge at filter loadings of 3.5 to 6 lb/hr-ft^2 (17.1 to 29.3 kg/h-m^2) has produced a cake with 14% to 22% solids (EPA, 1979).

Centrifugation of chemically conditioned digested primary sludge has produced a cake of 28% to 35% solids, and centrifugation of a chemically conditioned digested mixture of primary and secondary sludge has produced a cake of 15% to 30% solids. Filter presses dewatering a chemically conditioned digested mixture of primary and secondary sludge have produced a cake of 45% to 50% solids (EPA, 1979).

When sludge lagoons are used for dewatering, the digested sludge is added to about a 2.0-ft (0.61-m) depth and allowed to dry; then the filling and drying are repeated. Once the lagoon is filled with dry sludge, the sludge is removed. The lagoon loading is usually from 2.2 to 2.4 lb/yr-ft^3 (35.3 to 38.5 kg/yr-m^3) volume (EPA, 1979).

REFERENCES

American Society of Civil Engineers (ASCE). 1982. *Wastewater Treatment Plant Design*. ASCE Manual of Practice No. 36. New York: ASCE.

Barnes, D.; Bliss, P. J.; Gould, B. W.; and Vallentine, H. R. 1981. *Water and Wastewater Engineering Systems*. London: Pittman Books Limited.

Brisbin, S. G. 1957. Flow of Concentrated Raw Sewage Sludges in Pipes. *Proc. SED*, ASCE 83: SA3.

Conway, R. A., and Ross, R. D. 1980. *Handbook of Industrial Waste Disposal*. New York: Van Nostrand Reinhold.

Eckenfelder, W. W., Jr. 1989. *Industrial Water Pollution Control*. 2nd ed. New York: McGraw-Hill.

Eckenfelder, W. W., Jr. 1980. *Principles of Water Quality Management*. Boston: CBI Publishing.

Eckenfelder, W. W., Jr., and O'Connor, D. J. 1961. *Biological Waste Treatment*. London: Pergamon Press.

Eckenfelder, W. W., Jr., and Weston, R. F. 1956. Kinetics of Biological Oxidation. In *Biological Treatment of Sewage and Industrial Wastes*. Vol. 1, edited by J. McCabe and W. W. Eckenfelder, Jr. New York: Reinhold.

Environmental Protection Agency (EPA). 1979. *Sludge Treatment and Disposal*. EPA Process Design Manual, Washington, D.C.

Environmental Protection Agency (EPA). 1974. *Upgrading Existing Wastewater Treatment Plants*. EPA Process Design Manual, Washington, D.C.

Fair, G. M.; Geyer, J. C.; and Okun, D.A. 1968. *Water Purification and Wastewater Treatment, and Disposal*. Vol. 2. *Water and Wastewater Engineering*. New York: Wiley.

Fair, G. M., and Moore, E. W. 1932. Heat and Energy Relations in the Digestion of Sewage Solids. *Sewage Works Jour.* 4 (February):242; (May):248; (July):589; (September):728.

Fair, G. M., and Moore, E. W. 1937. Observations on the Digestion of Sewage Sludge over a Wide Range of Temperatures. *Sewage Works Jour.* 9, no. 1:3.

Ford, D. L. 1970. General Sludge Characteristics. In *Advances in Water Quality Improvement by Physical and Chemical Processes*, edited by E. F. Gloyna and W. W. Eckenfelder, Jr. Austin, Tex.: University of Texas Press.

Great Lakes–Upper Mississippi River Board of State Sanitary Engineers. 1978. Recommended Standards for Sewage Works. Ten state standards. Albany, N.Y.

Great Lakes–Upper Mississippi River Board of State Public Health and Environmental Managers. 1990. Recommended Standards for Wastewater Facilities. Ten state standards. Albany, N.Y.

Hardenbergh, W. A., and Rodie, E. B. 1963. *Water Supply and Waste Disposal*. Scranton, Pa.: International Textbook Co.

Heukelekian, H. 1956. Basic Principles of Sludge Digestion. In *Biological Treatment of Sewage and Industrial Wastes*. Vol. 2, edited by J. McCabe and W. W. Eckenfelder, Jr. New York: Reinhold.

Imhoff, K., and Fair, G. M. 1956. *Sewage Treatment*. New York: Wiley.

Imhoff, K.; Mueller, W. J.; and Thistlethwayte, D. K. B. 1971. *Disposal of Sewage and Other Waterborne Wastes*. Ann Arbor, Mich.: Ann Arbor Science Publishers.

Kampelmacher, E. H., and van Noorde, J. L. M. 1972. Reduction of Bacteria in Sludge Treatment. *Jour. WPCF* 44, no. 2:309.

Langford, L. L. 1956. Mesophilic Anaerobic Digestion: Design and Operating Considerations. In *Biological Treatment of Sewage and Industrial Wastes*. Vol. 2, edited by J. McCabe and W. W. Eckenfelder, Jr. New York: Reinhold.

Lawrence, A. W., and McCarty, P. L. 1969. Kinetics of Methane Fermentation in Anaerobic Treatment. *Jour. WPCF* 41, no. 2, part 2:R1.

Malina, J. F., Jr., and Mihotits, E. M. 1968. New Developments in the Anaerobic Digestion of Sludges. In *Advances in Water Quality Improvement*, edited by E. F. Gloyna and W. W. Eckenfelder, Jr. Austin, Tex.: University of Texas Press.

McCarty, P. L. 1968. Anaerobic Treatment of Soluble Wastes. In *Advances in Water Quality Improvement*, edited by E. F. Gloyna and W. W. Eckenfelder, Jr. Austin, Tex.: University of Texas Press.

McCarty, P. L. 1964. Anaerobic Waste Treatment Fundamentals. *Public Works Magazine* 95, no. 9:107; 95, no. 10:123; 95, no. 11:91; and 95, no. 12:95.

McGhee, T. J. 1991. *Water Supply and Sewage*. 6th ed. New York: McGraw-Hill.

McKinney, R. E. 1962. *Microbiology for Sanitary Engineers*. New York: McGraw-Hill.

Metcalf & Eddy, Inc. 1991. *Wastewater Engineering: Treatment, Disposal and Reuse*. 3rd ed. New York: McGraw-Hill.

Metcalf & Eddy, Inc. 1979. *Wastewater Engineering: Treatment, Disposal and Reuse.* 2nd ed. New York: McGraw-Hill.

Metcalf & Eddy, Inc., 1972. *Wastewater Engineering: Collection, Treatment and Disposal.* New York: McGraw-Hill.

Novotny, V.; Imhoff, K.; Olthof, M.; and Krenkel, P. 1989. *Karl Imhoff's Handbook of Urban Drainage and Wastewater Disposal.* New York: Wiley.

Qasim, S. R. 1985. *Wastewater Treatment Plant.* New York: Holt, Rinehart and Winston.

Ramalho, R. S. 1977. *Introduction to Wastewater Treatment Processes.* New York: Academic Press.

Sawyer, C. N. 1956. An Evaluation of High-Rate Digestion. In *Biological Treatment of Sewage and Industrial Wastes.* Vol. 2, edited by J. McCabe and W. W. Eckenfelder, Jr. New York: Reinhold.

Schroeder, E. D. 1977. *Water and Wastewater Treatment.* New York: McGraw-Hill.

Sundstrom, D. W., and Klei, H. E. 1979. *Wastewater Treatment.* Englewood Cliffs, N.J.: Prentice-Hall.

Tebbutt, T. H. Y. 1992. *Principles of Water Quality Control.* 4th ed. Oxford, England: Pergamon Press Limited.

Texas State Department of Health. 1976. Design Criteria for Sewage Systems. Austin, Tex.

Water Environment Federation (WEF). 1992. *Design of Municipal Wastewater Treatment Plants.* Vols. I and II, WEF Manual of Practice No. 8.

Water Pollution Control Federation (WPCF). 1968. *Anaerobic Sludge Digestion.* WPCF Manual of Practice No. 16, Washington, D.C.

Water Pollution Control Federation (WPCF). 1977. *Wastewater Treatment Plant Design.* WPCF Manual of Practice No. 8, Washington, D.C.

PROBLEMS

19.1 A heated low-rate digester is to be designed for an activated sludge plant treating the wastewater from 20,000 persons. The flowsheet is shown in Figure 19.15. The waste activated sludge flow, Q_w, is returned ahead of the primary clarifier, and the secondary sludge solids settle along with the influent solids that are removed. Both primary and secondary sludge solids are removed by the primary sludge pipe from the bottom of the unit — that is, the underflow from the unit. The primary sludge flow amounts to 1586 ft³/day (44.9 m³/d), and the digested sludge flow amounts to 564 ft³/day (16.0 m³/d). The digestion time at 98°F (35°C) is 23 days, and the storage time provided is 45 days. When the contents are stratified, the digesting and digested sludge occupies the bottom half of the side water depth. Determine, using USCS units:

a. The average volume of the digesting sludge.

b. The volume of the digesting sludge.

c. The volume of the stored digested sludge.

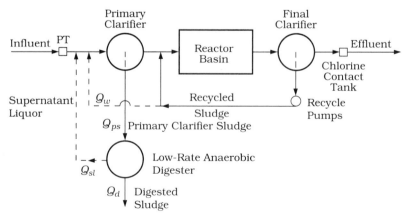

FIGURE 19.15 *Flowsheet for an Activated Sludge Plant with a Low-Rate Anaerobic Digester*

d. The volume of the digester.

19.2 Repeat Problem 19.1 using SI units.

19.3 A heated high-rate digester is to be designed for an activated sludge plant treating the wastewater from 20,000 persons. The flowsheet is shown in Figure 19.16. The waste activated sludge flow, Q_w, is returned ahead of the primary clarifier, and the secondary sludge solids settle along with the influent solids that are removed. Both primary and secondary sludge solids are removed by the primary sludge pipe from the bottom of the unit — that is, the underflow from the unit. The primary sludge flow amounts to 1586 ft^3/day (44.9 m^3/d), and the operating temperature of the digester is 98°F (35°C). Determine, using USCS units:

a. The minimum mean cell residence time, θ_c^{Min}.

b. The design mean cell residence time, θ_c, if a factor of safety of 2.5 is used.

c. The volume of the digester.

19.4 Repeat Problem 19.3 using SI units.

19.5 A heated low-rate digester is to be designed for an activated sludge plant treating the wastewater from 16,000 persons. The flowsheet is shown in Figure 19.15. The fresh sludge in the underflow pipe from the primary clarifier has 0.24 lb dry solids/cap-day (0.109 kg/cap-d), volatile solids are 72% of the dry solids, dry solids are 4.5% of the sludge, and the wet specific gravity is 1.01. Sixty-five percent of the volatile solids are destroyed by digestion, and the fixed solids remain unchanged. The digested sludge has 7.0% dry solids, and the wet specific gravity is 1.03. The operating temperature is 98°F (35°C), the digestion time is 23 days, and the storage time provided for the digested sludge is 45 days. When the contents are stratified, the sludge occupies the lower half of the side water depth, and the supernatant liquor occupies the top half. The gas production is 15 ft^3 per lb of volatile solids destroyed (0.935 m^3/kg). Determine, using USCS units:

a. The total digester volume.

b. The volume per capita.

c. The gas production per day.

19.6 Repeat Problem 19.5 using SI units.

19.7 A high-rate digester is to be designed for an activated sludge plant treating the wastewater from 16,000 persons. The flowsheet is shown in Figure 19.16. The fresh sludge in the underflow pipe from the primary clarifier has 0.24 lb dry solids/cap-day (0.109 kg/cap-day), the volatile solids are 72% of the dry solids, the dry solids are 4.5% of the sludge, and the wet specific gravity of the sludge is 1.01. The digester operating temperature is 98°F (35°C). Sixty-five percent of the volatile solids are destroyed during digestion, and the gas production is 15 ft^3/lb

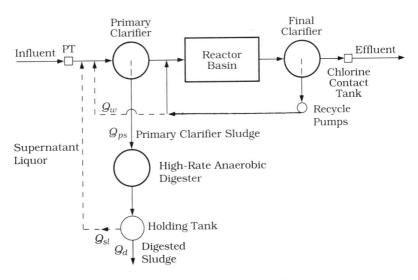

FIGURE 19.16 *Flowsheet for an Activated Sludge Plant with a High-Rate Anaerobic Digester*

volatile solids destroyed (0.935 m³/kg). Determine, using USCS units:

a. The design mean cell residence time, θ_c, if a safety factor of 2.5 is used.
b. The total digester volume.
c. The gas produced per day.

19.8 Repeat Problem 19.7 using SI units.

19.9 A heated low-rate digester with a floating cover is to be designed for an activated sludge plant treating the wastewater from 20,000 persons. The flowsheet is shown in Figure 19.15. The fresh sludge in the underflow pipe from the primary clarifier has 0.23 lb dry solids/cap-day (0.104 kg/cap-d), volatile solids are 70% of the dry solids, dry solids are 4.0% of the sludge, and the wet specific gravity of the sludge is 1.01. Sixty-five percent of the volatile solids are destroyed by digestion, and the fixed solids remain unchanged. The digested sludge has 7.0% dry solids, and the wet specific gravity is 1.03. The operating temperature of the digester is 98°F (35°C), the digestion time is 23 days, and the storage time provided for the digested sludge is 45 days. When the contents are stratified, the sludge occupies the bottom half of the side water depth, and the supernatant occupies the top half. The gas production is 15 ft³/lb volatile solids destroyed (0.935 m³/kg). Determine, using USCS units:

a. The total digester volume.
b. The theoretical diameter of the digester floating cover if the side water depth is between 28 ft and 30 ft. Assume a 29 ft depth for calculations.
c. The design diameter of the digester floating cover if covers are available in 5-ft increments in diameter.
d. The side water depth.
e. The volume per capita.
f. The percent volatile solids in the digested sludge.
g. The gas production per day.
h. The fuel value of the gas per day if the sludge gas is 65% methane, and methane has a fuel value of 963 Btu per ft³.

19.10 Repeat Problem 19.9 using SI units. Determine:

a. The total digester volume.
b. The theoretical diameter of the digester if the side water depth is 8.84 m.
c. The volume per capita.
d. The percent volatile solids in the digested sludge.
e. The gas production per day.
f. The fuel value of the gas per day if the sludge gas

is 65% methane, and methane has a fuel value of 35,850 kJ/m³.

19.11 A primary sludge is 4.5% dry solids and the volatile solids are 70%. A digested sludge has 7.0% dry solids and the volatile solids are 40%. Determine:

a. The specific gravities of the dry solids.
b. The specific gravities of the wet sludges.

19.12 A low-rate trickling filter plant treats the wastewater from 10,000 persons. The plant gets 95% BOD₅ removal, 33% BOD₅ removal by the primary clarifier, and 65% suspended solids removal by the primary clarifier. The load on the plant is 0.17 lb BOD₅/cap-day (0.077 kg/cap-d) and 0.20 lb suspended solids/cap-day (0.091 kg/cap-d). The primary sludge has 4.0% dry solids and a wet specific gravity of 1.01. The secondary sludge has 6.0% dry solids and a wet specific gravity of 1.02. The biological solids produced by the trickling filters amount to 0.30 lb solids/lb BOD₅ removed (kg/kg). Determine, using USCS units:

a. The primary sludge flowrate per day.
b. The secondary sludge flowrate per day.

19.13 Repeat Problem 19.12 using SI units.

19.14 An activated sludge treatment plant treats the wastewater from 20,000 persons and has a low-rate anaerobic digester. The mixture of primary and secondary sludge amounts to 0.23 lb/cap-day (0.104 kg/cap-d), the fresh sludge has 4.5% dry solids, and the digester operating temperature is 98°F (35°C). During the coldest month, January, the fresh sludge temperature is 52°F (11.1°C). The specific heat constant is 1.0 Btu/lb-°F (4200 J/kg-°C). Determine, using USCS units, the heat required to raise the temperature of the fresh incoming sludge to the temperature in the digester.

19.15 Repeat Problem 19.14 using SI units.

19.16 The low-rate digester for the activated sludge plant in Problem 19.14, which serves 20,000 persons, has a floating cover and is 65 ft (19.8 m) in diameter with a side water depth of 30 ft (9.14 m). The digester is recessed 3 ft (0.91 m) in the earth. For insulation, the earth is mounded around the digester up to a height of 14 ft (4.27 m). A wall height of 30 ft − (3 ft + 14 ft) = 13 ft [9.14 m − (0.91 m + 4.27 m) = 3.96 m] is exposed to air on the outside and to digester contents on the inside. The average monthly temperature in January is 43°F (6.1°C). The digester operating temperature is 98°F (35°C), and the dry earth mounded around the digester has a temperature equal to the average of the digester operating temperature and the average monthly temperature. The earth temperature below the digester floor is 98°F (35°C), and the temperature of the earth around the wall at the 3-ft (0.91-m) recess is 98°F (35°C). Determine, using USCS units, the heat required to make up for heat losses from the digester.

19.17 Repeat Problem 19.16 using SI units.

19.18 The fresh sludge solids for the 20,000-person municipality in Problems 19.14 and 19.16 have 70% volatile solids, and 65% of the volatile solids are destroyed during digestion. The sludge gas produced amounts to $15 \, \text{ft}^3/\text{lb}$ of volatile solids destroyed $(0.935 \, \text{m}^3/\text{kg})$ and is 65% methane. The fuel value of methane at standard conditions is $963 \, \text{Btu/ft}^3$ $(35,850 \, \text{kJ/m}^3)$. The sludge heater that heats the incoming sludge and makes up for digester heat losses has a thermal efficiency of 60%. Determine, using USCS units, what percent of the fuel value produced is required for the sludge heater.

19.19 Repeat Problem 19.18 using SI units.

19.20 A centrifugal sludge pump with a capacity of 350 gpm (1325 ℓpm) pumps a mixture of primary and secondary sludge from the primary clarifier to the anaerobic digester at a municipal wastewater treatment plant. The flowsheet for the plant is shown in Figure 19.15. The secondary sludge is mixed with the influent to the primary clarifier and allowed to resettle with the primary settlings in order to thicken the secondary sludge. The piping from the primary clarifier to the sludge pump and from the sludge pump to the digester is 8 in. (200 mm) in diameter. The discharge pipe in the digester runs vertically up to the gas dome of the floating cover, and the difference in elevation between the water surface of the primary clarifier and the end of the vertical discharge pipe is 31.0 ft (9.45 m). The gas is under a pressure of 8 in. (200 mm) of water column. The sludge mixture is 5.0% dry solids. The

length of the 8-in. (200-mm) pipe is 265 ft (80.8 m), and the fittings and form losses are equivalent to 125 ft (38.1 m) of pipe. The pump is operated by a time clock, and it runs for a preset time interval once every hour. Determine, using USCS units, the head at which the pump operates.

19.21 Repeat Problem 19.20 using SI units.

19.22 A heated low-rate digester is to be designed for an activated sludge plant treating the wastewater from 16,000 persons. The flowsheet is shown in Figure 19.15. The fresh sludge in the underflow pipe from the primary clarifier has 0.24 lb dry solids/cap-day (0.109 kg/cap-d), volatile solids are 72% of the dry solids, dry solids are 4.5% of the sludge, and the wet specific gravity is 1.01. Sixty-five percent of the volatile solids are destroyed by digestion, and 25% of the volatile solids destroyed are converted to fixed solids. The digested sludge has 7.0% dry solids, and the wet specific gravity is 1.03. The operating temperature is 98°F (35°C), the digestion time is 23 days, and the storage time provided for the digested sludge is 45 days. When the contents are stratified, the sludge occupies the lower half of the side water depth, and the supernatant liquor occupies the top half. Determine, using USCS units:

a. The total digester volume.

b. The volume per capita.

c. How do these values compare to those of Problem 19.5?

19.23 Repeat Problem 19.22 using SI units. How do the values compare to those of Problem 19.6?

19.24 A digester gas is 70% methane, is 30% carbon dioxide, and has trace amounts of other gases. Using USCS units, determine the pseudomolecular weight and the specific volume in ft^3/lb at STP.

19.25 A digester gas is 70% methane, is 30% carbon dioxide, and has trace amounts of other gases. Using SI units, determine the pseudomolecular weight and specific volume in m^3/kg at STP.

20

AEROBIC DIGESTION

Aerobic digestion may be defined as the biological oxidation of organic sludges under aerobic conditions. It closely resembles the activated sludge process since the aeration equipment and tanks are similar. Most of the microbes used in the process are facultative; however, some are obligate aerobes, such as the nitrifying bacteria. The following types of sludge have been successfully treated by aerobic digestion: (1) waste activated sludge, (2) primary and waste activated sludge, (3) trickling filter secondary sludge (humus), and (4) primary and trickling filter secondary sludge (humus). The most widespread use of aerobic digestion has been in the treatment of waste activated sludge. Most aerobic digesters that have been built have been for small- to medium-size plants, and most package plants have aerobic digestion systems.

Some advantages of aerobic digestion compared to anaerobic digestion are fewer operational problems, thus less laboratory control and daily maintenance required, much lower BOD concentrations in the supernatant liquor, and lower capital costs. The disadvantages of aerobic digestion compared to anaerobic digestion are as follows: Energy requirements are higher because an appreciable amount of aeration and mixing is required, methane that is a usable by-product is not produced, and the digested sludge has a lower solids content, and thus the volume of sludge to be dewatered is larger. The volatile solids loading and percent solids destruction are about the same for both aerobic and anaerobic digestion.

Aerobic digesters can be operated in either a batchwise or a continuous-flow manner, as shown in Figures 20.1, 20.2, and 20.3; however, the majority of the systems are continuous-flow. Not only do continuous-flow systems have lower operational costs, but they also provide relatively constant environmental conditions that aid in rapid digestion. If a batch digester is used, it is operated on a fill-and-draw basis, as shown in Figure 20.1. The continuous-flow systems may be operated with or without a thickener, but in nearly all cases, a thickener is provided. Details of a gravity-type sludge thickener are shown in Chapter 21. The feed sludge enters and settles to the bottom half of the tank, and most of the liquid or supernatant occupies the top half of the tank. The sludge rakes have vertical pickets mounted to them, and as the rakes move, the pickets rake through the sludge mass. The sludge mass will contain pockets of entrained water, and as the pickets pass through them, the water pockets are broken up and the water moves vertically upward to the supernatant liquid above the sludge mass. As the water pockets are released, the solids content of the sludge increases and the

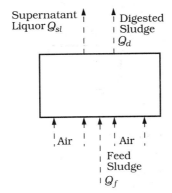

Supernatant Liquor Q_{sl}

Digested Sludge Q_d

Air   Air

Feed Sludge Q_f

(a) Batch-Operated Aerobic Digester

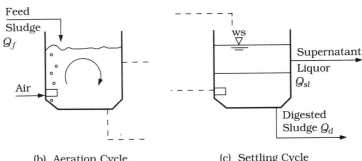

Feed Sludge Q_f

Air

ws

Supernatant Liquor Q_{sl}

Digested Sludge Q_d

FIGURE 20.1
Batch-Operated Aerobic Digester

(b) Aeration Cycle
(Sludge Feeding and Digestion)

(c) Settling Cycle
(Digested Sludge and Supernatant Withdrawal)

thickened sludge is removed at the bottom of the tank. Usually the thickened sludge has a solids content that is 2 to 5 times that of the feed sludge. The supernatant leaves the top portion of the tank and is returned to the head of the plant. If a digester does not have a thickener, the solids concentration in the digester will be the same as in the feed sludge, and the mean cell residence time will be equal to the hydraulic detention time. When a thickener is used, the solids concentration in the digester will be several orders of magnitude larger than in the feed sludge; this has the desirable result of producing a long mean cell residence time.

Aerobic digesters usually require a thickener either upstream or downstream from the digester, as shown in Figures 20.2 and 20.3. If the secondary sludge is wasted to the influent ahead of the primary clarifier, the solids content will be from 4% to 6% and a thickener is not required. In this case, the primary clarifier will have to be larger than otherwise, but the extra cost may offset the cost of a thickener. The usual case has a thickener ahead of the digester, as shown in Figure 20.2. The fresh sludge is thickened prior to digestion, which makes the digester volume smaller than without a thickener if the design is based on the detention time of the sludge flow. If the

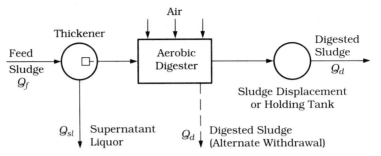

FIGURE 20.2 *Continuous-Flow Aerobic Digester System with Thickener Prior to Digester*

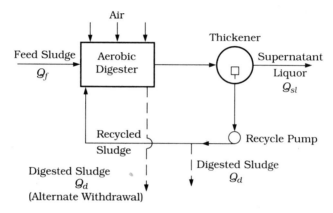

FIGURE 20.3 *Continuous-Flow Aerobic Digester System with Thickener Downstream from Digester*

thickener doubles the sludge solids content, the digester volume will be one-half of the volume without a thickener. This layout has a displacement or holding tank into which the digested sludge continuously overflows. The sludge in the holding tank has a fluctuating sludge depth so solids do not have to be continuously dewatered, which makes this arrangement very flexible. Digested sludge may be withdrawn directly from the digester, as shown in Figures 20.2 and 20.3, and for this case the aeration may be discontinued so that the sludge will settle and thicken prior to removal. This decreases the volume of sludge to be dewatered and decreases the costs of dewatering. Sludge can be withdrawn with the aeration system operating as shown in Figures 20.2 and 20.3. In Figure 20.3, the withdrawal is usually from the recycle line and the withdrawal is usually continuous.

It is essential to estimate accurately the sludge flows that are to be treated and their solids concentrations, because these significantly affect the design of the digester, the thickener-clarifier, and other digester facilities. If the digester system is inadequate in size, the solids will not be properly digested, and the thickener-clarifier will not give the proper solid-liquid separation.

When primary sludge is mixed with waste activated sludge or trickling filter humus and the combination is aerobically digested, there will be both oxidation of the organic matter in the primary sludge and endogenous oxidation of the cell mass produced from the biological oxidation and from the activated sludge or filter humus.

The generalized biochemical equation for the aerobic digestion of primary sludge solids is

$$\frac{\text{Organic}}{\text{matter}} + O_2 \xrightarrow[\text{microbes}]{\text{Aerobic}} \frac{\text{New}}{\text{cells}} + \frac{\text{Energy}}{\text{for cells}} + CO_2 + H_2O + \frac{\text{Other end}}{\text{products}}$$

(20.1)

Some of the other end products include NH_4^+, NO_2^-, NO_3^-, and PO_4^{-3}. During the biological oxidation, an appreciable amount of nitrogen is converted to the nitrate ion. If the $\text{BOD}_5 = 0.68\,\text{BOD}_u$, the oxygen required is $1/0.68$ or $1.47\,\text{lb}\,O_2/\text{lb}\,\text{BOD}_5$ removed ($1.47\,\text{kg/kg}$). Studies have shown that up to $1.9\,\text{lb}\,O_2/\text{lb}\,\text{BOD}_5$ is usually required ($1.9\,\text{kg/kg}$). The excess above $1.47\,\text{lb}$ ($1.47\,\text{kg}$) is the amount of oxygen required for nitrification.

The amount of living cell mass originally present from waste activated sludge or trickling filter humus, and the living cell mass produced by oxidation of the primary solids, is gradually reduced by endogenous decay until it is almost completely stabilized. The biochemical equation for the endogenous decay of cells with the nitrogen terminating in ammonia is

$$C_5H_7NO_2 + 5O_2 \rightarrow 5CO_2 + 2H_2O + NH_3 \qquad (20.2)$$

For this equation, the amount of oxygen required per pound of cell mass auto-oxidized ($MW = 113$) is $5 \times 32/113$ or $1.42\,\text{lb}$ of oxygen per pound of cells ($1.42\,\text{kg/kg}$). An appreciable amount of ammonia produced from this auto-oxidation is subsequently oxidized to the nitrate ion as the digestion proceeds. The biochemical equation for the overall endogenous decay of cell mass to form NO_3^- as one of the end products is

$$C_5H_7NO_2 + 7O_2 \rightarrow 5CO_2 + 3H_2O + H^+ + NO_3^- \qquad (20.3)$$

In this reaction, the cell nitrogen terminates as NO_3^-. For this reaction, the amount of oxygen required per pound of cell mass is $7 \times 32/113$ or $1.98\,\text{lb}$ of oxygen per pound of cell mass oxidized ($1.98\,\text{kg/kg}$). Research on the aerobic digestion of waste activated sludge has shown that $1.86\,\text{lb}$ of oxygen is required per pound of cell mass destroyed ($1.86\,\text{kg/kg}$). Approximately two-thirds of the cell mass is auto-oxidized during aerobic digestion. The remaining cell material consists of organic compounds, such as peptidoglycans, that are not readily biodegradable.

The pH of the digesting sludge is dependent upon the buffering capacity of the system and may drop as low as 5 to 6 at long hydraulic detention times; however, this does not inhibit the microbial action. The pH drop is possibly due to carbon dioxide production, which reduces the pH of the system, and possibly due to the increase in the nitrate ion concentration.

KINETICS OF AEROBIC BIOLOGICAL OXIDATION

The rate of aerobic oxidation of solid organic materials frequently has been found to be represented by the following pseudo–first-order biochemical equation:

$$-\frac{dX}{dt} = K_d X \qquad (20.4)$$

where

dX = change in the biodegradable organic matter

dt = time interval

K_d = reaction rate or degradation constant

X = concentration of biodegradable materials at any time t

Equation (20.4) may be arranged for integration between definite limits as follows:

$$\int_{X_0}^{X_t} \frac{dX}{X} = -K_d \int_0^t dt \qquad (20.5)$$

where X_t and X_0 represent the biodegradable matter at the respective times $t = t$ and $t = 0$. Integration gives

$$\ln \frac{X_t}{X_0} = -K_d t \qquad (20.6)$$

This equation may be rearranged to give

$$\frac{X_t}{X_0} = e^{-K_d t} \qquad (20.7)$$

Equation (20.6) will plot a straight line on semilog paper with the X_t/X_0 values on the y-axis and the t values on the x-axis. The slope will be $-K_d/2.303$.

DESIGN CONSIDERA-TIONS

The most important design and operational parameters for aerobic digesters are shown in Tables 20.1 and 20.2. The hydraulic detention time is equal to the digester volume divided by the feed sludge flowrate. If sludge recycle is used, the recycled sludge flowrate is not incorporated in the computations. The hydraulic detention time depends on the nature of the sludge and on the operational temperature. Waste activated sludge is more readily degradable than a mixture of primary and waste activated sludge; thus the

TABLE 20.1 *Aerobic Digester Design Parameters*

PARAMETER	VALUE
Hydraulic Detention Time, days at 20°C	
Primary and waste activated sludge or trickling filter sludge	18–22[l]
Waste activated sludge from a biosorption or contact stabilization plant (no primary settling)	16–18[l]
Waste activated sludge	12–16[l]
Minimum Design Mean Cell Residence Time, θ_c, days	
Primary and waste activated sludge	15–20[a,h,k,l]
Waste activated sludge	10–15[c,j,n]
Maximum Design Mean Cell Residence Time, θ_c, days	45–60
Solids Concentration, mg/ℓ	Up to 50,000
Organic Loading	
lb volatile solids per ft^3-day	0.04–0.20[c,g,i,k,l,n]
(kg volatile solids per m^3-day)	(0.64–3.20)
Volume Loading	
ft^3 per capita	1.5–4[d,i,n,q]
(m^3 per capita)	(0.042–0.113)
Operating Temperature	>15°C[f]
Volatile Solids Destruction, %	40–75[a,n,o]
Typical, %	65
Solids Destruction, %	35–55[a,n,o]
Oxygen Requirements	
Primary sludge, lb O_2/lb BOD$_5$ destroyed (kg/kg)	1.9[l]
Waste activated sludge, lb O_2/lb solids destroyed (kg/kg)	2.0[l,n]
Trickling filter humus, lb O_2/lb solids destroyed (kg/kg)	2.0[l]
Minimum Dissolved Oxygen, mg/ℓ	2.0[l,m,n,o]
Mixing Requirements	
Diffused aeration for primary and waste activated sludge, scfm/1000 ft^3 (m^3/min-1000 m^3)	>60[c,n,p]
Diffused aeration for waste activated sludge, scfm/1000 ft^3 (m^3/min-1000 m^3)	20–40[b,n,p]
Mechanical aeration for primary and waste activated sludge or waste activated sludge,	
hp/1000 ft^3	0.75–1.50[r]
(kW/1000 m^3)	(19.8–39.5)[r]

hydraulic detention time will be less. Table 20.1 gives the usual detention times for various sludges at a 20°C operating temperature. The required hydraulic detention time at temperatures other than 20°C is given by the equation

$$\theta_{h_2} = \theta_{h_{20}} \cdot \theta^{20 - T_2} \tag{20.8}$$

TABLE 20.2 *Reaction Rate or Degradation Constants* (K_d)

TYPE SLUDGE	TEMPERATURE (°C)	SOLIDS CONCENTRATION (mg/ℓ)	K_d (days^{-1})
Primary and Waste Activated Sludges	15	32,000	0.017[h]
Primary and Waste Activated Sludges	20	32,000	0.180[h]
Primary and Waste Activated Sludges	35	32,000	0.177[h]
Primary and Waste Activated Sludges	—	—	0.30[a]
Waste Activated Sludges			
Municipal Wastes	25	7,800	0.71[n]
Municipal Wastes	25	12,400	0.62[n]
Municipal Wastes	25	15,050	0.51[n]
Municipal Wastes	25	21,260	0.44[n]
Municipal Wastes	25	22,700	0.34[n]
Municipal Wastes	—	—	0.28[a]
Municipal and Textile Waste	—	—	0.43[a]
Pharmaceutical Waste	—	—	0.46[a]
Spent Sulfite Liquor Waste	—	—	0.19[a]
Primary and Waste Activated Sludge, Pulp and Paper Waste	—	—	0.14[a]

a. E. Barnhart, "Application of Aerobic Digestion to Industrial Waste Treatment," *Proceedings of the 16th Annual Purdue Industrial Waste Conference*, May 1961, p. 612.

b. R. S. Burd, "A Study of Sludge Handling and Disposal," FWPCA Publication no. WP-20-4, May 1968.

c. D. E. Dreier, "Aerobic Digestion of Solids," *Proceedings of the 18th Annual Purdue Industrial Waste Conference*, May 1963, p. 123.

d. D. E. Dreier, "Discussion on Aerobic Sludge Digestion," by N. Jaworski, G. W. Lawton, and G. A. Rohlich. Paper presented at the Conference on Biological Waste Treatment, Manhattan College, New York City, April 1960.

e. Environmental Protection Agency, *Sludge Treatment and Disposal*, EPA Process Design Manual, September 1979.

f. Environmental Protection Agency, *Upgrading Existing Wastewater Treatment Plants*, EPA Process Design Manual, October 1974.

g. R. H. L. Howe, "What to Do with Supernatant," *Waste Engineering* 30, no. 1 (January 1959): 12.

h. N. Jaworski, G. W. Lawton, and G. A. Rohlich, "Aerobic Sludge Digestion." Paper presented at the Conference on Biological Waste Treatment, Manhattan College, New York City, April 1960.

i. C. E. Levis, C. R. Miller, and L. E. Bosburg, "Design and Operating Experiences Using Turbine Dispersion for Aerobic Diqestion," *Jour. WPCF* 43, no. 3 (March 1971):417.

j. R. C. Loehr, "Aerobic Digestion: Factors Affecting Design," *Water and Sewage Works* 112 (30 November 1965):R169.

k. J. F. Malina and H. M. Burton, "Aerobic Stabilization of Primary Wastewater Sludge," *Proceedings of the 19th Annual Purdue Industrial Waste Conference*, May 1964, p. 123.

l. Metcalf & Eddy, Inc., *Wastewater Engineering: Treatment, Disposal and Reuse*, 2nd ed., (New York: McGraw-Hill, 1979).

m. C. N. Randall and C. T. Koch, "Dewatering Characteristics of Aerobically Digested Sludge," *Jour. WPCF* 41, no. 5, part 2 (May 1969):R215.

n. T. D. Reynolds, "Aerobic Digestion of Thickened Waste Activated Sludge," *Proceedings of the 28th Annual Purdue Industrial Waste Conference*, part 1, May 1973, p. 12.

o. L. Ritter, "Design and Operating Experiences Using Diffused Aeration for Sludge Digestion," *Jour. WPCF* 42, no. 10 (October 1970):1782.

p. R. Smith, R. G. Eilers, and E. D. Hall, "A Mathematical Model for Aerobic Digestion," EPA, Office of Research and Monitoring, Advanced Waste Treatment Research Laboratory, Cincinnati, Ohio, February 1973.

q. Texas State Department of Health, "Design Criteria for Sewage Systems," Austin, Tex., 1976.

r. Metcalf & Eddy, Inc., *Wastewater Engineering: Treatment, Disposal and Reuse*, 3rd ed. (New York: McGraw-Hill, 1991).

where

θ_{h_2} = hydraulic detention time in days at T_2 (°C)

$\theta_{h_{20}}$ = hydraulic detention time at 20°C

T_2 = temperature, °C

The value of θ ranges from 1.02 to 1.11; however, most θ values are in the upper half of this range. This high θ range shows that the process is extremely temperature dependent at usual detention times. Because appreciable air-water contact occurs during aeration, it may be assumed that the temperature of the digester contents could approach the average monthly temperature.

The mean cell residence time, θ_c, or solids retention time, θ_s, is given by

$$\theta_c = \frac{X}{\Delta X} \qquad (20.9)$$

where

X = pounds (kilograms) of solids in the digester

ΔX = pounds (kilograms) of solids produced per day in the digested sludge

In Eq. (20.9) the solids may be total or volatile solids. If the digester is operated without a thickener, the solids in the digester will be at the same concentration as the solids in the feed sludge, and the solids retention time will be equal to the hydraulic detention time. However, if a thickener is used, the solids concentration in the digester will be greater than in the feed sludge, and the mean cell residence time will be greater than the hydraulic detention time.

The reaction or degradation constant, K_d (Table 20.2), depends on the nature of the sludge, its solids concentration, and the temperature. Waste activated sludge, which is readily degraded, will have a high K_d value. A mixture of primary and waste activated sludge is not as easily degraded and, as a result, will have a lower K_d value. As the concentration of a sludge is increased, the K_d value decreases. The K_d value and the fraction of the sludge that is biodegradable may be determined by laboratory or pilot-plant studies.

The solids concentration may be up to 50,000 mg/ℓ; however, the usual range is from 25,000 to 35,000 mg/ℓ. It is essential to have a proper thickener-clarifier design in order to ensure that solids are maintained at these levels. Mixing requirements to keep the solids in suspension must be determined in addition to oxygen requirements.

The organic loading, based on limited data, should be in the range of 0.04 to 0.20 lb of volatile solids per day-ft^3 (0.064 to 3.2 kg/day-m^3), which is similar to anaerobic digester loadings. Usually, aerobic digester design is based on hydraulic detention time and solids retention time; however, the organic loading should be checked. The volumetric loading expressed in terms of population equivalents is usually from 1.5 to 4 ft^3/cap equivalent (0.042 to 0.113 m^3/cap).

The degree of volatile solids reduction or destruction depends on the nature of the sludge, the hydraulic detention time, the solids retention time, and the operating temperature, provided that no toxic substances are present in inhibitory concentrations. For mixtures of primary and waste activated sludge at operating temperatures of 15°C to 35°C, the volatile solids destruction has approximated a pseudo–first-order reaction. The required time and degree of volatile solids destruction will vary with the characteristics of the sludge and the operating temperature. The maximum volatile solids destruction is usually from 40% to 75%, and the maximum total solids destruction is usually from 35% to 55%. A typical value is 65% volatile solids destruction.

The maximum oxygen required for aerobic digestion of primary sludge is 1.9 lb of O$_2$ per pound of BOD$_5$ destroyed (1.9 kg/kg BOD$_5$ destroyed). Normally BOD$_5$ destruction in aerobic digestion is from 80% to 90%. For waste activated sludge and trickling filter humus, a maximum of 2.0 lb of O$_2$ are required per pound of solids destroyed (2.0 kg/kg solids destroyed). The dissolved oxygen concentration in a digester should be at least 2 mg/ℓ so that aerobic conditions will be maintained and digestion will proceed normally. If the oxygen concentration drops below 1 mg/ℓ, the inner portions of the solids will become anaerobic and poor digestion will occur. This, in turn, results in a sludge that has poor dewatering characteristics.

The mixing requirements depend primarily upon the nature of the sludge, the solids concentration, the sludge temperature, and the tank depth. Usually, the air needed to satisfy oxygen requirements is sufficient to attain adequate mixing; however, mixing requirements should be evaluated. A rational approach to mixing is needed. However, data collected so far indicate that for diffused compressed air used for the aerobic digestion of waste activated sludge, the air required for mixing is from 20 to 40 scfm per 1000 ft^3 (20 to 40 m^3/min–1000 m^3). The higher air requirements are needed for high suspended solids concentrations, relatively low sludge temperatures, and relatively shallow tanks. The lower air requirements are for low suspended solids concentrations, relatively high sludge temperatures, and relatively deep tanks. For diffused compressed air used in the aerobic digestion of mixtures of primary and waste activated sludge, the air required for mixing is about 60 scfm per 1000 ft^3 (60 m^3/min-1000 m^3) or more. For mechanical aeration, mixing requirements are usually from 0.75 to 1.50 hp per 1000 ft^3 (19.8 to 39.5 kW/1000 m^3), and bottom mixers are required if the solids are greater than about 8000 mg/ℓ. Since the degree of mixing by a mechanical aerator decreases with tank depth, high power levels are required for relatively deep tanks, and lower power levels are used for relatively shallow tanks.

The tanks or basins used for aerobic digestion are similar to those used for the activated sludge process, and both diffused compressed air and mechanical aeration are used. The criteria for aeration tanks and aeration systems, presented in Chapter 16, may also be used for aerobic digester design.

The thickener-clarifier used for an aerobic digester system should have surface skimmers to remove any floating materials, such as grease.

EXAMPLE 20.1

Aerobic Digester

An aerobic digester is to be designed for an activated sludge plant treating the wastewater from 10,000 persons. The flowsheet is shown in Figure 20.4. As shown, the primary and secondary sludge are to be blended together in a blending tank and then thickened and sent to the digester. Pertinent data are: influent BOD_5 = 200 mg/ℓ, influent suspended solids = 250 mg/ℓ, average flow = 100 gal/cap-day, 33% BOD_5 removal and 62% suspended solids removal by the primary clarifier, primary sludge has 5% solids, sludge density index = 6000 mg/ℓ, secondary sludge = 0.068 lb/cap-day, volatile solids for primary and secondary sludge = 70%, and the specific gravity of primary, secondary, blended, and thickened sludge = 1.01. Determine:

1. The primary, secondary, and blended sludge flow in ft^3/day.
2. The solids concentration in the blended flow.
3. The thickened sludge flow if the thickener increases the solids by 2.5 times and the supernatant liquor has negligible suspended solids.
4. The supernatant liquor flow.
5. The digester volume if the winter operating temperature is 18°C.

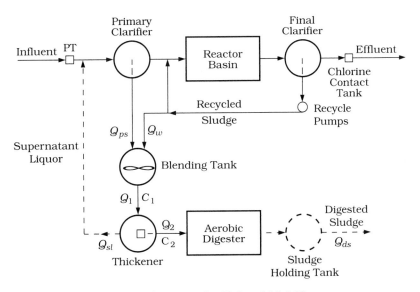

FIGURE 20.4 *Flowsheet for Examples 20.1 and 20.1 SI*

6. The oxygen requirements of the digester if 90% of the primary BOD$_5$ is destroyed and 65% of the waste activated sludge volatile solids are destroyed.

7. The air required if the air has $0.0175\,lb\,O_2/ft^3$ and the field transfer is 4%. Is this adequate for mixing?

8. The theoretical diameter of the thickener if the solids loading is $8\,lb/day\text{-}ft^2$ and the supernatant overflow rate is $600\,gal/day\text{-}ft^2$.

9. The design diameter of the thickener if scrapers are available in 5-ft increments in diameter.

SOLUTION   The primary sludge suspended solids are

$$lb/day = (10{,}000)\left(\frac{100\,gal}{cap\text{-}day}\right)\left(\frac{250}{10^6}\right)\left(\frac{8.34\,lb}{gal}\right)(0.62) = 1293\,lb/day$$

The primary sludge flow is

$$Q_{ps} = \left(\frac{1293\,lb}{day}\right)\left(\frac{100\,lb\ wet\ sludge}{5\,lb\ dry\ solids}\right)\left(\frac{ft^3}{62.4\,lb}\right)\left(\frac{1}{1.01}\right) = \boxed{410\,ft^3/day}$$

The waste activated sludge solids $= (10{,}000\,cap)(0.068\,lb/cap\text{-}day) = 680\,lb/day$. The waste activated sludge flow is

$$Q_w = \left(\frac{680\,lb}{day}\right)\left(\frac{10^6}{6000}\right)\left(\frac{ft^3}{62.4\,lb}\right)\left(\frac{1}{1.01}\right) = \boxed{1798\,ft^3/day}$$

The flow leaving the blending tank $= 410 + 1798 = \boxed{2208\,ft^3/day.}$ The solids content in the waste activated sludge $= 6000\,mg/\ell$ or 0.6%. A mass balance on the solids at the blending tank gives

$$Q_{ps} \cdot C_{ps} + Q_w \cdot C_w = (Q_{ps} + Q_w)C_1$$

or

$$(410)(5\%) + (1798)(0.6\%) = (410 + 1798)C_1$$

or

$$C_1 = \boxed{1.4\ percent}$$

The blended sludge, Q_1, is $410 + 1798 = 2208\,ft^3/day$. A mass balance on solids at the thickener gives

$$Q_1C_1 = Q_2C_2$$

Since the solids increase 2.5 times, $C_2 = (1.4)(2.5) = \boxed{3.5\%.}$ The thickened flow, Q_2, is

$$Q_2 = Q_1\frac{C_1}{C_2} = 2208\left(\frac{1.4}{3.5}\right) = \boxed{883\,ft^3/day}$$

A balance on the flows at the thickener gives

$$Q_1 = Q_2 + Q_{sl}$$

or

$$Q_{sl} = Q_1 - Q_2 = 2208 - 883 = \boxed{1325 \text{ ft}^3/\text{day}}$$

Assume $\theta_h = 20$ days at 20°C from Table 20.1. The correction for temperature is given by

$$\theta_{h_2} = \theta_{h_1} \cdot \theta^{20-T_2}$$

Assume θ is the mean value of 1.02 and 1.11, or 1.065. Thus,

$$\theta_{h_2} = 20 \cdot 1.065^{20-18}$$
$$= 22.7 \text{ days}$$

The digester volume $V = Q\theta_h$, or

$$V = (883 \text{ ft}^3/\text{day})(22.7 \text{ days})$$
$$= \boxed{20{,}000 \text{ ft}^3}$$

The oxygen required is for BOD$_5$ removal and waste activated sludge destruction. The BOD$_5$ removed is

$$\text{lb/day} = (10{,}000 \text{ cap})\left(\frac{100 \text{ gal}}{\text{cap-day}}\right)\left(\frac{200}{10^6}\right)\left(\frac{8.34 \text{ lb}}{\text{gal}}\right)(0.33)(0.90)$$
$$= 495 \text{ lb/day}$$

The waste activated sludge destruction is

$$\text{lb/day} = (10{,}000 \text{ cap})\left(\frac{0.068 \text{ lb}}{\text{cap-day}}\right)(0.70)(0.65) = 309 \text{ lb/day}$$

Assume, from Table 20.1, 1.9 lb O_2/lb BOD$_5$ removed and 2.0 lb O_2/lb VSS destroyed. The oxygen required is

$$O_2/\text{day} = (495)(1.9) + (309)(2.0)$$
$$= 941 + 618 = \boxed{1559 \text{ lb/day}}$$

The air required is

$$\text{ft}^3/\text{min} = \left(\frac{1560 \text{ lb}}{\text{day}}\right)\left(\frac{\text{day}}{1440 \text{ min}}\right)\left(\frac{\text{ft}^3}{0.0175 \text{ lb } O_2}\right)\left(\frac{1}{0.04}\right)$$
$$= \boxed{1550 \text{ cfm}}$$

The air per 1000 ft^3 is

$$\text{air} = \frac{1550}{20.0} = \boxed{78 \text{ cfm}/10^3 \text{ ft}^3}$$

Since 78 > 60 from Table 20.1, $\boxed{\text{the mixing is adequate.}}$

The dry solids of the sludge = 1293 + 680 = 1973 lb/day. The area required for thickening = 1973 lb/day/(8 lb/day-ft^2) = 246.6 ft^2. The area required for clarifying the supernatant = (1325 ft^3/day)(7.48 gal/ft^3) ÷ (600 gal/day-ft^2) = 16.5 ft^2. Since $(\pi/4)D^2$ = 246.6 ft^2, the theoretical diameter $D = \boxed{17.7\,\text{ft.}}$ For the next standard size, $D = \boxed{20\,\text{ft.}}$

(*Note*: In this problem, the design air required was determined by the field transfer percent to simplify the computations. Also, the primary and secondary sludges were blended prior to thickening, as is done in many small plants. In medium- to large-size activated sludge plants, the primary and secondary sludges are usually thickened separately and then blended. This is done because activated sludge usually thickens better by itself.)

EXAMPLE 20.1 SI   *Aerobic Digester*

An aerobic digester is to be designed for an activated sludge plant treating the wastewater from 10,000 persons. The flowsheet is shown in Figure 20.4. As shown, the primary and secondary sludge are to be blended together in a blending tank and then thickened and sent to the digester. Pertinent data are: influent BOD$_5$ = 200 mg/ℓ, influent suspended solids = 250 mg/ℓ, average flow = 380 ℓ/cap-d, 33% BOD$_5$ removal and 62% suspended solids removal by the primary clarifier, primary sludge has 5% solids, sludge density index = 6000 mg/ℓ, secondary sludge = 0.031 kg/cap-d, volatile solids for primary and secondary sludge = 70%, and the specific gravity of primary, secondary, blended, and thickened sludge = 1.01. Determine:

1. The primary, secondary, and blended sludge flow in m^3/d.
2. The solids concentration in the blended flow.
3. The thickened sludge flow if the thickener increases the solids by 2.5 times and the supernatant liquor has negligible suspended solids.
4. The supernatant liquor flow.
5. The digester volume if the minimum operating temperature is 18°C.
6. The oxygen requirements of the digester if 90% of the primary BOD$_5$ is destroyed and 65% of the volatile solids in the waste activated sludge are destroyed.
7. The air required if the air has 0.281 kg O$_2$/m^3 and the field transfer is 4%. Is the mixing adequate?
8. The diameter of the thickener if the solids loading is 39.1 kg/d-m^2 and the supernatant overflow rate is 24.4 m^3/d-m^2.

SOLUTION   The primary sludge suspended solids are

$$\text{kg/d} = (10{,}000\,\text{cap})\left(380\frac{\ell}{\text{cap-d}}\right)\left(\frac{250\,\text{mg}}{\ell}\right)\left(\frac{\text{gm}}{1000\,\text{mg}}\right)\left(\frac{\text{kg}}{1000\,\text{gm}}\right)(0.62)$$

$$= 589\,\text{kg/d}$$

The primary sludge flow is

$$Q_{ps} = \left(\frac{589 \text{ kg}}{d}\right)\left(\frac{100 \text{ kg wet sludge}}{5 \text{ kg dry solids}}\right)\left(\frac{m^3}{1000 \text{ kg}}\right)\left(\frac{1}{1.01}\right) = \boxed{11.7 \text{ m}^3/\text{d}}$$

The waste activated sludge solids = (10,000 cap)(0.031 kg/cap-d) = 310 kg/d.
The waste activated sludge flow is

$$Q_w = \left(\frac{310 \text{ kg}}{d}\right)\left(\frac{1000 \text{ gm}}{kg}\right)\left(\frac{1000 \text{ mg}}{gm}\right)\left(\frac{1}{6000 \text{ mg}}\right)\left(\frac{m^3}{1000 \text{ } \ell}\right)\left(\frac{1}{1.01}\right)$$

$$= \boxed{51.2 \text{ m}^3/\text{d}}$$

The flow leaving the blending tank = 11.7 + 51.2 = $\boxed{62.9 \text{ m}^3/\text{d.}}$ The solids content in the waste activated sludge = 6000 mg/ℓ or 0.6%. A mass balance on the solids at the blending tank gives

$$Q_{ps} \cdot C_{ps} + Q_w \cdot C_w = (Q_{ps} + Q_w)C_1$$

or

$$(11.7)(5\%) + (51.2)(0.6\%) = (11.7 + 51.2)C_1$$

or

$$C_1 = \boxed{1.4 \text{ percent}}$$

The blended sludge, Q_1, is 11.7 + 51.2 = 62.9 m^3/d. A mass balance on solids at the thickener gives

$$Q_1 C_1 = Q_2 C_2$$

Since the solids increase 2.5-fold, C_2 = (1.4)(2.5) = $\boxed{3.5\%.}$ The thickened flow, Q_2, is

$$Q_2 = Q_1 \frac{C_1}{C_2} = (62.9)\left(\frac{1.4}{3.5}\right) = \boxed{25.2 \text{ m}^3/\text{d}}$$

A balance on the flows at the thickener gives

$$Q_1 = Q_2 + Q_{sl}$$

or

$$Q_{sl} = Q_1 - Q_2 = 62.9 - 25.2 = \boxed{37.7 \text{ m}^3/\text{d}}$$

Assume θ_h = 20 d at 20°C from Table 20.1. The correction for temperature is given by

$$\theta_{h_2} = \theta_{h_1} \cdot \theta^{20-T_2}$$

Assume θ is the mean value of 1.02 and 1.11, or 1.065. Thus,

$$\theta_{h_2} = 20 \cdot 1.065^{20-18}$$
$$= 22.7 \text{ d}$$

The digester volume $V = Q\theta_h$, or

$$V = (25.2\,\text{m}^3/\text{d})(22.7\,\text{d}) = \boxed{572\,\text{m}^3}$$

The oxygen required is for BOD$_5$ removal and waste activated sludge destruction. The BOD$_5$ removed is

$$\text{kg/d} = (10{,}000\,\text{cap})\left(\frac{380\,\ell}{\text{cap-d}}\right)\left(\frac{200\,\text{mg}}{\ell}\right)\left(\frac{\text{gm}}{1000\,\text{mg}}\right)\left(\frac{\text{kg}}{1000\,\text{gm}}\right)(0.33)(0.90)$$

$$= 226\,\text{kg/d}$$

The waste activated sludge destruction is

$$\text{kg/d} = (10{,}000)\left(\frac{0.031\,\text{kg}}{\text{cap-d}}\right)(0.70)(0.65) = 141\,\text{kg/d}$$

Assume, from Table 20.1, 1.9 kg O$_2$/kg BOD$_5$ removed and 2.0 kg O$_2$/kg VSS destroyed. The oxygen required is

$$\text{O}_2/\text{day} = (226)(1.9) + (141)(2.0)$$

$$= 429 + 282 = \boxed{711\,\text{kg/day}}$$

The air required is

$$\text{m}^3/\text{min} = \left(\frac{711\,\text{kg}}{\text{d}}\right)\left(\frac{\text{d}}{1440\,\text{min}}\right)\left(\frac{\text{m}^3}{0.281\,\text{kg O}_2}\right)\left(\frac{1}{0.04}\right)$$

$$= \boxed{43.9\,\text{m}^3/\text{min}}$$

The air per 1000 m^3 is

$$\text{air} = \frac{43.9}{0.572} = \boxed{77\,\text{m}^3/\text{min-1000 m}^3}$$

Since 77 > 60 from Table 20.1, $\boxed{\text{the mixing is adequate.}}$

The dry solids in the sludge = 589 + 310 = 899 kg/d. The area required for thickening = 899 kg/d ÷ (39.1 kg/d-m^2) = 22.99 m^2. The area required for clarifying the supernatant = 37.7 m^3/d ÷ (24.4 m^3/d-m^2) = 1.55 m^2. Thus the area required = 22.99 m^2. Since $(\pi/4)D^2 = 22.99$, the diameter is $\boxed{5.41\,\text{m.}}$

(*Note*: In this problem, the design air required was determined by the field transfer percent to simplify the computations. Also, the primary and secondary sludges were blended prior to thickening, as is done in many small plants. In medium- to large-size activated sludge plants, the primary and secondary sludges are usually thickened separately and then blended. This is done because activated sludge usually thickens better by itself.)

REFERENCES

Ahlberg, N. R., and Boyko, B. I. 1972. Evaluation and Design of Aerobic Digesters. *Jour. WPCF* 44, no. 4:634.

American Society of Civil Engineers (ASCE). 1982. *Wastewater Treatment Plant Design*. ASCE Manual of Practice No. 36. New York: ASCE.

Aware, Inc. 1974. *Process Design Techniques for Industrial Waste Treatment*, edited by C. E. Adams and W. W. Eckenfelder, Jr. Nashville, Tenn.: Enviro Press.

Barnhart, E. 1961. Application of Aerobic Digestion to Industrial Waste Treatment. *Proceedings of the 16th Annual Purdue Industrial Waste Conference.*

Burd, R. S. 1968. A Study of Sludge Handling and Disposal. FWPCA Publication No. WP-20-4.

Cameron, J. W. 1972. Aerobic Digestion of Activated Sludge to Reduce Sludge Handling Costs. Paper presented at 45th Annual Conference, Water Pollution Control Federation, Atlanta, Ga.

Conway, R. A., and Ross, R. D. 1980. *Handbook of Industrial Waste Disposal*. New York: Van Nostrand Reinhold.

Dreier, D. E. 1963. Aerobic Digestion of Solids. *Proceedings of the 18th Annual Purdue Industrial Waste Conference.*

Dreier, D. E. 1960. Discussion on Aerobic Sludge Digestion, by Jaworski, N., Lawton, G. W., and Rohlich, G. A. Paper presented at the Conference on Biological Waste Treatment, Manhattan College, New York City.

Dreier, D. E., and Obma, C. A. 1963. *Aerobic Digestion of Solids*. Walker Process Equipment Co., bulletin no. 26-S-18194, Aurora, Ill.

Eckenfelder, W. W., Jr., 1989. *Industrial Water Pollution Control*. 2nd ed. New York: McGraw-Hill.

Eckenfelder, W. W., Jr., 1980. *Principles of Water Quality Management*. Boston: CBI Publishing.

Eckenfelder, W., W., Jr., and O'Connor, D. J. 1961. *Biological Waste Treatment*. New York: Pergamon Press.

Environmental Protection Agency (EPA). 1979. *Sludge Treatment and Disposal*. EPA Process Design Manual, Washington, D.C.

Environmental Protection Agency (EPA). 1974. *Upgrading Existing Wastewater Treatment Plants*. EPA Process Design Manual, Washington, D.C.

Great Lakes–Upper Mississippi River Board of State Sanitary Engineers. 1978. Recommended Standards for Sewage Works. Ten state standards. Albany, N.Y.

Great Lakes–Upper Mississippi River Board of State Public Health and Environmental Managers. 1990. Recommended Standards for Wastewater Facilities. Ten state standards. Albany, N.Y.

Howe, R. H. L. 1959. What to Do with Supernatant. *Waste Engineering* 30, no. 1:12.

Jaworski, N.; Lawton, G. W.; and Rohlich, G. A. 1960. Aerobic Sludge Digestion. Paper presented at the Conference on Biological Waste Treatment, Manhattan College, New York City.

Lawton, G. W., and Norman, J. D. 1964. Aerobic Digestion Studies. *Jour. WPCF* 36, no. 4:495.

Levis, C. E.; Miller, C. R.; and Bosburg, L. E. 1971. Design and Operating Experiences Using Turbine Dispersion for Aerobic Digestion. *Jour. WPCF* 43, no. 3:417.

Loehr, R. C. 1965. Aerobic Digestion: Factors Affecting Design. *Water and Sewage Works* 112:R169.

Malina, J. F., and Burton, H. M. 1964. Aerobic Stabilization of Primary Wastewater Sludge. *Proceedings of the 19th Purdue Industrial Waste Conference.*

Metcalf & Eddy, Inc. 1972. *Wastewater Engineering: Collection, Treatment and Disposal*. New York: McGraw-Hill.

Metcalf & Eddy, Inc. 1991. *Wastewater Engineering: Treatment, Disposal and Reuse*. 3rd ed. New York: McGraw-Hill.

Metcalf & Eddy, Inc. 1979. *Wastewater Engineering: Treatment, Disposal and Reuse*. 2nd ed., New York: McGraw-Hill.

Novotny, V.; Imhoff, K.; Olthof, M.; and Krenkel, P. 1989. *Karl Imhoff's Handbook of Urban Drainage and Wastewater Disposal*. New York: Wiley.

Qasim, S. R. 1985. *Wastewater Treatment Plant*. New York: Holt, Rinehart and Winston.

Ramalho, R. S. 1977. *Introduction to Wastewater Treatment Processes*. New York: Academic Press.

Randall, C. N., and Koch, C. T. 1969. Dewatering Characteristics of Aerobically Digested Sludge. *Jour. WPCF* 41, no. 5, part 2:R215. (Research Supplement)

Reynolds, T. D. 1973. Aerobic Digestion of Thickened Waste Activated Sludge. *Proceedings of the 28th Annual Purdue Industrial Waste Conference.*

Reynolds, T. D. 1967. Aerobic Digestion of Waste Activated Sludge. *Water and Sewage Works* 114, no. 22:37.

Ritter, L. 1970. Design and Operating Experiences Using Diffused Aeration for Sludge Digestion. *Jour. WPCF* 42, no. 10:1782.

Schroeder, E. D. 1977. *Water and Wastewater Treatment*. New York: McGraw-Hill.

Smith, A. R. 1971. Aerobic Digestion Gains Favor. *Water and Waste Engineering* 8, no. 2:24.

Smith, R.; Eilers, R. G.; and Hall, E. D. 1973. A Mathematical Model for Aerobic Digestion. EPA. Office of Research and Monitoring, Advanced Waste Treatment Research Laboratory, Cincinnati, Ohio.

Sundstrom, D. W., and Klei, H. E. 1979. *Wastewater Treatment*. Englewood Cliffs, N.J.: Prentice-Hall.

Texas Natural Resource Conservation Commission. 1993. Design Criteria for Sewage Systems. Austin, Texas.

Texas State Department of Health. 1976. Design Criteria for Sewage Systems. Austin, Tex.

Water Environment Federation (WEF). 1992. *Design of Municipal Wastewater Treatment Plants*. Vol. I and II. WEF Manual of Practice No. 8.

Water Pollution Control Federation (WPCF). 1977. *Wastewater Treatment Plant Design*. WPCF Manual of Practice No. 8, Washington, D.C.

PROBLEMS

20.1 An aerobic digester is to be designed for a population of 8100 persons. The flowsheet is shown in Figure 20.4. As shown, the primary and secondary sludges are blended together in a blending tank and then thickened and sent to the digester. Pertinent data are: influent BOD_5 = 205 mg/ℓ, influent suspended solids = 260 mg/ℓ, average flow = 95 gal/cap-day (360 ℓ/cap-d), 33% BOD_5 removal and 62% suspended solids removal by the primary clarifier, primary sludge has 4.5% dry solids, sludge density index = 6000 mg/ℓ, secondary sludge = 0.069 lb/cap-day (0.0313 kg/cap-d), volatile solids for the mixed primary and secondary sludge = 72%, and the specific gravity of the primary, secondary, blended, and thickened sludges = 1.01. Determine, using USCS units:

a. The primary, secondary, and blended sludge flows.

b. The solids concentration in the blended sludge flow, %.

c. The thickened sludge flow if the thickener increases the solids by 2.5 times and the supernatant flow has negligible suspended solids.

d. The supernatant flow.

e. The digester volume if the minimum winter operating temperature is 16°C.

f. The oxygen requirements for the digester if 90% of the primary BOD_5 is destroyed and 65% of the volatile solids is destroyed in the waste activated sludge.

g. The required air if the air has 0.0175 lb O_2/ft^3 and the field transfer is 4%. Is the mixing adequate?

h. The thickener theoretical diameter if the solids loading is 8 lb/day-ft^2 and the overflow rate is 600 gal/day-ft^2. What is the design diameter if scrapers are available in 5-ft increments in diameter?

20.2 Repeat Problem 20.1 using SI units. In part (g), there is 0.281 kg oxygen per m^3 of air. In part (h), determine the diameter of the thickener if the solids loading is 39.1 kg/d-m^2 and the overflow rate is 24.4 m^3/d-m^2.

20.3 An aerobic digester is to be designed to treat the primary and waste activated sludge (secondary sludge) from a municipal wastewater treatment plant serving 14,200 persons. The flowsheet is shown in Figure 20.5. The waste activated sludge is added to the plant influent flow ahead of the primary clarifier to thicken the solids, and as a result, the waste activated sludge solids leave with the underflow from the primary clarifier. Pertinent data are: primary and secondary solids in the underflow = 0.23 lb/cap-day (0.104 kg/cap-d) as dry solids, solids in the underflow = 5.0%, volatile solids in the underflow = 70%, design minimum operating temperature = 62°F (16.7°C), temperature correction coefficient θ = 1.06, specific gravity of the fresh sludge = 1.01, influent BOD_5 = 0.20 lb/cap-day (0.091 kg/cap-d), 33% of the influent BOD_5 is removed by the primary clarifier, aerobic digester destroys 90% of the BOD_5 in the underflow, 1.9 lb O_2 is required per lb BOD_5 removed (1.9 kg/kg), waste activated sludge or secondary sludge = 0.087 lb/cap-day (0.039 kg/cap-d), 70% volatile solids in the secondary sludge, 65% of the volatile solids in the secondary sludge are destroyed, and 2.0 lb O_2 are required per lb volatile solids destroyed (2.0 kg/kg). Determine, using USCS units:

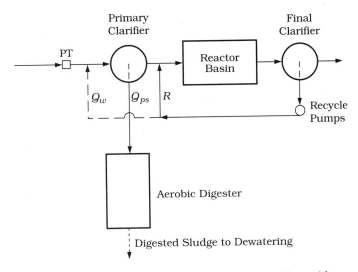

FIGURE 20.5 *Flowsheet of an Activated Sludge Plant with an Aerobic Digester*

a. The design hydraulic detention time.
b. The volume of fresh sludge produced.
c. The digester volume.
d. The volatile solids loading.
e. The volume per capita.
f. The oxygen requirements to stabilize the primary BOD$_5$.
g. The oxygen requirements to stabilize the waste activated sludge or secondary sludge.
h. The total oxygen requirements.
i. The total air requirements in scfm if there are $0.0175\,\text{lb}\,O_2/\text{ft}^3$ of air and the diffusers have a transfer efficiency of 4.0%.
j. The air required per $1000\,\text{ft}^3$.
k. Is the air required adequate for mixing?

20.4 Repeat Problem 20.3 using SI units. In part (i), there is 0.281 kg oxygen per m^3. In part (j), the air should be per $1000\,\text{m}^3$.

20.5 An aerobic digester is to be designed to treat the waste activated sludge or excess sludge from a completely mixed activated sludge plant without primary clarification. The flowsheet for the digester system is shown in Figure 20.6. The plant serves 12,100 persons, and the waste activated sludge amounts to 0.23 lb/cap-day (0.104 kg/cap-d). Pertinent data are: waste activated sludge is withdrawn from the recycle line, sludge density index = 6000 mg/ℓ (0.6%), volatile solids are 70%, spëific gravity of the wet sludge = 1.01, minimum design temperature = 62°F (16.7°C), temperature correction

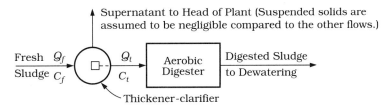

FIGURE 20.6 *Flowsheet of an Aerobic Digester and Its Thickener-Clarifier*

coefficient $\theta = 1.06$, 65% of the volatile solids are destroyed, and 2.0 lb O_2 are required per lb volatile solids destroyed (2.0 kg/kg). Determine, using USCS units:

a. The digester volume if the hydraulic detention time = 17 days at 20°C and the fresh sludge is thickened from 0.6% (6000 mg/ℓ) to 2.5% dry solids.

b. The oxygen required per day.

c. The air flow in scfm if the air contains 0.1075 lb O_2/ ft^3 and the oxygen transfer efficiency is 4.0%.

d. The volatile solids loading.

e. The volume per capita.

f. The air required for mixing. Is it adequate?

20.6 Repeat Problem 20.5 using SI units. In part (c), there is 0.281 kg oxygen per m^3.

21

SOLIDS HANDLING

In water treatment plants, such as the coagulation and softening type, and in wastewater treatment plants, sludges are produced that require disposal. Solids handling consists of the satisfactory processing of sludges for ultimate disposal to the environment, which is usually land disposal. In some cases, ultimate disposal may be by the air or water environment. In conventional wastewater treatment plants, the sludges are principally of an organic nature, such as primary sludge, excess (waste) activated sludge, or trickling filter humus; however, in advanced wastewater treatment, the sludges are mainly of a chemical nature because they result from coagulation or precipitation. Although the sludges in advanced wastewater treatment are considered chemical sludges, they do have some organic matter associated with them. In a wastewater treatment plant, which has both organic and chemical sludges, it is usually advantageous not to mix the two but, instead, to process each sludge separately. Mixing of the two types of sludges frequently results in a mixture that is difficult to process. In water treatment, the sludges from coagulation or precipitation are primarily of a chemical nature, although some organic material will be present.

WASTEWATER TREATMENT PLANT SLUDGES (ORGANIC)

Organic sludges are those produced by primary, secondary, or sludge treatment in biological wastewater treatment plants. Fresh or undigested sludges, such as primary or secondary sludges, are rather difficult to dewater as compared to digested sludges. The operations and processes used in solids handling in biological wastewater treatment plants may be classified as thickening, stabilization, conditioning, dewatering, incineration, and disposal. Primary sludges and secondary sludges, such as excess activated sludge and trickling filter humus, are frequently mixed by adding the secondary sludge to the incoming raw wastewater and resettling it in the primary clarifiers. This thickens, to a certain degree, the primary sludge and, in the case of excess activated sludge, the secondary sludge. The solids content of various municipal wastewater sludges is given in Table 19.4.

Typical solids handling systems for primary and secondary sludge mixtures mixed in this manner are (1) anaerobic digestion, sand-bed dewatering, and landfill disposal of the solids; (2) anaerobic digestion, chemical conditioning, vacuum filtration or centrifugation, and landfill disposal of the solids; (3) gravity thickening, chemical conditioning, vacuum filtration or centrifugation, multiple-hearth or fluidized-bed incineration, and landfill disposal of the ash; and (4) gravity thickening, chemical conditioning, vacuum filtration, flash drying, and land disposal of the solids as a fertilizer.

When primary sludge and excess activated sludge are thickened separately, typical solids handling systems are (1) gravity thickening of the primary sludge, air-flotation thickening of the excess activated sludge, anaerobic digestion, chemical conditioning, vacuum filtration or centrifugation, and landfill disposal of the solids, and (2) gravity thickening of the primary sludge, air-flotation thickening of the excess activated sludge, chemical conditioning, vacuum filtration or centrifugation, multiple-hearth or fluidized-bed incineration, and landfill disposal of the ash.

Excess activated sludge from activated sludge plants without primary clarification (extended aeration and contact stabilization plants) is frequently handled by the following system: aerobic digestion, sand-bed dewatering, and landfill disposal of the solids.

In general, the larger the treatment plant, the more complex will be the solids handling system because the most favorable disposal of the solids is desired.

Thickening   Thickening consists of increasing the solids content of a sludge; this reduces the volume of sludge to be processed by subsequent units. If, for example, a sludge having 3% solids content is thickened to 6% solids, the volume of sludge leaving the thickener is $\frac{3}{6}$ or $\frac{1}{2}$ the volume of the feed sludge.

Gravity thickening, which is the most common thickening method, can be used when the specific gravity of the solids is greater than 1. Figure 21.1 shows a gravity-type thickener. Note that it is very similar to a clarifier except that the floor slope is much greater and vertical pickets are mounted on the trusswork. The feed sludge enters the thickener at the inlet well in the center of the unit. As the flow goes radially outward from the inlet well, the solids settle into the sludge blanket that occupies the lower depths of the thickener. The thickened sludge leaves as the underflow, and the supernatant liquid leaves by discharging over the effluent weirs. The pickets mounted on the moving trusswork slowly rake through the sludge mass and break up sludge arching or bridging, thus releasing entrained water that rises to the surface, where it leaves as the supernatant. Also, the slow agitation releases gas bubbles that have been formed, and the scraper blades move the thickened sludge to the center of the unit where it is removed as the underflow. The thickener must serve two purposes: thickening of the sludge solids and clarification of the supernatant liquid. The design parameters for these two requirements are the solids loading in lb solids/day-ft^2 (kg/day-m^2) and the overflow rate in gal/day-ft^2 (m^3/day-m^2). These may be determined by the same methods of analysis as for activated sludge final clarifiers, presented in Chapter 9. The degree of thickening that may be accomplished, as expressed by the thickened sludge solids content, is a function of the nature of the sludge. Typical performance of gravity thickeners processing municipal sludges is shown in Table 21.1. The overflow rate is usually 400 to 800 gal/day-ft^2 (16.3 to 32.6 m^3/day-m^2) based on the supernatant flow. It is

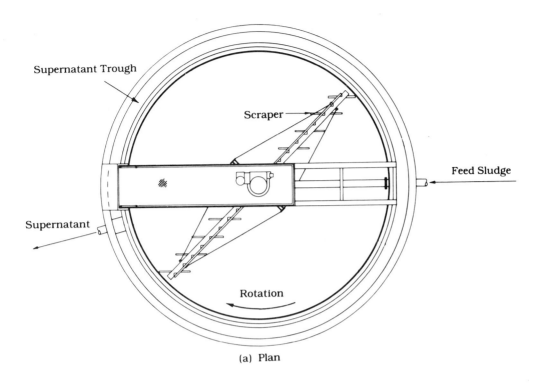

(a) Plan

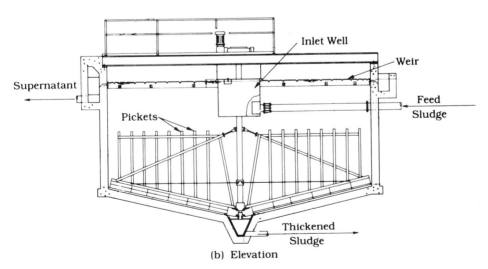

(b) Elevation

FIGURE 21.1 *Gravity-Type Sludge Thickener*

Courtesy of Infilco Degremont, Inc.

TABLE 21.1   *Gravity Thickener Performance for Processing Fresh Municipal Sludges*

TYPE OF SLUDGE	SOLIDS LOADING lb/day-ft² (kg/day-m²)	PERCENT SOLIDS	
		Unthickened	Thickened
Primary[a]	20–30 (100–150)	2.5–5	8–10
Primary[b]	20–30 (100–150)	2–7	5–10
Waste activated sludge[a]	5–6 (25–30)	0.5–1	2.5–3
Waste activated sludge[b]	4–8 (20–40)	0.5–1.5	2–3
Waste activated sludge, pure O_2	7–34 (35–170)	1.5–3.5	4–8
Waste activated sludge, pure O_2[b]	4–8 (20–40)	0.5–1.5	2–3
Trickling filter[a]	8–10 (40–50)	5–10	7–10
Trickling filter[b]	8–10 (40–50)	1–4	3–6
Rotating biological contactor[b]	7–10 (35–50)	2–6	5–8
Primary and waste activated sludge[a]	6–10 (30–50)	4–5	5–10
Primary and waste activated sludge[b]	5–14 (25–70)	0.5–1.5	4–6
	8–16 (40–80)	2.5–4	4–7
Primary and trickling filter[a]	10–12 (50–60)	3–6	7–9
Primary and trickling filter[b]	12–21 (60–100)	2–6	5–9
Primary and rotating biological contactor[b]	10–18 (50–90)	2–6	5–8

[a] D. Newton (1964).
[b] WPCF (1980).

important that the solids retention time in the thickener not be great enough to cause anaerobic conditions that will result in odors. The solids retention time is usually taken as the volume of the sludge blankert divided by the thickened sludge flowrate and is usually from 0.5 to 2 days (EPA, 1979). Low values should be used in warm climates because microbial action will be greater. The sidewater depth of a thickener is usually about 10 to 12 ft (3 to 3.7 m).

Thickening using dissolved **air flotation** may be used whenever the specific gravity of the solids is near unity. It has been very successfully used for thickening excess activated sludges. Figure 21.2 shows a schematic diagram of a typical flotation system, consisting of the flotation tank, the air dissolution tank, and the necessary pumps. Flotation uses the formation of air bubbles on the solid particles to buoy them to the surface, where they are skimmed from the flotation tank. It usually is accomplished by three methods. The first method, which is the most commonly used, is depicted in Figure 21.2. It consists of pressurizing the recycled flow in an air tank to dissolve air gases in the flow. The recycled flow then mixes with the feed sludge flow, and the mixture enters the flotation tank. Because of the release in pressure, the air gases come out of solution and form air bubbles around the sludge solids, which buoys them to the surface. They are skimmed off and leave as the thickened sludge flow. The air pressure in the dissolution tank is usually 40 to 70 psi (280 to 480 kPa), and the recycle flow is usually from 30% to 150% of the feed flow. The second method, depicted in Figure 21.3, consists of pressurizing the entire feed flow to dissolve air gases. The pressure is released just prior to the flow entering the flotation tank. The flotation action and removal of solids is the same as for the first method. Usually, pressurizing the entire flow is limited to small installations. The third method consists of aerating the feed sludge flow so that the liquid is saturated with dissolved air gases. The flow then goes to a covered flotation tank, where a vacuum is drawn upon the system. This causes the air gases to come out of solution and buoy the solids to the surface, where they are removed by skimming. Air-flotation thickening of waste activated sludge has an advantage over gravity thickening because the solids content in the thickened sludge is usually higher and the cost of the flotation system is usually less. Flotation aids, such as polymers, aid significantly, and nearly all installations employ them. Air-flotation thickening of waste activated

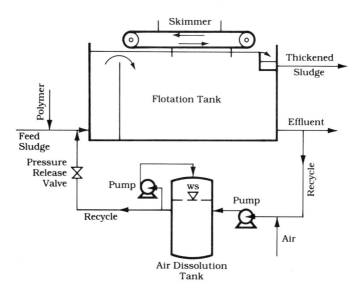

FIGURE 21.2
Air-Flotation System with Recycle

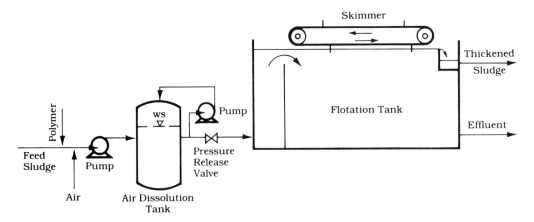

FIGURE 21.3   *Air-Flotation System without Recycle*

sludge with about 10,000 mg/ℓ suspended solids will result in a thickened sludge of 4% to 6% solids when flotation aids are used. Without flotation aids, the thickened sludge will be about 3% to 5% solids (Smith et al., 1972).

Although waste activated sludges from industrial wastewater treatment plants may flotate with more or less ease than waste activated sludges from municipal plants, nearly all waste activated sludges may be thickened by flotation. In the use of dissolved air flotation for the thickening of waste activated sludges, the air/solids ratio is usually from 0.005 to 0.06. The design procedure for dissolved air-flotation units, both with and without recycle, is presented in Chapter 23.

In **centrifugal thickening**, solid bowl, disc-nozzle, and basket-type centrifuges have been used to thicken waste activated sludges. The disc-nozzle type, shown in Figure 21.4, has been the most popular. In this centrifuge the feed sludge enters as shown in Figure 21.4 and passes upward through the discs, each disc acting as a single centrifuge. The centrifuged solids slide down each disc as a result of the centrifugal force and the angle of the discs; they then leave through the outlet port. The centrifugation of air and pure oxygen waste activated sludge with 4000 to 40,000 mg/ℓ suspended solids has thickened the sludges to 5% to 12% solids (WPCF, 1980). If the waste activated sludge is from a plant that does not have primary clarification (for example, the biosorption process), the sludge should be screened prior to centrifugal thickening. Centrifuges have appreciable maintenance and energy costs, so centrifugation is usually used when gravity or flotation thickening are not feasible because of such factors as space limitations.

Stabilization   Stabilization consists of treating the sludge so that future decomposition by biological action does not occur. It results in a sludge that will not undergo bacterial decomposition, has good dewatering characteristics, and has very

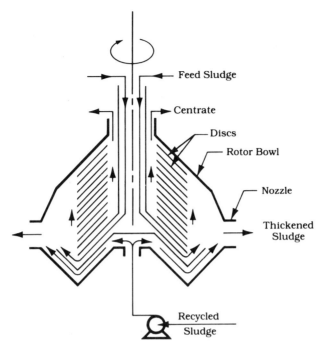

FIGURE 21.4 *Section through a Disc-Nozzle Centrifuge*

little odor. In the case of municipal sludges, it also results in a low pathogen content.

Anaerobic digestion, which was discussed in detail in Chapter 19, consists of the bio-oxidation of organic sludges under anaerobic conditions. One advantage that it has over aerobic digestion is the production of methane gas, which is a usable by-product.

Aerobic digestion, which was presented in Chapter 20, consists of the bio-oxidation of organic sludges under aerobic conditions. One advantage it has over anaerobic digestion is that less technical skill is required in operation because the process avoids many of the operational problems of anaerobic digestion.

Lagoons have found limited use in the digestion and disposal of waste activated sludges. In order to function as oxidation ponds and not merely as sludge storage lagoons, they must be designed properly. Oxidation ponds consist of earthen embankments with a pond depth of 3 to 5 ft (1 to 1.5 m). In their operation, the sludge settles to the bottom, where it undergoes anaerobic digestion. The top portion of the water depth remains aerobic, which prevents odors. If the organic loadings are too great, anaerobic conditions will occur throughout the water depth and odors will be produced. Oxidation ponds are limited to warm climates that have high sunlight intensities and to areas where land is relatively cheap and occasional odors are not objectionable.

Wet combustion uses chemical combustion of sludge with oxygen under wet conditions. One manufactured process uses pressures from 1000 to 1800 psi (6900 to 12,400 kPa). Injected steam raises the temperature of the sludge to a self-sustaining level, which is usually about 500°F (260°C). The chemical oxidation or combustion that occurs is not complete, however; the treated sludge is sterile and dewaters readily.

In **chemical treatment**, lime has been used to stabilize primary and excess activated sludges, temporarily preventing odors. If sufficient slaked lime is added to raise the pH above 11, almost all biological action ceases and a calcium carbonate precipitate is formed that greatly assists in the dewatering characteristics. Essentially all *E. coli* and *Salmonella typhosa* have been killed by high lime treatment at a pH of 11 to 11.5 and a 4-hr contact time at 15°C (59°F) (Riehl *et al.*, 1952). It has been found that lime-stabilized sludges disposed in lagoons have a gradual pH reduction and a gradual increase in biological action. Thus lime treatment should be considered as a temporary sludge stabilization method, because bio-oxidation will eventually occur. Chlorine has also been used for sludge stabilization. It differs from lime in that chemical oxidation occurs because chlorine is a strong oxidizing agent. Usually, about 2000 mg/ℓ chlorine concentration is employed, which produces a treated sludge that is stable and dewaters well on drying beds. The pH, however, will be about 2, and usually the sludge must be neutralized prior to dewatering by mechanical methods. Neutralization is necessary because of the corrosive nature of the sludge and also because the low pH interferes with any chemical conditioners that are required. The high concentration of chloramines in the water removed during dewatering must be considered in the treatment of such flows.

The **composting** of thickened and dewatered undigested primary and secondary sludges has been applied to a limited extent in the United States. Both soil and mechanical aeration systems have been employed, usually with the sludge being mixed with solid wastes. The stabilization is essentially an aerobic digestion process. The lack of use is primarily due to the lack of demand for the compost product.

Conditioning   Conditioning consists of treating a sludge prior to dewatering and sometimes thickening to enhance its dewatering characteristics.

In **chemical treatment**, both organic polymers and inorganic coagulants have been used for conditioning; however, the present trend is to use organic polymers (polyelectrolytes) because of their effectiveness. Sludges are stable suspensions, and the addition of polyelectrolytes causes coagulation and aggregation of the sludge solids and a release of the entrained water. Polyelectrolytes may have positive charges (cationic type), negative charges (anionic type), or neutral charges (nonionic type). The cationic type is the most commonly used in sludge conditioning because the sludge particles are slightly negatively charged. The coagulation of sludge particles occurs mainly as a result of bridging action between reactive groups on the polymer and the sludge particles; however, some coagulation by charge reduction does occur. Polymers may be obtained in powder, pellet, or liquid form; however, the liquid form is the easiest to mix with water. If the sludge is to be incinerated, polymers have an advantage over inorganic coagulant salts because inorganic coagulants decrease the fuel value of the treated sludge. Polymer requirements depend on the nature of the sludge and may be from 3 to 25 lb/ton (1.5 to 12.5 kg/1000 kg) of dry solids. Inorganic coagulants, particularly ferric chloride, have been used in sludge conditioning. Because of the high coagulant dosages used, the pH drop can be significant; ferric chloride is an acidic salt. The pH is maintained in the optimum coagulation range by the addition of hydrated lime, which also assists in the formation of the ferric hydroxide precipitate. Coagulation by inorganic salts is caused by charge reduction, by enmeshment in the hydroxide precipitate, and also by chemical bridging.

Elutriation consists of washing a sludge with water to remove components such as bicarbonates and fine solids that result in high chemical dosages when subsequent conditioning is done using inorganic salts. In particular, it has been used for anaerobically digested sludges that are to be chemically coagulated prior to dewatering. However, because of the widespread use of polymers, elutriation is rapidly decreasing in popularity.

Heat treatment has, in particular, been used to condition waste activated sludges. The solids are mainly microbial cells, so they contain significant amounts of water that are not removed by the usual dewatering methods. Heat treatment, by such processes as the Porteous and the Zimpro processes (EPA, 1979, 1988; WPCF MOP FD-14, 1988), causes the cells to rupture or lyse; the intercellular material, which has a high water content, is released. In the Porteous process, as shown in Figure 21.5, live steam is injected into

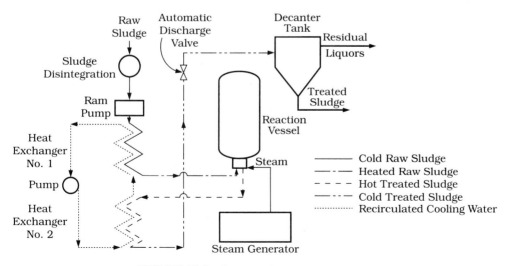

FIGURE 21.5 *Porteous Process*

Redrawn from WPCF, *Sludge Conditioning*, MOP No. FD-14 (1988), by permission.

a reaction vessel. The hot treated sludge is cooled in a water-to-sludge heat exchanger (exchanger no. 2). The heated water from exchanger no. 2 is used to preheat the raw sludge in another water-to-sludge heat exchanger (exchanger no. 1). In the Zimpro process, shown in Figure 21.6, live steam is injected into a reaction vessel. The hot treated sludge is used in a sludge-to-sludge heat exchanger to preheat the raw sludge. Air is injected into the raw sludge.

Heat treatment process design criteria (EPA, 1988) include the feed solids concentration, operating temperature and pressure, heat requirements, and treated sludge solids-liquid separation tank loading. Raw sludge or feed solids between 3% and 6% optimize capital and operating costs. Operating temperatures range from 350°F to 400°F (177°C to 204°C), with the higher temperature typically used for waste activated sludge. Sludge residence time in the reactor vessel ranges from 15 min for systems with air injection, such as the Zimpro process, to 30 min for systems without air injection, such as the Porteous process. Operating pressures at the high-pressure pump for a system with air injection range from 370 to 430 psi (2600 to 3000 kPa). For a system without air injection, operating pressures at the high-pressure pump range from 250 to 300 psi (1700 to 2100 kPa). The heat required for a system with air injection is lower than that required for a system without air injection because of the wet air oxidation of raw sludge COD achieved, typically 5%, with the air-injection systems. For systems with air injection, approximately 500 Btu of boiler fuel per gallon of sludge processed (140 kJ/ℓ) are required; for systems without air injection, approximately 800 Btu/gal

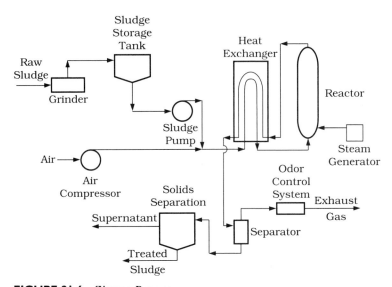

FIGURE 21.6 *Zimpro Process*

Redrawn from WPCF, *Sludge Conditioning*, MOP No. FD-14 (1988), by permission.

$(220 \, kJ/\ell)$ are required. Treated sludge solids loading to the solids-liquid separation tank are typically 40 to $60 \, lb/ft^2$-day (200 to $300 \, kg/m^2$-day). The hydraulic residence time in the solids-liquid separation tank is typically 6 to 18 hr. The dewatering characteristics of excess activated sludge are enhanced, and dewatering without chemical conditioning using vacuum filtration will produce a cake of 30% to 50% solids. The principal disadvantages of the process are the disposal of the supernatant liquor produced and the treatment of odorous sidestream gases. The supernatant is odorous, has a BOD_5 of 5000 to $15,000 \, mg/\ell$, and has a high nitrogen and phosphorus content. The supernatant has a pH of 4 to 5, and after neutralization, it may be treated by biological methods.

Ash obtained by sludge incineration may be added to sludges to increase their dewatering characteristics. The addition of ash to fresh primary and secondary sludges at a ratio of 0.25 to 0.5 lb (kg) of ash per pound (kilogram) of dry solids has resulted in a vacuum-filtered cake containing 33% solids (EPA, 1979).

Dewatering Dewatering consists of removing as much water from a sludge as possible so that the dewatered sludge volume to be subsequently processed is minimized. Fresh waste activated sludge is very difficult to dewater, and prior conditioning is nearly always required.

The most commonly used mechanical-type dewatering device is the **rotary vacuum filter**. It consists of a cylindrical drum having a filter medium, which may be cloth, wire mesh, or coil springs. The filters may be classified as drum, belt, and coil-spring type. Figure 21.7 shows a schematic section of the drum type that has a filter medium maintained on the drum throughout the filter cycle. The drum slowly rotates through the sludge vat, and the vacuum within the drum causes the sludge cake to form. Much of the water in the cake is drawn through the medium into the drum. The cake formed undergoes further dewatering in the sector shown in Figure 21.7. The piping inside the drum is arranged so that a vacuum exists in the sector shown for cake formation and dewatering. In the discharge sector, the cake is released

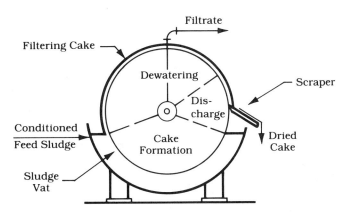

FIGURE 21.7
Rotary Vacuum Filter (Drum Type)

by a flow of compressed air blowing through the medium, and a scraper knife peels it from the drum. Then the medium may be spray washed prior to submergence in the sludge vat. Figure 21.8 shows a belt-type vacuum filter, and Figure 21.9 shows a schematic cross section of the belt-type filter. In the discharge sector, the medium or belt and the cake leave the drum, and as the belt passes over the first rollers, the cake falls from the belt. The belt is then washed and returns to the drum. The coil-spring type has a belt made of springs and operates in a manner similar to the belt type. The rotary vacuum filter requires appurtenances, such as a vacuum pump, a filtrate receiver and pump, and a sludge feed pump. For vacuum filters, the

FIGURE 21.8 *Vacuum Filter at a Wastewater Treatment Plant*

Courtesy of Envirex, Inc.

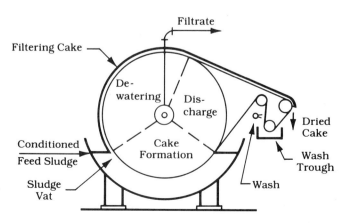

FIGURE 21.9
*Rotary Vacuum Filter
(Belt Type)*

performance is measured by the pounds (kilograms) of dry solids filtered per hour per square foot (square meter) of filter area, and also by the percent dry solids in the filter cake. The performance of vacuum filters processing municipal sludges is shown in Table 21.2. Generally, rotary vacuum filters have shown good results when filtering primary sludge or mixtures of primary and secondary sludges.

The basic equation for evaluating vacuum filtration is (Coackley, 1958)

$$\frac{dV}{dt} = \frac{\Delta p A^2}{\mu(wVR + AR_f)} \tag{21.1}$$

where

V = volume of filtrate

t = time

Δp = vacuum pressure differential

A = filter area

μ = absolute or dynamic viscosity of the filtrate

w = weight or mass of the dry sludge solids per unit volume of filtrate

R = specific resistance of the sludge cake

R_f = specific resistance of the filter medium

The specific resistance, R, of the sludge cake is the principal design parameter for vacuum filtration. The specific resistance is the reciprocal of the permeability. The specific resistance may be obtained in the laboratory using

TABLE 21.2 *Vacuum Filter Performance for Processing Municipal Sludges Conditioned by Ferric Chloride and Lime*

TYPE OF SLUDGE	SOLIDS LOADING lb/hr-ft² (kg/h-m²)	PERCENT SOLIDS OF CAKE
Fresh primary	6–8 (29–39)	25–38
Fresh primary and waste activated sludge	4–5 (20–24)	16–25
Fresh primary and waste activated sludge (pure oxygen)	5–6 (24–30)	20–28
Fresh primary and trickling filter humus	4–6 (20–29)	20–30
Digested primary (anaerobic)	5–8 (24–39)	25–32
Digested primary and waste activated sludge (anaerobic)	4–5 (20–24)	14–22

From the Environmental Protection Agency, *Sludge Treatment and Disposal*, EPA Process Design Manual, Washington, D.C., 1979.

the apparatus shown in Figure 21.10. For a constant vacuum pressure, Eq. (21.1) can be integrated and rearranged to give

$$\frac{t}{V} = \left(\frac{\mu w R}{2\Delta p A^2}\right)V + \frac{\mu R_f}{\Delta p A} \tag{21.2}$$

Using the laboratory apparatus, a batch of sludge can be dewatered, and the volume of the filtrate, V, versus filtration time, t, can be obtained. Equation (21.2) is of the straight-line form, $y = mx + b$; thus plotting the t/V values on the y-axis and the V values on the x-axis will result in a straight line. The slope of the line, m, will be equal to the term $(\mu w R/2\Delta p A^2)$ in Eq. (21.2). The specific resistance may then be obtained from

$$R = \left(\frac{2\Delta p A^2}{\mu w}\right)m \tag{21.3}$$

From the vacuum filtration test, the value of w may be obtained by drying the solids on the filter paper and dividing this weight by the volume of filtrate obtained during the test. The w value may also be computed using the dry solids concentration in the unfiltered sludge and in the sludge cake. The equation for w is

$$w = \frac{\gamma \text{ or } \rho}{[(1 - x)/x] - [(1 - x_c)/x_c]} \tag{21.4}$$

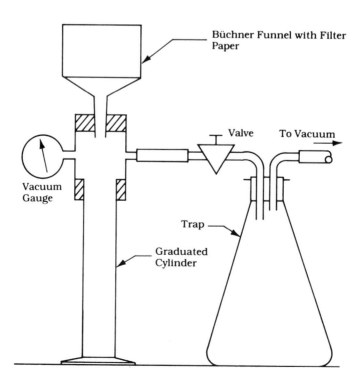

Büchner Funnel with Filter Paper

Valve   To Vacuum

Vacuum Gauge

Trap

Graduated Cylinder

FIGURE 21.10
Laboratory Apparatus for Vacuum Filtration Testing

where

> w = weight or mass of solids per unit volume of filtrate
>
> γ = specific weight of water
>
> x = dry solids content in the unfiltered sludge expressed as a fraction
>
> x_c = dry solids content in the cake expressed as a fraction
>
> ρ = density of water

If sludge conditioning with chemicals is to be done, the sludge must be conditioned prior to the vacuum filtration test. The effects of various chemical conditioners and their concentrations on vacuum filtration can be evaluated by a series of vacuum filtration tests. Conditioners will reduce the specific resistance of a sludge. The plot of specific resistance versus chemical dosage, shown in Figure 21.11, will give the optimum chemical dosage. Some organic sludges produce compressible sludge cakes, causing the specific resistance to vary with the vacuum level used. For these sludges, it is necessary to obtain a relationship giving specific resistance as a function of the vacuum level. A relationship for this is

$$R = r'\Delta p^s \tag{21.5}$$

where r' is the cake compressibility constant and s is the coefficient of compressibility. These constants are a characteristic of each sludge cake. To

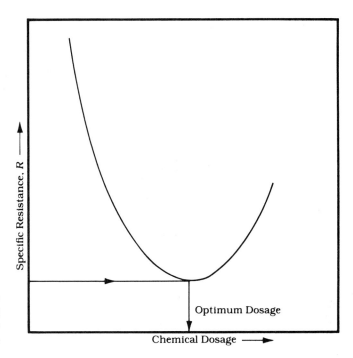

FIGURE 21.11

Specific Resistance versus Chemical Dosage Used for Conditioning

evaluate these constants, a series of vacuum filter tests may be conducted at different vacuum levels. Taking the log of Eq. (21.5) gives $\log R = \log r' + s \log \Delta p$. Thus plotting the specific resistance, R, on the y-axis of log-log paper and the vacuum levels, Δp, on the x-axis will result in a straight line with a slope equal to s. The y value when $\Delta p = 1$ equals the constant r'. If $s = 0$, the sludge cake is incompressible and the specific resistance is independent of the vacuum level. Beginning with Eq. (21.1), a derivation can be made to give the following performance equation for vacuum filters:

$$Y = \left(\frac{2\Delta p w \alpha}{\mu R \theta g}\right)^{1/2} \tag{21.6}$$

where

Y = filter yield

Δp = vacuum pressure differential, lb/ft^2 (N/m^2)

w = weight or mass of dry sludge solids per unit volume of filtrate, lb/ft^3 (kg/m^3)

α = ratio of form time to cycle time

μ = absolute or dynamic viscosity of the filtrate, lb-sec/ft^2 (N-s/m^2)

R = specific resistance of the sludge cake, sec^2/lb (s^2/kg)

θ = cycle time of the rotating drum — that is, time for one revolution, sec (s)

g = acceleration due to gravity, 32.17 ft/sec^2 (9.806 m/s^2)

In the derivation of Eq. (21.6), it is assumed that the specific resistance of the filter medium, R_f, is negligible. The filter yield is usually expressed in mass terms as lb/hr-ft^2 (kg/h-m^2). In inspecting Eq. (21.6), it can be seen that the filter yield can be increased by increasing Δp and α or by decreasing θ. Usually, the highest possible vacuum is maintained and operational control is accomplished by varying the drum rotational speed, which changes θ.

EXAMPLE 21.1   *Vacuum Filter Yield*

A conditioned digested sludge is to be dewatered on a rotary vacuum filter under a vacuum of 25 in. Hg. A vacuum filtration test has been done in the laboratory; it gave a specific resistance of 2.4×10^7 sec^2/gm, the unfiltered sludge solids were 4.5% by dry weight, and the sludge cake had 32% dry solids. The filtration temperature is 70°F, the cycle time is 6 min, and the form time is 40% of the cycle time. Determine the filter yield in lb/hr-ft^2.

SOLUTION   The dry solids per unit volume of filtrate is given by

$$w = \frac{62.4 \, \text{lb/ft}^3}{[(1 - 0.045)/0.045] - [(1 - 0.32)/0.32]}$$

$$= 3.27 \, \text{lb/ft}^3$$

The specific resistance is

$$R = (2.4 \times 10^7 \, \text{sec}^2/\text{gm})(454 \, \text{gm/lb})$$
$$= 1.0896 \times 10^{10} \, \text{sec}^2/\text{lb}$$

The vacuum differential is

$$\Delta p = \left(\frac{14.7 \, \text{psi}}{29.92 \, \text{in. Hg}} \right)(25 \, \text{in. Hg}) = 12.28 \, \text{lb/in}^2$$

The operating temperature of 70°F is 21.1°C. From the table in the Appendix giving the viscosity of water, a viscosity, μ, of 0.9820 centipoise is obtained by interpolating between 21°C and 22°C. The viscosity is

$$\mu = (0.9820 \, \text{centipoise})(2.088 \times 10^{-5})$$
$$= 2.05 \times 10^{-5} \, \text{lb-sec/ft}^2$$

The cycle time, θ, is $(6 \, \text{min})(60 \, \text{sec/min}) = 360 \, \text{sec}$. The α value is 0.40. Thus the yield is given by

$$Y = \left[\frac{2\left(12.28 \dfrac{\text{lb}}{\text{in}^2} \times \dfrac{144 \, \text{in}^2}{\text{ft}^2} \right)\left(3.27 \dfrac{\text{lb}}{\text{ft}^3} \right)(0.40)}{\left(2.05 \times 10^{-5} \dfrac{\text{lb-sec}}{\text{ft}^2} \right)\left(1.0896 \times 10^{10} \dfrac{\text{sec}^2}{\text{lb}} \right)(360 \, \text{sec})\left(32.17 \dfrac{\text{ft}}{\text{sec}^2} \right)} \right]^{1/2}$$

$$= \left[1.788 \times 10^{-6} \frac{\text{lb}^2}{\text{sec}^2\text{-ft}^4} \right]^{1/2} = 1.337 \times 10^{-3} \frac{\text{lb}}{\text{sec-ft}^2}$$

$$= \left(1.337 \times 10^{-3} \frac{\text{lb}}{\text{sec-ft}^2} \right)\left(3600 \frac{\text{sec}}{\text{hr}} \right)$$

$$= \boxed{4.81 \, \text{lb/hr-ft}^2}$$

EXAMPLE 21.1 SI *Vacuum Filter Yield*

A conditioned digested sludge is to be dewatered on a rotary vacuum filter under a vacuum of 640 mm Hg. A vacuum filtration test has been done in the laboratory; it gave a specific resistance of $2.4 \times 10^7 \, \text{s}^2/\text{gm}$, the unfiltered sludge solids were 4.5% by dry weight, and the sludge cake had 32% dry solids. The filtration temperature is 21.1°C, the cycle time is 6 min, and the form time is 40% of the cycle time. Determine the filter yield in kg/h-m².

SOLUTION The dry solids per unit volume of filtrate is given by

$$w = \frac{999.7 \, \text{kg/m}^3}{[(1 - 0.045)/0.045] - [(1 - 0.32)/0.32]}$$
$$= 52.35 \, \text{kg/m}^3$$

The specific resistance is

$$R = (2.4 \times 10^7 \, s^2/gm)(10^3 \, gm/kg)$$
$$= 2.4 \times 10^{10} \, s^2/kg$$

The vacuum differential is

$$\Delta p = \left(\frac{101.37 \, kPa}{760 \, mm \, Hg}\right)(640 \, mm \, Hg) = 85.36 \, kPa = 85.36 \times 10^3 \, N/m^2$$

From the table in the Appendix giving the viscosity of water, a viscosity, μ, of 0.9820 centipoise is obtained by interpolating between 21°C and 22°C. The viscosity is

$$\mu = (0.9820 \text{ centipoise})(10^{-3}) = 9.82 \times 10^{-4} \, N\text{-}s/m^2$$

The cycle time, θ, is $(6 \, min)(60 \, s/min) = 360 \, s$. The α value is 0.40. Thus the yield is given by

$$Y = \left[\frac{2\left(85.36 \times 10^3 \, \dfrac{N}{m^2}\right)\left(52.35 \, \dfrac{kg}{m^3}\right)(0.40)}{\left(9.82 \times 10^{-4} \, \dfrac{N\text{-}s}{m^2}\right)\left(2.4 \times 10^{10} \, \dfrac{s^2}{kg}\right)(360 \, s)\left(9.806 \, \dfrac{m}{s^2}\right)}\right]^{1/2}$$

$$= \left[4.297 \times 10^{-5} \, \frac{kg^2}{s^2\text{-}m^4}\right]^{1/2} = 6.55 \times 10^{-3} \, \frac{kg}{s\text{-}m^2}$$

$$= \left(6.55 \times 10^{-3} \, \frac{kg}{S\text{-}m^2}\right)(3600 \, s/hr)$$

$$= \boxed{23.6 \, kg/h\text{-}m^2}$$

EXAMPLE 21.2   *Filter Drum Area*

If the sludge in Example 21.1 has a flow of 85,000 gal/day, determine the required filter drum area.

SOLUTION   The pounds of solids per hour, assuming that the specific gravity is approximately 1.0, is

$$lb/hr = (85,000 \, gal/24 \, hr)(8.34 \, lb/gal)(4.5 \, lb/100 \, lb)$$
$$= 1329.2$$

Thus the required area is

$$A = (1329.2 \, lb/hr)(hr\text{-}ft^2/4.81 \, lb) = \boxed{276 \, ft^2}$$

EXAMPLE 21.2 SI   *Filter Drum Area*

If the sludge in Example 21.1 SI has a flow of 322,000 ℓ/d, determine the required filter drum area.

SOLUTION   The kilograms of solids per hour, assuming that the specific gravity is
approximately 1.0, is

$$kg/h = (322,000 \, \ell/24 \, h)(1 \, kg/\ell)(4.5 \, kg/100 \, kg) = 604$$

Thus the required area is

$$A = (604 \, kg/h)(h\text{-}m^2/23.6 \, kg) = \boxed{25.6 \, m^2}$$

The **pressure filter** is a batch-operated filter consisting of numerous
vertical filter plates mounted on a horizontal shaft, as depicted in Figure
21.12. The filter plates have recesses covered with filter cloth and filtrate
drain holes that discharge into an outlet port. The vertical plates are movable
and are mounted on a horizontal shaft, as shown in Figures 21.13 and 21.14.
When the press is closed, the plates are pressed together with either a
mechanical screw-type ram or a hydraulic ram. The sealing pressure must be
sufficient to withstand the hydraulic pressure that exists during the filtration
cycle. During the filtration cycle, the sludge is pumped under pressure into
the space between the plates, as shown in Figure 21.12. The filtrate passes
through the filter cloth into the filtrate drain holes and passes from the press
by an outlet port that is in the axial direction. The pressures used during the
filtration cycle are usually from 100 to 250 psi (690 to 1700 kPa). The sludge
flow is maintained until all the spaces within the press are filled with sludge

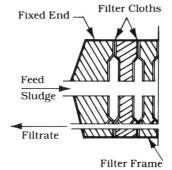

FIGURE 21.12
Section through a
Filter Press

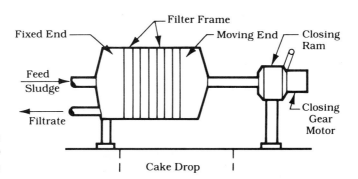

FIGURE 21.13
Elevation of a Filter
Press

cake, which usually is less than 1 hr. The fluid pressure is maintained until all the filterable water has passed from the cake. To remove the cake, the back pressure is released and the moving end is slid to the closing gear mechanism. Each individual plate then moves over the gap between the plates and the moving end, thus allowing the cake to fall from the press onto a conveyor or into a trailer. Frequently, presses are located at a sufficient height above ground level to allow the filtered cake to fall directly into trailers. Once the press has been emptied, it is closed again and the filtration cycle is repeated. The complete filtration cycle time is from about 1.5 to 2.5 hr. Filter presses are one of the most successful dewatering means used to dewater waste activated sludges. The following percentages of dry solids in the processed filter cake have been obtained for conditioned municipal sludges dewatered by filter presses: (1) raw primary sludge, 45% to 50%, (2) raw primary and fresh waste activated sludge, 45% to 50%, (3) fresh waste activated sludge, 50%, and (4) digested primary and waste activated sludge, 45% to 50% (Forster, 1972).

Belt presses have numerous designs; however, the design shown in Figure 21.15 is frequently used. The press consists of two converging belts mounted on rollers. The lower belt is made of fine wire mesh and is very porous. As conditioned sludge moves onto the belt, some of the entrained water drains through the belt in the gravity drain zone. The sludge passes through the press zone where the pressure provided by the converging belts and rollers causes dewatering. Finally, in the shear zone, the sludge must zigzag at changing directions, thus giving further dewatering. The following percentages of dry solids in the processed sludge cake have been obtained

FIGURE 21.14 *Sludge Filter Press at a Water Treatment Plant. The press dewaters sludge from alum coagulation.*

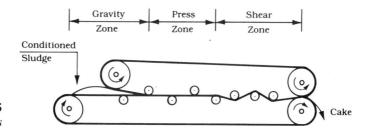

FIGURE 21.15
Belt Press

for municipal sludges chemically conditioned by polymers and dewatered by belt filter presses: (1) raw primary, 28% to 44%, (2) raw primary and waste activated sludge, 20% to 35%, (3) raw primary and trickling filter humus, 20% to 40%, (4) digested primary (anaerobic digestion), 26% to 36%, (5) digested primary and waste activated sludge (anaerobic digestion), 12% to 18%, and (6) digested primary and waste activated sludge (aerobic digestion), 12% to 18% (EPA, 1979). The belt press has low energy requirements and does not require a vacuum or pressure pump.

The main types of **centrifuges** that have been successfully used in dewatering sludges are the countercurrent solid bowl, shown in Figure 21.16, the concurrent solid-bowl, and the basket-type centrifuge shown in Figure 21.17.

In a countercurrent solid-bowl centrifuge, shown in Figure 21.16, both the bowl and the scroll conveyer rotate, and the speed of rotation of the scroll is slightly less than that of the bowl. The feed sludge, upon entering, is slung outward from the shaft, and as a result of the centrifugal force, the

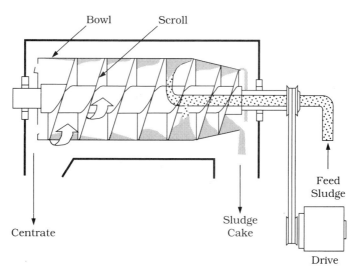

FIGURE 21.16 *Continuous Countercurrent Solid-Bowl Centrifuge*

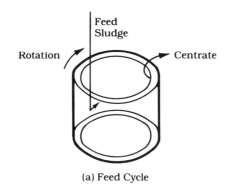

(a) Feed Cycle

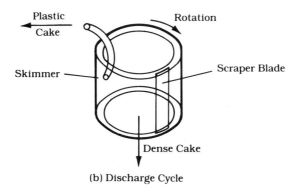

FIGURE 21.17
Operation of a Basket Centrifuge

(b) Discharge Cycle

solids collect on the inner wall of the bowl. The scroll conveyer, which spins at a slightly slower speed than the bowl, moves the collected solids down the bowl and up the inclined beach, where the cake is discharged. The clarified liquid moves countercurrent to the solids, and the centrate and the cake discharge at opposite ends of the machine. The concurrent solid-bowl centrifuge has a rotating bowl and scroll conveyer like the countercurrent type; however, the liquid and the collected solids move in the same direction. The centrate and the cake leave at the same end of the machine. The main advantage of the concurrent-type centrifuge is the concurrent movement of both the liquid and the collected solids. This tends to cause less disturbance to the solids than the countercurrent type. As a result, the centrate has less suspended solids content.

The following percentages of dry solids in the processed sludge cake have been obtained for municipal sludges chemically conditioned and dewatered by solid-bowl centrifugation: (1) raw primary sludge, 25% to 35%, (2) raw primary and fresh waste activated sludge, 15% to 23%, (3) raw primary and fresh trickling filter humus, 24% to 30%, (4) digested primary sludge (anaerobic digestion), 25% to 35%, (5) digested primary and waste activated sludge (anaerobic digestion), 14% to 18%, (6) digested primary and waste activated sludge (aerobic digestion), 10% to 14%, and (7) digested primary and trickling filter humus (anaerobic digestion), 20% to

28% (Bird Machine Co., 1971). In all of these cases, the solids capture was about 90%.

The basket-type centrifuge, shown in Figure 21.17, has a semi–batch-type operation. The sludge is introduced into the vertical mounted spinning bowl, and the solids are collected at the sides. The centrate overflows the bowl rim and is discharged. Once the solids have built up to a maximum thickness, the feed sludge is stopped, the rotation of the bowl is decreased, and a scraper blade peels the sludge cake from the walls and it discharges out of the bottom of the bowl. Once the cake is removed, the bowl rotation is increased and the feed sludge flow is resumed. The following percentages of dry solids in the processed sludge cake have been obtained for municipal sludges chemically conditioned by polymers and dewatered by basket-type centrifugation: (1) raw primary sludge, 25% to 30%, (2) raw primary and fresh waste activated sludge, 12% to 14%, and (3) digested primary and waste activated sludge, 12% to 14% (EPA, 1979). The solids capture was from 75% to 97%.

EXAMPLE 21.3

Sludge Centrifugation

A chemically conditioned digested sludge is to be dewatered using a centrifuge that has 90% solids capture. The digested primary and secondary sludge has 8% dry solids and a specific gravity of 1.03. The digested sludge flow is 20,000 gal/day, and the centrifuged cake has 18% dry solids. Determine the lb/day of cake and the gal/day centrate.

SOLUTION

Use one day of operation as a basis for calculation. The pounds of dry solids are $(20,000\,\text{gal/day})(8.34\,\text{lb/gal})(1.03)(8\,\text{lb}/100\,\text{lb})(1\,\text{day})$ or $13,744\,\text{lb}$. The pounds of dry solids in the cake are $(13,744\,\text{lb})(0.90)$ or $12,370\,\text{lb}$. The pounds of cake are $(12,370\,\text{lb})(100\,\text{lb}/18\,\text{lb})$ or $68,722\,\text{lb}$. A material balance on the total flows gives [Input] = [Output] or

$$(20,000\,\text{gal})(8.34\,\text{lb/gal})(1.03) = 68,722\,\text{lb} + (C)(8.34\,\text{lb/gal})$$

where C is the gallons of centrate. From the previous equation, $C = 12,360\,\text{gal}$. Thus the cake is $\boxed{68,700\,\text{lb/day}}$ and the centrate is $\boxed{12,400\,\text{gal/day.}}$

EXAMPLE 21.3 SI

Sludge Centrifugation

A chemically conditioned digested sludge is to be dewatered using a centrifuge that has 90% solids capture. The digested primary and secondary sludge has 8% dry solids and a specific gravity of 1.03. The digested sludge flow is 76,000 ℓ/d, and the centrifuged cake has 18% dry solids. Determine the kg/d of cake and the ℓ/d of centrate.

SOLUTION

Use one day of operation as a basis for calculation. The kilograms of dry solids are $(76,000\,\ell/\text{d})(\text{kg}/\ell)(1.03)(8\,\text{kg}/100\,\text{kg})(1\,\text{d})$ or $6262.4\,\text{kg}$. The kilograms of dry solids in the cake are (6262.4 kg)(0.90) or 5636.2 kg. The kilograms of cake are (5636.2 kg)(100 kg/18 kg) or 31,312 kg. A material balance on the total flows gives [Input] = [Output] or

$$(76,000\,\ell)(kg/\ell)(1.03) = 31,312\,kg + (C)(kg/\ell)$$

where C is the liters of centrate. From the previous equation, $C = 46,968\,\ell$. Thus the cake is $\boxed{31,300\,kg/d}$ and the centrate is $\boxed{47,000\,\ell/d.}$

Sludge-drying beds, shown in Figures 21.18 and 21.19, are the most commonly used dewatering method for digested sludges from small- to medium-size installations. If the population is greater than about 25,000 persons, other methods of dewatering should be considered because of the space required and the labor costs involved. Drying beds usually are 20 to 30 ft (6.9 to 9.1 m) wide, are 25 to 125 ft (7.6 to 38.1 m) long, and consist of a 6- to 10-in. (15- to 25-cm) coarse sand layer above a 6- to 12-in. (15- to 30-cm) graded gravel layer. The subgrade is sloped to a drain tile line that is beneath the center of the bed and extends the full length of the bed. At least two beds must be provided. Sludge is usually applied in an 8- to 12-in. (20-

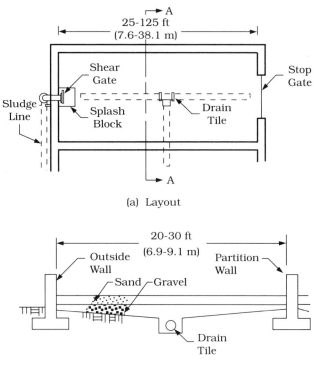

(a) Layout

(b) Section A-A

FIGURE 21.18
Sludge-Drying Bed

FIGURE 21.19 *Sludge-Drying Beds at a Wastewater Treatment Plant. Note concrete splash block at shear gate inlet and concrete pads for truck wheels.*

to 30-cm) depth. Dewatering occurs by both drainage from the bed and evaporation of water to the atmosphere. Sludge drainage occurs up to about 2 days after application of the sludge to the bed. Usually, the drainage is returned to the head of the plant. The final dewatering is by evaporation to the atmosphere. In dry weather, sludge may dewater sufficiently for removal after about 2 to 4 weeks, with the solids content of the dried sludge being about 30% to 40%. The amount of bed area required per population equivalent depends primarily on the type of digestion and the rainfall. The area required per capita for digested primary and secondary sludge mixtures is given by the equation (Texas State Department of Health, 1976)

$$\hat{A} = K(C_1 R + C_2) \tag{21.7}$$

where

$\hat{A}$ = area per capita, ft^2/cap (m^2/person)

K = factor depending on type of digestion; $K = 1.0$ for anaerobic digestion, $K = 1.6$ for aerobic digestion

$C_1 = 0.01$ for USCS units and 0.0366 for SI units

$C_2 = 1.0$ for USCS units and 0.0929 for SI units

R = annual rainfall, in. (m)

Only digested sludges may be dewatered on drying beds, because fresh sludges will decompose anaerobically and create severe odor problems.

The main dewatering action in **drying lagoons** is atmospheric evaporation. Lagoons are a simple, low-cost method of dewatering; however, they can occupy considerable land space in humid and rainy climates. Lagoons are usually used in pairs and must be surrounded by dikes to prevent surface runoff from entering. In operation, a lagoon is filled with about 2 ft (0.6 m) of digested sludge and allowed to dry; then the filling and drying are repeated. Once the lagoon is filled with dried sludge, the sludge is removed. The solids loading is usually from 2.2 to 2.4 lb/yr-ft^3 (35 to 38 kg/yr-m^3) (Great Lakes–Upper Mississippi River Board of State Sanitary Engineers, 1978). Only digested sludges should be dewatered by lagoons if severe odor problems are to be avoided.

Heat Drying   Heat drying is used when fresh sludge is to be processed to produce a fertilizer. Heat drying is accomplished by flash drying, kiln drying, or multiple-hearth furnaces. In flash drying, the fresh sludge is mixed with some previously dried sludge, and then the mixture is dried by a stream of hot combustion gases from a fuel-fired furnace. The dried sludge is separated from the gases by a cyclone. A portion of the dried material is mixed with incoming fresh sludge; the remainder is removed as the dried product. The vapors from the cyclone are returned to the furnace for deodorization. In the kiln dryer, the fresh sludge enters a sloped rotating kiln and moves through it countercurrent to hot combustion gases that dry the material. The dried sludge is discharged from the lower end of the kiln. In the multiple-hearth furnace, the sludge is dried as it passes downward through the hearths. The temperature used is usually about 700 to 900°F (371 to 482°C) and is not sufficient to cause sludge combustion. In all the drying methods, a fuel must be employed that involves considerable operational costs. The dried sludge has about 10% moisture, is sterile, and is a good fertilizer and soil conditioner. Although fertilizer production from sludge is not profit making, it does reduce handling costs by about 30% to 40%. It is limited to large operations because of the complexity of the process and the fact that the product demands are for large quantities.

Incineration   Incineration consists of dry combustion of a sludge to produce an inert ash. The ash is usually disposed in a sanitary landfill. The fuel requirements depend on the fuel value of the sludge solids and the water content. Raw primary and undigested secondary sludges will have fuel values from about 7000 to 8000 Btu/lb (16,000 to 19,000 kJ/kg) dry solids, and if they are dewatered to about 25% solids, incineration will be self-sustaining. Auxiliary fuel is required only during the startup of the incinerator. Digested sludge nearly always requires an auxiliary fuel, because the fuel value is only about half that of fresh sludge. The two types of incinerators used are the multiple-hearth type, shown in Figure 21.20, and the fluidized-bed incinerator, shown in Figure 21.21.

The **multiple-hearth incinerator**, depicted in Figure 21.20, consists of a hearth-type furnace in which the dewatered sludge is fed to the first hearth at the top. In the incinerator, the solids move downward by the raking

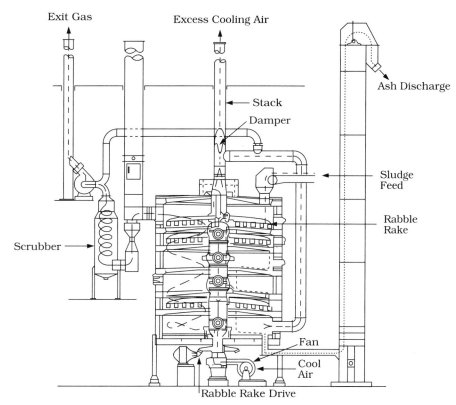

FIGURE 21.20
*Multiple-Hearth
Incinerator*

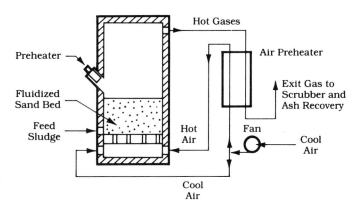

FIGURE 21.21
*Fluidized-Bed
Incinerator*

action of the rabble arms. In the upper hearths, the water content is vaporized and the sludge solids are dried at a temperature of about 900°F to 1200°F (482°C to 649°C). In the middle hearths, the sludge solids are ignited and burned at a temperature ranging from about 1200°F to 1500°F (649°C to 816°C). In the lower hearths, the slow-burning material is burned and the ash undergoes cooling at a temperature of about 600°F (316°C). The ash that is produced leaves at the bottom of the incinerator. The excess air required by this type of incinerator is from 50% to 100% of the theoretical amount. The efficiency of multiple-hearth furnaces is about 55%.

EXAMPLE 21.4   *Sludge Incineration*

A fresh sludge having a fuel value of 7000 Btu/lb dry solids is to be dewatered and then incinerated in a multiple-hearth furnace that has an efficiency of 55% — that is, 55% of the fuel value is available to evaporate water. The sludge has a winter temperature of 45°F. Determine the required dewatering if the furnace is to operate without auxiliary fuel.

SOLUTION   The heat required to raise the temperature of the sludge from 45°F to 212°F is $212 - 45$ or 167 Btu/lb. From steam tables, the heat required to evaporate 1 lb of water at 212°F is 970.3 Btu/lb. Thus the heat required to raise the temperature to 212°F and then to evaporate the water is equal to $167 + 970.3 = 1137.3$ Btu/lb. If x is the lb dry solids/lb wet sludge, the heat available is $(x)(7000)$(furnace efficiency as a fraction), and the heat required is $(1 - x)(1137.3)$. Thus,

$$(x)(7000)(0.55) = (1 - x)(1137.3)$$

From this equation, $x = 0.228$. Therefore, the sludge must be dewatered to obtain at least $\boxed{22.8\% \text{ dry solids}}$ in order to incinerate without auxiliary fuel.

EXAMPLE 21.4 SI   *Sludge Incineration*

A fresh sludge having a fuel value of 16,300 kJ/kg dry solids is to be dewatered and then incinerated in a multiple-hearth furnace that has an efficiency of 55% — that is, 55% of the fuel value is available to evaporate water. The sludge has a winter temperature of 7°C. Determine the required dewatering if the furnace is to operate without auxiliary fuel.

SOLUTION   The heat required to raise the temperature of the sludge from 7°C to 100°C is $(100 - 7)$ kg-cal/kg $\times$ 4.184 kJ/kg-cal or 389 kJ/kg. From steam tables, the heat required to evaporate 1 kg of water at 100°C is 2257.0 kJ/kg. Thus the heat required to raise the temperature to 100°C and then to evaporate the water is equal to $389 + 2257.0 = 2646$ kJ/kg. If x is the kg dry solids/kg wet sludge, the heat available is $(x)(16,300)$(furnace efficiency as a fraction), and the heat required is $(1 - x)(2646)$. Thus,

$$(x)(16,300)(0.55) = (1 - x)(2646)$$

From this equation, $x = 0.228$. Therefore, the sludge must be dewatered to obtain at least $\boxed{22.8\% \text{ dry solids}}$ in order to incinerate without auxiliary fuel.

The **fluidized-bed incinerator**, shown in Figure 21.21, consists of a combustion reactor containing a bed of sand above a grid. To start the furnace, the preheater is ignited and the fluidizing air is passed upward through the bed to suspend the sand. Once the sand temperature has reached about 1400°F to 1500°F (760°C to 816°C), the sludge feed to the incinerator is begun. The water is vaporized, and the sludge solids are burned in the fluidized sand bed. The intense agitation caused by the water vaporization and the solids combustion allows the furnace to be operated with only about 20% to 25% excess air. The ash created is carried from the reactor by the exit combustion gases and is subsequently removed by a cyclone or a scrubber.

Ultimate Disposal   The solids produced from sludge dewatering or the ash from incineration are usually disposed on land. Digested, air-dried municipal sludge may be spread on agricultural land and plowed under, thus serving as both a fertilizer and a soil conditioner. Wet digested sludge may also be spread on land, and once dried, it may be plowed under. Air-dried digested sludge may be spread on lawns for both fertilization and soil conditioning. Heat-dried sludge is sold as a fertilizer because it is sterile and has fertilizer value. Dewatered sludge and incinerator ash frequently are disposed in sanitary landfills.

WASTEWATER TREATMENT PLANT SLUDGES (CHEMICAL)   The chemical coagulants generally used for coagulation and precipitation in tertiary treatment of secondary effluents are lime and alum, lime being the more common. The coagulant generally used for the coagulation of raw municipal wastewaters is lime. Both alum and lime coagulants are effective in removing suspended solids and phosphorus.

Lime Sludges   Coagulation with lime requires large chemical dosages, and as a result, the amount of sludge produced is relatively large. Typical lime dosages are 100 to 500 mg/ℓ calcium oxide. Although the sludge may be disposed on land, it is generally recalcined to recover the lime because this greatly reduces the amount of new lime required and also minimizes the problem of sludge disposal. The operations and processes used for solids handling of lime sludges consist of thickening, dewatering, and lime recalcining.

Lime sludges are very dense and may be easily **thickened** using a gravity-type thickener. For a feed sludge of 5000 to 10,000 gm/ℓ solids, a thickener with a solids loading of about 200 lb/day-ft^2 (980 kg/day-m^2) and an overflow rate of about 1000 gal/day-ft^2 (41 m^3/day-m^2) has produced a thickened sludge with 8% to 20% solids (Culp and Culp, 1978; South Tahoe Public Utility District, 1971). Pumps for lime sludges are usually recessed-impeller centrifugals and should be located near the thickener to minimize the length of suction lines. Provisions should be made for rodding the lines.

Thickened lime sludges **dewater** readily. Concurrent flow centrifuges have dewatered a sludge with 8% to 20% solids to a cake of about 50% to

55% solids and a solids capture of about 90% (Culp and Culp, 1978; South Tahoe Public Utility District, 1971). Lime sludges will contain both calcium carbonate and calcium phosphate. Since the two precipitates have different densities, it is possible to operate a centrifuge so that the calcium phosphate is separated from the sludge to be recalcined. When this is done, two centrifuges are required. The first is operated to produce a cake of about 40% solids for recalcining, with the calcium phosphate leaving in the centrate. The centrate is then sent to the second centrifuge, which produces a cake containing the calcium phosphate precipitate. The calcium phosphate cake from the second centrifuge is usually disposed in a sanitary landfill.

Lime recalcining or lime recovery is usually accomplished by multiple-hearth furnaces similar to those used for sludge incineration. Usually six hearths are employed, and the hearth temperatures are about 800°F (427°C), 1250°F (677°C), 1850°F (1010°C), 1850°F (1010°C), 1850°F (1010°C), and 750°F (399°C) for the respective hearths from the top of the furnace to the bottom. As the lime sludge passes down through the furnace, the water is evaporated and the decomposition of calcium carbonate occurs as follows:

$$CaCO_3 \xrightarrow{\Delta} CaO + CO_2 \qquad \textbf{(21.8)}$$

The solids from the bottom hearth are passed through a grinder to break up large particles, and the recalcined lime is then cooled and conveyed to storage. Since some inerts will be present in lime, it is necessary to waste some lime sludge periodically to remove the inerts from the system. Lime recalcining may also be accomplished by fluidized-bed furnaces and rotary kilns. In lime recalcining, the organic solids in the sludge are incinerated during the recovery process.

Alum Sludges   Alum coagulation removes both suspended solids and phosphorus, requiring an alum dosage much less than that required for lime. Alum sludges are difficult to dewater, and where possible, they are frequently mixed with sewage sludges and sent to anaerobic digestion. In the digestion process, the aluminum hydroxide and the aluminum phosphate precipitates remain as solids and are disposed along with the digested sludge. In some cases, alum in sludges has been recovered by either an alkaline or an acid treatment (Culp and Culp, 1978).

WATER TREATMENT PLANT SLUDGES   In water coagulation, sludges are produced from the clarifier operations and the backwashing of the filters. The aluminum or iron salt coagulants create a gelatinous sludge that will contain the organic and inorganic materials that are coagulated and also the hydroxide precipitate. Dewatering of these sludges is very difficult. In the past, at most plants the clarifier sludge and the filter backwash water were returned to the water supply source, which was usually a river or lake. The present trend is to process the sludge from clarifiers for ultimate disposal and to catch the backwash water in basins and gradually return it to the treatment plant for reprocessing.

In water softening, the sludges produced are mainly calcium carbonate and magnesium hydroxide precipitates, although some organic and inorganic materials are present that have been removed from the water. These sludges dewater rather easily, and it is common to process them for ultimate disposal. The backwash water is usually caught in basins and slowly returned to the treatment plant for reprocessing.

The unit operations and processes used for solids handling in water treatment are similar to those employed in wastewater treatment. The main operations and processes include thickening, conditioning, dewatering, ultimate disposal, and, in some cases, coagulant or lime recovery.

Discharge to Sanitary Sewers　The gelatinous sludge from the clarifiers in a coagulation plant may be discharged to the sanitary sewage system if the primary clarifiers and solids handling facilities at the wastewater treatment plant have adequate capacity to process this additional sludge. Where this is practiced, the sludge is frequently discharged to the sewer system during the night when the wastewater flow is relatively low. Softening sludges should not be disposed in this manner because they are of a large quantity, may readily fill the digesters, and will produce encrustations on weirs, channels, piping, and so on.

Thickening　Gravity thickening of lime softening sludges has increased the solids content from 1% to about 30% when thickener loadings of 12.5 lb/day-ft^2 (61 kg/day-m^2) have been used (AWWA, 1969). Gravity thickening of alum sludges from water coagulation has increased the solids content from about 1% to 2% when thickener loadings of 4.0 lb/day-ft^2 (20 kg/day-m^2) have been used (AWWA, 1969).

Conditioning　The sludge from alum or iron salt coagulation may be conditioned to improve its dewatering characteristics.

Gelatinous hydroxide sludges may be **heated** in reactors at elevated temperatures and pressures to cause the bound water to be released. One study showed that vacuum filtration of heat-treated sludges at 10 to 20 lb/hr-ft^2 (49 to 98 kg/h-m^2) produced a cake having about 21% solids (Schroeder, 1970).

The **freezing and thawing** of gelatinous hydroxide sludges causes the bound water to be released, thus improving the dewatering characteristics. An alum sludge thickened to about 2% solids, and then frozen and thawed, created a sludge with about 20% solids that was subsequently dewatered by vacuum filtration to 34% solids (AWWA, 1969).

Dewatering　The sludges from water softening plants dewater readily; however, the metallic hydroxide sludges from water coagulation are difficult to dewater.

Vacuum filtration of gravity-thickened lime-softening sludges containing about 30% solids has produced a cake of about 65% solids at a vacuum filter loading of 40 lb/hr-ft^2 (195 kg/h-m^2) (AWWA, 1969). Vacuum filtration of coagulation sludges requires the use of a precoated **rotary vacuum filter**. In this operation, a filter aid is added to create a 2- to 4-in. (5- to 10-cm) thick precoat on the rotary filter drum prior to sludge addition. Tests have shown that a cake of 29% to 32% solids could be obtained (Mahoney and Duensing, 1972).

Centrifugation has been widely used to dewater lime softening sludges. Most installations use a gravity thickener to thicken the feed sludge to about 15% to 25% solids. Subsequent centrifugation produces a cake having about 55% to 60% solids (AWWA, 1969). In one installation, the underflow from the clarifier had 5% to 10% solids, and centrifugation of the underflow produced a cake of 55% to 60% solids (AWWA, 1969). Most centrifuges are the countercurrent solid-bowl type; however, the recently designed concurrent solid-bowl type gives a centrate with less solids. Recent installations usually employ this type. Centrifuges may also be used to classify softening sludges where recalcining is practiced. A lime softening sludge contains both calcium carbonate and magnesium hydroxide, and since these precipitates have different densities, centrifuges can be operated to separate them. The calcium carbonate will leave in the cake and the magnesium hydroxide in the centrate of the first centrifuge. A second centrifuge handling the centrate from the first can be used to remove the magnesium hydroxide. The solids capture by centrifuges handling softening sludges is usually from 70% to 90% without polymer aids and up to 95% when polymer aids are used. The centrifugation of alum sludges has not been very successful.

Filter presses have been used to process lime softening sludges to yield a cake having 60% to 65% solids. The filter press is the only mechanical dewatering device that is highly successful in dewatering coagulation sludges. Untreated alum sludges containing 1.5% to 2% solids have been processed to produce a cake of 15% to 20% solids (AWWA, 1969). If alum sludge is conditioned with lime or polymers, cakes are produced that have much higher percent solids. Filter precoating with diatomaceous earth and lime conditioning of an alum sludge at one installation have produced cakes containing 40% to 50% solids (Weir, 1972).

Lagoons are the most widely used method to handle softening and coagulation sludges, and although the operating costs are rather low, the land requirements are appreciable. They should be used in pairs and should have dike embankments to exclude surface runoff. The sludge is added until the lagoon is full; then it is taken out of service and allowed to dry. Once drying is sufficient, the sludge is removed; thus the lagoons serve for both thickening and dewatering. Lime softening sludges are effectively handled by lagoons, and the solids content of the dewatered sludge is usually about 50% (AWWA, 1969). Lagoons have not been nearly as successful in dewatering alum sludges. Usually alum sludges can be dewatered to about 1% to 10% solids (AWWA, 1969), and it is common to remove the partially dewatered sludge by drag lines, place it in trucks, and then spread it on land for ultimate disposal. A climatic condition of alternate freezing and thawing assists in releasing the bound water in coagulation sludges.

Ultimate Disposal

The final disposal of dewatering sludges from water treatment is land disposal either by placing the sludge in a sanitary landfill or by spreading it on land. The sanitary landfill is the more widely used.

Lime or Coagulant Recovery

The lime sludge produced from water softening may be calcified by centrifuges to separate the calcium carbonate from the magnesium hydroxide. The lime can then be recalcined and the quicklime that is produced reused.

Alum recovery, though it is not widely used, can be accomplished using acidification with sulfuric acid as follows:

$$2\,Al(OH)_3\downarrow\ +\ 3H_2SO_4 \rightarrow Al_2(SO_4)_3 + 6H_2O \qquad (21.9)$$

After acidification, the supernatant, which contains the alum, is separated from the solids, and the recovered solution is used as a liquid coagulant.

A magnesium carbonate and lime coagulation system has been developed that allows recovery of one or both of the coagulants. The coagulation reaction is

$$MgCO_3 + Ca(OH)_2 \rightarrow Mg(OH)_2\downarrow\ +\ CaCO_3\downarrow \qquad (21.10)$$

The sludge will contain calcium carbonate, magnesium hydroxide, and the coagulated matter. The sludge is carbonated to dissolve the magnesium hydroxide as follows:

$$Mg(OH)_2\downarrow\ +\ 2CO_2 \rightarrow Mg(HCO_3)_2 \qquad (21.11)$$

The sludge is filtered and the filtrate, which contains the magnesium bicarbonate, is returned as a liquid coagulant solution. When added with lime, the solution coagulates as follows:

$$Mg(HCO_3)_2 + 2Ca(OH)_2 \rightarrow Mg(OH)_2\downarrow\ +\ 2CaCO_3\downarrow\ +\ 2H_2O \quad (21.12)$$

It is also possible to separate the calcium carbonate in the sludge and recalcine it to produce quicklime, which may be reused.

REFERENCES

American Water Works Association (AWWA). 1971. *Water Quality and Treatment.* New York: McGraw-Hill.

American Water Works Association (AWWA) Research Foundation. 1969. Disposal of Wastes from Water Treatment Plants, Parts 1, 2, 3, and 4. *Jour. AWWA* 61, no. 10:541; 61, no. 11:619; 61, no. 12:681.

Bird Machine Co. 1971. *Bird Machine Company Product Manual.*

Coackley, P. 1958. Laboratory Scale Filtration Experiments and Their Application to Sewage Sludge Dewatering. In *Biological Treatment of Sewage and Industrial Wastes.* Vol. 2, edited by J. McCabe and W. W. Eckenfelder, Jr. New York: Reinhold.

Conway, R. A., and Ross, R. D. 1980. *Handbook of Industrial Waste Disposal.* New York: Van Nostrand Reinhold.

Culp, R. L., and Culp, G. L. 1978. *Handbook of Advanced Wastewater Treatment.* 2nd ed. New York: Van Nostrand Reinhold.

Culp, G. L., and Culp, R. L. 1974. *New Concepts in Water Purification.* New York: Van Nostrand Reinhold.

Dahlstrom, D. A., and Cornell, C. F. 1958. Improved Sludge Conditioning and Vacuum Filtration. In *Biological Treatment of Sewage and*

Industrial Wastes. Vol. 2, edited by J. McCabe and W. W. Eckenfelder, Jr. New York: Reinhold.

Dick, R. I., and Ewing, B. B. 1967. Evaluation of Activated Sludge Thickening Theories. *Jour. SED* 93, SA4:9.

Eckenfelder, W. W., Jr. 1980. *Principles of Water Quality Management.* Boston: CBI Publishing.

Environmental Protection Agency (EPA). 1987. *Dewatering Municipal Wastewater Sludges.* EPA Design Manual, Washington, D.C.

Environmental Protection Agency (EPA). 1979. *Sludge Treatment and Disposal.* EPA Process Design Manual, Washington, D.C.

Environmental Protection Agency (EPA). 1974. *Upgrading Existing Wastewater Treatment Plants.* EPA Process Design Manual, Washington, D.C.

Fair, G. M.; Geyer, J. C.; and Okun, D. A. 1968. *Water and Wastewater Engineering.* Vol. 2. *Water Purification and Wastewater Treatment and Disposal.* New York: Wiley.

Forster, H. W. 1972. Sludge Dewatering by Pressure Filtration. Paper presented at the AIChE Annual Meeting, New York City.

Great Lakes–Upper Mississippi River Board of State Sanitary Engineers. 1978. Recommended Standards for Sewage Works. Ten state standards. Albany, N.Y.

Krasauskas, J. W. 1969. Review of Sludge Disposal Practices. *Jour. AWWA* 61, no. 5:225.

Mahoney, P. F., and Duensing, W. J. 1972. Precoat Vacuum Filtration and Natural-Freeze Dewatering of Alum Sludge. *Jour. AWWA* 64, no. 10: 655.

Metcalf & Eddy, Inc. 1979. *Wastewater Engineering: Treatment, Disposal and Reuse.* 2nd ed. New York: McGraw-Hill.

Metcalf & Eddy, Inc. 1991. *Wastewater Engineering: Treatment, Disposal and Reuse.* 3rd ed. New York: McGraw-Hill.

Newton, D. 1964. Thickening by Gravity and Mechanical Means. In *Sludge Concentration, Filtration and Incineration.* University of Michigan. School of Public Health, Continuing Education Series 113, no. 4.

Ramalho, R. S. 1977. *Introduction to Wastewater Treatment Processes.* New York: Academic Press.

Rich, L. G. 1961. *Unit Operations of Sanitary Engineering.* New York: Wiley.

Riehl, M. L.; Weiser, H. H.; and Rheims, B. T. 1952. Effect of Lime-Treated Water on Survival of Bacteria. *Jour. AWWA* 44, no. 5:466.

Sanks, R. L. 1978. *Water Treatment Plant Design.* Ann Arbor, Mich.: Ann Arbor Science Publishers.

Schroeder, E. D. 1977. *Water and Wastewater Treatment.* New York: McGraw-Hill.

Schroeder, R. P. 1970. *Alum Sludge Disposal — Report no. 1.* R&D Project no. DP-6551, Eimco Corporation.

Smith, J. E.; Hathaway, S. W.; Farrell, J. B.; and Dean, R. B. 1972. Sludge Conditioning with Incinerator Ash. Paper presented at 27th Annual Purdue Industrial Waste Conference.

South Tahoe Public Utility District. 1971. *Advanced Wastewater Treatment as Practiced at South Tahoe.* Technical Report for the EPA, Project 17010 ELQ (WPRD 52-01-67).

Sundstrom, D. W., and Klei, H. E. 1979. *Wastewater Treatment.* Englewood Cliffs, N.J.: Prentice-Hall.

Texas State Dept. of Health. 1976. Design Criteria for Sewage Systems. Austin, Tex.

Thompson, C. G.; Singley, J. E.; and Black, A. P. 1972. Magnesium Carbonate: A Recycled Coagulant. *Jour. AWWA* 64, no. 1:11; 64, no. 2:93.

Vesilind, P. A. 1979. *Treatment and Disposal of Wastewater Sludges.* Ann Arbor, Mich.: Ann Arbor Science Publishers.

Water Environment Federation (WEF). 1992. *Sludge Digest.* WEF Digest Series, Alexandria, Va.

Water Environment Federation (WEF). 1992. *Sludge Incineration: Thermal Destruction of Residues.* WEF Manual of Practice FD-19, Alexandria, Va.

Water Environment Federation (WEF). 1993. *Sludge Stabilization.* WEF Manual of Practice No. FD-9, Alexandria, Va.

Water Environment Federation (WEF) and American Society of Civil Engineers (ASCE). 1991. *Design of Municipal Wastewater Treatment Plants.* WEF Manual of Practice No. 8, ASCE Manual and Report on Engineering Practice No. 76. Brattleboro, Vt: Book Press, Inc.

Water Pollution Control Federation (WPCF). 1988. *Incineration.* WPCF Manual of Practice No. OM-11, Alexandria, Va.

Water Pollution Control Federation (WPCF). 1988. *Sludge Conditioning.* WPCF Manual of Practice No. FD-14, Alexandria, Va.

Water Pollution Control Federation (WPCF). 1969. *Sludge Dewatering.* WPCF Manual of Practice No. 20. Washington, D.C.

Water Pollution Control Federation (WPCF). 1983. *Sludge Dewatering.* WPCF Manual of Practice No. 20. Alexandria, Va.

Water Pollution Control Federation (WPCF). 1980. *Sludge Thickening.* WPCF Manual of Practice No. FD-1. Washington, D.C.

Water Pollution Control Federation (WPCF). 1977. *Wastewater Treatment Plant Design.* WPCF Manual of Practice No. 8, Washington, D.C.

Water Pollution Control Federation (WPCF) Research Foundation. 1990. *Innovative Process Assessment: Sludge Processing, Disposal and Reuse.* WPCF Research Foundation Report 90-4, Alexandria, Va.

Weir, P. 1972. Research Activities by Water Utilities — Atlanta Water Dept. *Jour. AWWA* 64, no. 10:634.

Yoshioka, N.; Hotta, Y.; Tanaka, S.; Naito, S.; and Tsugami, S. 1957. Continuous Thickening of Homogeneous Flocculated Slurries. *Chem. Eng.* 21, Tokyo, Japan (in Japanese with English abstract).

PROBLEMS

21.1 A thickener is to thicken waste activated sludge from an industrial wastewater treatment plant. Pertinent data are: sludge flow = 250,000 gal/day, solids in the waste activated sludge = 10,000 mg/ℓ, specific gravity of the sludge = 1.002, solids in the thickened sludge = 25,000 mg/ℓ, specific gravity of the thickened sludge = 1.005, solids in the supernatant = 800 mg/ℓ, overflow rate = 600 gal/day-ft^2, and solids loading = 5 lb/day-ft^2. Determine:
a. The thickened sludge flow, gal/day.
b. The diameter of the thickener.

21.2 A thickener is to thicken waste activated sludge from an industrial wastewater treatment plant. Pertinent data are: sludge flow = 950,000 ℓ/d, solids in the waste activated sludge = 10,000 mg/ℓ, specific gravity of the sludge = 1.002, solids in the thickened sludge = 25,000 mg/ℓ, specific gravity of the thickened sludge = 1.005, solids in the supernatant = 800 mg/ℓ, overflow rate = 24.5 m^3/d-m^2, and solids loading = 24.4 kg/d-m^2. Determine:
a. The thickened sludge flow, ℓ/d.
b. The diameter of the thickener, m.

21.3 Vacuum filters are to be used to dewater a raw primary and waste activated sludge mixture from a municipal wastewater treatment plant. Pertinent data are: sludge flow = 200,000 gal/day, solids in the sludge = 4.5%, specific gravity of the sludge = 1.008, cake solids = 20%, solids in the filtrate = 1000 mg/ℓ, and solids loading = 4.5 lb/hr-ft^2. Determine:
a. The filtrate flow, gal/day.
b. The cake produced, tons/day.
c. The area of the filters, ft^2.

21.4 Vacuum filters are to be used to dewater a raw primary and waste activated sludge mixture from a municipal wastewater treatment plant. Pertinent data are: sludge flow = 760,000 ℓ/d, solids in the sludge = 4.5%, specific gravity of the sludge = 1.008, cake solids = 20%, solids in the filtrate = 1000 mg/ℓ, and solids loading = 22 kg/h-m^2. Determine:
a. The filtrate flow, ℓ/d.
b. The cake produced, kg/d.
c. The area of the filters, m^2.

21.5 The solids in the primary and waste activated sludge from an industrial wastewater treatment plant treating a pulp and paper mill wastewater are to be incinerated using a fluidized-bed incinerator. The solids have a fuel value of 6990 Btu/lb, and the efficiency of the incinerator is 55%; that is, 55% of the heat released during combustion is available to evaporate the water in the sludge. The remaining heat is lost, mainly through the stack. Assume the heat required to evaporate the water is 1120 Btu/lb water. Determine the minimum solids content in the dewatered sludge if the incinerator is to operate without auxiliary fuel.

21.6 The solids in the primary and waste activated sludge from an industrial wastewater treatment plant treating a pulp and paper mill wastewater are to be incinerated using a fluidized-bed incinerator. The solids have a fuel value of 16,260 kJ/kg, and the efficiency of the incinerator is 55%; that is, 55% of the heat released during combustion is available to evaporate the water in the sludge. The remaining heat is lost, mainly through the stack. Assume the heat required to evaporate the water is 2610 kJ/kg water. Determine the minimum solids content in the dewatered sludge if the incinerator is to operate without auxiliary fuel.

21.7 A centrifuge is to be used to dewater a sludge from a municipal lime-soda softening plant. The maximum plant capacity is 31 MGD, and 800 lb of dry sludge solids are produced per million gal treated. The solids content in the sludge is 10%, the cake from the centrifuge is 50% solids, and the solids capture is 90%. The specific gravity of the dry solids (s_s) is 2.50, and the specific gravity of an aqueous slurry is given by

$$s = \frac{p + (100 - p)}{p + (100 - p)/s_s}$$

where

s = specific gravity of the aqueous slurry

p = percent water

s_s = specific gravity of the dry solids

Determine:

a. The gallons of wet sludge and centrate produced per day.

b. The pounds of cake produced per day.

c. The solids content of the centrate, mg/ℓ.

21.8 A centrifuge is to be used to dewater a sludge from a municipal lime-soda softening plant. The maximum plant capacity is $117{,}000\,m^3/d$, and $0.10\,kg$ of dry sludge solids is produced per cubic meter of water treated. The solids content in the sludge is 10%, the cake from the centrifuge is 50% solids, and the solids capture is 90%. The specific gravity of the dry solids (s_s) is 2.50, and the specific gravity of an aqueous slurry is given by the equation in Problem 21.7. Determine:

a. The liters of wet sludge and centrate produced per day.

b. The kilograms of cake produced per day.

c. The solids content of the centrate, mg/ℓ.

21.9 A rotary vacuum filter is to dewater a digested sludge from a plant treating a wastewater from 200,000 persons. The digested sludge produced is 0.22 lb solids per person per day, and the solids content of the digested sludge is 10%. Six percent of the solids are lost in the filtrate, the sludge cake is 25% solids, and the filter is loaded at 2.5 lb/hr-ft^2 on a dry solids basis. Determine:

a. The pounds of cake produced per day.

b. The gallons of filtrate produced per day and the solids content of the filtrate.

c. The size of the filter if the length is twice the diameter.

21.10 A rotary vacuum filter is to dewater a digested sludge from a plant treating a wastewater from 200,000 persons. The digested sludge produced is

0.10 kg solids per person per day, and the solids content of the digested sludge is 10%. Six percent of the solids are lost in the filtrate, the sludge cake is 25% solids, and the filter is loaded at 12 kg/h-m^2 on a dry solids basis. Determine:

a. The kilograms of cake produced per d.

b. The liters of filtrate produced per day and the solids content of the filtrate.

c. The size of the filter if the length is twice the diameter.

21.11 A centrifuge is to dewater digested sludge from an anaerobic digester in an activated sludge plant. The sludge solids in the feed sludge to the digester amount to 60,000 lb/day, and the digested sludge solids produced from the digester are 45% of the sludge solids that enter. The feed sludge to the centrifuge contains 5.0% solids. The cake from the centrifuge is 30% solids, and the solids recovery is 90%. Determine:

a. The pounds of cake produced per d.

b. The gallons of centrate produced per d.

c. The solids content of the centrate, mg/ℓ.

21.12 A centrifuge is to dewater digested sludge from an anaerobic digester in an activated sludge plant. The sludge solids in the feed sludge to the digester amount to 27,000 kg/d, and the digested sludge solids produced from the digester are 45% of the sludge solids that enter. The feed sludge to the centrifuge contains 5.0% solids. The cake from the centrifuge is 30% solids, and the solids recovery is 90%. Determine:

a. The kilograms of cake produced per d.

b. The liters of centrate produced per d.

c. The solids content of the centrate, mg/ℓ.

21.13 A conditioned digested sludge is to be dewatered on a rotary vacuum filter under a vacuum of 25 in. Hg. A vacuum filtration test done in the laboratory gave a specific resistance of 3.9×10^7 sec^2/gm. The unfiltered solids were 4.9% by dry weight, and the sludge cake had 22% dry solids. The filtration temperature is 65°F, the cycle time is 5 min, and the form time is 42% of the cycle time. Determine the filter yield in lb/hr-ft^2.

21.14 A conditioned digested sludge is to be dewatered on a rotary vacuum filter under a vacuum of 635 mm Hg. A vacuum filtration test done in the laboratory gave a specific resistance of $3.9 \times 10^7\,s^2$/gm. The unfiltered solids were 4.9% by dry weight, and the sludge cake had 22% dry solids. The filtration temperature is 18°C, the cycle time is 5 min, and the form time is 42% of the cycle time. Determine the filter yield in kg/h-m^2.

21.15 For the conditions described in Problem 21.13, determine the filter drum area if the sludge flow is 62,000 gal/day and the volatile solids are 35% of the dry solids.

21.16 For the conditions described in Problem 21.14, determine the filter drum area if the sludge flow is 235,000 ℓ/d and the volatile solids are 35% of the dry solids.

21.17 A laboratory vacuum filtration test was performed using a conditioned, anaerobically digested sludge. The filtrate volumes shown in Table 21.3 were obtained at the indicated filtration times. Other pertinent data from the test are: vacuum = 24.5 in. Hg (622 mm Hg), dry solids content in the conditioned sludge = 8.2%, dry solids in the cake = 21%, temperature = 23°C, and the diameter of the Buchner funnel = 9.5 cm. Determine the specific resistance in sec^2/gm (s^2/gm).

TABLE 21.3

t (sec)	V (mℓ)
20	68
40	91
60	118
80	129
100	148

22

LAND TREATMENT OF MUNICIPAL WASTEWATER AND SLUDGES

Land application of sewage was practiced in ancient Athens, and municipal wastewater was used for farmland irrigation in Germany in the sixteenth century. The practice of **sewage farming** continued, with both successes and failures, in continental Europe and England, and during colonization the practice was brought to South Africa, Australia, and Mexico. Land application of sewage effluent began in the United States for irrigation purposes in the late 1800s and for groundwater recharge in the early 1900s. The simplicity of the practice for sewage treatment and disposal, the natural fertilization of vegetation derived from the practice, and water recycling and reuse were recognized as benefits in these early uses of land application of wastewaters (EPA, 1973). Sludge generated by municipal wastewater treatment plants has also been applied to land for many years in many countries. Estimates are that as much as 40% of the sewage sludge generated in the United States is potentially being applied to both food chain and non–food chain croplands. In addition to cropland application, municipal sludge is applied to forested lands, disturbed lands for land reclamation, and dedicated land disposal sites other than landfills (EPA, 1983).

WASTEWATER TREATMENT

Wastewater treatment systems referred to as natural treatment systems may be divided into two categories: soil-based systems and aquatic systems (Kruzik, 1994). The soil-based systems include the processes termed slow rate land treatment, rapid infiltration land treatment, and overland flow land treatment. In some cases, one or more of the processes may be combined to achieve a higher degree of treatment. The soil-based systems are discussed in this chapter. Aquatic systems are introduced in Chapter 23.

In the United States, land application of wastewaters has legal implications and potential use limitations that are embodied in the various states' water rights laws. These potential limitations are in addition to restrictions imposed by federal and state laws and regulations related to water quality. Because treatment by land application could divert waters away from the original source, and thereby potentially violate water use laws in a particular state, the U.S. Environmental Protection Agency sponsored research to define the legal questions associated with land application of wastewater

(EPA, 1978a). The research concluded that states' water use laws, in general, did not prevent use of land application as a treatment process; however, significant variability was found among the state laws and associated administrative regulations that influenced the use of land application. In some states, water rights laws and regulations impose more restrictions on the use of land application as a wastewater treatment alternative than in other states. Not only is treatment of wastewater by land application restricted by laws and regulations regarding water quality; it may also be restricted by laws and regulations regarding water use and reuse.

Wastewater treatment by any of the natural systems is accomplished by a complex mixture of physical, chemical, and biological mechanisms that occur in the water-soil-vegetation matrix (EPA, 1973; Metcalf & Eddy, 1991). The principle mechanisms include a combination of uptake by vegetation, precipitation, adsorption, oxidation, ion exchange, filtration, and microbial actions.

The pollutants in wastewater (including suspended solids, organic material, nitrogen, phosphorus, heavy metals, and microorganisms) are all reduced, some more than others. Suspended solids are removed by sedimentation and filtration. Biodegradable organic material is removed by a combination of mechanisms. Organic suspended solids separated from the wastewater by filtration are oxidized by microorganisms. This microbial mechanism occurs at or just beneath the surface of the soil-water interface. Dissolved organic material is removed both by direct biodegradation and by initial adsorption on soil and vegetative litter and subsequent biodegradation. Most systems are designed and operated as aerobic systems, so most of the microbial action in these systems is aerobic. In such systems, the odors that would result from anaerobic microbial action are not prevalent. Organic material that is more resistant to biodegradation (such as pesticides, cellulose, detergents, and phenols) is degraded slowly; however, excess amounts of phenols and similar organic material can be toxic and may destroy the microbial population (EPA, 1973).

Nitrogen applied to land systems is commonly in the organic and ammonia/ammonium forms unless the wastewater applied is nitrified effluent, in which case nitrate nitrogen will be present. Nitrogen in the nitrite form is rarely present in significant concentration because of the rapid oxidation of nitrite to nitrate. The nitrogen conversion and removal mechanisms in a land treatment system are shown schematically in Figure 22.1.

Organic nitrogen in the suspended material is removed from the water by filtration through the soil, where it is biologically converted to ammonium. Because of the normal pH range of domestic wastewater, relatively small amounts of ammonia exist, and volatilization of ammonia does not account for appreciable nitrogen removal in a land treatment system. The ammonium ion is removed from the water by adsorption and ion exchange processes in the soil. The adsorbed ammonium may be directly used by some vegetation and is used by microorganisms for cell maintenance, or it may be nitrified to nitrate. Ammonium ions adsorbed in anaerobic soil zones remain adsorbed

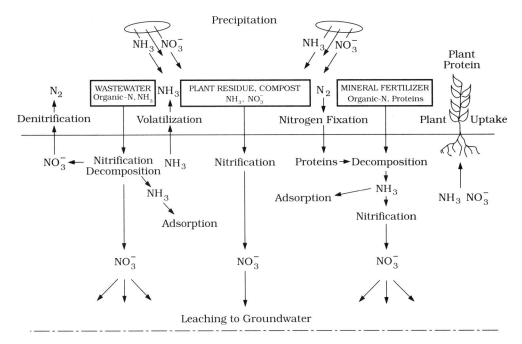

FIGURE 22.1 *Nitrogen Conversion and Removal in Soil*

Adapted from *Wastewater Treatment and Reuse by Land Application*, Vol. II, 1973. Environmental Protection Agency (EPA).

as long as the anaerobic conditions exist (Lance, 1972). Nitrate nitrogen is removed by vegetative uptake and chemical or biological denitrification. Biological denitrification does not usually occur, because land treatment systems are generally operated under aerobic conditions. However, biological denitrification can be achieved with appropriate operational control. Nitrate, unlike ammonium, is not adsorbed in the soil and readily moves through the soil, or leaches, as water movement through the soil occurs (Sepp, 1970).

Phosphorus is removed by adsorption, ion exchange, vegetative uptake, and incorporation in biological solids. Heavy metals are removed primarily by adsorption and ion exchange. Significant amounts of chromium and zinc can be removed by ion exchange (Wentink and Etzel, 1972). Some precipitation of heavy metals may occur, and small amounts are used by microorganisms for cell maintenance and by vegetation. Removal of microorganisms occurs by a number of mechanisms, including natural die-off, filtration, sedimentation, predation, desiccation, and adsorption. Virus removal is primarily by adsorption and natural die-off (Metcalf & Eddy, 1991). In irrigation systems, the spread of bacteria by drifting water mist or aerosol is a concern. Studies have indicated that bacteria may travel from 100 to 600 ft (30 to 200 m) in water mist or aerosol (Sorber *et al.*, 1976).

Slow Rate Process Wastewater treatment by the slow rate process is achieved mainly by percolation and evapotranspiration, as shown in Figure 22.2. Table 22.1 gives typical site characteristics, design parameters, and expected performance for slow rate land treatment. The slow rate process is the most widely used land treatment method (Metcalf & Eddy, 1991). Wastewater is applied to vegetated land for treatment of the wastewater and to irrigate the vegetation, which may include grains, fiber crops, oilseed crops, hay, fruit and nut orchards, vineyards, grass sod, nursery stock, woodlands, pastures, golf courses, and parklands (Dinges, 1982). The wastewater is applied by a variety of fixed or moving sprinklers or by flooding the land using ridge-and-furrow or border strip techniques. As shown in Figure 22.2, the wastewater percolates through the soil, and some is removed by plant use and evapotranspiration. The system is designed to avoid surface runoff, and provisions are usually made to collect surface runoff that may occur and return it to the system. Water generally percolates downward through the soil to the groundwater. As shown in Figure 22.3, the water also may move laterally to natural streams or lakes or may be recovered with an underdrain system or wells. The process may be designed and operated primarily as a wastewater treatment alternative, or it may also meet any of several objectives, including production of marketable crops, water conservation by use of the wastewater for irrigation, and preservation and enlargement of green belts or open spaces (EPA, 1981).

Land treatment of wastewaters by the slow rate process is limited by several factors. Site characteristics are the primary controlling or limiting factor. The slow rate process is land intensive. Depending on the primary objective of the process application, as much as 200 to 690 ac per MGD of wastewater (21.4 to 74 ha per 1000 m³/day) is required, not including the area required for buffer areas, roads, and ditches (Metcalf & Eddy, 1991; EPA, 1981). Geological characteristics of a site also may be a primary limiting factor for systems designed primarily for wastewater treatment. Shallow,

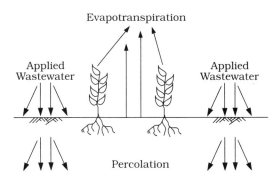

FIGURE 22.2 *Slow Rate Process, Application Hydraulic Pathway*

Adapted from *Land Treatment of Municipal Wastewater*, 1981. Environmental Protection Agency (EPA).

TABLE 22.1   *Slow Rate Land Treatment System Site Characteristics, Typical Design Features, and Expected Treated Water Quality*[a]

Grade	Less than 20% on cultivated land; less than 40% on noncultivated land
Soil permeability	Moderately slow to moderately rapid
Depth to groundwater	2–3.5 ft (0.6–1 m) (underdrains can be used to maintain this level at sites with high groundwater table)
Climatic restrictions	Storage often needed for cold weather and during heavy precipitation
Application technique	Sprinkler, ridge-and-furrow, border strip
Loading rates	
Annual	1.5–20 ft/yr (0.5–6 m/yr)
Weekly	0.5–4 in/wk (1.3–10 cm/wk)
Field area	57–690 ac/MGD (6–74 ha/1000 m^3/day), (not including buffer area, roads, or ditches)
Minimum pretreatment in United States	Preliminary treatment and primary sedimentation (with restricted public access; crops not for direct human consumption)
Disposition of applied wastewater	Evapotranspiration and percolation
Vegetation	Required
BOD_5	<2 to <5 mg/ℓ
Suspended solids	<1 to <5 mg/ℓ
Ammonia nitrogen	<0.5 to <2 mg/ℓ as N
Total nitrogen	3 to <8 mg/ℓ as N (depends on loading rate and crop)
Total phosphorus	<0.1 to <0.3 mg/ℓ as P
Fecal coliforms	0 to <10 per 100 mℓ

[a] Treated water quality values given are average to upper range expected with loading rates at the mid to low end of the loading rate range given, and with percolation of primary or secondary effluent through 5 ft (1.5 m) of unsaturated soil.
Adapted from EPA, *Land Treatment of Municipal Wastewater*, EPA Process Design Manual, 1981.

impermeable soils such as dense clays, underlying rock formations, and high water tables may prevent adequate percolation depth. Under these conditions, underdrain systems are required. Nitrogen loading may be a limiting factor because of nitrate leaching into the groundwater. In arid regions, the slow rate process may be limited on the basis of the concentrations of chloride and total dissolved solids that are acceptable for crop production. Use of the slow rate process for irrigation of golf courses, parks, and other areas with general public access requires positive control of pathogenic organisms. Irrigation of forested lands may be limited by the low-water requirements and tolerances of several types of trees, by nitrogen removal (which is lower unless the forested land is young and developing), or by shallow or rocky soils. Usually, fixed sprinkler distribution systems

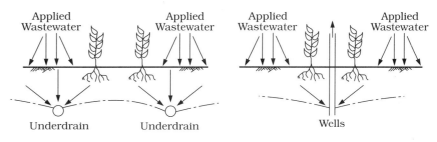

(a) Recovery Pathways

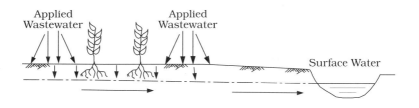

(b) Subsurface Drainage to Surface Water

FIGURE 22.3 *Slow   Rate   Process,   Subsurface   Hydraulic Pathway*

Adapted from *Land Treatment of Municipal Wastewater*, 1981. Environmental Protection Agency (EPA).

must be used for forest applications, and the added expense may be a limiting factor.

The wastewater treatment performance of the slow-rate method is excellent. The slow rate process can achieve the best level of treatment of all land treatment systems and usually has the capability to meet very strict nitrogen, phosphorus, BOD_5, suspended solids, pathogen, metals, and trace organic removal requirements (EPA, 1981). The slow rate process can achieve maximum levels of treatment without pretreatment of domestic wastewater that does not contain toxic materials from industrial sources. However, varying levels of pretreatment are required, depending on the objective of the slow rate application. In the United States, the EPA has set pretreatment guidelines to protect public health (EPA, 1981). Primary treatment — sedimentation — is considered acceptable for slow rate application on land with restricted public access or with crops that are not for direct human consumption. Higher levels of wastewater pretreatment, including disinfection, are needed when the land receiving slow rate application is used for crops produced for direct human consumption or for land with public access such as golf courses or parks. Short-term public health effects and potential long-term effects of slow rate irrigation are a concern. Dinges (1982) summarizes several studies, conducted by other researchers, of sites that have been successfully irrigated with wastewater for long periods of time, at which various chemical and biological parameters were measured and compared to control sites that received no wastewater irrigation. The book *Irrigation with Reclaimed Municipal Wastewater — A Guidance Manual*, edited by Pettygrove and Asano (1985), also contains information on potential short-term and long-term effects. These potential effects must be considered, systems must be well designed and properly operated, and each specific site should be monitored to ensure adequate protection of the public health.

The design of slow rate land systems for municipal wastewater treatment or reuse, as shown in Figure 22.4, is an iterative procedure that requires various input data not normally associated with confined wastewater treatment plants. The most important step in the design procedure is determination of the design hydraulic loading rate, which is used to determine the land area required for the slow rate system. The design hydraulic loading rate for domestic wastewater is either the allowable loading rate based on soil permeability or the allowable loading rate based on nitrogen limitations, whichever is smaller. Both the soil loading rate and the nitrogen loading rate

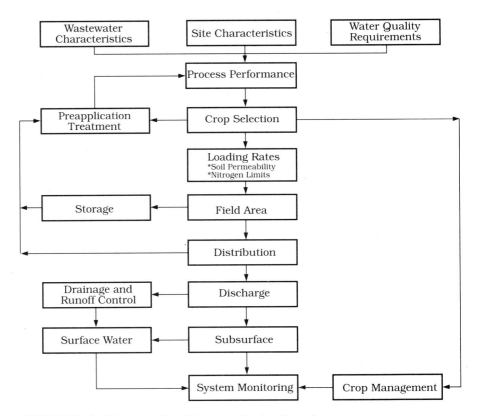

FIGURE 22.4 *Slow Rate Land Treatment Design Procedure*

Adapted from *Land Treatment of Municipal Wastewater*, 1981.
Environmental Protection Agency (EPA).

are influenced, iteratively, by crop selection, field area and associated storage requirements, the distribution method, and pretreatment, all of which are interrelated. The equations used to determine the design hydraulic loading rate and the required field area will be discussed, and their use illustrated by examples, in this text. Procedures for all of the design steps for slow rate systems may be found in *Land Treatment of Municipal Wastewater* (EPA, 1981) and *Irrigation with Reclaimed Municipal Wastewater — A Guidance Manual* (Pettygrove and Asano, 1985).

The hydraulic loading rate is the volume of wastewater applied per unit area of land per loading cycle. The allowable soil loading rate is essentially a water balance equation. Inflow (precipitation) and outflow (evaporation and percolation) rates on a monthly basis are used to determine allowable monthly soil loading rates. The monthly rates are summed to yield an allowable yearly rate. Runoff is assumed to be zero or is collected and returned. The monthly water balance is (EPA, 1981)

$$L_{w(s)} = ET - P_r + P_w \tag{22.1}$$

where

$L_{w(s)}$ = soil hydraulic loading rate, in./mo (cm/mo)

ET = evapotranspiration rate, in./mo (cm/mo)

P_r = precipitation rate, in./mo (cm/mo)

P_w = percolation rate, in./mo (cm/mo)

The allowable nitrogen loading rate is based on limiting the concentration of nitrate in underlying groundwater at the project site boundary to 10 mg/ℓ as nitrogen, in accordance with Primary Drinking Water Standards in the United States. The allowable annual nitrogen loading rate is determined from (EPA, 1981)

$$L_{w(n)} = \frac{(C_p)(P_r - ET) + (U)(K)}{(1 - f)(C_n) - C_p} \tag{22.2}$$

where

$L_{w(n)}$ = nitrogen hydraulic loading rate, in./yr (cm/yr)

C_p = nitrogen concentration in percolating water, mg/ℓ

P_r = precipitation rate, in./yr (cm/yr)

ET = evapotranspiration rate, in./yr (cm/yr)

U = nitrogen uptake by crop, lb/ac-yr (kg/ha-yr); values for selected crops are given in Table 22.2

K = unit conversion constant, 4.4 mg-in./ℓ-yr per lb/ac-yr for USCS units (10 mg-cm/ℓ-yr per kg/ha-yr for SI units)

C_n = nitrogen concentration in applied wastewater, mg/ℓ

f = fraction of applied nitrogen removed by denitrification and volatilization, 15% to 25% (EPA, 1981)

TABLE 22.2 *Nutrient Uptake Rates for Selected Crops*

CROP	NUTRIENT UPTAKE lb/ac-yr (kg/ha-yr)		
	Nitrogen	Phosphorus	Potassium
Forage crops			
Alfalfa[a]	201–481 (225–540)	20–31 (22–35)	156–201 (175–225)
Bromegrass	116–201 (130–225)	36–49 (40–55)	218 (245)
Coastal bermudagrass	357–602 (400–675)	31–40 (35–45)	201 (225)
Kentucky bluegrass	178–241 (200–270)	40 (45)	178 (200)
Quackgrass	209–250 (253–280)	27–40 (30–45)	245 (275)
Reed canarygrass	298–401 (335–450)	36–40 (40–45)	281 (315)
Ryegrass	178–250 (200–280)	53–76 (60–85)	241–290 (270–325)
Sweet clover[a]	156 (175)	18 (20)	89 (100)
Tall fescue	134–290 (150–325)	27 (30)	267 (300)
Orchardgrass	223–312 (250–350)	18–45 (20–50)	201–281 (225–315)
Field crops			
Barley	111 (125)	13 (15)	18 (20)
Corn	156–178 (175–200)	18–27 (20–30)	98 (110)
Cotton	67–98 (75–110)	13 (15)	36 (40)
Grain sorghum	120 (135)	13 (15)	62 (70)
Potatoes	205 (230)	18 (20)	218–290 (245–325)
Soybeans	223 (250)	9–18 (10–20)	27–49 (30–55)
Wheat	143 (160)	13 (15)	18–40 (20–45)

[a] Legumes also fix nitrogen from the atmosphere.

Adapted from *Land Treatment of Municipal Wastewater*, EPA Process Design Manual, 1981.

The required field area is determined by using the design hydraulic loading rate, $L_{w(d)}$, which is the smaller of the allowable soil hydraulic loading rate, $L_{w(s)}$, and the allowable nitrogen loading rate, $L_{w(n)}$. The equation for required area is (EPA, 1981)

$$A_w = \frac{(Q)\left(365\,\dfrac{\text{days}}{\text{yr}}\right) + \Delta V_s}{C(L_{w(d)})} \tag{22.3}$$

where

A_w = field area, ac (ha)

Q = average daily wastewater flow (annual basis), ft³/day (m³/day)

ΔV_s = net loss or gain in stored wastewater volume due to precipitation, evaporation, and seepage at storage pond, ft³/yr (m³/yr)

C = unit conversion constant, 3630 ft³/ac-in. for USCS units (100 m³/ ha-cm for SI units)

$L_{w(d)}$ = design hydraulic loading rate, in./yr (cm/yr)

The net loss or gain in stored wastewater volume, ΔV_s, is based on a water balance of inflow (precipitation) and outflow (evaporation and seepage) and may be determined from (EPA, 1981)

$$\Delta V_s = (P_r - E - S)A_s \tag{22.4}$$

where

ΔV_s = storage gain or loss, ac-ft/mo or ft³/mo (m³/mo)

P_r = precipitation, in./mo (cm/mo)

E = evaporation, in./mo (cm/mo)

S = seepage, in./mo (cm/mo)

A_s = storage surface area, ac (ha)

Appropriate conversion values must be used in Eq. 22.4: 1 ac = 43,560 ft² (1 ha = 10⁴ m²).

Because the storage area is determined after the initial field area is computed, as shown in Figure 22.4, the first calculation of required field area must be made assuming $\Delta V_s = 0$. This initial field area is an estimate that is revised in subsequent calculations. If the estimated field area and an assumed depth for the storage pond that is compatible with site conditions are used, the surface area for storage can be determined. By using the surface area, storage gains or losses due to precipitation, evaporation, and seepage may be determined on a monthly basis. An adjusted field area is then calculated to account for annual net gains or losses in storage. Next, the monthly volume of applied wastewater is calculated using the design monthly hydraulic loading rate and the adjusted field area. The monthly net change in storage is then determined using applied wastewater, available wastewater, and net gains or losses due to precipitation, evaporation, and seepage. The results are tabulated and the cumulative storage volume for each month is determined. The largest monthly cumulative volume is the storage volume required in the design. Finally, the storage pond depth that was assumed is adjusted to give the required storage volume, holding the required surface area constant. If the depth cannot be adjusted, and the surface area must be changed to provide the required storage volume, then another iteration must be performed because storage losses or gains will be different for the new surface area. Computer programs are available to aid in determining storage requirements for land treatment systems (EPA, 1981). Also, computer spreadsheet software is well suited to the calculations required for establishing the design hydraulic loading rate and determining storage requirements.

EXAMPLE 21.1 *Slow Rate Land Treatment System: Design Hydraulic Loading Rate*

A slow rate land treatment system is to be used to treat a municipal wastewater from a population of 10,000 persons. The wastewater flow is 100 gal/cap-day or 1.0 MGD. The wastewater will be pretreated using preliminary treatment and primary sedimentation. Site characteristics, vegetative cover characteristics, and wastewater data are:

1. Table 22.3 gives monthly operating days, Column (B); estimated evapotranspiration, ET, for the selected crop and climate, Column (C); monthly precipitation based on a 5-yr return period frequency analysis, Column (D); and nitrogen uptake by the selected vegetative cover, Column (E), distributed monthly by the same ratio as monthly to total annual evapotranspiration.

2. Field testing has yielded variable results for soil permeability. The average minimum permeability is 0.24 in./hr.

3. The system will have no runoff.

4. Maximum depth for wastewater storage is 15 ft, and seepage from storage will not occur.

5. Total nitrogen in the wastewater, C_n, is 25 mg/ℓ.

6. Allowable nitrogen in percolating water, C_p, is 10 mg/ℓ.

7. The denitrification-volatilization nitrogen removal fraction, f, is 0.18.

8. The annual nitrogen uptake rate for the selected vegetative cover, U, is 294 lb/ac-year.

9. In the absence of definitive data, evaporation is assumed to be equal to evapotranspiration.

Determine the annual design hydraulic loading rate, $L_{w(d)}$, which is the smaller of the allowable soil permeability loading rate, $L_{w(s)}$, and the allowable nitrogen loading rate, $L_{w(n)}$.

TABLE 22.3 *Tabulated Data for Examples 22.1 and 22.2*

(A) MONTH	(B) OPERATING DAYS	(C) ET (in/mo)	(D) P_r (in/mo)	(E) U (lb/ac-ft-mo)	(F) P_w (in/mo)	(G) $L_{w(s)}$ (in/mo)	(H) $L_{w(n)}$ (in/mo)	(I) $L_{w(d)}$ (in/mo)	(J) V_w (ac-ft/mo)	(K) W_a (ac-ft/mo)	(L) STORAGE (ac-ft/mo) (+)	(M) STORAGE (ac-ft/mo) (−)	(N) ΔV_s (ac-ft/mo)	(O) V_w (ac-ft/mo)	(P) STORAGE (ac-ft/mo) (+)	(Q) STORAGE (ac-ft/mo) (−)
Jan	10	0.9	1.2	4.5	2.3	2.0	2.2	2.0	30.6	93.4	62.8		0.5	28.4	65.4	
Feb	20	2.0	1.1	9.9	4.6	5.5	3.3	3.3	50.7	93.4	42.7		−1.4	47.1	45.0	
Mar	27	3.6	1.1	17.9	6.2	8.7	5.1	5.1	78.4	93.4	15.0		−3.9	72.8	16.7	
Apr	30	5.1	0.9	25.4	6.9	11.1	6.6	6.6	101.5	93.4		−8.1	−6.5	94.2		−7.3
May	31	7.1	0.3	35.3	7.1	13.9	8.3	8.3	127.4	93.4		−34.0	−10.5	118.3		−35.4
Jun	30	8.7	0.1	43.3	6.9	15.5	9.9	9.9	152.4	93.4		−58.8	−13.3	141.3		−61.2
Jul	31	9.4	trace	46.8	7.1	16.5	10.6	10.6	162.9	93.4		−69.5	−14.5	151.2		−72.3
Aug	31	9.1	0.1	45.3	7.1	16.1	10.4	10.4	159.1	93.4		−65.7	−13.9	147.7		−68.2
Sep	30	5.5	0.2	27.4	6.9	12.2	6.4	6.4	98.2	93.4		−4.8	−8.2	91.2		−6.0
Oct	31	4.4	0.3	21.9	7.1	11.2	5.3	5.3	80.6	93.4	12.8		−6.3	74.8	12.2	
Nov	28	2.4	0.6	11.9	6.4	8.2	3.3	3.3	50.3	93.4	43.1		−2.8	46.7	43.9	
Dec	9	0.9	1.1	4.5	2.1	1.9	2.1	1.9	28.6	93.4	64.8		0.3	26.6	67.1	
Annual		59.1	7.0	294.0	70.8	122.9	73.6	73.2	1120.5	1120.8	241.1	−240.9	−80.5	1040.4	250.4	−250.5

SOLUTION *Allowable Soil Permeability Hydraulic Loading Rate*

1. Determine a maximum daily percolation rate not to exceed 4% to 10% of the minimum soil permeability. (Use 4% because of the variable field permeability.)

$$\text{Daily } P_w = (0.24 \text{ in./hr})(24 \text{ hr/day})0.04 = 0.23 \text{ in./day}$$

2. Calculate monthly percolation rates, P_w, based on the number of operating days in each month. For January,

$$\text{Monthly } P_w \text{ (Jan)} = (10 \text{ days})(0.23 \text{ in./day}) = 2.3 \text{ in.}$$

Monthly P_w values are tabulated in Table 22.3, Column (F).

3. Calculate monthly soil permeability hydraulic loading rates, $L_{w(s)}$, using Eq. (22.1), which is

$$L_{w(s)} = ET - P_r + P_w$$

For January,

$$L_{w(s)} \text{ (Jan)} = (0.9 \text{ in./mo}) - (1.2 \text{ in./mo}) + (2.3 \text{ in./mo}) = 2.0 \text{ in./mo}$$

Monthly $L_{w(s)}$ values are tabulated in Table 22.3, Column (G).

4. The monthly hydraulic loading rates, $L_{w(s)}$, are summed to give the annual allowable soil permeability hydraulic loading rate, $L_{w(s)} = 122.9 \text{ in./yr}$, as shown in Table 22.3, Column (G), Annual.

Allowable Nitrogen Hydraulic Loading Rate

1. Calculate the allowable annual nitrogen loading rate, $L_{w(n)}$, using Eq. (22.2), which is

$$\begin{aligned}
L_{w(n)} &= \frac{(C_p)(P_r - ET) + (U)(K)}{(1 - f)(C_n) - C_p} \\
&= \frac{(10 \text{ mg/}\ell)(7.0 - 59.1) \text{ in./yr} + (294 \text{ lb/ac-yr})(4.4)}{(1 - 0.18)(25 \text{ mg/}\ell) - 10 \text{ mg/}\ell} \\
&= 73.58 \text{ in./yr}
\end{aligned}$$

2. Compare $L_{w(s)}$ and $L_{w(n)}$.

$$L_{w(n)} = 73.58 \text{ in./yr} < L_{w(s)} = 122.9 \text{ in./yr}$$

Therefore, the nitrogen loading rate, $L_{w(n)}$, controls. If the soil permeability loading rate, $L_{w(s)}$, were smaller (usually for humid regions), then the annual $L_{w(s)}$ would be used for design and the following monthly analysis would not be required. In this example, since the annual nitrogen loading rate, $L_{w(n)}$, controls, it is necessary to determine monthly values for $L_{w(n)}$, compare monthly values of $L_{w(s)}$ and $L_{w(n)}$, and use the smaller of the two monthly values for the allowable monthly design hydraulic loading rates. For January,

$$L_{w(n)} \text{ (Jan)} = \frac{(10 \text{ mg}/\ell)(1.2 - 0.9) \text{ in./mo} + (4.5 \text{ lb/ac-mo})(4.4)}{(1 - 0.18)(25 \text{ mg}/\ell) - 10 \text{ mg}/\ell}$$

$$= 2.17 \text{ in./mo}$$

Monthly $L_{w(n)}$ values are tabulated in Table 22.3, Column (H); monthly design loading rates, $L_{w(d)}$, appear in Column (I). The design monthly loading rates are summed to give a design annual hydraulic loading rate, $L_{w(d)} = \boxed{73.2 \text{ in./yr}}$, as shown in Table 22.3, Column (I), Annual.

EXAMPLE 21.1 SI   *Slow Rate Land Treatment System: Design Hydraulic Loading Rate*

A slow rate land treatment system is to be used to treat a municipal wastewater from a population of 10,000 persons. The wastewater flow is 380 ℓ/cap-d or 3800 m^3/d. The wastewater will be pretreated using preliminary treatment and primary sedimentation. Site characteristics, vegetative cover characteristics, and wastewater data are:

1. Table 22.4 contains monthly operating days, Column (B); estimated evapotranspiration, ET, for the selected crop and climate, Column (C); monthly precipitation based on a 5-yr return period frequency analysis, Column (D); and nitrogen uptake by the selected vegetative cover, Column (E), distributed monthly by the same ratio as monthly to total annual evapotranspiration.

2. Field testing has yielded variable results for soil permeability. The average minimum permeability is 0.6 cm/h.

3. The system will have no runoff.

4. Maximum depth for wastewater storage is 4.5 m, and seepage from storage will not occur.

5. Total nitrogen in the wastewater, C_n, is 25 mg/ℓ.

6. Allowable nitrogen in percolating water, C_p, is 10 mg/ℓ.

7. The denitrification-volatilization nitrogen removal fraction, f, is 0.18.

8. The annual nitrogen uptake rate for the selected vegetative cover, U, is 329 kg/ha-yr.

9. In the absence of definitive data, evaporation is assumed to be equal to evapotranspiration.

Determine the annual design hydraulic loading rate, $L_{w(d)}$, which is the smaller of the allowable soil permeability loading rate, $L_{w(s)}$, and the allowable nitrogen loading rate, $L_{w(n)}$.

TABLE 22.4   Tabulated Data for Examples 22.1 SI and 22.2 SI

(A) MONTH	(B) OPERATING DAYS	(C) ET (cm/mo)	(D) P_r (cm/mo)	(E) U (kg/ha-mo)	(F) P_w (cm/mo)	(G) $L_w(s)$ (cm/mo)	(H) $L_w(n)$ (cm/mo)	(I) $L_w(d)$ (cm/mo)	(J) V_w (10^3 m^3/mo)	(K) W_a (10^3 m^3/mo)	(L) STORAGE (10^3 m^3/mo) (+)	(M) STORAGE (10^3 m^3/mo) (−)	(N) ΔV_s (10^3 m^3/mo)	(O) V_w' (10^3 m^3/mo)	(P) STORAGE (10^3 m^3/mo) (+)	(Q) STORAGE (10^3 m^3/mo) (−)
Jan	10	2.3	3.0	5.0	5.8	5.1	5.5	5.1	38.1	115.6	77.5		0.5	35.4	80.7	
Feb	20	5.0	2.8	10.9	11.5	13.7	8.3	8.3	62.1	115.6	53.5		−1.6	57.6	56.3	
Mar	27	9.2	2.7	20.1	15.6	22.1	13.0	13.0	97.2	115.6	18.4		−4.9	90.2	20.5	
Apr	30	12.9	2.2	28.2	17.3	28.0	16.7	16.7	124.9	115.6		−9.3	−8.0	115.9		−8.3
May	31	18.1	0.7	39.6	17.9	35.3	21.1	21.1	157.8	115.6		−42.2	−13.0	146.4		−43.6
Jun	30	22.2	0.2	48.6	17.3	39.3	25.3	25.3	189.2	115.6		−73.6	−16.4	175.6		−76.4
Jul	31	24.0	trace	52.5	17.9	41.9	27.1	27.1	202.6	115.6		−87.0	−17.9	188.1		−90.4
Aug	31	23.1	0.2	50.5	17.9	40.8	26.3	26.3	196.6	115.6		−81.1	−17.1	182.5		−84.0
Sep	30	13.9	0.4	30.4	17.3	30.8	16.1	16.1	120.4	115.6		−4.8	−10.1	111.7		−6.2
Oct	31	11.1	0.7	24.3	17.9	28.3	13.2	13.2	98.7	115.6	16.9		−7.8	91.6	16.2	
Nov	28	6.2	1.4	13.6	16.1	20.9	8.3	8.3	62.1	115.6	53.5		−3.6	57.6	54.4	
Dec	9	2.4	2.7	5.3	5.2	4.9	5.3	4.9	36.6	115.6	78.9		0.2	34.0	81.8	
Annual		150.4	17.0	329.0	177.4	310.8	186.3	185.5	1386.2	1387.0	298.7	−298.0	−99.6	1286.7	310.0	−309.3

SOLUTION *Allowable Soil Permeability Hydraulic Loading Rate*

1. Determine a maximum daily percolation rate not to exceed 4% to 10% of the minimum soil permeability. (Use 4% because of the variable field permeability.)

$$\text{Daily } P_w = (0.6 \text{ cm/hr})(24 \text{ hr/d})0.04 = 0.576 \text{ cm/d}$$

2. Calculate monthly percolation rates, P_w, based on the number of operating days in each month. For January,

$$\text{Monthly } P_w \text{ (Jan)} = (10 \text{ days})(0.576 \text{ cm/d}) = 5.8 \text{ cm}$$

Monthly P_w values are tabulated in Table 22.4, Column (F).

3. Calculate monthly soil permeability hydraulic loading rates, $L_{w(s)}$, using Eq. (22.1), which is

$$L_{w(s)} = ET - P_r + P_w$$

For January,

$$L_{w(s)} \text{ (Jan)} = (2.3 \text{ cm/mo}) - (3.0 \text{ cm/mo}) + (5.8 \text{ cm/mo})$$
$$= 5.1 \text{ cm/mo}$$

Monthly $L_{w(s)}$ values are tabulated in Table 22.4, Column (G).

4. The monthly hydraulic loading rates, $L_{w(s)}$, are summed to give the annual allowable soil permeability hydraulic loading rate, $L_{w(s)} = 310.8 \text{ cm/yr}$, as shown in Table 22.4, Column (G), Annual.

Allowable Nitrogen Hydraulic Loading Rate

1. Calculate the allowable annual nitrogen loading rate, $L_{w(n)}$, using Eq. (22.2), which is

$$L_{w(n)} = \frac{(C_p)(P_r - ET) + (U)(K)}{(1 - f)(C_n) - C_p}$$
$$= \frac{(10 \text{ mg/}\ell)(17.0 - 150.4) \text{ cm/yr} + (329 \text{ kg/ha-yr})(10)}{(1 - 0.18)(25 \text{ mg/}\ell) - 10 \text{ mg/}\ell}$$
$$= 186.3 \text{ cm/yr}$$

2. Compare $L_{w(s)}$ and $L_{w(n)}$.

$$L_{w(n)} = 186.3 \text{ cm/yr} < L_{w(s)} = 310.8 \text{ cm/yr}$$

Therefore, the nitrogen loading rate, $L_{w(n)}$, controls. If the soil permeability loading rate, $L_{w(s)}$, where smaller (usually for humid regions), then the annual $L_{w(s)}$ would be used for design and the following monthly analysis would not be required. In this example, since the annual nitrogen loading rate, $L_{w(n)}$, controls, it is necessary to determine monthly values for $L_{w(n)}$, compare monthly values of

$L_{w(s)}$ and $L_{w(n)}$, and use the smaller of the two monthly values for the allowable monthly design hydraulic loading rates. For January,

$$L_{w(n)} \text{ (Jan)} = \frac{(10\,\text{mg}/\ell)(3.0 - 2.3)\,\text{cm/mo} + (5.0\,\text{kg/ha-mo})(10)}{(1 - 0.18)(25\,\text{mg}/\ell) - 10\,\text{mg}/\ell}$$

$$= 5.5\,\text{cm/mo}$$

Monthly $L_{w(n)}$ values are tabulated in Table 22.4, Column (H); monthly design loading rates, $L_{w(d)}$, appear in Column (I). The design monthly loading rates are summed to give a design annual hydraulic loading rate, $L_{w(d)} = \boxed{185.5\,\text{cm/yr,}}$ as shown in Table 22.4, Column (I), Annual.

EXAMPLE 22.2 | *Slow Rate Land Treatment System: Field Area and Wastewater Storage Volume*

For the conditions of Example 22.1, determine the field area and wastewater storage volume required for the population of 10,000 persons.

SOLUTION | Estimate the required field area.

Field area is determined using Eq. (22.3), which is

$$A_w = \frac{(Q)\left(365\,\dfrac{\text{days}}{\text{yr}}\right) + \Delta V_s}{C(L_{w(d)})}$$

From Example 22.1, $L_{w(d)} = 73.2\,\text{in./yr}$.

Because the storage volume and therefore the storage surface area are not yet known, the net loss or gain in stored wastewater volume, ΔV_s, due to precipitation, evaporation, and seepage must be assumed to be zero. The estimated field area is

$$A_w = \frac{\left(1 \times 10^6\,\dfrac{\text{gal}}{\text{day}} \times \dfrac{1\,\text{ft}^3}{7.48\,\text{gal}}\right)\left(365\,\dfrac{\text{day}}{\text{yr}}\right) + 0}{\left(3630\,\dfrac{\text{ft}^3}{\text{ac-in.}}\right)\left(73.2\,\dfrac{\text{in.}}{\text{yr}}\right)}$$

$$= 183.64\,\text{ac}$$

Estimate the required volume for wastewater storage.

1. Convert monthly design hydraulic loading rates, $L_{w(d)}$, to equivalent monthly volume loading rates, V_w.

$$V_w \text{ (Jan)} = (183.64\,\text{ac})\left(2.0\,\dfrac{\text{in.}}{\text{mo}}\right)\left(\dfrac{1\,\text{ft}}{12\,\text{in.}}\right)$$

$$= 30.6\,\text{ac-ft/mo}$$

Monthly V_w values are tabulated in Table 22.3, Column (J).

2. Calculate the monthly volume of available wastewater, W_a.

$$W_a \text{ (Jan)} = \left(1 \times 10^6 \frac{\text{gal}}{\text{day}}\right)\left(365\frac{\text{day}}{\text{yr}}\right)\left(\frac{1\,\text{yr}}{12\,\text{mo}}\right)\left(\frac{1\,\text{ft}^3}{7.48\,\text{gal}}\right)\left(\frac{1\,\text{ac}}{43{,}560\,\text{ft}^2}\right)$$

$$= 93.35\,\text{ac-ft/mo}$$

Monthly W_a values are tabulated in Table 22.3, Column (K).

3. Calculate the monthly change in storage volume by subtracting the volume that can be applied to the field, V_w, from the volume of available wastewater, W_a. If the result is positive, enter in Column (L); if negative, enter in Column (M). This procedure is similar to the mass tabulation commonly used to determine the required operating storage for water distribution systems. Round-off error may occur and can be minimized in computer spreadsheet calculations by using sufficient decimal places for field area and available wastewater volume in the calculations. This procedure has the advantage that it does not require rearranging the table by month to determine required storage volume. In this example, since there is only one cycle of addition to and withdrawal from storage during the year, summing either Column (L) or Column (M) yields the estimated required storage volume of 241 ac-ft. If conditions are such that two or more separate cycles of addition to and withdrawal from storage occur during the year, then summing each separate cycle and using the largest sum will yield the required storage. Applying this procedure for two or more storage cycles to either the positive or the negative column will yield the same result, within round-off error.

Determine the final storage volume and field area required to include net storage gain or loss, ΔV_s.

1. Using an assumed storage depth compatible with site subsurface conditions, calculate a required storage surface area.

$$A_s = 241\,\text{ac-ft}/13\,\text{ft (assumed)}$$

$$= 18.54\,\text{ac}$$

2. Calculate the monthly net storage gain or loss, ΔV_s, using

$$\Delta V_s = (P_r - E - S)A_s$$

For January,

$$\Delta V_s \text{ (Jan)} = (1.2\,\text{in./mo} - 0.9\,\text{in./mo} - 0)(1\,\text{ft}/12\,\text{in.})18.54\,\text{ac}$$

$$= 0.46\,\text{ac-ft/mo}$$

Monthly ΔV_s values are tabulated in Table 22.3, Column (N).

3. Calculate an adjusted field area to account for annual net gain or loss, ΔV_s, using

$$A'_w = (\Sigma \Delta V_s + \Sigma W_a)/L_{w(d)}$$
$$= (-80.5 + 1120.8)\,\text{ac-ft/yr} \,/\, (73.2\,\text{in./yr} \times 1\,\text{ft/12 in.})$$
$$= 170.5\,\text{ac}$$

4. Calculate the adjusted monthly volume of applied wastewater, V'_w, using the design monthly hydraulic loading rate, $L_{w(d)}$, and the adjusted field area, A'_w. For January,

$$V'_w\,(\text{Jan}) = (2.0\,\text{in./mo})(1\,\text{ft/12 in.})(170.5\,\text{ac})$$
$$= 28.4\,\text{ac-ft/mo}$$

Monthly V'_w values are tabulated in Table 22.3, Column (O).

5. Calculate the net monthly storage change, using

$$\Delta S_{\text{net}} = W_a + \Delta V_s - V'_w$$

For January,

$$\Delta S_{\text{net}}\,(\text{Jan}) = 93.35\,\text{ac-ft/mo} + 0.46\,\text{ac-ft/mo} - 28.4\,\text{ac-ft/mo}$$
$$= 65.4\,\text{ac-ft/mo}$$

Positive monthly values for ΔS_{net} are tabulated in Table 22.3, Column (P), negative values in column (Q). Again, round-off error may occur and can be minimized in computer spreadsheet calculations by using sufficient decimal places for storage surface area, net storage gain or loss, field area, and wastewater volume in the calculations. In this example, summing either Column (P) or Column (Q) yields the final design required storage of 250.4 ac-ft.

6. Adjust the assumed value of storage depth to provide the required storage volume.

$$d_s = 250.4\,\text{ac-ft/18.54 ac} = 13.5\,\text{ft}$$

If the adjusted depth exceeded site limitations, then the surface area would require adjustment. In that case, steps 1 through 6 would require reiteration.

Thus, for the population of 10,000 persons, the required field area is $\boxed{170.5\,\text{ac}}$ and the required wastewater storage volume is $\boxed{250.4\,\text{ac-feet}}$ with a storage depth of $\boxed{13.5\,\text{ft.}}$

EXAMPLE 22.2 SI   *Slow Rate Land Treatment System: Field Area and Wastewater Storage Volume*

For the conditions of Example 22.1 SI, determine the field area and wastewater storage volume required for the population of 10,000 persons.

SOLUTION Estimate the required field area.

Field area is determined using Eq. (22.3), which is

$$A_w = \frac{(Q)\left(365\frac{\text{days}}{\text{yr}}\right) + \Delta V_s}{C(L_{w(d)})}$$

From Example 22.1 SI, $L_{w(d)} = 185.5$ cm/yr.

Because the storage volume and therefore the storage surface area are not yet known, the net loss or gain in stored wastewater volume, ΔV_s, due to precipitation, evaporation, and seepage must be assumed to be zero. The estimated field area is

$$A_w = \frac{\left(3800\frac{\text{m}^3}{\text{d}}\right)\left(365\frac{\text{d}}{\text{yr}}\right) + 0}{\left(185.5\frac{\text{cm}}{\text{yr}}\right)\left(\frac{1\,\text{m}}{10^2\,\text{cm}}\right)\left(\frac{10^4\,\text{m}^2}{\text{ha}}\right)}$$

$$= 74.771\,\text{ha}$$

Estimate the required volume for wastewater storage.

1. Convert monthly design hydraulic loading rates, $L_{w(d)}$, to equivalent monthly volume loading rates, V_w.

$$V_w\,(\text{Jan}) = \left(74.771\,\text{ha} \times 10^4\frac{\text{m}^2}{\text{ha}}\right)\left(5.1\frac{\text{cm}}{\text{mo}}\right)\left(\frac{1\,\text{m}}{10^2\,\text{cm}}\right)$$

$$= 38.1 \times 10^3\,\text{m}^3$$

Monthly V_w values are tabulated in Table 22.4, Column (J).

2. Calculate the monthly volume of available wastewater, W_a.

$$W_a\,(\text{Jan}) = \left(3800\frac{\text{m}^3}{\text{d}}\right)\left(\frac{365\,\text{d}}{\text{yr}}\right)\left(\frac{1\,\text{yr}}{12\,\text{mo}}\right)$$

$$= 115.6 \times 10^3\,\text{m}^3/\text{mo}$$

Monthly W_a values are tabulated in Table 22.4, Column (K).

3. Calculate the monthly change in storage volume by subtracting the volume that can be applied to the field, V_w, from the volume of available wastewater, W_a. If the result is positive, enter in Column (L); if negative, enter in Column (M). This procedure is similar to the mass tabulation commonly used to determine the required operating storage for water distribution systems. Round-off error may occur and can be minimized in computer spreadsheet calculations by using sufficient decimal places for field area and available wastewater volume in the calculations. This procedure has the advantage that it does not require rearranging the table by month to determine required storage volume. In this example, since there is

only one cycle of addition to and withdrawal from storage during the year, summing either Column (L) or Column (M) yields the estimated required storage volume of $298.7 \times 10^3 \, m^3$. See the discussion at this design calculation step in Example 22.2.

Determine the final storage volume and field area required to include net storage gain or loss, ΔV_s.

1. Using an assumed storage depth, compatible with site subsurface conditions, of 4 m, calculate a required storage surface area, A_s.

$$A_s = (298.7 \times 10^3 \, m^3/4 \, m) \times 1 \, ha/10^4 \, m^2$$
$$= 7.4675 \, ha$$

2. Calculate the monthly net storage gain or loss, ΔV_s, using

$$\Delta V_s = (P_r - E - S)A_s$$

For January,

$$\Delta V_s \, (Jan) = (3.0 \, cm/mo - 2.3 \, cm/mo - 0)(1 \, m/10^2 \, cm)74.675$$
$$\times 10^3 \, m^2$$
$$= 0.52 \times 10^3 \, m^3/mo$$

Monthly ΔV_s values are tabulated in Table 22.4, Column (N).

3. Calculate an adjusted field area to account for annual net gain or loss, ΔV_s, using

$$A'_w = (\Sigma \Delta V_s + \Sigma W_a)/L_{w(d)}$$

$$= \frac{(-99.6 + 1387.0) \times 10^3 \dfrac{m^3}{yr}}{\left(185.5 \dfrac{cm}{yr}\right)\left(\dfrac{1 \, m}{10^2 \, cm}\right)} \times \frac{1 \, ha}{10^4 \, m^2}$$

$$= 69.402 \, ha$$

4. Calculate the adjusted monthly volume of applied wastewater, V'_w, using the design monthly hydraulic loading rate, $L_{w(d)}$, and the adjusted field area, A'_w. For January,

$$V'_w = (5.1 \, cm/mo \times 1 \, m/10^2 \, cm)(69.402 \, ha \times 10^4 \, m^2/ha)$$
$$= 35.40 \times 10^3 \, m^3/mo$$

Monthly V'_w values are tabulated in Table 22.4, Column (O).

5. Calculate the net monthly storage change, using

$$\Delta S_{net} = W_a + \Delta V_s - V'_w$$

For January,

$$\Delta S_{net} \, (Jan) = (115.6 + 0.52 - 35.40) \times 10^3 \, m^3/mo$$
$$= 80.72 \, m^3/mo$$

Positive monthly values for ΔS_{net} are tabulated in Table 22.4, Column (P), negative values in Column (Q). Again, round-off error may occur and can be minimized in computer spreadsheet calculations by using sufficient decimal places for storage surface area, net storage gain or loss, field area, and wastewater volume in the calculations. In this example, summing either Column (P) or Column (Q) yields the final design required storage of $310.0 \times 10^3\,m^3$.

6. Adjust the assumed value of storage depth to provide the required storage volume.

$$d_s = 310.0 \times 10^3\,m^3/74.675 \times 10^3\,m^2 = 4.2\,m$$

If the adjusted depth exceeded site limitations, then the surface area would require adjustment. In that case, steps 1 through 6 would require reiteration.

Thus, for the population of 10,000 persons, the required field area is $\boxed{69.4\,ha}$ and the required wastewater storage volume is $\boxed{310,000\,m^3}$ with a storage depth of $\boxed{4.2\,m.}$

Rapid Infiltration Wastewater treatment by the rapid infiltration process is achieved mainly by percolation through soil, as shown in Figure 22.5. Table 22.5 gives typical site characteristics, design parameters, and expected performance by the rapid infiltration process. A rapid infiltration system typically consists of earthen basins that are used for flooding, infiltration of the wastewater, and drying in repetitive loading cycles. The treated wastewater may be used for groundwater recharge, recovered for reuse or discharge, used for recharge of surface streams by interception of groundwater, or used to help protect an existing fresh groundwater from salt water intrusion. Wastewater is

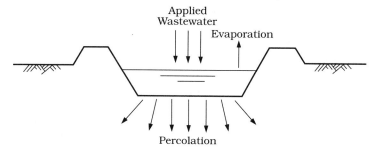

FIGURE 22.5 *Rapid Infiltration Process, Application Hydraulic Pathway*

Adapted from *Land Treatment of Municipal Wastewater*, 1981. Environmental Protection Agency (EPA).

TABLE 22.5 *Rapid Infiltration Land Treatment System Site Characteristics, Typical Design Features, and Expected Treated Water Quality*[a]

Grade	Not critical; excessive grades require much earthwork
Soil permeability	Rapid (sands, sandy loam)
Depth to groundwater	3.5 ft (1 m) during flood cycle (underdrains can be used to maintain this level at sites with high groundwater table); 5–10 ft (1.5–3 m) during drying cycle
Climatic restrictions	None (possibly modify operation during cold weather)
Application technique	Usually surface
Loading rate	
Annual	20–410 ft/yr (6–125 m/yr)
Weekly	4–95 in./wk (10–240 cm/wk)
Field area	7.5–57 ac/MGD (0.8–6.1 ha/1000 m³/d), (not including buffer area, roads, or ditches)
Minimum pretreatment in United States	Preliminary treatment and primary sedimentation (with restricted public access)
Disposition of applied wastewater	Mainly percolation
Vegetation	Optional
BOD$_5$	<5 to <10 mg/ℓ
Suspended solids	<2 to <5 mg/ℓ
Ammonia nitrogen	<0.5 to <2 mg/ℓ as N
Total nitrogen	10 to <20 mg/ℓ as N
Total phosphorus	1 to <5 mg/ℓ as P
Fecal coliforms	10 to <200 per 100 mℓ

[a] Treated water quality values are average to upper range expected with loading rates at the mid to low end of the loading rate range given, and with percolation of primary or secondary effluent through 15 ft (4.5 m) of unsaturated soil; phosphorus and fecal coliform removals increase with distance traveled vertically or horizontally.

Adapted from EPA, *Land Treatment of Municipal Wastewater*, EPA Process Design Manual, 1981.

applied to suitably permeable soils such as sand, loamy sand, sandy loam, and gravel. Soil permeability, or effective hydraulic conductivity, for successfully operating rapid infiltration systems ranges from 2 in./hr (5 cm/hr) to 6 in./hr (15 cm/hr) (EPA, 1984). Soils such as very coarse sand or gravel with very rapid permeability, greater than 8 in./hr (20 cm/hr), provide less effective treatment of applied wastewater. This is because percolation occurs too rapidly through the upper, relatively short distance of soil where the major biological and chemical actions take place (Metcalf & Eddy, 1991). Wastewater is applied to the soil by spreading in basins or by sprinkling. As shown in Figure 22.5, the wastewater percolates through the soil, and some evaporates. Evaporation is usually small in comparison to loading rates and ranges from about 2 ft/yr (0.6 m/yr) in cool climates to about 6 ft/yr (2 m/yr) under hot, arid conditions. Although vegetation is usually not included in the design, emergent weed or grass growth that may occur does not cause problems. As shown in Figure 22.6, water generally percolates downward into the ground water or laterally into surface streams, or it may be recovered by wells or underdrains for reuse.

Land treatment of wastewater by the rapid infiltration process is limited by several factors, some of which are the same as for the slow rate process. Site characteristics are the primary controlling or limiting factor. Suitable soils must be present and should be relatively uniform; otherwise, soil investigations become very intensive, complex, and expensive. A thorough knowledge of subsurface geologic conditions is necessary. These include the normal and seasonal variation in groundwater depth; location and depth of soil or bedrock, which influence groundwater movement; and quality of groundwater, not only at the application site, but also away from the site

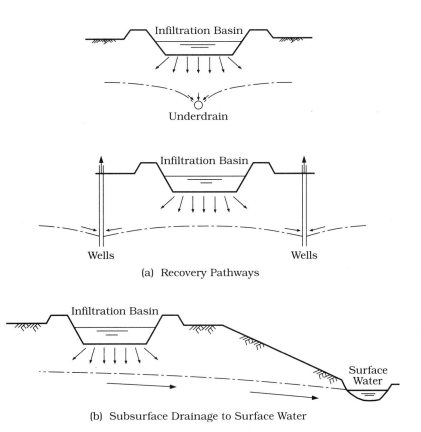

(a) Recovery Pathways

(b) Subsurface Drainage to Surface Water

FIGURE 22.6 *Rapid Infiltration Process, Subsurface Hydraulic Pathway*

Adapted from *Land Treatment of Municipal Wastewater*, 1981. Environmental Protection Agency (EPA).

where groundwater may travel. Nitrogen loadings may be a limiting factor because of groundwater quality requirements.

The performance of the rapid infiltration process is excellent. Suspended solids, BOD_5, and fecal coliform bacteria are almost completely removed by the biological, chemical, and physical actions occurring in the soil matrix. Excellent nitrification can be achieved with appropriate loading cycles. Nitrogen removal is about 50% and can be increased to about 80% by operational management of application cycles, recycling nitrate-rich treated water, reducing the infiltration rate, and providing an additional carbon source to promote denitrification. Phosphorus removal occurs primarily by adsorption and ranges from 70% to 99%, depending on soil characteristics, residence time, and travel distance of the wastewater in the soil (EPA, 1981).

The general design procedure for rapid infiltration systems is shown in Figure 22.7. The most important step in the design procedure is determination of the allowable hydraulic loading rate. The hydraulic loading rate is site specific and depends on the soil permeability or hydraulic conductivity, the loading cycle, the quality of the wastewater to be treated, and the required

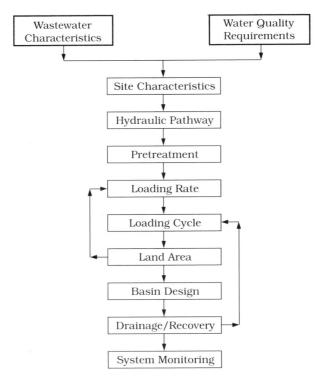

FIGURE 22.7 *Rapid Infiltration Land Treatment Design Procedure*

Adapted from *Land Treatment of Municipal Wastewater*, 1981. Environmental Protection Agency (EPA).

quality of the renovated wastewater. Determination of the long-term infiltration rate is critical to the design of rapid infiltration systems. Several methods for determining infiltration rate are available; discussion of these methods is beyond the scope of this text. Detailed discussion and procedures relevant to all aspects of rapid infiltration system design may be found in two EPA documents (1981) and (1984). Equations and procedures used to determine the design hydraulic loading rate, loading cycle, and land requirements will be discussed, and their use illustrated by examples, in this text.

The allowable hydraulic loading rate is primarily a function of the site-specific hydraulic capacity, which is often referred to as the hydraulic conductivity, permeability, or infiltration rate. After thorough field investigation (and laboratory testing as required) to measure the hydraulic capacity, the annual hydraulic loading rate may be determined from

$$L_{w(d)} = HC_m \times L_f/100 \tag{22.5}$$

where

$L_{w(d)}$ = design annual hydraulic loading rate, ft/yr (m/yr)

HC_m = minimum measured hydraulic capacity, in./hr (cm/hr)

L_f = appropriate loading factor, %

Recommended loading factors, L_f, are given in Table 22.6. Appropriate unit conversions must be used in Eq. (22.5) to yield the desired units of ft/yr (m/yr) for the loading rate, $L_{w(d)}$.

Other considerations that may limit the hydraulic loading rate are precipitation; wastewater constituent loading rates for BOD_5, suspended solids, nitrogen, and phosphorus; and climate. Precipitation adds to the hydraulic loading rate and must be considered. BOD_5 and suspended solids loadings do not usually limit the hydraulic loading rate for municipal wastewater. Typical BOD loadings at operating systems range from 14,600 to 57,600 lb/ac-yr (16,400 to 64,600 kg/ha-yr) (EPA, 1981). Typical suspended solids loadings

TABLE 22.6 *Rapid Infiltration Process — Recommended Annual Hydraulic Loading Factors*

FIELD MEASUREMENT METHOD	ANNUAL LOADING FACTOR
Rapid infiltration basin test method	10%–15% of minimum measured infiltration rate
Cylinder infiltrometer and air entry permeameter methods	2%–4% of minimum measured infiltration rate
Vertical hydraulic conductivity test method	4%–10% of conductivity of the most restrictive soil layer

Taken from EPA, *Land Treatment of Municipal Wastewater*, EPA Process Design Manual, 1981.

range from 11,000 to 36,500 lb/ac-yr (12,300 to 40,900 kg/ha-yr) (WPCF, 1990). Nitrogen loading and removal may influence both the hydraulic loading rate and the loading cycle. Under optimum conditions, the theoretical amount of nitrogen removed by denitrification may be determined from (EPA, 1981)

$$\Delta N = \frac{\text{TOC} - K}{2} \tag{22.6}$$

where

ΔN = change in total nitrogen concentration, mg/ℓ

TOC = total organic carbon in the applied wastewater, mg/ℓ

K = TOC remaining in renovated wastewater, which is assumed to equal 5 mg/ℓ

Data on which Eq. (22.6) is based indicate that the ratio of available carbon to available nitrogen in the wastewater to be treated, C:N, must be 2:1 or more (EPA, 1981). The equation can be used to estimate the potential amount of nitrogen that could be removed by denitrification. If the desired nitrogen removal can be achieved, a loading cycle favoring denitrification can be selected. Typical nitrogen loadings range from 1100 to 13,500 lb/ac-yr (1200 to 15,200 kg/ha-yr) (WPCF, 1990). The phosphorus removal potential for site-specific soils may be estimated by an empirical model that is, among other parameters, a function of infiltration rate and the subsurface distance through which wastewater moves both vertically and horizontally (EPA, 1981). Under conditions in which desired or required phosphorus removal is to be achieved, the potential phosphorus removal of the site must be analyzed. Typical phosphorus loadings range from 330 to 4300 lb/ac-yr (370 to 4800 kg/ha-yr) (WPCF, 1990). Climate also influences wastewater loading rates. Oxidation of organics, nitrification, denitrification, and drying rates all decrease with temperature. Under cold conditions, longer application and drying periods are required to achieve desired treatment results. The resulting decreased application rate and allowable annual loading rate require increased infiltration area based on cold-weather loading, or cold-weather storage. A combination of both may be appropriate.

Wastewater is not applied continuously but is applied to an individual infiltration basin in repetitive loading cycles of flooding, infiltration of the wastewater, and drying. Thus an application rate, which is dependent on the loading cycle, must be determined. Loading cycles consist of an application period and a drying period and are selected to maximize wastewater infiltration rates, nitrogen removal, or nitrification. Suggested loading cycles are given in Table 22.7. In general, to maximize infiltration rates, it is necessary to provide drying periods of sufficient length for soil reaeration and drying and for oxidation of filtered solids. To maximize nitrogen removal, drying periods of sufficient length to achieve anaerobic conditions must be provided. To maximize nitrification, short application periods followed by longer drying periods must be provided; these are essentially the same as the application

TABLE 22.7 *Rapid Infiltration Process — Recommended Loading Cycles*

LOADING CYCLE OBJECTIVE	APPLIED WASTEWATER	SEASON	APPLICATION PERIOD[a] (days)	DRYING PERIOD (days)
Maximize infiltration rates	Primary	Summer	1–2	5–7
		Winter	1–2	7–12
	Secondary	Summer	1–3	4–5
		Winter	1–3	5–10
Maximize nitrogen removal	Primary	Summer	1–2	10–14
		Winter	1–2	12–16
	Secondary	Summer	7–9	10–15
		Winter	9–12	12–16
Maximize nitrification	Primary	Summer	1–2	5–7
		Winter	1–2	7–12
	Secondary	Summer	1–3	4–5
		Winter	1–3	5–10

[a] As indicated, application periods should be limited to 1 to 2 days for wastewater receiving only primary treatment, regardless of season or cycle objective, to prevent excessive soil clogging.
Taken from EPA, *Land Treatment of Municipal Wastewater*, EPA Process Design Manual, 1981.

and drying periods recommended for maximizing the infiltration rate. It should be understood that the application and drying periods in Table 22.7 are only guidelines; the shorter drying periods given should be used for mild climates, the longer drying periods should be used for cooler climates, and longer drying periods than those given in Table 22.7 should be used for very cold climates (EPA, 1981). Once the loading cycle has been selected, the application rate may be determined using (EPA, 1981)

$$A_r = \frac{L_{w(d)}(A_t + D_t)}{365A_t} \tag{22.7}$$

where

A_r = application rate, ft/day (m/day)

$L_{w(d)}$ = hydraulic loading rate, ft/yr (m/yr)

A_t = application time period, days

D_t = drying time period, days

The maximum depth of an infiltration basin must be determined to permit design of the basin and to limit clogging and algal growth. Wastewater depth in a basin is determined by subtracting the measured minimum hydraulic capacity, HC_m, from the application rate, A_r. If the result is negative, the application rate should not result in standing wastewater in the basin at the end of the application period. If the result is positive, the maximum design wastewater depth at the end of the application period should not exceed 18 in. (46 cm), preferably not more than 12 in. (30 cm). If the calculated depth at the end of the application period is more than 18 in. (46 cm), then the application period must be increased or the loading rate decreased. From this discussion, the importance of accurately determining the hydraulic capacity or infiltration rate becomes readily apparent. If the hydraulic capacity is overestimated, then the design wastewater depth for a basin will be too shallow and operational difficulties will occur.

The total land area required for a rapid infiltration system, not including land required for dikes or land required between and around basins, is a function of wastewater flowrate and the annual hydraulic loading rate. The area required for infiltration also depends on whether wastewater flow equalization is provided. The infiltration area required may be determined from (EPA, 1981)

$$A_i = Q/L_{w(d)} \tag{22.8}$$

where

A_i = infiltration area required, ac (ha)

Q = average annual wastewater flow or largest average seasonal flow, gal/day (m^3/day)

$L_{w(d)}$ = design annual hydraulic loading rate, ft/yr (m/yr)

Appropriate conversion units must be used in Eq. (22.8) to yield the desired units of ac (ha) for infiltration area, A_i. In using Eq. (22.8), if wastewater flow equalization is provided, then the wastewater flow, Q, is the average annual flow. If wastewater flow varies seasonally and flow equalization is not provided, then the wastewater flow, Q, is the largest average seasonal flow.

Infiltration basin layout and individual basin dimensions are dependent on site topology, distribution hydraulics, and the wastewater loading rate. The number of basins is influenced by the loading cycle and the required total infiltration area. Table 22.8 gives the recommended minimum number of basins required for continuous wastewater application for various loading cycles. Individual basin size ranges from 0.5 to 5 ac (0.2 to 2 ha) for small to medium-size rapid infiltration systems and from 5 to 20 ac (2 to 8 ha) for large systems. In practice, to provide for basin maintenance between loadings, a sufficient number of basins should be available so that at least one basin is loaded at any time.

Adequate subsurface drainage is required for rapid infiltration systems. As shown in Figure 22.6 and Figure 22.8, two subsurface flow conditions may occur. Each requires analysis to ensure that both infiltration rates and treatment efficiencies are realized. For subsurface drainage to surface waters (Figure 22.6), the width of the infiltration area must be limited, which influences the dimensions of individual infiltration basins. Further, natural subsurface drainage conditions may not be feasible for drainage to a surface water, in which case engineered subsurface drainage is required. In rapid infiltration systems, wastewater initially travels in a general vertical direction

TABLE 22.8 *Recommended Minimum Number of Rapid Infiltration Basins for Continuous Application*

LOADING APPLICATION PERIOD (days)	CYCLE DRYING PERIOD (days)	MINIMUM NUMBER OF BASINS
1	5–7	6–8
2	5–7	4–5
1	7–12	8–13
2	7–12	5–7
1	4–5	5–6
2	4–5	3–4
3	4–5	3
1	5–10	6–11
2	5–10	4–6
3	5–10	3–5
1	10–14	11–15
2	10–14	6–8
1	12–16	13–17
2	12–16	7–9
7	10–15	3–4
8	10–15	3
9	10–15	3
7	12–16	3–4
8	12–16	3
9	12–16	3

Taken from EPA, *Land Treatment of Municipal Wastewater*, EPA Process Design Manual, 1981.

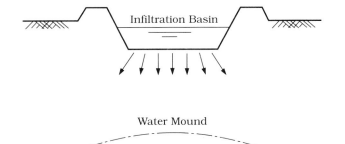

FIGURE 22.8 *Rapid Infiltration System Groundwater Mounding*

Adapted from *Land Treatment of Municipal Wastewater*, 1981. Environmental Protection Agency (EPA).

toward the groundwater and a groundwater mounding effect results, as shown in Figure 22.8. If natural drainage is not adequate and excessive mounding occurs, engineered subsurface drainage is required. Procedures for determining the adequacy of natural subsurface drainage, the need for constructed subsurface drainage, and underdrain design are presented in several references, including EPA (1981), EPA (1984), and WPCF (1990).

EXAMPLE 22.3   *Rapid Infiltration Land Treatment System: Design Hydraulic Loading Rate*

A rapid infiltration land treatment system is to be used to treat a municipal wastewater flow from 10,000 persons. The wastewater flow is 100 gal/cap-day or 1.0 MGD. The wastewater will be pretreated using preliminary treatment and primary sedimentation. Site characteristics, treatment objectives, and wastewater data are:

1. The minimum soil infiltration rate measured during extensive field study by the basin infiltration method is 1.6 in./hr.

2. The climate is mild; however, winter operation will control.

3. A well-nitrified renovated water is desired.

4. Seasonal variation in wastewater flow is negligible.

5. After wastewater pretreatment, $BOD_5 = 163$ mg/ℓ, suspended solids = 88 mg/ℓ, total nitrogen as N = 26 mg/ℓ, and total phosphorus as P = 5 mg/ℓ.

Determine the design annual hydraulic loading rate, the wastewater application rate, and the required infiltration area, and compare annual BOD_5, suspended solids, nitrogen, and phosphorus loadings to typical ranges for each.

SOLUTION   The design annual hydraulic loading rate, $L_{w(d)}$, is determined using Eq. (22.5), which is

$$L_{w(d)} = HC_m \times L_f/100$$

From Table 22.6, the loading factor, L_f, for infiltration basin testing ranges from 10% to 15%. Using $L_f = 10\%$,

$$L_{w(d)} = \left(1.6\frac{\text{in.}}{\text{hr}}\right)\left(\frac{1\,\text{ft}}{12\,\text{in.}}\right)\left(\frac{24\,\text{hr}}{\text{day}}\right)\left(\frac{365\,\text{days}}{\text{yr}}\right)\left(\frac{10}{100}\right)$$

$$= \boxed{116.8\,\text{ft/yr}}$$

The application rate, A_r, is determined using Eq. (22.7), which is

$$A_r = \frac{L_{w(d)}(A_t + D_t)}{365A_t}$$

From Table 22.7, an application period, A_t, of 1 day and a drying period, D_t, of 7 days are selected to maximize nitrification and with consideration for the pretreatment provided and the controlling winter conditions. The application rate is

$$A_r = \frac{\left(116.8\frac{\text{ft}}{\text{yr}}\right)(1 \text{ day} + 7 \text{ days})}{\left(365\frac{\text{days}}{\text{yr}}\right)(1 \text{ day})}$$

$$= \boxed{2.56\,\text{ft/day}}$$

The maximum depth of applied wastewater is analyzed by subtracting the minimum measured infiltration rate or hydraulic capacity, HC_m, from the wastewater application rate, A_r.

$$2.56\,\text{ft/day} - 1.6\,\text{in./hr} \times 1\,\text{ft/12 in.} \times 24\,\text{hr/day} = -0.64\,\text{ft/day}$$

The negative value indicates that the application rate of 2.56 ft/day should not result in standing wastewater in the infiltration basin at the end of the application period.

The required total infiltration area, for infiltration only, is determined using Eq. (22.8), which is

$$A_i = Q/L_{w(d)}$$

Since seasonal variation in wastewater flow is negligible, use the average annual wastewater flow. The area for infiltration is

$$A_i = \frac{\left(1.0 \times 10^6 \frac{\text{gal}}{\text{day}}\right)\left(365\frac{\text{day}}{\text{yr}}\right)\left(\frac{1\,\text{ft}^3}{7.48\,\text{gal}}\right)}{\left(116.8\frac{\text{ft}}{\text{yr}}\right)\left(43,560\frac{\text{ft}^2}{\text{ac}}\right)}$$

$$= \boxed{9.59\,\text{ac}}$$

Wastewater constituent loadings are:

For BOD₅,

$$\left(1.0 \times 10^6 \frac{\text{gal}}{\text{day}}\right)\left(365\frac{\text{day}}{\text{yr}}\right)\left(163\frac{\text{lb BOD}}{10^6\,\text{lb}}\right)\left(8.34\frac{\text{lb}}{\text{gal}}\right)\left(\frac{1}{9.59\,\text{ac}}\right) = 51,700\frac{\text{lb BOD}}{\text{ac-yr}}$$

The other wastewater constituent loadings are determined in the same manner and are

For suspended solids, 27,900 lb SS/ac-yr

For nitrogen, 8300 lb N/ac-yr

For phosphorus, 1600 lb P/ac-yr

All constituent loadings are within typical ranges.

Thus, for a population of 10,000 persons, the required infiltration area is 9.59 ac.

EXAMPLE 22.3 SI *Rapid Infiltration Land Treatment System: Design Hydraulic Loading Rate*

A rapid infiltration land treatment system is to be used to treat a municipal wastewater flow from 10,000 persons. The wastewater flow is 380 ℓ/cap-d or 3800 m^3/d. The wastewater will be pretreated using preliminary treatment and primary sedimentation. Site characteristics, treatment objectives, and wastewater data are:

1. The minimum soil infiltration rate measured during extensive field study by the basin infiltration method is 4.1 cm/h.
2. The climate is mild; however, winter operation will control.
3. A well-nitrified renovated water is desired.
4. Seasonal variation in wastewater flow is negligible.
5. After wastewater pretreatment, BOD$_5$ = 163 mg/ℓ, suspended solids = 88 mg/ℓ, total nitrogen as N = 26 mg/ℓ, and total phosphorus as P = 5 mg/ℓ.

Determine the design annual hydraulic loading rate, the wastewater application rate, and the required infiltration area, and compare annual BOD, suspended solids, nitrogen, and phosphorus loadings to typical ranges for each.

SOLUTION The design annual hydraulic loading rate, $L_{w(d)}$, is determined using Eq. (22.5), which is

$$L_{w(d)} = HC_m \times L_f/100$$

From Table 22.6, the loading factor, L_f, for infiltration basin testing ranges from 10% to 15%. Using $L_f = 10\%$,

$$L_{w(d)} = \left(4.1 \frac{\text{cm}}{\text{h}}\right) \times \left(\frac{1\,\text{m}}{100\,\text{cm}}\right) \times \left(\frac{24\,\text{h}}{\text{day}}\right) \times \left(\frac{365\,\text{days}}{\text{yr}}\right) \times \left(\frac{10}{100}\right)$$

$$= \boxed{35.9\,\text{m/yr}}$$

The application rate, A_r, is determined using Eq. (22.7), which is

$$A_r = \frac{L_{w(d)}(A_t + D_t)}{365A_t}$$

From Table 22.7, an application period, A_t, of 1 day and a drying period, D_t, of 7 days are selected to maximize nitrification and with consideration for the pretreatment provided and the controlling winter conditions. The application rate is

$$A_r = \frac{\left(35.9 \frac{\text{m}}{\text{yr}}\right)(1\,\text{d} + 7\,\text{d})}{\left(365 \frac{\text{d}}{\text{yr}}\right)(1\,\text{d})}$$

$$= \boxed{0.79\,\text{m/d}}$$

The maximum depth of applied wastewater is analyzed by subtracting the minimum measured infiltration rate or hydraulic capacity, HC_m, from the wastewater application rate, A_r.

$$(0.79 \, \text{m/d} - 4.1 \, \text{cm/h})(1 \, \text{m}/100 \, \text{cm})(24 \, \text{h/day}) = -0.19 \, \text{m/d}$$

The negative value indicates that the application rate of 0.79 m/d should not result in standing wastewater in the infiltration basin at the end of the application period.

The required total infiltration area, for infiltration only, is determined using Eq. (22.8), which is

$$A_i = Q/L_{w(d)}$$

Since seasonal variation in wastewater flow is negligible, use the average annual wastewater flow. The area for infiltration is

$$A_i = \frac{\left(3800 \, \dfrac{\text{m}^3}{\text{d}}\right)\left(365 \, \dfrac{\text{d}}{\text{yr}}\right)}{\left(35.9 \, \dfrac{\text{m}}{\text{yr}}\right)\left(10^4 \, \dfrac{\text{m}^2}{\text{ha}}\right)}$$

$$= \boxed{3.86 \, \text{ha}}$$

Wastewater constituent loadings are as follows

For BOD,

$$\left(3800 \, \frac{\text{m}^3}{\text{d}}\right)\left(365 \, \frac{\text{d}}{\text{yr}}\right)\left(163 \, \frac{\text{mg BOD}}{\ell}\right)\left(\frac{10^3 \, \ell}{\text{m}^3}\right)\left(\frac{1 \, \text{kg}}{10^6 \, \text{mg}}\right)\left(\frac{1}{3.86 \, \text{ha}}\right) = 58,600 \, \frac{\text{kg BOD}_5}{\text{ha-yr}}$$

The other wastewater constituent loadings are determined in the same manner and are:

For suspended solids, 31,600 kg SS/ha-yr

For nitrogen, 9340 kg N/ha-yr

For phosphorus, 1800 kg P/ha-yr

All constituent loadings are within typical ranges.

Thus, for a population of 10,000, the required infiltration area is $\boxed{3.86 \, \text{ha.}}$

Overland Flow Wastewater treatment by the overland flow process is achieved mainly by percolation and evapotranspiration, as shown in Figure 22.9. Table 22.9 gives typical site characteristics, design parameters, and expected performance by the overland flow process. The wastewater is applied at the upper region of sloping, vegetated fields termed terraces. Wastewater is distributed onto the terraces by gated piping operated at pressures of 2 to 5 psi (14 to 35 kPa), low-pressure fan spray devices mounted on fixed risers and operated

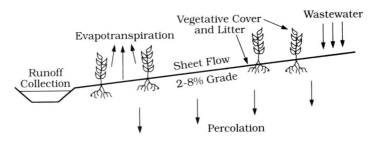

FIGURE 22.9 *Overland Flow Process, Hydraulic Pathway*

Adapted from *Land Treatment of Municipal Wastewater*, 1981.
Environmental Protection Agency (EPA).

TABLE 22.9 *Overland Flow Land Treatment System Site Characteristics, Typical Design Features, and Expected Treated Water Quality[a]*

Grade	Finished slopes 2%–8% (steeper slopes might be feasible at reduced hydraulic loading)
Soil permeability	Slow (clays, silts, and soils with impermeable barriers)
Depth to ground water	Not critical (impact on groundwater should be considered for more permeable soils)
Climatic restrictions	Storage usually needed for cold weather
Application technique	Sprinkler or surface
Loading rate[b]	
Annual	10–65 ft/yr (3–20 m/yr)
Weekly	2.5–16 in./wk (6–40 cm/wk); (range includes raw wastewater to secondary treatment, higher rates for higher level of pretreatment)
Field area	16–109 ac/MGD (1.7–11.6 ha/1000 m^3/d), not including buffer area, roads, or ditches
Minimum pretreatment in United States	Comminution and grit removal (with restricted public access)
Disposition of applied wastewater	Surface runoff and evapotranspiration with some percolation
Vegetation	Required
BOD_5	<10 to <15 mg/ℓ
Suspended solids	<10 to <20 mg/ℓ
Ammonia nitrogen	<4 to <8 mg/ℓ as N
Total nitrogen	5 to <10 mg/ℓ as N (higher values expected when operating through a moderately cold winter or when using secondary effluent at high rates)
Total phosphorus	4 to <6 mg/ℓ as P
Fecal coliforms	200 to <2000 per 100 ml

[a] Treated water quality values are average to upper range expected with loading rates at the mid to low end of the loading rate range given, and with treated comminuted, screened wastewater using a terrace length of 100–120 ft (30–36 m).

[b] See Table 22.10 for loading rates based on more recent data.

Adapted from EPA, *Land Treatment of Municipal Wastewater*, EPA Process Design Manual, 1981.

at 5 to 20 psi (35 to 138 kPa), or high-pressure impact sprinklers at 20 to 80 psi (138 to 550 kPa). Although sprinkler systems can distribute wastewater more uniformly over a terrace, higher system treatment performance with the use of sprinkler systems for municipal wastewater has not been demonstrated, and there is a potential risk of aerosol-borne bacteria. Sites with natural land slopes of up to 8% are ideal, and with terraced construction or reduced hydraulic loading, sites with slopes up to 12% may be suitable. Because the development of continual sheet flow over a terrace is necessary for proper system performance, final terrace grades should be within a tolerance of 0.05 ft (1.5 cm). The overland flow process can be used conservatively at sites with surface soils of low permeability, less than 0.2 in./hr (0.5 cm/hr), or where an underlying soil such as clay, at about 1 to 2 ft (0.3 to 0.6 m) from the surface, restricts downward flow. The overland flow process may be suitable for sites with higher soil permeabilities, up to 2 in./hr (5 cm/hr), and such sites should not be eliminated from consideration. Most of the applied wastewater flows over a terrace at a shallow depth and is collected in drainage channels. Some of the applied wastewater, usually less than 20%, may percolate into the underlying soil; however, percolation usually decreases with time as soil clogging by wastewater particulate material occurs. Groundwater is generally affected very little, because the small amount of wastewater that does percolate is treated by interaction with the soil. However, precautions must be taken to ensure that groundwaters are protected. The collected runoff is discharged to receiving waters and must meet appropriate discharge quality standards (EPA, 1981, 1984; Metcalf & Eddy, 1991).

The overland flow process is capable of treating screened raw wastewater, primary effluent, and treatment pond effluent to secondary treatment effluent quality. Algae are not readily removed by the overland flow process (Witherow and Bledsoe, 1983); thus pretreatment processes, such as ponds, that generate algae should be avoided (EPA, 1984). The process also can be used as an upgrade to existing conventional secondary treatment for high levels of BOD_5, nitrogen, and suspended solids removal. Generally, BOD_5 removal occurs similarly to that which occurs in trickling filters. Biological oxidation of soluble BOD_5 by the microbial population in the soil and surface organic layer accounts for much of the BOD_5 removal. Suspended and colloidal solids are removed by sedimentation and filtration as the wastewater moves over the soil and through the vegetative cover. Biological oxidation of the organic fraction of the particulate material by the microbial population accounts for the remainder of BOD_5 removal. Nitrogen removal occurs as a result of nutrient uptake by the vegetative cover, biological nitrification and denitrification, and volatilization of ammonia nitrogen. Nitrogen removals of 75% to 90% usually can be achieved and are dependent on temperature and on the application rate, time period, and frequency. Phosphorus removal occurs by adsorption and precipitation and ranges from 40% to 60%. Removal of phosphorus is limited because of the short contact time of the wastewater with soil adsorption sites (EPA, 1981).

The general design procedure for overland flow systems is shown in Figure 22.10. The major design parameters are the application rate, the hydraulic loading rate, the application period and frequency, and the terrace length and grade. The land area required is dependent on or influenced by each of the design parameters listed. Three design methods exist for overland flow systems. The method recommended by the EPA (1984) and WPCF (1990) is an empirical method based on design parameter values used at successfully operated overland flow systems. Two alternative, rational methods that may be used are the U.S. Army Cold Regions Research and Engineering Laboratory (CRREL) method, which is based on detention time on a terrace, and the University of California, Davis (UCD) method, which is based on terrace length and application rate. Each of the three methods is discussed in several references, including EPA (1981), EPA (1984), and WPCF (1990). The empirical design method equations and procedures for determining the application and hydraulic loading rates, and the land area requirements will be discussed, and their use illustrated by examples, in this text. Application

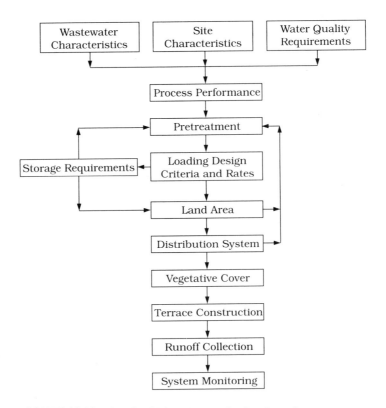

FIGURE 22.10 *Overland Flow Process Design Procedure*
Adapted from *Land Treatment of Municipal Wastewater*, 1981. Environmental Protection Agency (EPA).

periods, application frequency, terrace length and grade, and storage will be discussed briefly.

Suggested overland flow design guidelines are given in Table 22.10. Most of these guidelines have been developed at existing overland flow facilities, although some have been developed at research and demonstration facilities. For the existing overland flow facilities, the guidelines have produced spring, summer, and fall effluent concentrations of less than 20 mg/ℓ BOD$_5$, less than 20 mg/ℓ suspended solids, less than 10 mg/ℓ total nitrogen, and less than 6 mg/ℓ total phosphorus (EPA, 1981). In warm climates, mid- to upper-range values can be used, and in cold climates, lower- to middle-range values are recommended (WPCF, 1990). The empirical design method makes use of design parameters used at successfully operating overland flow systems.

The application rate, R_a, is the flowrate applied to a terrace per unit width of the terrace, expressed as gal/min-ft (m^3/h-m). The hydraulic loading rate, L_w, is the average flowrate per unit area of terrace, expressed as in./day (cm/day). The hydraulic loading rate, the application rate and application period, and the terrace length are related as shown in (EPA, 1981)

$$L_w = \frac{(R_a)(P)}{S}(C) \qquad (22.9)$$

where

L_w = hydraulic loading rate, in./day (cm/day)

R_a = application rate, gal/min-ft (m^3/h-m)

TABLE 22.10 *Suggested Overland Flow Design Guidelines*

PREAPPLICATION TREATMENT	HYDRAULIC LOADING RATE in./day (cm/day)	APPLICATION RATE gal/min-ft (m³/hr-m)	APPLICATION PERIOD hr/day	APPLICATION FREQUENCY day/wk	TERRACE LENGTH ft (m)
Screening	0.79–2.8 (2.0–7.0)	0.09–0.16 (0.07–0.12)	8–12	5–7	118–148 (36–45)
Primary sedimentation	0.79–2.8 (2.0–7.0)	0.09–0.16 (0.07–0.12)	8–12	5–7	98–118 (30–36)
Stabilization pond[a]	1.0–3.5 (2.5–9.0)	0.12–0.20 (0.09–0.15)	8–18	5–7	148 (45)
Complete secondary biological[b]	1.2–3.9 (3.0–10.0)	0.15–0.23 (0.11–0.17)	8–12	5–7	98–118 (30–36)

[a] Does not include removal of algae.

[b] Recommended only for upgrading existing secondary treatment.

Adapted from EPA, *Land Treatment of Municipal Wastewater*, and *Land Treatment of Municipal Wastewater, Supplement on Rapid Infiltration and Overland Flow*, EPA Process Design Manuals, 1981 and 1984.

P = application period, hr/day

S = terrace length, ft (m)

C = (ft^3/7.48 gal × 60 min/hr × 12 in./ft) for USCS units
   = (100 cm/m) for SI units

There are two ways to approach using Eq. (22.9). In one, the application rate, application period, and terrace length are selected from design guidelines, and the hydraulic loading rate is calculated. Alternatively, the application period, terrace length, and hydraulic loading rate are selected from the guidelines, and the application rate is calculated. The application period, P, is selected to provide operational flexibility, and 6 to 12 hr has been used most frequently. For example, by using an 8-hr application period in a system designed for 24-hr operation, wastewater could be applied to one-third of the total area during any 8-hr period. When maintenance or vegetation harvest was required, one-third of the total system could be taken out of service, and wastewater could be applied with an application period of 12 hrs per day to the remaining system. Under these conditions, the application rate would not be increased.

The total terrace area required for wastewater application, A_s, may be determined from (EPA, 1981)

$$A_s = \frac{Q\left(\dfrac{365 \text{ days}}{\text{yr}} \times C_1\right) + \Delta V_s}{(D_a)(L_w \times C_2)} \times C_3 \qquad \textbf{(22.10)}$$

where

A_s = total terrace area, ac (ha)

Q = average daily wastewater flow, gal/day (m^3/day)

ΔV_s = net gain or loss of stored wastewater, ft^3/year (m^3/yr)

D_a = operating days per year

L_w = design hydraulic loading rate, in./day (cm/day)

C_1 = 1 ft^3/7.48 gal (1)

C_2 = 1 ft/12 in. (1 m/100 cm)

C_3 = 1 ac/43,560 ft^2 (1 ha/10^4 m^2)

The net gain or loss of stored wastewater, ΔV_s, due to precipitation, evaporation, or seepage can be determined only after the storage requirement and dimensions of the storage reservoir have been determined. Storage of wastewater may be required for cold-weather operation, for storm water runoff, or for wastewater flow equalization. Equation 22.10 can be used for systems operated through the year in warmer climates. By decreasing the application rate and increasing the application period during the winter, compensation for reduced treatment efficiency due to reduced soil temperature can be achieved without reducing the hydraulic loading rate. Thus terrace area is minimized and no winter storage will be required.

For systems operated throughout the year, but with reduced hydraulic loading during the winter, two alternatives may used. The first alternative requires no winter storage, and the total terrace area is determined as follows (EPA, 1981):

$$A_s = \frac{Q_w \times C_1}{L_{ww} \times C_2} \times C_3 \tag{22.11}$$

where

A_s = total terrace area, ac (ha)

Q_w = average daily winter wastewater flow, gal/day (m³/day)

L_{ww} = winter hydraulic loading rate, in./day (cm/day)

The unit conversion constants, C_1, C_2, and C_3 are as defined for Eq. (22.10). This alternative eliminates the need for winter storage but results in increased terrace area and low warm-weather loading. The second alternative requires winter storage, and the total terrace area is determined as follows (EPA, 1981):

$$A_s = \frac{Q\left(\dfrac{365 \text{ days}}{\text{yr}} \times C_1\right) + \Delta V_s}{(L_{ww}D_{aw} + L_{ws}D_{as})(C_2)} \times C_3 \tag{22.12}$$

where

A_s = total terrace area, ac (ha)

Q = average daily wastewater flow, gal/day (m³/day)

ΔV_s = net gain or loss of stored wastewater, ft³/yr (m³/yr)

L_{ww} = winter hydraulic loading rate, in./day (cm/day)

D_{aw} = operating days at winter hydraulic loading rate

L_{ws} = nonwinter hydraulic loading rate, in./day (cm/day)

D_{as} = operating days at nonwinter hydraulic loading rate

The unit conversion constants C_1, C_2, and C_3 are as previously defined. This alternative minimizes terrace area but requires winter storage. Procedures for determining storage requirements, determining the net gain or loss of storage, and revising an initial estimated terrace area are similar to those previously discussed for the slow rate process. These procedures may be found in an EPA document (EPA, 1981).

The application frequency generally used is 7 days/week. This frequency minimizes terrace area requirements and reduces, or may eliminate, storage requirements. Constituent loading rates at existing facilities have not been found to affect system performance. System performance has been observed to be directly related to terrace length and inversely related to application rate.

Terrace length ranges are given in Table 22.10. The longer lengths should be used with higher application rates. Shorter lengths may be used with lower application rates to achieve an equal level of treatment. The shorter terrace lengths have been used with gated pipe and low-pressure spray distribution systems. With high-pressure, part-circle sprinkler distribution systems, the recommended minimum terrace length is 150 ft (45 m), and for full-circle systems, the recommended minimum length is the sprinkler circle diameter plus about 65 ft (20 m). The optimum terrace grade ranges from 2% to 8%. Within this range, system performance has been shown to be unaffected by grade. With grades less than 2%, ponding may occur, and with grades greater than 8%, erosion, short circuiting, or channeling may occur. Grade changes should be closely controlled to ensure that sheet flow of wastewater is maintained over the terrace (EPA, 1981).

EXAMPLE 22.4   *Overland Flow Land Treatment System: Terrace Area Requirement*

An overland flow land treatment system is to be used to treat a municipal wastewater flow from 10,000 persons. The wastewater flow is 100 gal/cap-day or 1.0 MGD. The wastewater will be pretreated using preliminary treatment and primary sedimentation. The system will be located in a warm climate and will be operated year-round. An application rate of 0.25 gal/min-ft for 8 hr per day has been selected. Use a terrace length of 120 ft, and assume that no winter storage is required. Determine the required terrace area in acres.

SOLUTION   1. Determine the hydraulic loading rate, L_w, using Eq. (22.9), which is

$$L_w = \frac{(R_a)(P)}{S}(C)$$

$$= \frac{\left(0.25\dfrac{\text{gal}}{\text{min-ft}}\right)\left(8\dfrac{\text{hr}}{\text{day}}\right)}{120\,\text{ft}}\left(\frac{1\,\text{ft}^3}{7.48\,\text{gal}} \times \frac{60\,\text{min}}{\text{hr}} \times \frac{12\,\text{in.}}{\text{ft}}\right)$$

$$= 1.6\,\text{in./day}$$

2. Determine the terrace area, A_s, using Eq. (22.10), which is

$$A_s = \frac{Q\left(\dfrac{365\,\text{days}}{\text{yr}} \times C_1\right) + \Delta V_s}{(D_a)(L_w \times C_2)} \times C_3$$

Since no storage is required, $\Delta V_s = 0$, and

$$A_s = \frac{\left(1 \times 10^6\,\dfrac{\text{gal}}{\text{day}}\right)\left(\dfrac{365\,\text{days}}{\text{yr}} \times \dfrac{1\,\text{ft}^3}{7.48\,\text{gal}}\right) + 0}{(365\,\text{days})\left(1.6\dfrac{\text{in.}}{\text{day}} \times \dfrac{1\,\text{ft}}{12\,\text{in.}}\right)}\left(\frac{1\,\text{ac}}{43,560\,\text{ft}^2}\right)$$

$$= \boxed{23.0\,\text{ac}}$$

EXAMPLE 22.4 SI   *Overland Flow Land Treatment System: Terrace Area Requirement*

An overland flow land treatment system is to be used to treat a municipal wastewater flow from 10,000 persons. The wastewater flow is 380 ℓ/cap-d or 3800 m³/d. The wastewater will be pretreated using preliminary treatment and primary sedimentation. The system will be located in a warm climate and will be operated year-round. An application rate of 0.19 m³/h-m for 8 h per day has been selected. Use a terrace length of 37 m, and assume that no winter storage is required. Determine the required terrace area in hectares.

SOLUTION   1. Determine the hydraulic loading rate, L_w, using Eq. (22.9), which is

$$L_w = \frac{(R_a)(P)}{S}(C)$$

$$= \frac{\left(0.19\frac{m^3}{h\text{-}m}\right)\left(8\frac{h}{d}\right)}{37\,m}\left(\frac{100\,cm}{m}\right)$$

$$= 4.1\,cm/d$$

2. Determine the terrace area, A_s, using Eq. (22.10), which is

$$A_s = \frac{Q\left(\dfrac{365\ \text{days}}{yr}\times C_1\right) + \Delta V_s}{(D_a)(L_w \times C_2)} \times C_3$$

Since no storage is required, $\Delta V_s = 0$, and

$$A_s = \frac{3800\dfrac{m^3}{d}\left(\dfrac{365\,d}{yr}\times 1\right) + 0}{(365\,d)\left(4.1\dfrac{cm}{d}\times\dfrac{1\,m}{100\,cm}\right)}\left(\frac{1\,ha}{10^4\,m^2}\right)$$

$$= \boxed{9.27\,ha}$$

Combination Systems   Land treatment systems may be combined in series when a high-quality effluent is required or to increase treatment reliability. The combination of an overland flow system followed by either a slow rate system or a rapid infiltration system could be used to produce an effluent quality better than that produced by the overland flow system alone, particularly for BOD, suspended solids, nitrogen, and phosphorus. Overland flow has been used prior to rapid infiltration to reduce nitrogen levels. The rapid infiltration process followed by slow rate land treatment may be able to produce an irrigation water that would meet or exceed the restrictions for irrigation of food crops. Examples of two potential land treatment combinations are shown in Figure 22.11.

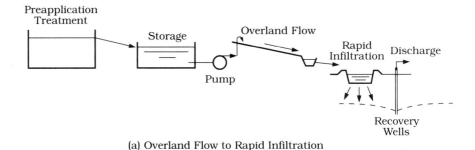

(a) Overland Flow to Rapid Infiltration

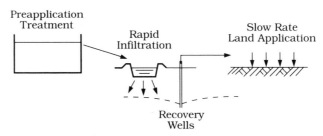

(b) Rapid Infiltration to Slow Rate Land Application

FIGURE 22.11 *Combined Land Treatment Systems*

Adapted from *Land Treatment of Municipal Wastewater*, 1981. Environmental Protection Agency (EPA).

SLUDGE TREATMENT

Depending on the quality of the sludge, wastewater sludge can be considered a valuable resource. Sludge applied to land receives treatment by several mechanisms. Soil microorganisms use biodegradable material in the sludge as a food and energy source. Natural drying, exposure to ultraviolet radiation in sunlight, adsorption in the soil, and nutrient use by vegetation are other mechanisms by which treatment of the sludge occurs.

Agricultural use of sludge for its fertilizer value and as a soil amendment, both of which applications enhance crop production, is the predominant form of land treatment. Sludge may be applied to forest land to enhance forest productivity. Application of sludge to unproductive or disturbed land assists in reclamation and in the establishment of vegetation. Application of sludge to dedicated land disposal sites, other than landfills, is accomplished for the primary purpose of sludge disposal. In dedicated land disposal, use of the sludge as a valuable resource is a minor consideration or is not considered at all. These alternatives may be combined to achieve more than one beneficial use of the sludge at a single site. Each of the alternatives has advantages, disadvantages, and limitations. In the planning and design of land treatment of sludge, the considerations that must be addressed include the chemical, biological, and physical characteristics of the sludge; the applicable federal, state, and local regulations; the selection of the land treatment alternative; the estimation of land area required; and the evaluation of available sludge transport options. Discussion of the advantages, disadvantages, and limitations of each alternative, and of the planning considerations, may be found in several references, including Reed and Crites (1984); Reed *et al.* (1988); and EPA (1983). The basics for estimation of the land area required for a sludge land treatment system are discussed, and related examples are provided, in the remainder of this chapter.

A preliminary estimate, for planning purposes, of the required land area may be made by using the guidelines provided in Table 22.11. Using typical values from Table 22.11, for agricultural use of 1 ton, dry weight, of sludge generated annually, 5 ac of land would be required. Each year, the same 5 ac of agricultural land could receive the sludge generated. For agricultural use of 1 metric ton (tonne), dry weight, of sludge generated annually, 11 ha of land would be required. Each year, the same 11 ha of agricultural land could receive the sludge generated. For land reclamation use of 1 ton, dry weight, of sludge generated annually, 50 ac would be required. An additional 50 acres of land to be reclaimed would be needed for each succeeding year. For land reclamation use of 1 metric ton, dry weight, of sludge generated annually, 112 ha of land would be required, and an additional 112 ha would be needed for each succeeding year. These land requirements are for sludge application only and may be used only for planning purposes. Detailed calculations are needed for final design.

Design sludge application rates are limited by sludge constituent regulatory criteria or by nutrient needs of an agricultural crop, usually its need for nitrogen. Sludge application rates on nonagricultural land are usually

TABLE 22.11 *Sludge Application Rates[a] for Preliminary Estimation of Land Area Requirements*

DISPOSAL OPTION	APPLICATION PERIOD	RANGE ton/ac (mt/ha)	TYPICAL RATE ton/ac (mt/ha)
Agricultural use	Annual	1–30 (2–70)	5 (11)
Forest use	One time, or at 3- to 5-yr intervals	4–100 (10–220)	20 (44)
Land reclamation use	One time	3–200 (7–450)	50 (112)
Dedicated disposal site	Annual	100–400 (220–900)	150 (340)

[a] Rates given are for sludge application area only, not including area required for buffer zone, sludge storage, or other site requirements.
Note: 1 ton = 2000 lb, 1 mt = 1000 kg
    1 ac = 43,560 ft^2, 1 ha = 10^4 m^2
Adapted from EPA, *Land Application of Municipal Sludge*, EPA Process Design Manual, 1983.

limited by sludge constituent regulatory criteria. The determination of design application rates assumes that the sludge will be used as a replacement for commercial fertilizers on agricultural lands. The nitrogen and phosphorus needs of a particular crop grown in a particular soil must be determined. Application rates on agricultural land may be limited by regulatory criteria on sludge constituents such as pathogens, metals, and organics. Before allowable rates are determined, the amount of total solids, total nitrogen, ammonia nitrogen, nitrate nitrogen, total phosphorus, total potassium, and total pcbs, lead, zinc, copper, nickel, and cadmium must be known to ensure good design (EPA, 1983). In addition, *current* local, state, and federal regulations must be known.

Application rates based on regulatory criteria for a particular sludge constituent may be determined using (EPA, 1983)

$$S_m = \frac{L}{C_m \times C} \qquad (22.13)$$

where

S_m = allowable amount of sludge that can be applied for a particular sludge constituent over a selected time interval on a dry weight basis, ton/ac (mt/ha)

L = maximum amount of sludge constituent that can be applied for the selected time interval based on the regulatory limit, lb/ac (kg/ha)

C_m = concentration of the sludge constituent expressed as a decimal (for example, 10 ppm = 0.00001)

C = conversion constant, 2000 lb/ton (1000 kg/mt)

Sludge application rates based on the nitrogen needs of a crop are determined on a yearly basis with consideration for the amount of nitrogen remaining as available nitrogen from previous years. For any year of application, nitrogen available from sludge applied in that year is estimated from (Metcalf & Eddy, 1991; Reed and Crites, 1984; Reed *et al.*, 1988)

$$N_a = (C)[NO_3 + k_v(NH_4) + f_n(N_o)] \qquad (22.14)$$

where

N_a = plant-available nitrogen in sludge for application year on dry solids basis, lb N/ton-yr (kg N/mt-yr)

C = conversion constant, 2000 lb/ton (1000 kg/mt)

NO_3 = percent nitrate in sludge expressed as a decimal

k_v = volatilization factor for ammonia
  = 0.5 for surface- or sprinkler-applied liquid sludge
  = 1.0 for incorporated liquid sludge or dewatered sludge

NH_4 = percent ammonia in sludge expressed as a decimal

f_n = mineralization factor for year $n = 1$ from Table 22.12 expressed as a decimal

N_o = percent organic N in sludge expressed as a decimal

The mineralization of organic nitrogen available from previous-year applications is determined using the following equation (Metcalf & Eddy, 1991; Reed, *et al.*, 1984 and 1988).

$$N_{ap} = (C)\Sigma[f_2(N_o)_2 + f_3(N_o)_3 + \cdots + f_n(N_o)_n] \qquad (22.15)$$

where

N_{ap} = plant-available mineralized organic nitrogen from sludge applied in the previous n years on a dry weight basis, lb N/ton-yr (kg N/mt-yr)

C = conversion constant 2000 lb/ton (1000 kg/mt)

$f_{2,3,n}$ = mineralization factor from Table 22.12 expressed as a decimal; subscript is year under consideration

$(N_o)_n$ = decimal fraction of organic nitrogen remaining in sludge from year n.

The annual nitrogen sludge application rate is given by (Metcalf & Eddy, 1991; Reed and Crites, 1984; Reed *et al.*, 1988)

$$R_n = \frac{U_n}{N_a + N_{ap}} \qquad (22.16)$$

where

R_n = annual sludge application rate in year n on a dry weight basis, ton/ac-yr (kg/ha-yr)

U_n = annual nitrogen uptake by vegetation from Table 22.2, lb N/ac-yr (kg N/ha-yr)

TABLE 22.12 *Estimated Mineralization Percentages[a] for Organic Nitrogen after Sludge Application to Soil*

TIME AFTER APPLICATION (YEARS)	UNSTABILIZED PRIMARY AND WASTE ACTIVATED	AEROBICALLY DIGESTED	ANAEROBICALLY DIGESTED	COMPOSTED
0–1	40	30	20	10
1–2	20	15	10	5
2–3	10	8	5	3
3–4	5	4	3	3
4–10	3	3	3	3

[a] Percent of initial organic N present mineralized *per year* during time interval shown.
Adapted from Sommers, Parker, and Meyers, 1981. Volatilization, Plant Uptake and Mineralization of Nitrogen in Soils Treated with Sewage Sludge, Purdue Univ. Water Resources Research Technical Report 133. In EPA, *Land Application of Municipal Sludge*, EPA Process Design Manual, 1983.

N_a and N_{ap} are as previously defined.

The sludge application rate based on phosphorus as the limiting parameter is (Metcalf & Eddy, 1991; Reed and Crites, 1984; Reed *et al.*, 1988)

$$S_p = \frac{U_p}{C_p \times C} \qquad (22.17)$$

where

> S_p = sludge application rate based on phosphorus limitation, ton P/ac-yr (mt P/ha-yr)
>
> U_p = annual phosphorus uptake rate by vegetation from Table 22.2, lb P/ac-yr (kg P/ha-yr)
>
> C_p = concentration of phosphorus in the sludge expressed as a decimal (for example, 10 ppm = 0.00001)
>
> C = conversion constant, 2000 lb/ton (1000 kg/mt)

Normally, about 50% of the phosphorus content in sludge is available for crop uptake (Metcalf & Eddy, 1991).

After all loading rates have been determined on the basis of nutrient and regulatory limiting criteria, the final design sludge application rate R_d will be the smallest of the loading rates. The required land area may then be determined from (Reed and Crities, 1984; Reed *et al.*, 1988)

$$A = \frac{Q_s}{R_d} \qquad (22.18)$$

where

> A = application area required, ac (ha)
>
> Q_s = total sludge to be applied on a dry weight basis, ton/yr (mt/yr)
>
> R_d = design sludge application rate on a dry weight basis, ton/ac-yr (mt/ha-yr)

EXAMPLE 22.5 *Sludge Treatment by Land Application for Agricultural Use: Application Rates and Land Area Required*

Anaerobically digested sludge from a wastewater treatment facility is to be treated by land application to an agricultural site at which corn will be grown. The wastewater treatment plant serves a population of 10,000 persons and produces 220 tons of dry sludge per year. The sludge will be applied as a liquid by surface distribution. The design is to be based on the nitrogen needs of the corn crop and is to be limited by a cumulative cadmium loading of 10 lb Cd/ac for the length of time the site can be used. The sludge contains 100 mg Cd/kg of dry sludge, 1% ammonia nitrogen, 2.2% organic nitrogen, and no nitrate nitrogen. Determine the allowable annual dry sludge application rate, the land area required, and the useful life of the site for sludge application.

SOLUTION Determine the nitrogen loading rate for the first year due to direct application using Eq. (22.14), which is

$$N_a = (2000)[NO_3 + k_v(NH_4) + f_n(N_o)]$$

From Table 22.12, the mineralization rate, f_1, is 0.20, and

$$N_a = \left(2000\frac{lb}{ton}\right)[0 + 0.5(0.01) + 0.2(0.022)]$$

$$= 18.8\,lb\,N/ton\,dry\,sludge$$

Determine the additional nitrogen available in the second year and in subsequent years from mineralization of the organic nitrogen fraction from previous years of application. Use Eq. (22.15), which is

$$N_{ap} = \left(2000\frac{lb}{ton}\right)\Sigma[f_2(N_o)_2 + f_3(N_o)_3 + \cdots + f_n(N_o)_n]$$

1. Determine the decimal fraction of organic nitrogen, $(N_o)_n$, remaining each year after application, using mineralization factor, f_n, values from Table 22.12.

$$(N_o)_1 = (N_o) = 0.022$$
$$(N_o)_2 = (N_o)_1 - f_1(N_o)_1 = (0.022) - 0.2(0.022) = 0.0176$$
$$(N_o)_3 = (N_o)_2 - f_2(N_o)_2 = (0.0176) - 0.1(0.0176) = 0.0158$$
$$(N_o)_4 = (N_o)_3 - f_3(N_o)_3 = (0.0158) - 0.05(0.0158) = 0.0150$$
$$(N_o)_5 = (N_o)_4 - f_4(N_o)_4 = (0.0150) - 0.03(0.0150) = 0.0146$$
$$(N_o)_6 = (N_o)_5 - f_5(N_o)_5 = (0.0146) - 0.03(0.0146) = 0.0142$$
$$(N_o)_7 = (N_o)_6 - f_6(N_o)_6 = (0.0142) - 0.03(0.0142) = 0.0138$$
$$(N_o)_8 = (N_o)_7 - f_7(N_o)_7 = (0.0138) - 0.03(0.0138) = 0.0134$$
$$(N_o)_9 = (N_o)_8 - f_8(N_o)_8 = (0.0134) - 0.03(0.0134) = 0.0130$$

2. Using Eq. (22.15), determine the available nitrogen from mineralization. For the purpose of this example, assume that the nitrogen available from mineralization is negligible after the ninth year.

$$N_{ap} = 2000\,lb/ton\,[0.1(0.0176) + 0.05(0.0158) + 0.03(0.0150)$$
$$+ 0.03(0.0146) + 0.03(0.0142) + 0.03(0.0138) + 0.03(0.0134)$$
$$+ 0.03(0.0130)]$$
$$= 10.1\,lb\,N/ton\,dry\,sludge$$

Determine the nitrogen-loading-based allowable annual sludge application rate using Eq. (22.16), which is

$$R_n = \frac{U_n}{N_a + N_{ap}}$$

From Table 22.2, use a nitrogen uptake rate for corn, U_n, of 160 lb N/ac-yr.

$$R_n = \frac{160 \dfrac{\text{lb-N}}{\text{ac-yr}}}{(18.8 + 10.1)\dfrac{\text{lb-N}}{\text{ton dry sludge}}}$$

$$= \boxed{5.5 \text{ ton dry sludge/ac-yr}}$$

Determine the cadmium-limited allowable amount of sludge that can be applied during the useful life of the site using Eq. (22.13), which is

$$S_m = \frac{L}{C_m \times C}$$

$$= \frac{10 \dfrac{\text{lb Cd}}{\text{acre}}}{0.000100 \times 2000 \dfrac{\text{lb}}{\text{ton}}}$$

$$= 50 \text{ ton/ac}$$

Determine the land area required for sludge application using Eq. (22.18), which is

$$A = \frac{Q_s}{R_d}$$

$$= \frac{220 \dfrac{\text{ton}}{\text{yr}}}{5.5 \dfrac{\text{ton}}{\text{ac-yr}}}$$

$$= \boxed{40.0 \text{ acres}}$$

Determine the design life of the site on the basis of the cadmium limitation.

$$\text{Design life} = \frac{50 \dfrac{\text{ton}}{\text{ac}}}{5.5 \dfrac{\text{ton}}{\text{ac-yr}}}$$

$$= \boxed{9.1 \text{ yr}}$$

EXAMPLE 22.5 SI   *Sludge Treatment by Land Application for Agricultural Use: Application Rates and Land Area Required*

Anaerobically digested sludge from a wastewater treatment facility is to be treated by land application to an agricultural site at which corn will be grown. The wastewater treatment plant serves a population of 10,000 persons and produces 200 metric tons of dry sludge per year. The sludge will be applied as a liquid by surface distribution. The design is to be based on the nitrogen needs of the corn crop and is to be limited by a cumulative cadmium loading of 11 kg Cd/ha for the length of time the site can be used. The sludge contains 100 mg Cd/kg of dry sludge, 1% ammonia nitrogen, 2.2% organic nitrogen, and no nitrate nitrogen. Determine the allowable annual dry sludge application rate, the land area required, and the useful life of the site for sludge application.

SOLUTION Determine the nitrogen loading rate for the first year due to direct application using Eq. (22.14), which is

$$N_a = (1000)[NO_3 + k_v(NH_4) + f_n(N_o)]$$

From Table 22.12, the mineralization rate, f_1, is 0.20, and

$$N_a = \left(1000\frac{kg}{mt}\right)[0 + 0.5(0.01) + 0.2(0.022)]$$

$$= 9.4 \text{ kg N/mt dry sludge}$$

Determine the additional nitrogen available in the second year and in subsequent years from mineralization of the organic nitrogen fraction from previous years of application. Use Eq. (22.15), which is

$$N_{ap} = \left(1000\frac{kg}{mt}\right)\Sigma[f_2(N_o)_2 + f_3(N_o)_3 + \cdots + f_n(N_o)_n]$$

1. Determine the decimal fraction of organic nitrogen, $(N_o)_n$, remaining each year after application using mineralization factor, f_n, values from Table 22.12.

   $(N_o)_1 = (N_o) = 0.022$
   $(N_o)_2 = (N_o)_1 - f_1(N_o)_1 = (0.022) - 0.2(0.022) = 0.0176$
   $(N_o)_3 = (N_o)_2 - f_2(N_o)_2 = (0.0176) - 0.1(0.0176) = 0.0158$
   $(N_o)_4 = (N_o)_3 - f_3(N_o)_3 = (0.0158) - 0.05(0.0158) = 0.0150$
   $(N_o)_5 = (N_o)_4 - f_4(N_o)_4 = (0.0150) - 0.03(0.0150) = 0.0146$
   $(N_o)_6 = (N_o)_5 - f_5(N_o)_5 = (0.0146) - 0.03(0.0146) = 0.0142$
   $(N_o)_7 = (N_o)_6 - f_6(N_o)_6 = (0.0142) - 0.03(0.0142) = 0.0138$
   $(N_o)_8 = (N_o)_7 - f_7(N_o)_7 = (0.0138) - 0.03(0.0138) = 0.0134$
   $(N_o)_9 = (N_o)_8 - f_8(N_o)_8 = (0.0134) - 0.03(0.0134) = 0.0130$

2. Using Eq. (22.15), determine the available nitrogen from mineralization. For the purpose of this example, assume that the nitrogen available from mineralization is negligible after the ninth year.

$$N_{ap} = 1000 \text{ kg/mt}[0.1(0.0176) + 0.05(0.0158) + 0.03(0.0150)$$
$$+ 0.03(0.0146) + 0.03(0.0142) + 0.03(0.0138) + 0.03(0.0134)$$
$$+ 0.03(0.0130)]$$
$$= 5.1 \text{ kg N/mt dry sludge}$$

Determine the nitrogen-loading-based allowable annual sludge application rate using Eq. (22.16), which is

$$R_n = \frac{U_n}{N_a + N_{ap}}$$

From Table 22.2, use a nitrogen uptake rate for corn, U_n, of 180 kg-N/ha-yr.

$$R_n = \frac{180\dfrac{\text{kg-N}}{\text{ha-yr}}}{(9.4 + 5.1)\dfrac{\text{kg-N}}{\text{mt dry sludge}}}$$

$$= \boxed{12.4 \text{ mt dry sludge/ha-yr}}$$

Determine the cadmium-limited allowable amount of sludge that can be applied during the useful life of the site using Eq. (22.13), which is

$$S_m = \frac{L}{C_m \times C}$$

$$= \frac{11\dfrac{\text{kg Cd}}{\text{ha}}}{0.000100 \times 1000\dfrac{\text{kg}}{\text{mt}}}$$

$$= 110 \text{ mt/ha}$$

Determine the land area required for sludge application using Eq. (22.18), which is

$$A = \frac{Q_s}{R_d}$$

$$= \frac{200\dfrac{\text{mt}}{\text{yr}}}{12.4\dfrac{\text{mt}}{\text{ha-yr}}}$$

$$= \boxed{16.1 \text{ ha}}$$

Determine the design life of the site on the basis of the cadmium limitation.

$$\text{Design life} = \frac{110\dfrac{\text{mt}}{\text{ha}}}{12.4\dfrac{\text{mt}}{\text{ha-yr}}}$$

$$= \boxed{8.9 \text{ yr}}$$

COMPOSTING

Sludge composting is the biological decomposition of sludge under conditions that allow development of thermophilic temperatures resulting from biologically produced heat. The final product is sufficiently stable for storage and application to land without adverse environmental effects (Haug, 1980). Composting may be accomplished either aerobically or anaerobically. Aerobic systems are used most often because of the typical end products of anaerobic decomposition, such as methane and low-molecular-weight organic acids, and the associated potential for offensive odors. Aerobic biological composting of municipal sludge can produce a safe, aesthetically acceptable product with high potential use in areas where a market is available or as a product provided at minimal or no cost as a public service (Vesilind, 1980).

In composting systems, moisture must be controlled. Too much moisture in the sludge reduces composting temperatures and the efficiency of the composting process. Dewatered municipal sludge may contain as much as 70% to 80% water and lacks sufficient porosity for adequate aeration. Continuous agitation of such sludge must be used to provide sufficient aeration. Oxygen transfer has been enhanced by blending previously composted material with the dewatered sludge, by amending the dewatered sludge with an organic material such as sawdust, straw, peat, rice hulls, or tree and lawn trimmings, or by adding an organic or inorganic material such as wood chips or peanut shells as a bulking agent (Haug, 1980). Wood chips, the most commonly used bulking agent, can be recovered and reused.

Composting systems have been classified by Haug (1980) as reactor systems or nonreactor systems. In reactor systems, the material to be composted is contained within a vessel. In nonreactor systems, a vessel is not used. Both reactor systems and nonreactor systems may employ mechanical equipment and may be enclosed under or in protective facilities. Reactor systems may employ either vertical flow or horizontal and inclined flow of material to be composted. Vertical-flow reactors may be moving agitated bed reactors or moving packed bed reactors. In either type of vertical-flow reactor, sludge may be supplied continuously or intermittently. Horizontal and inclined reactor systems may be tumbling solids bed reactors, which employ rotating drums, or agitated solids bed reactors, which use various bin geometry and agitation methods. Nonreactor systems may be of either the agitated solids bed type or the static solids bed type. Agitated solids bed systems employ turning, tumbling, or other means of agitating the composting material. An example is the windrow system, in which material to be composted is placed in rows and turned periodically by mechanical equipment. Either oxygen is supplied naturally during the turning process or its introduction is aided by forced or induced aeration. Static solids bed systems employ either forced aeration or natural aeration without agitation during the composting cycle. During composting, the temperature may reach 120°F to 160°F (50°C to 70°C), and significant pathogen reduction occurs. The composted material can be used as a low-grade fertilizer and soil conditioner for agricultural and horticultural applications. However, it should not be used on crops eaten raw by humans, such as tomatoes, peppers, and similar vegetables.

The number of municipal sludge composting projects in the United States under consideration in planning, design, or construction or in operation increased from 90 projects in 1983 to 219 projects in 1988 (Goldstein, 1988). More than 400 sewage sludge composting facilities have been constructed in North America (La Trobe and Ross, 1991). For additional reading on composting, see *Compost Engineering, Principles and Practice* by R. T. Haug (1980); the Water Environment Federation (WEF) publication WEF Manual of Practice FD-9, *Sludge Stabilization* (1993b); and a series of four papers by Finstein *et al.* (1987) in *Biocycle*.

REFERENCES

Abernathy, A. R.; Zirschky, J.; and Borup, M. B. 1985. Overland Flow Wastewater Treatment at Easley, S.C. *Jour. WPCF* 57, no. 4:291.

Bruggeman, A. C., and Mostaghimi, S. 1993. Sludge Application Effects on Runoff, Infiltration, and Water Quality. *Water Resources Bulletin* 29, no. 1:15.

Buchberger, S. G., and Maidment, D. R. 1989. Design of Wastewater Storage Ponds at Land Treatment Sites I: Parallels with Applied Reservoir Theory. *Jour. of Environmental Engineering* 115, no. 4:689.

Buchberger, S. G., and Maidment, D. R. 1989. Design of Wastewater Storage Ponds at Land Treatment Sites II: Equilibrium Storage Performance Functions. *Jour. of Environmental Engineering* 115, no. 4:704.

Bouwer, H., and Rice, R. C. 1984. Renovation of Wastewater at the 23rd Avenue Rapid Infiltration Project. *Jour. WPCF* 56, no. 1:76.

Demirjian, Y. A.; Westman, T. R.; Joshi, A. M.; Rop, D. J.; Buhl, R. V.; and Clark, W. R. 1984. Land Treatment of Contaminated Sludge with Wastewater Irrigation. *Jour. WPCF* 56, no. 4:370.

Dinges, R. 1982. *Natural Systems for Water Pollution Control.* New York: Van Nostrand Reinhold.

Environmental Protection Agency (EPA). 1982. *Dewatering Municipal Wastewater Sludge.* EPA Process Design Manual, Washington, D.C.

Environmental Protection Agency (EPA). 1983. *Land Application of Municipal Sludge.* EPA Process Design Manual, Washington, D.C.

Environmental Protection Agency (EPA). 1978a. *Land Application of Wastewater and State Water Law, State Analysis*, Vol. II. Environmental Protection Technology Series EPA-600/2-78-175, Ada, Ok.

Environmental Protection Agency (EPA). 1981. *Land Treatment of Municipal Wastewater.* EPA Process Design Manual, Washington, D.C.

Environmental Protection Agency (EPA). 1984. *Land Treatment of Municipal Wastewater, Supplement on Rapid Infiltration and Overland Flow.* EPA Process Design Manual, Washington, D.C.

Environmental Protection Agency (EPA). 1978b. *Municipal Sludge Landfills.* EPA Process Design Manual. NTIS PB-279 675, Springfield, Va.

Environmental Protection Agency (EPA). 1979. *Municipal Wastewater Treatment by the Overland Flow Method of Land Application.* Environmental Protection Technology Series EPA-600/2-79-178, Ada, Okla.

Environmental Protection Agency (EPA). 1979. *Sludge Treatment and Disposal.* EPA Process Design Manual, Washington, D.C.

Environmental Protection Agency (EPA). 1979. *Treatment of Secondary Effluent by Infiltration-Percolation.* Environmental Protection Technology Series EPA-600/2-79-174, Ada, Okla.

Environmental Protection Agency (EPA). 1979. *Wastewater Irrigation at Tallahassee, Florida.* Environmental Protection Technology Series EPA-600/2-79-151, Ada, Okla.

Environmental Protection Agency (EPA). 1973. *Wastewater Treatment and Reuse by Land Application*, Vol. II. Environmental Protection Technology Series EPA-660/2-73-006b, Washington, D.C.

Finstein, M. S.; Miller, F. C.; Hogan, J. A.; and Strom, P. F. 1987. Analysis of EPA Guidance on Composting Sludge, Part I — Biological Heat Generation and Temperature. *Biocycle* 28, no. 1:21.

Finstein, M. S.; Miller, F. C.; Hogan, J. A.; and Strom, P. F. 1987. Analysis of EPA Guidance on Composting Sludge, Part II — Biological Process Control. *Biocycle* 28, no. 2:42.

Finstein, M. S.; Miller, F. C.; Hogan, J. A.; and Strom, P. F. 1987. Analysis of EPA Guidance on Composting Sludge, Part III — Oxygen, Moisture, Odor, Pathogens. *Biocycle* 28, no. 3:38.

Finstein, M. S.; Miller, F. C.; Hogan, J. A.; and Strom, P. F. 1987. Analysis of EPA Guidance on Composting Sludge, Part IV — Facility Design and Operation. *Biocycle* 28, no. 4:56.

Gan, D. R., and Berthouex, P. M. 1994. Disappearance and Crop Uptake of PCBs from Sludge-Amended Farmland. *Water Environment Research*, 66, no. 1:54.

Goldstein, N. 1988. Steady Growth for Sludge Composting. *Biocycle* 29, no. 4:27.

Haug, R. T. 1980. *Compost Engineering, Principles and Practice.* Ann Arbor, Mich.: Ann Arbor Science.

Kruzik, A. P. 1994. Natural Treatment Systems. *Water Environment Research* 66, no. 4:357.

Lance, J. C. 1972. Nitrogen Removal by Soil Mechanisms. *Jour. WPCF* 44, no. 7:1352.

La Trobe, B. E., and Ross, W. R. 1991. Forced Aeration Co-composting of Domestic Refuse and Night Soil. Presented at the 64th Annual Conference, Water Pollution Control Federation, Toronto, Canada. In Water Environment Federation (WEF) 1993. *Natural Systems Digest.* Alexandria, Va.

Leach, L. E., and Enfield, C. G. 1983. Nitrogen Control in Domestic Wastewater Rapid Infiltration Systems. *Jour. WPCF* 55, no. 9:1150.

Loehr, R. C., and Overcash, M. R. 1985. Land Treatment of Waste: Concepts and General Design. *Jour. of Environmental Engineering* 111, no. 2:141.

Meo, M.; Day, J. W., Jr.; and Ford, T. B. 1975. *Overland Flow in the Louisiana Coastal Zone.* Center for Wetland Resources, Louisiana State University, Pub. no. LSUSG-T-75-04, Baton Rouge, La.

Metcalf & Eddy, Inc. 1991. *Wastewater Engineering: Treatment, Disposal and Reuse.* 3rd. ed. New York: McGraw-Hill.

Overcash, M. R., and Pal, D. 1979. *Design of Land Treatment Systems for Industrial Wastes: Theory and Practice.* Ann Arbor, Mich.: Ann Arbor Science.

Pardue, J. H.; DeLaune, R. D.; and Patrick, W. H., Jr. 1988. Removal of PCBs from Wastewater in a Simulated Overland Flow Treatment System. *Water Research* 22, no. 8:1011.

Pettygrove, G. S., and Asano, T., editors. 1985. *Irrigation with Reclaimed Municipal Wastewater — A Guidance Manual.* Chelsea, Mich.: Lewis Publishers.

Reed, S. C., and Crites, R. W. 1984. *Handbook of Land Treatment Systems for Industrial and Municipal Wastes.* Park Ridge, N.J.: Noyes Data Corp.

Reed, S. C.; Middlebrooks, E. J.; and Crites, R. W. 1988. *Natural Systems for Waste Management and Treatment.* New York: McGraw-Hill.

Sanks, R. L., and Asano, T. 1976. *Land Treatment and Disposal of Municipal and Industrial Wastewater.* Ann Arbor, Mich.: Ann Arbor Science.

Sepp, E. 1970. Nitrogen Cycle in Groundwater. Bureau of Sanitary Engineering, California State Department of Public Health, Berkeley.

Smith, R. G., and Schroeder, E. D. 1985. Field Studies of the Overland Flow Process for the Treatment of Raw and Primary Treated Municipal Wastewater. *Jour. WPCF* 57, no. 7:785.

Smith, R. G. and Schroeder, E. D. 1983. Physical Design of Overland Flow Systems. *Jour. WPCF* 55, no. 3:255.

Sommers, L. E.; Parker, C. F.; and Meyers, G. J. 1981. Volatilization, Plant Uptake and Mineralization of Nitrogen in Soils Treated with Sewage Sludge. Purdue University Water Resources Research Center Technical Report 133. In *Land Application of Municipal Sludge*, EPA, 1983.

Sopper, W. E., and Kardos, L. T., editors. 1973. *Recycling Treated Municipal Wastewater Through Forest and Cropland, Symposium Proceedings*, August 1972. Pennsylvania State University, College of Agriculture and the Institute for Research on Land and Water Resources, University Park, Penn.: Pennsylvania University Press.

Sorber, C. A.; Bausum, H. T.; Schaub, S. A.; and Small, M. J. 1976. A Study of Bacterial Aerosols at a Wastewater Irrigation Site. *Jour. WPCF* 48, no. 10:2367.

Suzuki, T.; Katsuno, T.; and Yamaura, G. 1992. Land Application of Wastewater Using Three Types of Trenches Set in Lysimeters and Its Mass Balance of Nitrogen. *Water Research* 26, no. 11:1433.

Suzuki, T., and Yamaura, G. 1989. Natural Recovery of Chemical Components Accumulated in Soil by Wastewater Application. *Water Research*, 23 no. 10:1285.

Vesilind, P. A. 1980. *Treatment and Disposal of*

Wastewater Sludges. Ann Arbor, Mich.: Ann Arbor Science.

Water Environment Federation (WEF). 1993a. *Natural Systems Digest.* Alexandria, Va.

Water Environment Federation (WEF). 1993b. *Sludge Stabilization.* WEF Manual of Practice FD-9. Alexandria, Va.

Water Environment Research Foundation (WERF). 1993. *Document Long-Term Experience of Biosolids Land Application Programs.* WERF Project 91-ISP-4, Final Report, Alexandria, Va.

Water Pollution Control Federation (WPCF). 1989. *Beneficial Use of Waste Solids.* WPCF MOP No. FD-15, Alexandria, Va.

Water Pollution Control Federation (WPCF). 1990.

Natural Systems for Wastewater Treatment. WPCF Manual of Practice No. FD-16, Alexandria, Va.

Water Pollution Control Federation (WPCF). 1971. *Utilization of Municipal Wastewater Sludge.* WPCF Manual of Practice No. 2, Washington, D.C.

Wentink, G. R., and Etzel, J. E. 1972. Removal of Metal Ions by Soil. *Jour. WPCF* 44, no. 8:1561.

Witherow, J. L., and Bledsoe, B. E. 1983. Algae Removal by the Overland Flow Process. *Jour. WPCF* 55, no. 10:1256.

Zirschky, J., and Crawford, D. 1989. Effect of Wastewater Application Device on Ammonia Volatilization. *Jour. of Environmental Engineering* 115, no. 6:1258.

PROBLEMS

22.1 A slow rate land treatment system is to be used to treat a municipal wastewater from a population of 15,000 persons. The wastewater flow is 100 gal/cap-day. The wastewater will be pretreated using preliminary treatment and primary sedimentation. Site characteristics, vegetative cover characteristics, and wastewater data are:

1. Table 22.13 gives monthly operating days, estimated monthly evapotranspiration for the selected crop and climate, and monthly precipitation.

2. Field testing has yielded relatively consistent results with minor variability for soil permeability. The average minimum permeability is 0.26 in./hr,

and the maximum daily percolation rate should not exceed 6% of the minimum permeability.

3. The system will be constructed to prevent runoff.

4. The maximum depth for wastewater storage is 20 ft, and seepage from storage will not occur.

5. Total nitrogen in the wastewater, C_n, is 25 mg/ℓ.

6. Allowable nitrogen in percolating water, C_p, is 10 mg/ℓ.

7. Use a denitrification-volatilization nitrogen removal fraction, $f = 0.18$.

8. The annual nitrogen uptake rate for the selected vegetative cover, U, is 300 lb/ac-year.

TABLE 22.13 *Data for Problem 22.1*

MONTH	OPERATING DAYS/MONTH	EVAPOTRANSPIRATION (in./mo)	PRECIPITATION (in./mo)
Jan	11	1.0	1.4
Feb	21	1.9	1.2
Mar	26	3.5	1.2
Apr	30	5.0	1.0
May	31	7.2	0.5
Jun	30	8.6	0.2
Jul	31	9.3	0.1
Aug	31	9.0	0.2
Sep	30	5.6	0.2
Oct	30	4.5	0.4
Nov	27	2.5	0.8
Dec	8	0.8	1.2

9. Assume that evaporation is equal to evapotranspiration.
Determine:
a. The annual design hydraulic loading rate, in./yr.
b. The field area required, ac.
c. The wastewater storage volume in acre-ft, the storage surface area in acres, and the required storage depth in feet.

22.2 A slow rate land treatment system is to be used to treat a municipal wastewater from a population of 15,000 persons. The wastewater flow is 380 ℓ/cap-d. The wastewater will be pretreated using preliminary treatment and primary sedimentation. Site characteristics, vegetative cover characteristics, and wastewater data are:
1. Table 22.14 gives monthly operating days, estimated monthly evapotranspiration for the selected crop and climate, and monthly precipitation.
2. Field testing has yielded relatively consistent results with minor variability for soil permeability. The average minimum permeability is 0.7 cm/h, and the maximum daily percolation rate should not exceed 6% of the minimum permeability.
3. The system will be constructed to prevent runoff.
4. The maximum depth for wastewater storage is 6 m, and seepage from storage will not occur.
5. Total nitrogen in the wastewater, C_n, is 25 mg/ℓ.
6. Allowable nitrogen in percolating water, C_p, is 10 mg/ℓ.
7. Use a denitrification-volatilization nitrogen removal fraction, $f = 0.18$.

8. The annual nitrogen uptake rate for the selected vegetative cover, U, is 340 kg/ha-yr.
9. Assume that evaporation is equal to evapotranspiration.
Determine:
a. The annual design hydraulic loading rate, cm/yr.
b. The field area required, in ha.
c. The wastewater storage volume in cubic meters, the required surface area in hectares, and the required depth in meters.

22.3 A rapid infiltration land treatment system is to be used to treat a municipal wastewater flow from 15,000 persons. The wastewater flow is 100 gal/cap-day. The wastewater will be pretreated using preliminary treatment and primary sedimentation. Site characteristics, treatment objectives, and wastewater data are:
1. The minimum soil infiltration rate measured by the basin infiltration method is 1.5 in./hr. Use a loading factor, L_f of 15%.
2. The climate is mild; however, winter operation will control.
3. A well-nitrified renovated water is desired, so use an application period of 1 day and a drying period of 7 days.
4. Seasonal variation in wastewater flow is negligible.
5. After wastewater pretreatment, BOD$_5$ = 160 mg/ℓ, suspended solids = 85 mg/ℓ, total nitrogen as N = 25 mg/ℓ, and total phosphorus as P = 5 mg/ℓ.
Determine:
a. The design annual hydraulic loading rate, m/yr.

TABLE 22.14   *Data for Problem 22.2*

MONTH	OPERATING DAYS/MONTH	EVAPOTRANSPIRATION (cm/mo)	PRECIPITATION (cm/mo)
Jan	11	2.5	3.6
Feb	21	4.8	3.0
Mar	26	8.9	3.0
Apr	30	12.7	2.5
May	31	18.3	1.3
Jun	30	21.8	0.5
Jul	31	23.6	0.3
Aug	31	22.9	0.5
Sep	30	14.2	0.5
Oct	30	11.4	1.0
Nov	27	6.4	2.0
Dec	8	2.0	3.0

b. The wastewater application rate, ft/day.

c. The required infiltration area, ac.

d. The annual loadings of BOD_5, suspended solids, nitrogen, and phosphorus, lb/ac-yr.

22.4 A rapid infiltration land treatment system is to be used to treat a municipal wastewater flow from 15,000 persons. The wastewater flow is $380\,\ell$/cap-day. The wastewater will be pretreated using preliminary treatment and primary sedimentation. Site characteristics, treatment objectives, and wastewater data are:

1. The minimum soil infiltration rate measured by the basin infiltration method is 4 cm/h. Use a loading factor, L_f of 15%.

2. The climate is mild; however, winter operation will control.

3. A well-nitrified renovated water is desired, so use an application period of 1 day and a drying period of 7 days.

4. Seasonal variation in wastewater flow is negligible.

5. After wastewater pretreatment, $BOD_5 = 160$ mg/ℓ, suspended solids = 85 mg/ℓ, total nitrogen as N = 25 mg/ℓ, and total phosphorus as P = 5 mg/ℓ.

Determine:

a. The design annual hydraulic loading rate, m/yr.

b. The wastewater application rate, m/d.

c. The required infiltration area, ha.

d. The annual loadings of BOD_5, suspended solids, nitrogen, and phosphorus, kg/ha-yr.

22.5 An overland flow land treatment system is to be used to treat a municipal wastewater flow from 15,000 persons. The wastewater flow is 100 gal/cap-day. The wastewater will be pretreated using preliminary treatment and primary sedimentation. The system is to be operated year-round. Use an application rate of 0.15 gal/min-ft for 8 hr per day, use a terrace length of 115 ft, and assume that no winter storage will be required. Determine the required terrace area in acres.

22.6 An overland flow land treatment system is to be used to treat a municipal wastewater flow from 15,000 persons. The wastewater flow is $380\,\ell$/cap-d. The wastewater will be pretreated using preliminary

treatment and primary sedimentation. The system is to be operated year-round. Use an application rate of $0.11\,m^3$/h-m for 8 h per day, use a terrace length of 35 m, and assume that no winter storage will be required. Determine the required terrace area in hectares.

22.7 Anaerobically digested sludge from a wastewater treatment facility is to be treated by land application to an agricultural site at which potatoes will be grown. The wastewater treatment plant serves a population of 15,000 persons and produces 330 tons of dry sludge per year. The sludge will be applied as a liquid by surface distribution. The design is to be based on the nitrogen needs of the potato crop and is to be limited by a cumulative cadmium loading of 8 lb Cd/ac for the useful life of the site. The sludge contains 90 mg Cd/kg of dry sludge, 1.5% ammonia nitrogen, 1.5% organic nitrogen, and no nitrate nitrogen. Determine:

a. The allowable annual dry sludge application rate, tons dry sludge/ac-yr.

b. The land area required for sludge application, ac.

c. The useful life of the site, in yr.

22.8 Anaerobically digested sludge from a wastewater treatment facility is to be treated by land application to an agricultural site at which potatoes will be grown. The wastewater treatment plant serves a population of 15,000 persons and produces 300 metric tons of dry sludge per year. The sludge will be applied as a liquid by surface distribution. The design is to be based on the nitrogen needs of the potato crop and is to be limited by a cumulative cadmium loading of 10 kg Cd/ha for the useful life of the site. The sludge contains 90 mg Cd/kg of dry sludge, 1.5% ammonia nitrogen, 1.5% organic nitrogen, and no nitrate nitrogen. Determine:

a. The allowable annual dry sludge application rate, mt dry sludge/ha-yr.

b. The land area required for sludge application, ha.

c. The useful life of the site, yr.

23

OTHER UNIT OPERATIONS AND PROCESSES

This chapter presents a brief coverage of several water or wastewater treatment unit operations and processes that are not discussed in other chapters or that are only mentioned. This chapter is intended to be introductory, and only some of the most common treatments are presented.

WATER TREATMENT

Fluoridation, defluoridation, iron and manganese removal, and removal of volatile materials from groundwater are unit operations and processes that are not presented in other chapters. These operations and processes are introduced here.

Fluoridation

A fluoride content of 0.7 to 1.2 mg/ℓ in drinking water has been found to be beneficial because it results in the development of a hard, decay-resistant enamel during the formation of permanent teeth. Fluoride is usually added in the form of sodium fluoride salt, although hydrofluorosilicic acid and sodium silico-fluoride salt have been used.

Defluoridation

Many waters, in particular groundwaters, may have excessive amounts of the fluoride ion, F^-. It is usually present as the free ion, but sometimes it is complexed with polyvalent cations. Fluoride concentrations greater than about 4 mg/ℓ may result in the mottling of tooth enamel during the formation of permanent teeth. Although several methods have been tried for fluoride removal, the most successful has been adsorption by calcined alumina. Regeneration of the adsorbent requires the use of a strong base, such as sodium hydroxide, followed by a sulfuric acid rinse. This process has resulted in reduction of the fluoride ion from about 8 to 1 mg/ℓ at Bartlett, Texas (Steel and McGhee, 1979).

Iron and Manganese Removal

Iron and manganese may be present in groundwaters, particularly from deep aquifers, as divalent ions. The most commonly practiced removal involves oxidation of the iron and manganese to form precipitates that are removed by settling or filtration. Sufficient time must be provided for oxidation, and lime is often used for pH adjustment to reduce the time required for adequate oxidation. Oxidation may be accomplished by aeration, chlorination, ozonation, or the addition of a potassium permanganate solution. Other removal methods use sodium or hydrogen cation exchangers or filtration through a potassium permanganate medium (Hampel and Hawley, 1973; Montgomery, 1985).

Iron present in groundwaters having zero dissolved oxygen usually will be in the ferrous state. After aeration it will be oxidized to the ferric state and will precipitate as ferric oxide. Thus treatment usually consists of aeration, settling, and filtration. Manganese can be oxidized to an insoluble state by passing through beds of potassium permanganate. Thus treatment consists of oxidation, settling, and filtration. Iron present as the ferrous ion in low concentrations in groundwaters having zero dissolved oxygen may be complexed by certain polyphosphates, such as sodium hexametaphosphate, so that it will not oxidize and precipitate. For this to be successful, the complexing agent must be added at the well prior to aeration or chlorination. Iron and manganese in surface waters can be removed by rapid sand filtration plants. Iron removal and manganese removal are discussed in *Water Quality and Treatment* (AWWA, 1990) and *Water Treatment Principles and Design* (Montgomery, 1985).

Removal of Volatile Contaminants from Groundwater

Volatile organic compounds (VOCs) found in groundwater may be removed by pumping the water from wells to aeration facilities where air stripping is employed to remove the volatile components from the water. Some of the VOCs occurring in groundwater are dichloroethene, trichloroethene, tetrachloroethene, benzene, toluene, ethyl benzene, and *m*-, *o*-, and *p*-xylene. The combination of benzene, toluene, ethyl benzene, and the xylene isomers is often referred to as BTEX. Technology evaluations by Joel and Gerbasi (1992) and Thomas *et al.* (1992) for the treatment of contaminated groundwaters have included consideration of air stripping for removal of volatile components. The evaluations concluded that air stripping was effective for volatile component removal; however, these evaluations noted that off-gases produced by air stripping must meet ambient air guideline concentrations and may require treatment to reduce emissions.

Air stripping, discussed in Chapter 11 for ammonia removal, is based on equilibrium relationships for a gas-liquid system that are given by Henry's Law and on the fundamentals of gas mass transfer. These basic principles are applicable to air stripping for VOC removal. Numerous air stripping techniques to accomplish removal of gases from water exist. Among the various methods are diffused aeration in contact tanks, coke tray aerators, countercurrent packed towers, and cross-flow towers (Kavanaugh and Trussell, 1989). Packed towers are commonly used to remove VOCs from groundwaters serving as sources of drinking water. VOCs can also be removed by granular activated carbon columns or beds.

WASTEWATER TREATMENT

Flotation as a means of thickening sludge is presented in Chapter 21; its application is expanded, and the design procedure presented, in this chapter. Also presented here are the anaerobic contact process, the submerged anaerobic filter, biological phosphorus removal (mentioned in Chapter 11), aquatic systems, and septic tank–soil adsorption systems.

Flotation Low-density solid or liquid particles may be separated from a liquid by flotation. In this operation, fine air bubbles are introduced into the liquid, resulting in the attachment of the bubbles to the particles. The attached bubbles cause the particles to rise to the liquid surface, where they are removed by skimming.

Flotation may be accomplished by three methods of air introduction. The first is the injection of compressed air through diffusers in the flotation tank to buoy the particles to the surface. This has not been highly successful except for flotating grease in wastewaters having an unusually high grease content. The second method consists of supersaturating the wastewater flow with air gases and then sending the flow to an open flotation tank, thus releasing the pressure. The air gases come out of solution and form minute bubbles around the solid and liquid particles, thus causing them to float to the surface where they are removed by skimming. For small installations, the entire flow is injected with air, as shown in Figure 23.1 (a) and then pumped under a pressure of 30 to 60 psi (21 to 42 kPa) to a closed retention tank with several minutes detention time. From there the flow goes to

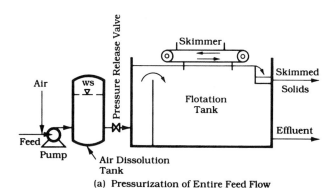

(a) Pressurization of Entire Feed Flow

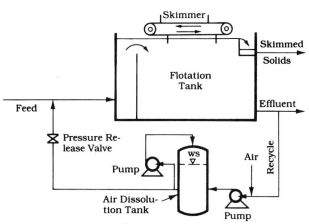

FIGURE 23.1
Air Flotation Systems

(b) Pressurization of Recycled Flow

the flotation tank. For large installations, 30% to 150% of the effluent is recycled and pressurized to become supersaturated with air gases, as in Figure 23.1 (b). The recycle is then mixed with the main wastewater flow, and the mixture goes to the flotation tank. The third method consists of saturating the main wastewater flow with air gases by air injection as the flow goes to a covered tank where a small vacuum is drawn by a vacuum pump. This causes the air gases to come out of solution, flotation occurs, and particle removal is effected by skimming. Frequently, chemical coagulants are added ahead of the flotation operation to enhance the efficiency. A dissolved air flotation unit employing pressurized recycle is shown in Figure 23.2.

Flotation has been used to remove oil emulsions from wastewaters, thicken biological sludges, and remove suspended solids from secondary effluents, in addition to numerous applications in industrial wastewater treatment. The use of flotation to thicken sludges is presented in Chapter 21.

The principal parameter in the design of a dissolved air flotation system is the air-to-solids ratio, A/S, which is expressed as mass/mass. To determine the optimum A/S ratio, bench-scale batch flotation experiments are used; the procedures are outlined by Eckenfelder and Ford (1970). The air-to-solids ratio for a system in which all the wastewater flow is pressurized is

$$\frac{A}{S} = \frac{1.3a_s(fP - 1)}{S_s} \qquad (23.1)$$

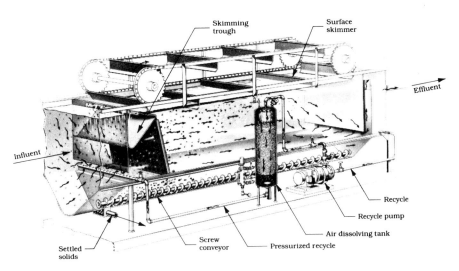

FIGURE 23.2 *Dissolved Air Flotation Unit with Pressurized Recycle*

Courtesy of Envirex Inc.

where

A/S = air to solids ratio, mg/mg

a_s = air solubility, mℓ/ℓ

f = fraction of air dissolved at a given pressure, usually 0.5 to 0.8

P = absolute pressure in atmospheres

S_s = suspended solids concentration, mg/ℓ

For a system where pressurized recycle is used, Eq. (23.1) is modified as follows:

$$\frac{A}{S} = \frac{1.3a_s(fP - 1)R}{S_sQ} \qquad (23.2)$$

where

R = recycle flowrate

Q = wastewater flowrate

The solubility of air at 0°C, 10°C, 20°C, and 30°C is 29.2, 22.8, 18.7, and 15.7 mℓ/ℓ, respectively. Figure 23.3 shows a schematic of the laboratory equipment required. For a nonrecycle system, the air dissolving chamber is partially filled with the wastewater, and the vessel is pressurized with compressed air, shaken, and then allowed to sit to absorb air. The pressure is then released, causing flotation of the solids. The floated sludge and the clarified liquid are withdrawn, the suspended solids are determined, and the air-to-solids ratio is computed. The test is repeated at several different air pressures to give a curve of A/S versus effluent suspended solids concentration. For a recycle system, the air dissolving chamber is filled partly with effluent, pressurized with air, shaken, and allowed to sit to dissolve the air.

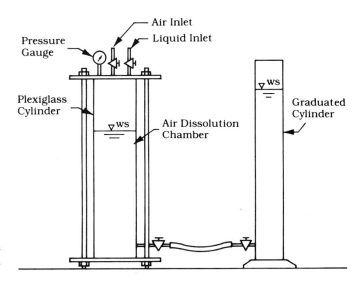

FIGURE 23.3
Flotation Test Equipment

The saturated liquid is released by a tube in the bottom of a graduated cylinder containing the wastewater. Flotation occurs, and the floated sludge and clarified liquid are withdrawn and the suspended solids are determined. The experiment is repeated at several pressures to obtain a curve of A/S versus effluent suspended solids. The amount of pressurized effluent used depends on the recycle ratio. If R/Q is $\frac{1}{4}$ and 800 mℓ of wastewater are used, then the saturated effluent volume to be used is (800) $\frac{1}{4}$ or 200 mℓ. In design, the area of the flotation tank is based on an overflow rate of 0.2 to 4.0 gal/min-ft² (12 to 235 m³/day-m²).

EXAMPLE 23.1
AND 23.1 SI

Air Flotation Without Recycle

A wastewater flow of 0.50 MGD (1900 m³/d) has 200 mg/ℓ suspended solids. Air flotation tests show that 0.05 mg air/mg solids gives optimum flotation. The design temperature is 20°C, and $a_s = 18.7$ mℓ/ℓ at this temperature. The fraction of absorption is 0.5, and the overflow rate is 2.0 gal/min-ft² (117 m³/d-m²). Determine the required pressure and flotation tank area.

SOLUTION

Substituting into Eq. (23.1) gives

$$0.05 = \frac{(1.3)(18.7)(0.5P - 1)}{200}$$

or

$$\boxed{P = 2.82 \, \text{atm}}$$

$$\text{Plan area} = (500,000 \, \text{gal/day})(\text{day}/1440 \, \text{min})$$
$$\times \, (\text{min-ft}^2/2.0 \, \text{gal})$$
$$= \boxed{174 \, \text{ft}^2}$$

The SI solution for flotation tank area is

$$\text{Plan area} = (1900 \, \text{m}^3/\text{d})(\text{d-m}^2/117 \, \text{m}^3)$$
$$= \boxed{16.2 \, \text{m}^2}$$

EXAMPLE 23.2
AND 23.2 SI

Air Flotation with Recycle

For the data in Example 23.1, determine the recycle ratio if the operating pressure is 3.00 atm. Also determine the flotation tank area.

$$0.05 = \frac{(1.3)(18.7)(0.5 \times 3.00 - 1)R}{200Q}$$

or

$$\boxed{R/Q = 0.82}$$

$$\text{Plan area} = (500{,}000\,\text{gal/day})(\text{day}/1440\,\text{min})$$
$$\times (1 + 0.82)(\text{min-ft}^2/2.0\,\text{gal})$$
$$= \boxed{316\,\text{ft}^2}$$

The SI solution for flotation tank area is

$$\text{Plan area} = (1900\,\text{m}^3/\text{d})(1 + 0.82)(\text{d-m}^2/117\,\text{m}^3)$$
$$= \boxed{29.6\,\text{m}^2}$$

Oil Separation Since oils are relatively insoluble in water, they frequently may be separated from a wastewater by gravity separators, such as the American Petroleum Institute (API) separator. If oils are present as an emulsion, their effective removal usually requires coagulation followed by gravity separation or air flotation.

Anaerobic Contact Process This is essentially an anaerobic activated sludge treatment process, as shown in Figure 23.4. The influent wastewater and the recycled biological solids go to a covered, continuously mixed anaerobic reactor where most of the organic matter is biologically oxidized. The mixed liquor goes to a covered clarifier where the biological solids are separated from the supernatant, which leaves as the effluent. The settled biological solids are recycled and mixed with the influent flow. The off-gases of CH_4 and CO_2 are vented from the system. Because of the low cell synthesis in anaerobic systems, the waste sludge volume is much smaller than from an activated sludge system. This process has been mainly used for high-BOD_5 wastewaters.

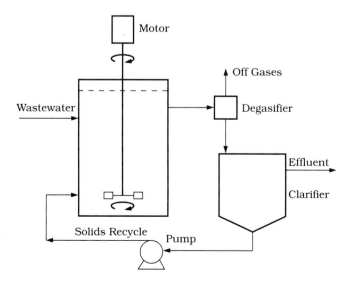

FIGURE 23.4
Anaerobic Contact Process

Submerged
Anaerobic Filter

This consists of an anaerobic filter, as shown in Figure 23.5, with a packing, usually of stone. The filter is totally enclosed to maintain anaerobic conditions. The wastewater enters the bottom of the filter and passes upward through the media, which has fixed-microbial growths. The organic matter is biologically oxidized by anaerobic biochemical action by the fixed growths. The effluent leaves from the top of the unit and the off-gases of CH_4 and CO_2 are vented from the filter. This process has been used on low-BOD$_5$ wastewaters.

Biological
Phosphorus
Removal

Under usual conditions of secondary biological treatment, microorganisms use between 10% and 30% of the influent phosphorus for cell maintenance and synthesis during normal metabolism. Biological phosphorus removal beyond the normal amount used may be accomplished by providing alternating aerobic, anoxic, and anaerobic conditions in an appropriate sequence of reactors. *Acinetobacter*, one of the primary microorganisms responsible for phosphorus removal, release stored phosphorus under anaerobic conditions in the presence of volatile fatty acids. When subjected to an aerobic environment following anaerobic conditions, the *Acinetobacter* remove more phosphorus from the wastewater than is required for normal cell metabolism. The phenomenon of additional phosphorus uptake beyond that required for normal cell metabolism is termed **luxury uptake**. Phosphorus is ultimately removed from the wastewater either by wasting the sludge that contains the higher levels of phosphorus or by removing phosphorus from the sludge by treatment in a side stream. When the sludge is subjected to anaerobic conditions in a side stream, the phosphorus is released, or stripped, from the sludge; this process is termed **side stream stripping**. After side stream stripping, the phosphorus is removed from the wastewater by chemical treatment, usually with lime, to precipitate the phosphorus. Several biological treatment processes take advantage of the excess uptake and release of phosphorus by microorganisms that occurs when the microbes are stressed under alternating anaerobic and aerobic

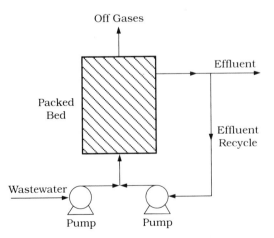

FIGURE 23.5
Submerged Anaerobic
Filter Process

conditions, as shown in Figure 23.6. Effluent total phosphorus concentrations of 1 to 2 mg/ℓ or less can be achieved by these processes (EPA, 1987; Metcalf & Eddy, 1991).

The proprietary five-stage Bardenpho process, discussed in Chapter 11, combines both nitrogen removal and phosphorus removal with carbonaceous BOD$_5$ removal. As shown in Figure 23.6 (a), anaerobic conditions exist in the first and second stage (anoxic), followed by an aerobic third stage, an anaerobic fourth stage (anoxic), and an aerobic fifth stage. The alternating anaerobic and aerobic conditions stress the microorganisms, which promotes luxury uptake of phosphorus. The biological sludge, containing additional phosphorus from luxury uptake, is wasted, resulting in phosphorus reduction

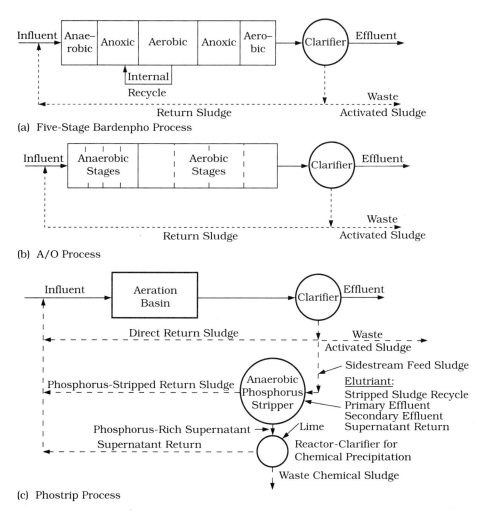

(a) Five-Stage Bardenpho Process

(b) A/O Process

(c) Phostrip Process

FIGURE 23.6 *Biological Phosphorus Removal Processes (EPA, 1987).*

in the clarified effluent. The five-stage Bardenpho process is usually designed for relatively low overall loading rates because of the longer detention time required for nitrification and denitrification (EPA, 1987). Another proprietary process, the A/O process, shown in Figure 23.6 (b), is similar to the Bardenpho process and also takes advantage of luxury uptake of phosphorus by subjecting the microorganisms to alternating anaerobic and aerobic conditions. As shown in Figure 23.6(b), anaerobic conditions followed by aerobic conditions are maintained in a reactor, and sludge is returned to the head of the reactor. The microorganisms are stressed by the alternating anaerobic and aerobic conditions, resulting in luxury uptake of phosphorus. Phosphorus is released in the anaerobic stages, and luxury uptake occurs in the aerobic stages of the reactor. Phosphorus is removed from the system by wasting sludge. The A/O process is usually designed as a high-rate activated sludge system. Phosphorus removal in these processes is dependent on the BOD_5-to-phosphorus ratio of the wastewater influent. Total BOD_5 to total phosphorus ratios greater than $20:1$ to $25:1$ have been recommended to achieve treated effluent soluble phosphorus concentrations of $1\,mg/\ell$ or less (EPA, 1987).

Other processes used for biological phosphorus removal also take advantage of luxury uptake of phosphorus by subjecting the microorganisms to alternating anaerobic and aerobic conditions but accomplish ultimate phosphorus removal by side stream stripping rather than only by wasting sludge. In one such process, the proprietary Phostrip process shown in Figure 23.6 (c), anaerobic conditions are provided in a side stream reactor, and alternating anaerobic and aerobic conditions to promote luxury uptake of phosphorus are achieved by returning sludge from the side stream anaerobic reactor. After phosphorus has been stripped from the sludge under anaerobic conditions, the supernatant from the side stream reactor is chemically treated, usually with lime, to precipitate phosphorus for ultimate removal of phosphorus. The phosphorus removal performance of the Phostrip process is controlled more by the side stream stripping and chemical treatment operation and is not as dependent as the five-stage Bardenpho process and the A/O process on the ratio of total BOD_5 to total phosphorus or on the organic loading rates. Effluent total phosphorus concentrations of less than $1\,mg/\ell$ can be achieved, and the Phostrip process produces a high-phosphorus-content lime sludge for which handling is of less concern than for a high-phosphorus-content biological sludge. However, the side stream stripping and chemical treatment operations may require more control and operator skill, and lime storage and handling can be a significant problem (EPA, 1987).

Aquatic Systems Centralized wastewater treatment facilities for both larger and smaller populations have traditionally consisted of tanks connected by pipes or open channels constructed in a relatively small area. Because of both construction and operation costs for the traditional wastewater treatment facilities, less costly, environmentally effective alternatives are evolving.

The U.S. Environmental Protection Agency (EPA, 1988) indicates that one such alternative is the use of aquatic treatment systems and discusses three aquatic treatment system categories: natural wetlands, aquatic plant systems, and constructed wetlands. **Natural wetlands** are, generally, part of or contiguous with natural waters and are regulated with the same, or with more stringent, criteria as the natural waters. Natural wetlands are, therefore, limited for wastewater treatment. **Aquatic plant systems** are shallow ponds with floating aquatic plants such as duckweed, water hyacinth, and pennywort or with submerged aquatic plants such as waterweed, water milfoil, and watercress. **Constructed wetlands** either free water surface systems or subsurface flow systems. The free water surface systems consist of a shallow basin or channel with an impervious layer to prevent seepage and a medium capable of supporting vegetative growth. Wastewater flows at a low velocity over the medium supporting the vegetation and through the vegetation stalks. In long, narrow channels, flow is essentially plug flow. The subsurface flow systems, often referred to as the root-zone method or rock-reed-filter, consist of an inclined trench or bed with an impervious layer to prevent seepage and a permeable medium, supporting vegetative growth, through which wastewater flows. Systems with permeable media of sand or rock have received the most study in the United States.

In the aquatic systems described, the principal treatment mechanisms are biological metabolism and sedimentation. The aquatic plants serve primarily to support an aquatic environment conducive to the conventional treatment mechanisms; there is little direct treatment by the plants. Some of the functions of the aquatic plants are to provide surfaces on which bacteria grow, to provide a medium for filtration and adsorption of solids, and to attenuate sunlight, which prevents the growth of algae. Not only are wetland systems effective for BOD_5, suspended solids, and nitrogen removal, but they can also effectively reduce metals, trace organics, and pathogens. Constructed wetlands technology has also been successfully used for treatment of acid mine drainage (EPA, 1993). For further reading, see *Natural Systems for Waste Management and Treatment* (1988), a textbook by S. C. Reed, E. J. Middlebrooks, and R. W. Crites.

Septic Tank–Soil Absorption Systems

Wastewater treatment by a conventional, centralized treatment facility for small suburban communities, and particularly in rural areas, is in most cases neither practical nor economically feasible. In recent years, migration from large centers of population in the United States to rural and suburban areas has increased the number of houses that are not served by conventional, centralized wastewater treatment facilities. Estimates place from 25% to 30% of the houses in the United States in this category, and the numbers are increasing (EPA, 1980; Kaplan, 1987). Properly designed, constructed, and maintained on-site treatment and disposal systems are an effective and economically feasible means of managing wastewater in these areas. The most common type of on-site wastewater treatment system is the septic tank–soil absorption system. This system consists of a septic tank, which receives raw wastewater from the source, and a soil absorption drainfield for disposal of liquid from the septic tank.

For an individual home or dwelling, the septic tank is usually a reinforced concrete or fiberglass tank with a capacity of 500 to 1500 gal (1900 to 5700 ℓ). The tank is usually a single-compartment tank, as shown in Figure 23.7, or, in some cases, a two-compartment tank. The two-compartment tank, which generally is used for extremely large tanks, has a second compartment in series with the first one to clarify the effluent better before it goes to the soil absorption system. The detention time required is usually 2 days. For homes with modern conveniences, such as garbage grinders, dish washers, and clothes washers, typical minimum recommended capacities are 750 gal (2840 ℓ) for a one- or two-bedroom house, 1000 gal (3790 ℓ) for a three-bedroom house, 1250 gal (4730 ℓ) for a four-bedroom house, and 250 gal (945 ℓ) more for each additional bedroom (GLUMRB, 1980). The function of the septic tank is to provide the initial treatment by receiving the household wastewater, separating and storing solids from the liquid, providing digestion of organic matter, and allowing the clarified liquid to exit the tank for further treatment and disposal. As shown in Figure 23.7, settleable solids accumulate at the tank bottom where decomposition takes place under anaerobic conditions. Floatable solids (fats and greases) rise to

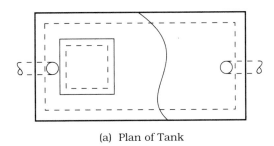

(a) Plan of Tank

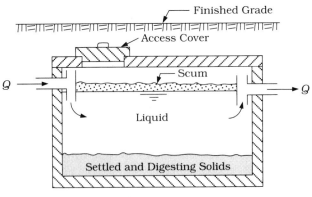

FIGURE 23.7
Septic Tank

(b) Section through Tank

the surface where limited microbial decomposition results in the formation of a scum layer. Between the sludge at the bottom and the scum layer at the top, a partially clarified liquid remains. The partially clarified liquid exits the septic tank through an outlet located just below the scum layer and proceeds to the soil absorption system for further treatment.

Proper operation and maintenance of the septic tank are essential. Because a septic tank has no moving parts, very little routine maintenance is required. A well-designed concrete, fiberglass, or plastic tank can be expected to last about 50 years. The major cause of failure of a septic tank to function properly is the failure to remove accumulated solids. Excessive solids accumulation reduces the effective volume and decreases the detention time, which reduces treatment efficiency and increases the amount of solids leaving the tank. Increased solids exiting the septic tank adversely affect performance of the soil absorption system.

There are several different types of soil absorption systems, including trenches and beds, seepage pits, mounds, fills, and artificially drained systems. All of these are soil excavations that are backfilled with porous media and covered with soil. The most commonly used disposal systems are trenches, as shown in Figure 23.8, and beds, as shown in Figure 23.9.

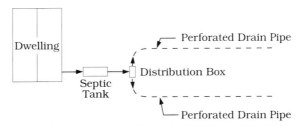

(a) Layout of Septic Tank and Trench Soil Absorption System

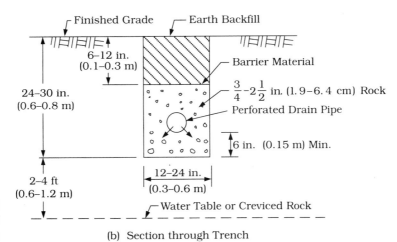

FIGURE 23.8
Typical Trench Soil Absorption System

(b) Section through Trench

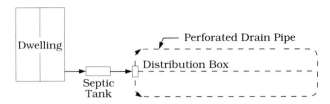

(a) Layout of Septic Tank and Bed Soil Absorption System

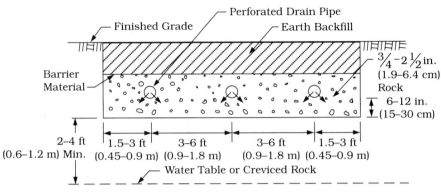

FIGURE 23.9
Typical Bed Soil Absorption System

(b) Section through Bed Soil Absorption System

Seepage pits are used when land area is limited or where soil underlying the upper 3 to 4 ft (0.9 to 1.2 m) is more suitable. Mound disposal systems are used when soil permeability or high water tables prevent use of the more commonly used bed or trench systems. Fill disposal systems may be used when unsuitable surface soils overlay sands or sandy loams and the highest water table or bedrock surface is within 1 ft (0.3 m) of the suitable soil surface. The upper, unsuitable soil is removed and replaced with a suitable fill material in which a trench or bed system is installed. Where seasonally high water tables prevent the use of beds, trenches, or pits, the water table may be artificially lowered, if possible, by use of curtain drains or vertical drains. Perforated piping is used to distribute the septic tank effluent into and throughout the porous media. The function of the soil absorption component is to receive the clarified effluent from the septic tank, to provide an environment in which additional decomposition of organic material and removal of pathogenic organisms occur, and to allow the liquid to disperse by percolation, evaporation, plant uptake, and transpiration.

Knowledge of the soil characteristics of the site is essential to designing the soil absorption component. Soil profile, permeability or hydraulic conductivity, proximity to groundwater, and variations in water table are critical factors that must be investigated. It is estimated that only about 32% of the land area in the United States is suitable for on-site disposal (EPA, 1980). The percolation test is used to determine the hydraulic capacity of soil to be used for on-site disposal. For trench-type soil absorption systems, GLUMRB

(1980) recommends the following drainfield areas per bedroom on the basis of the time required for the water surface to drop 1 in. (2.54 cm) in the percolation test: 0 to 5 min, 125 ft² (12 m²); 6 to 10 min, 165 ft² (15 m²); 11 to 30 min, 250 ft² (23 m²); 31 to 45 min, 300 ft² (28 m²); and 46 to 60 min, 330 ft² (31 m²). For bed-type soil absorption systems, GLUMRB (1980) recommends the following drainfield areas per bedroom on the basis of the time required for the water surface to drop 1 in. (2.54 cm) in the percolation test: 0 to 5 min, 250 ft² (23 m²); 6 to 10 min, 330 ft² (31 m²); and 11 to 30 min, 500 ft² (46 m²). For details on the septic tank and drainfield system, see *Recommended Standards for Individual Sewage Systems*, by the Great Lakes–Upper Mississippi River Board of State Sanitary Engineers (GLUMRB, 1980), and *Environmental Engineering and Sanitation* (1992), a textbook by J. A. Salvato.

REFERENCES

American Water Works Association (AWWA). 1989. *Air Stripping for Volatile Organic Contaminant Removal*. Denver, Colo.

American Water Works Association (AWWA). 1990. *Water Quality and Treatment*. 4th ed. Denver, Colo.

AWARE, Inc. 1974. *Process Design Techniques for Industrial Waste Treatment*, edited by C. E. Adams. Jr., and W. W. Eckenfelder, Jr. Nashville, Tenn.: Enviro Press.

Barth, E. F.; Brenner, R. C.; and Lewis, R. F. 1968. Chemical-Biological Control of Nitrogen and Phosphorus in Wastewater Effluents. *Jour. WPCF* 40:2040.

Culp, G. L. 1975. *Treatment Processes for Wastewater Reclamation for Groundwater Recharge*. Preliminary Report for the State of California, Department of Water Resources.

Culp, R. L.; Westner, G. M.; and Culp, G. L. 1978. *Handbook of Advanced Wastewater Treatment*. 2nd ed. New York: Van Nostrand Reinhold.

Eckenfelder, W. W., Jr. 1966. *Industrial Water Pollution Control*. New York: McGraw-Hill.

Eckenfelder, W. W., Jr. 1989. *Industrial Water Pollution Control*. 2nd ed. New York: McGraw-Hill.

Eckenfelder, W. W., Jr. 1980. *Principles of Water Quality Management*. Boston: CBI Publishing.

Eckenfelder, W. W., Jr. 1970. *Water Quality Engineering for Practicing Engineers*. New York: Barnes & Noble.

Eckenfelder, W. W., Jr., and Ford, D. L. 1970. *Water Pollution Control*. Austin, Tex.: Pemberton Press.

Eliassen, R., and Tchoganoglous, G. 1969. Removal of Nitrogen and Phosphorus from Waste Water. *Environmental Science and Technology* 3, no. 6:536.

Environmental Protection Agency (EPA). 1977a. Alternatives for Small Wastewater Treatment Systems, On-Site Disposal/Septage Treatment and Disposal. EPA Technology Transfer Seminar Publication, Washington, D.C.

Environmental Protection Agency (EPA). 1977b. Alternatives for Small Wastewater Treatment Systems, Pressure Sewers/Vacuum Sewers. EPA Technology Transfer Seminar Publication, Washington, D.C.

Environmental Protection Agency (EPA). 1977c. Alternatives for Small Wastewater Treatment Systems, Cost/Effectiveness Analysis. EPA Technology Transfer Seminar Publication, Washington, D.C.

Environmental Protection Agency (EPA). 1988. *Constructed Wetlands and Aquatic Plant Systems for Municipal Wastewater Treatment*. EPA Design Manual, Washington, D.C.

Environmental Protection Agency (EPA). 1993. *Handbook for Constructed Wetlands Receiving Acid Mine Drainage*. EPA Superfund Innovative Technology Evaluation Emerging Technology Summary, Cincinnati, Ohio.

Environmental Protection Agency (EPA). 1980. *Onsite Wastewater Treatment and Disposal Systems*. EPA Design Manual, Washington, D.C.

Environmental Protection Agency (EPA). 1987. *Phosphorus Removal*. EPA Design Manual, Washington, D.C.

Environmental Protection Agency (EPA). 1979. Small

Wastewater Treatment Facilities, EPA Design Seminar Handout, Washington, D.C.

Envronmental Protection Agency (EPA). 1979. Small Wastewater Treatment Facilities, EPA Design Seminar Handout, Washington, D.C.

Environmental Protection Agency (EPA). 1974. *Upgrading Existing Wastewater Treatment Plants.* EPA Process Design Manual, Washintong, D.C.

Environmental Protection Agency (EPA). 1992. *Wastewater Treatment/Disposal for Small Communities.* EPA Manual, Washington, D.C.

Fair, G. M.; Geyer, J. C.; and Okun, D. A. 1968. *Water and Wastewater Engineering.* Vol. 2. *Water Purification and Wastewater Treatment and Disposal.* New York: Wiley.

Great Lakes–Upper Mississippi River Board of State Public Health and Environmental Managers. 1990. Recommended Standards for Wastewater Facilities. Albany, N.Y.: Health Education Services.

Great Lakes–Upper Mississippi River Board of State Sanitary Engineers (GLUMRB). 1980. Recommended Standards for Individual Sewage Systems. Albany, N.Y.

Hampel, C. A., and Hawley, G. G. 1973. *The Encyclopedia of Chemistry*, 3rd ed. New York: Van Nostrand Reinhold.

Joel, A. R., and Gerbasi, P. 1992. Gasoline Contaminated Ground Water Treatment Study. *Proceedings of the 47th Purdue Industrial Waste Conference*, West Lafayette, Ind.

Kaplan, O. B. 1987. *Septic Systems Handbook.* Chelsea, Mich.: Lewis Publishers.

Karpiscak, M. M.; Foster, K. E.; Hopf, S. B.; Bancroft, J. M.; and Warshall, P. J. 1994. Using Water Hyacinth to Treat Municipal Wastewater in the Desert Southwest. *Water Resources Bulletin* 30, no. 2:219.

Kavanaugh, M. C., and Trussell, R. R. 1989. Design of Aeration Towers to Strip Volatile Contaminants from Drinking Water. *Air Stripping for Volatile Organic Contaminant Removal*, 1989. Denver, Colo.: AWWA, p. 43.

Metcalf & Eddy, Inc. 1979. *Wastewater Engineering: Treatment, Disposal and Reuse.* 2nd ed. New York: McGraw-Hill.

Metcalf & Eddy, Inc. 1991. *Wastewater Engineering: Treatment, Disposal and Reuse.* 3rd ed. New York: McGraw-Hill.

Montgomery, J. M., Consulting Engineers, Inc. 1985. *Water Treatment Principles and Design.* New York: Wiley.

Piluk, R. J., and Oliver, J. H. 1989. Evaluation of On-Site Waste Disposal System for Nitrogen Reduction. *Jour. of Environmental Engineering* 115, no. 4:725.

Ramadori, R., ed. 1987. *Biological Phosphate Removal from Wastewaters.* New York: Pergamon Press.

Reed, S. C., and Brown, D. S. 1992. Constructed Wetlands Design — The First Generation. *Water Environment Research* 64, no. 6:776.

Reed, S. C.; Middlebrooks, E. J.; and Crites, R. W. 1988. *Natural Systems for Waste Management and Treatment.* New York: McGraw-Hill.

Salvato, J. A. 1992. *Environmental Engineering and Sanitation*, 4th ed. New York: Wiley.

Sanks, R. L. 1978. *Water Treatment Plant Design.* Ann Arbor, Mich.: Ann Arbor Science Publishers.

Schroeder, E. D. 1977. *Water and Wastewater Treatment.* New York: McGraw-Hill.

Stark, L. R.; Brooks, R. P.; Williams, F. M.; Stevens, S. E., Jr.; and Davis, L. K. 1994. Water Quality During Storm Events from Two Constructed Wetlands Receiving Mine Drainage. *Water Resources Bulletin* 30, no. 4:639.

Steel, E. W., and McGhee, T. J. 1979. *Water Supply and Sewerage*, 6th ed. New York: McGraw-Hill.

Thomas, A. J.; Adams, C. E., Jr.; Falco, R. J.; Campaigne, R. T.; and Hurd, S. R. 1992. Remediation Evaluation of Innovative Technologies for Low Organic Strength Groundwater. *Proceedings of the 47th Purdue Industrial Waste Conference*, West Lafayette, Ind.

Vesilind, P. A.; Peirce, J. J.; and Weiner, R. F. 1990. *Environmental Pollution and Control.* 3rd ed. Stoneham, Mass.: Butterworth-Heinemann.

Wanielista, M. P., and Eckenfelder, W. W., Jr. 1979. *Advances in Water and Wastewater Treatment, Biological Nutrient Removal.* Ann Arbor, Mich.: Ann Arbor Science.

Water Environment Federation (WEF) and American Society of Civil Engineers (ASCE). 1992. *Design of Municipal Wastewater Treatment Plants.* WEF MOP No. 8 and ASCE Manual and Report on Engineering Practice No. 76. Brattleboro, Vt: Book Press.

Water Pollution Control Federation (WPCF). 1977. *Wastewater Treatment Plant Design.* WPCF Manual of Practice No. 8. Washington, D.C.

Winneberger, J. H. T. 1984. *Septic-Tank Systems: A Consultant's Toolkit.* Vol. I. *Subsurface Disposal of Septic-Tank Effluents.* Boston: Butterworth.

Wright, F. B. 1977. *Rural Water Supply and Sanitation.* 3rd ed. Huntington, N.Y.: Robert E. Krieger Publishing Co.

Zachritz, W. H., II, and Fuller, J. W. 1993. Performance of an Artificial Wetlands Filter Treating Facultative Lagoon Effluent at Carville, Louisiana. *Water Environment Research* 65, no. 1:46.

PROBLEMS

23.1 The maintenance shops at a large international airport have a process wastewater flow of 0.4 MGD and a free and emulsified oil content of 180 mg/ℓ as oil and grease. Air flotation using pressurized recycle and a polymer coagulant is to be employed to lower the oil and grease to 50 mg/ℓ. The flotation effluent is to be mixed with the sanitary wastewater and treated at an activated sludge plant owned by the airport authority. Laboratory experiments have been done to find the optimum coagulant dosage. Using this dosage, laboratory batch air flotation studies have been performed to determine the air required. The data obtained are shown in Table 23.1. The fraction of air dissolved in a retention tank having a 1-min detention time is 0.5. Other pertinent data are: average operating temperature = 20°C, air pressure = 45 psia, and overflow rate based on $Q + R$ = 3.0 gpm/ft². Determine:

a. The recycle flow R, MGD.
b. The area of the flotation tank.
c. The pounds of air required per day.
d. The cubic feet of air required per day.

23.2 The maintenance shops at a large international airport have a process wastewater flow of 1500 m³/d and a free and emulsified oil content of 180 mg/ℓ as oil and grease. Air flotation using pressurized recycle and a polymer coagulant is to be employed to lower the oil and grease to 50 mg/ℓ. The flotation effluent is to be mixed with the sanitary wastewater and treated at an activated sludge plant owned by the airport authority. Laboratory experiments have been done to find the optimum coagulant dosage. Using this dosage, laboratory batch air flotation studies have been performed to determine the air required. The data obtained are

shown in Table 23.1. The fraction of air dissolved in a retention tank having a 1-min detention time is 0.5. Other pertinent data are: average operating temperature = 20°C, air pressure = 310 kPa absolute, and overflow rate based on $Q + R$ = 176 m³/d-m². Determine:

a. The recycle flow R, ℓ/d.
b. The area of the flotation tank.
c. The kilograms of air required per day.
d. The cubic meters of air required per day.

23.3 A design of a dissolved air flotation unit without recycle is to be made for the airport and conditions described in Problem 23.1. Determine:

a. The area of the flotation tank.
b. The air pressure required in atmospheres and psia.

23.4 A design of a dissolved air flotation unit without recycle is to be made for the airport and conditions described in Problem 23.2. Determine:

a. The area of the flotation tank.
b. The air pressure required in atmospheres and kPa absolute.

TABLE 23.1

AIR/SOLIDS (gm/gm)	EFFLUENT OIL AND GREASE (mg/ℓ)
0.056	12
0.042	27
0.028	47
0.015	77
0.008	113
0.003	155

24

DISINFECTION

Disinfection is the destruction of pathogenic microorganisms. It does not apply to nonpathogenic microorganisms or to pathogens that might be in the spore state. The term that applies to the destruction of all living organisms, and especially to microorganisms including spores, is **sterilization** (McCarthy and Smith, 1974).

In the United States, water and wastewater disinfection is accomplished almost solely by chlorination. Very little use is made of other disinfection techniques. Findings of recent years, however, relating to the generation of undesirable trihalomethanes (Rook, 1977) and other chlorinated organic products (Jolley, 1975) by the chlorination process suggest that future disinfection practice in the United States might well be modified to lessen the magnitude of these undesirable consequences.

The chlorination of drinking water supplies was introduced in the United States in 1908, and by 1914 "the greater part of the water supplied in cities in the U.S. [was] treated in this [or an equivalent] way . . ." (National Academy of Sciences, 1977). As a result, a marked decrease in the incidence of water-borne diseases occurred. The Mills-Reincke theorem holds that for every death from water-borne typhoid, there are several deaths from other diseases for which the causal agents are transmitted by water. The National Academy of Sciences Safe Drinking Water Committee (1977) observes that the theorem appears to have considerable merit, but although disinfection to levels based on coliform criteria has ensured freedom from typhoid fever, it does not give a similar guarantee of freedom from other infections.

Disinfection is described by Chang (1971) as a complex rate process dependent upon:

1. the physico-chemistry of the disinfectant
2. the cyto-chemical nature and physical state of the pathogens
3. the interaction of 1 and 2 above
4. quantitative effects of factors in the reaction medium, such as temperature, pH, electrolytes, and interfering substances.

He also classified disinfectants as follows:

1. oxidizing agents (ozone, halogens, halogen compounds)
2. cations of heavy metals (silver, gold, mercury)
3. organic compounds
4. gaseous agents
5. physical agents (heat, UV and ionizing radiation, pH).

The cyto-chemical nature and physical state of the pathogens encompasses a broad range of organisms categorized by Engelbrecht and Hendricks (1982) as including bacteria, viruses, protozoa, and helminths. Bacterial (and fungal) spores are much more resistant to disinfection than are vegetative forms. Spore-formers, however, are usually not important in water and wastewater disinfection. The same might be said of some bacterial pathogens, such as the tuberculosis organism, which are more resistant than the Gram-negative coliform cells that serve as our criterion of water and wastewater disinfection. Finally, cysts and ova of protozoa and helminths are also more resistant to disinfection than vegetative bacterial cells, but other treatment processes, notably flocculation, sedimentation, and filtration, are usually quite effective in their removal, when they are working properly.

Among the more important differences that exist between various microorganisms and their resistance to disinfection processes are those relating to various viruses on the one hand and coliforms on the other. The former, in general, are much more resistant to chloramines and are also more resistant under many naturally occurring circumstances, such as when they are embedded or enmeshed in tissues or suspended material or when they are aggregated in clumps. A technique often used to measure the effectiveness of various disinfection processes has been to seed with substantial numbers of viruses, but these seeded viruses have been found to be considerably more vulnerable than the viruses that occur naturally.

The rate of disinfection by a chemical agent may in many cases conform to Chick's law of disinfection (1908), which is presented by the following pseudo–first-order reaction:

$$-\frac{dN}{dt} = kN \tag{24.1}$$

where

dN/dt = rate of cell destruction, number/time

k = rate constant

N = number of living cells remaining at time t

The constant, k, depends on the species and form of the microorganisms being destroyed, the disinfectant and its nature, the concentration of the disinfectant, and environmental factors such as pH and temperature. Chick's law is limited in application, since in many cases, it has been found that the rate of kill in the latter duration of the disinfection process may be more or less than indicated by Eq. (24.1).

An empirical equation that relates the concentration of a disinfectant and the contact time is

$$C^n t_c = K \tag{24.2}$$

where

C = concentration of disinfectant at time $t = 0$

t_c = time of contact required to kill a given percentage of the microbes

K, n = experimental constants

The value of n depends on the nature of the disinfectant used. If n is greater than 1, the disinfecting action is greatly dependent on the concentration of the disinfectant. Conversely, if n is less than 1, the disinfectant action is primarily dependent on the time of contact. The value of the constant, K, depends on the types of microorganisms being destroyed and on environmental factors such as pH and temperature. For the usual range of microbe concentration encountered in water treatment, microbial concentration usually has very little effect on Eq. (24.2). Equation (24.2) plots as a straight line on log-log paper, as shown in Figures 24.1 and 24.2. Figure 24.1 illustrates the effectiveness of the various forms of chlorine for *Escherichia coli* destruction. It can be seen that the free chlorine forms are much more effective than combined chlorine in the form of monochloramine. Figure 24.2 shows the effectiveness of free chlorine for the destruction of *E. coli* and three enteric viruses. As depicted, two of the viruses were more resistant than *E. coli*.

Temperature affects disinfecting action because an increase in temperature results in a more rapid kill. If organic matter is present, chemical disinfecting reagents may react with it, thus reducing the effective concentration of the disinfectant. In chlorination particularly, pH is important because it influences the relative distribution of agents of varying effectiveness, as will be discussed. The currently used disinfection models are

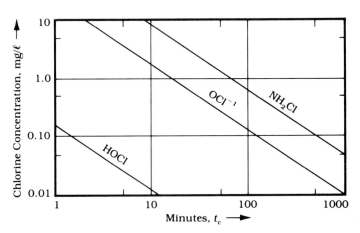

FIGURE 24.1 *Concentration versus Contact Time for 99% Kill of* **E. coli** *by Various Forms of Chlorine at 2°C to 6°C*

Adapted from "Influence of pH and Temperature on the Survival of Coliforms and Enteric Pathogens When Exposed to Free Chlorine" by C. T. Butterfield, E. Wattie, S. Megregian, and C. W. Chambers in *U.S. Public Health Reports* 58 (1943):1837.

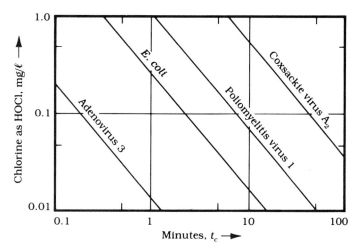

FIGURE 24.2 *Concentration versus Contact Time for 99% Kill of* **E. coli** *and Three Enteric Viruses by HOCl at 0°C to 6°C*

Adapted from "The Virus Hazard in Water Supplies" by G. Berg in *Journal of New England Water Works Association* 78 (1964):79. Reprinted by permission.

particularly deficient when applied to long contact periods (Engelbrecht and Hendricks, 1982).

The generation of undesirable organo-chlorines in the chlorination process and the increasing need to treat used waters (that is, wastewaters) for subsequent reuse is resulting in study and consideration of other disinfection methods. Most prominent among these are ozonation, chlorine dioxide, uv irradiation, high-pH treatment, and the use of other halogens, such as iodine and bromine. The latter two are not discussed in the following text.

CHLORINATION

Chlorine is the most widely used disinfectant because it is effective at low concentration, is cheap, and forms a residual if applied in sufficient dosage. It may be applied as a gas or as a hypochlorite, the gas form being more common. The gas is liquified at 5 to 10 atm and shipped in steel cylinders. Pressurized liquid chlorine (99.8% Cl_2) is available in cylinders containing 100, 150, or 2000 lb (45, 68, or 910 kg) of the liquefied gas. The disinfecting ability of chlorine is due to its powerful oxidizing properties, which oxidize those enzymes of microbial cells that are essential to the cells' metabolic processes (Butterfield *et al.*, 1943).

Reaction

Chlorine gas reacts readily with water to form hypochlorous acid, HOCl, and hydrochloric acid.

$$Cl_2 + H_2O \rightarrow HOCl + HCl \qquad \textbf{(24.3)}$$

In dilute solution and with pH greater than 3, the reaction is appreciably displaced to the right and very little molecular chlorine gas will remain dissolved and unreacted. The hypochlorous acid produced then dissociates to yield hypochlorite ion.

$$HOCl \rightleftharpoons H^+ + OCl^- \qquad (24.4)$$

The relative distribution of HOCl and OCl$^-$ is a function of pH, as shown in Figure 24.3.

Hypochlorite salts are available in dry (calcium hypochlorite) or liquid (sodium hypochlorite) form. The dry form is cheaper but must be dissolved in water.

$$Ca(OCl)_2 \xrightarrow{\text{H}_2\text{O}} Ca^{+2} + 2OCl^- \qquad (24.5)$$

The OCl^{-1} will then seek an equilibrium with the hydrogen ions as indicated in Eq. (24.4), and therefore when hypochlorites are used in such applications as swimming pools, it is often necessary to add acid.

Although both hypochlorous acid and hypochlorite ion are excellent disinfecting agents, the acid form is the more effective (Engelbrecht and

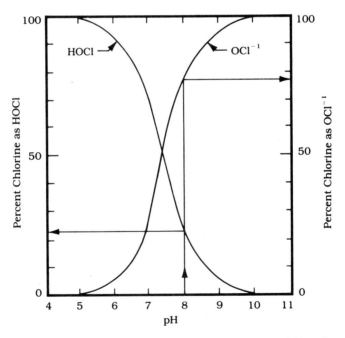

FIGURE 24.3 *Relative Amounts of Chlorine as HOCl and OCl^{-1} at 20°C versus pH*

Adapted from "Behavior of Chlorine as a Disinfectant" by G. M. Fair *et al.*, *Journal of the American Water Works Association* 40, no. 10 (October 1948):1051. By permission. Copyright 1948, the American Water Works Association.

Hendricks, 1982). They also react with certain inorganic and organic materials in water. One of the important reactions is with ammonia:

$$NH_3 + HOCl \rightarrow NH_2Cl + H_2O$$
monochloramine   (24.6)

$$NH_2Cl + 2HOCl \rightarrow NHCl_2 + 2H_2O$$
dichloramine   (24.7)

$$NHCl_2 + 3HOCl \rightarrow NCl_3 + 3H_2O$$
nitrogen trichloride   (24.8)

The relative amounts of the various chloramines that are formed are mainly a function of the hypochlorous acid present and the pH. The monochloramine form predominates at a pH greater than about 6.0, the dichloramine at about pH 5.

Reactions also occur with reduced substances and with organic materials. Dissolved chlorine gas will react with hydrogen sulfide to produce sulfuric and hydrochloric acids. It will react with other inorganic reducing ions or substances such as Fe^{+2}, Mn^{+2}, and NO_2^-. Dissolved chlorine will also react with organic compounds, particularly unsaturated compounds. Two of the organo reactions are particularly important — those that result in chlorophenols and those that produce trihalomethanes. Chlorophenols, formed from the reaction of chlorine with phenols, impart undesirable tastes and odors to water that are detectable at phenol concentrations less than one microgram per liter. Reaction of chlorine with innocuous humic substances results in the formation of trihalomethanes, including

$CHCl_3$	chloroform
$CHCl_2Br$	bromodichloromethane
$CHClBr_2$	dibromochloromethane

These compounds, including bromoform, are limited by drinking water regulations to a total of 0.1 milligram per liter because of tumorigenic properties.

Chloramines are effective compounds against bacteria but are not nearly so effective against viruses. The difference in effectiveness of chloramines was illustrated in a bench study (Figure 24.4) by Kruse *et al.* (1970), who utilized a synthetic waste, *Escherichia coli*, and F2 coliphage. A plot of the data of Durham and Wolf (1973) shows the same type of results on effluents from two trickling filter plants using total coliforms and any phages accepted by *E. coli* $K12(f^+)$ cells. The extrapolation of these findings on coliphages to all animal viruses is not warranted, but the results convey some inherent differences that can exist between bacteria and viruses in their susceptibility to chloramines.

Dosages, Demands, and Residuals   The **dosage** is the amount of chlorine added, the **demand** is the amount used for oxidation of materials present, and the **residual** is the amount remaining after oxidation. The relationships among them are shown in Figure 24.5. The residual equals the dosage minus the demand.

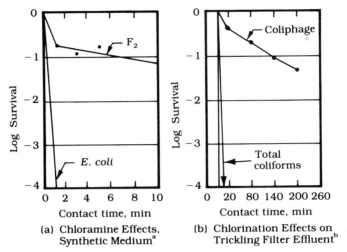

FIGURE 24.4 *Chloramine Effects on Bacteria and Viruses*

(a) C. W. Kruse, Y-C. Hsu, A. C. Griffiths, and R. Stringer, "Halogen Action on Bacteria, Viruses, and Protozoa." Proceedings of the National Specialty Conference on Disinfection. American Society of Civil Engineers, 1970, p. 113. Reproduced by permission of ASCE. (b) D. Durham, and H. W. Wolf, "Wastewater Chlorination: Panacea or Placebo?", *Water and Sewage Works* 120, no. 10:67.

Contact time is very important in the disinfection process. In chlorination, increased time of contact results not only in greater destruction of micro-organisms but also in an increased demand and, if appropriate precursors and free chlorine are present, in an increased amount of various chlorinated by-products.

Chlorine gas, hypochlorous acid, and the hypochlorite ion remaining after the demand is satisfied are collectively termed **free chlorine residuals**. The chloramines and other reactive chlorine forms remaining after the demand is satisfied are referred to as **combined chlorine residuals**. Free chlorine residuals are faster acting than combined residuals, and for the same concentration and time, free chlorine residuals have much greater disinfecting capacity than combined residuals, especially for viruses.

As depicted in Figure 24.5, an increase in chlorine dosage results in an equivalent increase in the residual up to a molar ratio of chlorine to ammonia nitrogen of 1:1. The residual formed is predominantly mono- and dichloramine. If the chlorine dosage is increased above this ratio, some nitrogen trichloride will be formed; however, as the dosage is increased, most of the chloramines will be oxidized to nitrogen gases. The oxidation reaction is essentially complete, for any particular point in time, at the minimum dip in the residual curve, which is termed the **breakpoint**. The breakpoint occurs at a chlorine dosage of about $1\frac{1}{2}$ to 2 moles of chlorine per mole of ammonia nitrogen and represents the dosage when the chloramines

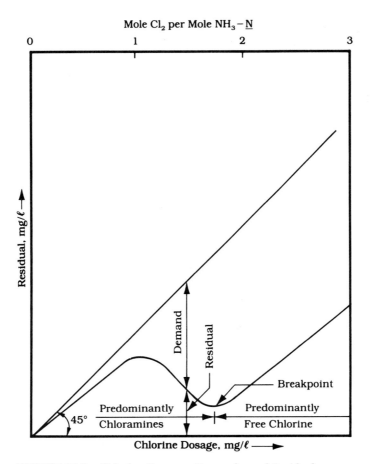

FIGURE 24.5 *Chlorine Dosages, Demands, and Residuals*

Adapted from *Chemistry for Environmental Engineers* by C. N. Sawyer and P. L. McCarty. Copyright © 1978 by McGraw-Hill, Inc. Reprinted by permission.

have been converted to the nitrogen gases. Some of the gases have been identified as free nitrogen, nitrous oxide, and nitrogen trichloride, free nitrogen being the most predominant. Continued addition of chlorine beyond the breakpoint gives a residual that is predominantly free chlorine.

The breakpoint dosage is very much dependent upon water quality, but for many drinking water supplies it ranges over 4 to 10 mg/ℓ. The desirable residual to be maintained at the farthest tap on the distribution system is at least 0.2 mg/ℓ, all free chlorine.

Wastewater chlorination practice varies with states' policies. Some states utilize a fecal coliform criterion such as 200/100 mℓ. Other states specify a type of residual after a specific contact period. Texas specifies 1.0 mg/ℓ total residual after 20 min of contact. Breakpoint chlorination of wastewater is seldom practiced. Dosages of 50 to 70 mg/ℓ may be necessary to reach breakpoint in many wastewaters, and this renders the effluents highly toxic to much of the aquatic life (Brungs, 1973).

Chlorine Application

Gaseous chlorine may be dissolved in water using any one of a variety of proprietary chlorinators, and the concentrated solution is then piped to the water stream to be disinfected. Hypochlorites can be added using solution-type feeders. Dry hypochlorites are first dissolved in water in a plastic or clay vessel; the liquid is then decanted off by a solution-type feeder. Some hypochlorites are packaged in polymers to render them amenable to dry feed equipment, but for the usual size of installation, these cannot compete economically with gaseous chlorine.

In water and wastewater treatment, prechlorination is the application of chlorine prior to any treatments, whereas postchlorination is chlorination after all treatments. Prechlorination is practiced to control undesirable growth such as might occur in a pipeline aqueduct. Similarly, prechlorination in wastewater treatment might be applied to sewers to control odors that develop as a result of undesirable growth. Postchlorination is sometimes called **terminal disinfection**.

Dechlorination

The toxic effects of chlorinated effluents on receiving waters has received significant attention since about 1970 (wpcf, 1986). Haugh (1993) reports that the removal of chlorine residuals prior to discharge is becoming a common requirement. Haugh lists sulfur dioxide, sodium sulfite, sodium bisulfite, sodium thiosulfate, hydrogen peroxide, and ammonia as chemicals used for dechlorination; of these chemicals, sulfur dioxide (SO_2) is the most widely used in municipal wastewater treatment. Sulfur dioxide is supplied commercially in steel cylinders or tanks of 100, 150, 2000, or 3000 lb (45, 68, 907, or 1365 kg) and is also available in bulk quantity in railroad tank cars (Metcalf & Eddy, 1991; epa, 1986). Activated carbon adsorption, although well established as a method for removal of chlorine residuals from potable water and some industrial waters, has received limited application in waste-water treatment. The process is expensive and requires a relatively long contact time for monochloramine removal. However, because activated carbon for dechlorination may simultaneously achieve reduction of organic material to low levels, its application in wastewater treatment for combined dechlorination and organic material removal may increase (epa, 1986; Metcalf & Eddy, 1991; wpcf, 1976 and 1986). Dechlorination by carbon adsorption will not be discussed here.

Some of the physical, chemical, and toxicity characteristics of sulfur dioxide, and chemical equations related to the use of sulfur dioxide for dechlorination of wastewater effluents, are presented in the following discussion (wpcf, 1986). Sulfur dioxide gas is colorless and has a strong, pungent odor. Sulfur dioxide gas has a solubility in water of about 18.6% at 32°F (0°C), and its reaction with water forms a weak solution of sulfurous acid, H_2SO_3. Dissociation of the sulfurous acid occurs as follows:

$$H_2SO_3 \rightleftharpoons H^+ + HSO_3^- \qquad (24.9)$$

$$HSO_3^- \rightleftharpoons H^+ + SO_3^{-2} \qquad (24.10)$$

Liquid or gaseous sulfur dioxide in the absence of moisture is not corrosive to steel or other common metals materials; however, it is corrosive to galvanized metals. With sufficient moisture, it is corrosive to most metals. Neither the gaseous nor the liquid form is flammable or explosive. Sulfur dioxide gas may cause irritation to mucous membranes and is extremely irritating to the respiratory system.

Both free and combined forms of chlorine react readily with the sulfite ion (SO_3^{-2}) produced in Eq. (24.10) as follows:

$$SO_3^{-2} + HOCl \rightarrow SO_4^{-2} + Cl^- + H^+ \qquad (24.11)$$

$$SO_3^{-2} + NH_2Cl + H_2O \rightarrow SO_4^{-2} + Cl^- + NH_4^+ \qquad (24.12)$$

The reactions in Eqs. (24.11) and (24.12) indicate that the stoichiometric mass ratio of sulfur dioxide to chlorine is $0.9:1$; however, experience indicates that the required mass ratio is closer to $1.1:1$.

OZONATION

Ozone is an allotrope of oxygen. It is a powerful oxidant and is more powerful than hypochlorous acid. In aqueous solution it is relatively unstable, having a half-life of 20 to 30 min in distilled water at 20°C. The presence of oxidant-demanding materials in solution will render the half-life even shorter (Rice *et al.*, 1979).

Ozone is widely used in drinking water treatment practice in Europe. Its first application was in 1893 at Oudshoorn, Netherlands, where it was used for settled and filtered Rhine River water. Today more than 1000 plants throughout the world use ozone. Canada has 22 plants, and Montreal has probably the world's largest (Rice *et al.*, 1979). Ozone use in wastewater treatment has significantly increased in recent years. In the United States, more than 40 full-scale facilities have been constructed. Extensive effort in research and development has been made to develop ozonation for wastewater disinfection (EPA, 1986).

Ozone must be produced on-site because it cannot be stored as chlorine can. This is not necessarily bad; serious accidents have happened with chlorine because of breaks in storage systems. Ozone is produced by passing air between oppositely charged plates or through tubes in which a core and the tube walls serve as the oppositely charged surfaces. Air is refrigerated to below the dew point to remove much of the atmospheric humidity and then is passed through desiccants, such as silica gel, activated alumina, to dry the air to a dew point of $-40°C$ to $-60°C$. The use of dry, clean air results in less frequent ozone generator maintenance, long-life units, and more ozone production per unit of power used (Jolley, 1975).

Gomella and co-workers observed complete destruction of poliovirus samples in distilled water at a residual of $0.3 \, mg/\ell$ at the end of 3 min of exposure. They then observed the same effectiveness when the viruses were suspended in Seine River water and recommended the use of $0.4\,mg/\ell$ after a contact of 4 min (Coin *et al.*, 1964). It is important that ozone demand be completely satisfied before this disinfecting step.

Usual French practice uses two contactors. In the first, with a contact time of 8 to 12 min, the ozone demand is satisfied and a residual of $0.4\,mg/\ell$ obtained. In the second, with a contact time of 4 to 8 min, the $0.4\,mg/\ell$ residual is maintained (Rice *et al.*, 1979).

German use of ozone often exploits its ability to render some refractory organics biodegradable. Sontheimer uses a 20-min contact period in this application. The dose of ozone to be used is determined from pilot studies that are conducted in Europe for 1-year periods and even longer. Ozone-treated water is then passed through granular activated carbon, which serves as a fixed-bed biological contactor allowing saprophytic organisms to decompose the biodegradable materials. Such a contactor biologically regenerates itself and is called a BAC, or biological activated carbon, process. Ozone is never used as a terminal treatment because experience has shown that organisms can under certain circumstances proliferate in distribution systems, causing all types of problems. Hence, many European plants utilize the desirable residual action of chlorine as a terminal disinfectant, but the dose is very low, such as 0.1 to $0.3\,mg/\ell$.

Ozone disinfection of wastewater has several notable environmental advantages compared to chlorination (WPCF, 1984; EPA, 1986). Because ozone rapidly degrades to oxygen, toxic residuals are not present and oxygen levels in the effluent are often at saturation. Ozone disinfection does not result in increased total dissolved solids in the effluent. Although ozonation systems require higher capital and operational costs than chlorination, wastewater treatment plants using pure oxygen activated sludge treatment can realize a reduction in costs for ozone disinfection. Ozone has been applied for the removal of phenols, cyanides, and heavy metals in industrial processes (WPCF, 1986). The ability of ozone to render some refractory organics biodegradable further underscores its potential use in wastewater treatment. McCarthy and Smith (1974) observed a marked affinity of ozone for suspended matter. Thus, for good results, the suspended solids should not be too high. Buys (1980) observed that for a petrochemical wastewater, a dose of $12\,mg\,O_3/\ell$ had no effect on the raw wastewater but that on the treated wastewater, BOD_5 values were increased by about 50%. This observation has been reported for municipal wastewater effluents. He also noted that the nonadsorbable COD was decreased by about 50%. The use of ozone in wastewater treatment is increasing and is likely to continue to increase in the future.

It should be noted that there are uncertainties about the reaction of ozone with organic materials in wastewater. Current research is directed toward identifying byproducts of the reaction of ozone with organic materials (WPCF, 1986). Ozone's reaction with organics destroys the original molecule of organic material, often producing a more biodegradable material. On the other hand, ozone's reaction with pesticides may produce a more toxic material. The formation of several persistent, potentially dangerous epoxides has been predicted by ozone reaction models. These byproducts may have significant human health and environmental consequences that will influence the use of ozone for wastewater disinfection.

CHLORINE DIOXIDE

Chlorine dioxide (ClO_2) was discovered in 1811 but was not used in water treatment until 1944 at Niagara Falls, New York. Its application was primarily to alleviate taste and odor problems that arose from chlorination of phenolic contaminants in the water. A 1958 survey indicated that 56 water treatment plants were using ClO_2 — most for taste and odor control, some for iron or manganese removal, and others for disinfection.

ClO_2 is a more powerful oxidant than chlorine. It does not react with water as does chlorine gas, is easily removed from water by aeration, is readily decomposed by exposure to ultraviolet irradiation, does not react with ammonia as does chlorine, and persists to maintain a stable residual.

There are at least four general methods of preparing ClO_2:

1. Acid and sodium chlorite

$$NaClO_2 + HCl \rightarrow ClO_2 + NaCl + H^+ \qquad \textbf{(24.13)}$$

2. Chlorine gas and sodium chlorite (excess chlorine)

$$(Cl_2 + H_2O \rightarrow HOCl + HCl)$$

$$HOCl + HCl + 2\,NaClO_2 \rightarrow 2ClO_2 + 2\,NaCl + H_2O \qquad \textbf{(24.14)}$$

3. Sodium hypochlorite and sodium chlorite

$$2\,NaClO_2 + NaOCl + 2\,HCl \rightarrow 2\,ClO_2 + 3\,NaCl + H_2 \qquad \textbf{(24.15)}$$

4. Sodium chlorate
   Four different processes are used to produce ClO_2 from sodium chlorate, but none are used in water treatment.

The process most widely used in water treatment is the excess chlorine method. The excess chlorine is to insure that all of the chlorite ions will produce chlorine dioxide. Some uncertainties exist regarding the health effects of the chlorite ion, ClO_2^-, which might also be introduced after the oxidizing reaction.

The ClO_2 residual is longer-lasting than HOCl. Terminal disinfection is practiced using only 0.10 to 3.0 mg/ℓ ClO_2. However, taste and odor control applications may see dosages up to 10 mg/ℓ (Miller *et al.*, 1978).

Applications of chlorine dioxide in wastewater treatment (wPCF, 1986; EPA, 1986) have been limited to phenolic waste treatment and to the control of sulfide in wastewater collection systems. Unlike chlorine, however, chlorine dioxide does not produce measurable amounts of THMs (trihalomethanes) or TOXs (total organic halogens), X representing chlorine, bromine, or iodine. The high expense of both the equipment and the sodium chlorite used in generating chlorine dioxide are major disadvantages of its application in wastewater disinfection. However, recent technological advances in generation equipment and sodium chlorite production may reduce the cost for applications in wastewater disinfection, and since chlorine dioxide is an effective disinfectant, its use in wastewater treatment may become a viable alternative.

Ultraviolet (UV) irradiation has been used for the disinfection of drinking water supplies aboard ships for many years. Pilot plant studies by numerous researchers (WPCF, 1984) have indicated that the UV process is a most viable alternative for wastewater disinfection. Disinfection of treated wastewater effluent by the UV process is both effective and economical as an alternative to chlorination or ozonation (WPCF, 1986). A summary list, compiled in 1984, indicated that 117 installations of UV disinfection were in the design stage, under construction, or in operation at wastewater treatment plants in the United States (EPA, 1986). The design size of these facilities ranged from less than 0.1 MGD (380 m^3/day) to greater than 50 MGD (190,000 m^3/day). There have been 18 reports discussing the effectiveness, key wastewater parameters, repair mechanisms (photoreactivation), pilot studies, process design, and operational considerations of UV irradiation for effluent disinfection (WEF, 1993). Scheible (1993) estimates that between 500 and 600 UV installations are now in the planning, design, construction, or operational phase in the United States.

UV can effectively disinfect both water and wastewater. The lack of a residual is a major disadvantage in water treatment, and this is recognized in the Centers for Disease Control regulations, which recommend the use of chlorine aboard ship even when UV systems are installed and functioning. If chlorine must be used, there does not appear to be any justification for using UV also. The advantages of the UV irradiation process for wastewater disinfection are its effectiveness in pathogen inactivation and ability to achieve disinfection goals, its viable application to a wide range of wastewater qualities, its cost effectiveness, its relative simplicity, and the absence of residuals and chemical intermediates (EPA, 1986).

Two different types of UV installations were tested in Dallas, where it was found that the submergence of the UV bulb encased in a quartz tube was superior to an overhead bulb radiating downward through a shallow 1- to 2-in. (2.5 to 5.1-cm) depth of water. The major factor in achieving good microorganism kill is the ability of the radiation to pass through the water and get to the target organism. Interestingly, this transmittance was found to be markedly less a function of turbidity or suspended solids, but very much a function of the COD or TOC. Since lower COD or TOC values are associated with long mean cell residence times accompanying highly nitrified operation, an association of good transmittance — and hence good kill — with lower NH$_3$-N values was observed.

Comparative dose-response equations for three organisms showed only a little spread and similar slopes:

$$y = 1.48x - 3.21 \quad \text{(fecal coliforms)} \tag{24.16}$$

$$y = 1.62x - 4.48 \quad \text{(polioviruses)} \tag{24.17}$$

$$y = 1.59x - 4.68 \quad \text{(coliphages)} \tag{24.18}$$

where

$y = $ log reduction

$x = $ log uv dose

Unlike chloramines, uv performs very well against viruses.

A slime buildup occurs on the quartz sleeves housing the uv lamps. A proprietary cleaning solution was effective in cleaning the sleeves. A cleaning frequency of once every 2 weeks, or perhaps every 3 weeks, is necessary to keep the problem under control.

The mean intensity of radiation, measured by a radiometer in microwatts per square centimeter (μw/cm^2), was multiplied by the detention time (in seconds) of the flow in the irradiation chamber to yield the indicated dose, D_I, in watt-sec/cm^2. For total and fecal coliforms, the relationship of D_I to log reduction was determined to be

$$D_I = 24{,}800x - 60{,}000 \quad \text{(fecal coliforms)} \tag{24.19}$$

$$D_I = 20{,}000x - 48{,}200 \quad \text{(total coliforms)} \tag{24.20}$$

where $x = $ log reduction.

Figure 24.6 is a plot of the log of the fecal coliform density whose probability of exceedence is equal to or greater than 5% versus the log of

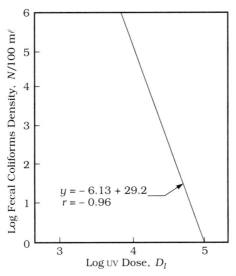

FIGURE 24.6 *Fecal Coliform Densities with an Exceedence Probability Greater than or Equal to 5% versus Indicated Dose*

D_I. The curve indicates that a fecal coliform density of $200/100\,m\ell$ can be expected to be exceeded 5% of the time if the indicated dose is $24,400\,\mu\text{watt-sec/cm}^2$ (Petrasek *et al.*, 1980).

HIGH-pH TREATMENT

The use of lime in treating the contents of privies, dead animals, and battlefield mortalities to alleviate nuisance conditions is historical. Its effectiveness in destroying bacteria in water treatment application at high pH has been recognized since the 1920s. Even so, it is not relied on as the sole disinfectant, and chlorine is always employed as well, because after neutralization of the high pH, there remains no residual to protect the water.

Berg, Dean, and Dahling (1968) studied virus removal from secondary effluents by lime flocculation and observed higher removals at higher pH values. Dallas studies using seeded polio I and coliphages showed very high removals of both viruses (neither was detected in the effluents) and of vegetative bacteria from secondary effluents, an observation reinforced by South African studies (Nupen *et al.*, 1974). Prior work in Dallas on coliphages alone had indicated the critical pH to contact time relationships to be in the pH range of 11.2 to 11.3 and a contact time of 1.56 to 2.40 hours. In the studies showing no surviving viruses, a pH of 11.0 or greater and a theoretical contact time of 5 hours and 10 minutes were employed. It should be pointed out that a residence time distribution function study showed peak dye concentrations in the effluent occurring only 2 hours after the addition of a slug load. The pH was neutralized as water left the high-pH tank, and growth of surviving or contaminating bacteria was observed.

High-pH treatment for disinfection of drinking water supplies, like ozone or UV, would be highly effective against viruses but still requires the use of an additional residual disinfectant. However, the additional disinfectant need only be added in minimal amount. In wastewater treatment, high-pH treatment for disinfection could be coupled with the additional objectives of ammonia and phosphorus removal. Recarbonation prior to discharge would also be necessary unless discharge were to sodic or acidic soils.

High-pH treatment of wastewater sludges, called **lime stabilization**, is a recommended procedure. Lime is added to pH 12 and maintained at that value for 30 min. Noland, Edwards, and Kipp (1978) observed on sludge what Berg, Dean, and Dahling (1968) had observed on wastewater — that is, the higher the pH, the greater the bacterial reduction. However, a pH of 12 or greater appears to be required by sludges. Unfortunately, virus evaluations were not undertaken and parasite data indicate little effect of the high pH; however, the time of exposure to the high pH in the study appears to have been limited to only 30 min (Wolf *et al.*, 1974).

REFERENCES

Archer, A. J., and Juven, B. J. 1977. Destruction of Coliforms in Water and Sewage Water by Dye-Sensitized Photooxidation. *Applied and Environmental Microbiology* 33:1019.

Bauer, R. C., and Snoeyink, V. L. 1973. Reactions of Chloramines with Active Carbon. *Jour. WPCF* 45, no. 11:2290.

Berg, G. 1964. The Virus Hazard in Water Supplies. *Jour. New England Water Works Assoc.* 78:79.

Berg, G.; Dean, R. B.; and Dahling, D. R. 1968. Removal of Poliovirus I from Secondary Effluents by Lime Flocculation and Rapid Sand Filtration. *Jour. AWWA* 60, no. 2:193.

Brungs, W. A. 1973. Effects of Residual Chlorine on Aquatic Life. *Jour. WPCF* 45, no. 10:2180.

Butterfield, C. T.; Wattie, E.; Megregian, S.; and Chambers, C. W. 1943. Influence of pH and Temperature on the Survival of Coliforms and Enteric Pathogens When Exposed to Free Chlorine. *U.S. Pub. Health Repts.* 58:1837.

Buys, R. E. 1980. The Effect of Solids Retention Time on Tertiary Ozonation and Carbon Adsorption of Petrochemical Wastewaters. M.S. thesis, Texas A&M University.

Chang, S. L. 1971. Modern Concept of Disinfection. *Jour. SED* 97, no. SA5:689.

Chick, H. 1908. Investigation of the Laws of Disinfection. *Jour. Hygiene* 8:92.

Clark, R. M.; Adams, J. Q.; and Lykins, B. W., Jr. 1994. DBP Control in Drinking Water: Cost and Performance. *Jour. of Environmental Engineering* 120, no. 4:759.

Coin, L.; Gomella, C.; and Hannoun, C. 1964. Inactivation of Poliomyelitis Virus by Ozone in the Presence of Water. *la Presse Med.* 72, no. 37:2153.

Durham, D., and Wolf, H. W. 1973. Wastewater Chlorination: Panacea or Placebo? *Water and Sewage Works* 120:67.

Engelbrecht, R. S., and Hendricks, C. W. 1982. Microbiological Criteria and Standards for Potable Reuse. *Proceedings of the Water Reuse Symposium II*, Washington, D.C., August 23–28, 1981, Vol. 3:2123. EPA.

Environmental Protection Agency (EPA). 1986. *Municipal Wastewater Disinfection.* EPA Design Manual. Washington, D.C.

Evans, F. L., III. 1972. *Ozone in Water and Wastewater Treatment.* Ann Arbor, Mich.: Ann Arbor Science.

Fair, G. M.; Morris, J. C.; Chang, S. L.; Weil, I.; and Burden, R. P. 1948. Behavior of Chlorine as a Disinfectant. *Jour. AWWA* 40. no. 10:1051.

Gerba, C. P.; Wallis, C.; and Melnick, J. L. 1977. Disinfection of Wastewater by Photodynamic Oxidation. *Jour. WPCF* 49, no. 4:575.

Gomella, C. Personal communication.

Green, D. E., and Stumpf, P. K. 1946. The Mode of Action of Chlorine. *Jour. AWWA* 38, no. 11:1301.

Harris, D. H.; Adams, V.; Sorrensen, D. L.; and Dupont, R. R. 1987. The Influence of Photoreactivation and Water Quality on Ultraviolet Disinfection of Secondary Municipal Wastewater. *Jour. WPCF* 59, no. 8:781.

Haugh, R. S. 1993. Instrumentation and Control of Sulfur Dioxide Dechlorination: An Update on State-of-the-Art. *Proceedings, Planning, Design and Operations of Effluent Disinfection Systems.* Water Environment Federation (WEF) Specialty Conference Series, Whippany, N.J. Alexandria, Va: WEF.

Johnson, J. D.; Aldrich, J.; Francisco, D. E.; Wolff, T.; and Elliott, M. 1978. UV Disinfection of Secondary Effluent. *Progress in Wastewater Disinfection Technology*, edited by A. D. Venosa. Environmental Protection Agency, Washington, D.C.

Jolley, R. L. 1975. Chlorine-Containing Organic Constituents in Chlorinated Effluents. *Jour. WPCF* 47, no. 3:601.

Kruse, C. W.; Hsu, Y-C.; Griffiths, A. C.; and Stringer, R., 1970. Halogen Action on Bacteria, Viruses, and Protozoa. *Proceedings of the National Specialty Conference on Disinfection.* American Society of Civil Engineers.

Longley, K. E.; Moore, B. L.; and Sorber, C. A. 1980. Comparison of Chlorine and Chlorine Dioxide as Disinfectants. *Jour. WPCF* 52, no. 8:2098.

Lykins, B. W., Jr., and Clark, R. M. 1994. U.S. Drinking-Water Regulations: Treatment Technologies and Cost. *Jour. of Environmental Engineering* 120, no. 4:783.

Lykins, B. W., Jr.; Koffskey, W. E.; and Patterson, K. S. 1990. Alternative Disinfectants for Drinking Water Treatment. *Jour. of Environmental Engineering* 120, no. 4:745.

Metcalf & Eddy, Inc. 1991. *Wastewater Engineering: Treatment, Disposal and Reuse.* 3rd ed. New York: McGraw-Hill.

McCarthy, J. J., and Smith, C. H. 1974. The Use of Ozone in Advanced Wastewater Treatment, *Jour. AWWA* 66, no. 12:718.

Miller, G. W.; Rice, R. G.; Robson, C. M.; Kuhr, W.; and Wolf, H. 1978. *An Assessment of*

Ozone and Chlorine Dioxide Technologies for Treatment of Municipal Water Supplies. Environmental Protection Agency Grant R 804385-01, Washington, D.C.

National Academy of Sciences. 1977. *Drinking Water and Health*. Washington, D.C.

Noland, R. F.; Edwards, J. D.; and Kipp, M. 1978. Full Scale Demonstration of Lime Stabilization. Environmental Protection Agency Report 600/2-68-171, Washington, D.C.

Nupen, E. M.; Bateman, B. W.; and McKenny, N. C. 1974. The Reduction of Virus by the Various Unit Processes Used in the Reclamation of Sewage to Potable Waters. *Virus Survival in Water and Wastewater Systems*. Austin, Tex.: University of Texas Press.

Oliver, B. G., and Carey, J. H. 1976. Ultraviolet Disinfection, an Alternative to Chlorination. *Jour. WPCF* 48, no. 11:2619.

Petrasek, A. C., Jr.; Wolf, H. W.; Esmond, S. E.; and Andrews, D. C. 1980. *Ultraviolet Disinfection of Municipal Wastewater Effluents*. Environmental Protection Agency Report 600/2-80-102, Washington, D.C.

Rice, R. G.; Miller, G. W.; Robson, C. M.; and Hill, A. G. 1979. Ozone Utilization in Europe. AIChE, 8th Annual Meeting, Houston, Tex.

Rook, J. J. 1977. Chlorination Reactions of Fulvic Acids in Natural Waters. *Environmental Sci. and Tech.* 11:478.

Sawyer, C. N., and McCarty, P. L. 1978. *Chemistry for Environmental Engineers*. 3rd ed. New York: McGraw-Hill.

Scheible, O. K. 1993. Current Assessment of Design and O&M Practices for UV Disinfection. *Proceedings, Planning, Design and Operations of Effluent Disinfection Systems*. Water Environment Federation (WEF) Speciality Conference Series, Whippany, N.J. Alexandria, Va: WEF.

Scheible, O. K. 1987. Development of a Rationally Based Design Protocol for the Ultraviolet Light Disinfection Process. *Jour. WPCF* 59, no. 1:25.

Severin, B. F. 1980. Disinfection of Municipal Waste Effluents with Ultraviolet Light. *Jour. WPCF* 52, no. 7:2007.

Singer, P. C. 1994. Control of Disinfection By-Products in Drinking Water. *Jour. of Environmental Engineering* 120, no. 4:727.

Sontheimer, H. Personal communication.

Water Environment Federation (WEF). 1993. *Proceedings, Planning, Design and Operations of Effluent Disinfection Systems*. Water Environment Federation (WEF) Specialty Conference Series, Whippany, N.J. Alexandria, Va: WEF.

Water Pollution Control Federation (WPCF). 1976 (reprinted 1986). *Chlorination of Wastewater*. WPCF Manual of Practice No. 4. Alexandria, Va.

Water Pollution Control Federation (WPCF). 1984. *Wastewater Disinfection: A State-of-the-Art Report*. WPCF Disinfection Committee, Alexandria, Va.

Water Pollution Control Federation (WPCF). 1986. *Wastewater Disinfection*. WPCF Manual of Practice FD-10. Alexandria, Va.

Webster's New Collegiate Dictionary. 1976. Springfield, Mass.: G. & C. Merriam.

White, G. C. 1978. *Disinfection of Wastewater and Water for Reuse*. New York: Van Nostrand Reinhold.

White, G. C. 1972. *Handbook of Chlorination*. New York: Van Nostrand Reinhold.

Wolf, H. W.; Safferman, R. S.; Mixson, A. R.; and Stringer, C. E. 1974. Virus Inactivation during Tertiary Treatment. *Jour. AWWA* 66, no. 9:526.

Appendix A

CONVERSIONS AND MEASURES

CONVERSIONS AND MEASURES

1. LENGTH

mi	yd	ft	in.	mm	cm	m	km
1	1760	5280	—	—	—	—	1.609
—	1	3	36	—	91.44	0.9144	—
—	—	1	12	—	30.48	0.3048	—
—	—	—	1	25.40	2.54	—	—
—	—	—	—	10	1	—	—
—	1.094	3.281	—	—	100	1	—
0.6214	—	—	—	—	—	1000	1

2. AREA

mi^2	ac	yd^2	ft^2	in^2	cm^2	m^2	km^2
1	640	—	—	—	—	—	—
—	1	—	43,560	—	—	—	—
—	—	1	9	—	—	—	—
—	—	—	1	144	—	—	—
—	—	—	—	1	6.45	—	—
—	—	1.196	10.764	—	10^2	1	—
0.3863	247	—	—	—	—	10^6	1

1 hectare (ha) = 10,000 m^2 = 2.47 ac

3. VOLUME

ft^3	Imp. gal	U.S. gal	U.S. qt	in^3	ℓ	cm^3	m^3
1	6.23	7.481	29.92	1728	28.32	—	—
—	1	1.2	4.8	277.4	4.536	—	—
—	—	1	4	231	3.785	—	—
—	—	—	1	57.75	0.946	—	—
—	—	0.264	1.057	61.02	1	1000	—
35.31	—	264.2	—	—	1000	10^6	1

4. WEIGHT OR MASS

ton	long ton	lb	grains	mg	gm	kgm	tonne
1	—	2000	—	—	—	—	0.908
—	1	2240	—	—	—	—	—
—	—	1	7000	—	454	—	—
—	—	—	15.43	1000	1	—	—
—	—	2.203	—	—	1000	1	—
—	—	—	—	—	—	1000	1

5. PRESSURE

psi	ft of water	in. of Hg	mm of Hg	atm	kPa
1	2.307	2.036	51.714	—	6.8957
0.4335	1	0.8825	22.416	—	—
0.4912	1.33	1	25.4	—	—
14.70	33.93	29.92	760	1	101.37

6. VELOCITY

mi/hr	ft/sec	in./min	cm/s	km/h
1	1.467	1056	—	1.609
—	—	1	0.423	—
0.6215	—	—	—	1

7. DISCHARGE

ft^3/sec	MGD	gpm	ℓ/s	ℓ/min	m^3/sec
1	0.6464	448.9	28.32	1699	—
1.547	1	694.4	—	—	—
—	—	15.85	1	—	—
35.32	22.82	15,850	1000	60,000	1

8. POWER

kw	hp	ft-lb/sec	J/sec (watt)
1	1.341	737.6	1000
0.7457	1	550	745.7

9. WORK, ENERGY, HEAT

kw-hr	hp-hr	ft-lb	Btu	gm-cal	J
1	1.341	—	3412	—	—
0.7457	1	—	2544	—	—
—	—	1	—	—	1.356
—	—	777.5	1	252	1054
—	—	—	—	1	4.184
—	—	0.7376	—	—	1

10. MISCELLANEOUS

1 U.S. gal = 8.34 lb of water

1 Imp. gal = 10 lb of water

1 ft^3 = 62.43 lb of water

1 m^3 = 2205 lb of water

1 newton (N) = 0.2248 lb force = 1 kg-m/s^2

1 pascal (Pa) = 1 N/m^2

1 joule (J) = 1 N-m

1 joule/sec (J/sec) = 1 N-m/sec

1 watt (W) = 1 joule/sec = 1 N-m/sec

1 ppm = 1 mg/ℓ if the specific gravity = 1.00

°F = 9/5 (°C) + 32

°C = 5/9 (°F − 32)

°R = °F + 460

°K = °C + 273

g = 32.17 ft/sec^2 = 9.806 m/s^2

Appendix B

INTERNATIONAL ATOMIC WEIGHTS OF ELEMENTS

INTERNATIONAL ATOMIC WEIGHTS OF NUMEROUS ELEMENTS (1961), BASED ON CARBON 12

ELEMENT	SYMBOL	ATOMIC NUMBER	ATOMIC WEIGHT
Aluminum	Al	13	26.98
Antimony	Sb	51	121.75
Arsenic	As	33	74.92
Barium	Ba	56	137.34
Boron	B	5	10.81
Bromine	Br	35	79.91
Cadmium	Cd	48	112.40
Calcium	Ca	20	40.08
Carbon	C	6	12.01
Chlorine	Cl	17	35.45
Chromium	Cr	24	52.00
Cobalt	Co	27	58.93
Copper	Cu	29	63.54
Fluorine	F	9	19.00
Gold	Au	79	196.97
Helium	He	2	4.00
Hydrogen	H	1	1.008
Iodine	I	53	126.90
Iron	Fe	26	55.85
Lead	Pb	82	207.19
Magnesium	Mg	12	24.31
Manganese	Mn	25	54.94
Mercury	Hg	80	200.59
Molybdenum	Mo	42	95.94
Nickel	Ni	28	58.71
Nitrogen	N	7	14.01
Oxygen	O	8	16.00
Phosphorus	P	15	30.97
Platinum	Pt	78	195.09
Potassium	K	19	39.10

ELEMENT	SYMBOL	ATOMIC NUMBER	ATOMIC WEIGHT
Selenium	Se	34	78.96
Silicon	Si	14	28.09
Silver	Ag	47	107.87
Sodium	Na	11	22.99
Strontium	Sr	38	87.62
Sulfur	S	16	32.06
Tin	Sn	50	118.69
Zinc	Zn	30	65.37

The above is a partial listing. In particular, the radioactive elements are not included. For a complete listing, see *Lange's Handbook of Chemistry* (New York: McGraw-Hill, 1961).

Appendix C

DENSITY AND VISCOSITY OF WATER

DENSITY AND VISCOSITY OF WATER

TEMPERATURE (°C)	DENSITY (gms/cm^3, γ)	ABSOLUTE VISCOSITY (centipoise[a], μ)	KINEMATIC VISCOSITY (centistokes[b], ν)
0	0.99987	1.7921	1.7923
1	0.99993	1.7320	1.7321
2	0.99997	1.6740	1.6741
3	0.99999	1.6193	1.6193
4	1.00000	1.5676	1.5676
5	0.99999	1.5188	1.5188
6	0.99997	1.4726	1.4726
7	0.99993	1.4288	1.4288
8	0.99988	1.3872	1.3874
9	0.99981	1.3476	1.3479
10	0.99973	1.3097	1.3101
11	0.99963	1.2735	1.2740
12	0.99952	1.2390	1.2396
13	0.99940	1.2061	1.2068
14	0.99927	1.1748	1.1756
15	0.99913	1.1447	1.1457
16	0.99897	1.1156	1.1168
17	0.99880	1.0876	1.0888
18	0.99862	1.0603	1.0618
19	0.99843	1.0340	1.0356
20	0.99823	1.0087	1.0105
21	0.99802	0.9843	0.9863
22	0.99780	0.9608	0.9629
23	0.99757	0.9380	0.9403
24	0.99733	0.9161	0.9186
25	0.99707	0.8949	0.8975
26	0.99681	0.8746	0.8774
27	0.99654	0.8551	0.8581
28	0.99626	0.8363	0.8394
29	0.99597	0.8181	0.8214
30	0.99568	0.8004	0.8039
31	0.99537	0.7834	0.7870
32	0.99505	0.7670	0.7708

TEMPERATURE (°C)	**DENSITY (gms/cm³, γ)**	**ABSOLUTE VISCOSITY (centipoise[a], μ)**	**KINEMATIC VISCOSITY (centistokes[b], ν)**
33	0.99473	0.7511	0.7551
34	0.99440	0.7357	0.7398
35	0.99406	0.7208	0.7251
36	0.99371	0.7064	0.7109
37	0.99336	0.6925	0.6971
38	0.99299	0.6791	0.6839
39	0.99262	0.6661	0.6711

[a] 1 centipoise = 10^{-2} (gram mass)/(cm)(s). To express the absolute viscosity (μ) as (N)(s)/(m^2), multiply centipoise by 10^{-3}. To express the absolute viscosity (μ) as (lb mass)/(ft)(sec), multiply centipoise by 6.72×10^{-4}. To express the absolute viscosity (μ) as (lb force)(sec)/(ft^2), multiply centipoise by 2.088×10^{-5}.
[b] 1 centistoke = 10^{-2} cm^2/s. To express the kinematic viscosity (ν) as m^2/s, multiply centistokes by 10^{-6}. To express the kinematic viscosity (ν) as ft^2/sec, multiply centistokes by 1.075×10^{-5}.
From *International Critical Tables*, 1928 and 1929.

Appendix D

DISSOLVED OXYGEN SATURATION VALUES IN WATER

DISSOLVED OXYGEN SATURATION VALUES IN FRESH AND SEA WATER EXPOSED TO AN ATMOSPHERE CONTAINING 20.9% OXYGEN UNDER A PRESSURE OF 760 mm Hg

Temperature (°C)	DISSOLVED OXYGEN (mg/ℓ) FOR STATED CONCENTRATIONS OF CHLORIDE (mg/ℓ)				
	0	5,000	10,000	15,000	20,000
0	14.62	13.79	12.97	12.14	11.32
1	14.23	13.41	12.61	11.82	11.03
2	13.84	13.05	12.28	11.52	10.76
3	13.48	12.72	11.98	11.24	10.50
4	13.13	12.41	11.69	10.97	10.25
5	12.80	12.09	11.39	10.70	10.01
6	12.48	11.79	11.12	10.45	9.78
7	12.17	11.51	10.85	10.21	9.57
8	11.87	11.24	10.61	9.98	9.36
9	11.59	10.97	10.36	9.76	9.17
10	11.33	10.73	10.13	9.55	8.98
11	11.08	10.49	9.92	9.35	8.80
12	10.83	10.28	9.72	9.17	8.62
13	10.60	10.05	9.52	8.98	8.46
14	10.37	9.85	9.32	8.80	8.30
15	10.15	9.65	9.14	8.63	8.14
16	9.95	9.46	8.96	8.47	7.99
17	9.74	9.26	8.78	8.30	7.84
18	9.54	9.07	8.62	8.15	7.70
19	9.35	8.89	8.45	8.00	7.56
20	9.17	8.73	8.30	7.86	7.42
21	8.99	8.57	8.14	7.71	7.28
22	8.83	8.42	7.99	7.57	7.14
23	8.68	8.27	7.85	7.43	7.00
24	8.53	8.12	7.71	7.30	6.87
25	8.38	7.96	7.56	7.15	6.74
26	8.22	7.81	7.42	7.02	6.61

Temperature (°C)	DISSOLVED OXYGEN (mg/ℓ) FOR STATED CONCENTRATIONS OF CHLORIDE (mg/ℓ)				
	0	5,000	10,000	15,000	20,000
27	8.07	7.67	7.28	6.88	6.49
28	7.92	7.53	7.14	6.75	6.37
29	7.77	7.39	7.00	6.62	6.25
30	7.63	7.25	6.86	6.49	6.13

Under any other barometric pressure, P(mm), the solubility, DO'_s (mg/ℓ) can be computed from the corresponding value in the table by the equation

$$DO'_s = DO_s \frac{P - p}{760 - p}$$

where DO_s is the solubility at 760 mm (29.92 in.) and p is the pressure (mm) of saturated water vapor at the temperature of the water. For elevations less than 3000 ft and temperatures less than 25°C, p can be ignored. The equation then becomes

$$DO'_s = DO_s \frac{P}{760}$$

From Whipple and Whipple, *Jour. Am. Chem. Soc.* 33, (1911):362.

Appendix E

LABORATORY BIOLOGICAL TREATMENT PLANT STUDIES

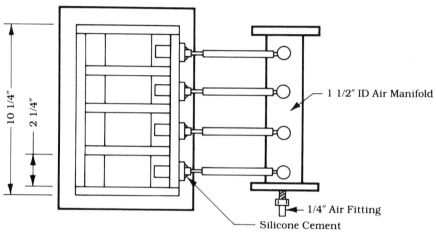

Plan View

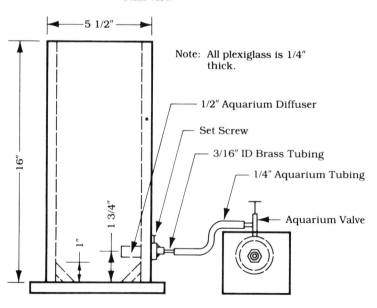

Elevation View

FIGURE A.1
Construction Details for a Batch Reactor Module of Four 2-ℓ Reactors
Note: 1 in. = 25.4 mm.

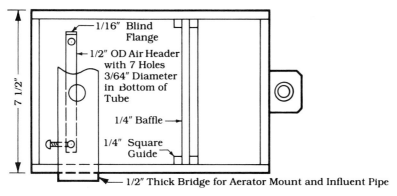

Plan View

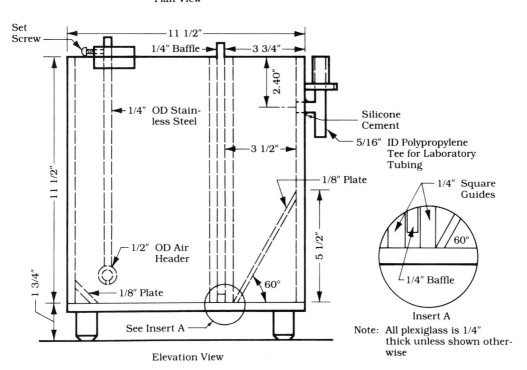

Elevation View

FIGURE A.2 *Construction Details for a 10-ℓ Completely Mixed Activated Sludge Reactor*

Note: 1 in. = 25.4 mm.

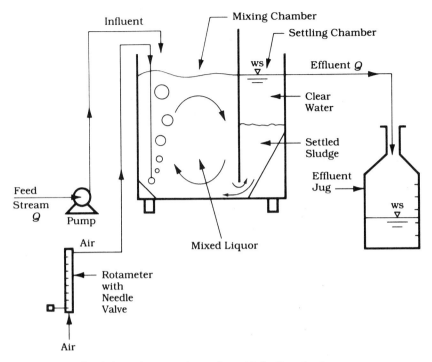

FIGURE A.3 *Schematic Drawing of a 10-ℓ Completely Mixed Activated Sludge Reactor in Operation*

Operational Notes:

1. To acclimate a sludge for a particular wastewater, fill the reactor to the desired MLSS concentration using activated sludge from a municipal plant. Start aeration and feed the reactor in a stepwise manner using 0.05 lb BOD₅/lb MLSS-day (kg/kg-d) increments for 2 to 3 days until the desired F/M ratio is attained. Waste sludge during the period to keep the desired MLSS concentration. If filamentous organisms occur, restart the reactor and use longer feed intervals.

2. For proper recycle, the space below the baffle should be from 1/4 to 3/8 in. (6 to 10 mm).

3. The baffle should be pulled and the contents allowed to mix when withdrawing samples for MLSS analyses, oxygen uptake rates, settling tests, and for wasting sludge and withdrawing sludge for batch reactor studies. The effluent pipe should be plugged prior to pulling the baffle.

4. The oxygen uptake rate can be determined with a BOD bottle fitted with a DO probe and stirrer. A mixed-liquor sample larger than 300 mℓ is aerated by shaking in a stoppered flask. The sample is poured into the BOD bottle, and the oxygen concentration versus time is obtained. The slope of the plot is the oxygen uptake rate.

5. The detention time, θ, is based on $\theta = V/Q$, where V is the total reactor volume (both aeration and settling chambers — that is, 10 ℓ).

GLOSSARY

acclimated culture or acclimated activated sludge A culture of microbes that has been developed to use a particular substrate or wastewater. To develop a culture, activated sludge from a municipal plant is gradually fed increasing amounts of the substrate or wastewater until the microbes that use these food substances have grown in abundance.

activated carbon Carbon particles made by carbonization of a cellulosic material in the absence of air. It has a high adsorptive capacity.

activated sludge The active biological solids in an activated sludge wastewater treatment plant.

activated sludge process A biological wastewater treatment process in which a mixture of the wastewater and activated sludge is aerated in a reactor basin or aeration tank. The active biological solids bio-oxidize the waste matter, and the biological solids are then removed by secondary clarification or final settling.

advanced waste treatment Consists of: (1) The use of physical and chemical means for treating raw wastewaters, such as municipal wastewaters, in lieu of primary and secondary treatment. (2) The use of physical, chemical, and biological means to upgrade the quality of a secondary effluent. (3) The use of flowsheets, treatment operations, and/or processes not routinely used in wastewater treatment.

aerated lagoon A wastewater treatment pond in which mechanical or diffused-air aeration is used to supplement the oxygen supply artificially.

aeration tank A tank or basin in which mixed liquor, wastewater, sludge, or other liquid is aerated.

Aerobacter aerogenes One of the species of bacteria in the coliform group.

aerobic Requiring the presence of free molecular oxygen.

aerobic bacteria Bacteria that require free molecular oxygen for their life processes.

aerobic digestion Digestion of suspended organic matter by aerobic microbial action.

algae Primitive microorganisms, single or multicellular, usually aquatic and capable of utilizing food materials by photosynthesis.

alkalinity The ability of a water to neutralize an acid. It is due to the presence of bicarbonate, carbonate, and hydroxide ions, although occasionally it is caused by the presence of borate, silicate, and phosphate ions. It is expressed as mg/ℓ of equivalent $CaCO_3$ or meq/ℓ.

allowable breakthrough concentration The maximum acceptable concentration of a solute in the effluent from an adsorption or ion exchange column.

amino acid Organic acid containing one or more amino groups $(-NH_2)$. Amino acids are the building blocks for proteins.

ammonia stripping The passage of a high-pH wastewater downward through a packed tower countercurrent to an induced air flow passing upward. Because of the favorable equilibrium, the ammonia is stripped from the water and leaves with the air.

anaerobic Requiring combined oxygen, such as SO_4^{-2}, PO_4^{-3}, NO_3^{-1}, and so on, and requiring the absence of free molecular oxygen.

anaerobic bacteria Bacteria that require combined oxygen and the absence of free molecular oxygen.

anaerobic digestion Digestion of suspended organic matter by anaerobic microbial action.

anion A negatively charged ion in an electrolyte solution that migrates to the anode when an electrical potential is applied.

autotrophic bacteria Bacteria that use inorganic materials for energy and growth.

available chlorine A measure of the oxidizing power of hypochlorous acid and the hypochlorite ion.

bacilli Rod-shaped or cylindrical bacterial cells.

bacteria A group of universally distributed, rigid, essentially unicellular micoscopic organisms lack-

Note: Some of the terms in the Glossary are reprinted from the publication *Glossary — Water and Wastewater Engineering*, by permission of the Water Environment Federation.

769

ing chlorophyll. Usually, they have a spheriod, rodlike, or spiral shape. Some use organic matter as a foodstuff; others use inorganic matter.

bar rack or bar screen   A screen consisting of parallel bars, either vertical or inclined, that is placed in a waterway to remove debris. The screenings are raked from the screen by manual or mechanical means.

batch activated carbon process   The use of activated carbon adsorption in a batchwise manner. That is, the activated carbon and water or wastewater are added to a vessel, mixing is provided, transfer of the organic material from the liquid to the carbon occurs, the contents are allowed to settle, and then the treated liquid is withdrawn.

batch activated sludge process   The use of the activated sludge process in a batchwise manner. That is, the wastewater and activated sludge are placed in a reaction vessel, aeration is provided until the biochemical reaction is essentially complete, the contents are allowed to settle, and then the treated wastewater is withdrawn.

batch operation or process   An operating technique that is batchwise in manner — that is, there are no continuous flows in or out of the operation or process.

batch reactor   A reactor that does not have continuous streams entering or leaving. The reactants are added, the reaction occurs, and then the products are discharged.

bench-scale   A laboratory scale operation, process, or combination thereof.

binary fission   The manner in which most bacteria multiply. The parent cell divides into two daughter cells.

biochemical action   A chemical change resulting from the metabolism of living organisms.

biochemical oxidation   An oxidation caused by biological activity resulting in a chemical combination of oxygen with organic matter to produce relatively stable end products. Same as bio-oxidation and biological oxidation.

biochemical oxygen demand (BOD)   The amount of oxygen required by microbes in the stabilization of a decomposable waste under aerobic conditions.

biodegradation (biodegradability)   The biological oxidation of natural or synthetic organic materials by soil microorganisms in soils, waterbodies, or wastewater treatment plants.

biological oxidation   An oxidation caused by biological activity resulting in a chemical combination of oxygen with organic matter to produce relatively stable end products. Same as biochemical oxidation and bio-oxidation.

biomass   Living biological matter.

bio-oxidation   Same as biochemical oxidation or biological oxidation.

biosorption activated sludge   See contact stabilization activated sludge process.

blowdown   (1) Water discharged from a boiler or cooling tower to dispose of accumulated salts. (2) Removal of a portion of flow to maintain constituents of flow within desired levels.

BOD₅   See five-day biochemical oxygen demand.

BODᵤ   See ultimate biochemical oxygen demand.

boiler feedwater   Water to be used in a boiler.

brackish water   Water having a dissolved solids content between that of fresh water and that of sea water.

breakpoint chlorination   Addition of chlorine to a water or wastewater until the chlorine demand has been satisfied. Further additions result in a residual that is directly proportional to the amount added beyond the breakpoint.

breakthrough curve   A performance curve for a test or design column used for carbon adsorption and ion exchange. The solute concentration in the effluent is plotted on the y-axis versus the effluent throughput volume on the x-axis.

Brownian movement   The random movement of microscopic particles suspended in a liquid medium.

brush aerator   A surface aerator consisting of a rotating horizontal axle with protruding steel bristles partially submerged in the still water surface. Oxygen is transferred by air entrainment in the vicinity of the rotating bristles and also by the spray and impingement area. Known as the Kessener brush in Europe.

buffer action   The action of certain ions in solution to oppose a change in pH.

bulking sludge   An activated sludge that settles poorly because of a floc with a low bulk density.

cake   Sludge that has been dewatered.

carbohydrates   Organic compounds containing carbon, hydrogen, and oxygen. Common carbohydrates are sugars, starches, and cellulose.

carbon adsorption   The use of activated carbon to remove dissolved organic matter from a water or wastewater.

carbonate hardness (CH) That part of the total hardness that is chemically equivalent to the bicarbonate plus carbonate alkalinity present in the water. Usually expressed as mg/ℓ of equivalent $CaCO_3$ or meq/ℓ. See hardness.

cathode A negatively charged electrode in an electrolytic cell.

cation A positively charged ion in an electrolyte solution that migrates to the cathode when an electrical potential is applied.

centrate The liquid flow leaving a centrifuge.

centrifuge A mechanical device that uses a centrifugal force to separate solids from a liquid.

chemical analysis Analysis by chemical methods to determine the composition and concentration of substances present.

chemical coagulation The destabilization and initial aggregation of colloidal and finely suspended matter by the addition of a floc-forming chemical coagulant.

chemical oxygen demand (COD) The amount of oxygen required to oxidize chemically the organic and sometimes inorganic matter in a water or wastewater. Usually expressed in mg/ℓ. The COD test does not measure the oxygen required to convert ammonia to nitrites and nitrites to nitrates; thus COD is frequently assumed to be equal to the ultimate first-stage biochemical oxygen demand.

chemical precipitation Precipitation resulting from the addition of a chemical.

chemical sludge Sludge produced by chemical coagulation or chemical precipitation.

chemical treatment A process involving the addition of chemicals to obtain a desired result.

chemically coagulated raw wastewater A raw wastewater that has been chemically coagulated. Usually, it has been settled and filtered.

chemically conditioned sludge A sludge that has had chemicals added to enhance its dewatering characteristics.

chemically treated secondary effluent A secondary effluent that has been chemically treated, usually by coagulation along with other processes or operations.

chloramines Compounds of organic or inorganic nitrogen and chlorine.

chlorination The addition of chlorine to a water or wastewater, usually for disinfection purposes;

however, sometimes it is for accomplishing other biological or chemical results.

chlorine contact tank A tank that allows added chlorine sufficient time to disinfect or accomplish other desired results.

chlorine demand The difference between the amount of chlorine added to a water or wastewater and the amount of residual chlorine remaining after contact of a specific duration.

clarification The removal of settleable suspended solids from a water or wastewater by gravity settling in a quiescent tank or basin. Also called *sedimentation* or *settling*.

clarified wastewater A wastewater that has had most of the settleable solids removed by clarification. Also called *settled wastewater*.

clarifier Same as sedimentation basin.

clear well A ground storage reservoir for filtered water that is of sufficient volume to allow the filtration plant to operate at a constant flowrate on the day of maximum demand.

clinoptilolite A naturally occurring zeolite that has ion exchange ability. It is used in removing the ammonium ion.

coagulant A compound that causes coagulation — that is, a floc-forming agent.

coagulant aid A chemical or substance used to assist in coagulation.

coagulation In water or wastewater treatment, the destabilization and initial aggregation of colloidal and finely divided suspended solids by the addition of a floc-forming chemical.

coarse rack A rack with relatively wide spaces between the bars, usually 1 in. (2.54 cm) or more.

coarse screen A mesh or bar screen with openings of 1 in. (2.54 cm) or more.

cocci Spherical bacterial cells.

COD See chemical oxygen demand.

coliform bacteria A group of bacteria predominantly living in the intestines of humans and other warm-blooded animals but also found elsewhere, such as in soils. It includes all aerobic and facultative anaerobic, Gram-negative, non–spore-forming bacilli that ferment lactose with gas production.

coliphage A virus pathogenic to coliforms.

colloids Fine suspended solids that will not settle by gravity but may be removed by coagulation, biological action, or membrane filtration.

combined chlorine The concentration of chlorine

that is combined with ammonia as chloramines or as other chloroderivatives, yet is still available for chemical oxidation. Same as combined available chlorine.

combined chlorine residual That part of the total residual chlorine remaining in a water or wastewater at the end of a contact of specific duration. It will react chemically or biologically as chloramines or organic chloramines. Same as combined available chlorine residual.

comminution The cutting and screening of solids contained in a wastewater prior to pumping or any treatments at a wastewater treatment plant.

complete treatment A wastewater treatment that uses both primary and secondary treatment.

completely mixed activated sludge An activated sludge process with a completely mixed reactor basin. The usual basin is square, circular, or slightly rectangular in plan view, and the influent, on entering, is almost immediately dispersed throughout the reactor basin.

completely mixed reactor A reactor where the fluid elements, on entering, are dispersed almost immediately throughout the reactor volume. Usually, it is circular, square, or slightly rectangular in plan view.

contact stabilization activated sludge process An activated sludge process with a contact tank where sorption of the organic materials occurs and a sludge stabilization tank where the sludge biooxidizes the sorbed organic matter. Same as biosorption.

continuous-flow operation or process An operating technique that is continuous in manner; that is, there are continuous flows into and out of the operation or process.

continuous-flow reactor A reactor that has a continuous stream of reactants entering and a continuous stream of products leaving.

conventional activated sludge process An activated sludge plant with rectangular reactor basin and air diffusers or aerators spaced uniformly along the basin length.

conventional digester A low-rate anaerobic digester.

conventional wastewater treatment The use of primary and secondary treatment.

cooling water Water used to reduce the temperature of fluids and gases by use of industrial condensers.

countercurrent An operation or process that has two streams of materials in contact with each other and moving in opposite directions. For example, in countercurrent ion exchange, the ion exchange resin moves continuously down the column or bed, while the water or wastewater continuously moves upward.

Crenothrix A genus of bacteria that occur as filaments having a sheath of deposited iron. They cause color, odor, and objectionable tastes in a water.

culture Microbial growth that has been developed by provision of suitable nutrients and environment.

decomposition of wastewater The breakdown of organic matter in wastewater by microbial action. It may be under aerobic or anaerobic conditions.

degradation The breakdown of substances by biological oxidation.

demineralization The removal of all salts from a water.

desalting Removal of salts from a brackish water or a tertiary effluent.

dewater To remove part of the water present in a sludge.

dewatered sludge A sludge that has had some of its water content removed.

diatomaceous-earth filter A filter usually used in water treatment that utilizes a built-up layer of diatomaceous earth as a filter medium.

diffused air aeration Aeration produced in a liquid by the use of compressed air passed through air diffusers.

digested sludge Sludge digested by aerobic or anaerobic action to the degree that the volatile content is low enough for the sludge to be stable.

digester A tank used for sludge digestion. See sludge digestion.

digestion The biological oxidation of organic matter in sludge, resulting in stabilization. See sludge digestion.

disinfection The killing or inactivation of most of the microorganisms in or on a substance with the probability that all pathogenic bacteria are killed by the disinfecting agent used.

dispersed plug-flow activated sludge An activated sludge process with a dispersed plug-flow reactor basin. The basin is rectangular in plan view and has significant longitudinal or axial dispersion of fluid elements throughout its length.

dispersed plug-flow reactor A reactor that is rectangular in plan view and has significant longitudinal mixing of fluid elements throughout its length.

dissolved oxygen　The oxygen dissolved in a liquid, usually expressed in mg/ℓ. Abbreviated DO.

dissolved solids　Dissolved substances in a water or wastewater.

DO　See dissolved oxygen.

domestic wastewater　Wastewater mainly from dwellings, business buildings, institutions, and so on.

dry suspended solids　The suspended matter in water and, in particular, wastewater, that is removed by laboratory filtration and is dried for one hour at 103°C.

dry-weather flow　The flow of wastewater in a sanitary sewer during dry weather. It is wastewater and dry-weather infiltration.

effluent　Wastewater or other liquid, partially or completely treated or in its natural state, flowing out of a basin, reservoir, treatment plant, or industrial treatment plant or parts thereof.

electrolyte　A substance that dissociates into ions when it is dissolved in water.

endocellular　Inside a microbial cell — for example, endocellular enzymes.

endogenous decay　The continuous process in which microbial cell tissue decays.

end products　Substances created by microbial metabolism and growth.

enteric bacteria　Bacteria that inhabit the intestines of humans and other animals.

enzymes　Organic catalysts that are proteins and are produced by living cells.

equalization basin　A holding basin where variations in flowrate and liquid composition are averaged. An equalization basin provides an effluent of reasonably uniform composition and flowrate to subsequent treatment operations or processes.

Escherichia coli (E. coli)　A species of bacteria in the coliform group. Its presence is considered indicative of fresh fecal contamination.

essential amino acid　An amino acid that is required by a living cell but that it cannot synthesize.

excess activated sludge　Same as waste activated sludge.

exocellular　Outside a microbial cell — for example, exocellular enzymes.

extended aeration activated sludge process　An activated sludge process with a detention time long enough to allow the amount of cells synthesized to be endogenously decayed.

facultative anaerobic bacteria　Bacteria that use either free molecular oxygen, if available, or combined oxygen. Also called *facultative bacteria*.

facultative stabilization pond　A basin for retention of a wastewater, in which biological oxidation of the organic matter occurs by bacterial action. A significant amount of molecular oxygen used by the microbes is supplied by algal photosynthesis.

fats　Triglyceride esters of fatty acids. Mistakenly used as synonymous with grease.

fermentation　(1) A biochemical change caused by a ferment, such as yeast enzymes. (2) A biochemical change in organic matter or organic wastes caused by anaerobic biological action.

fill and draw reactor　A batch-operated activated sludge reactor. See batch activated sludge process.

filter　A tanklike structure with a granular-bed and underdrain system that is used to remove fine suspended solids and colloids from a water or wastewater. The separation occurs as the liquid passes through the bed.

filter bed　A tank used for water or wastewater filtration that has an underdrain system covered by a granular filter medium.

filter operating table　A table set on the operating floor of a rapid sand filtration plant. It is placed in front of the filter that it operates and supports all the equipment for control and operation of the filter. It has the controls or all valves, the rate-of-flow gauge, the loss-of-head gauge, and so on.

filtration plant　A water treatment plant consisting of all operations and processes, structures, and appurtenances required for the filtration of water. Also called rapid sand filtration plant.

filter press　A mechanical press used to dewater sludges.

filter run　The time interval between the backwashings of a rapid sand filter.

filter underdrains　The system of underdrains for a granular medium filter. It is used for collecting the filtrate and for distributing the washwater.

filtration　The unit operation that consists of passing a liquid through a granular medium for the removal of suspended and colloidal matter.

final clarifier　The last settling basin or settling tank at a wastewater treatment plant. In the activated sludge process, it separates the biological solids from the final effluent. In the trickling filter process, it separates the trickling filter humus (that is, sloughed growths) from the final effluent.

final effluent The effluent from the final clarifier, final sedimentation basin, or final settling tank at a wastewater treatment plant.

fine rack A bar rack with clear spaces of about 1 in. (2.54 cm) or less between the bars.

fine screen In water treatment, a screen with openings less than 1 in. (2.54 cm). In wastewater treatment, a screen with openings less than 1/16 in. (1.6 mm).

first-stage biochemical oxygen demand That part of the biochemical oxygen demand that results from the biological oxidation of carbonaceous materials, as distinct from nitrogenous materials. Generally, the major portion of the carbonaceous materials are bio-oxidized before the bio-oxidation of nitrogenous materials (the second-stage biochemical oxygen demand) begins.

five-day biochemical oxygen demand (BOD$_5$) The oxygen required by microbes in the stabilization of a decomposable waste under aerobic conditions for a period of five days at 20°C and under specified conditions. It represents the breakdown of carbonaceous materials as distinct from nitrogenous materials.

fixed bed In carbon adsorption or ion exchange treatments using columns or open beds, a bed that is stationary in the column or in the structure for the open bed.

fixed solids The residue remaining after the ignition of suspended or dissolved solids at 500°C.

floc In water treatment, the small, gelatinous masses formed in the water by the adding of a coagulant. In wastewater treatment, the small, gelatinous biological solids formed at an activated sludge treatment plant.

flocculation In water and wastewater treatment, the slow stirring of a coagulated water or wastewater to aggregate the destabilized particles and form a rapidly settling floc. In biological wastewater treatment where a coagulant is not used, the aggregation may be accomplished biologically.

flocculator A basin in which flocculation is done or a mechanical device to enhance the formation of floc in a liquid.

flotation The raising of suspended matter to the surface of a liquid, where it is removed by skimming.

flowsheet A diagrammatic representation of unit operations and processes used in a water or wastewater treatment scheme.

fluidized bed (1) In carbon adsorption or ion exchange treatment, a bed in which the particles are not in continuous contact because of the upward flow of the water or wastewater. (2) In sludge incineration, an incineration bed of sand particles that are not in continuous contact because of the upward flow of the combustion air.

free chlorine The amount of chlorine available as dissolved gas, hypochlorous acid, or hypochlorite ion.

free chlorine residual The portion of the total residual chlorine remaining in a water or wastewater at the end of contact of a specific duration, which will react chemically as hypochlorous acid or hypochlorite ion. Same as free available chlorine residual.

fresh sludge Undigested organic sludge.

fresh wastewater Raw wastewater containing some dissolved oxygen.

fungi Small, multicellular, nonphotosynthetic, microorganisms that lack chlorophyll, roots, stems, or leaves and that feed on organic matter. Their decomposition after death may cause disagreeable tastes and odors in a water. They are found in water, wastewater, wastewater effluents, and soil.

gas stripping An operation in which a solute gas is stripped from a liquid.

gasification The conversion of soluble organic matter into gas during anaerobic bio-oxidation.

Gram-negative See Gram stain.

Gram-positive See Gram stain.

Gram stain A staining procedure used in the identification of a bacterial species. A microscopic slide is made of the culture, and a purple stain, such as crystal violet, is added to the slide. Then the slide is washed using special solutions, and a red stain, such as safranin, is added. The slide is then washed with water. Once dry, the slide is viewed under a microscope. A species stained purple is Gram-positive and a species stained red is Gram-negative. A species always has the same staining characteristics — for example, *E. coli* is always Gram-negative.

granular medium A granular material, such as sand or crushed anthracite coal, that serves as the filter bed.

granular medium filtration Filtration through a bed of granular material.

gravity filters Filters that have gravity flow of the water through the filter bed.

gravity thickening Thickening of a sludge using gravity settling in a tank. Pickets, usually mounted on the trusswork for the sludge, scrapers, rake through the sludge, releasing entrained water. This allows the sludge to subside and concentrate.

grease In wastewater treatment, a group of substances such as fats, waxes, free fatty acids, calcium and magnesium soaps, mineral oils, and other fatlike materials.

grit The dense, mineral, suspended matter present in a water or wastewater, such as silt and sand.

grit chamber In wastewater treatment, a settling chamber to remove grit from organic solids.

grit removal The removal of heavy suspended mineral matter present in a wastewater, such as sand and silt.

groundwater Subsurface water from which springs and wells are fed.

hard water A water with significant hardness. Waters with hardness greater than $75 \, mg/\ell$ as $CaCO_3$ are usually considered hard. See hardness.

hardness The ability of a water to consume excessive amounts of soap prior to forming a lather and to produce scale in hot-water heaters, boilers, or other units in which the temperature of the water is significantly increased. It is due to the presence of polyvalent metallic ions, mainly calcium and magnesium. Occasionally strontium, ferrous, and manganous ions are present. Usually expressed as mg/ℓ of equivalent $CaCO_3$ or meq/ℓ.

heavy metals Metals that can be precipitated by hydrogen sulfide in an acid solution — for example, lead, mercury, cadmium, zinc, silver, gold, bismuth, and copper. Above certain concentrations, they can inhibit microbial action in wastewater treatment plants and are toxic to humans.

helminth An intestinal worm.

herbicide An agent used to inhibit or kill plant growth.

heterotrophs Bacteria that feed on preformed organic matter.

high-rate digester An anaerobic digester with continuous mixing, continuous feeding, and digester heating.

high-rate trickling filter A trickling filter with continuous recycle, an organic loading greater than $800 \, lb \, BOD_5/ac\text{-}ft\text{-}day$ $(0.29 \, kg/m^3\text{-}day)$, and a hydraulic loading greater than $10 \, MGD/ac$ $(9.35 \, m^3/m^2\text{-}day)$.

high service pumps Pumps that pump from a clear well to a distribution system.

hydrocarbons Organic compounds consisting of carbon and hydrogen.

Imhoff tank A deep, two-storied wastewater treatment tank. The upper story is a continuous-flow primary settling chamber, and the lower story is an anaerobic digestion chamber. The sludge from the upper chamber passes through trapped slots to the lower chamber.

incineration The combustion of organic sludges to produce CO_2, H_2O, and other stable forms. The material remaining will be ash.

industrial wastewater The liquid wastes from industrial processes.

industrial water Water used in an industrial water system for process water, cooling water, and boiler water.

infiltration water Water that has migrated from the ground into a sewer system.

influent Water, wastewater, or other liquid flowing into a reservoir, treatment plant, or any unit thereof.

inoculate To introduce viable cells into a medium.

inorganic industrial wastewater An industrial wastewater that has inorganic ions or compounds as the objectionable constituents.

insecticide An agent used to kill insects.

institutional wastewater Wastewater from institutions such as hospitals, prisons, or charitable institutions.

ion A charged atom, molecule, or radical.

ion exchange A chemical process involving the reversible exchange of ions between a liquid and a solid.

ion exchange softening A process in which calcium and magnesium ions are removed from water by ion exchange. The exchanger exchanges sodium ions for the calcium and magnesium ions that are removed.

ion exchange treatment The use of ion exchangers such as resins to remove unwanted ions from a liquid and substitute more desirable ions.

iron bacteria Bacteria that assimilate iron and excrete its compounds in their life processes.

isoelectric point The pH at which electroneutrality occurs.

lime or slaked lime Calcium hydroxide, $Ca(OH)_2$.

lime recalcination The heat treatment of a sludge resulting from lime coagulation. It converts the calcium carbonate precipitate to calcium oxide.

lime slaking The reacting of quicklime, CaO, with water to produce calcium hydroxide, $Ca(OH)_2$.

lime-soda softening A process in which calcium and magnesium ions are precipitated from a water by reaction with lime and soda ash.

liquefaction The changing of organic solids to soluble forms.

low-rate digester An anaerobic digester with intermittent mixing and intermittent feeding. It may or may not be heated. Same as conventional digester.

low-rate trickling filter A trickling filter with recycle only during periods of low flow, an organic loading less than 800 lb BOD_5/ac-ft-day (0.29 kg/m^3-day), and a hydraulic load less than 6 MGD/ac (5.61 m^3/m^2-day). Same as standard-rate trickling filter.

make-up water Water added to a circulating water system to replace water lost by leakage, evaporation, or blowdown.

manganese bacteria Bacteria that utilize dissolved manganese and deposit it as hydrated manganic hydroxide.

mass transfer The transfer of a substance from one phase to another. For example, in diffused air aeration, there is transfer of oxygen from the diffused compressed air into the liquid.

mean cell residence time (θ_c) The average time a microbial cell remains in an activated sludge or sludge digestion system. It is equal to the mass of cells divided by the rate of cell wastage from the system.

mechanical aeration (1) The transfer of oxygen from the atmosphere into a liquid by the mechanical action of a turbine or other mechanisms. (2) The mixing by mechanical means of the mixed liquor in the reactor basin or aeration tank of an activated sludge treatment plant.

media For trickling filters, the packing, such as stone, in the filter bed.

medium (1) A porous material used in filters. (2) A nutrient material used to cultivate bacteria.

mesophilic digestion Anaerobic digestion by biological oxidation by anaerobic action at or below 45°C (110°F).

microbes Same as microorganisms.

microbial activity Chemical changes resulting from biochemical action — that is, the metabolism of living organisms.

microorganisms Minute organisms visible only by means of a microscope. Same as microbe.

microscopic Visible only by magnification with an optical microscope. The size range is from 0.5 to 100 μ.

microscreen A rotating drum with a screen mesh surrounding the drum. The wastewater enters the interior of the drum and flows outward. A backwash at the top of the drum continuously removes the screenings.

milliequivalent (meq) The weight in milligrams of a substance that combines with or displaces 1 mg of hydrogen. It is equal to the formula weight in mg divided by its valence.

mixed culture A microbial culture consisting of two or more species.

mixed liquor A mixture of wastewater and activated sludge undergoing aeration in a reactor basin or aeration tank in an activated sludge wastewater treatment plant.

mixed-liquor suspended solids (MLSS) The suspended solids in the mixed liquor — that is, the mixture of wastewater and activated sludge undergoing aeration at an activated sludge wastewater treatment plant.

mixed-liquor volatile suspended solids (MLVSS) The volatile fraction of the mixed-liquor suspended solids. The MLVSS is usually considered to be more representative of the active biological solids than is the MLSS.

mixing tank A tank to provide thorough mixing of chemicals added to a liquid. Also called *mixing basin*, *mixing chamber*, or *rapid-mix tank*.

MLSS See mixed-liquor suspended solids.

MLVSS See mixed-liquor volatile suspended solids.

moisture content The amount of water present in wastewater sludge, industrial wastewater sludge, water treatment sludge, or soil. Usually expressed as a percentage of the wet weight.

morphology The shape and grouping of bacterial cells. The three basic shapes are bacilli (rod or cylindrical shapes), cocci (spherical shape), and spirilli (spiral shape). Cells usually appear as single cells, bunches, or chains, although some other groupings do occur. It is a characteristic of a species to have the same morphology. For example, *E. coli* is always rod-shaped and occurs as single cells.

most probable number (MPN) The statistical number of organisms in a 100-mℓ sample.

moving bed In carbon adsorption or ion exchange treatments using columns or open beds, a bed that moves downward through the column or the structure for the open bed. The bed depth is a constant value because the carbon or exchanger material continuously enters above the bed.

MPN See most probable number.

multimedia Two or three granular media used for filter beds, such as crushed coal and sand, or crushed coal, sand, and garnet.

multimedia filtration The filtration of a water or wastewater through a granular bed containing two or more filter media.

multiple-hearth furnace A furnace that consists of numerous hearths and is used to incinerate organic sludges or recalcinate lime.

municipal wastewater Wastewater derived principally from dwellings, business buildings, institutions, and the like. It may or may not contain groundwater, surface water, storm water, and industrial wastewater.

nematode A worm that feeds on minute organisms.

nitrification The conversion of nitrogenous matter into nitrates by bacterial action.

Nitrobacter A genus of bacteria that oxidize nitrites to nitrates.

Nitrosomonas A genus of bacteria that oxidize ammonia to nitrites.

noncarbonate hardness (NCH) That amount of the hardness equal to the total hardness minus the carbonate hardness. Usually expressed as mg/ℓ of equivalent $CaCO_3$ or meq/ℓ. See hardness.

nonsettleable solids Suspended solids in a wastewater that will not settle by gravity means within a given duration of time, usually considered 1 hour for laboratory testing.

operating floor The floor in a rapid sand filter building on which the operating and indicating devices are generally installed.

organic industrial wastewater An industrial wastewater that has organic compounds as the objectionable constituents.

organic matter Chemical substances of animal and vegetable origin that are carbon compounds.

organic-matter degradation The conversion of organic matter to inorganic forms by biological action.

organic nitrogen Organic compounds or molecules containing nitrogen, such as proteins, amines, and amino acids.

organic sludges Sludges that have a high organic content. Usually, these are primary or secondary sludges at a wastewater treatment plant.

oxidation ditch An extended aeration process with a racetrack-shaped aeration basin.

oxidation pond A pond where the organic matter in a wastewater is biologically oxidized. Same as a waste stabilization pond or, if artificial aeration is provided, an aerated lagoon.

ozone Oxygen in molecular form consisting of three atoms of oxygen forming each molecule (O_3).

parasitic bacteria Bacteria that require a living host organism but do not harm the host.

pathogen Pathogenic or disease-producing organism.

pathogenic bacteria Bacteria that require a living host organism and harm the host by causing disease.

pesticide Any substance or chemical applied to kill or control pests, such as weeds, insects, algae, rodents, and so on.

petrochemicals Products or compounds produced by the chemical processing of petroleum and natural gas hydrocarbons.

pH The reciprocal of the logarithm of the hydrogen ion concentration in gram moles per liter. Neutral water has a pH of 7.

phosphorus The phosphorus radicals found in a wastewater. Usually, the major portion is orthophosphate, PO_4^{-3}.

photosynthesis The synthesis of complex organic materials from carbon dioxide, water, and inorganic salts using sunlight as the energy source and a catalyst such as chlorophyll.

physical analysis The examination of a water or wastewater to determine physical characteristics, such as temperature, turbidity, color, odor, and so on.

physical-chemical treatment The use of physical and chemical means for treating a raw wastewater or a secondary effluent.

pilot scale A pilot-scale operation, process, or combination thereof. It is operated in the field.

pipe gallery A gallery provided in a rapid sand filtration plant that houses the pipes going to and from the filters, along with their valves and other appurtenances. It also serves as an access passageway.

plain sedimentation The gravity settling of suspended solids in a water or wastewater without the aid of chemical coagulants.

plug-flow reactor A reactor in which all fluid elements that enter the reactor at the same time flow through it with the same velocity and leave at the same time. The travel time of the fluid elements is equal to the theoretical detention time, and there is no longitudinal mixing. It is approached in long, narrow tanks.

polishing ponds Stabilization lagoons or oxidation ponds that have detention times less than 24 hours and are used as a finishing treatment. They are frequently used for trickling filter effluents.

pollution A conditon caused by the presence of harmful or objectionable material in a water.

polyelectrolytes Organic polymers used as coagulant aids or coagulants.

polymers Organic polyelectrolytes used as coagulant aids or coagulants.

precipitate To separate from solution as a precipitate.

precipitation The phenomenon that occurs when a substance held in solution in a liquid passes out of solution into solid form.

precursor A substance from which another substance is formed.

preliminary treatment In a wastewater treatment plant, unit operations like screening, comminution, and grit removal that prepare the wastewater for subsequent major operations.

pressure filters Filters that operate under pressure. Usually the flow is pumped to and through the filter.

primary clarifier In a wastewater treatment plant, this is the first clarifier used. It is for removal of settleable suspended solids. Also called *primary settling tank*.

primary sludge The sludge from a primary clarifier.

primary treatment The treatment of a wastewater by sedimentation to remove a substantial amount of the suspended solids.

process water Industrial water used in making or manufacturing a product.

proteins Complex nitrogenous compounds formed in living organisms. The compounds consist of amino acids bound together by the peptide linkage.

protozoa Small, single-celled, animal-like microbes including amoebas, ciliates, and flagellants.

pulsed bed A countercurrent bed with intermittent bed movement.

pure culture A microbial culture consisting of only one species.

pure oxygen activated sludge process An activated sludge process that uses pure molecular oxygen, instead of atmospheric oxygen, for microbial respiration.

purification The removal of objectionable material from a water or wastewater by natural or artificial means.

quicklime A calcined material that is mainly calcium oxide, CaO.

rack A device consisting of parallel bars evenly spaced that is placed in a waterway to remove suspended or floating solids from a wastewater.

rapid-mix basins Basins or tanks with agitation that are used to disperse and dissolve one or more chemicals in a water or wastewater.

rapid sand filter A granular-medium filter used in water treatment in which the water is passed downward through the filter medium. Typical media are sand and/or crushed anthracite coal. The water is pretreated by coagulation, flocculation, and sedimentation.

rapid sand filtration plant A water treatment plant consisting of coagulation, flocculation, sedimentation, rapid sand filtration, and disinfection. The rapid sand filters used contain a granular medium or multimedia.

rate-of-flow controller An automatic device that controls the rate of flow of a fluid, usually flowing in a pipe.

rate-of-flow indicator A device that indicates the rate of flow of a fluid at any time. It may or may not have a recorder.

raw sludge A settled sludge that is undigested. Also called *undigested sludge* and *fresh sludge*.

raw wastewater A wastewater before it receives any treatments.

raw water Untreated water. Usually the water entering the first treatment unit at a water treatment plant.

reactor basin An aeration tank at an activated sludge plant.

recarbonation The diffusion of carbon dioxide in a water or wastewater after lime coagulation to lower the pH.

receiving body of water A natural watercourse, lake, or ocean into which a treated or untreated wastewater is discharged.

recycle ratio For an activated sludge or trickling

filter plant, the rate of recycled flow divided by the rate of influent flow.

residual chlorine Chlorine remaining in a water or wastewater at the end of a specified contact time as free or combined chlorine. Same as chlorine residual.

resins Synthetic organic resins that have ion exchange capability.

respiration The furnishing of energy for microbial metabolism and growth.

returned sludge Settled activated sludge that is returned to mix with the raw or primary settled wastewater.

rotary distributor A revolving distributor with long arms with jets that are used to spray wastewater on a trickling filter bed.

rotary biological contactor A biological contactor consisting of circular discs mounted on a horizontal rotating axle. Fixed biological growths are on the discs, and the discs are partially submerged in a vat containing the wastewater. As the discs rotate, the fixed biological growths sorb the organic matter and bio-oxidize the materials. Oxygen is supplied by absorption from the atmosphere as the discs are partially exposed during their rotation.

rotifer A minute, usually microscopic, multicellular invertebrate animal that feeds on organic matter.

sanitary landfill A landfill for disposing of solid wastes.

sanitary wastewater (1) Domestic wastewater without storm and surface run-off. (2) Wastewater from the sanitary conveniences in dwellings, office buildings, industrial plants, and institutions. (3) The water supply of a community after it has been used and discharged to a sewer.

saprophytic bacteria Bacteria that feed on dead or nonliving organic matter.

SDI See sludge density index.

secondary clarifier The final settling basin at a wastewater treatment plant. Also called *secondary settling tank*, *secondary settling basin*, or *final clarifier*.

secondary effluent The effluent leaving the secondary or final clarifier at a wastewater treatment plant.

secondary sludge The sludge from the final clarifier at a wastewater treatment plant. For the activated sludge process, it is the sludge to be recycled. For the trickling filter process, it is the trickling

filter growths that have sloughed off — that is, the trickling filter humus.

secondary treatment The treatment of a wastewater by biological oxidation after primary treatment by sedimentation.

second-stage biochemical oxygen demand That part of the biochemical oxygen demand that results from the biological oxidation of nitrogenous materials. This includes the bio-oxidation of ammonia to nitrites and of nitrites to nitrates. The oxidation of nitrogenous materials usually does not begin until a significant portion of the carbonaceous material has been bio-oxidized in the first stage.

sedimentation The removal of settleable suspended solids from a water or wastewater by gravity settling in a quiescent tank or basin. Also called *clarification* or *settling*.

sedimentation basin A basin or tank through which a water or wastewater is passed to remove settleable suspended solids by gravity settling. Also called *sedimentation tank*, *settling tank*, *settling basin*, or *clarifier*.

sedimentation tank Same as sedimentation basin.

settleable solids Suspended solids removed by gravity settling.

settled wastewater Wastewater that has been treated by sedimentation. Also called *clarified wastewater*.

settling Same as sedimentation and clarification.

settling basin Same as sedimentation basin.

settling tank Same as sedimentation basin.

sewage See wastewater.

shock load See slug load.

side water depth The depth of water in a tank measured from the bottom of the tank to the water surface at an exterior wall.

skimmings The grease, solids, and scum skimmed from a wastewater settling tank or basin.

slake To mix with water so that a chemical combination takes place, such as the slaking of lime where CaO reacts with H_2O to produce $Ca(OH)_2$.

slow sand filter A filter for water purification in which water without previous treatment is passed through a sand bed. It is characterized by a slow rate of filtration, such as 3 to 6 MGD/ac (0.032 to 0.065 ℓ/m^2-s) of area.

sludge (1) The accumulated solids removed from a sedimentation basin, settling tank, or clarifier in a water or wastewater treatment plant. (2) The precipitate resulting from the chemical coagulation, flocculation, and sedimentation of a water or wastewater.

sludge age   Same as mean cell residence time.

sludge bed   An area consisting of natural or artificial layers of porous materials on which digested wastewater sludge is dewatered by drainage and evaporation.

sludge blanket   (1) The depth of activated sludge in a final clarifier or thickener. (2) The accumulation of sludge hydrodynamically suspended in a solids-contact unit.

sludge bulking   A phenomenon that occurs in activated sludge treatment plants in which the sludge does not settle and concentrate readily.

sludge cake   Sludge that has been dewatered to increase the solids content to 15% to 45% based on dry solids.

sludge collector   A mechanical device for scraping the sludge along the bottom of a settling tank to a hopper where it can be withdrawn.

sludge conditioning   Treatment of a sludge, usually by chemical means, to enhance its dewatering characteristics.

sludge density index (SDI)   The concentration of an activated sludge after settling one liter of the mixed liquor for 30 minutes in a one-liter graduated cylinder. It is the reciprocal of the sludge volume index with appropriate conversions. Usually expressed as mg/ℓ.

sludge dewatering   The removal of part of the water in a sludge by any method such as centrifugation, filter pressing, vacuum filtration, or passing through a belt press.

sludge digestion   The biological oxidation of organic or volatile matter in sludges to produce more stable substances. See anaerobic and aerobic digestion.

sludge digestion gas   Gas produced from the biological oxidation of organic matter by anaerobic digestion.

sludge digestion tank   Tank used for the anaerobic digestion of organic sludges.

sludge dryer   A device that utilizes heat for the removal of a large portion of the water in a sludge.

sludge drying   The removal of a large portion of the moisture in a sludge by drainage or evaporation.

sludge gas utilization   The use of digester gas from anaerobic digesters for beneficial purposes such as heating and fueling engines.

sludge moisture content   The amount of water in a sludge, usually expressed as percentage of dry weight.

sludge processing   The collection, treatment, and disposal of sludge.

sludge rakes   Rakes used in the collection of settled sludge from a clarifier used in water or wastewater treatment.

sludge thickener   A tank or a piece of equipment that concentrates the solids in a sludge.

sludge thickening   Increasing the solids content of a sludge.

sludge treatment   The processing of wastewater sludges to render them innocuous. Common methods include anaerobic or aerobic digestion followed by sludge dewatering.

sludge volume index (SVI)   The volume in mℓ occupied by one gram of mixed liquor after settling for 30 minutes in a liter graduate cylinder. Expressed as mℓ/gm.

slug load   A sudden load, either organic or hydraulic, to a wastewater treatment plant. Same as shock load.

soft water   Water having a low concentration of calcium and magnesium ions. Usually considered to have less than 75 mg/ℓ as equivalent $CaCO_3$.

softening   The removal of hardness — that is, the polyvalent metallic ions such as calcium and magnesium — from a water.

solids handling system   In water or wastewater treatment, the system for the collection, treatment, and dewatering of sludges.

solute   A substance dissolved in a fluid.

***Sphaerotilus* bulking**   Sludge bulking caused by the presence of the filamentous bacteria *Sphaerotilus* in large numbers.

spirilla   Spiral-shaped bacterial cells.

spore   A reproductive body created by some bacteria and capable of development into a new bacterial cell under proper environmental conditons.

spore formers   Bacteria capable of spore formation.

stabilization   In lime-soda softening or lime coagulation, any process that minimizes or eliminates scale-forming tendencies.

stabilization pond   An oxidation pond where the organic matter in a wastewater is biologically oxidized without artificial aeration.

stale wastewater   Raw wastewater that does not contain any dissolved oxygen.

standard biochemical oxygen demand   The biochemical oxygen demand determined by standard laboratory procedure for five days at 20°C. Usually expressed in mg/ℓ. See five-day biochemical oxygen demand.

Standard Methods Methods for the Examination of Water and Wastewater, which are jointly published by the American Public Health Association, the American Water Works Association, and the Water Environment Federation.

standard-rate trickling filter Same as low-rate trickling filter.

substrate The food substances for microorganisms in a liquid solution.

sulfur bacteria Bacteria that use dissolved sulfur or dissolved sulfur compounds for their metabolism or growth.

supernatant liquor The liquid released during anaerobic or aerobic digestion. In a stratified tank, it lies above the sludge and beneath the scum layer.

surface water Water that appears on the surface of the earth as distinguished from groundwater.

suspended matter Solids in suspension in a water or wastewater that can be removed by laboratory filtration techniques, such as membrane filtration.

suspended solids Solids in suspension in a water or wastewater that can be removed by laboratory filtration techniques, such as membrane filtration. Same as suspended matter.

SVI See sludge volume index.

synthesis The forming of new cell tissue in microbial metabolism and growth.

tapered aeration activated sludge process An activated sludge plant with a rectangular reactor basin and air diffusers or aerators spaced along the reactor length in accordance with the oxygen demand.

tertiary treatment The use of physical, chemical, or biological means to upgrade a secondary effluent.

theoretical oxygen demand (TOD) The amount of oxygen stoichiometrically required to convert organic matter to stabilized substances such as CO_2, H_2O, NO_3^-, and so on.

thermophilic digestion Anaerobic digestion at a temperature within the thermophilic range, which is usually considered to be between 110° and 145°F (43.3 and 62.8°C).

θ_c See mean cell residence time.

thickened sludge Sludge that has had some of its water removed.

thickener-clarifier A clarifier that settles and thickens the solids in the influent, while at the same time it clarifies the effluent leaving the tank.

thickening of sludge See sludge thickening.

TOC See total organic carbon.

TOD See theoretical oxygen demand.

total hardness (TH) The sum of the polyvalent metallic ion concentration in a water. It is principally due to calcium and magnesium ions in the water, although occasionally strontium, ferrous, and manganous ions are present. Usually expressed as mg/ℓ of equivalent $CaCO_3$ or meq/ℓ. See hardness.

total organic carbon (TOC) A measure of the organic matter in a water or wastewater in terms of the organic carbon content. Usually reported as mg/ℓ.

total solids The sum of the dissolved and suspended solids in a water or wastewater. Usually expressed as mg/ℓ.

traveling screen A rotating trash screen.

treated wastewater A wastewater that has undergone complete treatment or partial treatment.

trickling filter A biological filter consisting of a bed of coarse material, such as stone, over which wastewater is distributed by a spray from a moving distributor or other device. The wastewater trickles through the bed to the underdrains, giving the microbial slimes an opportunity to absorb the organic material and clarify the wastewater.

trickling filter media The packing in a trickling filter.

turbidity Suspended matter in a water or wastewater that causes the scattering or absorption of light rays.

ultimate biochemical oxygen demand (BOD$_u$) (1) Generally, the amount of oxygen required to completely satisfy the first-stage BOD. (2) More strictly, the oxygen required to completely satisfy the first- and second-stage BOD.

ultimate first-stage biochemical oxygen demand (BOD$_u$) The total amount of oxygen required to satisfy the firststage biochemical oxygen demand. The relationship between the BOD (y) at any day during the first stage and the BOD$_u$ is $y = \text{BOD}_u[1 - \exp(-kt)]$, where k is the rate constant to base e and t is the time in days.

undigested sludge Untreated, settled sludge that has been removed from a sedimentation basin or tank. Also called *fresh* or *raw sludge*.

unsaturated hydrocarbons Hydrocarbons having at least one double bond between two carbon atoms.

vacuum filter A sludge dewatering filter consisting of a cylindrical drum mounted on a horizontal axle, covered with a filter cloth, and rotating partly submerged in a sludge vat. A vacuum is maintained within the drum for the major part of the revolution to form the sludge cake on the drum and to dewater the cake. The cake is continuously scraped off or removed by other means.

viable cells Living cells.

virus The smallest form capable of producing disease in humans, other animals, and plants, being 10 to 300 mμ in diameter.

volatile acids Fatty acids with six or fewer carbon atoms that are water soluble. Usually expressed as equivalent acetic acid in mg/ℓ.

volatile matter Matter within a residue that is lost at 500°C ignition temperature. The ignition time must be sufficient to reach a constant weight of residue, usually 15 minutes. See volatile solids.

volatile solids The quantity of solids, either suspended or dissolved, lost in ignition at 500°C.

waste activated sludge The excess activated sludge produced by the microbial solids in an activated sludge plant. This amount has to be wasted from the system at the rate it is produced. Same as excess activated sludge.

waste treatment Any operation or process that removes objectionable constituents from a wastewater and renders it less offensive or dangerous. Same as wastewater treatment.

wastewater The spent water that consists of a combination of liquid and water-carried wastes from dwellings, business buildings, industrial plants, and institutions, along with any surface or groundwater infiltration. In recent years, the term *wastewater* has taken precedence over the word *sewage*.

wastewater analysis The determination of the physical, chemical, and biological characteristics of a wastewater or treatment plant effluent.

wastewater composition (1) The relative quantities of the various solids, liquids, and gases in a wastewater. (2) The physical, chemical, and biological constituents of a wastewater.

wastewater renovation The treatment of a wastewater for reuse.

wastewater treatment Any operation or process that removes objectionable constituents from a wastewater and renders it less offensive or dangerous. Same as waste treatment.

water analysis The determination of the physical, chemical, and biological characteristics of a water.

water-borne disease A disease caused by organisms or toxic materials transported by water. The most common water-borne diseases are typhoid fever, cholera, dysentery, and other intestinal disturbances.

water quality The physical, chemical, and biological characteristics of a water in regard to its suitability for a particular use.

water softening The process of removing calcium and magnesium ions from a water. The removal may be partial or total.

water treatment The treatment of a water by operations and processes to make it acceptable for a specific use.

water treatment plant A plant for removal of objectionable constituents from a water to make it satisfactory for a particular use.

zeolite Natural minerals or synthetic resins that have ion exchange capabilities.

zooglea The gelatinous material resulting from the attrition of bacterial slime layers. It is an important constituent of activated sludge floc and trickling filter growths.

GLOSSARY REFERENCES

Glossary — Water and Wastewater Engineering. 1969. American Water Works Association, Water Pollution Control Federation, American Society of Civil Engineers, and American Public Health Association.

McKinney, R. E. 1962. *Microbiology for Sanitary Engineers.* New York: McGraw-Hill.

Sawyer, C. N.; McCarty, P. L.; and Parkin G. F. 1994. *Chemistry for Environmental Engineering.* 4th ed. New York: McGraw-Hill.

Standard Methods for the Examination of Water and Wastewater. 1992. 18th ed. American Water Works Association, Water Environment Federation, and American Public Health Association.

Webster's Third New International Dictionary. 1968. Vols. I, II, and III. Copyright 1966, G. & C. Merriam Co. Encyclopaedia Brittannica, Inc., London, England.

ANSWERS TO SELECTED PROBLEMS

CHAPTER 1

1.1 $Na^+ = 23$, $Ca^{+2} = 40$
$Na^+ = 23$ mg/meq, $Ca^{+2} = 20$ mg/meq
1.3 $Na^+ = 168$ mg/ℓ
1.5 An insoluble product is formed.
1.7 3.70 meq/ℓ as $CaCO_3$
1.9 **a.** Second
   b. Zero
   c. First
1.11 **a.** Second order
   b. $K = 0.011575$ ℓ/mg-hr
1.13 8.67 mg/ℓ
1.15 Volume = 37, 100 ft^3 or 1050 m^3
1.17 Ratio = 0.60

CHAPTER 2

2.1 **a.** $dS/dt = 90$ mg/ℓ-hr
   b. $dS/dt = 2.5$ mg/ℓ-hr
2.3 **a.** $K = 0.506$ ℓ/gm-hr
   b. $K = 0.377$ ℓ/gm-hr

CHAPTER 3

3.1 **a.** First order, $k = 0.0633$/hr
   b. $C = 13.7$ mg/ℓ
3.3 **a.** $\theta = 5.04$ hr,
   volume = 30,200 gal = 115,000 liters
   b. Plot as required
3.5 **a.** $\theta_{dpf} = 12.7$ hr,
   volume = 76,200 gal = 290,000 liters
   b. $\theta_{dpf} = 10.4$ hr,
   volume = 62,400 gal = 237,000 liters
3.7 **a.** Graph as required
   b. $\theta = 2.90$ hr,
   volume = 17,400 gal each = 66,100 liters each
   $C_{A_1} = 70$ mg/ℓ, $C_{A_2} = 32$ mg/ℓ,
   $C_{A_3} = 15$ mg/ℓ
   $r_{A_1} = 28$ mg/ℓ-hr, $r_{A_2} = 13$ mg/ℓ-hr,
   $r_{A_3} = 6.5$ mg/ℓ-hr
   $C_{A_1} = 33$ mg/ℓ, 78% conversion
3.9 **a.** Graph as required.
   b. $\theta = 2.17$ hr, volume = 13,000 gal each = 49,500 liters each

$C_{A_1} = 72$ mg/ℓ, $C_{A_2} = 43$ mg/ℓ,
$C_{A_3} = 30$ mg/ℓ
$r_{A_1} = 35.6$ mg/ℓ-hr, $r_{A_2} = 13$ mg/ℓ-hr,
$r_{A_3} = 6$ mg/ℓ-hr
3.11 **a.** $\theta = 2.89$ hr,
   volume = 17,300 gal = 65,900 liters
   b. $C_{A_1} = 70$ mg/ℓ, $C_{A_2} = 32$ mg/ℓ,
   $C_{A_3} = 15$ mg/ℓ
   c. $\theta = 8.67$ hr,
   volume = 51,900 gal = 98,000 liters
   d. $\theta = 22.5$ hr,
   volume = 135,000 gal = 513,000 liters

CHAPTER 4

4.1 **a.** 1996 = 34,038
   b. 2006 = 44,900
   2011 = 50,800
   2016 = 57,200
   2021 = 64,000
4.3 Demand = 166 gal/cap-day
4.5 **a.** Suspended solids = 22 mg/ℓ
   b. Dissolved solids = 618 mg/ℓ
4.7 **a.** Ca^{+2} hardness = 0.55 meq/ℓ = 27.5 mg/ℓ as $CaCO_3$
   Mg^{+2} hardness = 0.292 meq/ℓ = 14.6 mg/ℓ as $CaCO_3$
   Fe^{+2} hardness = 0.004 meq/ℓ = 0.2 mg/ℓ as $CaCO_3$
   b. Total hardness = 0.846 meq/ℓ = 42 mg/ℓ as $CaCO_3$
   c. Alkalinity = 0.361 meq/ℓ = 18.1 mg/ℓ as $CaCO_3$
   d. Nitrate nitrogen (N) = 0.045 mg/ℓ
   e. $[H^+] = 10^{-6.7}$ moles/ℓ
   $[OH^-] = 10^{-7.3}$ moles/ℓ
4.9 **a.** Ca^{+2} hardness = 0.460 meq/ℓ = 23 mg/ℓ as $CaCO_3$
   Mg^{+2} hardness = 0.217 meq/ℓ = 11 mg/ℓ as $CaCO_3$
   Fe^{+2} hardness = 0.0007 meq/ℓ = 0.35 mg/ℓ as $CaCO_3$
   b. Total hardness = 0.678 meq/ℓ = 34 mg/ℓ as $CaCO_3$

c. Alkalinity = 6.72 meq/ℓ = 336 mg/ℓ as CaCO$_3$

d. Nitrate nitrogen (N) = 0

e. [H$^+$] = 10$^{-7.6}$ moles/ℓ
[OH$^-$] = 10$^{-6.4}$ moles/ℓ

4.11 CO$_3^{-2}$ and OH$^-$ are in negligible concentrations in the neutral pH range.

CHAPTER 5

5.1 a. L_0 = 233 mg/ℓ
b. y/L_0 = 0.858 = 85.8%

5.3 Total suspended solids = 316 mg/ℓ;
fixed suspended solids = 84 mg/ℓ;
volatile suspended solids = 232 mg/ℓ or 73.4%

CHAPTER 7

7.1 Approach velocity = 2.31 fps (0.771 m/s),
head loss = 0.08 ft (0.0157 m)

7.2 a. Head = 3.50 ft
b. The coordinates for the parabola are h = 0.51 ft, w = 2.13 ft; h = 1.04 ft, w = 2.96 ft; h = 2.10 ft, w = 4.14 ft; h = 3.16 ft, w = 5.06 ft; and h = 3.70 ft, w = 5.45 ft. The design section is a rectangular area over a trapezoidal area. The rectangular area is 4 ft-6 in. wide and 1.60 ft high. The trapezoidal area is 4 ft-6 in. wide at the top and 1 ft-6 in. wide at the bottom, and the height is 2.10 ft.
c. t = 53.7 sec
d. Length = 53.7 ft
e. Design length = 72 ft-6 in.
f. Head = 2.07 ft, velocity = 0.93 fps
g. Head = 1.31 ft, velocity = 0.91 fps

7.3 a. Head = 1.08 m
b. The coordinates for the parabola are h = 0.153 m, w = 0.644 m; h = 0.312 m, w = 0.893 m; h = 0.631 m, w = 1.249 m; and h = 1.135 m, w = 1.678 m. The design section is a rectangular area over a trapezoidal area. The rectangular area is 1.678 m wide and 0.48 m high. The trapezoidal area is 1.678 m wide at the top and 0.450 m wide at the bottom, and the height is 0.655 m.
c. t = 54.0
d. Length = 16.2 m
e. Design length = 21.9 m
f. Head = 0.639 m, velocity = 0.28 m/s
g. Head = 0.402 m, velocity = 0.27 m/s

7.6 a. Width = 11.6 ft, depth = 7.73 ft, and length = 46.4 ft
b. Air flow = 148 ft^3/min

7.7 a. Width = 3.53 m, depth = 2.35 m, and length = 14.12 m
b. Air flow = 4.23 m^3/min

7.10 Volume = 1,074,000 gal or 144,000 ft^3

7.11 Volume = 3,930,000 liters

CHAPTER 8

8.1 a. Width = 5 ft-4 in., depth = 6 ft-8 in.
b. Power = 7.62 hp
c. Speed = 1.23 rps = 73.7 rpm

8.2 a. Width = 1.61 m, depth = 2.01 m
b. Power = 5578 W
c. Speed = 1.42 rps = 85.2 rpm

8.5 a. Width = 10 ft-11 in., length = 21 ft-10 in., depth = 9 ft-10 in.
b. Air flow = 173 cfm
c. 44 diffusers

8.6 a. Width = 3.32 m, length = 6.64 m, depth = 3.0 m
b. Air flow = 4.90 m^3/min
c. 44 diffusers

8.9 a. CaO = 2570 lb/MG, Na$_2$CO$_3$ = 1080 lb/MG
b. CaO = 5780 tons/month, Na$_2$CO$_3$ = 2430 tons/month

8.10 a. CaO = 308 kg/ML, Na$_2$CO$_3$ = 129 kg/ML
b. CaO = 5250 tonnes/month, Na$_2$CO$_3$ = 2200 tonnes/month

CHAPTER 9

9.1 a. Diameter = 52.5 ft
b. Diameter = 55 ft
c. Depth = 10 ft
d. Head over weirs = 0.153 ft
e. Depth of weirs = 2.84 in.
f. Depth (d) = 0.90 ft
g. Depth (H_0) = 1.29 ft
h. Depth = 19.5 in.

9.2 a. Diameter = 16.0 m
b. Depth = 3.05 m
c. Head over weirs = 47.5 mm
d. Depth of weirs = 72.5 mm
e. Depth (d) = 0.275 m
f. Depth (H_0) = 0.391 m
g. Depth = 0.493 m

9.5 a. Effluent box ws = El. 321.52
b. Primary clarifier ws = El. 322.39

9.6 **a.** Effluent box ws = El. 97.983 m
b. Primary clarifier ws = El. 98.244 m
9.9 **a.** Overflow rate = 972 gal/day-ft^2
b. Settling velocity = 0.0458 cm/sec
c. Diameter = 0.047 cm
9.10 **a.** Overflow rate = 39.62 m^3/d-m^2
b. Settling velocity = 0.0459 cm/s
c. Diameter = 0.047 cm
9.13 **a.** Settling velocity = 0.0943 cm/sec
b. Diameter = 0.067 cm
9.16 **a.** Diameter = 54.0 ft
b. Use 55 ft diameter.
c. Side water depth = 11 ft
d. Peak weir load = 29,200 gal/day-ft. It is acceptable.
e. BOD$_5$ removal = 33%
f. Suspended solids removal = 62%
9.17 **a.** Diameter = 16.5 m
b. Depth = 3.35 m
c. Peak wier load = 368 m^3/d-m
d. BOD$_5$ removal = 33%
e. Suspended solids removal = 61%
9.20 **a.** Width = 84.5 ft, length = 169 ft
b. Width = 85 ft, length = 170 ft
c. Depth = 19.3 ft
d. Channel length = 227 ft
e. Downstream length = 75.0 ft and channel upstream length (x) = 76.0 ft
f. No
9.21 **a.** Width = 25.8 m, length =51.6 m
b. Depth = 5.94 m
c. Channel length = 69.32 m
d. Downstream length = 22.76 m and channel upstream length (x) = 23.28 m
e. No
9.24 **a.** Graph as required
b. Graph as required
c. G_L (Design) = 0.573 lb/hr-ft^2
d. Diameter = 85.7 ft
e. Design diameter = 90 ft
9.25 **a.** Graph as required
b. Graph as required
c. G_L (Design) = 2.48 kg/h-m^2
d. Diameter = 27.7 m

CHAPTER 10

10.1 Head loss = 3.51 ft
10.2 Head loss = 1.07 m
10.5 Head loss = 1.60 ft
10.6 Head loss = 0.521 m

10.9 **a.** Depth = 0.0968 ft
b. Depth (y_c) = 0.64 ft
c. Depth (H_0) = 1.18 ft
d. Trough depth = 18 in.
10.10 **a.** Depth = 0.0293 m
b. Depth (y_c) = 0.195 m
c. Depth (H_0) = 0.360 m
d. Trough depth = 435 mm

CHAPTER 11

11.1 **a.** Theoretical air flow = 2480 lb air/hr-ft^2
b. Design air = 4340 lb air/hr-ft^2
c. Design air = 1070 cfm/ft^2 or 535 ft^3/gal
d. Tower plan dimensions = 32.5 ft × 32.5 ft
11.2 **a.** Theoretical air flow = 12,000 kg air/h-m^2
b. Design air = 21,000 kg air/h-m^2
c. Design air = 18,300 m^3/h-m^2 or 3.77 m^3/ℓ
d. Tower plan dimensions = 9.9 m × 9.9 m
11.5 **a.** Minimum solids retention time = 3.76 days
b. Design solids retention time = 3.87 days
c. Design nitrate removal rate = 0.332 lb NO$_3^-$-N/lb MLVSS-day
d. Hydraulic detention time = 0.031 day or 0.74 h
e. Sludge wasting = 635 lb MLVSS/day
f. Menthanol = 3060 lb methanol/day
11.6 **a.** Minimum solids retention time = 3.76 d
b. Design solids retention time = 3.87 d
c. Design nitrate removal rate = 0.332 kg NO$_3^-$-N/kg MLVSS-d
d. Hydraulic detention time = 0.031 d or 0.74 h
e. Sludge wasting = 289 kg MLVSS/d
f. Methanol = 1400 kg methanol/d

CHAPTER 12

12.1 **a.** Liquid flowrate = 2.35 BV/hr, volume treated/mass carbon = 44.9 gal/lb
b. Kinetic constants k_1 = 55.7 gal/hr-lb and q_0 = 0.171 lb/lb
12.2 **a.** Liquid flowrate = 2.36 BV/h, volume treated/mass carbon = 378 ℓ/kg
b. Kinetic constants k_1 = 0.129 ℓ/s-kg and q_0 = 0.170 kg/kg
12.5 **a.** Diameter = 4 ft-0 in.; height = 14.02 ft
b. Two columns in series
12.6 **a.** Diameter = 1.18 m; height = 8.5 m
b. Two columns in series
12.9 **a.** Volume = 2090 ft^3
b. Diameter = 10 ft-6 in.; height = 24.4 ft, 3 in.

c. Mass = 52,300 lb
d. Carbon exhausted = 584 lb/day
e. On-line time = 89.9 days
12.10 a. Volume = 59.4 m³
b. Diameter = 3.09 m; height = 7.92 m
c. Mass = 23,800 kg
d. Carbon exhausted = 266 kg/d
e. On-line time = 89.3 d

CHAPTER 13

13.3 a. Resin mass flowrate = 5.50 lb/hr-ft²
b. Wastewater treated/lb resin = 49.1 gal/lb
c. Diameter = 5 ft-6 in.
13.4 a. Resin mass flowrate = 26.4 kg/h-m²
b. Wastewater treated/kg resin = 409 ℓ/kg
c. Diameter = 1.64 m
13.7 Moist resin = 2910 lb; dry resin = 1690 lb
13.8 Moist resin = 1320 kg; dry resin = 766 kg

CHAPTER 14

14.1 a. Efficiency = 50%; number of membranes = 202; power = 134,000 watts; power cost = $0.80/1000 gal
b. Brine flow = 1750 galday; produced water flow = 98,250 gal/day
14.2 a. Efficiency = 50%; number of membranes = 204; power = 132,000 watts; power cost = $0.21/1000 liters
b. Brine flow = 6.63 m³/d; produced water flow = 373 m³/d

CHAPTER 15

15.1 Nondegradable COD = 92 mg/ℓ, pseudo–first-order reaction, K = 0.224 ℓ/gm MLSS-hr or K = 0.268 ℓ/gm MLVSS-hr.
15.3 a. θ = 9.30 h, volume = 389,000 ft³
b. θ_{dpf} = 13.2 hr, volume = 551,000 liters
15.4 a. θ = 9.30 h, volume 11,000 m³
b. θ_{dpf} = 13.2 hr, volume = 15,600 m³
15.7 a. θ = 6.0 h
b. Volume = 460,000 ft³
c. F/M = 0.17 lb BOD₅/lb MLVSS-day
d. Volumetric load = 33.8 lb BOD₅/day-10³ ft³
15.8 a. θ = 6.0 h
b. Volume = 13,100 m³
c. F/M = 0.17 kg BOD₅/kg MLVSS-d
d. Volumetric load = 0.54 kg BOD₅/d-m³
15.11 a. θ = 29.7 h
b. Volume = 695,000 ft³

15.12 a. θ = 29.7 h
b. Volume = 19,700 m³
15.15 a. Volume = 127,000 ft³
b. Operating MLSS = 2470 mg/ℓ
c. Length = 1058 ft
15.16 a. Volume = 3600 m³
b. Operating MLSS = 2470 mg/ℓ
c. Length = 324 m
15.19 Volume = 14,900 ft³
15.20 Volume = 425 m³
15.23 a. Volume = 300,000 ft³
b. Waste sludge = 4250 lb MLSS/day
c. Waste sludge = 50,500 gal/day
d. θ_c = 11.0 days
15.24 a. Volume = 8500 m³
b. Waste sludge = 1930 kg/d
c. Waste sludge = 193,000 ℓ/d
d. θ_c = 11.0 d
15.27 a. Volume = 537,000 ft³
b. Waste sludge = 5380 lb MLSS/day
c. Waste sludge = 63,900 gal/day
d. θ_c = 15.6 days
e. Oxygen required = 21,500 lb/day
f. θ_c (Design) = 2.68 days. No.
g. Fraction of nitrifiers = 0.0350 = 3.50%
h. Rate of nitrification = 80.9 mg/ℓ-day
i. θ = 5.62 h. No.
j. Oxygen demand by quarters: 1st, 314 lb/hr; 2nd, 233 lb/hr; 3rd, 179 lb/hr; 4th, 170 lb/hr
15.28 a. Volume = 15,200 m³
b. Waste sludge = 2450 kg/d
c. Waste sludge = 245,000 ℓ/d
d. θ_c = 15.5 d
e. Oxygen required = 7510 kg/d
f. θ_c (Design) = 2.68 d. No.
g. Fraction of nitrifiers = 0.035 = 3.50%
h. Rate of nitrification = 80.9 mg/ℓ-d
i. θ = 5.62 h. No.
j. Oxygen demand by quarters: 1st, 109.5 kg/h; 2nd, 81.4 kg/h; 3rd, 62.6 kg/h; 4th, 59.5 kg/h
15.31 a. θ_{dpf} = 6.46 hr, volume = 48,000 ft³
b. Waste sludge = 385 lb/MLSS/day
c. Waste sludge = 4570 gal/day
d. Oxygen demand = 1870 lb/day, 78.0 lb/hr or 26.1 mg/ℓ-hr
e. Oxygen demand per quarter: 1st, 27.3 lb/hr; 2nd, 20.3 lb/hr; 3rd, 15.6 lb/hr; 4th, 14.8 lb/hr.

Oxygen demand: 1st, 36.5 mg/ℓ-hr;
2nd, 27.2 mg/ℓ-hr; 3rd, 20.9 mg/ℓ-hr;
4th, 19.8 mg/ℓ-hr
  f. Standard air per quarter: 1st, 543 scfm;
2nd, 323 scfm; 3rd, 249 scfm; 4th, 236 scfm
  g. Air is 1780 ft^3/lb BOD$_5$ applied.
15.32 a. $\theta_{dpf} = 6.46$ h, volume $= 1360$ m^3
  b. Waste sludge $= 176$ kg/d
  c. Waste sludge $= 17,400$ ℓ/d
  d. Oxygen demand $= 852$ kg/d, 35.5 kg/h, or
26.1 mg/ℓ-h
  e. Oxygen demand per quarter:
1st, 12.4 kg/h; 2nd, 9.23 kg/h,
3rd, 7.10 kg/h; 4th, 6.75 kg/h.
Oxygen demand: 1st, 36.4 mg/ℓ-h;
2nd, 27.1 mgℓ-h; 3rd, 20.9 mg/ℓ-h;
4th, 19.8 mg/ℓ-h
  f. Standard air per quarter: 1st, 12.3 m^3/min;
2nd, 9.22 m^3/min; 3rd, 7.06 m^3/min;
4th, 6.71 m^3/min
  g. Air is 111 m^3/kg BOD$_5$ applied
15.35 a. $\theta_i = 6.60$ h, volume $= 297,000$ ft^3
  b. Waste sludge $= 3940$ lb MLSS/day
  c. Waste sludge $= 59,200$ gal/day
  d. $F/M = 0.12$ lb BOD$_5$/day-lb MLSS
  e. Space loading $= 30.6$ lb/day-10^3 ft^3
15.36 a. $\theta = 6.60$ h, volume $= 8420$ m^3
  b. Waste sludge $= 1810$ kg/d
  c. Waste sludge $= 226,000$ ℓ/d
  d. $F/M = 0.12$ kg BOD$_5$/kg MLSS-d
  e. Space loading $= 0.491$ kg/d-m^3
15.38 a. $\theta_c = 6.60$ days
  b. $\theta_i = 26.4$ hr
  c. Volume $= 36,800$ ft^3
  d. Waste sludge $= 2610$ lb MLSS/day, or
25,000 gal/day if wasted from recycle line
15.39 a. $\theta_c = 6.60$ d
  b. $\theta = 26.4$ h
  c. Volume $= 1040$ m^3
  d. Waste sludge $= 1180$ kg/d or
94,600 ℓ/d if wasted from recycle line
15.42 $K = 0.0404$ ℓ/gm MLSS-hr
  $Y = 0.949$ mg MLSS/mg BOD$_5$
  $k_e = 0.099$ day^{-1}
15.43 a. $\theta = 30.0$ hr
  b. Volume $= 41,800$ ft^3
  c. Waste sludge $= 2370$ lb MLSS/day. If wasted
from reactor, $Q_w = 37,900$ gal/day.
15.44 a. $\theta = 30.0$ h
  b. Volume $= 1880$ m^3

  c. Waste sludge $= 1080$ kg/d. If wasted
from reactor, $Q_w = 144,000$ ℓ/d.
15.47 $Y = 0.48$ mg MLSS/mg COD $= 0.42$ mg MLVSS/
mg COD
  $Y' = 0.40$ mg O$_2$/mg COD
  $k_e = 0.181$ day^{-1}
  $k'_e = 0.26$ mg O$_2$/mg MLVSS-day
15.50 a. Diameter $= 15.6$ in.
  b. Diameter $= 16$ in.
  c. Head on weirs $= 0.11$ ft
  d. Depth (d) $= 0.89$ ft; depth (H_0) $= 1.15$ ft
  e. Head losses are as follows: entrance,
0.14 ft; friction, 0.39 ft; change in direction
and expansion, 0.36 ft; friction in center
column $= 0.01$ ft; exit, 0.03 ft.
  f. $\Delta Z_1 = 0.70$ ft
  g. $\Delta Z_2 = 0.93$ ft
  h. $\Delta Z_1 + \Delta Z_2 = 1.63$ ft
15.51 a. Diameter $= 396$ mm
  b. Diameter $= 400$ mm
  c. Head on weirs $= 0.033$ m
  d. Depth (d) $= 0.273$ m, depth (H_0) $=$
0.399 m
  e. Head losses are as follows: entrance,
0.046 m; friction $= 0.113$ m; change in
direction and expansion $= 0.101$ m; friction
in center column $= 0.002$ m; exit, 0.009 m.
  f. $\Delta Z_1 = 0.261$ m
  g. $\Delta Z_2 = 0.271$ m
  h. $\Delta Z_1 + \Delta Z_2 = 0.532$ m

CHAPTER 16

16.1 a. 0.660 lb O$_2$/hr
  b. 4.2%
16.2 a. 0.300 kg O$_2$/h
  b. 4.2%
16.5 1270 cfm
16.6 36.0 m^3/min
16.9 a. 0.775 lb O$_2$/hr per diffuser or 1.55 lb O$_2$/hr
per pair
  b. Oxygen demand per quarter:
1st, 115.5 lb/hr; 2nd, 85.8 lb/hr;
3rd, 66.0 lb/hr; 4th, 62.7 lb/hr
  c. Number of pairs and spacing per quarter:
1st, 75, 1.33 ft; 2nd, 56, 1.79 ft;
3rd, 43, 2.33 ft; 4th, 41, 2.44 ft
  d. Standard air $= 3010$ scfm
  e. Actual air $= 3190$ cfm
  f. Power $= 93.5$ hp

16.10 a. 0.352 kg/h per diffuser or 0.704 kg/h per pair
b. Oxygen demand per quarter:
1st, 52.5 kg/h; 2nd, 39.0 kg/h; 3rd 30.0 kg/h; 4th, 28.5 kg/h
c. Number of pairs and spacing per quarter:
1st, 75, 0.407 m; 2nd, 56, 0.545 m; 3rd, 43, 0.709 m; 4th, 41, 0.744 m
d. Standard air = 85.6 m³/min
e. Actual air = 90.9 m³/min
f. Power = 70.8 kW

16.13 a. 1.08 lb O_2/hp-hr
b. 56.3 hp

16.14 a. 0.656 kg O_2/kW-h
b. 42.3 kW

16.17 a. 0.618 lb O_2/hr
b. Number of diffusers per quarter:
1st, 107; 2nd, 80; 3rd, 61; 4th, 58
c. Spacing per quarter: 1st, 0.925 ft; 2nd, 1.24 ft; 3rd, 1.62 ft; 4th, 1.70 ft

16.18 a. 0.281 kg/h
b. Number of diffusers per quarter: 1st, 107; 2nd, 79; 3rd, 61; 4th, 58
c. Spacing per quarter:
1st, 0.280 m; 2nd, 0.380 m; 3rd, 0.492 m; 4th, 0.517 m

16.21 a. α = 0.82
b. β = 0.95

16.22 a. Free air = 1810 cfm
b. Compressor power = 63.0 hp
c. Motor power = 66.3 hp

16.23 a. Free air = 51.4 m³/min
b. Compressor power = 47.2 kW
c. Motor power = 49.7 kW

16.26 2.21 lb O_2/hp-hr

16.27 1.34 kg O_2/kW-h

16.30 a. Free air = 1280 cfm
b. Motor power = 47.0 hp

16.31 a. Free air = 35.77 m³/min
b. Motor power = 45.7 kW

CHAPTER 17

17.1 a. Diameter = 119 ft
b. Design diameter = 120 ft
c. Depth = 5.44 ft

17.2 Diameter = 36.4 m

17.5 a. Diameter = 23.7 ft
b. Design diameter = 25 ft
c. Use 20 ft of packing.

d. Hydraulic load = 114 MG/ac-day
e. Organic load = 45.8 lb BOD_5/day-10³ ft³ or 2000 lb BOD_5/day-ac-ft

17.6 a. Diameter = 10.6 m
b. Hydraulic load = 104 m³/d-m²
c. Organic load = 0.801 kg BOD_5/d-m³

17.9 a. Diameter = 79.6 ft
b. Design diameter = 80 ft
c. Depth = 5.94 ft

17.10 Diameter = 24.3 m

17.13 a. Organic load = 2664 lb BOD_5/day-ac-ft
b. Diameter = 62.5 ft
c. Design diameter = 65 ft
d. Depth = 5.54 ft

17.14 a. Organic load = 0.980 kg BOD_5/d-m³
b. Diameter = 19.0 m

17.17 Secondary sludge = 0.454 lb dry solids/lb BOD_5 removed

17.18 Secondary sludge = 0.456 kg dry solids/kg BOD_5 removed

17.21 a. Hydraulic load = 1.99 gal/day-ft²
b. S_1 = 85 mg/ℓ, S_2 = 52 mg/ℓ, S_3 = 32 mg/ℓ, S_4 = 20 mg/ℓ
c. Area per stage = 271,000 ft²
d. Number of shafts = 2.71; use 3. Total shafts = 12

17.23 a. Hydraulic load = 81.3 ℓ/d-m²
b. S_1 = 85 mg/ℓ, S_2 = 52 mg/ℓ, S_3 = 32 mg/ℓ, S_4 = 20 mg/ℓ
c. Area per stage = 25,100 m²
d. Number of shafts = 2.70; use 3. Total shafts = 12

17.25 a. Hydraulic load = 1.725 gal/day-ft²
b. S_1 = 80 mg/ℓ, S_2 = 48 mg/ℓ, S_3 = 29 mg/ℓ, S_4 = 17 mg/ℓ

17.26 a. Hydraulic load = 70.6 ℓ/d-m²
b. S_1 = 80 mg/ℓ, S_2 = 48 mg/ℓ, S_3 = 29 mg/ℓ, S_4 = 17 mg/ℓ

CHAPTER 18

18.1 a. 6.12 ac or 267,000 ft²
b. 43.1 lb BOD_5/day-ac
c. Width = 258 ft, length = 516 ft, depth = 6 ft

18.2 a. 2.48 ha
b. 48.2 kg/d-ha
c. Width = 78.7 m, length = 157 m, depth = 1.52 m

18.5 a. θ = 3.76 days
b. Volume = 355,000 ft³

c. Width = 94.2 ft, length = 189 ft, depth = 10 ft

d. Oxygen required = 1290 lb/day, power = 32.6 hp

e. $P/V = 0.092$ hp/10^3 ft^3. It is not sufficient; use 53.3 hp.

18.6 a. $\theta = 3.76$ d

  b. Volume = 10,100 m^3

  c. Width = 29.0 m, length = 58.0 m, depth = 3.0 m

  d. Oxygen required = 586 kg/d, power = 23.5 kW

  e. $P/V = 2.33$ kW/10^3-m^3. It is not sufficient; use 39.9 kW.

18.9 a. $K = 0.550$ hr^{-1}

  b. $Y = 0.475$ mg MLSS/mg COD
  $Y = 0.418$ mg MLVSS/mg COD
  $k_e = 0.375$ day^{-1}

  c. $Y' = 0.406$ mg O$_2$/mg COD
  $k'_e = 0.533$ mg O$_2$/mg MLVSS-day

  d. $\theta = 86.1$ hr or 3.59 days

  e. Width = 150 ft, length = 300 ft

  f. $S_t = 27.9$ mg/ℓ total COD

  g. 65%

  h. Oxygen required = 7400 lb/day per lagoon

  i. 206 hp

  j. $P/V = 0.57$ hp/10^3 ft^3; mixing is adequate.

  k. 2.81 lb O$_2$/lb BOD$_5$

18.10 a. $K = 0.550$ h^{-1}

  b. $Y = 0.475$ mg MLSS/mg COD
  $Y = 0.418$ mg MLVSS/mg COD
  $k_e = 0.375$ d^{-1}

  c. $Y' = 0.406$ mg O$_2$/mg COD
  $k'_e = 0.533$ mg O$_2$/mg MLVSS-d

  d. $\theta = 86.1$ h = 3.59 d

  e. Width = 45.7 m, length = 91.4 m

  f. $S_t = 27.9$ mg/ℓ total COD

  g. 65%

  h. Oxygen required = 3360 kg/d per lagoon

  i. 154 kW per lagoon

  j. $P/V = 12.4$ kW/10^3 m^3; mixing is adequate.

  k. 2.81 kg O$_2$/kg BOD$_5$

18.13 a. $\theta_i = 13.7$ d

  b. 1,530,000 ft^2

18.14 a. $\theta_i = 13.7$ d

  b. 142,000 m^2

CHAPTER 19

19.1 a. $V_{avg} = 905$ ft^3/day

  b. V (digesting sludge) = 20,800 ft^3

c. V (stored sludge) = 25,400 ft^3

d. V (total) = 92,400 ft^3

19.2 a. $V_{avg} = 25.6$ m^3/d

  b. V (digesting sludge) = 610 m^3

  c. V (stored sludge) = 720 m^3

  d. V (total) = 26,600 m^3

19.5 a. V (total) = 75,600 ft^3

  b. 4.73 ft^3/cap

  c. Gas produced = 27,000 ft^3/day

19.6 a. V (total) = 2170 m^3

  b. 0.136 m^3/cap

  c. Gas produced = 763 m^3/d

19.9 a. V (total) = 95,200 ft^3

  b. Diameter = 64.7 ft

  c. Design diameter = 65 ft

  d. Side water depth = 28.7 ft

  e. 4.76 ft^3/cap

  f. 45.0% volatile solids

  g. Gas produced = 31,400 ft^3/day

  h. Fuel value = 19,700,000 Btu/day

19.10 a. V (total) = 2730 m^3

  b. Diameter = 19.8 m

  c. 0.137 m^3/cap

  d. 45.0% volatile solids

  e. Gas produced = 885 m^3/d

  h. Fuel value = 20,600,000 kJ/d

19.12 a. Primary sludge = 3860 gal/d

  b. Secondary sludge = 619 gal/d

19.13 a. Primary sludge = 14,600 ℓ/d

  b. Secondary sludge = 2340 ℓ/d

19.16 79,300 Btu/hr

19.17 79,900,000 J/h

19.20 Head = 37.2 ft

19.21 Head = 11.5 m

19.24 Pseudomolecular weight = 24.40 lb/lb mole, specific volume = 14.7 ft^3/lb

19.25 Pseudomolecular weight = 24.40 kg/kg mole, specific volume = 0.918 ℓ/gm

CHAPTER 20

20.1 a. Primary sludge = 365 ft^3/day
  Secondary sluge = 1480 ft^3/day
  Blended sludge = 1845 ft^3/day

  b. Solids = 1.4%

  c. Thickened sludge = 738 ft^3/day

  d. Supernatant flow = 1107 ft^3/day

  e. Volume = 19,000 ft^3

  f. Oxygen required = 1270 lb/day

  g. Air = 1260 cfm, air/10^3 ft^3 = 66.3 cfm; mixing is adequate.

h. Diameter = 15.9 ft, design diameter = 20 ft

20.2 a. Primary sludge = 10.3 m^3/d, secondary sludge = 41.7 m^3/d, blended sludge = 52.0 m^3/d

b. Solids = 1.4%

c. Thickened sludge = 20.8 m^3/d

d. Supernatant liquor = 31.2 m^3/d

e. Volume = 535 m^3

f. Oxygen required = 566 kg/d

g. Air = 35.0 m^3/min, air/10^3 m^3 = 65.4 m^3/min; mixing is adequate.

20.5 a. Volume = 36,400 ft^3

b. Oxygen required = 2530 lb/day

c. Air = 2510 cfm

d. Volatile solids load = 0.054 lb VS/ft^3-day

e. 3.00 ft^3/cap

f. Air for mixing = 69.0 cfm/10^3 ft^3
It is adequate.

20.6 a. Volume = 10.40 m^3

b. Oxygen required = 1144 kg/d

c. Air = 70.7 m^3/min

d. Volatile solids load = 0.847 kg VS/d-m^3

e. 0.086 m^3/cap

f. Air for mixing = 68.1 m^3/min-10^3 m
It is adequate.

CHAPTER 21

21.1 a. Thickened sludge flow = 99,700 gal/day

b. Diameter = 75 ft (for standard size)

21.2 a. Thickened sludge flow = 379,000 ℓ/d

b. Diameter = 22.3 m

21.5 Minimum solids content = 22.6%

21.6 Minimum solids content = 22.6%

21.9 a. Cake produced = 165,000 lb/day

b. Filtrate produced = 32,900 gal/day; solids content = 9620 mg/ℓ

c. Diameter = 8 ft; length = 16 ft

21.10 a. Cake produced = 75,200 kg/d

b. Filtrate produced = 125,000 ℓ/d; solids content = 9620 mg/ℓ

c. Diameter = 3.54 m; length = 7.08 m

21.13 Filter yield = 4.49 lb/hr-ft^2

21.14 Filter yield = 21.9 kg/h-m^2

21.17 Specific resistance = 3.10 × 10^7 sec^2/gm

CHAPTER 22

22.1 a. Design hydraulic loading rate = 77.6 in./yr

b. Field area = 247 acres

c. Volume = 360 acre-ft; area = 19.4 acres; depth = 18.6 ft

22.2 a. Design hydraulic loading rate = 202 cm/yr

b. Field area = 97.6 ha

c. Volume = 451,000 m^3; area = 8.72 ha; depth = 5.17 m

22.5 Terrace area = 55.0 acres

22.6 Terrace area = 22.7 ha

CHAPTER 23

23.1 a. Recycle flow = 0.145 MGD

b. Flotation tank area = 126 ft^2

c. Air required = 15.6 lb/day

d. Air required = 208 ft^3/day

23.2 a. Recycle flow = 545,000 ℓ/d

b. Flotation tank area = 11.6 m^2

c. Air required = 7.02 kg/d

d. Air required = 5.84 m^3/d

INDEX